Collins
BIG ROAD ATLAS
BRITAIN

Contents

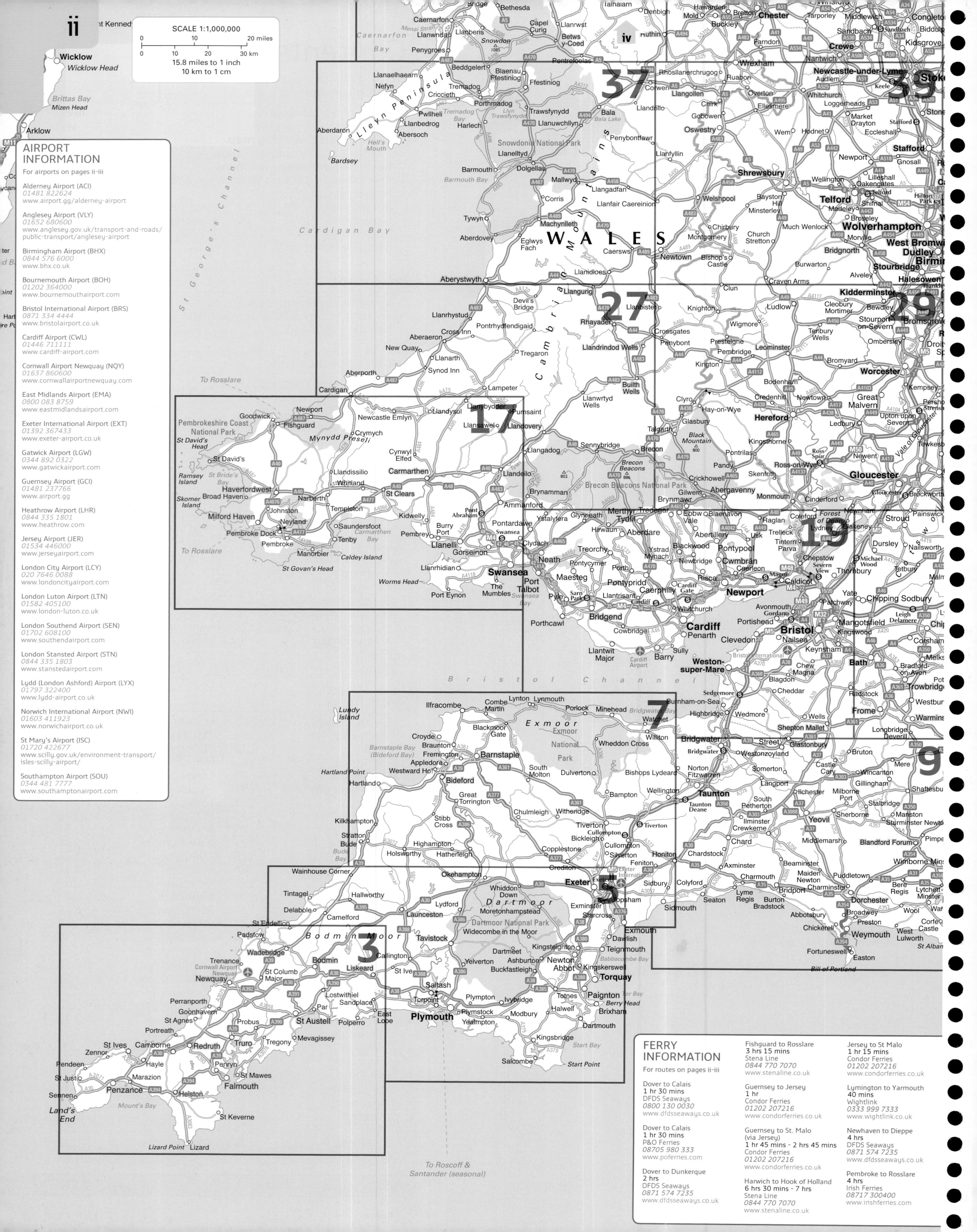

AIRPORT INFORMATION
For airports on pages ii-iii

Alderney Airport (ACI)
01481 822624
www.airport.gg/alderney-airport

Anglesey Airport (VLY)
01652 680600
www.anglesey.gov.uk/transport-and-roads/public-transport/anglesey-airport/

Birmingham Airport (BHX)
0844 576 6000
www.bhx.co.uk

Bournemouth Airport (BOH)
01202 364000
www.bournemouthairport.com

Bristol International Airport (BRS)
0871 334 4444
www.bristolairport.co.uk

Cardiff Airport (CWL)
01446 711111
www.cardiff-airport.com

Cornwall Airport Newquay (NQY)
01637 860600
www.cornwallairportnewquay.com

East Midlands Airport (EMA)
0800 083 8759
www.eastmidlandsairport.com

Exeter International Airport (EXT)
01392 367433
www.exeter-airport.co.uk

Gatwick Airport (LGW)
0344 892 0322
www.gatwickairport.com

Guernsey Airport (GCI)
01481 237766
www.guernsey-airport.gov.gg

Heathrow Airport (LHR)
0844 335 1801
www.heathrow.com

Jersey Airport (JER)
01534 446000
www.jerseyairport.com

London City Airport (LCY)
020 7646 0088
www.londoncityairport.com

London Luton Airport (LTN)
01582 405100
www.london-luton.co.uk

London Southend Airport (SEN)
01702 608100
www.southendairport.com

London Stansted Airport (STN)
0844 335 1803
www.stanstedairport.com

Lydd (London Ashford) Airport (LYX)
01797 322400
www.lydd-airport.co.uk

Norwich International Airport (NWI)
01603 411923
www.norwichairport.co.uk

St Mary's Airport (ISC)
01720 422677
www.scilly.gov.uk/environment-transport/isles-scilly-airport/

Southampton Airport (SOU)
0344 481 7777
www.southamptonairport.com

FERRY INFORMATION
For routes on pages ii-iii

Dover to Calais
1 hr 30 mins
DFDS Seaways
0800 130 0030
www.dfdsseaways.co.uk

Dover to Calais
1 hr 30 mins
P&O Ferries
08705 980 333
www.poferries.com

Dover to Dunkerque
2 hrs
DFDS Seaways
0871 574 7235

Fishguard to Rosslare
3 hrs 15 mins
Stena Line
0844 770 7070
www.stenaline.co.uk

Guernsey to Jersey
1 hr
Condor Ferries
01202 207216
www.condorferries.co.uk

Guernsey to St. Malo
(via Jersey)
1 hr 45 mins - 2 hrs 45 mins
Condor Ferries
01202 207216
www.condorferries.co.uk

Harwich to Hook of Holland
6 hrs 30 mins - 7 hrs
Stena Line
0844 770 7070
www.stenaline.co.uk

Jersey to St Malo
1 hr 15 mins
Condor Ferries
01202 207216
www.condorferries.co.uk

Lymington to Yarmouth
40 mins
Wightlink
0333 999 7333
www.wightlink.co.uk

Newhaven to Dieppe
4 hrs
DFDS Seaways
0871 574 7235
www.dfdsseaways.co.uk

Pembroke to Rosslare
4 hrs
Irish Ferries
08717 300400
www.irishferries.com

SCALE 1:1,000,000
15.8 miles to 1 inch
10 km to 1 cm

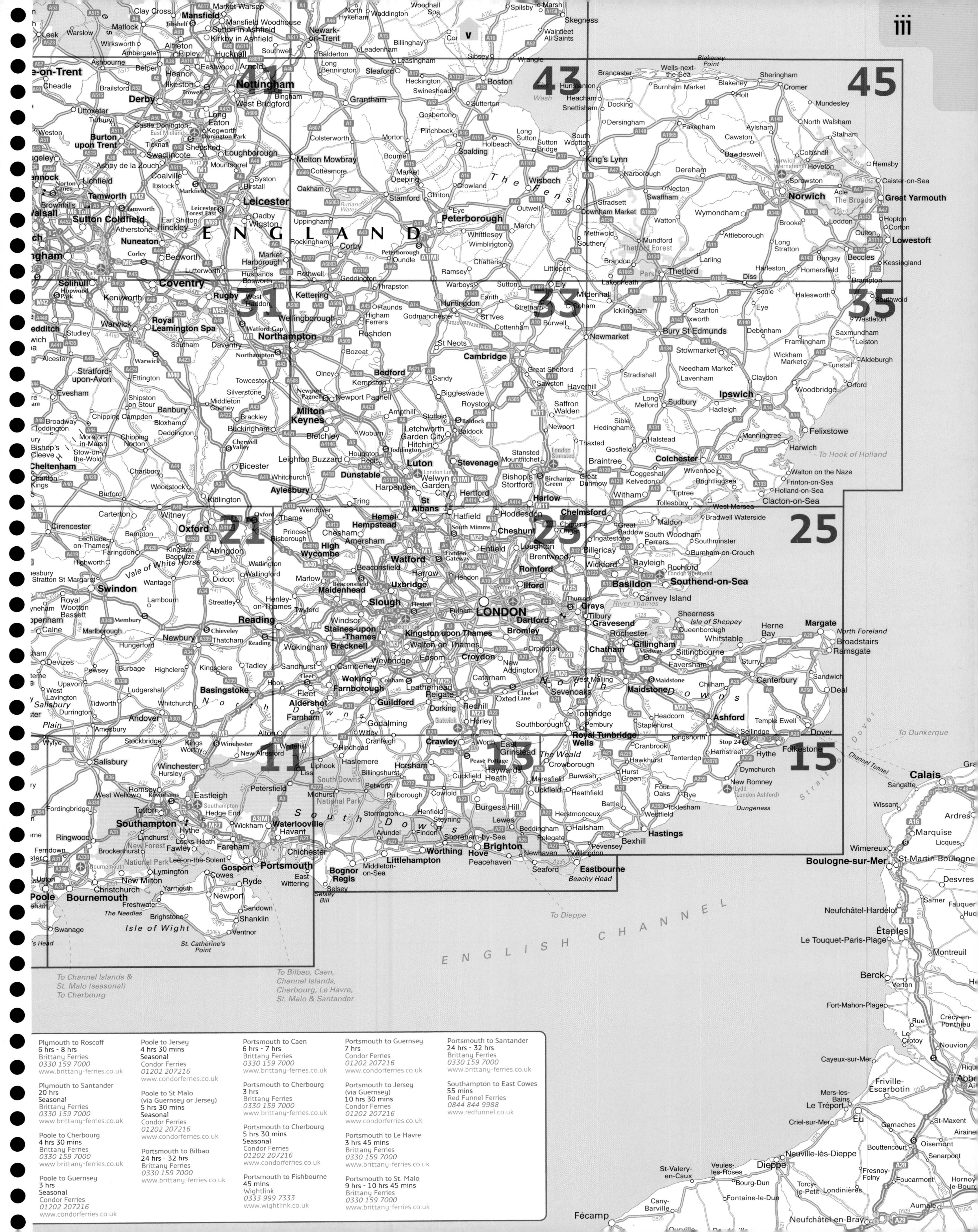
ENGLAND
ENGLISH CHANNEL
11
13
15
21
23
25
31
33
35
41
43
45
LONDON
Nottingham
Leicester
Coventry
Northampton
Cambridge
Norwich
Ipswich
Colchester
Oxford
Reading
Southampton
Portsmouth
Brighton
Hastings
Canterbury
Dover
Calais
Boulogne-sur-Mer
Dieppe
The Fens
The Wash
Isle of Wight
South Downs
North Downs
Isle of Sheppey
Channel Tunnel
To Hook of Holland
To Dunkerque
To Dieppe
To Channel Islands & St. Malo (seasonal) To Cherbourg
To Bilbao, Caen, Channel Islands, Cherbourg, Le Havre, St. Malo & Santander
Plymouth to Roscoff
6 hrs - 8 hrs
Brittany Ferries
0330 159 7000
www.brittany-ferries.co.uk
Plymouth to Santander
20 hrs
Seasonal
Brittany Ferries
0330 159 7000
www.brittany-ferries.co.uk
Poole to Cherbourg
4 hrs 30 mins
Brittany Ferries
0330 159 7000
www.brittany-ferries.co.uk
Poole to Guernsey
3 hrs
Seasonal
Condor Ferries
01202 207216
www.condorferries.co.uk
Poole to Jersey
4 hrs 30 mins
Seasonal
Condor Ferries
01202 207216
www.condorferries.co.uk
Poole to St Malo
(via Guernsey or Jersey)
5 hrs 30 mins
Seasonal
Condor Ferries
01202 207216
www.condorferries.co.uk
Portsmouth to Bilbao
24 hrs - 32 hrs
Brittany Ferries
0330 159 7000
www.brittany-ferries.co.uk
Portsmouth to Caen
6 hrs - 7 hrs
Brittany Ferries
0330 159 7000
www.brittany-ferries.co.uk
Portsmouth to Cherbourg
3 hrs
Brittany Ferries
0330 159 7000
www.brittany-ferries.co.uk
Portsmouth to Cherbourg
5 hrs 30 mins
Seasonal
Condor Ferries
01202 207216
www.condorferries.co.uk
Portsmouth to Fishbourne
45 mins
Wightlink
0333 999 7333
www.wightlink.co.uk
Portsmouth to Guernsey
7 hrs
Condor Ferries
01202 207216
www.condorferries.co.uk
Portsmouth to Jersey
(via Guernsey)
10 hrs 30 mins
Condor Ferries
01202 207216
www.condorferries.co.uk
Portsmouth to Le Havre
3 hrs 45 mins
Brittany Ferries
0330 159 7000
www.brittany-ferries.co.uk
Portsmouth to St. Malo
9 hrs - 10 hrs 45 mins
Brittany Ferries
0330 159 7000
www.brittany-ferries.co.uk
Portsmouth to Santander
24 hrs - 32 hrs
Brittany Ferries
0330 159 7000
www.brittany-ferries.co.uk
Southampton to East Cowes
55 mins
Red Funnel Ferries
0844 844 9988
www.redfunnel.co.uk

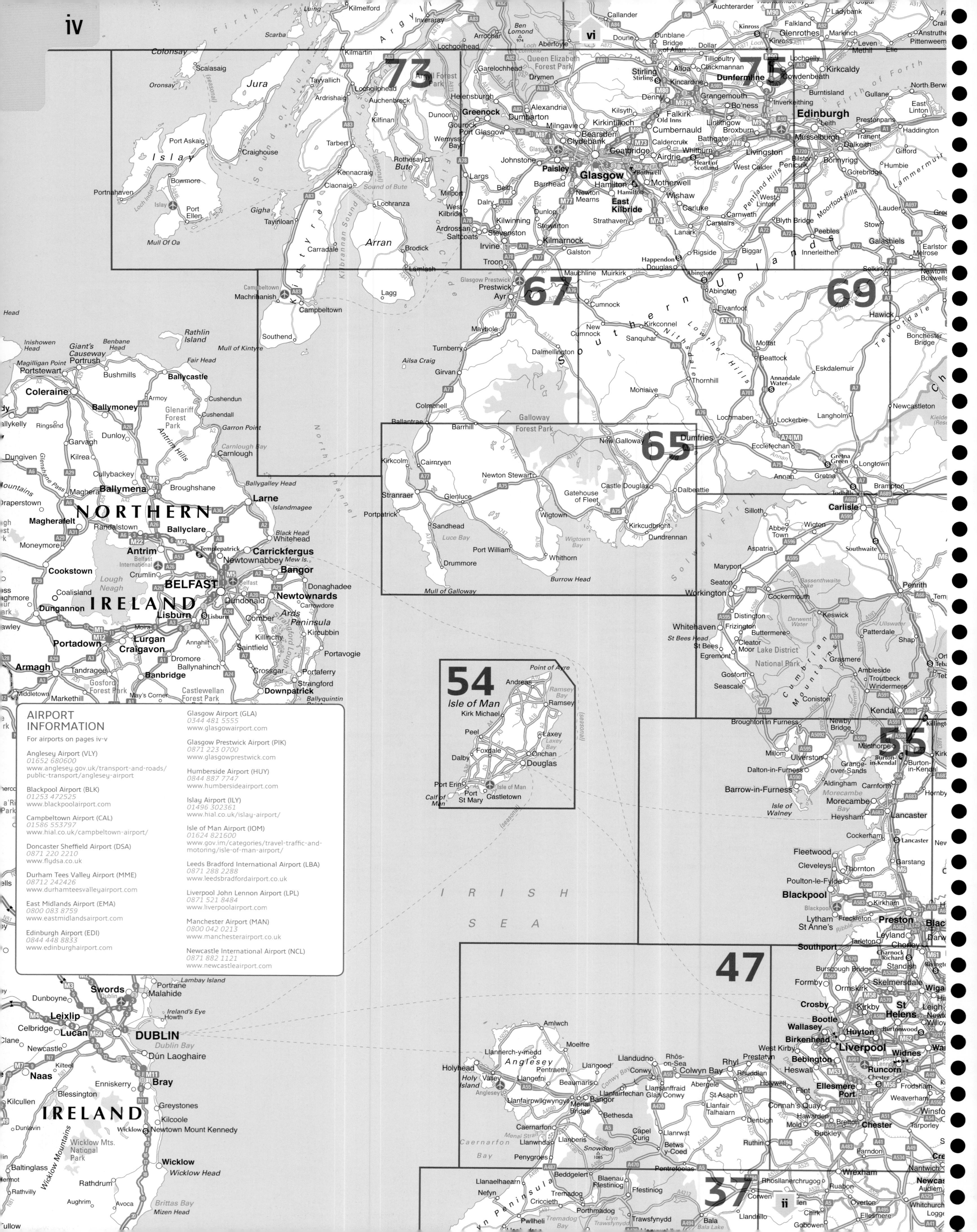

AIRPORT INFORMATION

For airports on pages iv-v

Anglesey Airport (VLY)
01652 680600
www.anglesey.gov.uk/transport-and-roads/
public-transport/anglesey-airport

Blackpool Airport (BLK)
01253 472525
www.blackpoolairport.com

Campbeltown Airport (CAL)
01586 553797
www.hial.co.uk/campbeltown-airport/

Doncaster Sheffield Airport (DSA)
0871 220 2210
www.flydsa.co.uk

Durham Tees Valley Airport (MME)
08712 242426
www.durhamteesvalleyairport.com

East Midlands Airport (EMA)
0800 083 8759
www.eastmidlandsairport.com

Edinburgh Airport (EDI)
0844 448 8833
www.edinburghairport.com

Glasgow Airport (GLA)
0344 481 5555
www.glasgowairport.com

Glasgow Prestwick Airport (PIK)
0871 223 0700
www.glasgowprestwick.com

Humberside Airport (HUY)
0844 887 7747
www.humbersideairport.com

Islay Airport (ILY)
01496 302361
www.hial.co.uk/islay-airport/

Isle of Man Airport (IOM)
01624 821600
www.gov.im/categories/travel-traffic-and-
motoring/isle-of-man-airport/

Leeds Bradford International Airport (LBA)
0871 288 2288
www.leedsbradfordairport.co.uk

Liverpool John Lennon Airport (LPL)
0871 521 8484
www.liverpoolairport.com

Manchester Airport (MAN)
0800 042 0213
www.manchesterairport.co.uk

Newcastle International Airport (NCL)
0871 882 1121
www.newcastleairport.com

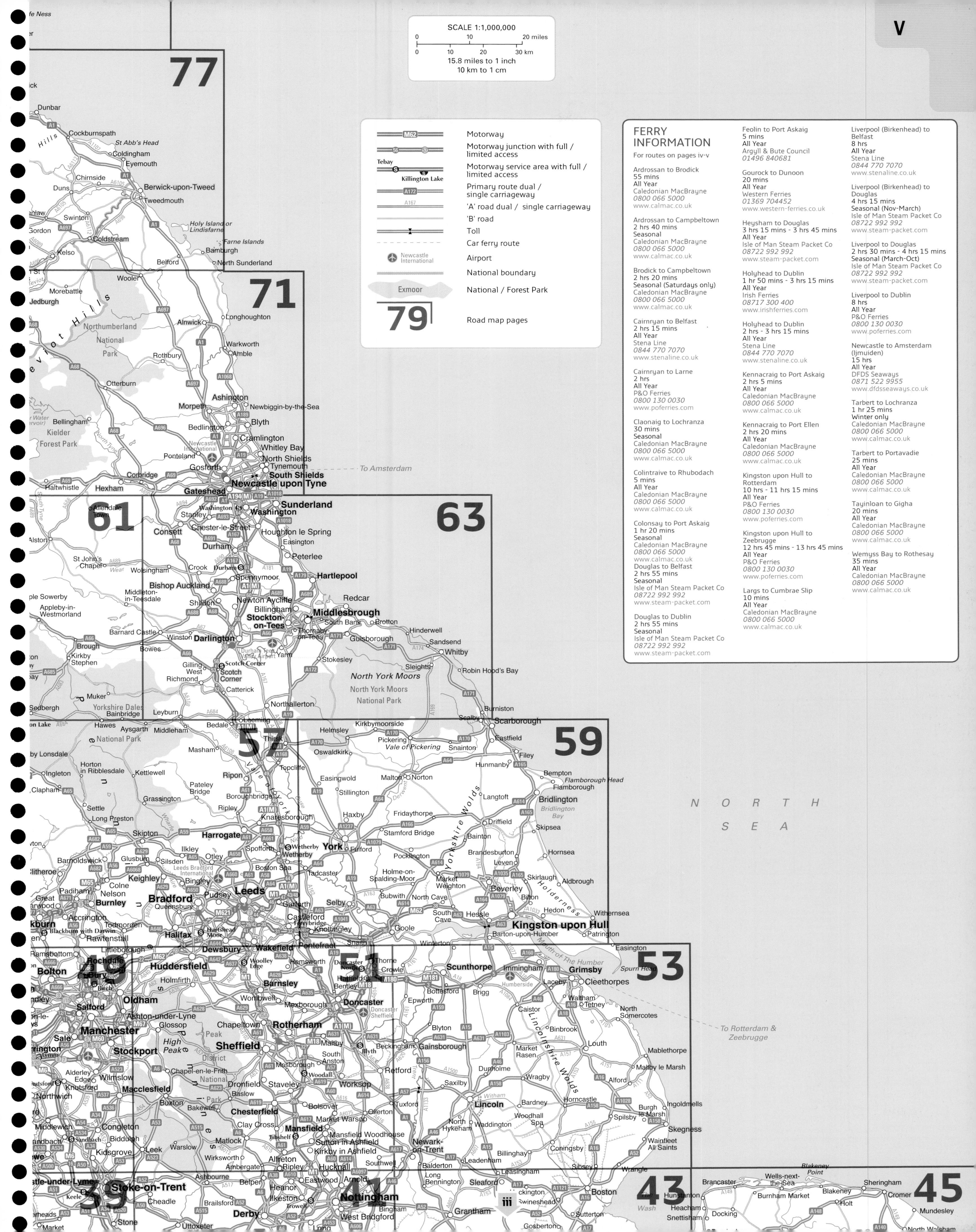

SCALE 1:1,000,000

0   10   20 miles
0   10   20   30 km
15.8 miles to 1 inch
10 km to 1 cm

Legend

M62	Motorway
	Motorway junction with full / limited access
Tebay S / S Killington Lake	Motorway service area with full / limited access
A172	Primary route dual / single carriageway
A167	'A' road dual / single carriageway
	'B' road
T	Toll
	Car ferry route
Newcastle International ✈	Airport
	National boundary
Exmoor	National / Forest Park
79	Road map pages

FERRY INFORMATION

For routes on pages iv-v

Ardrossan to Brodick
55 mins
All Year
Caledonian MacBrayne
0800 066 5000
www.calmac.co.uk

Ardrossan to Campbeltown
2 hrs 40 mins
Seasonal
Caledonian MacBrayne
0800 066 5000

Brodick to Campbeltown
2 hrs 20 mins
Seasonal (Saturdays only)
Caledonian MacBrayne
0800 066 5000
www.calmac.co.uk

Cairnryan to Belfast
2 hrs 15 mins
All Year
Stena Line
0844 770 7070
www.stenaline.co.uk

Cairnryan to Larne
2 hrs
All Year
P&O Ferries
0800 130 0030
www.poferries.com

Claonaig to Lochranza
30 mins
Seasonal
Caledonian MacBrayne
0800 066 5000
www.calmac.co.uk

Colintraive to Rhubodach
5 mins
All Year
Caledonian MacBrayne
0800 066 5000
www.calmac.co.uk

Colonsay to Port Askaig
1 hr 20 mins
Seasonal
Caledonian MacBrayne
0800 066 5000
www.calmac.co.uk

Douglas to Belfast
2 hrs 55 mins
Seasonal
Isle of Man Steam Packet Co
08722 992 992
www.steam-packet.com

Douglas to Dublin
2 hrs 55 mins
Seasonal
Isle of Man Steam Packet Co
08722 992 992
www.steam-packet.com

Feolin to Port Askaig
5 mins
All Year
Argyll & Bute Council
01496 840081

Gourock to Dunoon
20 mins
All Year
Western Ferries
01369 704452
www.western-ferries.co.uk

Heysham to Douglas
3 hrs 15 mins - 3 hrs 45 mins
All Year
Isle of Man Steam Packet Co
08722 992 992
www.steam-packet.com

Holyhead to Dublin
1 hr 50 mins - 3 hrs 15 mins
All Year
Irish Ferries
08717 300 400
www.irishferries.com

Holyhead to Dublin
2 hrs - 3 hrs 15 mins
All Year
Stena Line
0844 770 7070
www.stenaline.co.uk

Kennacraig to Port Askaig
2 hrs 5 mins
All Year
Caledonian MacBrayne
0800 066 5000
www.calmac.co.uk

Kennacraig to Port Ellen
2 hrs 20 mins
All Year
Caledonian MacBrayne
0800 066 5000
www.calmac.co.uk

Kingston upon Hull to Rotterdam
10 hrs - 11 hrs 15 mins
All Year
P&O Ferries
0800 130 0030
www.poferries.com

Kingston upon Hull to Zeebrugge
12 hrs 45 mins - 13 hrs 45 mins
All Year
P&O Ferries
0800 130 0030
www.poferries.com

Largs to Cumbrae Slip
10 mins
All Year
Caledonian MacBrayne
0800 066 5000
www.calmac.co.uk

Liverpool (Birkenhead) to Belfast
8 hrs
All Year
Stena Line
0844 770 7070
www.stenaline.co.uk

Liverpool (Birkenhead) to Douglas
4 hrs 15 mins
Seasonal (Nov-March)
Isle of Man Steam Packet Co
08722 992 992
www.steam-packet.com

Liverpool to Douglas
2 hrs 30 mins - 4 hrs 15 mins
Seasonal (March-Oct)
Isle of Man Steam Packet Co
08722 992 992
www.steam-packet.com

Liverpool to Dublin
8 hrs
All Year
P&O Ferries
0800 130 0030
www.poferries.com

Newcastle to Amsterdam (Ijmuiden)
15 hrs
All Year
DFDS Seaways
0871 522 9955
www.dfdsseaways.co.uk

Tarbert to Lochranza
1 hr 25 mins
Winter only
Caledonian MacBrayne
0800 066 5000
www.calmac.co.uk

Tarbert to Portavadie
25 mins
All Year
Caledonian MacBrayne
0800 066 5000
www.calmac.co.uk

Tayinloan to Gigha
20 mins
All Year
Caledonian MacBrayne
0800 066 5000
www.calmac.co.uk

Wemyss Bay to Rothesay
35 mins
All Year
Caledonian MacBrayne
0800 066 5000
www.calmac.co.uk

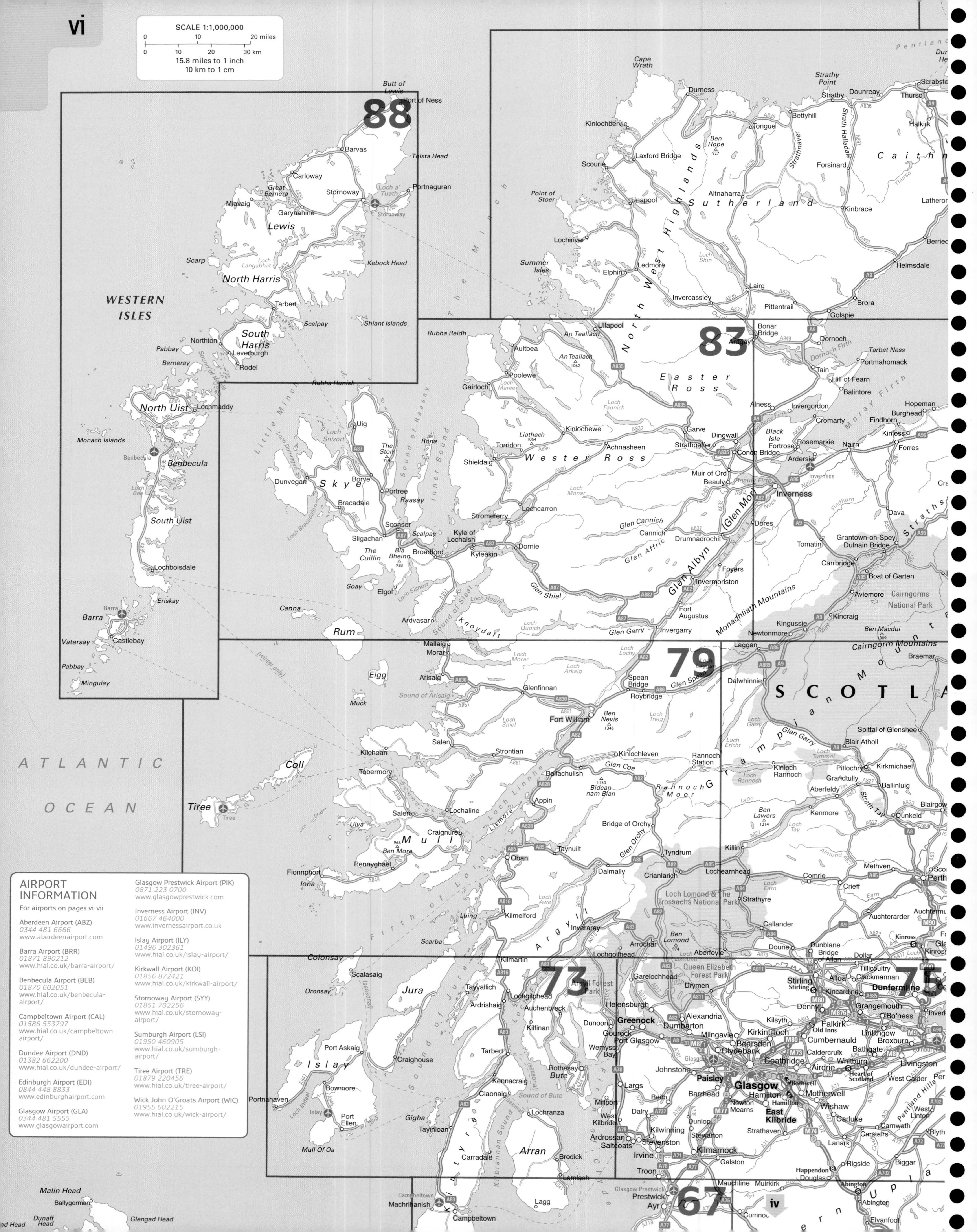

vi

88

83

79

73

75

67

SCALE 1:1,000,000

0   10   20 miles
0   10   20   30 km
15.8 miles to 1 inch
10 km to 1 cm

AIRPORT INFORMATION

For airports on pages vi–vii

Aberdeen Airport (ABZ)
0344 481 6666
www.aberdeenairport.com

Barra Airport (BRR)
01871 890212
www.hial.co.uk/barra-airport/

Benbecula Airport (BEB)
01870 602051
www.hial.co.uk/benbecula-airport/

Campbeltown Airport (CAL)
01586 553797
www.hial.co.uk/campbeltown-airport/

Dundee Airport (DND)
01382 662200
www.hial.co.uk/dundee-airport/

Edinburgh Airport (EDI)
0844 448 8833
www.edinburghairport.com

Glasgow Airport (GLA)
0344 481 5555
www.glasgowairport.com

Glasgow Prestwick Airport (PIK)
0871 223 0700
www.glasgowprestwick.com

Inverness Airport (INV)
01667 464000
www.invernessairport.co.uk

Islay Airport (ILY)
01496 302361
www.hial.co.uk/islay-airport/

Kirkwall Airport (KOI)
01856 872421
www.hial.co.uk/kirkwall-airport/

Stornoway Airport (SYY)
01851 702256
www.hial.co.uk/stornoway-airport/

Sumburgh Airport (LSI)
01950 460905
www.hial.co.uk/sumburgh-airport/

Tiree Airport (TRE)
01879 220456
www.hial.co.uk/tiree-airport/

Wick John O'Groats Airport (WIC)
01955 602215
www.hial.co.uk/wick-airport/

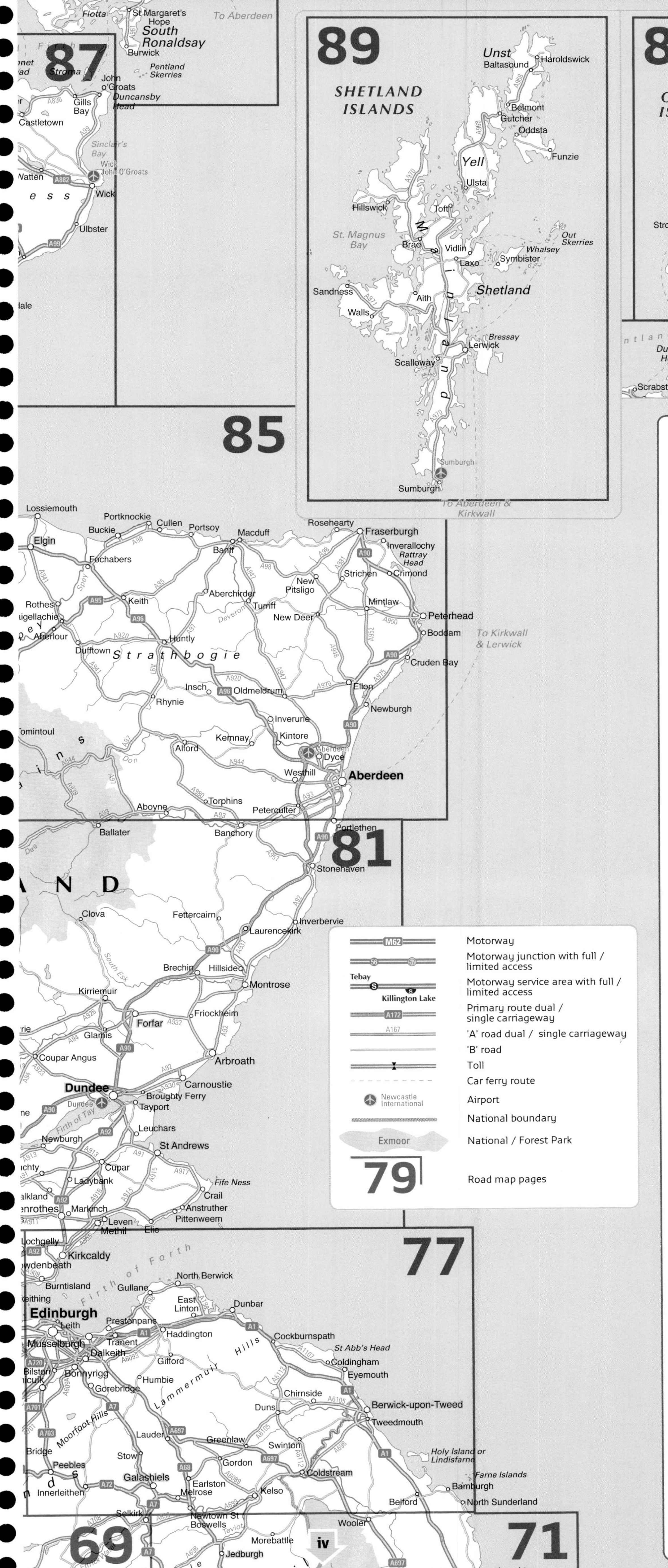

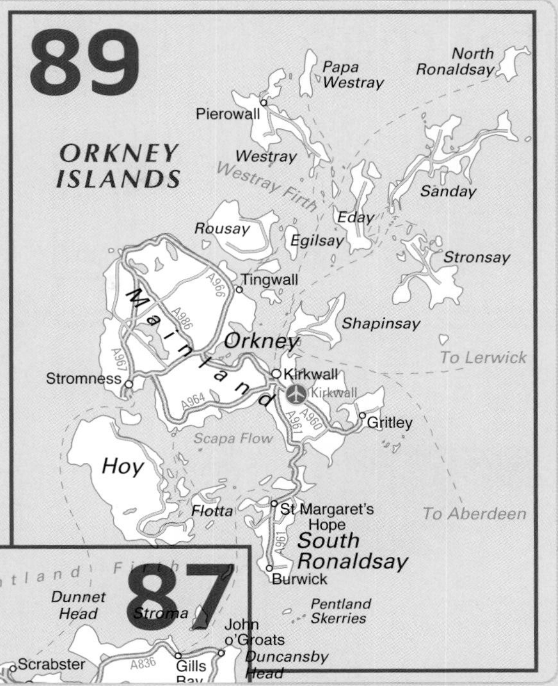

87

89

SHETLAND ISLANDS

89

ORKNEY ISLANDS

87

85

81

77

79

69

iv

71

Aberdeen

Dundee

Edinburgh

	Motorway
M62	Motorway junction with full / limited access
	Motorway service area with full / limited access
A172	Primary route dual / single carriageway
A167	'A' road dual / single carriageway
	'B' road
	Toll
	Car ferry route
Newcastle International	Airport
	National boundary
Exmoor	National / Forest Park
79	Road map pages

FERRY INFORMATION

For routes on pages vi-vii

Aberdeen to Kirkwall
6 hrs - 7 hrs 15 mins
All Year
North Link Ferries
0845 6000 449
www.northlinkferries.co.uk

Aberdeen to Lerwick
12 hrs 30 mins
All Year
North Link Ferries
0845 6000 449
www.northlinkferries.co.uk

Ardrossan to Brodick
55 mins
All Year
Caledonian MacBrayne
0800 066 5000
www.calmac.co.uk

Ardrossan to Campbeltown
2 hrs 40 mins
Seasonal
Caledonian MacBrayne
0800 066 5000
www.calmac.co.uk

Barra to Eriskay
40 mins
All Year
Caledonian MacBrayne
0800 066 5000
www.calmac.co.uk

Belmont to Gutcher
10 mins
All Year
Shetland Islands Council
01806 244200
www.shetland.gov.uk/ferries/

Belmont to Hamars Ness
30 mins
All Year
Shetland Islands Council
01806 244200
www.shetland.gov.uk/ferries/

Brodick to Campbeltown
2 hrs 20 mins
Seasonal (Saturdays only)
Caledonian MacBrayne
0800 066 5000
www.calmac.co.uk

Claonaig to Lochranza
30 mins
Seasonal
Caledonian MacBrayne
0800 066 5000
www.calmac.co.uk

Coll to Tiree
55 mins - 1 hr
All Year
Caledonian MacBrayne
0800 066 5000
www.calmac.co.uk

Colonsay to Port Askaig
1 hr 20 mins
Seasonal
Caledonian MacBrayne
0800 066 5000
www.calmac.co.uk

Cromarty to Nigg
5 mins
Seasonal
Highland Ferries
07468 417137

Eday to Sanday
20 mins
All Year
Orkney Ferries
01856 872044
www.orkneyferries.co.uk

Eday to Stronsay
35 mins
All Year
Orkney Ferries
01856 872044
www.orkneyferries.co.uk

Egilsay to Rousay
20 mins
All Year
Orkney Ferries
01856 872044
www.orkneyferries.co.uk

Egilsay to Wyre
15 mins
All Year
Orkney Ferries
01856 872044
www.orkneyferries.co.uk

Gill's Bay to St. Margaret's Hope
1 hr
All Year
Pentland Ferries
01856 831226
www.pentlandferries.co.uk

Glenelg to Kylerhea
5 mins
Seasonal
Skye Ferry
01599 522273
www.skyeferry.co.uk

Gourock to Dunoon
20 mins
All Year
Western Ferries
01369 704452
www.western-ferries.co.uk

Gutcher to Hamars Ness
25 mins
All Year
Shetland Islands Council
01806 244200
www.shetland.gov.uk/ferries/

Houton to Flotta
35 mins
All Year
Orkney Ferries
01856 872044
www.orkneyferries.co.uk

Houton to Lyness
35 mins
All Year
Orkney Ferries
01856 872044
www.orkneyferries.co.uk

Kennacraig to Port Askaig
2 hrs 5 mins
All Year
Caledonian MacBrayne
0800 066 5000
www.calmac.co.uk

Kennacraig to Port Ellen
2 hrs 20 mins
All Year
Caledonian MacBrayne
0800 066 5000
www.calmac.co.uk

Kirkwall to Eday
1 hr 15 mins
All Year
Orkney Ferries
01856 872044
www.orkneyferries.co.uk

Kirkwall to Lerwick
7 hrs 45 mins
All Year
North Link Ferries
0845 6000 449
www.northlinkferries.co.uk

Kirkwall to North Ronaldsay
2 hrs 40 mins
All Year
Orkney Ferries
01856 872044
www.orkneyferries.co.uk

Kirkwall to Papa Westray
1 hr 50 mins
All Year
Orkney Ferries
01856 872044
www.orkneyferries.co.uk

Kirkwall to Sanday
1 hr 25 mins
All Year
Orkney Ferries
01856 872044
www.orkneyferries.co.uk

Kirkwall to Shapinsay
45 mins
All Year
Orkney Ferries
01856 872044
www.orkneyferries.co.uk

Kirkwall to Stronsay
1 hr 35 mins
All Year
Orkney Ferries
01856 872044
www.orkneyferries.co.uk

Kirkwall to Westray
1 hr 25 mins
All Year
Orkney Ferries
01856 872044
www.orkneyferries.co.uk

Largs to Cumbrae Slip
10 mins
All Year
Caledonian MacBrayne
0800 066 5000
www.calmac.co.uk

Laxo to Symbister
30 mins
All Year
Shetland Islands Council
01806 244200
www.shetland.gov.uk/ferries/

Lerwick to Bressay
5 mins
All Year
Shetland Islands Council
01806 244200
www.shetland.gov.uk/ferries/

Lerwick to Kirkwall
5 hrs 30 mins
All Year
North Link Ferries
0845 6000 449
www.northlinkferries.co.uk

Lerwick to Skerries
2 hrs 30 mins
All Year
Shetland Islands Council
01806 244200
www.shetland.gov.uk/ferries/

Leverburgh to Berneray
1 hr
All Year
Caledonian MacBrayne
0800 066 5000
www.calmac.co.uk

Lochaline to Fishnish
15 mins
All Year
Caledonian MacBrayne
0800 066 5000
www.calmac.co.uk

Longhope to Flotta
30 mins
All Year
Orkney Ferries
01856 872044
www.orkneyferries.co.uk

Longhope to Lyness
30 mins
All Year
Orkney Ferries
01856 872044
www.orkneyferries.co.uk

Luing to Seil
5 mins
All Year
Argyll and Bute Council
01852 300382

Lyness to Flotta
20 mins
All Year
Orkney Ferries
01856 872044
www.orkneyferries.co.uk

Mallaig to Armadale
30 mins
All Year
Caledonian MacBrayne
0800 066 5000
www.calmac.co.uk

Mallaig to Lochboisdale
3 hrs 30 mins
All Year
Caledonian MacBrayne
0800 066 5000
www.calmac.co.uk

Oban to Castlebay
4 hrs 45 mins
All Year
Caledonian MacBrayne
0800 066 5000
www.calmac.co.uk

Oban to Coll
2 hrs 45 mins
All Year
Caledonian MacBrayne
0800 066 5000
www.calmac.co.uk

Oban to Colonsay
2 hrs 20 mins
All Year
Caledonian MacBrayne
0800 066 5000
www.calmac.co.uk

Oban to Craignure
45 mins
All Year
Caledonian MacBrayne
0800 066 5000
www.calmac.co.uk

Oban to Lismore
55 mins
All Year
Caledonian MacBrayne
0800 066 5000
www.calmac.co.uk

Oban to Lochboisdale
5 hrs 20 mins
Winter only
Caledonian MacBrayne
0800 066 5000
www.calmac.co.uk

Oban to Tiree
3 hrs 30 mins - 4 hrs 15 mins
All Year
Caledonian MacBrayne
0800 066 5000
www.calmac.co.uk

Rousay to Wyre
5 mins
All Year
Orkney Ferries
01856 872044
www.orkneyferries.co.uk

Sconser to Raasay
25 mins
All Year
Caledonian MacBrayne
0800 066 5000

Scrabster to Stromness
2 hr 15 mins
All Year
North Link Ferries
0845 6000 449
www.northlinkferries.co.uk

Tarbert to Lochranza
1 hr 25 mins
Winter only
Caledonian MacBrayne
0800 066 5000
www.calmac.co.uk

Tarbert to Portavadie
25 mins
All Year
Caledonian MacBrayne
0800 066 5000
www.calmac.co.uk

Tayinloan to Gigha
20 mins
All Year
Caledonian MacBrayne
0800 066 5000
www.calmac.co.uk

Tingwall to Rousay
25 mins
All Year
Orkney Ferries
01856 872044
www.orkneyferries.co.uk

Tobermory to Kilchoan
35 mins
All Year
Caledonian MacBrayne
0800 066 5000
www.calmac.co.uk

Toft to Ulsta
20 mins
All Year
Shetland Islands Council
01806 244200
www.shetland.gov.uk/ferries/

Uig to Lochmaddy
1 hr 45 mins
All Year
Caledonian MacBrayne
0800 066 5000
www.calmac.co.uk

Uig to Tarbert
1 hr 40 mins
All Year
Caledonian MacBrayne
0800 066 5000
www.calmac.co.uk

Ullapool to Stornoway
2 hrs 30 mins
All Year
Caledonian MacBrayne
0800 066 5000
www.calmac.co.uk

Vidlin to Skerries
1 hr 30 mins
All Year
Shetland Islands Council
01806 244200
www.shetland.gov.uk/ferries/

Vidlin to Symbister
45 mins
All Year
Shetland Islands Council
01806 244200
www.shetland.gov.uk/ferries/

Wemyss Bay to Rothesay
35 mins
All Year
Caledonian MacBrayne
0800 066 5000
www.calmac.co.uk

Westray to Papa Westray
40 mins - 1 hr 45 mins
All Year
Orkney Ferries
01856 872044
www.orkneyferries.co.uk

Wyre to Tingwall
45 mins
All Year
Orkney Ferries
01856 872044
www.orkneyferries.co.uk

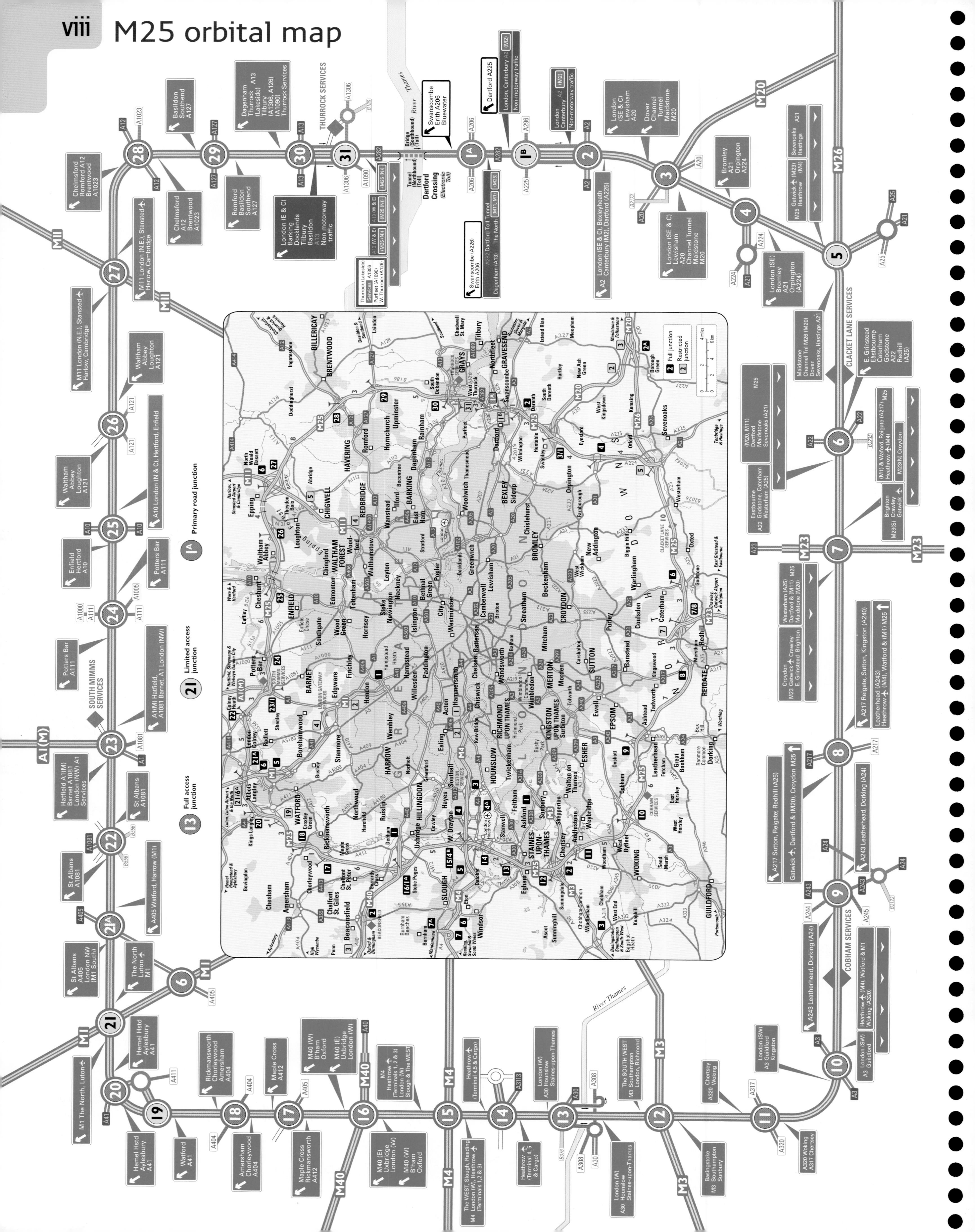

1A Primary road junction
21 Limited access junction
13 Full access junction
THURROCK SERVICES
SOUTH MIMMS SERVICES
CLACKET LANE SERVICES
COBHAM SERVICES
Dartford Crossing (Electronic Toll)
River Thames
Full junction
Restricted junction

Birmingham motorways map

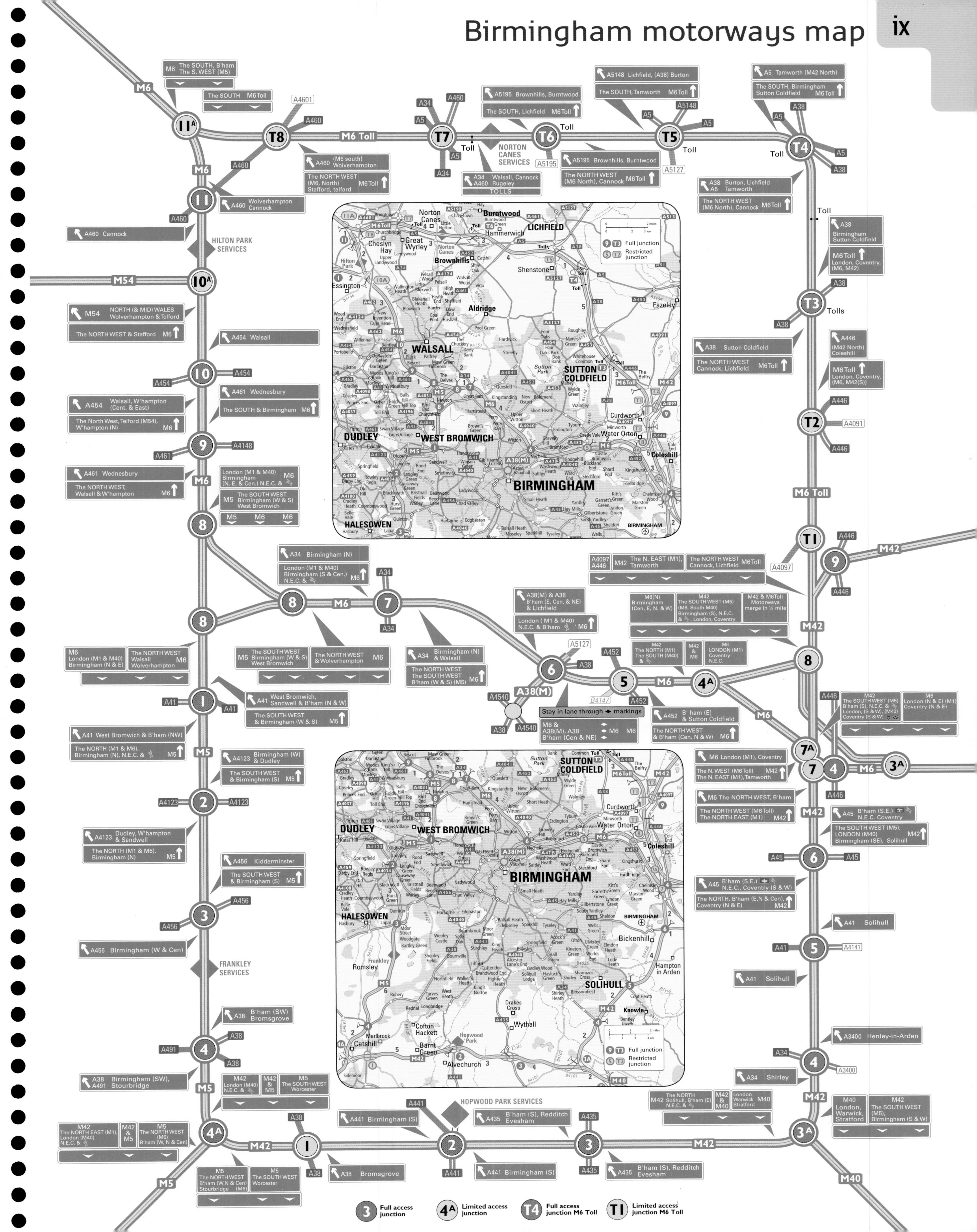

M6
M6 Toll
M54
M5
M42
M40
A38(M)
NORTON CANES SERVICES
TOLLS
HILTON PARK SERVICES
FRANKLEY SERVICES
HOPWOOD PARK SERVICES
The SOUTH, B'ham The S. WEST (M5)
The SOUTH M6Toll
A460 (M6 south) Wolverhampton
The NORTH WEST (M6, North) Stafford, telford M6Toll
A5195 Brownhills, Burntwood
The SOUTH, Lichfield M6Toll
A34 Walsall, Cannock A460 Rugeley
A5148 Lichfield, (A38) Burton
The SOUTH, Tamworth M6Toll
A5195 Brownhills, Burntwood
The NORTH WEST (M6 North), Cannock M6Toll
A5 Tamworth (M42 North)
The SOUTH, Birmingham Sutton Coldfield M6Toll
A38 Burton, Lichfield A5 Tamworth
The NORTH WEST (M6 North), Cannock M6Toll
A460 Wolverhampton Cannock
A460 Cannock
M54 NORTH (& MID) WALES Wolverhampton & Telford
The NORTH WEST & Stafford M6
A454 Walsall
A454 Walsall, W'hampton (Cent. & East)
The North West, Telford (M54), W'hampton (N) M6
A461 Wednesbury
The SOUTH & Birmingham M6
A461 Wednesbury
The NORTH WEST, Walsall & W'hampton M6
London (M1 & M40) Birmingham (N. E. & Cen.) N.E.C. M6
M5 The SOUTH WEST Birmingham (W & S) West Bromwich
A38 Birmingham Sutton Coldfield
M6Toll London, Coventry, (M6, M42)
A38 Sutton Coldfield
The NORTH WEST Cannock, Lichfield M6Toll
A446 (M42 North) Coleshill
M6Toll London, Coventry, (M6, M42(S))
A34 Birmingham (N)
London (M1 & M40) Birmingham (S & Cen.) N.E.C. M6
A38(M) & A38 B'ham (E, Cen, & NE) & Lichfield
London (M1 & M40) N.E.C. & B'ham M6
A4097 A446 M42 The N. EAST (M1), Tamworth
The NORTH WEST Cannock, Lichfield M6Toll
M6(N) Birmingham (Cen, E, N. & W)
M42 The SOUTH WEST (M5) (M6, South M40) Birmingham (S), N.E.C. London, Coventry
M42 & M6Toll Motorways merge in ¼ mile
M6 London (M1 & M40) Birmingham (N & E)
The NORTH WEST Walsall Wolverhampton M6
M5 The SOUTH WEST Birmingham (W & S) West Bromwich
The NORTH WEST & Wolverhampton M6
A34 Birmingham (N) & Walsall
The NORTH WEST The SOUTH WEST B'ham (W & S) (M5) M6
M42 The NORTH (M1) The SOUTH (M40)
M42 & M6
M6 LONDON (M1) Coventry N.E.C.
Stay in lane through markings
M6 & A38(M), A38 B'ham (Cen & NE) M6
A452 B'ham (E) & Sutton Coldfield
The NORTH WEST & B'ham (Cen. N & W) M6
M42 The SOUTH WEST (M5) B'ham (S), N.E.C. & London, (S, W), (M40) Coventry (S & W)
M6 London (N & E) (M1) Coventry (N & E)
A41 West Bromwich, Sandwell & B'ham (N & W)
The SOUTH WEST & Birmingham (W & S) M5
A41 West Bromwich & B'ham (NW)
The NORTH (M1 & M6), Birmingham (N), N.E.C. M5
A4123 Birmingham (W) & Dudley
The SOUTH WEST & Birmingham (S) M5
A4123 Dudley, W'hampton & Sandwell
The NORTH (M1 & M6), Birmingham (N) M5
A456 Kidderminster
The SOUTH WEST & Birmingham (S) M5
A456 Birmingham (W & Cen)
A38 B'ham (SW) Bromsgrove
A38 Birmingham (SW), A491 Stourbridge
M42 London (M40) N.E.C.
M42 & M5
M5 The SOUTH WEST Worcester
M42 The NORTH EAST (M1), London (M40) N.E.C.
M42 & M5
M5 The NORTH WEST (M6) B'ham (W, N & Cen)
M5 The NORTH WEST B'ham (W,N & Cen) Stourbridge (M6)
M5 The SOUTH WEST Worcester
A441 Birmingham (S)
A435 B'ham (S), Redditch Evesham
A38 Bromsgrove
A441 Birmingham (S)
A435 B'ham (S), Redditch Evesham
M6 London (M1), Coventry
The N. WEST (M6Toll) The N. EAST (M1), Tamworth M42
M6 The NORTH WEST, B'ham
The NORTH WEST (M6Toll) The NORTH EAST (M1) M42
A45 B'ham (S.E.) N.E.C. Coventry
The SOUTH WEST (M5), LONDON (M40) Birmingham (SE), Solihull M42
A45 B'ham (S.E.) N.E.C., Coventry (S & W)
The NORTH, B'ham (E,N & Cen), Coventry (N & E) M42
A41 Solihull
A41 Solihull
A3400 Henley-in-Arden
A34 Shirley
M42 The NORTH Solihull, B'ham (E) N.E.C.
M42 & M40
London Warwick Stratford M40
M40 London, Warwick, Stratford
M42 The SOUTH WEST (M5), Birmingham (S & W)
BIRMINGHAM
WALSALL
DUDLEY
WEST BROMWICH
SUTTON COLDFIELD
HALESOWEN
LICHFIELD
SOLIHULL
Full junction
Restricted junction
Full access junction
Limited access junction
Full access junction M6 Toll
Limited access junction M6 Toll

Motorway services information

All motorway service areas have fuel, food, toilets, disabled facilities and free short-term parking

For further information on motorway services providers:
Moto www.moto-way.com
RoadChef www.roadchef.com
Welcome Break www.welcomebreak.co.uk
Euro Garages www.eurogarages.com
Extra www.extraservices.co.uk
Westmorland www.westmorland.com

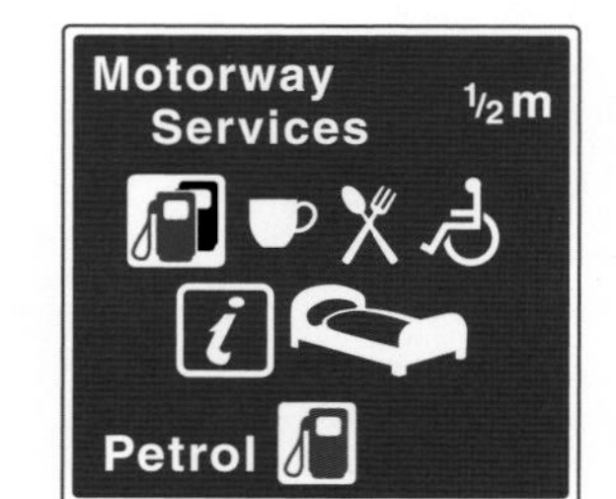

Facility columns (left to right): Information · Accommodation · Conference facilities · Showers · M&S Simply Food · Costa Coffee · Starbucks · Burger King · KFC · McDonalds · Wimpy

Motorway	Junction	Service provider	Service name	Fuel supplier
A1(M)	1	Welcome Break	South Mimms	BP
	10	Extra	Baldock	Shell
	17	Extra	Peterborough	Shell
	34	Moto	Blyth	Esso
	46	Moto	Wetherby	BP
	53	Moto	Scotch Corner	Esso
	61	RoadChef	Durham	Total
	64	Moto	Washington	BP
A74(M)	16	RoadChef	Annandale Water	BP
	22	Welcome Break	Gretna Green	BP
M1	2-4	Welcome Break	London Gateway	Shell
	11-12	Moto	Toddington	BP
	14-15	Welcome Break	Newport Pagnell	Shell
	15A	RoadChef	Northampton	BP
	16-17	RoadChef	Watford Gap	BP
	21-21A	Welcome Break	Leicester Forest East	BP
	22	Euro Garages	Markfield	BP
	23A	Moto	Donington Park	BP
	25-26	Moto	Trowell	BP
	28-29	RoadChef	Tibshelf	Shell
	30-31	Welcome Break	Woodall	Shell
	38-39	Moto	Woolley Edge	BP
M2	4-5	Moto	Medway	BP
M3	4A-5	Welcome Break	Fleet	Shell
	8-9	Moto	Winchester	Shell
M4	3	Moto	Heston	BP
	11-12	Moto	Reading	BP
	13	Moto	Chieveley	BP
	14-15	Welcome Break	Membury	BP
	17-18	Moto	Leigh Delamere	BP
	23A	RoadChef	Magor	Esso
	30	Welcome Break	Cardiff Gate	Total
	33	Moto	Cardiff West	Esso
	36	Welcome Break	Sarn Park	Shell
	47	Moto	Swansea	BP
	49	RoadChef	Pont Abraham	Texaco
M5	3-4	Moto	Frankley	BP
	8	RoadChef	Strensham (South)	BP
	8	RoadChef	Strensham (North)	Texaco
	11-12	Westmorland	Gloucester	Texaco
	13-14	Welcome Break	Michaelwood	BP
	19	Welcome Break	Gordano	Shell
	21-22	RoadChef	Sedgemoor (South)	Total
	21-22	Welcome Break	Sedgemoor (North)	Shell
	24	Moto	Bridgwater	BP
	25-26	RoadChef	Taunton Deane	Shell
	27	Moto	Tiverton	Shell
	28	Extra	Cullompton	Shell
	29-30	Moto	Exeter	BP
M6 Toll	T6-T7	RoadChef	Norton Canes	BP
M6	3-4	Welcome Break	Corley	Shell
	10-11	Moto	Hilton Park	BP
	14-15	RoadChef	Stafford (South)	Esso
	14-15	Moto	Stafford (North)	BP
	15-16	Welcome Break	Keele	Shell
	16-17	RoadChef	Sandbach	Esso
	18-19	Moto	Knutsford	BP
	20	Moto	Lymm	BP
	27-28	Welcome Break	Charnock Richard	Shell
	32-33	Moto	Lancaster	BP
	35A-36	Moto	Burton-in-Kendal (N)	BP
	36-37	RoadChef	Killington Lake (S)	BP
	38-39	Westmorland	Tebay	Total
	41-42	Moto	Southwaite	BP
	44-45	Moto	Todhills	BP/Shell
M8	4-5	BP	Heart of Scotland	BP
M9	9	Moto	Stirling	BP
M11	8	Welcome Break	Birchanger Green	Shell
M18	5	Moto	Doncaster North	BP
M20	8	RoadChef	Maidstone	Esso
	11	Stop 24	Stop 24	Shell
M23	11	Moto	Pease Pottage	BP
M25	5-6	RoadChef	Clacket Lane	Total
	9-10	Extra	Cobham	Shell
	23	Welcome Break	South Mimms	BP
	30	Moto	Thurrock	Esso
M27	3-4	RoadChef	Rownhams	Esso
M40	2	Extra	Beaconsfield	Shell
	8	Welcome Break	Oxford	BP
	10	Moto	Cherwell Valley	Esso
	12-13	Welcome Break	Warwick	BP
M42	2	Welcome Break	Hopwood Park	Shell
	10	Moto	Tamworth	Esso
M48	1	Moto	Severn View	BP
M54	4	Welcome Break	Telford	Shell
M56	14	RoadChef	Chester	Shell
M61	6-7	Euro Garages	Rivington	BP
M62	7-9	Welcome Break	Burtonwood	Shell
	18-19	Moto	Birch	BP
	25-26	Welcome Break	Hartshead Moor	Shell
	33	Moto	Ferrybridge	Esso
M65	4	Extra	Blackburn with Darwen	Shell
M74	4-5	RoadChef	Bothwell (South)	BP
	5-6	RoadChef	Hamilton (North)	BP
	11-12	Cairn Lodge	Happendon	Shell
	12-13	Welcome Break	Abington	Shell
M80	6-7	Shell	Old Inns	Shell
M90	6	Moto	Kinross	BP

There are a number of operators of motorway service areas in Britain; RoadChef, Welcome Break and Moto being the biggest three. All motorway service areas are required by law to provide fuel, free toilets and free short term parking 24 hours a day. Details of other facilities provided at each service area are shown opposite, although most of these will not be open 24 hours a day.

As part of its *Think, don't drive tired* road safety campaign the Government has the following tips for drivers:

- If you are feeling tired, opening the window or turning up the radio does not work, instead find a safe place to stop.

- On long journeys take a 15 minute break every 2 hours.

- If feeling tired, a 15 minute nap will help as will drinking 2 cups of coffee or other high caffeine drink. The most effective solution is to have some caffeine and then take a short sleep which gives the caffeine time to kick in.

- Avoid making long trips between midnight and 6am when you are most susceptible to sleepiness.

- Don't begin a journey if you are already feeling tired.

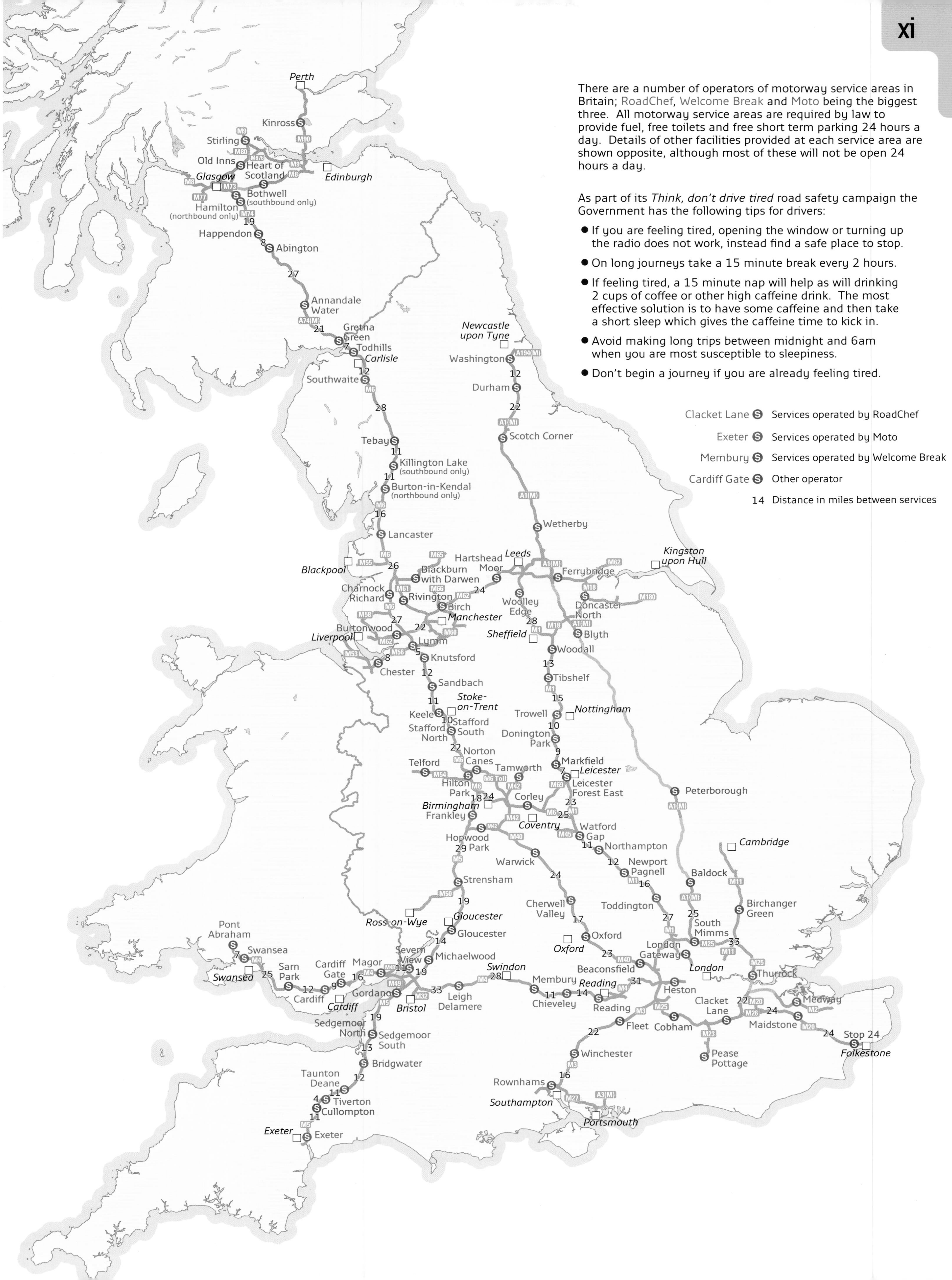

Clacket Lane Ⓢ Services operated by RoadChef

Exeter Ⓢ Services operated by Moto

Membury Ⓢ Services operated by Welcome Break

Cardiff Gate Ⓢ Other operator

14 Distance in miles between services

Perth
Kinross
M9 Stirling
M80 M90
Old Inns Heart of Scotland
M8 Glasgow M9 M8 Edinburgh
M73
M77 Bothwell (southbound only)
Hamilton (northbound only) M74
19
Happendon 8 Abington
27
Annandale Water
A74(M) 21 Gretha Green
7 Todhills
Carlisle Newcastle upon Tyne
12 Southwaite Washington A194(M)
M6 12
28 Durham
22
Tebay 11 A1(M)
Killington Lake (southbound only) Scotch Corner
11
Burton-in-Kendal (northbound only)
16
Lancaster A1(M) Wetherby
M6 M65 Hartshead Leeds Kingston upon Hull
Blackpool M55 26 Moor A1(M) M62
Blackburn with Darwen Ferrybridge
Charnock M61 M66 24 M18
Richard Rivington M62 Woolley Doncaster M180
M58 Birch Edge North
M6 27 Manchester 28 M18 A1(M) Blyth
Burtonwood 22 M60 Sheffield
Liverpool M62 Lymm M1 Woodall
M53 Knutsford 13
8 5
Chester 12 Tibshelf
Sandbach M1
11 15 Nottingham
Stoke-on-Trent Trowell
Keele 10 Donington 10
Stafford Stafford South Park 9 Markfield
North 22 Norton Leicester
Telford M6 Canes Tamworth 7 Leicester
M54 M6 Toll Leicester Forest East Peterborough
Hilton M6 M42 M69
Park 18 24 Corley 23 A1(M)
Birmingham M42 25
Frankley M1
Hopwood M42 Coventry Watford
Park 29 M40 Gap Northampton
Warwick M49 11
M5 12 Newport Baldock
Strensham Pagnell
24 16 Birchanger M11
M50 19 Toddington Green
Ross-on-Wye Cherwell 27 25 Cambridge
Gloucester Valley South A1(M)
Gloucester 17 Mimms M25 33
14 Oxford M11 M25
Pont Severn Oxford 23 London Birchanger
Abraham View Michaelwood London M40 Gateway Thurrock
7 Swansea Swindon Beaconsfield 31 Heston M25
M4 Magor 28 Membury Medway
Sarn Cardiff M48 M4 Reading Clacket M20
Swansea 25 Park Gate 16 11 14 Lane 22 M2
12 9 Gordano M4 Chieveley M3 M25 24
Cardiff M5 M32 33 Leigh Reading 22 Maidstone
Cardiff Bristol Delamere Fleet Cobham M20
19 M23 24 Stop 24
Sedgemoor Winchester Pease Folkestone
North Sedgemoor Pottage
13 South 16
Bridgwater
Taunton
Deane 12
4 Tiverton
11 Cullompton
11
Exeter M5 Exeter
Rownhams M27 A3(M)
Southampton
Portsmouth

Restricted motorway junctions

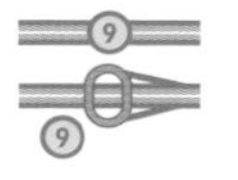

 Restricted motorway junctions are shown on the maps as

A1(M) LONDON TO NEWCASTLE

②
Northbound : No access
Southbound : No exit

③
Southbound : No access

⑤
Northbound : No exit
Southbound : No access
: No exit

㊵
Northbound : No exit to M62 Eastbound

㊸
Northbound : No exit to M1 Westbound

Dishforth
Southbound : No access from A168 Eastbound

㊼
Northbound : No access
: Exit only to A66(M) Northbound
Southbound : Access only from A66(M) Southbound
: No exit

�65
Northbound : No access from A1
Southbound : No exit to A1

A3(M) PORTSMOUTH

①
Northbound : No exit
Southbound : No access

④
Northbound : No access
Southbound : No access

A38(M) BIRMINGHAM

Victoria Road
Northbound : No exit
Southbound : No access

A48(M) CARDIFF

Junction with M4
Westbound : No access from M4 ㉙ Eastbound
Eastbound : No exit to M4 ㉙ Westbound

㉙A
Westbound : No exit to A48 Eastbound
Eastbound : No access from A48 Westbound

A57(M) MANCHESTER

Brook Street
Westbound : No exit
Eastbound : No access

A58(M) LEEDS

Westgate
Southbound : No access

Woodhouse Lane
Westbound : No exit

A64(M) LEEDS

Claypit Lane
Eastbound : No access

A66(M) DARLINGTON

Junction with A1(M)
Northbound : No access from A1(M) Southbound
: No exit
Southbound : No access
: No exit to A1(M) Northbound

A74(M) LOCKERBIE

⑱
Northbound : No access
Southbound : No exit

A167(M) NEWCASTLE

Campden Street
Northbound : No exit
Southbound : No access
: No exit

M1 LONDON TO LEEDS

②
Northbound : No exit
Southbound : No access

④
Northbound : No exit
Southbound : No access

⑥A
Northbound : Access only from M25 ㉑
: No exit
Southbound : No access
: Exit only to M25 ㉑

⑦
Northbound : Access only from A414
: No exit
Southbound : No access
: Exit only to A414

⑰
Northbound : No access
: Exit only to M45
Southbound : Access only from M45
: No exit

⑲
Northbound : Exit only to M6
Southbound : Access only from M6

㉑A
Northbound : No access
Southbound : No exit

㉓A
Northbound : No access from A453
Southbound : No exit to A453

㉔A
Northbound : No access
Southbound : No exit

�35A
Northbound : No access
Southbound : No exit

㊸
Northbound : No access
: Exit only to M621
Southbound : No access
: Access only from M621

㊽
Northbound : No exit to A1(M) Southbound
: Access only from A1(M) Northbound
Southbound : Exit only to A1(M) Southbound

M2 ROCHESTER TO CANTERBURY

①
Westbound : No exit to A2 Eastbound
Eastbound : No access from A2 Westbound

M3 LONDON TO WINCHESTER

⑧
Westbound : No access
Eastbound : No exit

⑩
Northbound : No access
Southbound : No exit

⑬
Southbound : No exit to A335 Eastbound
: No access

⑭
Westbound : No access
Eastbound : No exit

M4 LONDON TO SWANSEA

①
Westbound : No access from A4 Eastbound
Eastbound : No exit to A4 Westbound

②
Westbound : No access from A4 Eastbound
: No exit to A4 Eastbound
Eastbound : No access from A4 Westbound
: No exit to A4 Westbound

㉑
Westbound : No access from M48 Eastbound
Eastbound : No exit to M48 Westbound

㉓
Westbound : No exit to M48 Eastbound
Eastbound : No access from M48 Westbound

㉕
Westbound : No access
Eastbound : No exit

㉕A
Westbound : No access
Eastbound : No exit

㉙
Westbound : No access
Eastbound : No exit
: Exit only to A48(M)

㊳
Eastbound : Access only from A48(M) Eastbound
: No access

㊴
Westbound : No access
Eastbound : No access
: No exit

㊶
Westbound : No access
Eastbound : No access
: No exit

㊷
Westbound : No exit to A48
Eastbound : No access from A48

M5 BIRMINGHAM TO EXETER

⑩
Northbound : No exit
Southbound : No access

⑪A
Northbound : No access from A417 Eastbound
Southbound : No exit to A417 Westbound

M6 COVENTRY TO CARLISLE

Junction with M1
Northbound : No access from M1 ⑲ Southbound
Southbound : No exit to M1 ⑲ Northbound

③A
Northbound : No access from M6 Toll
Southbound : No exit to M6 Toll

④
Northbound : No exit to M42 Northbound
: No access from M42 Southbound
Southbound : No exit to M42
: No access from M42 Southbound

④A
Northbound : No access from M42 ⑧ Northbound
: No exit
Southbound : No access
: Exit only to M42 ⑧

⑤
Northbound : No access
Southbound : No exit

⑩A
Northbound : No access
: Exit only to M54
Southbound : Access only from M54
: No exit

⑪A
Northbound : No exit to M6 Toll
Southbound : No access from M6 Toll

㉔
Northbound : No exit
Southbound : No access

㉕
Northbound : No access
Southbound : No exit

㉚
Northbound : Access only from M61 Northbound
: No exit
Southbound : No access
: Exit only to M61 Southbound

㉛A
Northbound : No access
Southbound : No exit

M6 Toll BIRMINGHAM

T1
Northbound : Exit only to M42
: Access only from A4097
Southbound : No exit
: Access only from M42 Southbound

T2
Northbound : No exit
: No access

T5
Northbound : No exit
Southbound : No access

T7
Northbound : No access
Southbound : No exit

T8
Northbound : No access
Southbound : No exit

M8 EDINBURGH TO GLASGOW

⑥A
Westbound : No exit
Eastbound : No access

⑦
Westbound : No exit
Eastbound : No access

⑦A
Westbound : No access
Eastbound : No exit

⑧
Westbound : No access from M73 ② Southbound
: No access from A8 Eastbound
: No access from A89 Eastbound
Eastbound : No access from A89 Westbound
: No exit to M73 ② Northbound

⑨
Westbound : No exit
Eastbound : No access

⑬
Westbound : Access only from M80
Eastbound : Exit only to M80

⑭
Westbound : No exit
Eastbound : No access

⑯
Westbound : No access
Eastbound : No exit

⑰
Eastbound : Access only from A82, not central Glasgow
: Exit only to A82, not central Glasgow

⑱
Westbound : No access
Eastbound : No access

⑲
Westbound : Access only from A814 Eastbound
Eastbound : Exit only to A814 Westbound, not central Glasgow

⑳
Westbound : No access
Eastbound : No exit

㉑
Westbound : No exit
Eastbound : No access

㉒
Westbound : No access
: Exit only to M77 Southbound
Eastbound : Access only from M77 Northbound
: No exit

㉓
Westbound : No access
Eastbound : No exit

㉕A
Eastbound : No access
Westbound : No access

㉘
Westbound : No access
Eastbound : No exit

㉘A
Westbound : No access
Eastbound : No exit

M9 EDINBURGH TO STIRLING

②
Westbound : No exit
Eastbound : No access

③
Westbound : No access
Eastbound : No exit

⑥
Westbound : No access
Eastbound : No access

⑧
Westbound : No access
Eastbound : No exit

M11 LONDON TO CAMBRIDGE

④
Northbound : No access from A1400 Westbound
: No exit
Southbound : No access
: No exit to A1400 Eastbound

⑤
Northbound : No access
Southbound : No exit

⑧A
Northbound : No access
Southbound : No exit

⑨
Northbound : No access
Southbound : No exit

⑬
Northbound : No access
Southbound : No exit

⑭
Northbound : No access from A428 Eastbound
: No exit to A428 Westbound
: No exit to A1307
Southbound : No access from A428 Eastbound
: No access from A1307
: No exit

M20 LONDON TO FOLKESTONE

②
Westbound : No exit
Eastbound : No access

③
Westbound : No access
: Exit only to M26 Westbound
Eastbound : Access only from M26 Eastbound
: No exit

⑪A
Westbound : No exit
Eastbound : No access

M23 LONDON TO CRAWLEY

⑦
Northbound : No exit to A23 Southbound
Southbound : No access from A23 Northbound

⑩A
Southbound : No access from B2036
Northbound : No exit to B2036

M25 LONDON ORBITAL MOTORWAY

①B
Clockwise : No access
Anticlockwise : No access

⑤
Clockwise : No access from M26 Eastbound
Anticlockwise : No access from M26 Westbound

Spur of M25 ⑤
Clockwise : No access from M26 Westbound
Anticlockwise : No exit to M26 Eastbound

⑲
Clockwise : No access
Anticlockwise : No exit

㉑
Clockwise : No access from M1 ⑥A Northbound
: No exit to M1 ⑥A Southbound
Anticlockwise : No access from M1 ⑥A Northbound
: No exit to M1 ⑥A Southbound

㉛
Clockwise : No exit
Anticlockwise : No access

M26 SEVENOAKS

Junction with M25 ⑤
Westbound : No exit to M25 Anticlockwise
: No exit to M25 spur
Eastbound : No access from M25 Clockwise
: No access from M25 spur

Junction with M20
Westbound : No access from M20 ③ Eastbound
Eastbound : No exit to M20 ③ Westbound

M27 SOUTHAMPTON TO PORTSMOUTH

④ West
Westbound : No exit

④ East
Westbound : No access
Eastbound : No exit

⑩
Westbound : No exit
Eastbound : No access

⑫ West
Westbound : No access

⑫ East
Westbound : No access from A3
Eastbound : No exit

M40 LONDON TO BIRMINGHAM

③
Westbound : No access
Eastbound : No exit

⑦
Westbound : No access
Eastbound : No exit

⑧
Northbound : No access
Southbound : No exit

⑬
Northbound : No access
Southbound : No exit

⑭
Northbound : No access
Southbound : No exit

⑯
Northbound : No access
Southbound : No exit

M42 BIRMINGHAM

①
Northbound : No exit
Southbound : No access

⑦
Northbound : No access
: Exit only to M6 Northbound
Southbound : Access only from M6 Northbound
: No exit

⑦A
Northbound : No access
: Exit only to M6 Eastbound
Southbound : No access
: No exit

⑧
Northbound : Access only from M6 Southbound
: No exit
Southbound : Access only from M6 Southbound
: Exit only to M6 Northbound

M45 COVENTRY

Junction with M1
Westbound : No access from M1 ⑰ Southbound
Eastbound : No exit to M1 ⑰ Northbound

Junction with A45
Westbound : No access
Eastbound : No access

M48 CHEPSTOW

M4
Westbound : No exit to M4 Eastbound
Eastbound : No access from M4 Westbound

M49 BRISTOL

⑱A
Northbound : No access from M5 Southbound
Southbound : No access from M5 Northbound

M53 BIRKENHEAD TO CHESTER

⑪
Northbound : No access from M56 ⑮ Eastbound
: No exit to M56 ⑮ Westbound
Southbound : No access from M56 ⑮ Eastbound
: No exit to M56 ⑮ Westbound

M54 WOLVERHAMPTON TO TELFORD

Junction with M6
Westbound : No access from M6 ⑩A Southbound
Eastbound : No exit to M6 ⑩A Northbound

M56 STOCKPORT TO CHESTER

①
Westbound : No access from M60 Eastbound
: No access from A34 Northbound
Eastbound : No access from M60 Westbound
: No exit to A34 Southbound

②
Westbound : No access
Eastbound : No exit

③
Westbound : No exit
Eastbound : No access

④
Westbound : No exit
Eastbound : No access

⑦
Westbound : No exit
Eastbound : No access

⑧
Westbound : No exit
Eastbound : No access

⑨
Westbound : No exit to M6 Southbound
Eastbound : No access from M6 Northbound

⑮
Westbound : No access
: No access from M53 ⑪
Eastbound : No exit
: No exit to M53 ⑪

M57 LIVERPOOL

③
Northbound : No exit
Southbound : No access

⑤
Northbound : Access only from A580 Westbound
: No exit
Southbound : No access
: Exit only to A580 Eastbound

M58 LIVERPOOL TO WIGAN

①
Westbound : No access
Eastbound : No exit

M60 MANCHESTER

②
Westbound : No exit
Eastbound : No access

③
Westbound : No access from M56 ①
: No access from A34 Southbound
: No exit to A34 Northbound
Eastbound : No access from A34 Southbound
: No exit to M56 ①
: No exit to A34 Northbound

④
Westbound : No access
Eastbound : No exit to M56

⑤
Westbound : No access from A5103 Southbound
: No exit to A5103 Southbound
Eastbound : No access from A5103 Northbound
: No exit to A5103 Northbound

⑭
Westbound : No access from A580
: No exit to A580 Eastbound
Eastbound : No access from A580 Westbound
: No exit to A580

⑯
Westbound : No access
Eastbound : No exit

⑳
Westbound : No access
Eastbound : No exit

㉒
Westbound : No access

㉕
Westbound : No access

㉖
Westbound : No access
: No exit

㉗
Westbound : No exit
Eastbound : No access

M61 MANCHESTER TO PRESTON

②
Northbound : No access from A580 Eastbound
: No access from A666
Southbound : No exit to A580 Westbound

③
Northbound : No access from A580 Eastbound
: No access from A666
Southbound : No exit to A580 Westbound

Junction with M6
Northbound : No exit to M6 ㉚ Southbound
Southbound : No access from M6 ㉚ Northbound

M62 LIVERPOOL TO HULL

㉓
Westbound : No access
Eastbound : No access

㉜A
Westbound : No exit to A1(M) Southbound

M65 BURNLEY

⑨
Westbound : No exit
Eastbound : No access

⑪
Westbound : No access
Eastbound : No exit

M66 MANCHESTER TO EDENFIELD

①
Northbound : No access
Southbound : No exit

Junction with A56
Northbound : Exit only to A56 Northbound
Southbound : Access only from A56 Southbound

M67 MANCHESTER

①
Westbound : No exit
Eastbound : No access

②
Westbound : No access
Eastbound : No exit

M69 COVENTRY TO LEICESTER

②
Northbound : No exit
Southbound : No access

M73 GLASGOW

①
Northbound : No access from A721 Eastbound
Southbound : No exit to A721 Eastbound

②
Northbound : No access from M8 ⑧ Eastbound
Southbound : No exit to M8 ⑧ Westbound

M74 GLASGOW

①
Westbound : No exit to M8 Kingston Bridge
Eastbound : No access from M8 Kingston Bridge

③
Westbound : No access
Eastbound : No exit

③A
Westbound : No exit
Eastbound : No access

⑦
Westbound : No access
Eastbound : No exit

⑨
Northbound : No exit
: No access

⑩
Northbound : No exit
Southbound : No access

⑪
Northbound : No access
Southbound : No exit

⑫
Northbound : Access only from A70 Northbound
Southbound : Exit only to A70 Southbound

M77 GLASGOW

Junction with M8
Northbound : No exit to M8 ㉒ Westbound
Southbound : No access from M8 ㉒ Eastbound

④
Northbound : No access
Southbound : No access

⑥
Northbound : No exit to A77
Southbound : No access from A77

⑦
Northbound : No access
: No exit

⑧
Northbound : No access
Southbound : No access

M80 STIRLING

④A
Northbound : No access
Southbound : No exit

⑥A
Northbound : No exit
Southbound : No access

⑧
Northbound : No access from M876
Southbound : No exit to M876

M90 EDINBURGH TO PERTH

①
Northbound : No exit to A90

②A
Northbound : No access
Southbound : No exit

⑦
Northbound : No exit
Southbound : No access

⑧
Northbound : No access
Southbound : No exit

⑩
Northbound : No access from A912
: No exit to A912 Southbound
Southbound : No access from A912 Northbound
: No exit to A912

M180 SCUNTHORPE

①
Westbound : No exit
Eastbound : No access

M606 BRADFORD

Straithgate Lane
Northbound : No access

M621 LEEDS

②A
Northbound : No exit
Southbound : No access

⑤
Northbound : No access
Southbound : No exit

⑥
Northbound : No exit
Southbound : No access

M876 FALKIRK

Junction with M80
Westbound : No exit to M80 ⑧ Northbound
Eastbound : No access from M80 ⑧ Southbound

Junction with M9
Westbound : No access
Eastbound : No exit

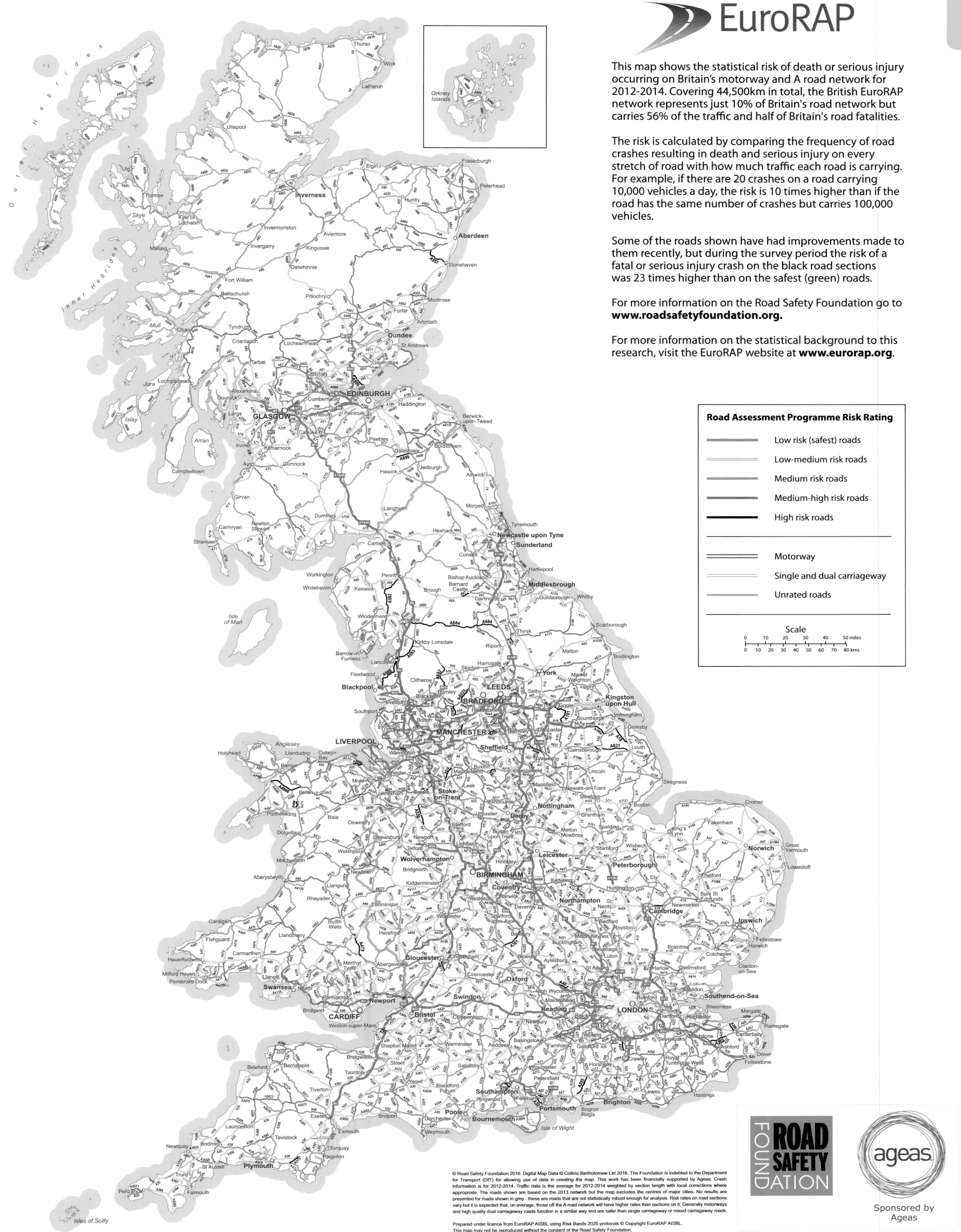

EuroRAP

This map shows the statistical risk of death or serious injury occurring on Britain's motorway and A road network for 2012-2014. Covering 44,500km in total, the British EuroRAP network represents just 10% of Britain's road network but carries 56% of the traffic and half of Britain's road fatalities.

The risk is calculated by comparing the frequency of road crashes resulting in death and serious injury on every stretch of road with how much traffic each road is carrying. For example, if there are 20 crashes on a road carrying 10,000 vehicles a day, the risk is 10 times higher than if the road has the same number of crashes but carries 100,000 vehicles.

Some of the roads shown have had improvements made to them recently, but during the survey period the risk of a fatal or serious injury crash on the black road sections was 23 times higher than on the safest (green) roads.

For more information on the Road Safety Foundation go to **www.roadsafetyfoundation.org**.

For more information on the statistical background to this research, visit the EuroRAP website at **www.eurorap.org**.

Road Assessment Programme Risk Rating

	Low risk (safest) roads
	Low-medium risk roads
	Medium risk roads
	Medium-high risk roads
	High risk roads
	Motorway
	Single and dual carriageway
	Unrated roads

Scale

0 10 20 30 40 50 miles
0 10 20 30 40 50 60 70 80 kms

ROAD SAFETY FOUNDATION

ageas

Sponsored by Ageas

Stansted

Manchester

Heathrow

Glasgow

Gatwick

Birmingham

Key to map symbols   🅿 Short stay car park   🅿 Mid stay car park   🅿 Long stay car park   🅿 Other car park   Airport terminal building

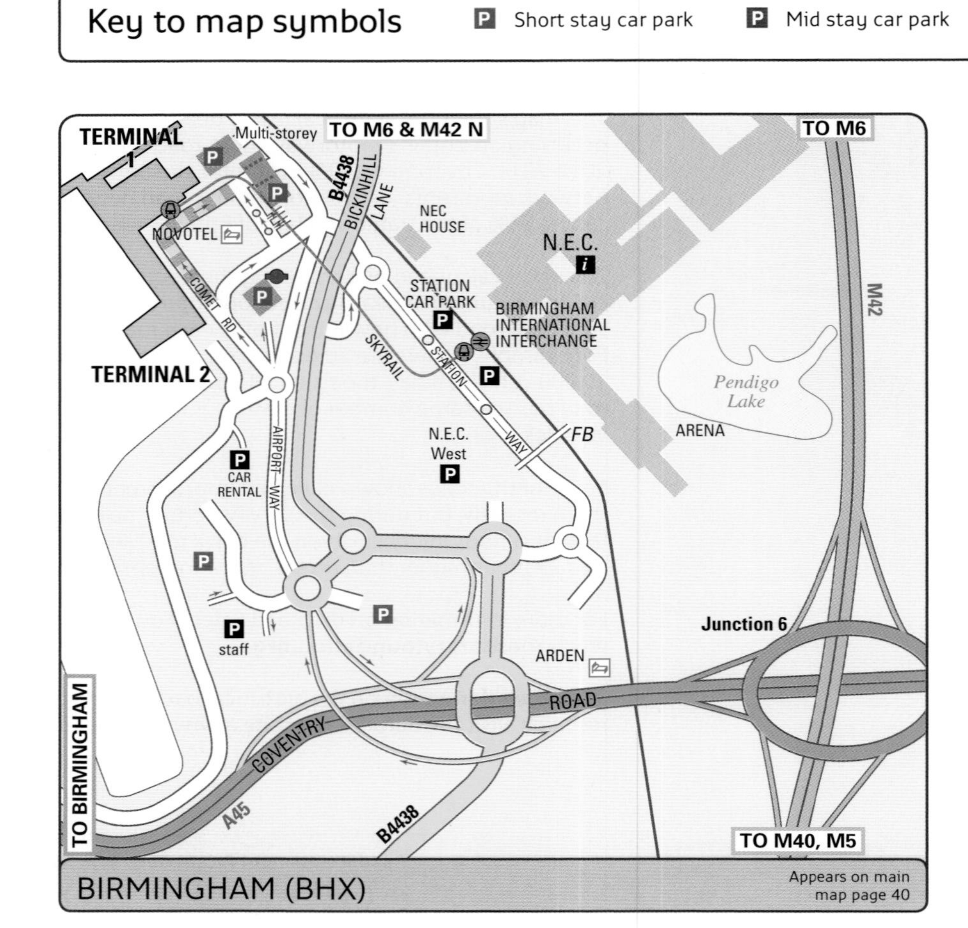

BIRMINGHAM (BHX)

Appears on main map page 40

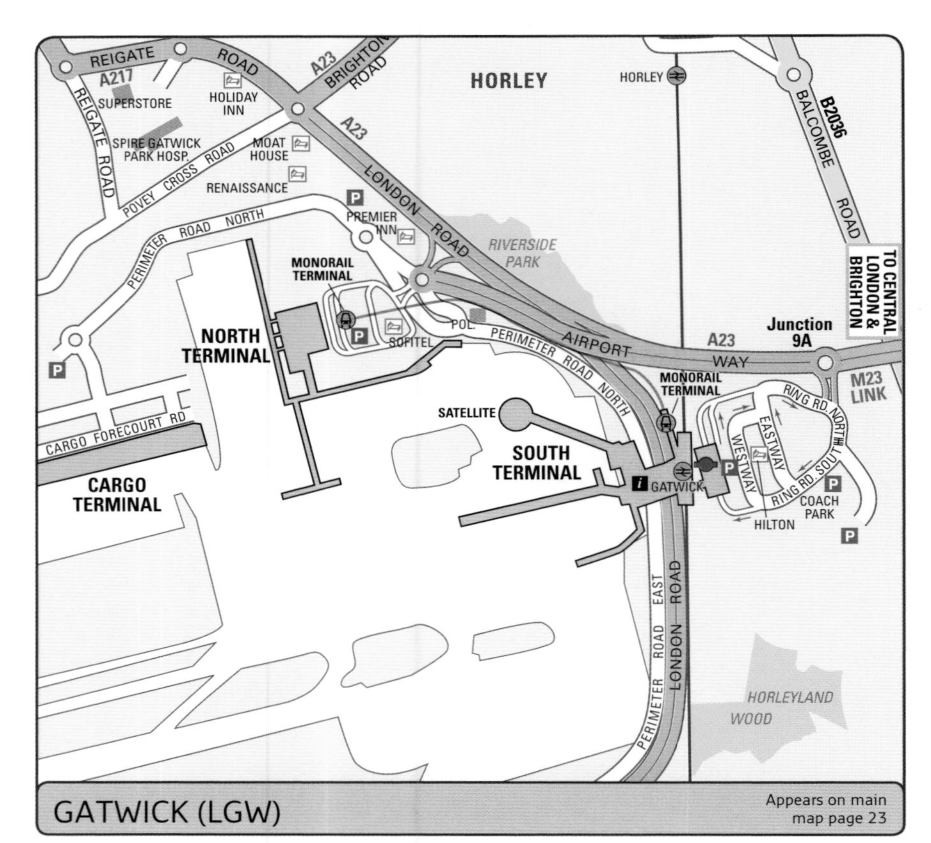

GATWICK (LGW)

Appears on main map page 23

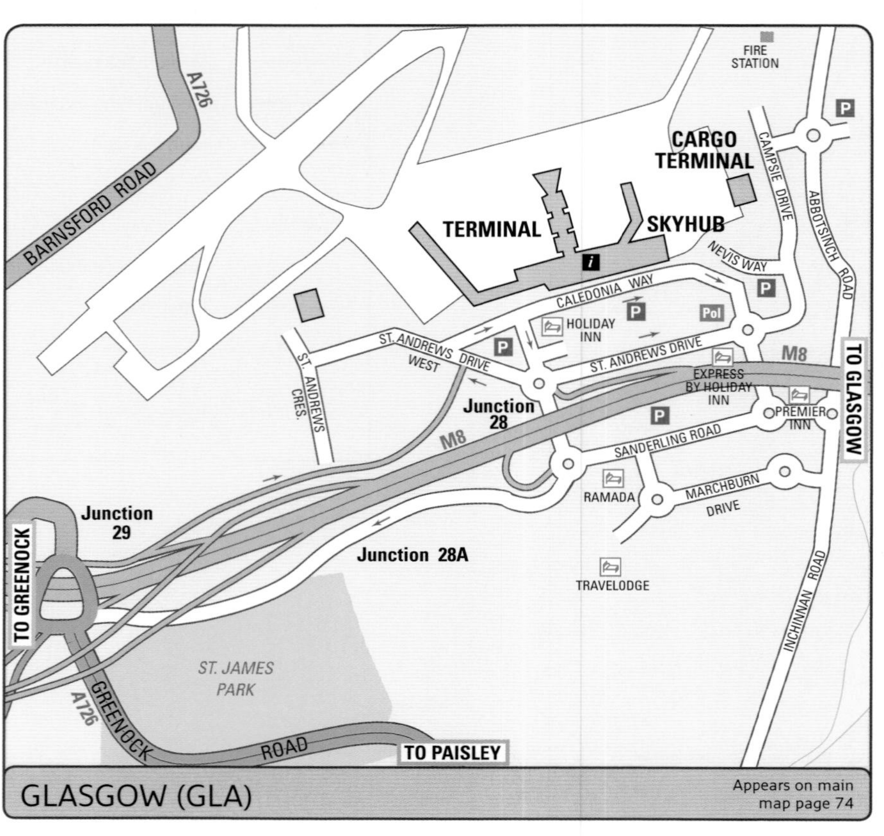

GLASGOW (GLA)

Appears on main map page 74

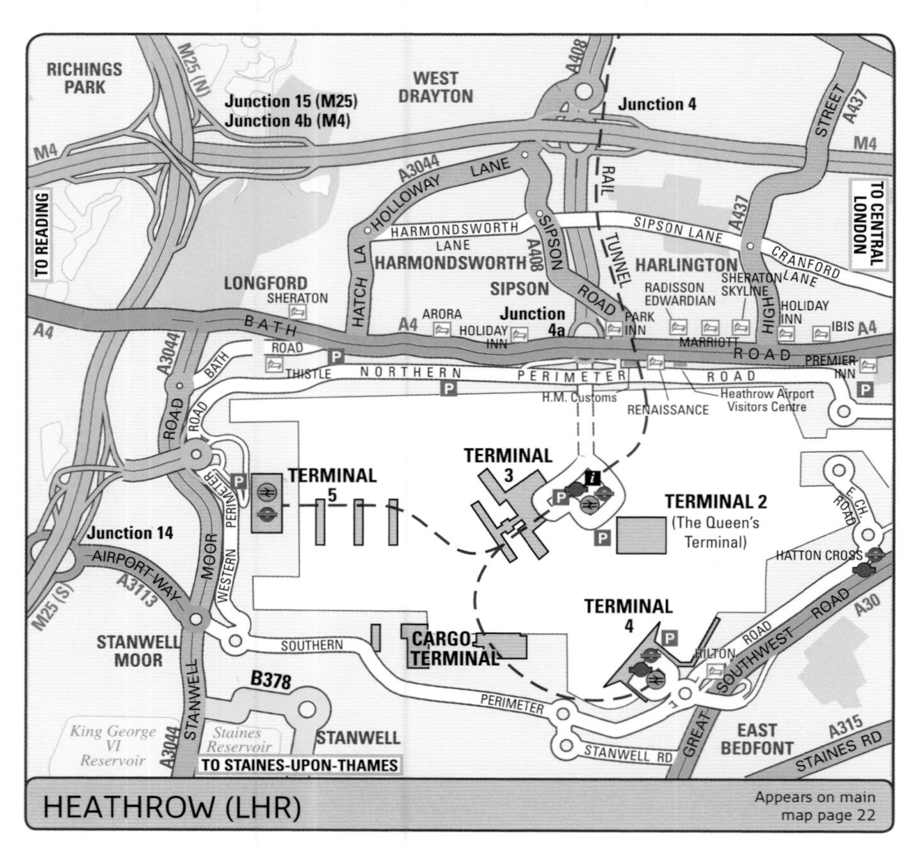

HEATHROW (LHR)

Appears on main map page 22

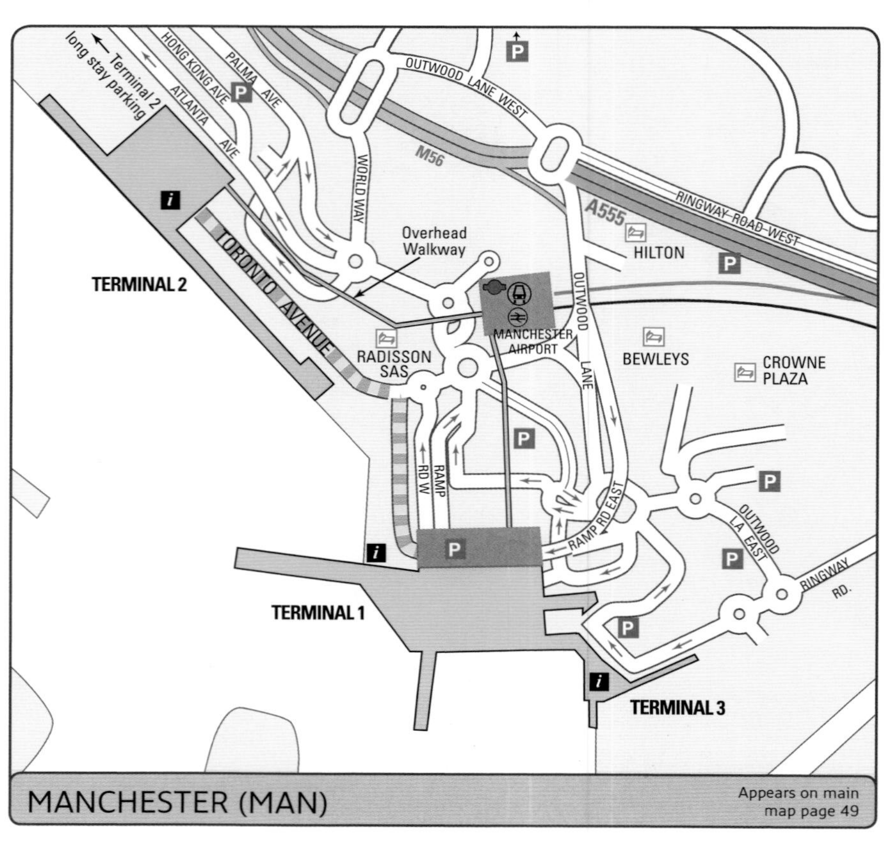

MANCHESTER (MAN)

Appears on main map page 49

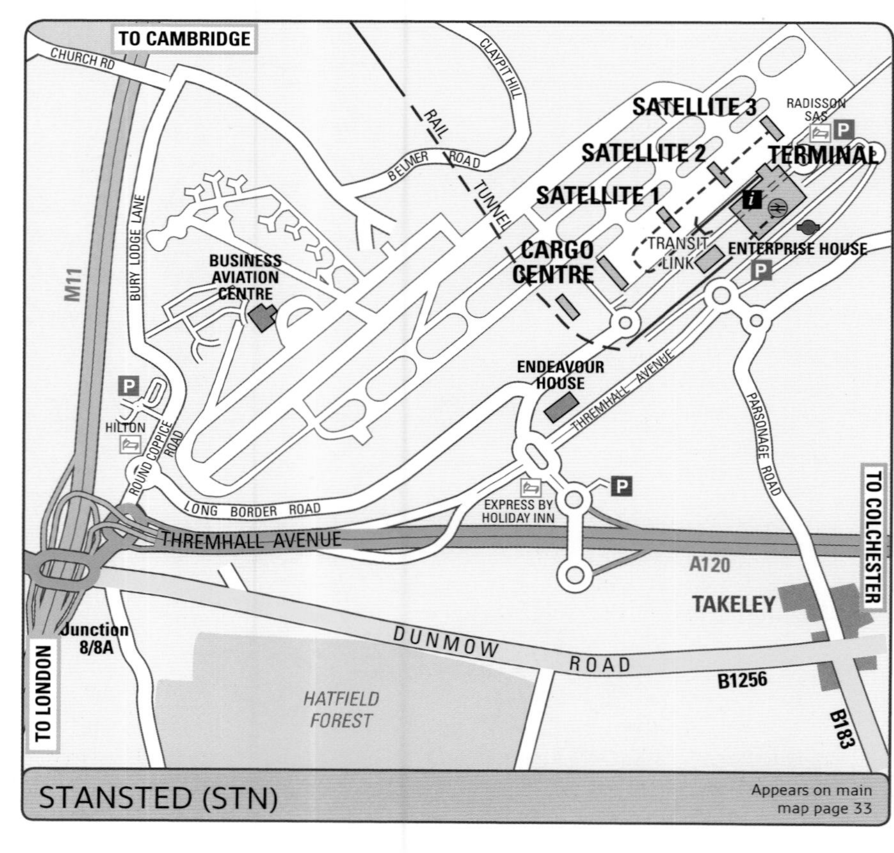

STANSTED (STN)

Appears on main map page 33

Map scale

A scale bar appears at the bottom of every page to help with measurements.

```
0        2        4        6 miles
0    2    4    6    8    10 km
```

England, Wales & Southern Scotland are at a scale of 1:200,000 or 3.2 miles to 1 inch
Northern Scotland is at a scale of 1:316,800 or 5 miles to 1 inch.
Orkney & Shetland are at a scale of 1:411,840 or 6.5 miles to 1 inch

Symbols used on the map

M5	Motorway			Fixed safety camera / fixed average-speed safety camera. Speed shown by number within camera, a V indicates a variable limit.
M6Toll	Toll motorway			
8 9	Motorway junction with full / limited access (in congested areas there is just a numbered symbol)		✈ ✈	Airport with / without scheduled services
Maidstone Birch Sarn	Motorway service area with off road / full / limited access		Ⓗ	Heliport
A556	Primary route dual / single carriageway		P&R P&R	Park and Ride site operated by bus / rail (runs at least 5 days a week)
S	24 hour service area on primary route			Built up area
Peterhead	Primary route destination Primary route destinations are places of major traffic importance linked by the primary route network. They are shown on a green background on direction signs.		▫ ▫ ▫	Town / Village / Other settlement
A30	'A' road dual / single carriageway		**Hythe**	Seaside destination
B1403	'B' road dual / single carriageway		-------	National boundary
	Minor road		**KENT**	County / Unitary Authority boundary and name
=========	Road with restricted access			Heritage Coast
▪▪▪▪▪	Roads with passing places			National Park
	Road proposed or under construction			Regional / Forest Park boundary
	Multi-level junction with full / limited access (with junction number)			Woodland
	Roundabout		Danger Zone	Military range
4	Road distance in miles between markers		468 ▲941	Spot / Summit height (in metres)
	Road tunnel			Lake / Dam / River / Waterfall
	Steep hill (arrows point downhill)			Canal / Dry canal / Canal tunnel
Toll / Electronic Toll	Toll / Electronic Toll			Lighthouse
×	Level crossing			Beach
St. Malo 8hrs	Car ferry route with journey times		SEE PAGE 91	Area covered by urban area map
	Railway line /station / tunnel		90	National Grid reference figures
Wales Coast Path	National Trail / Long Distance Route		**SY**	National Grid reference letters

Places of interest

A selection of tourist detail is shown on the mapping. It is advisable to check with the local tourist information centre regarding opening times and facilities available.

Any of the following symbols may appear on the map in maroon ★ which indicates that the site has World Heritage status.

ℹ	Tourist information centre (open all year)	⚽	Major football club
ℹ	Tourist information centre (open seasonally)	£	Major shopping centre / Outlet village
m	Ancient monument		Major sports venue
	Aquarium		Motor racing circuit
	Aqueduct / Viaduct		Mountain bike trail
	Arboretum	🏛	Museum / Art gallery
⚔ 1643	Battlefield		Nature reserve (NNR indicates a National Nature Reserve)
	Blue flag beach		Racecourse
⛺ 🚐	Camp site / Caravan site		Rail Freight Terminal
	Castle		Ski slope (artificial / natural)
	Cave		Spotlight nature reserve (Best sites for access to nature)
	Country park		Steam railway centre / preserved railway
	County cricket ground		Surfing beach
	Distillery		Theme park
✝	Ecclesiastical feature		University
	Event venue		Vineyard
	Farm park		Wildlife park / Zoo
❀	Garden		Wildlife Trust nature reserve
	Golf course	★	Other interesting feature
	Historic house	(NT) (NTS)	National Trust / National Trust for Scotland property
	Historic ship		

Reading our maps

Safety Camera The number inside the camera shows the speed limit at the camera location.

Multi-level junctions Non-motorway junctions where slip roads are used to access the main roads.

Distances Blue numbers give distances in miles between junctions shown with a blue marker.

Park & Ride Sites are shown that operate at least 5 days a week. Bus operated sites have a yellow symbol and rail operated sites a pink symbol.

Motorway service area

World Heritage site Places of interest defined by UNESCO as special on a world scale.

Places of interest Blue symbols indicate places of interest. See the section at the bottom of the page for the different types of feature represented on the map.

More detailed maps Green boxes indicate busy built-up-areas. More detailed mapping is available.

Map pages

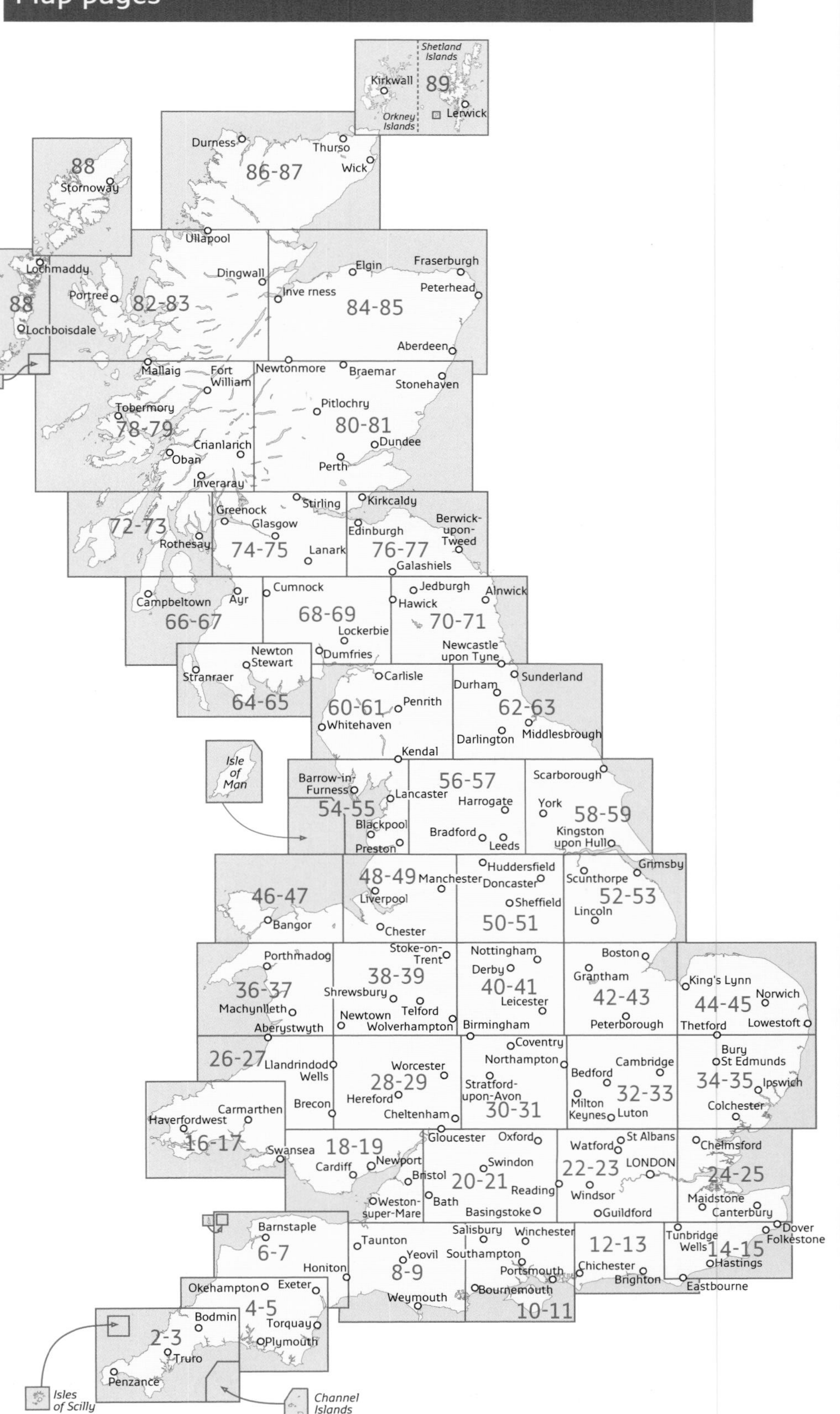

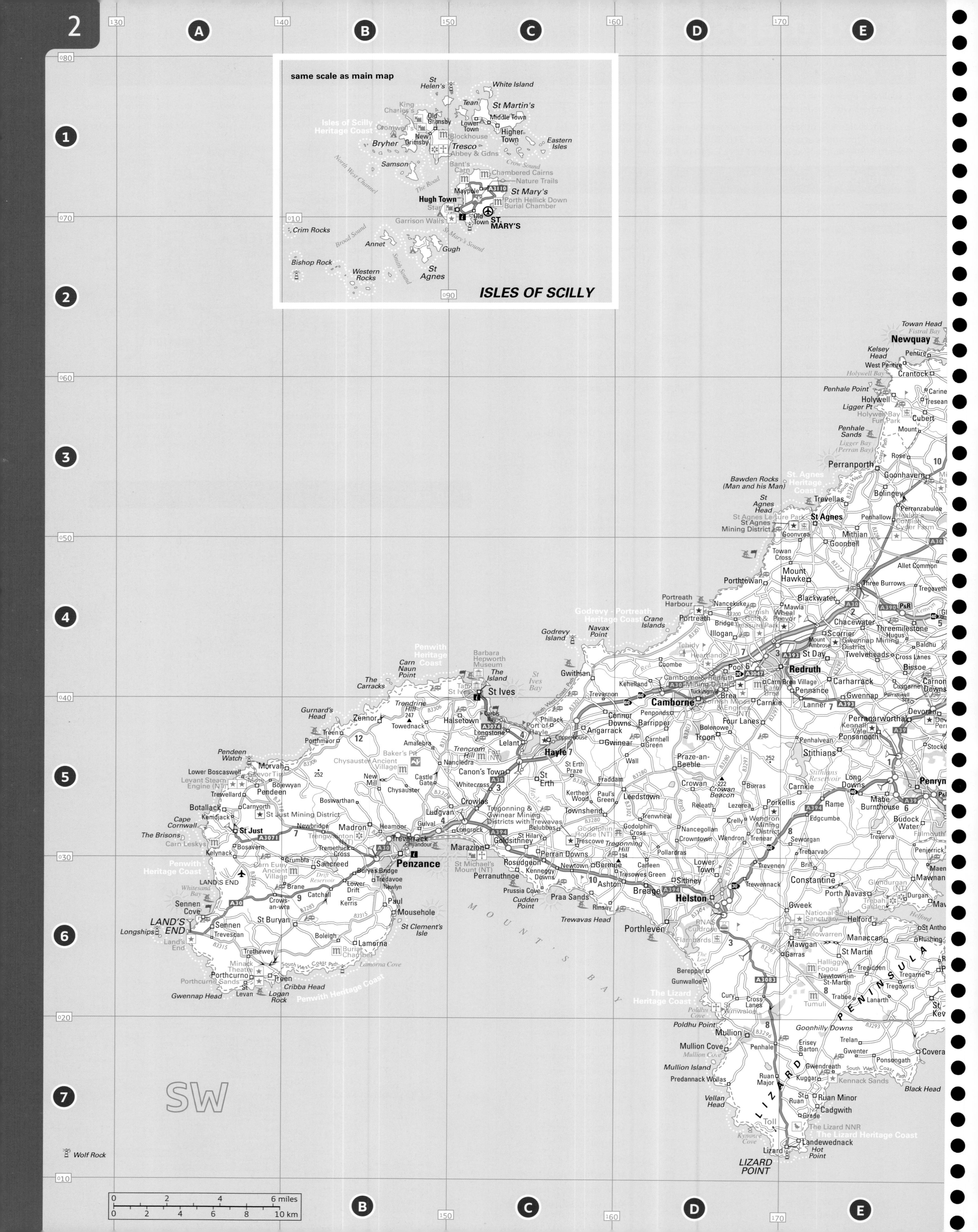

ISLES OF SCILLY (inset map)

same scale as main map

St Helen's
White Island
Tean
King Charles's
Isles of Scilly Heritage Coast
Old Grimsby
Cromwell's
St Martin's
Middle Town
Lower Town
Higher Town
New Grimsby
Blockhouse
Bryher
Tresco
Abbey & Gdns
Eastern Isles
Samson
Crow Sound
Bant's Carn
Chambered Cairns
Nature Trails
Maypole
St Mary's
Porth Hellick Down
Burial Chamber
Hugh Town
St Mary's
Star
ST. MARY'S
Garrison Walls
Old Town
Crim Rocks
Broad Sound
Bishop Rock
Annet
Gugh
Western Rocks
St Agnes
St Mary's Sound
South Sound

ISLES OF SCILLY

SW

Wolf Rock

Towan Head
Fistral Bay
Newquay
Kelsey Head
Pentire
West Pentire
Crantock
Holywell Bay
Penhale Point
Holywell
Ligger Pt
Tresean
Cubert
Holywell Bay Fun Park
Mount
Carine
Penhale Sands
Ligger Bay (Perran Bay)
Perranporth
Rose
Goonhavern
10
Bawden Rocks (Man and his Man)
St Agnes Heritage Coast
Trevellas
Bolingey
Perranzabuloe
Penhallow
Cornish Cyder Farm
St Agnes Head
St Agnes
St Agnes Leisure Park
St Agnes Mining District
Mithian
Goonbell
Goonvrea
Towan Cross
Allet Common
Mount Hawke
Blackwater
Three Burrows
Tregavethan
A30
Porthtowan
Nancekuke
Mawla
Wheal Peevor
Chacewater
Threemilestone
5
Godrevy - Portreath Heritage Coast
Crane Islands
Portreath
Cornish Gold & Treasure Park
Scorrier
Gwennap Mining District
Godrevy Island
Navax Point
Illogan
Pool
Baldhu
Redruth
St Day
Twelveheads
Cross Lanes
Bissoe
Gwithian
Tehidy Heartland
Camborne Redruth Mining District
Carn Brea Village
Carharrack
Carnon Downs
Barbara Hepworth Museum
Trevarnon
Keheland
Camborne
Pennance
Gwennap
Perranarworthal
The Island
Camborne Redruth
Connor Downs
Penponds
Four Lanes
Lanner
A393
Devoran
St Ives
Carrack Gladden
St Ives Bay
Phillack
Port of Hayle
Angarrack
Carnhell Green
Troon
Bolenowe
Penhalvean
Stithians
Ponsanooth
Trendrine Hill 247
Carbis Bay
Copperhouse
Gwinear
Wall
Praze-an-Beeble
Burras
Carnkie
Stithians Reservoir
Zennor
Halsetown
Longstone
Lelant
Hayle
St Erth Praze
Crowan
Crowan Beacon
Long Downs
Penryn
Gurnard's Head
Trendrine Hill
Towednack
Nancledra
St Erth
Fraddam
Kerthen Wood
Leedstown
Porkellis
Carnkie
Rame
Mabe Burnthouse
Pen
Penwith Heritage Coast
Carn Naun Point
The Carracks
Treen
Porthmeor
Amalebra
New Mill
Canon's Town
Paul's Green
Townshend
Releath
Lezerea
Wendron Mining District
Edgcumbe
Trenear
Treverva
Carnyorth
Castle Gate
Trenwheal
Nancegollan
Godolphin Cross
Sewargan
Treverbyn
Pendeen Watch
Morvah
Chysauster Ancient Village
Whitecross
Townshend
Crelly
Perranuthnoe
Lower Boscaswell
Levant Steam Engine (NT)
Bojewyan
252
Baker's Pit
Chysauster
Ludgvan
Trengwainton & Gwinear Mining Districts with Trewavas
Belubbus
Godolphin House (NT)
Crowntown
Wendron
Trenear
Trebarvah
Trevellard
Pendeen
Boswarthan
Gulval
Longrock
St Hilary
Carleen
Pollardras
Sithney
Trevenen
Brill
Penjerrick
Botallack
Kemidjack
St Just Mining District
Newbridge
Madron
Heamoor
Longrock
A394
Goldsithney
Rosudgeon
Newtown
Germoe
Tresowes Green
Lower Town
Constantine
Maen
Cape Cornwall
Trewellard
Trevega
Tremethick Cross
Gyllandour
Marazion
St Michael's Mount (NT)
Perranuthnoe
Kennegy Downs
10
Ashton
Breage
Helston
Trewennack
Porth Navas
The Brisons
Carn Leskys
St Just
Bosavern
Sancreed
Penzance
Prussia Cove
Cudden Point
Praa Sands
Rinsey
Gweek
Trebah Garden
Durgan
Penwith Heritage Coast
Kelynack
Grumbla
Crows-an-wra
Drift Reservoir
Buryas Bridge
Tredavoe
Newlyn
Paul
RNAS Culdrose
Mawgan
St Martin
Flushing
LAND'S END
Sennen Cove
Brane
Catchall
Kerris
Mousehole
St Clement's Isle
Mount's Bay
Trewavas Head
Porthleven
Flambards
Halliggye Fogou
Newtown-in-St-Martin
Manaccan
St Anthony
Longships
Sennen
Trevescan
St Buryan
Boleigh
Lamorna
Bereppa
Tregidden
Tregarne
Land's End
Trethewey
Lamorna Cove
Gunwalloe
St Keverne
Minack Theatre
Treen
Burial Chamber
Cury
Cross Lanes
Gunwalloe
Trelowarren
Garras
Tumuli
St Martin
Tregowris
Porthcurno
Cribba Head
Logan Rock
Penwith Heritage Coast
The Lizard Heritage Coast
Gwennap Head
Porthcurno Sands
St Levan
Poldhu Cove
Mullion
Goonhilly Downs
Trelan
Gwenter
Coverack
Poldhu Point
Mullion Cove
Penhale
Erisey Barton
Ponsongath
Mullion Island
Mullion
Predannack Wollas
Ruan Major
St Grade
Kuggar
Kennack Sands
Black Head
Vellan Head
Ruan Minor
Cadgwith
Lizard NNR
Kynance Cove
Landewednack
The Lizard Heritage Coast
Toll
Lizard Hot Point
LIZARD POINT

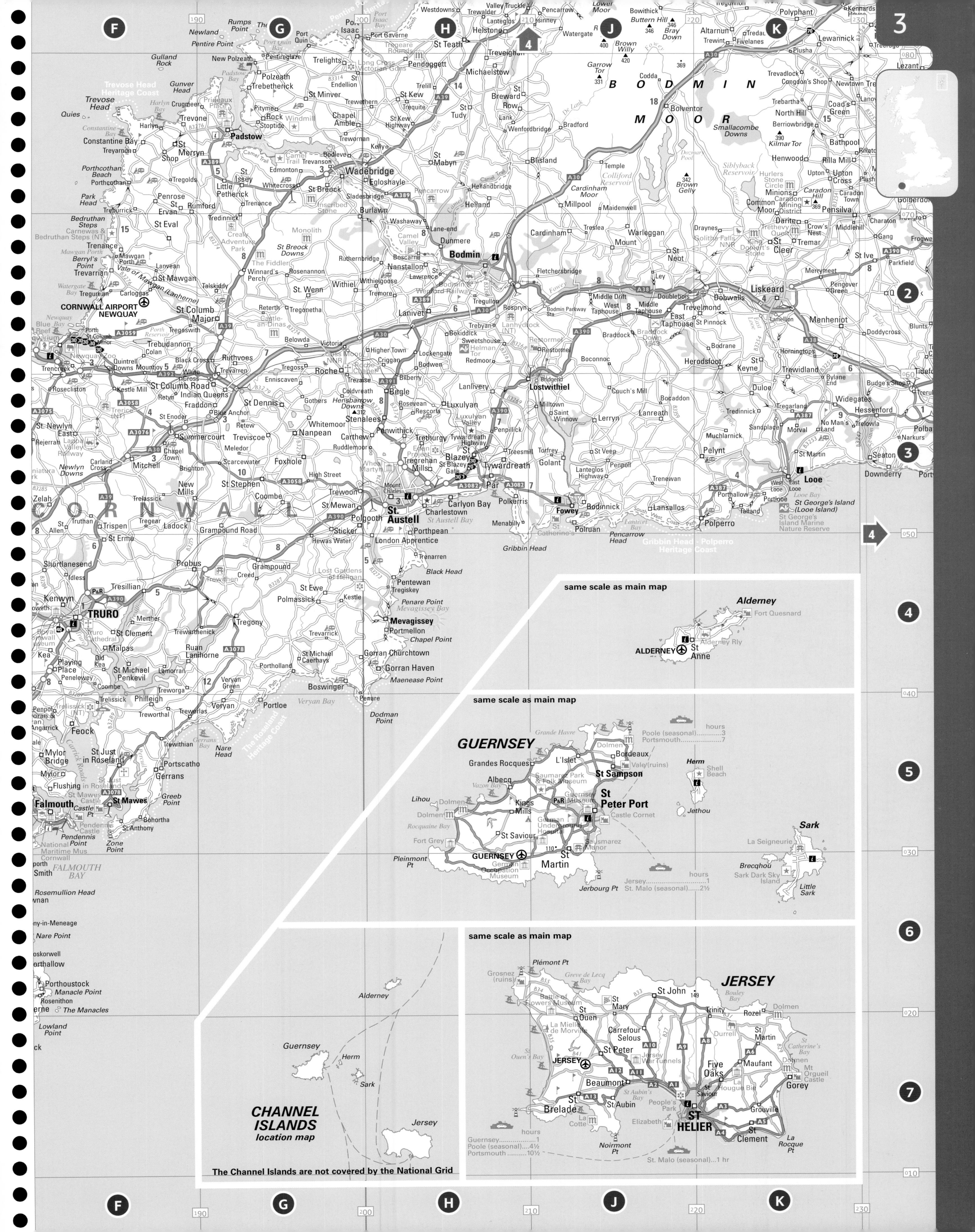
3
F
G
H
J
K
BODMIN MOOR
CORNWALL
Padstow
Wadebridge
Bodmin
Lostwithiel
Liskeard
Looe
Fowey
St. Austell
TRURO
Falmouth
St Mawes
Mevagissey
Newquay
CORNWALL AIRPORT NEWQUAY
St Columb Major
St Columb Road
Indian Queens
St Dennis
Roche
Bugle
Stenalees
Par
Tywardreath
Charlestown
Carlyon Bay
Polperro
Polruan
Camelford
Delabole
St Teath
Port Isaac
Polzeath
Rock
St Minver
Trevose Head
Trevose Head Heritage Coast
Gribbin Head - Polperro Heritage Coast
The Roseland Heritage Coast
Gribbin Head
Dodman Point
Nare Head
Black Head
Chapel Point
Gorran Haven
Gorran Churchtown
Veryan
Portloe
Tregony
Probus
Grampound
Grampound Road
St Mewan
Polgooth
Sticker
London Apprentice
Pentewan
Mitchell
Summercourt
St Enoder
Fraddon
New Mills
St Stephen
Ladock
Trispen
St Erme
Tresillian
St Clement
Malpas
Kea
Feock
Playing Place
Philleigh
St Just in Roseland
Portscatho
Gerrans
Flushing
Mylor Bridge
St Anthony
Pendennis Point
Zone Point
FALMOUTH BAY
Rosemullion Head
Nare Point
Porthoustock
Manacle Point
The Manacles
Lowland Point
Veryan Bay
Mevagissey Bay
St Austell Bay
St George's Island (Looe Island)
Looe Bay
Talland
Pelynt
Lanreath
Lerryn
Duloe
Menheniot
Dobwalls
Doublebois
East Taphouse
Middle Taphouse
Lanivet
Nanstallon
Lanhydrock (NT)
Restormel
Blisland
St Mabyn
St Tudy
St Breward
Helstone
Altarnun
Bolventor
North Hill
Upton Cross
Minions
Pensilva
St Cleer
St Neot
Warleggan
Cardinham
Mount
Temple
Millpool
Colliford Reservoir
Siblyback Reservoir
Dozmary Pool
Brown Willy
Kilmar Tor
Lewannick
Polyphant
Trewint
Fivelanes
Launceston
same scale as main map
Alderney
ALDERNEY
St Anne
Fort Quesnard
Alderney Rly
GUERNSEY
Grandes Rocques
L'Islet
Bordeaux
St Sampson
Albecq
Vazon Bay
Kings Mills
St Peter Port
Castle Cornet
St Saviour
St Martin
Jerbourg Pt
Pleinmont Pt
Lihou
Fort Grey
Rocquaine Bay
Herm
Shell Beach
Jethou
Sark
Little Sark
Brecqhou
La Seigneurie
Sark Dark Sky Island
Poole (seasonal)............3
Portsmouth......................7
hours
Jersey..................................1
St. Malo (seasonal)......2½
JERSEY
Plémont Pt
Grosnez (ruins)
Greve de Lecq Bay
St John
St Mary
St Ouen
La Mielle de Morville
Carrefour Selous
St Peter
Jersey War Tunnels
Trinity
Rozel
St Martin
Five Oaks
Maufant
Gorey
Mt Orgueil Castle
Beaumont
St Aubin
Brelade
St Brelade
People's Park
Elizabeth
ST HELIER
Grouville
St Clement
La Rocque Pt
Noirmont Pt
Guernsey.................1
Poole (seasonal).......4½
Portsmouth..........10½
St. Malo (seasonal)...1 hr
CHANNEL ISLANDS
location map
Alderney
Guernsey
Herm
Sark
Jersey
The Channel Islands are not covered by the National Grid
190
200
210
220
230
010
020
030
040
050
060
070
080
1
2
3
4
5
6
7

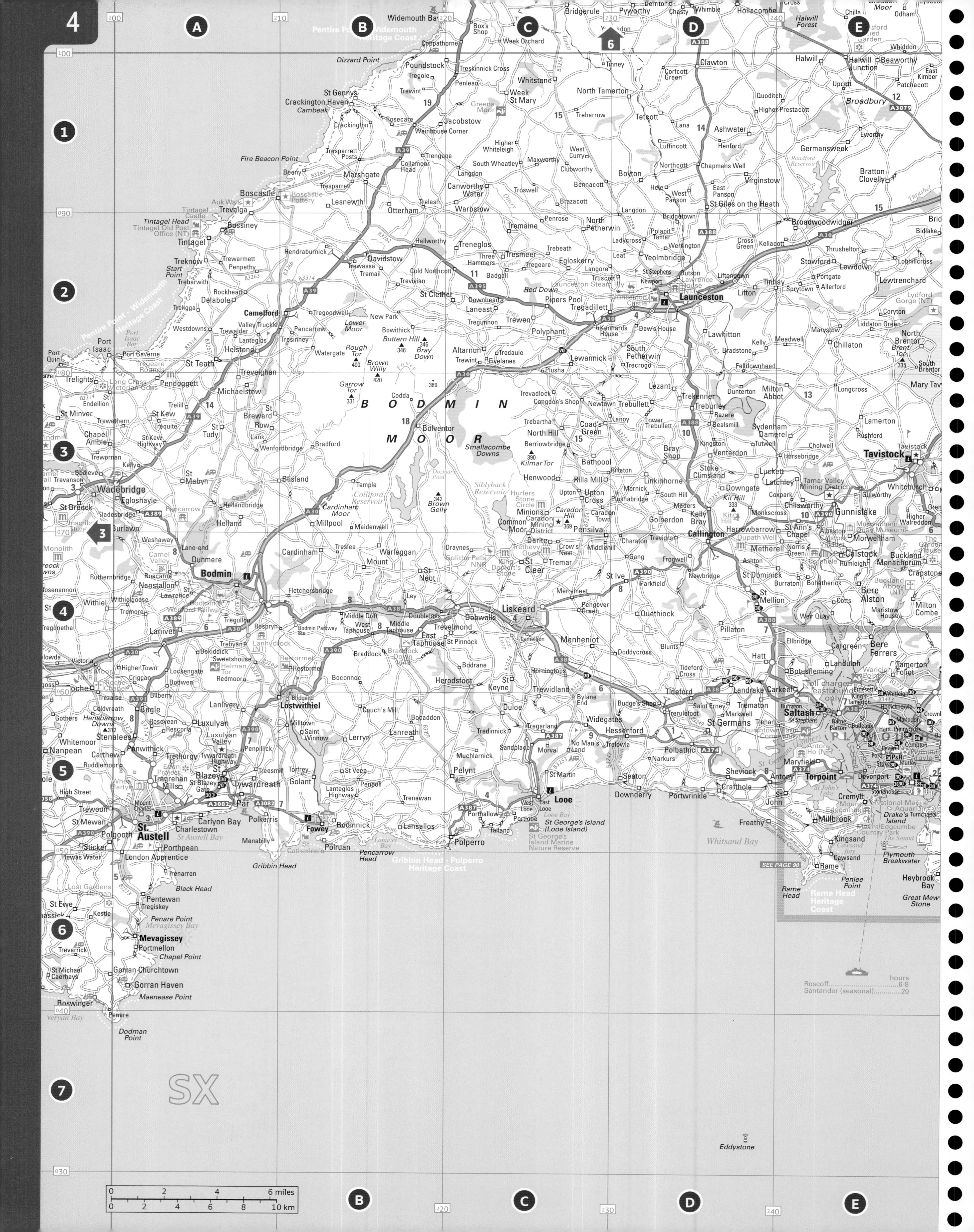
A
B
C
D
E
1
2
3
4
5
6
7
6
3
SX
Widemouth Bay
Pentire Point - Widemouth Heritage Coast
Box's Shop
Bridgerule
Pyworthy
Chasty
Whimble
Holsworthy
Halwill Forest
Chilla
Odham
Coppathorne
Week Orchard
A388
Dizzard Point
Poundstock
Treskinnick Cross
Tregole
Penlean
Whitstone
Tinney
Clawton
Corfcott Green
Halwill
Halwill Junction
Beaworthy
East Kimber
Upcott
Patchacott
St Gennys
Crackington Haven
Cambeak
Trewint
Week St Mary
North Tamerton
Quoditch
Broadbury
A3079
Higher Prestacott
Rosecare
Jacobstow
Wainhouse Corner
Greena Moor
Crackington
Trebarrow
Tetcott
Lana
Ashwater
Ewortny
Fire Beacon Point
Trengune
Higher Whiteleigh
Luffincott
Henford
Germansweek
Tresparrett Posts
A39
Collamoor Head
South Wheatley
Maxworthy
West Curry
Clubworthy
Northcott
Chapmans Well
Roadford Reservoir
Bratton Clovelly
Beeny
B3263
Marshgate
Langdon
Boyton
Canworthy Water
Troswell
Bennacott
Hele
West Panson
East Panson
Virginstow
Boscastle
Boscastle Pottery
Tresparrett
Lesnewth
Trelash
Brazacott
St Giles on the Heath
Auk Walk
Trevalga
Tintagel Castle
Otterham
Warbstow
Langdon
Penrose
North Petherwin
Bridgetown
Broadwoodwidger
Tintagel Head
Tintagel Old Post Office (NT)
Bossiney
Tremaine
Polapit Tamar
Werrington
Tintagel
Hallworthy
Treneglos
Ladycross
Cross Green
Kellacott
Thrushelton
Hendraburnick
Davidstow
Trebeath
Leat
Yeolmbridge
Treknow
Trewarmett
Penpethy
Three Hammers
Tresmeer
Tregeare
Egloskerry
Stowford
Lewdown
Lobhillcross
Start Point
Trewassa
Tremail
Cold Northcott
Langore
St Stephens
Dutson
Lawrence House
Liftondown
Portgate
Trebarwith
Badgall
Truscott
Newport
Tinhay
Lewtrencharrd
Rockhead
Trevivian
A395
Launceston Steam Rly
Launceston
Lifton
Sprytown
Allerford
Delabole
St Clether
Red Down
Launceston Castle
Lydford Gorge (NT)
Trekkgga
Camelford
Tregoodwell
Downhead
Laneast
Pipers Pool
Tregadillett
Coryton
New Park
Tregunnon
Trewen
Westdowns
Valley Truckle
Lower Moor
Bowithick
Polyphant
Kennards House
Daw's House
Lawhitton
Marystow
Liddaton Green
North Brentor
Port Isaac
Trewalder
Pencarrow
Buttern Hill
Brentor
Port Quin
Port Gaverne
Lanteglos
Tresinney
Watergate
Rough Tor
Bray Down
Altarnun
Tredaule
South Petherwin
Bradstone
Kelly
Meadwell
Chillaton
Helstone
Trewint
Fivelanes
Lewannick
Trecrogo
Felldownhead
South Brentor
Tregeare Rounds
St Teath
Brown Willy
Plusha
Trelights
Pendoggett
Treveighan
Garrow Tor
Codda
Trevadlock
Congdon's Shop
Lezant
Dunterton
Milton Abbot
Longcross
Mary Tavy
Long Cross Victorian Gdns
Michaelstow
Newtown
Trebullett
Trekenner
Treburley
Endellion
Trelill
St Breward
B O D M I N
Trebartha
Lanoy
Lower Trebullett
Rezare
Lamerton
St Minver
Trewethern
St Kew
Trequite
Row
Bolventor
North Hill
Coad's Green
Bealsmill
Sydenham Damerel
Chapel Amble
St Kew Highway
St Tudy
Lank
M O O R
Berriowbridge
Kingston
Tutwell
Rushford
Wenfordbridge
Bradford
Smallacombe Downs
Kilmar Tor
Bathpool
Bray Shop
Venterdon
Cholwell
Horsebridge
Tavistock
Treworran
Kelly
Bodieve
St Mabyn
Henwood
Rillaton
Stoke Climsland
Luckett
Latchley
Tamar Valley Mining District
Trevanson
Wadebridge
Blisland
Temple
Siblyback Reservoir
Rilla Mill
Linkinhorne
Downgate
Whitchurch
Egloshayle
Colliford Reservoir
Hurlers Stone Circle
Upton
Upton Cross
Mornick
Plushabridge
South Hill
Kit Hill
Coxpark
Gulworthy
St Breock
Pencarrow
Hellandbridge
Brown Gelly
Minions
Caradon Hill
Caradon Town
Maders
Chilsworthy
Sladesbridge
A389
Cardinham Moor
Monkscross
Gunnislake
Helland
Millpool
Common Moor
Caradon Mining District
Pensilva
Golberdon
Kelly Bray
Higher Walreddon
Burlawn
Maidenwell
Callington
Harrowbarrow
St Ann's Chapel
Albaston
Morwellham Quay Museum
Morwellham
Washaway
Darite
Charaton
Trevigro
Dupath Well
Calstock Viaduct
Monolith
Camel Valley
Cardinham
Treslea
Warleggan
Draynes
Golitha Falls NNR
Trethevy Quoit
Crow's Nest
Middlehill
Metherell
Norris Green
Calstock
Buckland Monachorum
Dunmere
Mount
Gang
Frogwell
Ashton
Cotehele
Rumleigh
Rosenannon
Ruthernbridge
Boscarne
Bodmin
St Neot
King Doniert's Stone
St Cleer
Tremar
St Ive
Parkfield
St Dominick
Crapstone
Nanstallon
Fletchersbridge
Merrymeet
Newbridge
Burraton
Bohetherick
Buckland Abbey (NT)
Withiel
Withielgoose
St Lawrence
Bodmin & Wenford Railway
Ley
Pengover Green
St Mellion
Bere Alston
Cotts
Milton Combe
Tremore
Middle Drift
Doublebois
Dobwalls
Liskeard
Quethiock
Weir Quay
Maristow House
Tregonetha
Lanivet
A30
Tregullon
Respryn
West Taphouse
Bodmin Parkway Sta
Middle Taphouse
Trevelmond
East Taphouse
St Pinnock
Menheniot
Pillaton
Cargreen
Bere Ferrers
Victoria
Trebyan
Lanhydrock (NT)
Braddock
Braddock Down 1643
Lamellion
Blunts
Ellbridge
Lockengate
Bokiddick
Sweetshouse
Helman Tor
Restormel
Bodrane
Doddycross
Hatt
Landulph
Tamerton Foliot
Higher Town
Criggan
Bodwen
Redmoor
Herodsfoot
St Keyne
Horningtops
Tideford Cross
Botusfleming
Roche
Trezaise
Bilberry
Boconnoc
Lanlivery
Lostwithiel
Bridgend
Trewidland
Bylane End
Tideford
Landrake
Carkeel
Whitleigh
Coldvreath
Bugle
Couch's Mill
Duloe
Budge's Shop
Trematon
Saltash
St Stephens
Gothers
Hensbarrow Downs
Bosevean
Rescorla
Luxulyan
Milltown
Bocaddon
Widegates
Trerulefoot
Saint Erney
Markwell
Honicknowle
Stenalees
Penwithick
Luxulyan Valley
Saint Winnow
Lerryn
Lanreath
Tredinnick
Tregarland
A387
Hessenford
St Germans
Barne Barton
Whitemoor
Nanpean
Carthew
Rudlemoor
Trethurgy
Tywardreath Highway
Penpillick
Torfrey
Sandplace
No Man's Land
Morval
Trelowla
Polbathic
A374
Antony House (NT)
Plymouth
Tregrehan Mills
St Blazey
St Blazey Gate
Tywardreath
Treesmill
Golant
St Veep
Penpoll
Muchlarnick
Pelynt
St Martin
Narkurs
Maryfield
Torpoint
Devonport
High Street
Par
A3082
Polkerris
Lanteglos Highway
Trenewan
Seaton
Downderry
Portwrinkle
Sheviock
Crafthole
Antony
St John
Cremyll
Trewoon
Charlestown
Carlyon Bay
St Austell
St Austell Bay
Fowey
Bodinnick
Lansallos
Porthallow
West Looe
East Looe
Looe
Looe Bay
Portlooe
Talland
St George's Island (Looe Island)
St George's Island Marine Nature Reserve
Mount Edgcumbe
Millbrook
Mount Edgcumbe Country Park
Drake's Island
National Marine Aquarium
St Mewan
Polgooth
Porthpean
Menabilly
St Catherine's
Polruan
Lantivet Bay
Pencarrow Head
Polperro
Freathy
Whitsand Bay
Kingsand
Cawsand
Cawsand Bay
The Sound
Plymouth Breakwater
Sticker
Hewas Water
London Apprentice
Gribbin Head
Gribbin Head - Polperro Heritage Coast
SEE PAGE 90
Rame
Penlee Point
Heybrook Bay
Trenarren
Black Head
Rame Head
Rame Head Heritage Coast
Great Mew Stone
Lost Gardens of Heligan
Pentewan
Tregiskey
St Ewe
Kestle
Penare Point
Mevagissey Bay
Mevagissey
Portmellon
Chapel Point
Trevarrick
Gorran Churchtown
St Michael Caerhays
Gorran Haven
Maenease Point
Boswinger
Penare
Veryan Bay
Dodman Point
hours
Roscoff.........................6-8
Santander (seasonal)..............20
Eddystone
0 2 4 6 miles
0 2 4 6 8 10 km

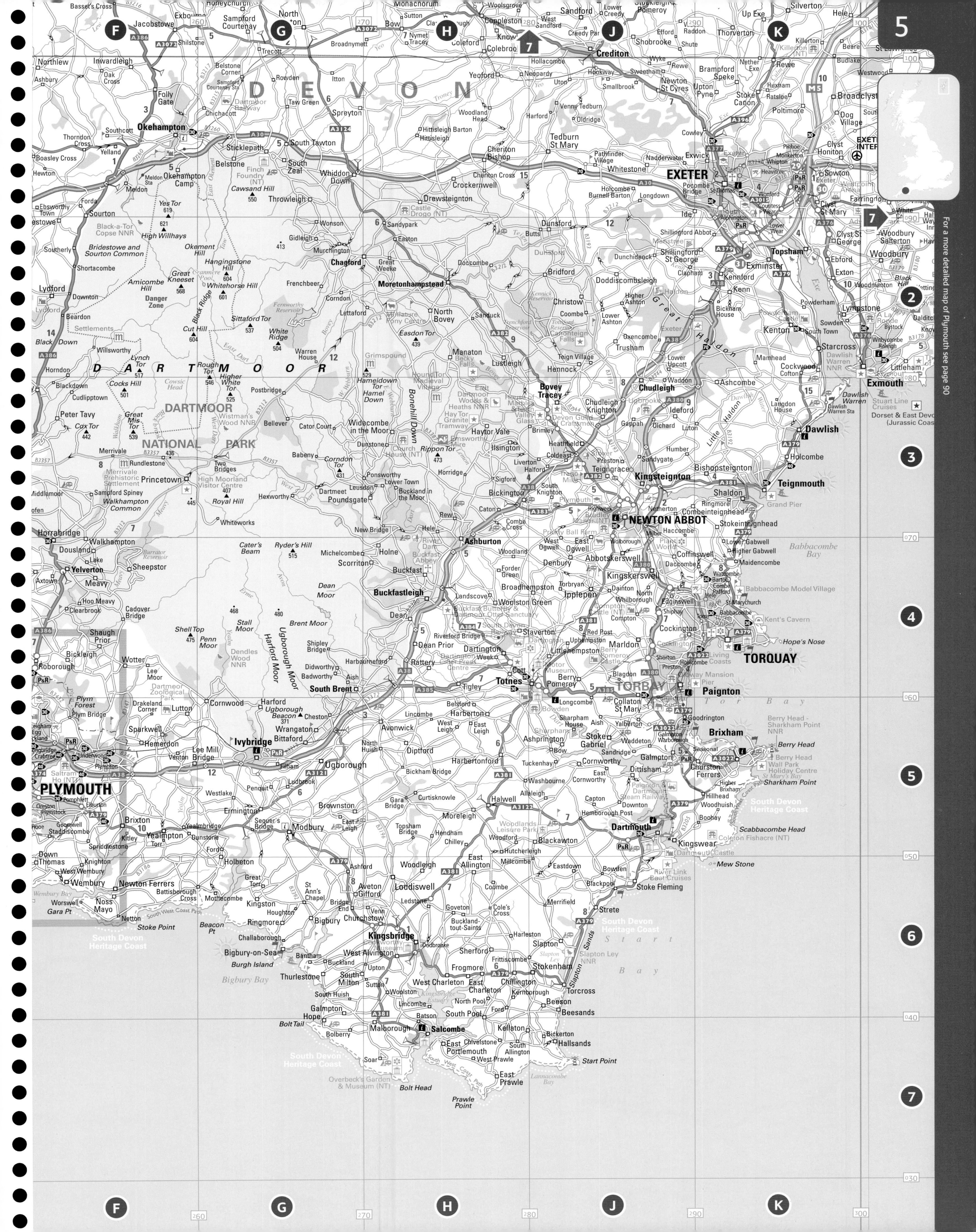

F
G
H
J
K
DEVON
DARTMOOR
DARTMOOR NATIONAL PARK
Okehampton
Crediton
EXETER
Exmouth
Topsham
Exminster
Dawlish
Teignmouth
Shaldon
Kingsteignton
NEWTON ABBOT
TORQUAY
TORBAY
Paignton
Brixham
Dartmouth
Kingswear
Totnes
Dartington
Ashburton
Buckfastleigh
South Brent
Ivybridge
PLYMOUTH
Yelverton
Princetown
Moretonhampstead
Chagford
Bovey Tracey
Chudleigh
Lydford
Modbury
Kingsbridge
Salcombe
Start Point
Bolt Head
Prawle Point
Bolt Tail
Burgh Island
Bigbury-on-Sea
Bigbury Bay
Start Bay
Tor Bay
Babbacombe Bay
Hope's Nose
Berry Head
South Devon Heritage Coast
Stoke Fleming
Slapton Sands
Torcross
Stokenham
Loddiswell
Modbury
Yealmpton
Newton Ferrers
Noss Mayo
Wembury
Ugborough
Harford
Cornwood
Holne
Widecombe in the Moor
Haytor Vale
Ilsington
Ipplepen
Kingskerswell
Marldon
Stoke Gabriel
Dittisham
Cockington
Kenton
Starcross
Dawlish Warren
Lympstone
Woodbury
Whiddon Down
Drewsteignton
Sticklepath
South Zeal
Belstone
Tedburn St Mary
Cheriton Bishop
Dunsford
Christow
Haldon
Great Haldon
Little Haldon
For a more detailed map of Plymouth see page 90
2
3
4
5
6
7

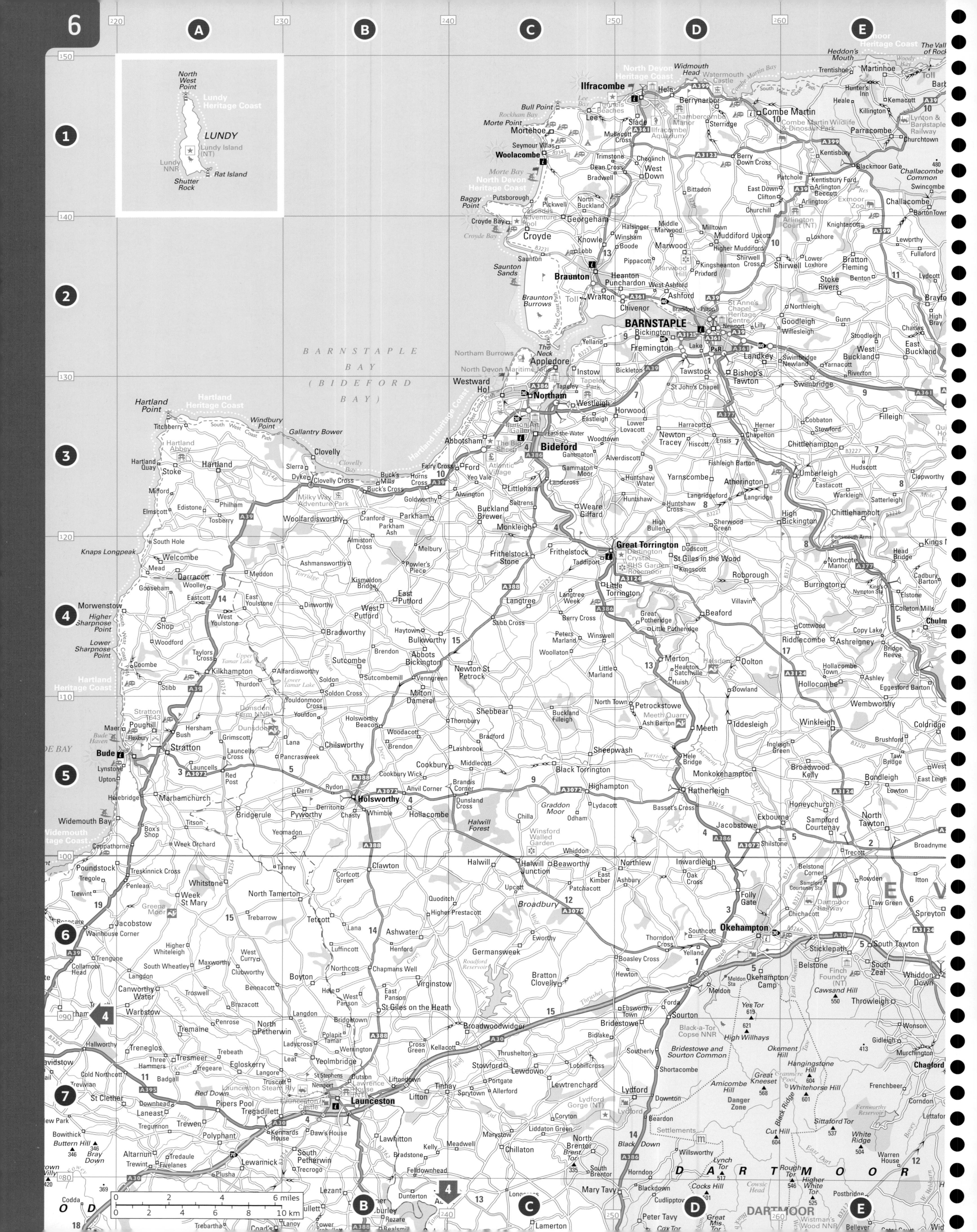
A
B
C
D
E
1
2
3
4
5
6
7
LUNDY
North West Point
Lundy Heritage Coast
Lundy Island (NT)
Lundy NNR
Rat Island
Shutter Rock
BARNSTAPLE BAY (BIDEFORD BAY)
Ilfracombe
Hele
Widmouth Head
Watermouth Castle
North Devon Heritage Coast
Berrynarbor
Combe Martin
Combe Martin Wildlife & Dinosaur Park
Lee
Slade
Ilfracombe Aquarium
Bull Point
Morte Point
Mortehoe
Woolacombe
Morte Bay
Baggy Point
Croyde Bay
Croyde
Georgeham
Braunton
Saunton
Saunton Sands
Braunton Burrows
Heanton Punchardon
Wrafton
Chivenor
Ashford
BARNSTAPLE
Bickington
Fremington
Instow
Yelland
Appledore
Northam Burrows
North Devon Maritime Museum
Westward Ho!
Northam
Westleigh
Bideford
Abbotsham
Landkey
Bishop's Tawton
Swimbridge
Goodleigh
Shirwell
Muddiford
Marwood
Parracombe
Martinhoe
Heddon's Mouth
Trentishoe
Kentisbury
Blackmoor Gate
Challacombe
Exmoor Zoo
Arlington Court (NT)
Bratton Fleming
Stoke Rivers
East Buckland
West Buckland
Filleigh
Chittlehampton
Umberleigh
Atherington
Yarnscombe
Chittlehamholt
High Bickington
Hartland Point
Hartland Heritage Coast
Hartland Abbey
Hartland
Hartland Quay
Stoke
Clovelly
Gallantry Bower
Windbury Point
Milky Way Adventure Park
Buck's Cross
Horns Cross
Fairy Cross
Parkham
Woolfardisworthy
Littleham
Buckland Brewer
Monkleigh
Weare Giffard
Great Torrington
Dartington Crystal
RHS Garden Rosemoor
St Giles in the Wood
Roborough
Little Torrington
Frithelstock
Frithelstock Stone
Langtree
Beaford
Merton
Dolton
Petrockstowe
Meeth
Iddesleigh
Winkleigh
Sheepwash
Black Torrington
Highampton
Hatherleigh
Monkokehampton
Jacobstowe
Exbourne
Sampford Courtenay
North Tawton
Knaps Longpeak
Welcombe
Morwenstow
Higher Sharpnose Point
Lower Sharpnose Point
Kilkhampton
Stibb
Bude
Poughill
Stratton
Marhamchurch
Widemouth Bay
Holsworthy
Bradworthy
Sutcombe
Abbots Bickington
Bulkworthy
Shebbear
Newton St Petrock
Milton Damerel
Halwill Junction
Beaworthy
Northlew
Okehampton
Sticklepath
South Tawton
Belstone
South Zeal
Whitstone
Week St Mary
North Tamerton
Clawton
Tetcott
Ashwater
Germansweek
Broadbury
Boyton
Virginstow
St Giles on the Heath
Broadwoodwidger
Bridestowe
Sourton
Lydford
Launceston
Lifton
Egloskerry
Tresmeer
Tregadillet
Polyphant
Altarnun
Lewannick
South Petherwin
Lezant
North Petherwin
Werrington
Yeolmbridge
Lewdown
Lewtrenchard
Mary Tavy
Peter Tavy
North Brentor
Brent Tor
High Willhays
Yes Tor
Great Kneeset
Cut Hill
Hangingstone Hill
Whitehorse Hill
Sittaford Tor
White Ridge
Chagford
Whiddon Down
Throwleigh
Cawsand Hill
DARTMOOR
D A R T M O O R
DEVON
Dartmoor Railway
Meldon Reservoir
Roadford Reservoir
Tamar Lake
Dunsdon Farm NNR
Halwill Forest
Lydford Gorge (NT)
Danger Zone
Black-a-Tor Copse NNR
Bridestowe and Sourton Common
Amicombe Hill
Hallworthy
Davidstow
St Clether
Trewen
Tregeare
A39
A361
A386
A388
A3072
A3079
A3124
A377
A399
A30
A395
A3123
0 2 4 6 miles
0 2 4 6 8 10 km
220
230
240
250
260
150
140
130
120
110
100
090
080
4

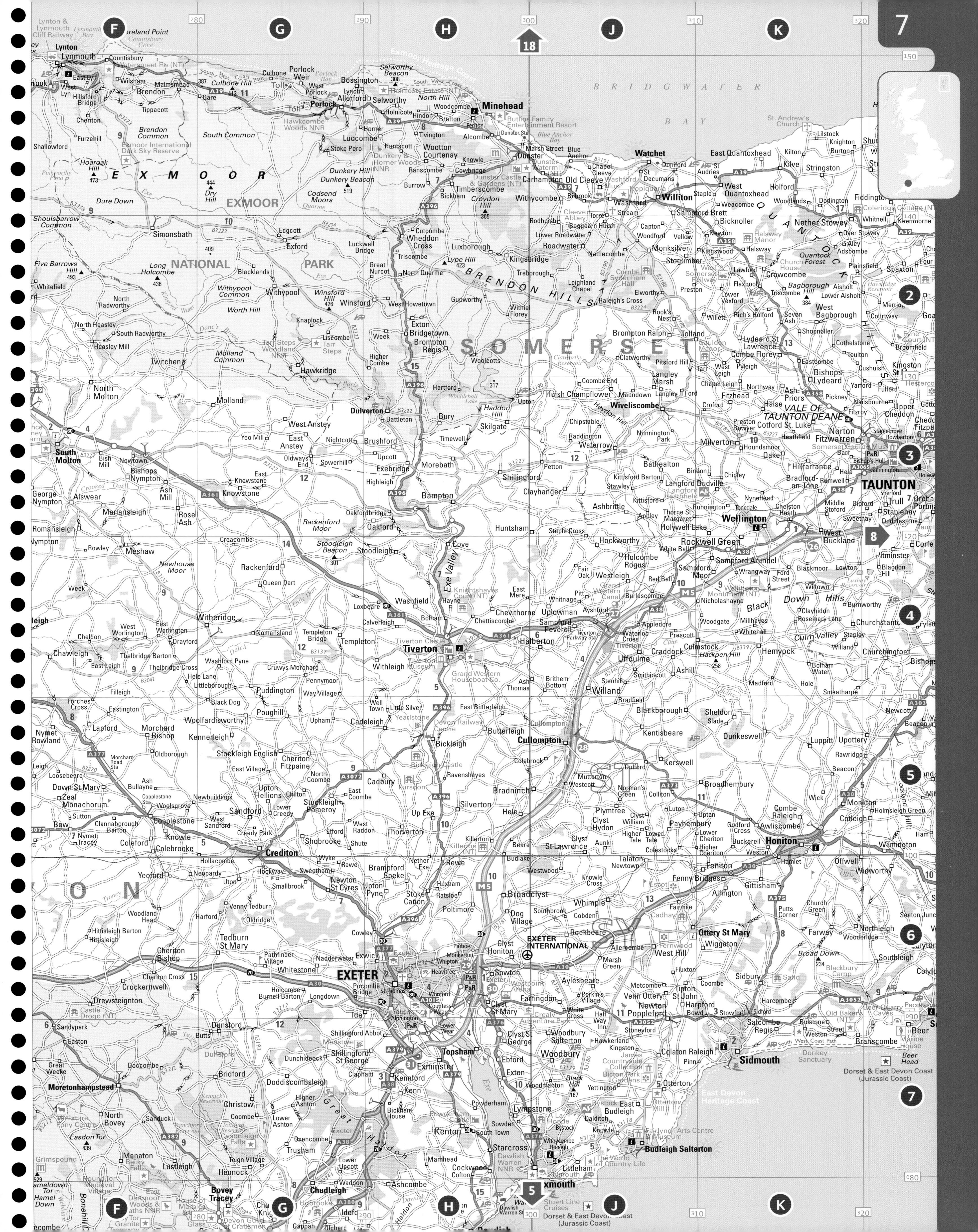
F
G
H
J
K
280
290
300
310
320
18
150
BRIDGWATER
BAY
Lynton & Lynmouth Cliff Railway
Lynton
Lynmouth
Countisbury
Watersmeet Ho (NT)
Brendon
Malmsmead
Culbone Hill
Culbone
Porlock Weir
Porlock
Porlock Bay
Bossington
Selworthy Beacon
308
Holnicote Estate (NT)
Lynch
Allerford
Selworthy
North Hill
Minehead
Woodcombe
Butlins Family Entertainment Resort
Exmoor Heritage Coast
South West Coast Path
Tippacott
Cheriton
Furzehill
Shallowford
Hoaroak Hill
473
Pinkworthy Pond
Brendon Common
Exmoor International Dark Sky Reserve
South Common
Hawkcombe Woods NNR
Horner
Luccombe
Stoke Pero
Hindon
Bratton
Tivington
Wootton Courtenay
Dunkery & Horner Woods NNR
Dunkery Hill
Dunkery Beacon
519
Ranscombe
Alcombe
Dunster Sta
Blue Anchor Bay
Marsh Street
Blue Anchor
Dunster
Knowle
Cowbridge
Dunster Castle & Gardens (NT)
Timberscombe
Carhampton
Old Cleeve
Chapel Cleeve
Washford
Watchet
Doniford
St Decumans
Williton
East Quantoxhead
Kilton
Lilstock
Knighton
Burton
St Audries
West Quantoxhead
Kilve
Stringston
Holford
St. Andrew's Church
EXMOOR
EXMOOR NATIONAL PARK
444
Dry Hill
Codsend Moors
Quarme
Burrow
Bickham
Croydon Hill
365
Withycombe
Bilbrook
Washford Mus
Cleeve Abbey
Torre
Stream
Sampford Brett
Bicknoller
Capton
Woodford
Vellow
Rodhuish
Beggearn Huish
Lower Roadwater
Roadwater
Nettlecombe
Monksilver
Stogumber
Kingswood
Halsway Manor
Halsway
Weacombe
Woodlands
Dodington
Nether Stowey
Over Stowey
Fiddington
Coleridge Cottage (NT)
Whitnell
Keenthorne
Dure Down
Shoulsbarrow Common
Simonsbath
Edgcott
Exford
Luckwell Bridge
Cutcombe
Wheddon Cross
Luxborough
Kingsbridge
Triscombe
Lype Hill
423
Treborough
Leighland Chapel
Combe Sydenham Hall
Elworthy
Five Barrows Hill
493
Long Holcombe
436
409
Blacklands
Great Nurcot
North Quarme
BRENDON HILLS
Raleigh's Cross
Whitefield
North Radworthy
Withypool Common
Withypool
Winsford Hill
426
Winsford
West Hawetown
Gupworthy
Withiel Florey
Worth Hill
North Heasley
South Radworthy
Heasley Mill
Knaplock
Liscombe
Tarr Steps Woodland NNR
Tarr Steps
Exton
Bridgetown
Brompton Regis
Week
Higher Combe
SOMERSET
Woolcotts
Clatworthy Reservoir
Clatworthy
Brompton Ralph
Tolland
Rook's Nest
Willett
Preston
Lawford
Flaxpool
Crowcombe
Lower Vexford
Rich's Holford
Seven Ash
Bagborough Hill
384
Triscombe
Aisholt
Lower Aisholt
QUANTOCK HILLS
Quantock Forest
Plainsfield
Spaxton
West Bagborough
Courtway
Lydeard St Lawrence
Combe Florey
Pitsford Hill
Shopnoller
Cothelstone
Fyne Court (NT)
Broomfield
Toulton
Cushuish
Kingston St Mary
Eastcombe
Bishops Lydeard
Twitchen
Molland Common
Hawkridge
North Molton
Molland
Dulverton
Battleton
Hartford
Wimbleball Lake
317
Haddon Hill
Upton
Bury
Skilgate
Huish Champflower
Coombe End
Maundown
Langley Marsh
Langley
Ford
Wiveliscombe
Heydon Hill
Chipstable
Raddington
Waterrow
Nunnington Park
Fitzhead
Croford
Halse
Northway
Ash Priors
Pickney
Yarford
Fulford
VALE OF TAUNTON DEANE
Cotford St. Luke
Preston Bowyer
Norton Fitzwarren
Milverton
Heathfield
Upper Cheddon
Cheddon Fitzpaine
Staplegrove
Rowbarton
Somerset Cricket Mus
Bishop's Hull
Galmington
TAUNTON
South Molton
Bish Mill
Newtown
Bishops Nympton
West Anstey
East Anstey
Yeo Mill
Oldways End
Nightcott
Brushford
Sowerhill
Upcott
Exebridge
Morebath
Timewell
Shillingford
Petton
Bathealton
Kittisford Barton
Stawley
Langford Budville
Bindon
Chipley
Bradford-on-Tone
Hillfarrance
Rumwell
Hele
Oake
Houndsmoor
Sherford
Trull
Orchard Portman
Staplehay
Dipford
Middle Stoford
Sweethay
Duddlestone
George Nympton
Alswear
Ash Mill
Knowstone
East Knowstone
Highleigh
Bampton
Clayhanger
Ashbrittle
Kittisford
Appley
Thorne St Margaret
Runnington
Nynehead
Tonedale
Chelston Heath
Wellington
Holywell Lake
West Buckland
Mariansleigh
Rose Ash
Rackenford Moor
Oakfordbridge
Oakford
Huntsham
Staple Cross
Hockworthy
Rockwell Green
White Ball
Sampford Arundel
Pitminster
Corfe
Blagdon Hill
Romansleigh
Nympton
Rowley
Meshaw
Creacombe
Stoodleigh Beacon
301
Stoodleigh
Cove
Holcombe Rogus
Fair Oak
Westleigh
Sampford Moor
Wrangway
Ford Street
Blackmoor
Lowton
Newhouse Moor
Rackenford
Queen Dart
Exe Valley
Knightshayes Court (NT)
East Mere
Grand Western Canal
Burlescombe
Red Ball
Wellington Monument (NT)
Nicholashayne
Black Down Hills
Wiltown
Burnworthy
Clayhidon
Rosemary Lane
Churchstanton
Week
Washfield
Hayne
Loxbeare
Calverleigh
Bolham
Chevithorne
Chettiscombe
Whitnage
Pitt
Uplowman
Ayshford
Sampford Peverell
Tiverton Parkway Sta
Appledore
Woodgate
Millhayes
Whitehall
Witheridge
West Worlington
East Worlington
Drayford
Nomansland
Templeton Bridge
Templeton
Tiverton Castle
Tiverton
Tiverton Museum
Halberton
Waterloo Cross
Prescott
Culmstock
Hackpen Hill
258
Hemyock
Culm Valley
Stapley
Willand
Churchingford
Cheldon
Chawleigh
Thelbridge Barton
Thelbridge Cross
East Leigh
Washford Pyne
Cruwys Morchard
Withleigh
Grand Western Houseboat Co.
Ash Thomas
Brithem Bottom
Uffculme
Craddock
Ashill
Smithincott
Bolham Water
Bishops
Hele Lane
Littleborough
Pennymoor
Way Village
Stenhill
Willand
Madford
Hole
Smeatharpe
Filleigh
Puddington
Black Dog
Well Town
Little Silver
Yearlstone
East Butterleigh
Bradfield
Blackborough
Sheldon
Slade
Forches Cross
Eastington
Poughill
Upham
Cadeleigh
Devon Railway Centre
Butterleigh
Cullompton
Kentisbeare
Dunkeswell
Luppitt
Upottery
Newcott
Nymet Rowland
Lapford
Morchard Bishop
Woolfardisworthy
Kennerleigh
Bickleigh
Stockleigh English
Cheriton Fitzpaine
Oldborough
Bickleigh Castle
Colebrook
Kerswell
Rawridge
Beacon
Leigh
Morchard Road Sta
Loosebeare
East Village
North Coombe
Cadbury
Fursdon
Ravenshayes
Dulford
Mutterton
Westcott
Broadhembury
Down St Mary
Zeal Monachorum
Ash Bullayne
Copplestone Sta
Woolsgrove
Newbuildings
Upton Hellions
Chilton
East Coombe
Stockleigh Pomeroy
Bradninch
Silverton
Hele
Norman's Green
Colliton
Wick
Monkton
Holmsleigh Green
Combe Raleigh
Awliscombe
Cotleigh
Bow
Sutton
Clannaborough Barton
Copplestone
West Sandford
Sandford
Lower Creedy
Efford
West Raddon
Up Exe
Thorverton
Plymtree
Clyst William
Clyst Hydon
Luton
Upton
Payhembury
Godford Cross
Buckerell
Honiton
Higher Tale
Lower Tale
Lower Cheriton
Higher Cheriton
Colestocks
Nymet Tracey
Coleford
Knowle
Colebrooke
Creedy Park
Shobrooke
Shute
Killerton (NT)
Beare
Clyst St Lawrence
Aunk
Weston
Hamlet
Wilmington
Hollacombe
Crediton
Wyke
Rewe
Nether Exe
Budlake
Talaton
Newtown
Feniton
Offwell
Widworthy
Yeoford
Neopardy
Hookway
Sweetham
Brampford Speke
Upton Pyne
Newton St Cyres
Uton
Smallbrook
Huxham
Stoke Canon
Ratsloe
Westwood
Knowle Cross
Fenny Bridges
Escot
Gittisham
Allington
DEVON
Venny Tedburn
Harford
Oldridge
Woodland Head
Poltimore
Broadclyst
Whimple
Dog Village
Southbrook
Cobden
Fairmile
Cadhay
Putts Corner
Church Green
Seaton Junc
Hittisleigh Barton
Hittisleigh
Tedburn St Mary
Cowley
Pinhoe
EXETER INTERNATIONAL
Clyst Honiton
Rockbeare
Allercombe
Ottery St Mary
Wiggaton
Farway
Broad Down
234
Northleigh
Woodbridge
Southleigh
Colyford
Cheriton Bishop
Pathfinder Village
Nadderwater
Exwick
Whitestone
Whipton
Monkerton
Heavitree
Exeter
EXETER
Sowton
Marsh Green
West Hill
Fernwood
Fluxton
Sidbury
Blackbury Camp
Cheriton Cross
Crockernwell
Pocombe Bridge
Holcombe Burnell Barton
Longdown
Wonford
Countess Wear
Westpoint Arena
Aylesbeare
Metcombe
Venn Ottery
Tipton St John
Harpford
Bowd
Coombe
Harcombe
Drewsteignton
Castle Drogo (NT)
Ide
Alphington
Clyst St Mary
Farringdon
Crealy Adventure Park
Perkin's Village
White Cross
Half Way Inn
Newton Poppleford
Stowford
Sidford
Salcombe Regis
Beer Quarry Caves
Old Bakery (NT)
Bulstone
Branscombe
Weston
Street
Beer
Beer Head
Sandypark
Easton
Butts
Dunsford
Shillingford Abbot
Lower Wear
Topsham
Clyst St George
Woodbury Salterton
Woodbury
Stoneyford
Hawkerland
Kingston
Colaton Raleigh
Pinn
Sidmouth
Donkey Sanctuary
Marine House
Dorset & East Devon Coast (Jurassic Coast)
Great Weeke
Doccombe
Bridford
Dunchideock
Shillingford St George
Exminster
Clapham
Kennford
Ebford
Exton
Woodmanton
Black Hill
167
Yettington
Countryside Collection
Bicton Park Gardens
Otterton
Moretonhampstead
Doddiscombsleigh
Kenn
Powderham
Lympstone
Bystock
East Budleigh
Dalditch
Ottertory Mill
East Devon Heritage Coast
Higher Ashton
Lower Ashton
Christow
Great Haldon
Bickham House
Kenton
Powderham Castle
Sowden
South Town
La Ronde
Knowle
North Bovey
Sandrick
Kennick Reservoir
Coombe
Canonteign Falls
Oxencombe
Trusham
Exeter Racecourse
Starcross
Dawlish Warren NNR
Withycombe Raleigh
Fairlynch Arts Centre & Museum
Budleigh Salterton
Easdon Tor
439
Manaton
Becky Falls
Lustleigh
Teign Village
Hennock
Lower Upcott
Mamhead
Cockwood
Cofton
Littleham
The World of Country Life
Grimspound
529
Hameldown Tor
Hamel Down
Hound Tor Medieval Village
Bovey Tracey
Chudleigh
Waddon
Ashcombe
Exmouth
Stuart Line Cruises
Dorset & East Devon Coast (Jurassic Coast)
Dawlish
Ugbrooke
Idefo
Gappah
Bonehill
Haldon
8
5
140
130
120
110
100
090
080
1
2
3
4
5
6
7

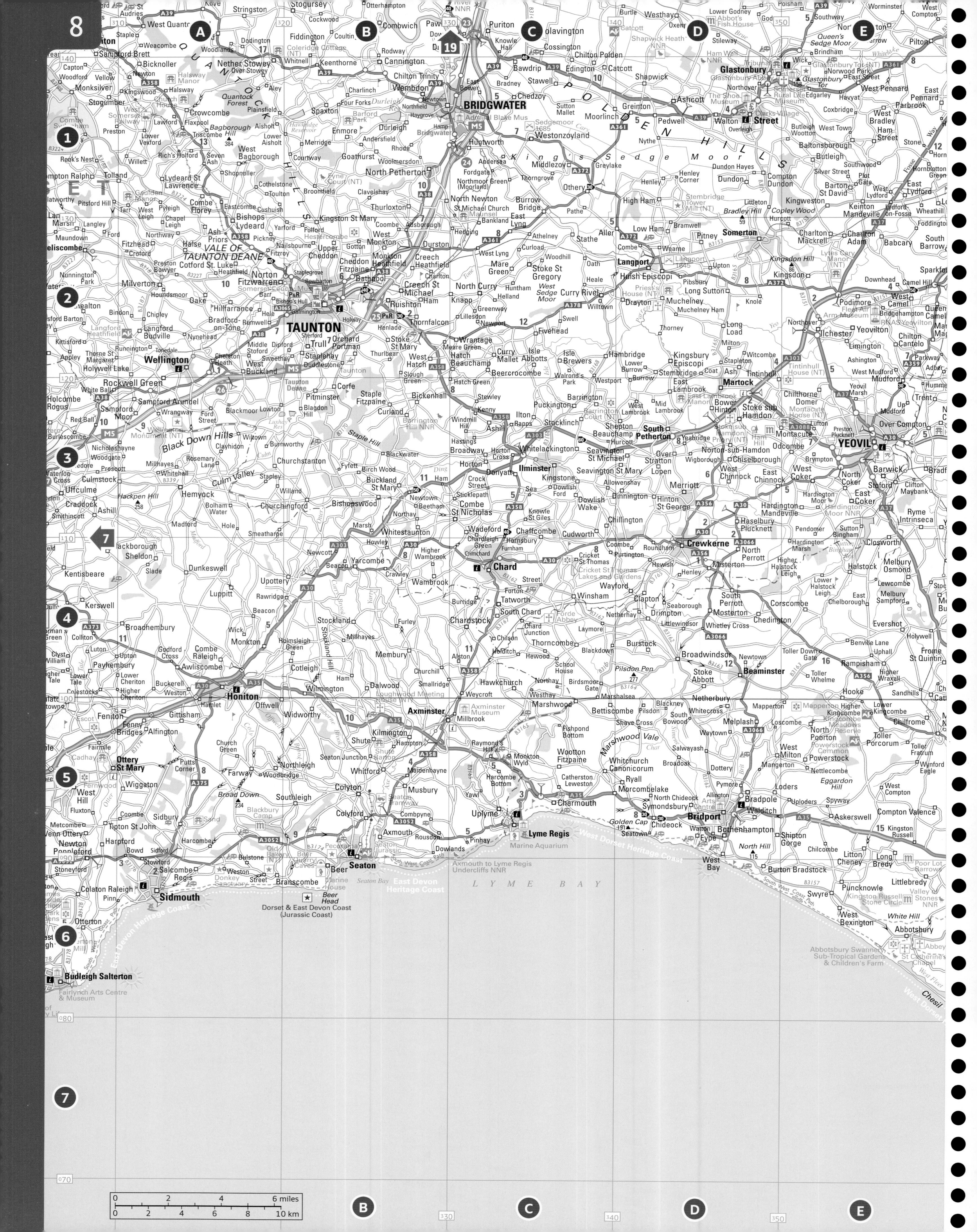
A
B
C
D
E
1
2
3
4
5
6
7
West Quantoxhead
Stringston
Stogursey
Puriton
Cossington
Glastonbury
Street
Quantock Hills
Nether Stowey
Bridgwater
Polden Hills
King's Sedge Moor
Cannington
Bawdrip
Edington
Catcott
Shapwick
Ashcott
Walton
North Petherton
Westonzoyland
Middlezoy
Othery
High Ham
Langport
Somerton
Kingsdon
Bishops Lydeard
Vale of Taunton Deane
Norton Fitzwarren
Taunton
Wellington
Milverton
Ruishton
Creech St Michael
North Curry
Curry Rivel
Muchelney
Martock
Ilchester
Yeovilton
Yeovil
Stoke sub Hamdon
South Petherton
Ilminster
Crewkerne
Chard
Black Down Hills
Culm Valley
Hemyock
Churchinford
Broadway
Horton
Donyatt
Hinton St George
East Coker
West Coker
Honiton
Axminster
Ottery St Mary
Sidmouth
Seaton
Beer
Colyton
Lyme Regis
Charmouth
Chideock
Bridport
West Bay
Beaminster
Broadwindsor
Netherbury
Burton Bradstock
Abbotsbury
Budleigh Salterton
Otterton
Lyme Bay
Dorset & East Devon Coast (Jurassic Coast)
West Dorset Heritage Coast
East Devon Heritage Coast
Marine Aquarium
Chesil
0 2 4 6 miles
0 2 4 6 8 10 km

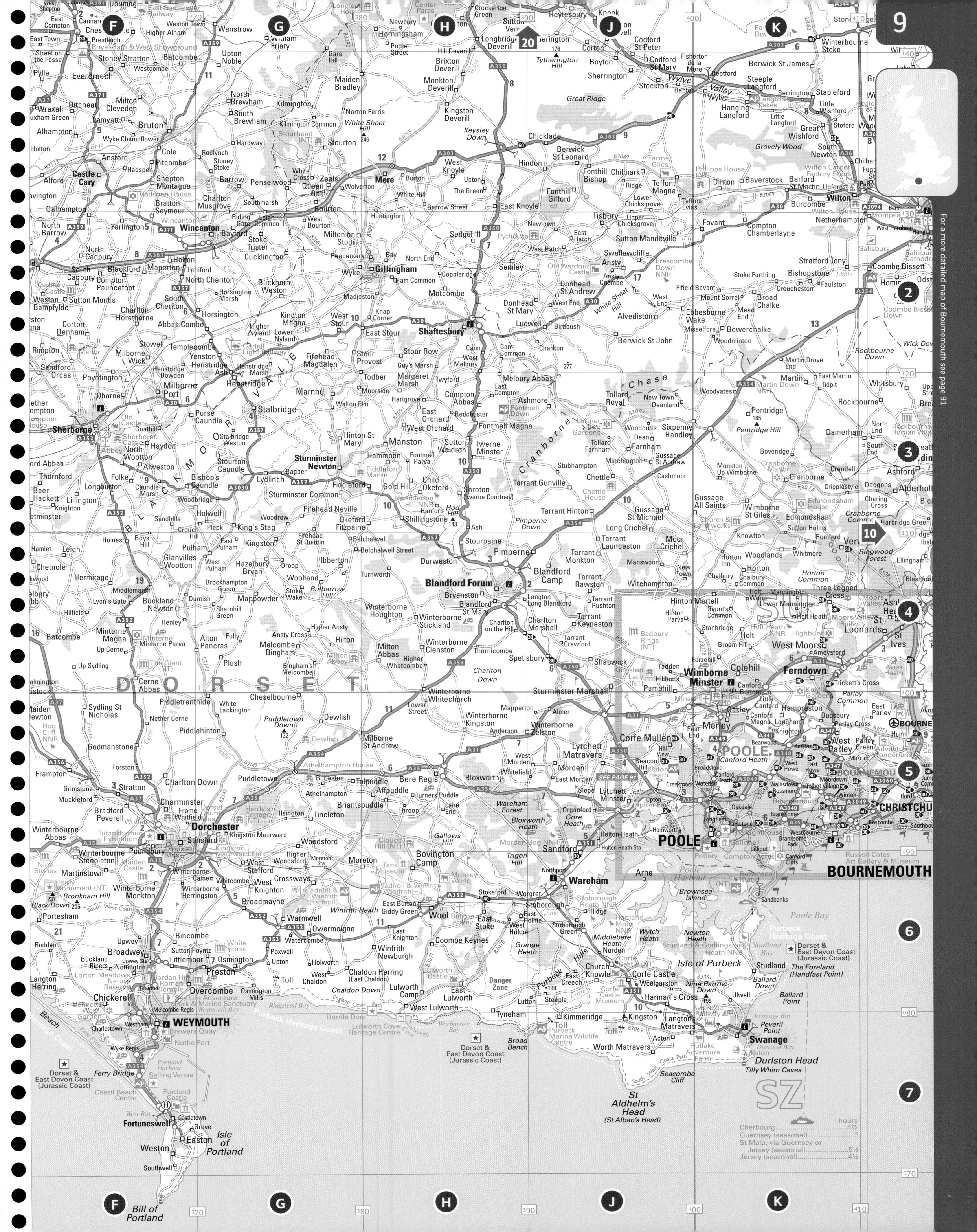

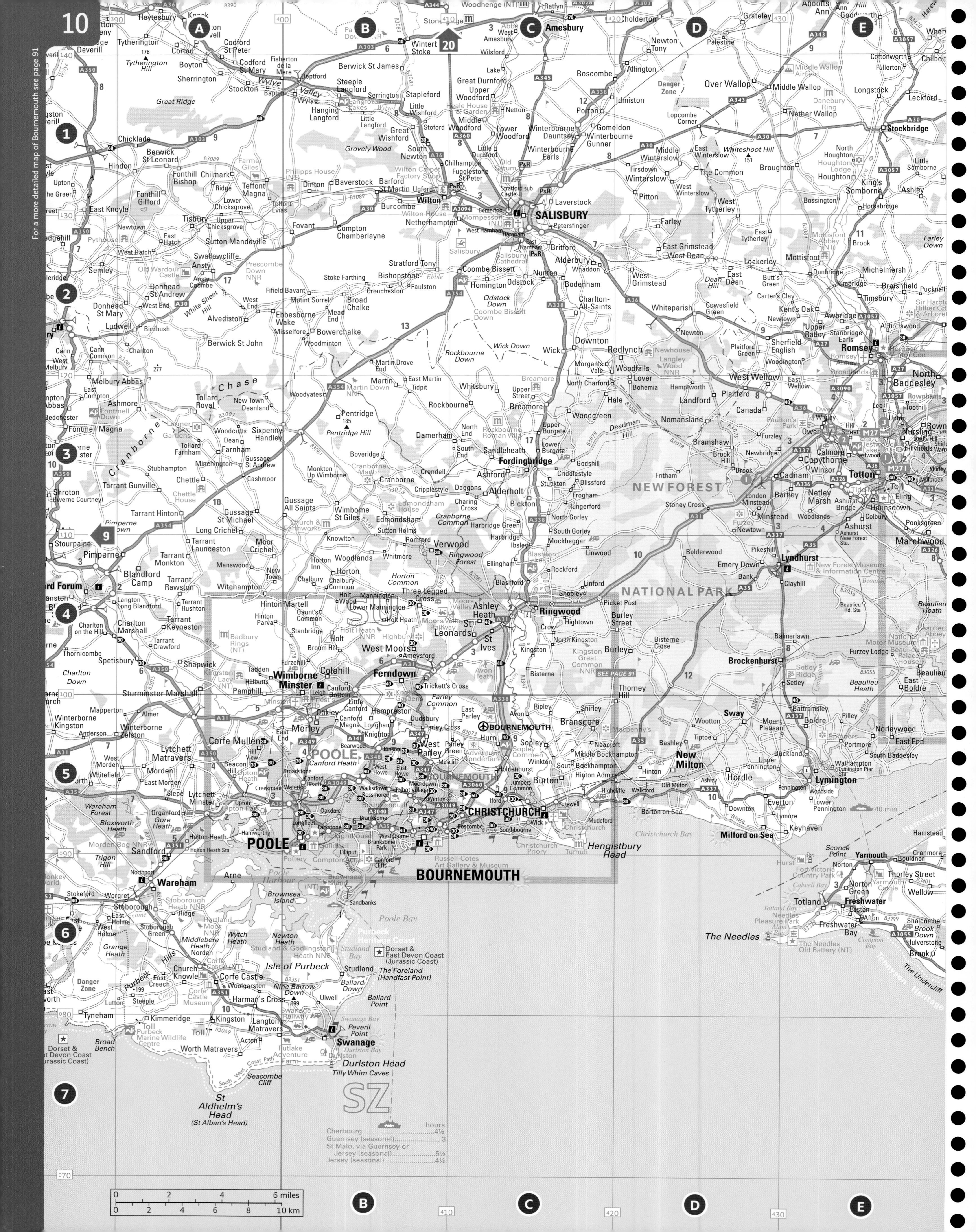
For a more detailed map of Bournemouth see page 91
A
B
C
D
E
1
2
3
4
5
6
7
20
9
SALISBURY
Amesbury
Stockbridge
Romsey
Totton
Lyndhurst
Lymington
Ringwood
Fordingbridge
Wimborne Minster
Ferndown
Verwood
Blandford Forum
Wareham
POOLE
BOURNEMOUTH
CHRISTCHURCH
New Milton
Milford on Sea
Swanage
Yarmouth
Freshwater
Totland
NEW FOREST
NATIONAL PARK
Wilton
Downton
Alderholt
Cranborne
Sixpenny Handley
Tisbury
Hindon
Mere
Corfe Mullen
Corfe Castle
Studland
Isle of Purbeck
Poole Bay
Poole Harbour
Brownsea Island
Christchurch Bay
Hengistbury Head
The Needles
Durlston Head
St Aldhelm's Head
(St Alban's Head)
Cranborne Chase
Great Ridge
Grovely Wood
Pentridge Hill
Brockenhurst
Sway
Burley
St Leonards
St Ives
West Moors
Colehill
Cadnam
Ashurst
Beaulieu
Old Sarum
Wilton House
Salisbury Cathedral
Kingston Lacy
Badbury Rings
SZ
Cherbourg 4½
Guernsey (seasonal) 3
St Malo, via Guernsey or Jersey (seasonal) 5½
Jersey (seasonal) 4½
hours
0 2 4 6 miles
0 2 4 6 8 10 km

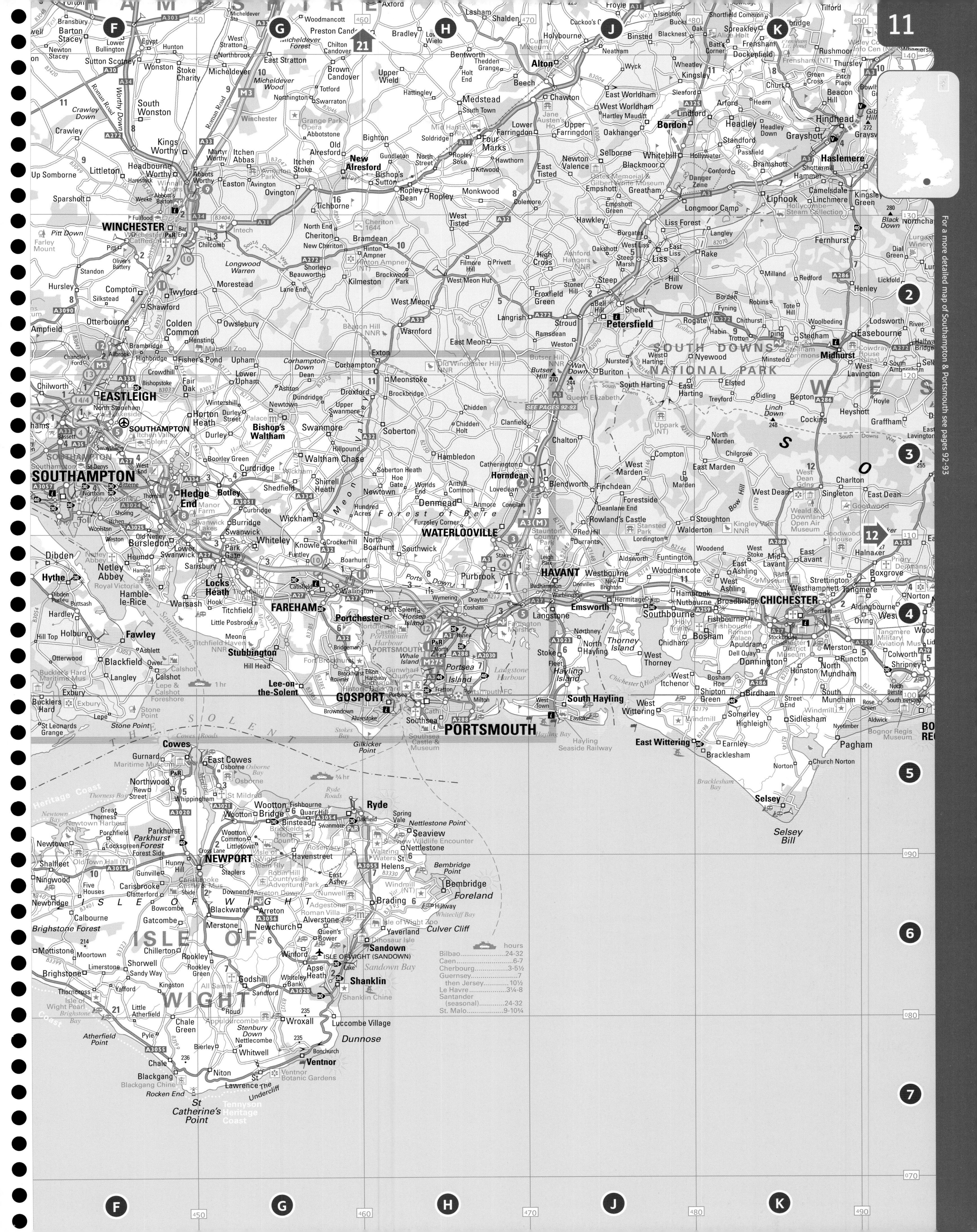

11
For a more detailed map of Southampton & Portsmouth see pages 92-93
HAMPSHIRE
SOUTH DOWNS NATIONAL PARK
WEST SUSSEX
ISLE OF WIGHT
THE SOLENT
WINCHESTER
EASTLEIGH
SOUTHAMPTON
Alton
Petersfield
Haslemere
Midhurst
Bordon
Hindhead
Liphook
Liss
New Alresford
Bishop's Waltham
Hedge End
Botley
Fareham
Gosport
PORTSMOUTH
WATERLOOVILLE
Horndean
HAVANT
Emsworth
CHICHESTER
Selsey
Selsey Bill
East Wittering
West Wittering
South Hayling
Hayling Island
Portsea Island
Lee-on-the-Solent
Hythe
Fawley
Cowes
East Cowes
NEWPORT
Ryde
Sandown
Shanklin
Ventnor
Bembridge
St Catherine's Point
Culver Cliff
Dunnose
Sandown Bay
Compton
Twyford
Swanmore
Wickham
Denmead
Purbrook
Rowland's Castle
Westbourne
Bosham
Pagham
SEE PAGES 92-93
hours
Bilbao........................24-32
Caen..............................6-7
Cherbourg.................3-5½
Guernsey.........................7
then Jersey..............10½
Le Havre................3¼-8
Santander
(seasonal)................24-32
St. Malo..................9-10¾
F
G
H
J
K
2
3
4
5
6
7
12
21

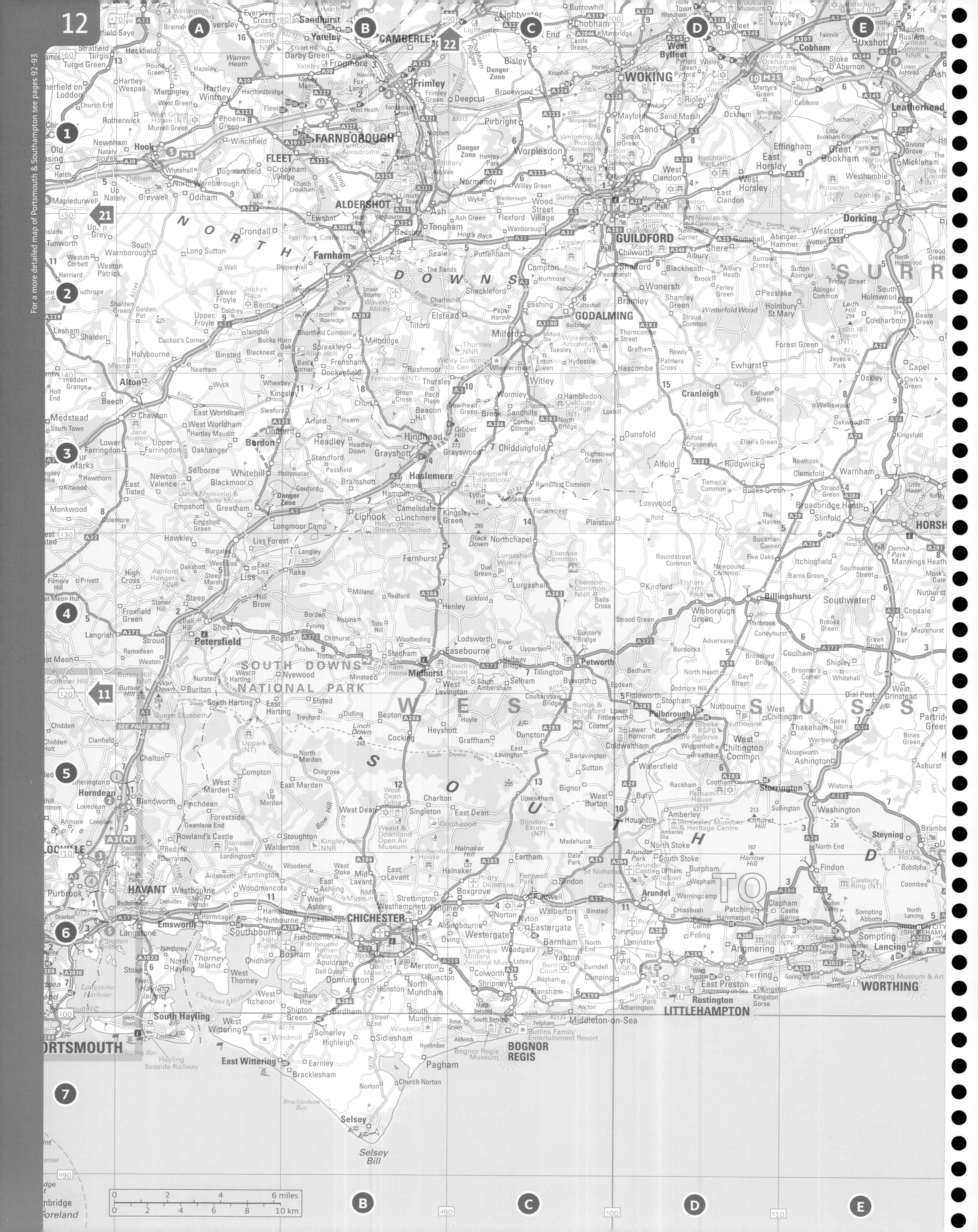

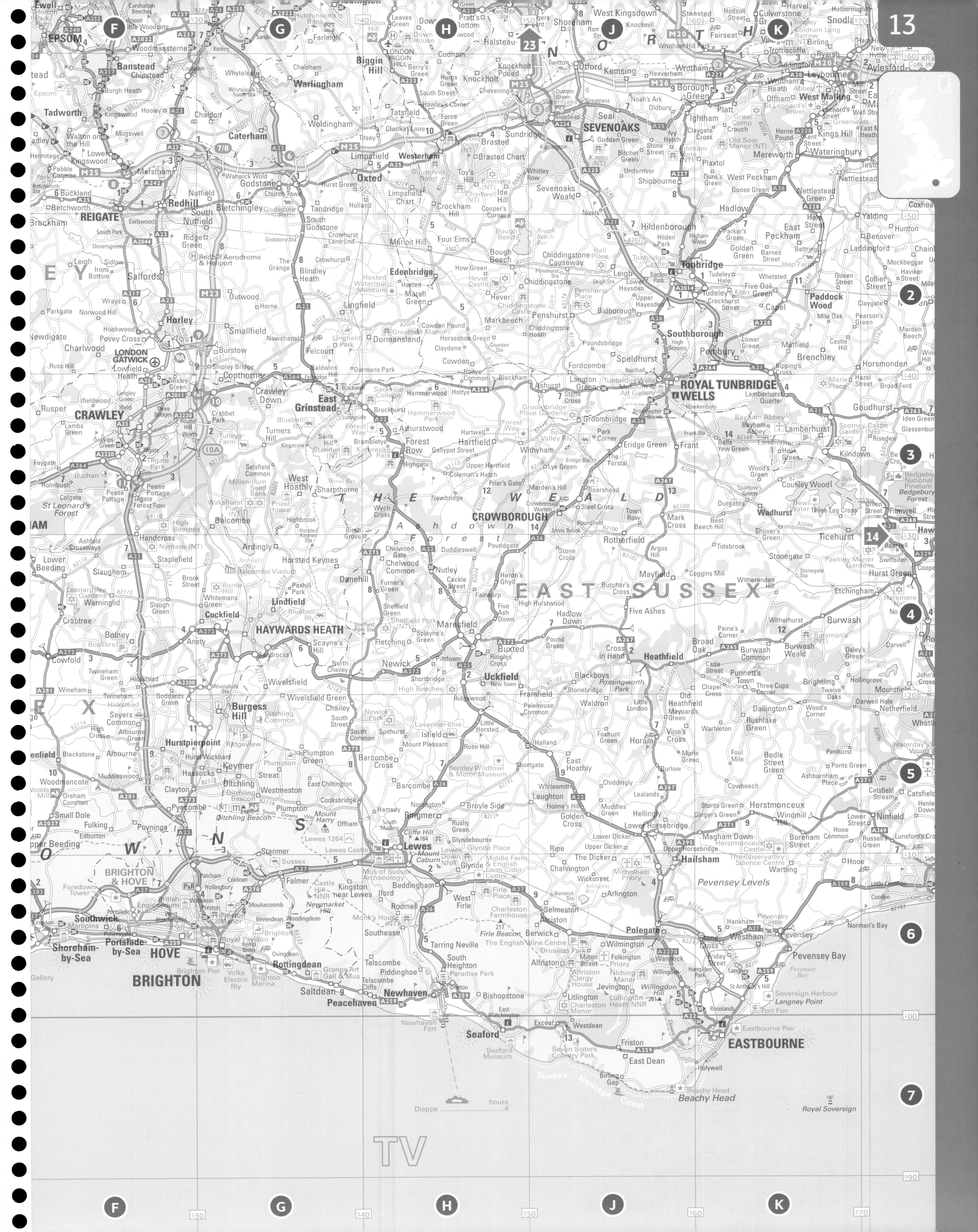
F
G
H
J
K
2
3
4
5
6
7
23
14
EPSOM
Banstead
Tadworth
REIGATE
Redhill
Caterham
Warlingham
Biggin Hill
Oxted
Westerham
SEVENOAKS
Borough Green
West Malling
Aylesford
Tonbridge
Paddock Wood
Horley
LONDON GATWICK
CRAWLEY
East Grinstead
Edenbridge
ROYAL TUNBRIDGE WELLS
Southborough
Pembury
Crowborough
Forest Row
Hartfield
Wadhurst
Ticehurst
Lamberhurst
Goudhurst
Mayfield
Heathfield
Uckfield
Haywards Heath
Cuckfield
Burgess Hill
Hurstpierpoint
Lindfield
EAST SUSSEX
THE WEALD
Ashdown Forest
Burwash
Hailsham
Polegate
Lewes
Ringmer
Seaford
Newhaven
Peacehaven
Rottingdean
BRIGHTON
HOVE
BRIGHTON & HOVE
Portslade-by-Sea
Shoreham-by-Sea
Southwick
EASTBOURNE
Beachy Head
Pevensey Bay
Pevensey Levels
Herstmonceux
Royal Sovereign
Sussex Heritage Coast
Dieppe
hours 4
TV
M25
M23
M20
M26
A22
A23
A26
A27
A259
A264
A21
A272
530
540
550
560
570
100
090
110
120
130
140
150
160

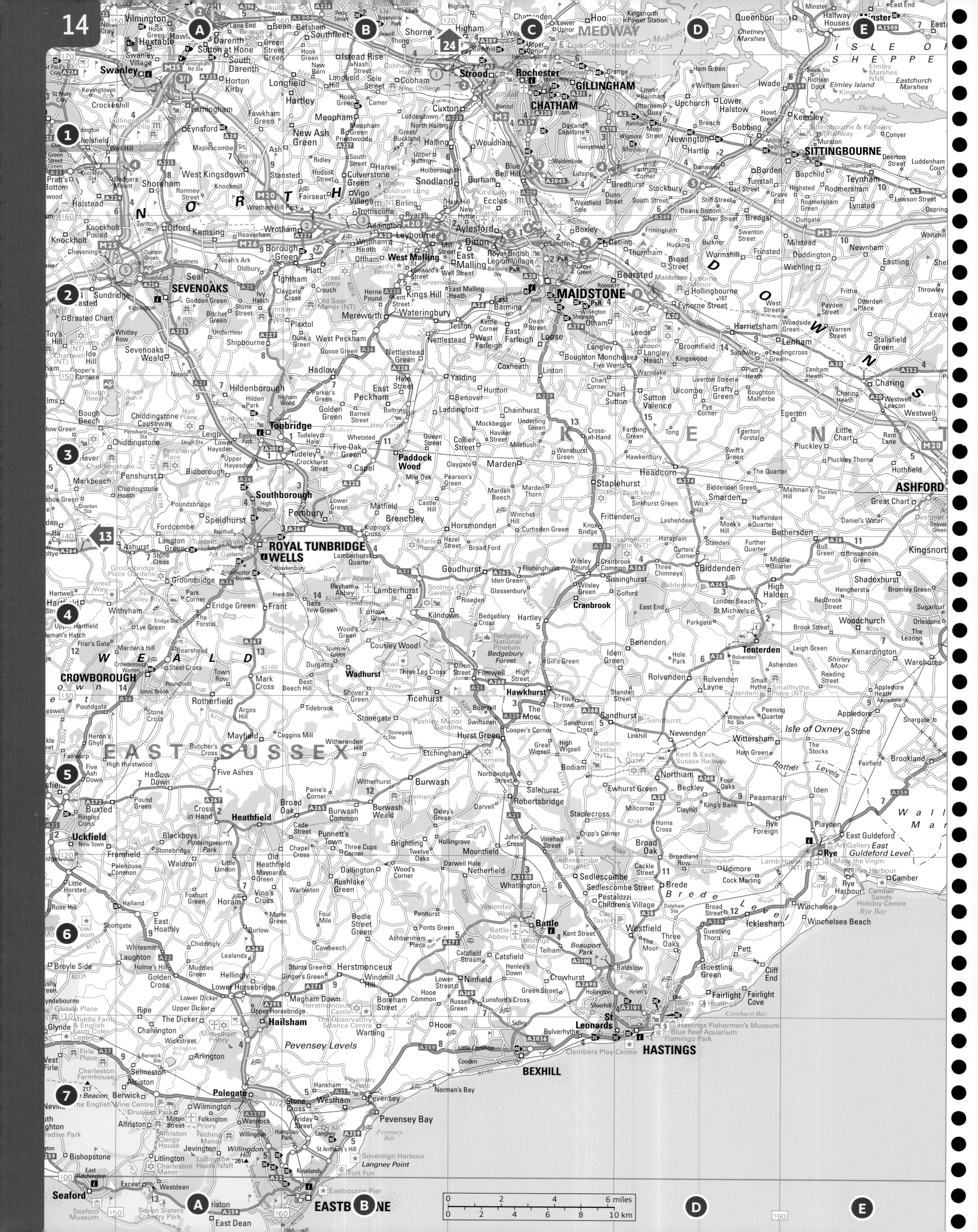
A
B
C
D
E
1
2
3
4
5
6
7
24
13
MEDWAY
ISLE OF SHEPPEY
NORTH DOWNS
KENT
WEALD
EAST SUSSEX
Swanley
Strood
Rochester
GILLINGHAM
CHATHAM
SITTINGBOURNE
SEVENOAKS
West Malling
MAIDSTONE
Tonbridge
Southborough
ROYAL TUNBRIDGE WELLS
Paddock Wood
ASHFORD
Cranbrook
Tenterden
CROWBOROUGH
Wadhurst
Hawkhurst
Heathfield
Uckfield
Hailsham
Battle
Rye
HASTINGS
BEXHILL
Polegate
Seaford
EASTBOURNE
Pevensey Levels
Bedgebury Forest
Isle of Oxney
Brede Level
Pevensey Bay
Langney Point
Rye Bay
Hastings Fishermen's Museum
Blue Reef Aquarium
Flamingo Park
Clambers Play Centre
Eastbourne Pier
0 2 4 6 miles
0 2 4 6 8 10 km

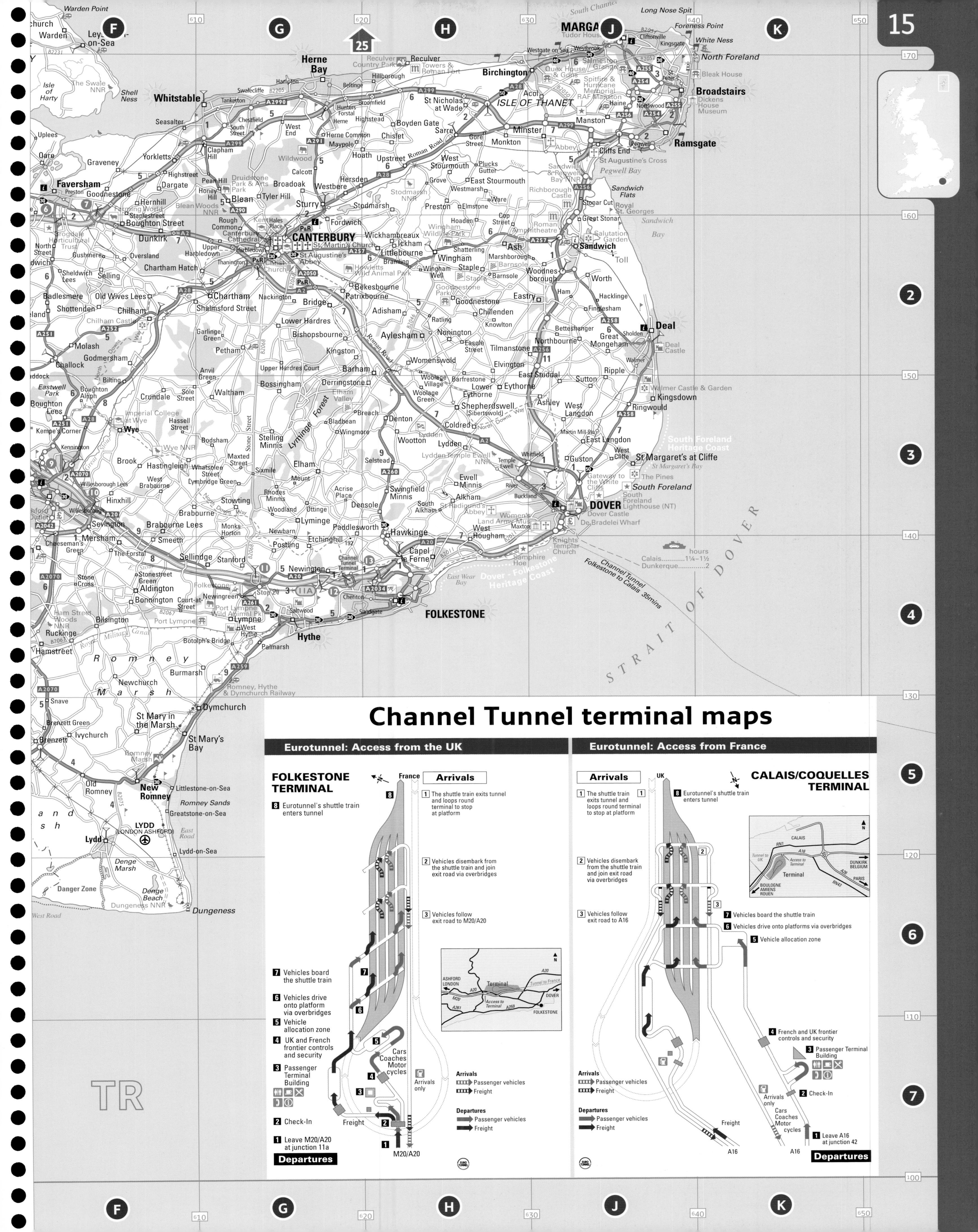
F
G
H
J
K
25
Warden Point
Warden
Leysdown-on-Sea
Isle of Harty
The Swale NNR
Shell Ness
Whitstable
Herne Bay
Reculver
Birchington
MARGATE
Long Nose Spit
Foreness Point
White Ness
North Foreland
Bleak House
Broadstairs
Dickens House Museum
ISLE OF THANET
Manston
Minster
Monkton
Sarre
St Nicholas at Wade
Ramsgate
Cliffs End
St Augustine's Cross
Pegwell Bay
Sandwich Flats
Royal St. Georges
Sandwich Bay
Sandwich
Seasalter
Yorkletts
Graveney
Faversham
Boughton Street
Dunkirk
Chartham Hatch
Blean
Tyler Hill
Sturry
Fordwich
CANTERBURY
Wickhambreaux
Ickham
Littlebourne
Wingham
Ash
Woodnesborough
Worth
Eastry
Deal
Deal Castle
Walmer
Walmer Castle & Garden
Kingsdown
Ringwould
Great Mongeham
Northbourne
Tilmanstone
Nonington
Aylesham
Adisham
Bekesbourne
Patrixbourne
Bridge
Chartham
Shalmsford Street
Chilham
Chilham Castle
Old Wives Lees
Selling
Badlesmere
Shottenden
Molash
Godmersham
Challock
Petham
Bishopsbourne
Kingston
Barham
Womenswold
Elvington
Eythorne
East Studdal
Sutton
Ripple
West Langdon
Shepherdswell (Sibertswold)
Coldred
Lydden
Whitfield
Guston
East Langdon
West Cliffe
St Margaret's at Cliffe
St Margaret's Bay
South Foreland Heritage Coast
The Pines
South Foreland
South Foreland Lighthouse (NT)
DOVER
Dover Castle
De Bradelei Wharf
River
Buckland
Alkham
Ewell Minnis
Temple Ewell
Lydden Temple Ewell NNR
Wootton
Denton
Derringstone
Bossingham
Waltham
Crundale
Wye
Wye NNR
Imperial College at Wye
Hastingleigh
Brook
Elham
Lyminge Forest
Stelling Minnis
Swingfield Minnis
Acrise Place
Densole
Hawkinge
Paddlesworth
Etchinghill
Lyminge
Newington
Postling
Capel le Ferne
West Hougham
Samphire Hoe
Dover - Folkestone Heritage Coast
East Wear Bay
Knights Templar Church
Channel Tunnel Folkestone to Calais 35mins
Calais 1¼–1½
Dunkerque 2
hours
STRAIT OF DOVER
FOLKESTONE
Cheriton
Sandgate
Saltwood
Hythe
West Hythe
Lympne
Port Lympne Wild Animal Pk
Palmarsh
Botolph's Bridge
Burmarsh
Dymchurch
Romney, Hythe & Dymchurch Railway
St Mary in the Marsh
St Mary's Bay
New Romney
Littlestone-on-Sea
Romney Sands
Greatstone-on-Sea
Lydd
LYDD (LONDON ASHFORD)
Lydd-on-Sea
Denge Marsh
Danger Zone
Denge Beach
Dungeness NNR
Dungeness
Romney Marsh
Old Romney
Ivychurch
Brenzett
Brenzett Green
Snave
Newchurch
Hamstreet
Ruckinge
Bilsington
Aldington
Bonnington
Mersham
Sevington
Smeeth
Sellindge
Stanford
Brabourne Lees
Brabourne
Stowting
Hinxhill
Kennington
TR
Channel Tunnel terminal maps
Eurotunnel: Access from the UK
FOLKESTONE TERMINAL
France
Arrivals
1 The shuttle train exits tunnel and loops round terminal to stop at platform
2 Vehicles disembark from the shuttle train and join exit road via overbridges
3 Vehicles follow exit road to M20/A20
8 Eurotunnel's shuttle train enters tunnel
7 Vehicles board the shuttle train
6 Vehicles drive onto platform via overbridges
5 Vehicle allocation zone
4 UK and French frontier controls and security
3 Passenger Terminal Building
2 Check-In
1 Leave M20/A20 at junction 11a
Cars Coaches Motor cycles
Arrivals only
Freight
M20/A20
Arrivals
Passenger vehicles
Freight
Departures
Passenger vehicles
Freight
Departures
ASHFORD LONDON
Terminal
Tunnel to France
Access to Terminal
DOVER
FOLKESTONE
Eurotunnel: Access from France
CALAIS/COQUELLES TERMINAL
UK
Arrivals
1 The shuttle train exits tunnel and loops round terminal to stop at platform
2 Vehicles disembark from the shuttle train and join exit road via overbridges
3 Vehicles follow exit road to A16
8 Eurotunnel's shuttle train enters tunnel
7 Vehicles board the shuttle train
6 Vehicles drive onto platforms via overbridges
5 Vehicle allocation zone
4 French and UK frontier controls and security
3 Passenger Terminal Building
2 Check-In
1 Leave A16 at junction 42
CALAIS
Tunnel to UK
Access to Terminal
Terminal
DUNKIRK BELGIUM
PARIS
BOULOGNE AMIENS ROUEN
Arrivals only
Cars Coaches Motor cycles
Freight
A16
Arrivals
Passenger vehicles
Freight
Departures
Passenger vehicles
Freight
Departures

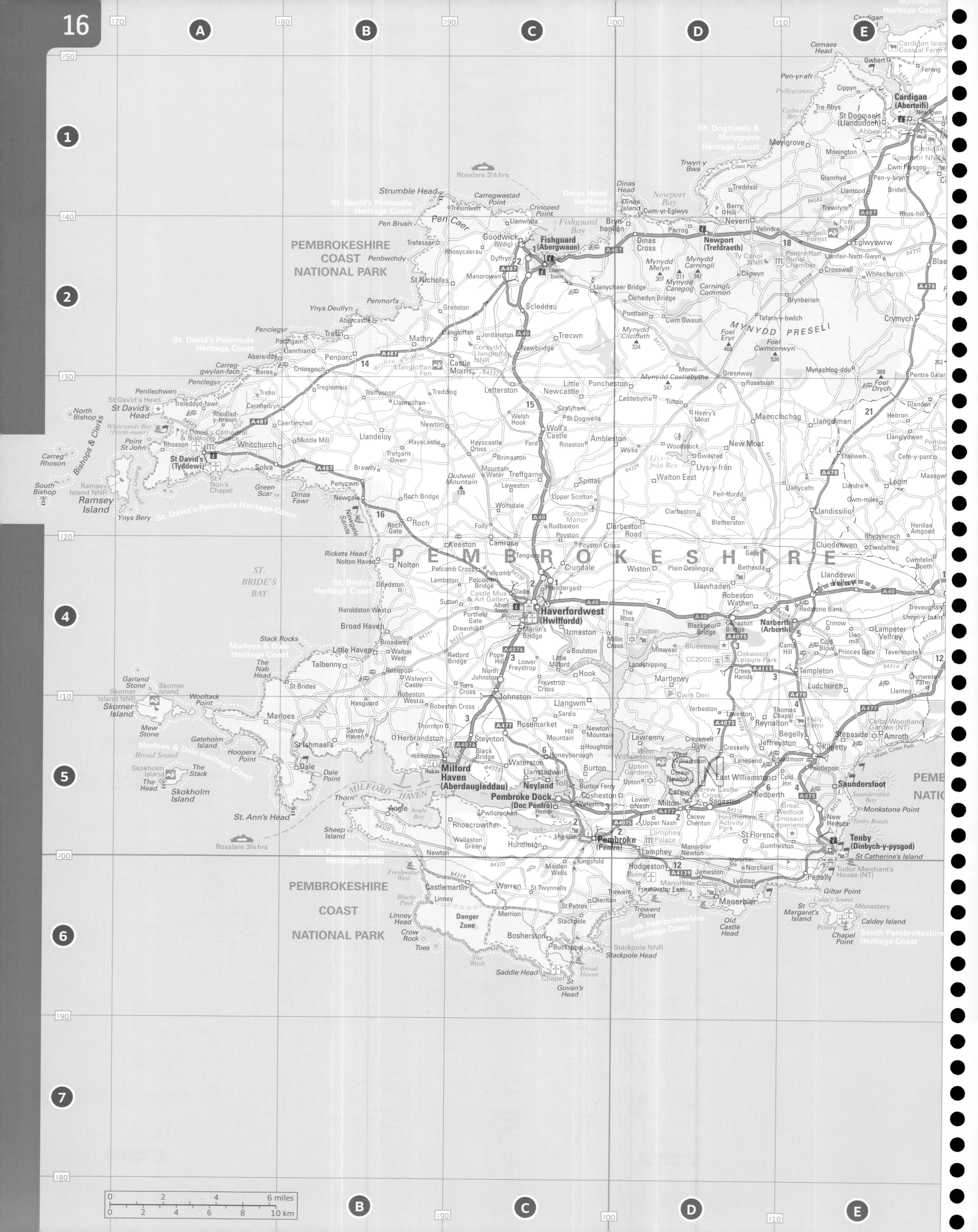
PEMBROKESHIRE
PEMBROKESHIRE COAST NATIONAL PARK
ST. BRIDE'S BAY
MILFORD HAVEN
MYNYDD PRESELI
Strumble Head
Pen Caer
Fishguard (Abergwaun)
Goodwick (Wdig)
Fishguard Bay
Newport (Trefdraeth)
Newport Bay
Dinas Head
Cardigan (Aberteifi)
St Dogmaels (Llandudoch)
Cemaes Head
Moylgrove
Eglwyswrw
Crymych
Mathry
St Nicholas
Trefin
Abereiddy
Porthgain
St David's Head
St David's (Tyddewi)
Whitesands Bay (Porth-mawr)
Ramsey Island
Ramsey Sound
North Bishop
South Bishop
Bishops & Clerks
Carreg Rhoson
Solva
Newgale
Newgale Sands
Roch
Letterston
Puncheston
Maenclochog
Wolf's Castle
Treffgarne
Spittal
Camrose
Keeston
Nolton Haven
Druidston
Broad Haven
Little Haven
Haverfordwest (Hwlffordd)
Merlin's Bridge
Johnston
Narberth (Arberth)
Templeton
Clunderwen
Llanddewi Velfrey
Kilgetty
Saundersfoot
Tenby (Dinbych-y-pysgod)
Penally
Caldey Island
Giltar Point
Manorbier
Lamphey
Pembroke (Penfro)
Pembroke Dock (Doc Penfro)
Neyland
Milford Haven (Aberdaugleddau)
Herbrandston
St Ishmael's
Dale
Marloes
St. Ann's Head
Skomer Island
Skokholm Island
Angle
Castlemartin
Bosherston
Stackpole Head
St Govan's Head
Saddle Head
Danger Zone
Rosslare 3¼ hrs
A487
A40
A478
A477
A4076
A4075
0 2 4 6 miles
0 2 4 6 8 10 km

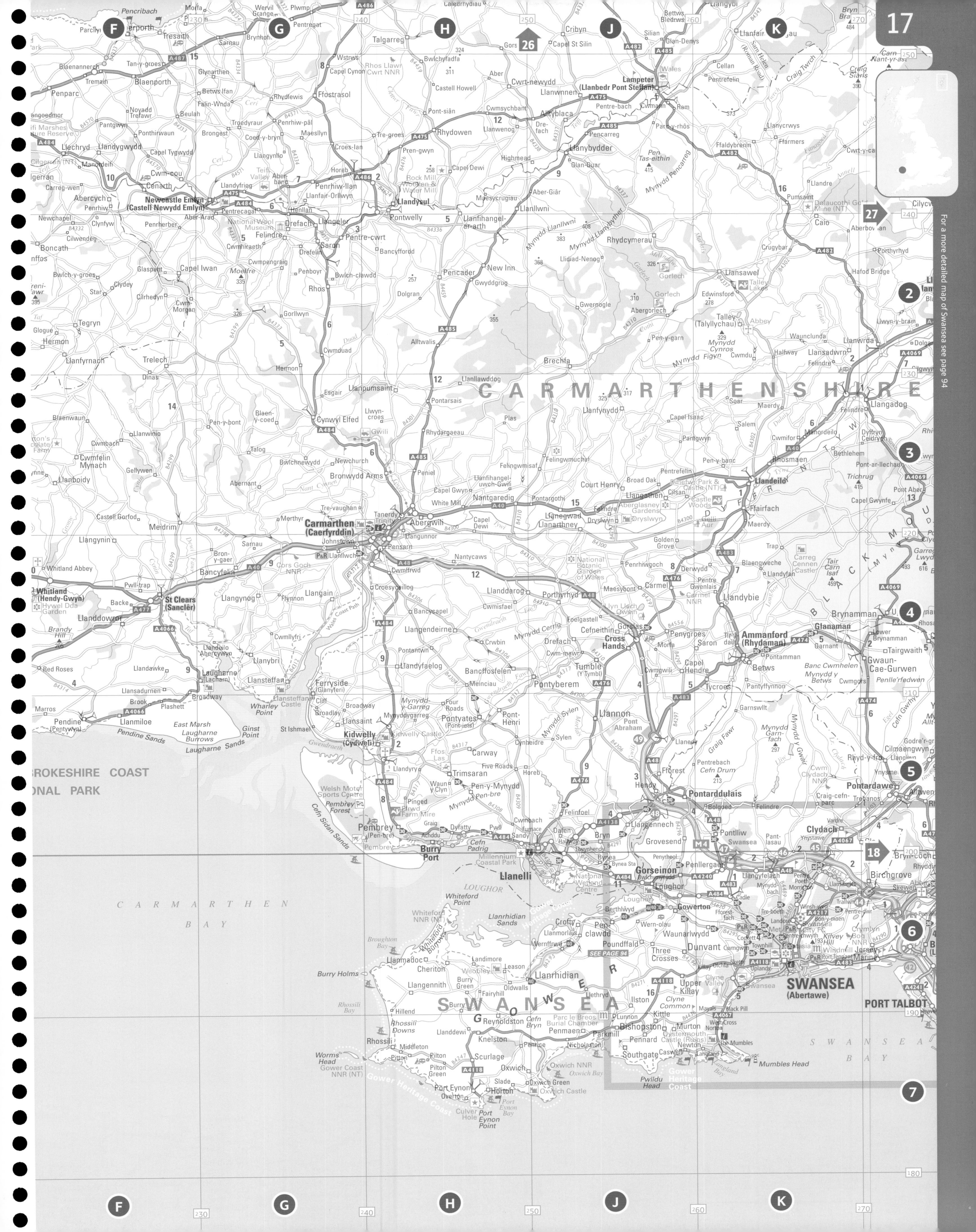

F
G
H
J
K
26
27
18
CARMARTHENSHIRE
SWANSEA
SWANSEA (Abertawe)
PORT TALBOT
Carmarthen (Caerfyrddin)
Llanelli
Lampeter (Llanbedr Pont Steffan)
Newcastle Emlyn (Castell Newydd Emlyn)
Llandysul
Llandeilo
Ammanford (Rhydaman)
Whitland (Hendy-Gwyn)
St Clears (Sanclêr)
Kidwelly (Cydweli)
Pembrey (Pen-bre)
Burry Port
Gorseinon
Loughor
Gowerton
Pontarddulais
Cross Hands
Llandybie
Brynamman
Glanaman
Gwaun-Cae-Gurwen
Pontardawe
Clydach
Birchgrove
Pontyberem
Tumble (Y Tymbl)
Laugharne (Lacharn)
Ferryside (Glanyferi)
Pendine (Pentywyn)
Pendine Sands
Laugharne Sands
Cefn Sidan Sands
Llanrhidian
Reynoldston
Port Eynon
Rhossili
Mumbles Head
Pwlldu Head
Worms Head
Burry Holms
Whiteford Point
LOUGHOR
CARMARTHEN BAY
SWANSEA BAY
PEMBROKESHIRE COAST NATIONAL PARK
BLACK MOUNTAIN
TYWI
Mynydd Llanllwni
Mynydd Pencarreg
Mynydd Sylen
Mynydd Cerrig
Gower Heritage Coast
Millennium Coastal Park
National Botanic Garden of Wales
SEE PAGE 94
For a more detailed map of Swansea see page 94
230
240
250
260
270
180
190
200
210
220
230
240
250
1
2
3
4
5
6
7

For a more detailed map of Swansea see page 94

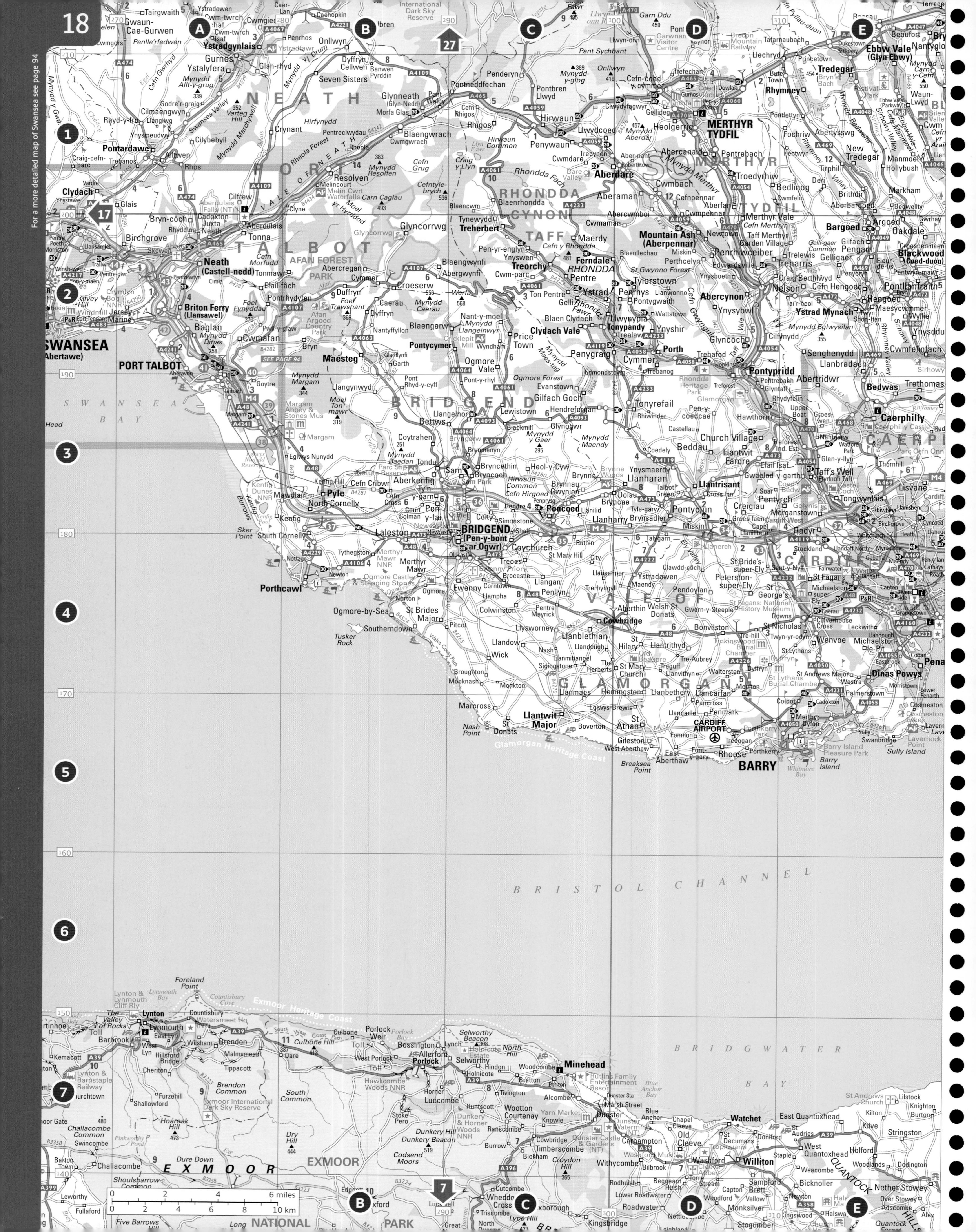

A
B
C
D
E
1
2
3
4
5
6
7
27
17
SEE PAGE 94
Gwaun-Cae-Gurwen
Tairgwaith
Ystradgynlais
Cwmgiedd
Gurnos
Ystalyfera
Onllwyn
Dyffryn Cellwen
Seven Sisters
Glan-rhyd
International Dark Sky Reserve
NEATH
PORT
TALBOT
Pontardawe
Alltwen
Rhos
Cilybebyll
Rhyd-y-fro
Cilmaengwyn
Ynysmeudwy
Clydach
Glais
Bryn-coch
Birchgrove
Neath (Castell-nedd)
Tonna
Aberdulais
Cadoxton-Juxta-Neath
Skewen
Briton Ferry (Llansawel)
Baglan
Cwmafan
SWANSEA (Abertawe)
PORT TALBOT
Aberavon
Margam
SWANSEA BAY
Kenfig
Pyle
North Cornelly
South Cornelly
Porthcawl
Sker Point
Nottage
Newton
Glynneath
Resolven
Blaengwrach
Rheola
Melincourt
Crynant
Pontneddfechan
Penderyn
Hirwaun
Penywaun
Llwydcoed
Aberdare
Cwmdare
Trecynon
Aberaman
Cwmbach
Mountain Ash (Aberpennar)
Penrhiwceiber
Abercynon
Ynysybwl
Pontypridd
MERTHYR TYDFIL
Merthyr Vale
Treharris
Nelson
Troedyrhiw
Rhymney
Tredegar
Ebbw Vale (Glyn Ebwy)
Bargoed
Blackwood (Coed-duon)
Ystrad Mynach
Caerphilly
RHONDDA
CYNON
TAFF
Treherbert
Treorchy
Ferndale
Maerdy
Tylorstown
Ystrad
Tonypandy
Porth
Penygraig
Clydach Vale
Cymmer
Tonyrefail
Blaengwynfi
Glyncorrwg
Croeserw
Cymmer
Caerau
Maesteg
Pontycymer
Price Town
Ogmore Vale
Nantymoel
BRIDGEND
BRIDGEND (Pen-y-bont ar Ogwr)
Bettws
Llangeinor
Tondu
Aberkenfig
Sarn
Bryncethin
Pencoed
Coychurch
Laleston
Ewenny
Ogmore-by-Sea
Southerndown
St Brides Major
Tusker Rock
Llanharan
Llantrisant
Pontyclun
Church Village
Beddau
Llanharry
Cowbridge
Llanblethian
Llantwit Major
St Donats
Nash Point
Marcross
Boverton
St Athan
Gileston
East Aberthaw
Breaksea Point
Rhoose
CARDIFF AIRPORT
BARRY
Barry Island
Sully
Dinas Powys
Wenvoe
St Nicholas
Peterston-super-Ely
St Fagans
Radyr
Creigiau
Pentyrch
Taff's Well
Tongwynlais
Lisvane
Bedwas
VALE OF GLAMORGAN
CARDIFF
Glamorgan Heritage Coast
BRISTOL CHANNEL
BRIDGWATER BAY
Foreland Point
Lynton
Lynmouth
Countisbury
Barbrook
Brendon
Exmoor Heritage Coast
Culbone Hill
Porlock Weir
Porlock
Bossington
Allerford
Selworthy
Minehead
Alcombe
Dunster
Blue Anchor
Watchet
Williton
Old Cleeve
Washford
Carhampton
Timberscombe
Cutcombe
Wheddon Cross
Luccombe
Wootton Courtenay
Dunkery Beacon
Exford
Simonsbath
EXMOOR
Dure Down
Brendon Common
Challacombe
Parracombe
St Audries
West Quantoxhead
Kilve
Stringston
Holford
Nether Stowey
Over Stowey
Bicknoller
Crowcombe
QUANTOCK HILLS
EXMOOR NATIONAL PARK
0 2 4 6 miles
0 2 4 6 8 10 km

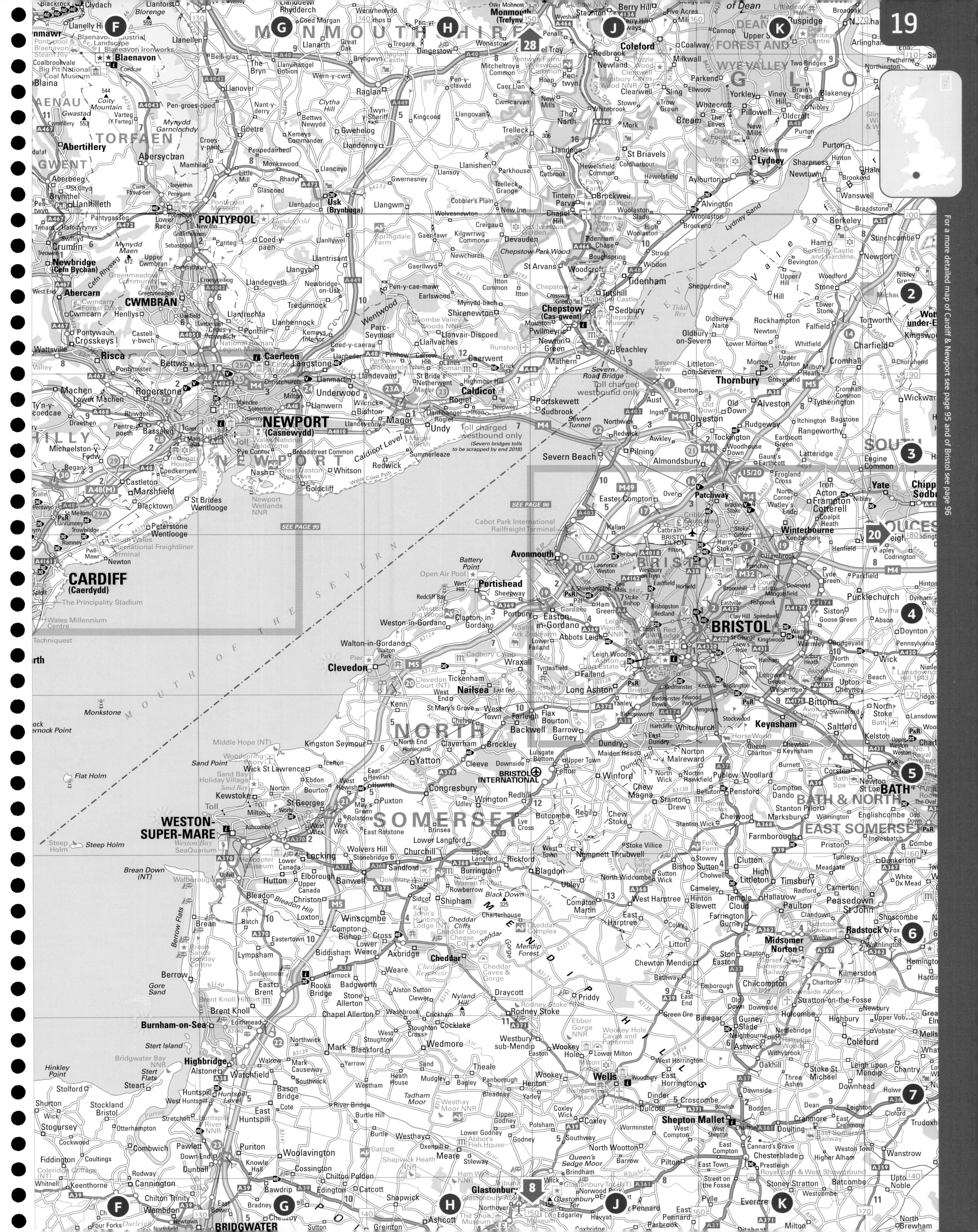
For a more detailed map of Cardiff & Newport see page 95 and of Bristol see page 96
MONMOUTHSHIRE
Monmouth (Trefynwy)
28
FOREST AND WYE VALLEY
DEAN
Coleford
Blaenavon
TORFAEN
Abertillery
Pontypool
CWMBRÂN
Usk (Brynbuga)
Raglan
Chepstow (Cas-gwent)
Lydney
Caerleon
NEWPORT (Casnewydd)
NEWPORT
Caldicot
Toll charged westbound only
Severn Road Bridge
Severn Tunnel
Toll charged westbound only (Severn bridges tolls to be scrapped by end 2018)
SEE PAGE 95
SEE PAGE 96
CARDIFF (Caerdydd)
The Principality Stadium
Wales Millennium Centre
Thornbury
Berkeley
Patchway
BRISTOL
Avonmouth
Portishead
Clevedon
Nailsea
Keynsham
BATH
BATH & NORTH EAST SOMERSET
NORTH SOMERSET
WESTON-SUPER-MARE
BRISTOL INTERNATIONAL
MOUTH OF THE SEVERN
Flat Holm
Steep Holm
Cheddar
MENDIP HILLS
Wells
Shepton Mallet
Midsomer Norton
Radstock
Burnham-on-Sea
Highbridge
Glastonbury
BRIDGWATER
Winterbourne
Yate
20
8
F G H J K
1 2 3 4 5 6 7

For a more detailed map of Bristol see page 96

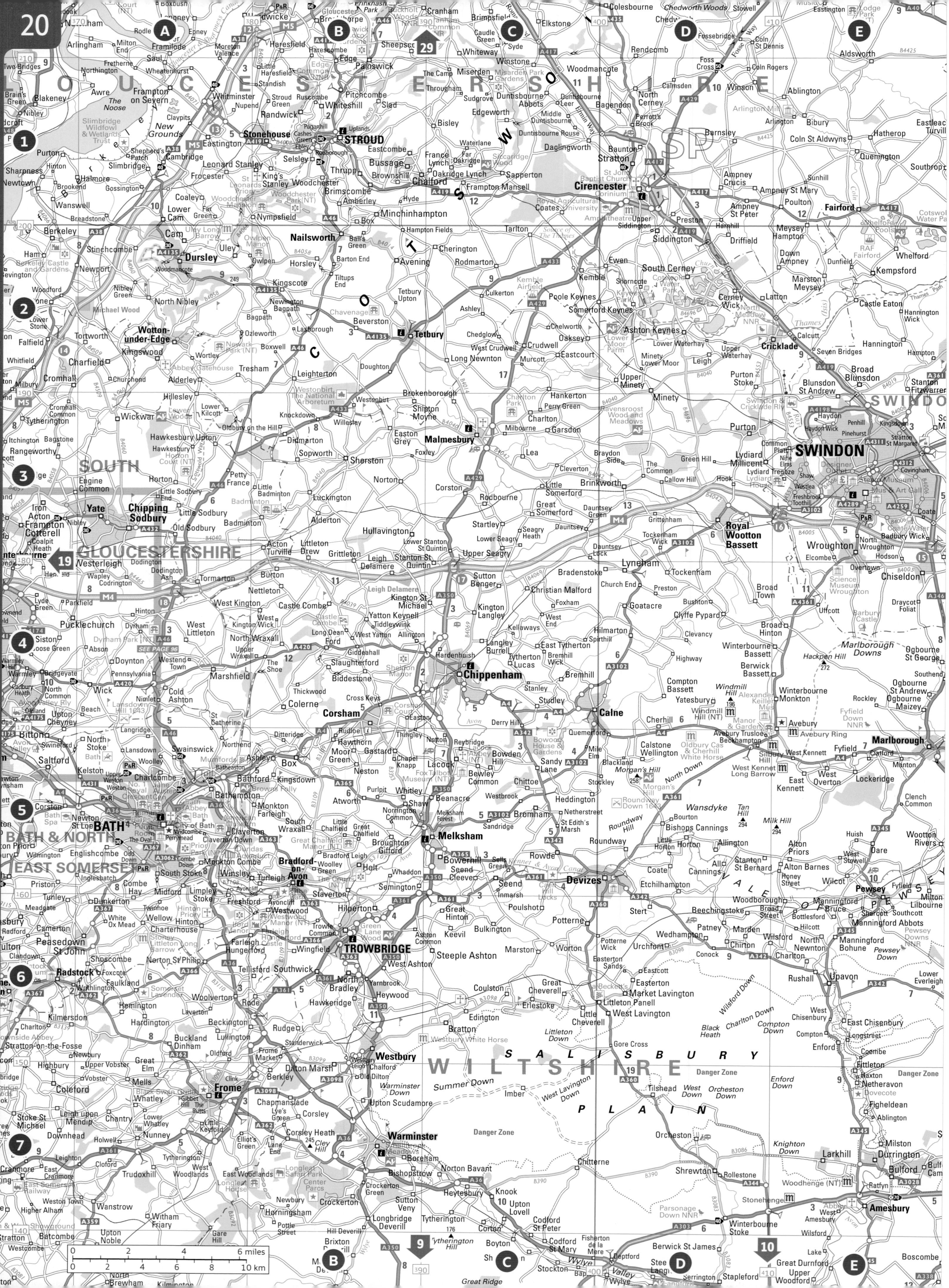

GLOUCESTERSHIRE

WILTSHIRE

SWINDON

SALISBURY PLAIN

BATH & NORTH EAST SOMERSET

SOUTH GLOUCESTERSHIRE

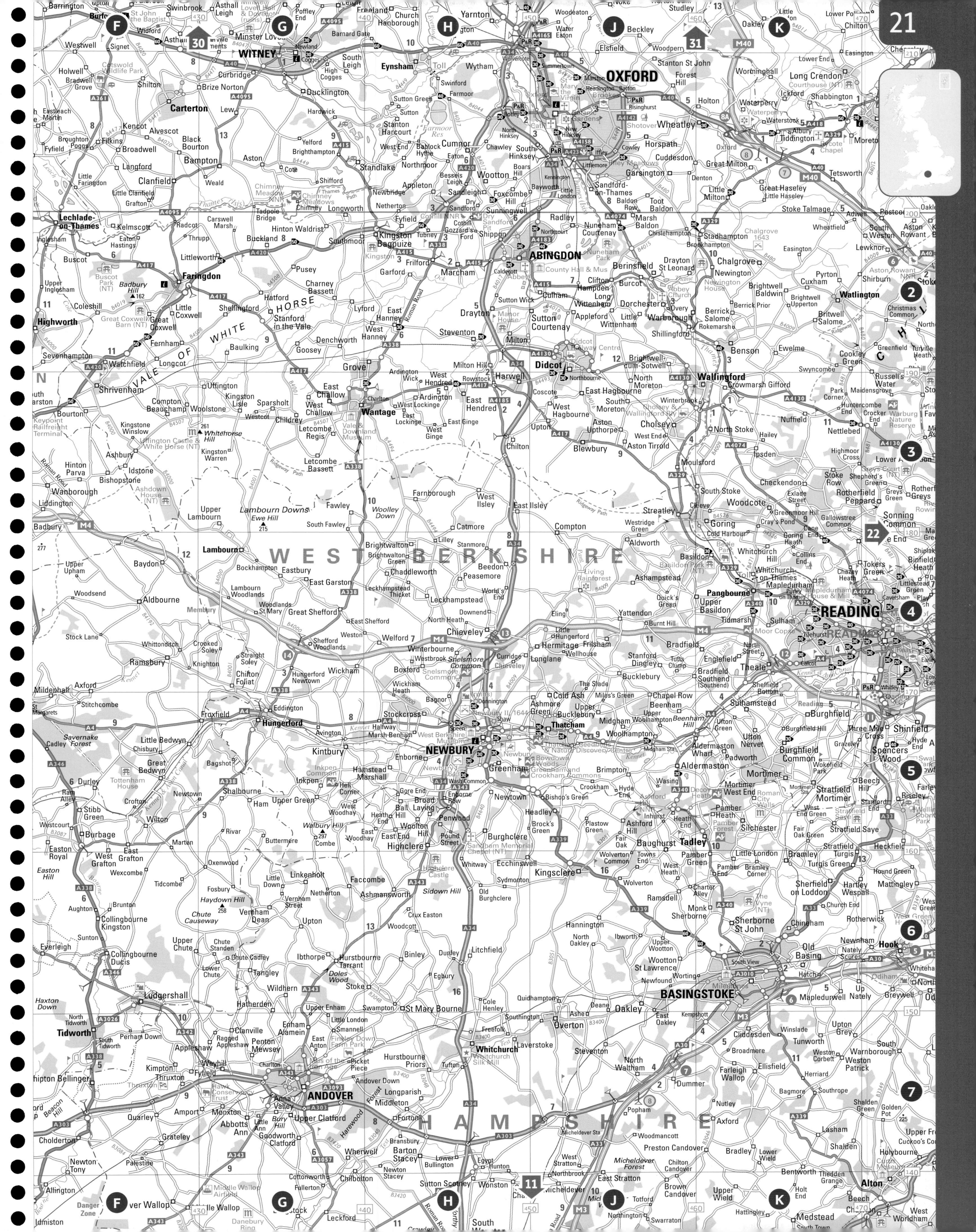
F
G
H
J
K
30
31
22
11
WITNEY
OXFORD
ABINGDON
Didcot
Wallingford
Wantage
Faringdon
Carterton
Highworth
Shrivenham
Lechlade-on-Thames
Bampton
Eynsham
Kingston Bagpuize
Wheatley
Thame
Watlington
Tetsworth
Benson
Goring
Pangbourne
READING
Theale
Shinfield
Lambourn
Hungerford
NEWBURY
Thatcham
Kintbury
Greenham
Kingsclere
Tadley
Mortimer
BASINGSTOKE
Hook
Whitchurch
Overton
Oakley
ANDOVER
Ludgershall
Tidworth
Alton
WEST BERKSHIRE
HAMPSHIRE
VALE OF WHITE HORSE
Lambourn Downs
Stockbridge
Micheldever
Sutton Scotney
Oakley
Sonning Common
Nettlebed
Cholsey
Streatley
Compton
Chieveley
Highclere
Burghclere
Kingsclere
Chilton
Harwell
Milton
Drayton
Sutton Courtenay
Steventon
Grove
Culham
Dorchester
Berinsfield
Garsington
Cumnor
Standlake
Ducklington
Minster Lovell
Burford
Asthall
Brize Norton
Clanfield
Kelmscott
Buckland
Uffington
Ashbury
Aldbourne
Ramsbury
Baydon
Great Shefford
Froxfield
Shalbourne
Burbage
Collingbourne Ducis
Enham Alamein
Andover Down
Longparish
Barton Stacey
Hurstbourne Priors
St Mary Bourne
Whitchurch
Laverstoke
Dummer
North Waltham
Cliddesden
Old Basing
Sherborne St John
Bramley
Silchester
Aldermaston
Burghfield
Woolhampton
Bucklebury
Hermitage
Bradfield
Pangbourne
Mapledurham
Woodcote
Checkendon
Crowmarsh Gifford
Stadhampton
Chalgrove
Great Haseley
Little Milton
Tiddington
Long Crendon
Forest Hill
Stanton St John
Beckley
Kidlington
Yarnton
Botley
Kennington
Radley
Nuneham Courtenay
Cowley
Headington
Marston

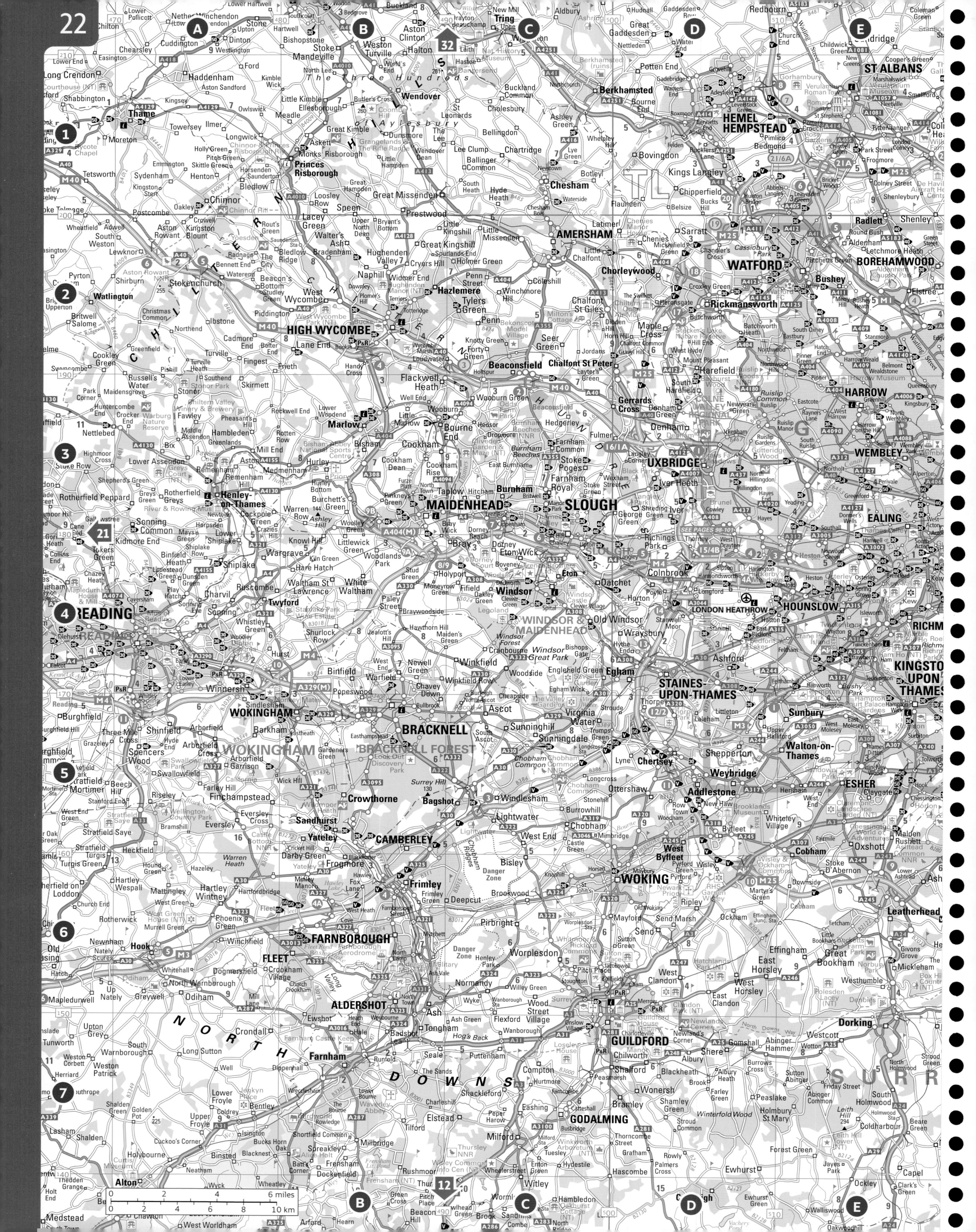
A
B
C
D
E
1
2
3
4
5
6
7
32
21
12
Thame
Princes Risborough
Wendover
Tring
Berkhamsted
HEMEL HEMPSTEAD
ST ALBANS
Chesham
Kings Langley
AMERSHAM
WATFORD
BOREHAMWOOD
Radlett
Chorleywood
Rickmansworth
Watlington
Stokenchurch
HIGH WYCOMBE
Hazlemere
Beaconsfield
Chalfont St Peter
Harefield
HARROW
WEMBLEY
Marlow
Henley-on-Thames
MAIDENHEAD
SLOUGH
UXBRIDGE
EALING
Windsor
HOUNSLOW
LONDON HEATHROW
READING
WOKINGHAM
BRACKNELL
Ascot
Egham
STAINES-UPON-THAMES
KINGSTON UPON THAMES
Sunbury
Walton-on-Thames
ESHER
Weybridge
Chertsey
Addlestone
Crowthorne
Sandhurst
CAMBERLEY
Bagshot
WOKING
West Byfleet
Cobham
Leatherhead
FARNBOROUGH
FLEET
Hook
ALDERSHOT
Farnham
GUILDFORD
GODALMING
Dorking
Alton
C H I L T E R N
N O R T H
D O W N S
S U R R
0 2 4 6 miles
0 2 4 6 8 10 km

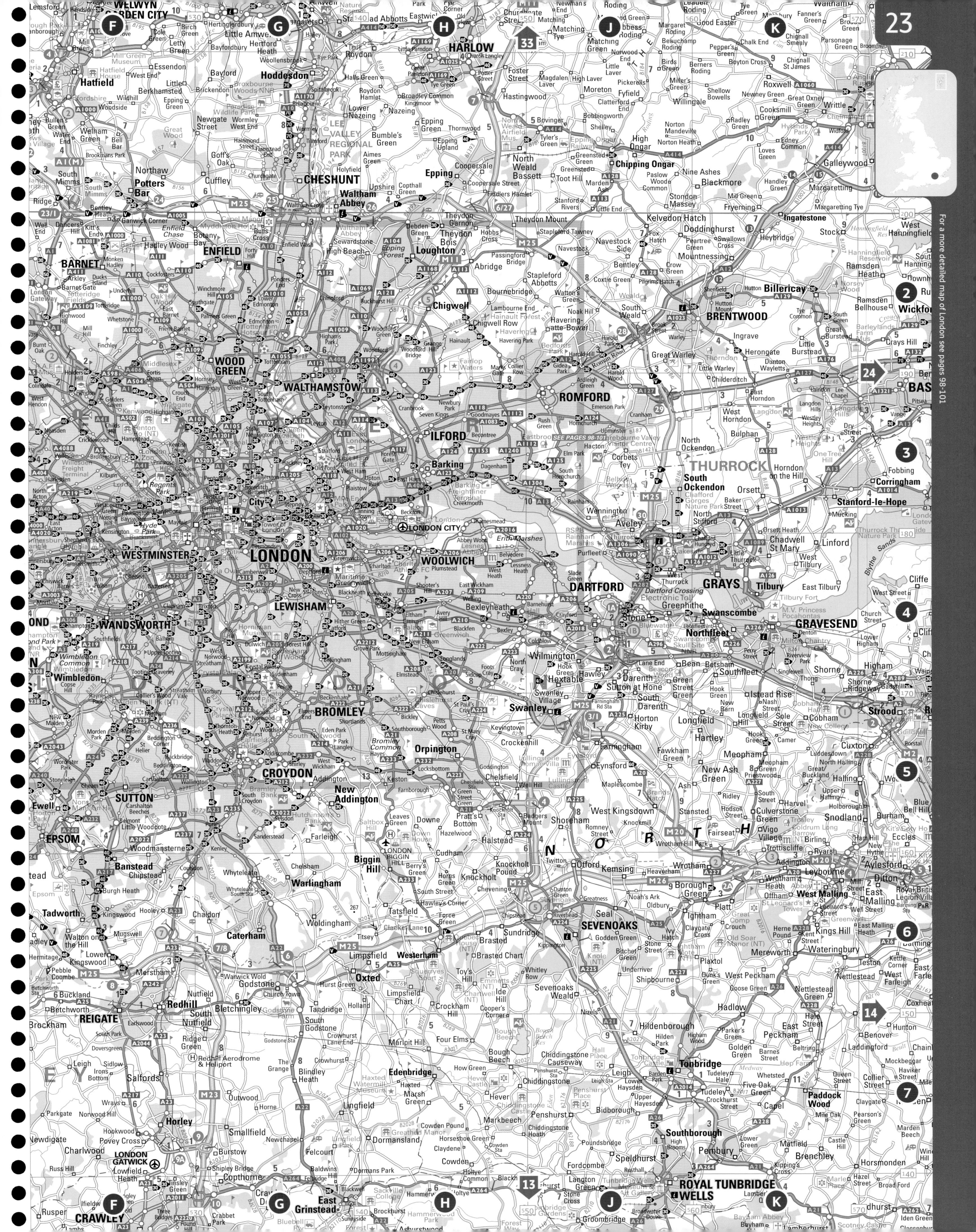
For a more detailed map of London see pages 98-101
33
24
14
13
F
G
H
J
K
2
3
4
5
6
7
WELWYN GARDEN CITY
Hatfield
HARLOW
Hoddesdon
CHESHUNT
Epping
Chipping Ongar
Potters Bar
Waltham Abbey
ENFIELD
BARNET
Loughton
Chigwell
BRENTWOOD
Billericay
WOOD GREEN
WALTHAMSTOW
ROMFORD
ILFORD
Barking
THURROCK
South Ockendon
Stanford-le-Hope
LONDON
WESTMINSTER
City
WOOLWICH
LONDON CITY
DARTFORD
GRAYS
Tilbury
LEWISHAM
WANDSWORTH
Wimbledon
GRAVESEND
Northfleet
Swanscombe
BROMLEY
Orpington
Swanley
CROYDON
New Addington
SUTTON
EPSOM
Strood
Biggin Hill
Warlingham
Caterham
Banstead
Tadworth
Westerham
Oxted
SEVENOAKS
Redhill
REIGATE
West Malling
Edenbridge
Tonbridge
Horley
LONDON GATWICK
CRAWLEY
East Grinstead
ROYAL TUNBRIDGE WELLS
Southborough
Paddock Wood
Dartford Crossing
M25
M20
M26
M23
M11

For a more detailed map of London see pages 98-101

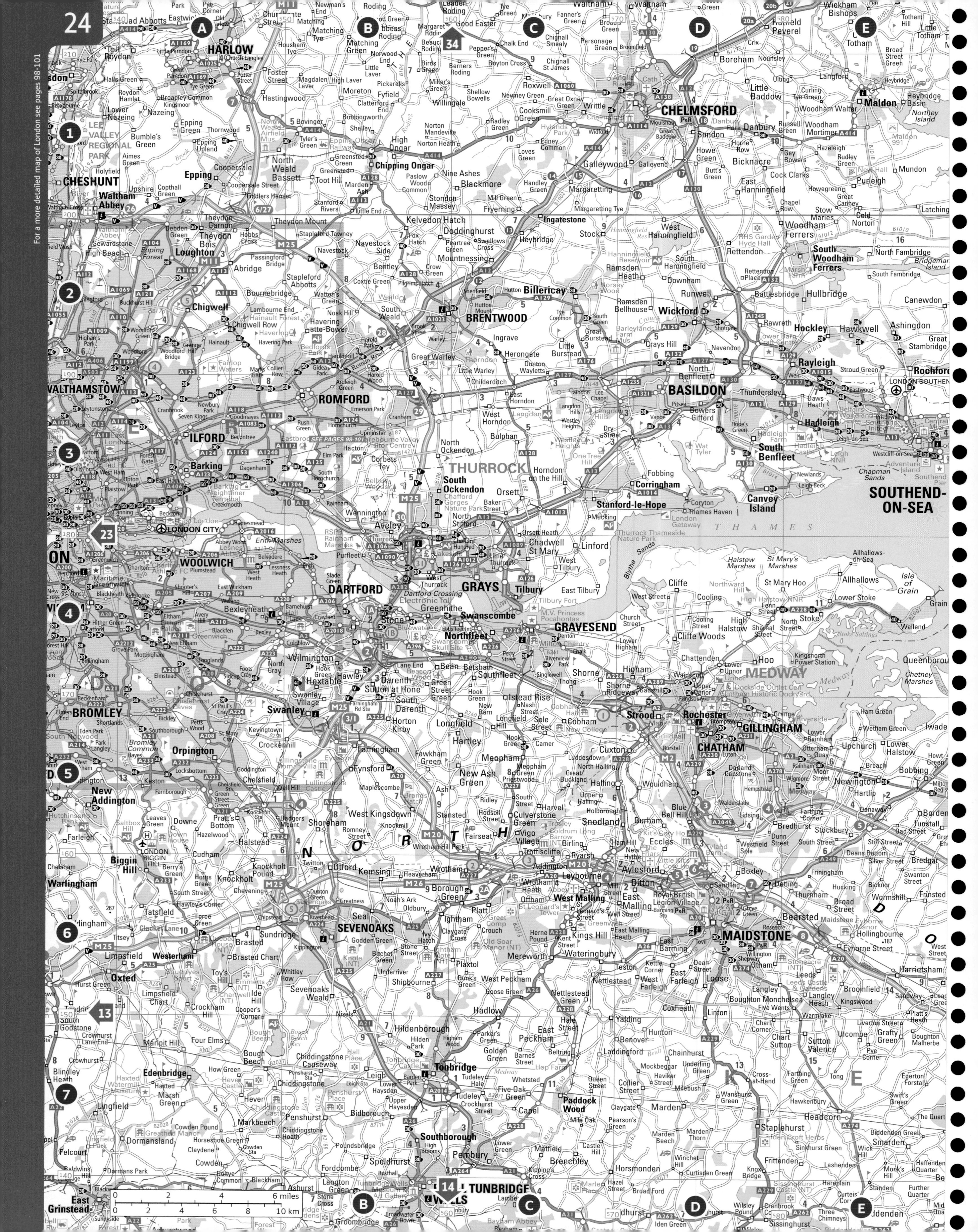
HARLOW
CHELMSFORD
Maldon
CHESHUNT
Waltham Abbey
Epping
Chipping Ongar
Loughton
Chigwell
BRENTWOOD
Billericay
Wickford
Hockley
Rayleigh
Rochford
WALTHAMSTOW
ROMFORD
BASILDON
ILFORD
Barking
THURROCK
South Ockendon
Stanford-le-Hope
Canvey Island
SOUTHEND-ON-SEA
LONDON CITY
WOOLWICH
DARTFORD
GRAYS
Tilbury
GRAVESEND
Swanscombe
Northfleet
Bexleyheath
MEDWAY
Strood
Rochester
GILLINGHAM
CHATHAM
BROMLEY
Orpington
Swanley
New Addington
Biggin Hill
Warlingham
Westerham
Oxted
SEVENOAKS
West Malling
MAIDSTONE
Edenbridge
Tonbridge
Southborough
TUNBRIDGE WELLS
East Grinstead
Lingfield
Paddock Wood
THAMES
SEE PAGES 98-101
0 2 4 6 miles
0 2 4 6 8 10 km

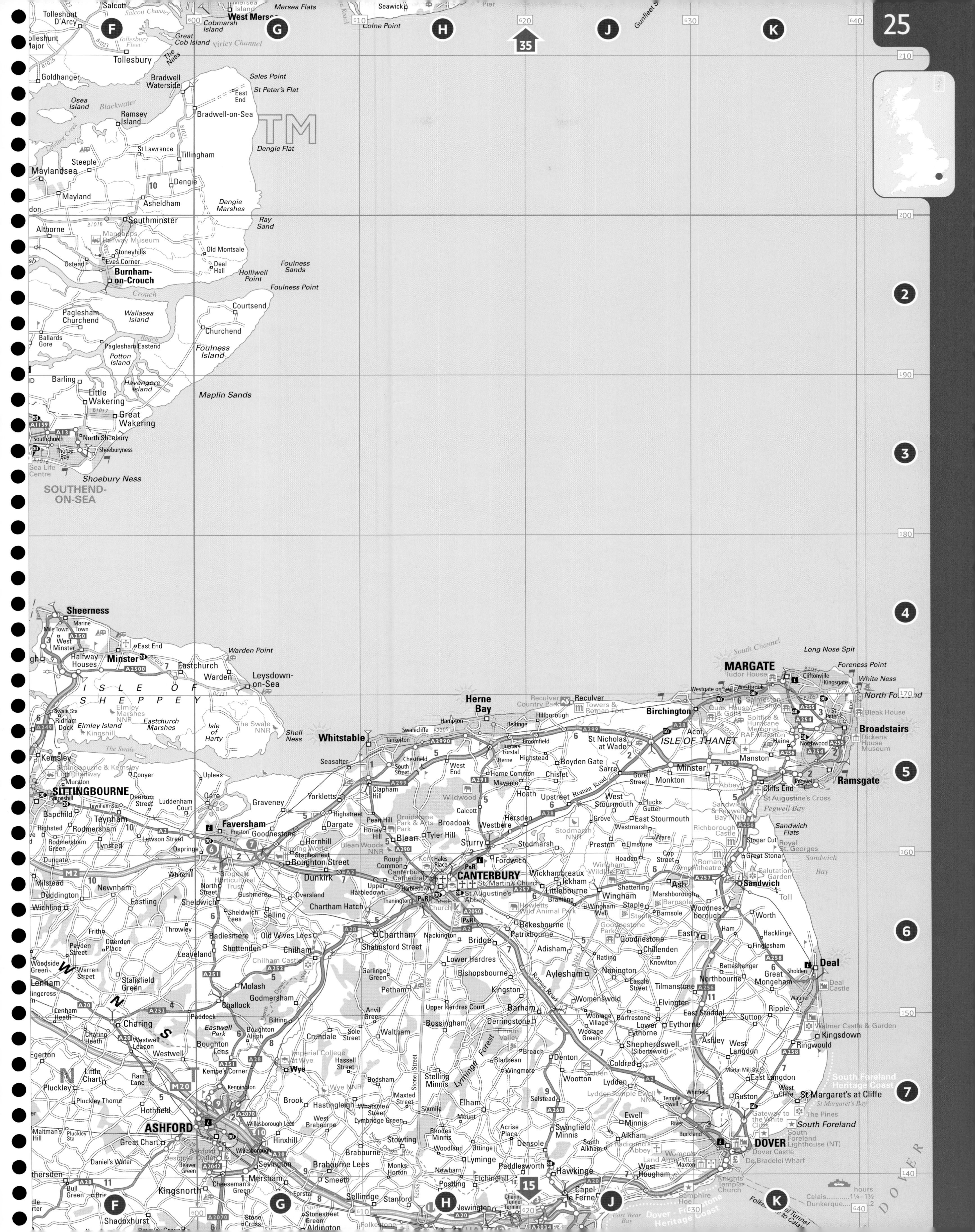

F
G
H
J
K
35
15
TM
Salcott
Tolleshunt D'Arcy
Tolleshunt Major
Tollesbury
Mersea Island
West Mersea
Mersea Flats
Cobmarsh Island
Great Cob Island
Virley Channel
Seawick
Colne Point
Gunfleet Sands
The Nass
Goldhanger
Bradwell Waterside
Sales Point
St Peter's Flat
East End
Osea Island
Blackwater
Ramsey Island
Bradwell-on-Sea
Steeple
St Lawrence
Tillingham
Dengie Flat
Maylandsea
Mayland
Dengie
Asheldham
Dengie Marshes
Southminster
Ray Sand
Althorne
Mangapps Railway Museum
Stoneyhills
Old Montsale
Ostend
Eves Corner
Burnham-on-Crouch
Deal Hall
Holliwell Point
Foulness Sands
Foulness Point
Crouch
Paglesham Churchend
Wallasea Island
Courtsend
Ballards Gore
Churchend
Paglesham Eastend
Foulness Island
Potton Island
Barling
Havengore Island
Little Wakering
Maplin Sands
Great Wakering
Southchurch
North Shoebury
Shoeburyness
Thorpe Bay
Sea Life Centre
Shoebury Ness
SOUTHEND-ON-SEA
Sheerness
Marine Town
Mile Town
West Minster
East End
Halfway Houses
Minster
Warden Point
Eastchurch
Warden
Leysdown-on-Sea
ISLE OF SHEPPEY
Elmley Marshes NNR
Eastchurch Marshes
Ridham Dock
Elmley Island
Kingshill
Isle of Harty
The Swale NNR
Shell Ness
The Swale
Kemsley
Sittingbourne & Kemsley Light Railway
Murston
Conyer
Uplees
SITTINGBOURNE
Oare
Deerton Street
Luddenham Court
Graveney
Bapchild
Teynham
Highsted
Rodmersham
Faversham
Preston
Goodnestone
Lewson Street
Ospringe
Rodmersham Green
Lynsted
Dungate
M2
Whitehill
Brogdale Horticultural Trust
North Street
Milstead
Doddington
Newnham
Eastling
Sheldwich
Wichling
Sheldwich Lees
Throwley
Badlesmere
Frith
Otterden Place
Payden Street
Leaveland
Woodside Green
Warren Street
Stalisfield Green
Lenham
Lenham Heath
Charing
Charing Heath
Westwell Leacon
Westwell
Egerton
Little Chart
Pluckley
Pluckley Thorne
Hothfield
Maltman's Hill
Pluckley Sta
Great Chart
ASHFORD
Daniel's Water
Bethersden
Bull Green
Brisley
Kingsnorth
Shadoxhurst
Whitstable
Seasalter
Swalecliffe
Tankerton
Chestfield
South Street
Herne Bay
Hampton
Reculver Country Park
Reculver
Towers & Roman Fort
Hillborough
Beltinge
Broomfield
Hunters Forstal
Herne
Herne Common
Highstead
Maypole
Hoath
Boyden Gate
Chislet
Upstreet
Roman Road
West Stourmouth
Plucks Gutter
Birchington
Acol
St Nicholas at Wade
Sarre
Gore Street
Monkton
Minster
ISLE OF THANET
Manston
MARGATE
Tudor House
Westgate on Sea
Westbrook
Cliftonville
Kingsgate
Salmestone Grange
Quex House & Gdns
Spitfire & Hurricane Memorial
RAF Manston
Haine
St Peter's
Northwood
Bleak House
Broadstairs
Dickens House Museum
Pegwell
Ramsgate
Cliffs End
St Augustine's Cross
Pegwell Bay
Sandwich Flats
Sandwich & Pegwell Bay NNR
Richborough Castle
Stonar Cut
Royal St. Georges
Sandwich Bay
Great Stonar
Roman Amphitheatre
Salutation Garden
Sandwich
Toll
Long Nose Spit
Foreness Point
White Ness
North Foreland
South Channel
Yorkletts
Highstreet
Dargate
Clapham Hill
Wildwood
West End
Calcott
Broadoak
Druidstone Park & Arts Park
Pean Hill
Honey Hill
Blean
Tyler Hill
Sturry
Westbere
Hersden
Stodmarsh NNR
Stodmarsh
Grove
Westmarsh
East Stourmouth
Ware
Preston
Elmstone
Hoaden
Cop Street
Ash
Hernhill
Mount Ephraim Gardens
Staplestreet
Boughton Street
Blean Woods NNR
Rough Common
Kent Hales Place
Fordwich
CANTERBURY
Canterbury Cathedral
St Martin's Church
St Augustine's Abbey
Harbledown
Upper Harbledown
Thanington
Wickhambreaux
Ickham
Wingham Wildlife Park
Littlebourne
Wingham
Shatterling
Marshborough
Barnsole
Staple
Dunkirk
Oversland
Chartham Hatch
Bushmere
Selling
Howletts Wild Animal Park
Bramling
Wingham Well
Goodnestone Park
Goodnestone
Bekesbourne
Patrixbourne
Nackington
Bridge
Chartham
Shalmsford Street
Chilham
Chilham Castle
Old Wives Lees
Shottenden
Molash
Godmersham
Challock
Lower Hardres
Bishopsbourne
Adisham
Aylesham
Ratling
Nonington
Chillenden
Knowlton
Woodnesborough
Worth
Ham
Eastry
Northbourne
Hacklinge
Finglesham
Betteshanger
Great Mongeham
Sholden
Deal
Deal Castle
Walmer
Tilmanstone
Easole Street
Kingston
Barham
Womenswold
Elvington
East Studdal
Sutton
Ripple
Walmer Castle & Garden
Kingsdown
Ringwould
Garlinge Green
Petham
Anvil Green
Upper Hardres Court
Bossingham
Derringstone
Elham Valley
Woolage Village
Woolage Green
Barfrestone
Lower Eythorne
Eythorne
Shepherdswell (Sibertswold)
Coldred
Ashley
West Langdon
Paddock
Eastwell Park
Boughton Aluph
Bilting
Crundale
Sole Street
Waltham
Boughton Lees
Kempe's Corner
Imperial College at Wye
Wye
Hassell Street
Wye NNR
Bodsham
Stone Street
Stelling Minnis
Lyminge Forest
Breach
Bladbean
Wingmore
Denton
Wootton
Lydden
Lydden Temple Ewell NNR
Temple Ewell
Whitfield
Martin Mill Sta
East Langdon
Guston
West Cliffe
St Margaret's at Cliffe
St Margaret's Bay
Gateway to the White Cliffs
The Pines
South Foreland
South Foreland Lighthouse (NT)
South Foreland Heritage Coast
Kennington
Brook
Hastingleigh
Whatsole Street
Maxted Street
Sixmile
Elham
Selstead
Mount
Ewell Minnis
Alkham
River
Buckland
DOVER
Dover Castle
De Bradelei Wharf
Willesborough Lees
West Brabourne
Lymbridge Green
Rhodes Minnis
Acrise Place
Swingfield Minnis
South Alkham
St Radigund's Abbey
Women's Land Army Museum
Maxton
Hinxhill
Stowting
Woodland
Ottinge
Densole
Willesborough
Designer Outlet
Sevington
Beaver Green
Brabourne
Brabourne Lees
Monks Horton
Lyminge
Paddlesworth
Hawkinge
West Hougham
Mersham
Cheeseman's Green
Smeeth
Newbarn
Postling
Etchinghill
Capel le Ferne
Forstal
Sellinge
Stanford
Channel Tunnel Terminal
Newington
Samphire Hoe
Knights Templar Church
Stonestreet Green
Aldington
Stone Cross
Folkestone
East Wear Bay
Dover - Folkestone Heritage Coast
Folkestone to Calais
Channel Tunnel
hours
Calais 1¼-1½
Dunkerque 2
DOVER

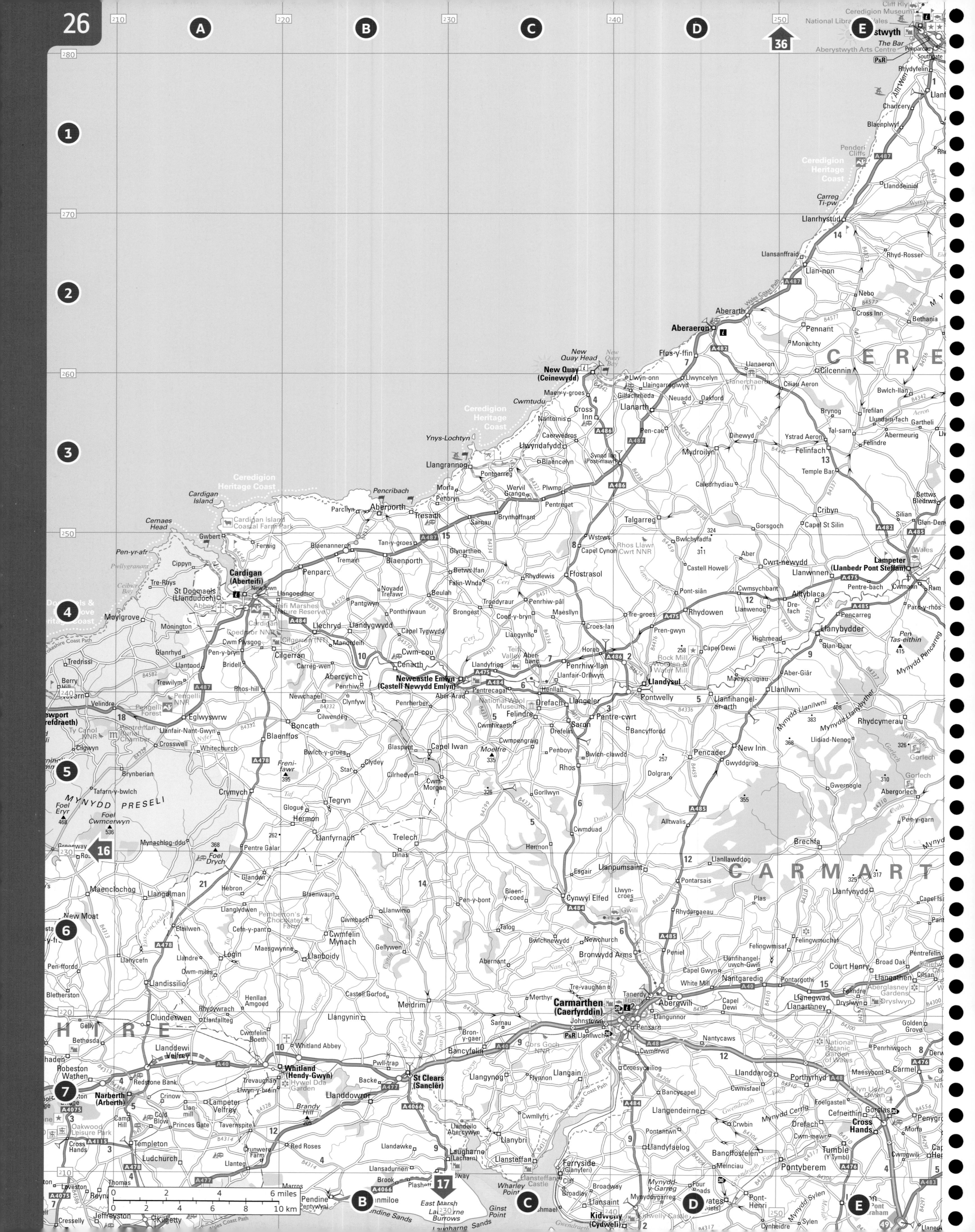
A
B
C
D
E
1
2
3
4
5
6
7
36
16
17
Aberystwyth
Ceredigion Museum
National Library of Wales
Aberystwyth Arts Centre
The Bar
Rhydyfelin
Chancery
Blaenplwyf
Penderi Cliffs
Ceredigion Heritage Coast
Llanddeiniol
Carreg Ti-pw
Llanrhystud
Llansanffraid
Llan-non
Rhyd-Rosser
Nebo
Cross Inn
Bethania
Aberarth
Aberaeron
Pennant
Monachty
Ffos-y-ffin
CERE
New Quay Head
New Quay (Ceinewydd)
Llwyncelyn
Llanerchaeron (NT)
Cilcennin
Ciliau Aeron
Llanarth
Neuadd
Oakford
Bwlch-llan
Cwmtudu
Maen-y-groes
Gilfachrheda
Cross Inn
Brynog
Trefilan
Llundain-fach
Gartheli
Ceredigion Heritage Coast
Pen-cae
Dihewyd
Ystrad Aeron
Tal-sarn
Abermeurig
Felindre
Ynys-Lochtyn
Caerwedros
Mydroilyn
Felinfach
Llwyndafydd
Synod Inn (Post-mawr)
Temple Bar
Llangrannog
Blaencelyn
Pontgarreg
Caerdrhydiau
Ceredigion Heritage Coast
Cardigan Island
Pencribach
Morfa
Penbryn
Wervil Grange
Plwmp
Pentregat
Bettws Bledrws
Cribyn
Silian
Parcllyn
Aberporth
Tresaith
Sarnau
Brynhoffnant
Talgarreg
Gorsgoch
Capel St Silin
Cemaes Head
Cardigan Island Coastal Farm Park
Gwbert
Ferwig
Blaenannerch
Tan-y-groes
Glynarthen
Wstrws
Capel Cynon
Rhos Llawr Cwrt NNR
Bwlchyfadfa
Aber
Cwrt-newydd
Lampeter (Llanbedr Pont Steffan)
Pen-yr-afr
Cippyn
Cardigan (Aberteifi)
Tremain
Blaenporth
Betws Ifan
Ffostrasol
Castell Howell
Llanwnnen
Pentre-bach
Cwmann
Tre-Rhys
St Dogmaels (Llandudoch)
Abbey
Penparc
Noyadd Trefawr
Rhydlewis
Falin-Wnda
Pont-siân
Cwmsychbant
Alltyblacca
Llangoedmor
Teifi Marshes Nature Reserve
Pantgwyn
Beulah
Brongest
Troedyraur
Penrhiw-pâl
Rhydowen
Llanwenog
Dre-fach
Pencarreg
Moylgrove
Monington
Llechryd
Llandygwydd
Ponthirwaun
Coed-y-bryn
Maesllyn
Tre-groes
Pren-gwyn
Llanybydder
Pen Tas-eithin
Glanrhyd
Pen-y-bryn
Cilgerran
Capel Tygwydd
Llangynllo
Croes-lan
Highmead
Glan-Duar
Tredrissi
Llantood
Bridell
Carreg-wen
Cwm-cou
Cenarth
Llandyfriog
Aber-banc
Horeb
Penrhiw-llan
Rock Mill Woollen & Water Mill
Capel Dewi
Mynydd Pencarreg
Berry Hill
Trewilym
Abercych
Newcastle Emlyn (Castell Newydd Emlyn)
Llanfair-Orllwyn
Llandysul
Maesycrugiau
Aber-Giâr
Newport (Trefdraeth)
Velindre
Pengelli Forest NNR
Rhos-hill
Newchapel
Clynfyw
Penrhiw
Penrherber
Aber-Arad
Pentrecagal
Henllan
Pontwelly
Llanfihangel-ar-arth
Llanllwni
Eglwyswrw
Pentre Ifan Burial Chamber
Cilwendeg
National Wool Museum
Felindre
Drefach
Llangeler
Pentre-cwrt
Mynydd Llanllwni
Rhydcymerau
Llanfair-Nant-Gwyn
Boncath
Cwmhiraeth
Saron
Bancyffordd
Crosswell
Whitechurch
Blaenffos
Cwmpengraig
Drefelin
Glaspant
Capel Iwan
Bwlch-y-groes
Moelfre
Penboyr
Bwlch-clawdd
Pencader
New Inn
Gwyddgrug
Llidiad-Nenog
Gorlech
Brynberian
Freni-fawr
Star
Clydey
Cilrhedyn
Cwm-Morgan
Rhos
Dolgran
Gwernogle
Tafarn-y-bwlch
MYNYDD PRESELI
Crymych
Tegryn
Gorllwyn
Abergorlech
Foel Eryr
Foel Cwmcerwyn
Glogue
Hermon
Cwmduad
Alltwalis
Pen-y-garn
Mynachlog-ddu
Foel Drych
Pentre Galar
Llanfyrnach
Trelech
Dinas
Hermon
Brechfa
Maenclochog
Llangolman
Glandwr
Hebron
Blaenwaun
Esgair
Llanpumsaint
Llanllawddog
Pontarsais
CARMARTHEN
Llanfynydd
New Moat
Llanglydwen
Pemberton's Chocolate Farm
Llanwinio
Pen-y-bont
Blaen-y-coed
Cynwyl Elfed
Llwyn-croes
Gwili
Plas
Capel Isaac
Efailwen
Cwmbach
Cwmfelin Mynach
Talog
Rhydargaeau
Cefn-y-pant
Bwlchnewydd
Newchurch
Felingwmuchaf
Llanycefn
Maesgwynne
Llanboidy
Gellywen
Bronwydd Arms
Peniel
Felingwmisaf
Pen-ffordd
Login
Llandre
Cwm-miles
Abernant
Llanfihangel-uwch-Gwili
Court Henry
Broad Oak
Llandissilio
Tre-vaughan
Capel Gwyn
Nantgaredig
Pontargothi
Llangathen
Cilsan
Bletherston
Henllan Amgoed
Castell Gorfod
Meidrim
Merthyr
White Mill
Aberglasney Gardens
Rhydywrach
Llanfallteg
Clunderwen
Llangynin
Carmarthen (Caerfyrddin)
Tanerdy
Trinity
Abergwili
Llangunnor
Capel Dewi
Llanegwad
Llanarthney
Dryslwyn
Gelly
Bethesda
Cwmfelin Boeth
Sarnau
Johnstown
Llanllwch
Pensarn
Nantycaws
Golden Grove
Llanddewi Velfrey
Whitland Abbey
Bron-y-gaer
Bancyfelin
Cors Goch NNR
Cwmffrwd
National Botanic Garden of Wales
Penrhiwgoch
Robeston Wathen
Whitland (Hendy-Gwyn)
Pwll-trap
St Clears (Sanclêr)
Llangynog
Llangain
Croesyceiliog
Llanddarog
Porthyrhyd
Maesybont
Carmel
Redstone Bank
Trevaughan
Hywel Dda Garden
Backe
Ffynnon
Bancycapel
Cwmisfael
Llyn Llech Owain
Narberth (Arberth)
Crinow
Lampeter Velfrey
Llwyn-y-brain
Llanddowror
Llandeilo Abercywyn
Cwmllyfri
Llangendeirne
Mynydd Cerrig
Foelgastell
Cefneithin
Gorslas
Cross Hands
Penygroes
Camp Hill
Cold Blow
Lan-mill
Brandy Hill
Crwbin
Drefach
Oakwood Leisure Park
Princes Gate
Tavernspite
Pontantwn
Cwm-mawr
Meinciau
Morfa
Templeton
Red Roses
Llandawke
Laugharne (Talacharn)
Llanybri
Llandyfaelog
Tumble (Y Tymbl)
Cross Hands
Ludchurch
Crunwere Farm
Llanteg
Llansadurnen
Llansteffan
Ferryside (Glanyferi)
Bancffosfelen
Pontyberem
Cwmgwili
Llansteffan Castle
Wharley Point
Thomas Chapel
Loveston
Reynalton
Marros
Brook
Plashett
Iway
Pendine
Broadlay
Broadway
Mynydd-y-Garreg
Four Roads
Pont-Henri
Mynydd Sylen
Ginst Point
East Marsh
Laugharne Burrows
Kidwelly (Cydweli)
Kidwelly Castle
Llansaint
Cynheidre
Sylen
Jeffreyston
Kilgetty
Cresselly
Pendine Sands
Laugharne Sands
Wales Coast Path
0 2 4 6 miles
0 2 4 6 8 10 km

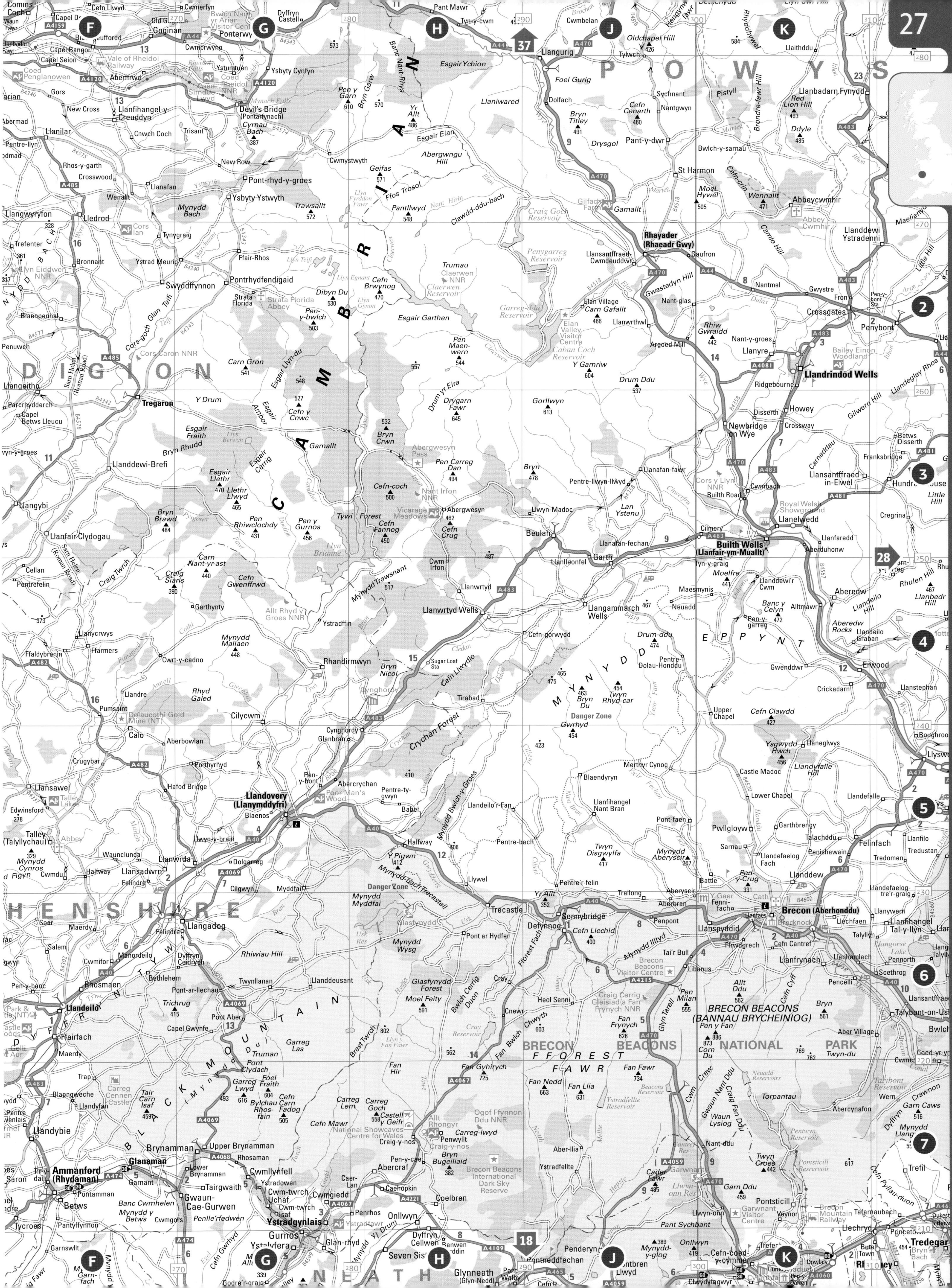

POWYS

DIGION

CAMBRIAN

EPPYNT

BRECON BEACONS (BANNAU BRYCHEINIOG) NATIONAL PARK

FFOREST FAWR

BLACK MOUNTAIN

NEATH

Rhayader (Rhaeadr Gwy)

Llandrindod Wells

Builth Wells (Llanfair-ym-Muallt)

Llanwrtyd Wells

Tregaron

Llandovery (Llanymddyfri)

Brecon (Aberhonddu)

Sennybridge

Ammanford (Rhydaman)

Llandeilo

Glanaman

Ystradgynlais

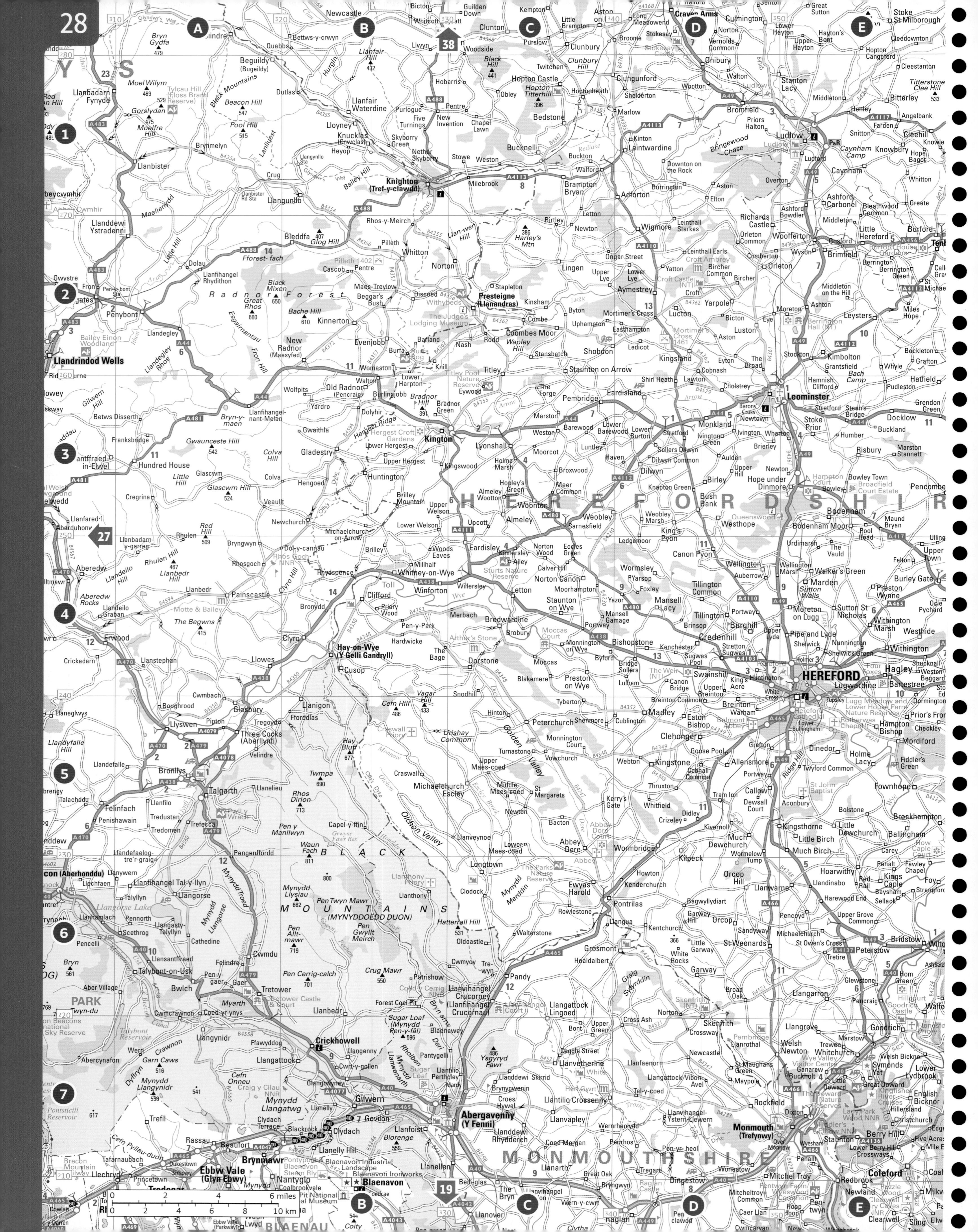

38
27
19
A
B
C
D
E
1
2
3
4
5
6
7
HEREFORDSHIRE
MONMOUTHSHIRE
Radnor Forest
BLACK MOUNTAINS
(MYNYDDOEDD DUON)
Knighton
(Tref-y-clawdd)
Llandrindod Wells
Presteigne
(Llanandras)
Kington
Leominster
Hay-on-Wye
(Y Gelli Gandryll)
HEREFORD
Ludlow
Craven Arms
Crickhowell
Abergavenny
(Y Fenni)
Monmouth
(Trefynwy)
Ebbw Vale
(Glyn Ebwy)
Brynmawr
Blaenavon
Coleford
Talgarth
Brecon (Aberhonddu)
Llanbister
New Radnor
Old Radnor
Pembridge
Weobley
Bromyard
Golden Valley
Olchon Valley
Longtown
Pontrilas
Skenfrith
Goodrich
Ross
0 2 4 6 miles
0 2 4 6 8 10 km

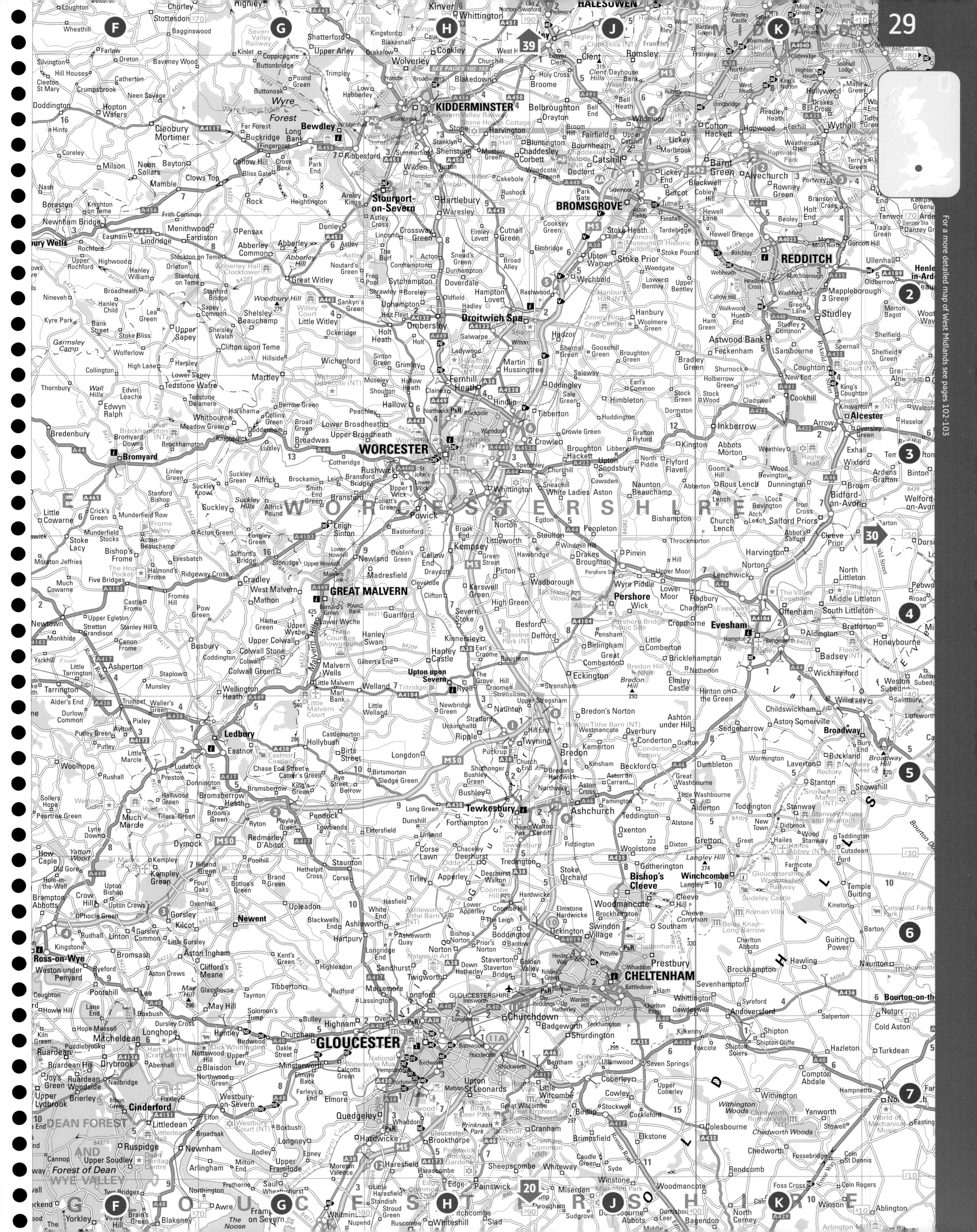

F
G
H
J
K
39
SEE PAGES 102-103
For a more detailed map of West Midlands see pages 102-103
30
20
HALESOWEN
KIDDERMINSTER
Wyre Forest
Bewdley
Cleobury Mortimer
Stourport-on-Severn
BROMSGROVE
REDDITCH
Droitwich Spa
WORCESTER
Bromyard
WORCESTERSHIRE
GREAT MALVERN
Pershore
Evesham
Alcester
Studley
Upton upon Severn
Ledbury
Tewkesbury
Ashchurch
Bredon
Broadway
Winchcombe
Bishop's Cleeve
CHELTENHAM
Newent
Ross-on-Wye
GLOUCESTER
Cinderford
DEAN FOREST
Forest of Dean
WYE VALLEY
Churchdown
Painswick
GLOUCESTERSHIRE
COTSWOLD HILLS
Vale of Evesham

For a more detailed map of West Midlands see pages 102-103 and of Coventry see page 104

SEE PAGES 102-103

SEE PAGE 104

40

29

20

0 2 4 6 miles

0 2 4 6 8 10 km

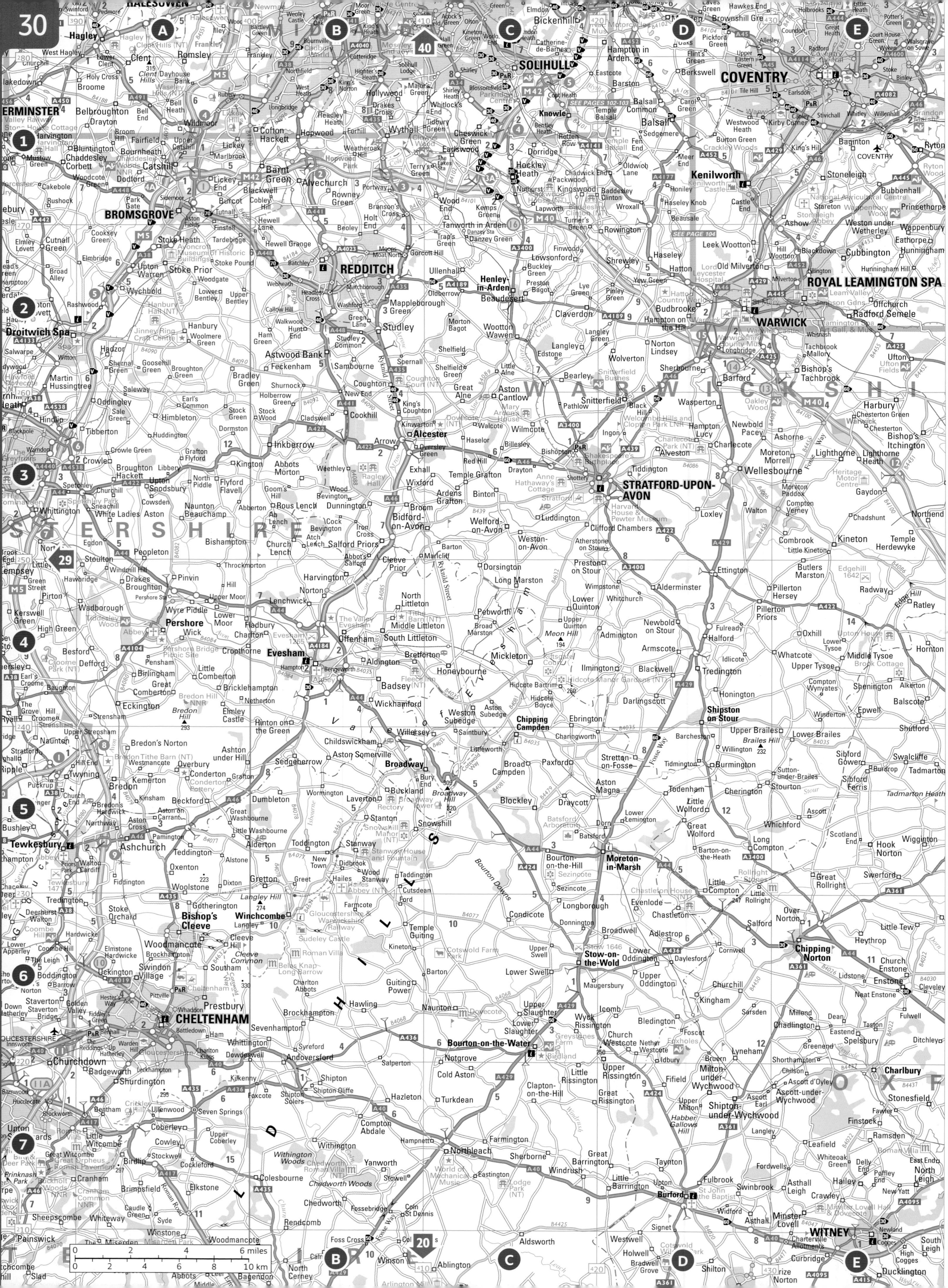

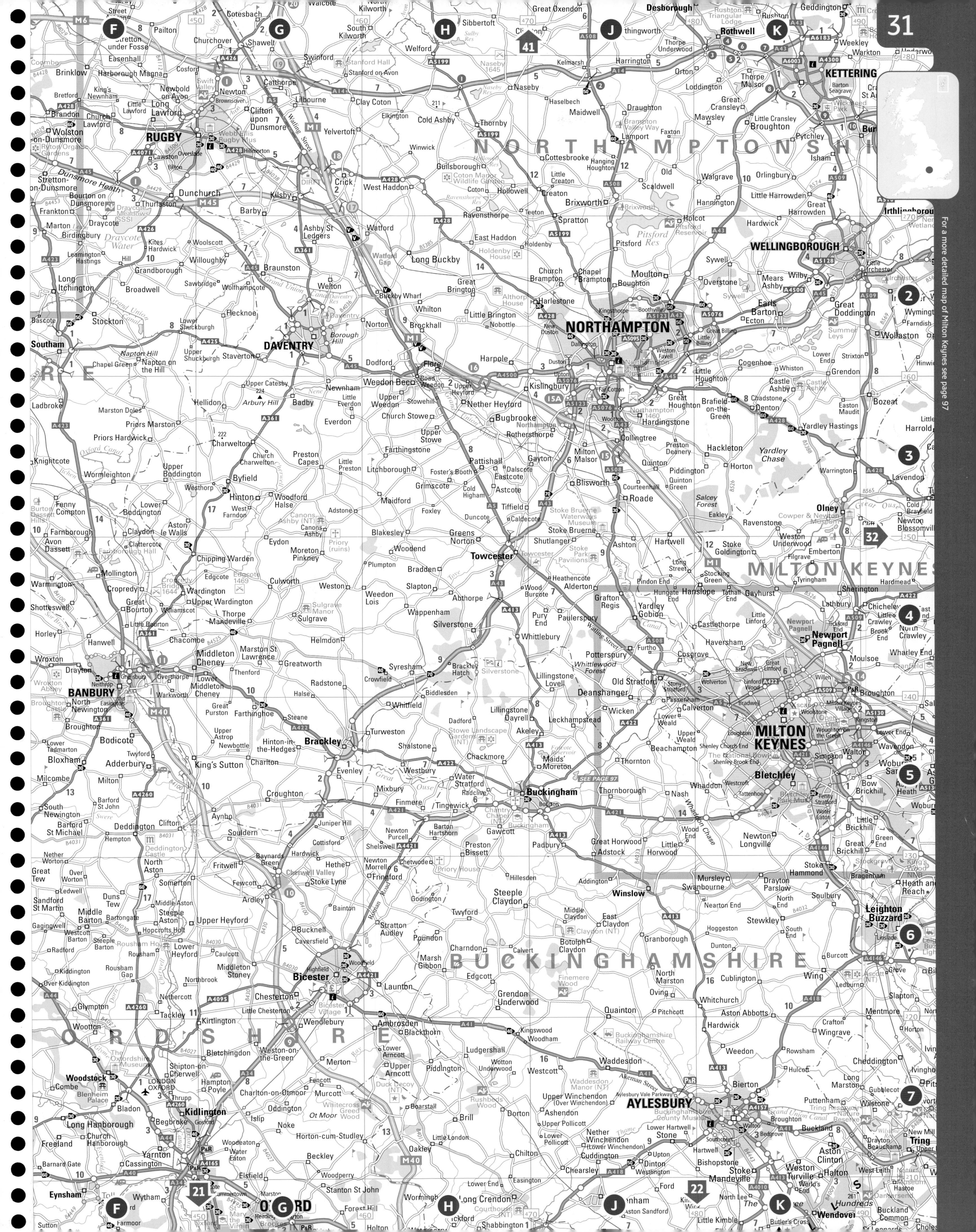

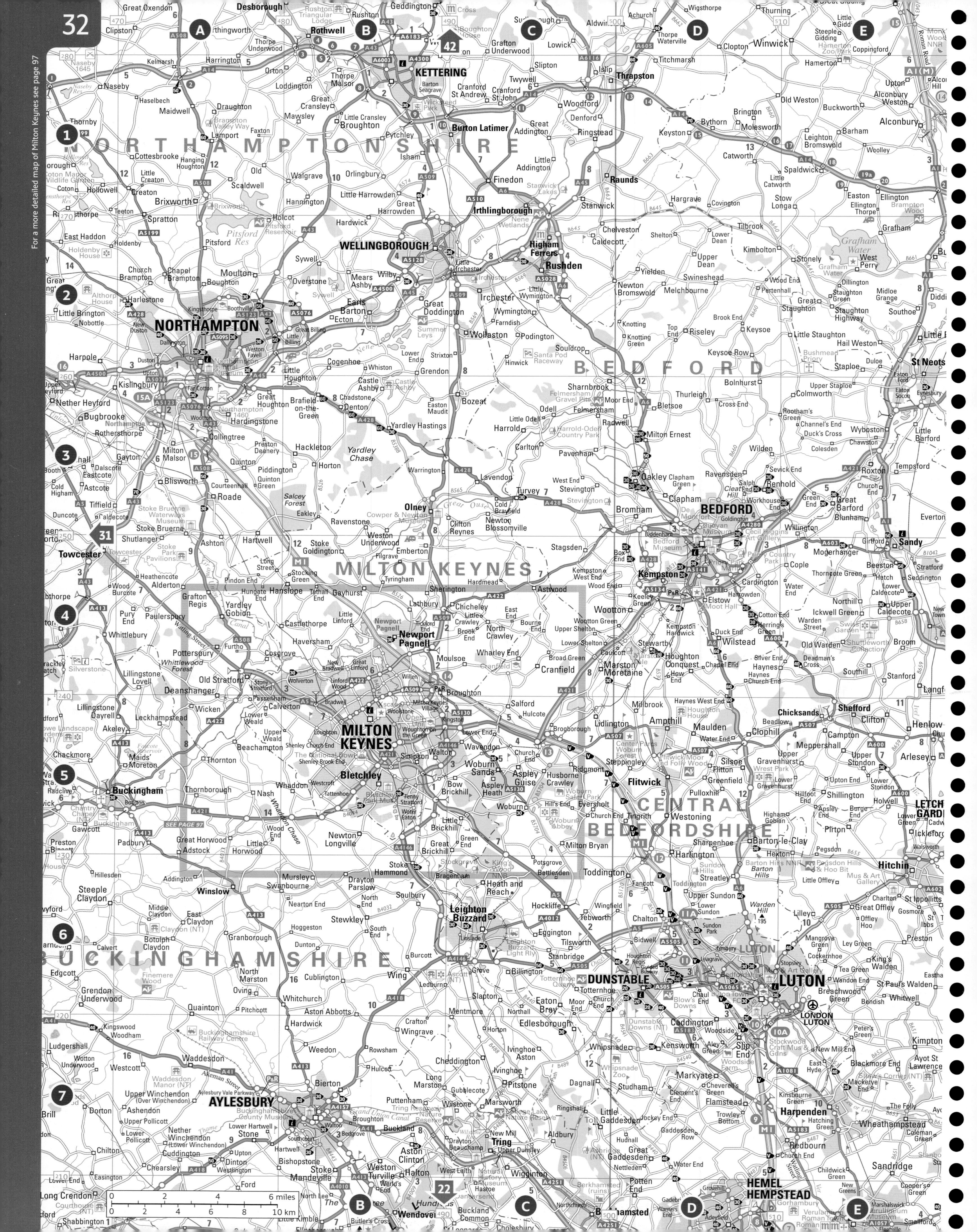

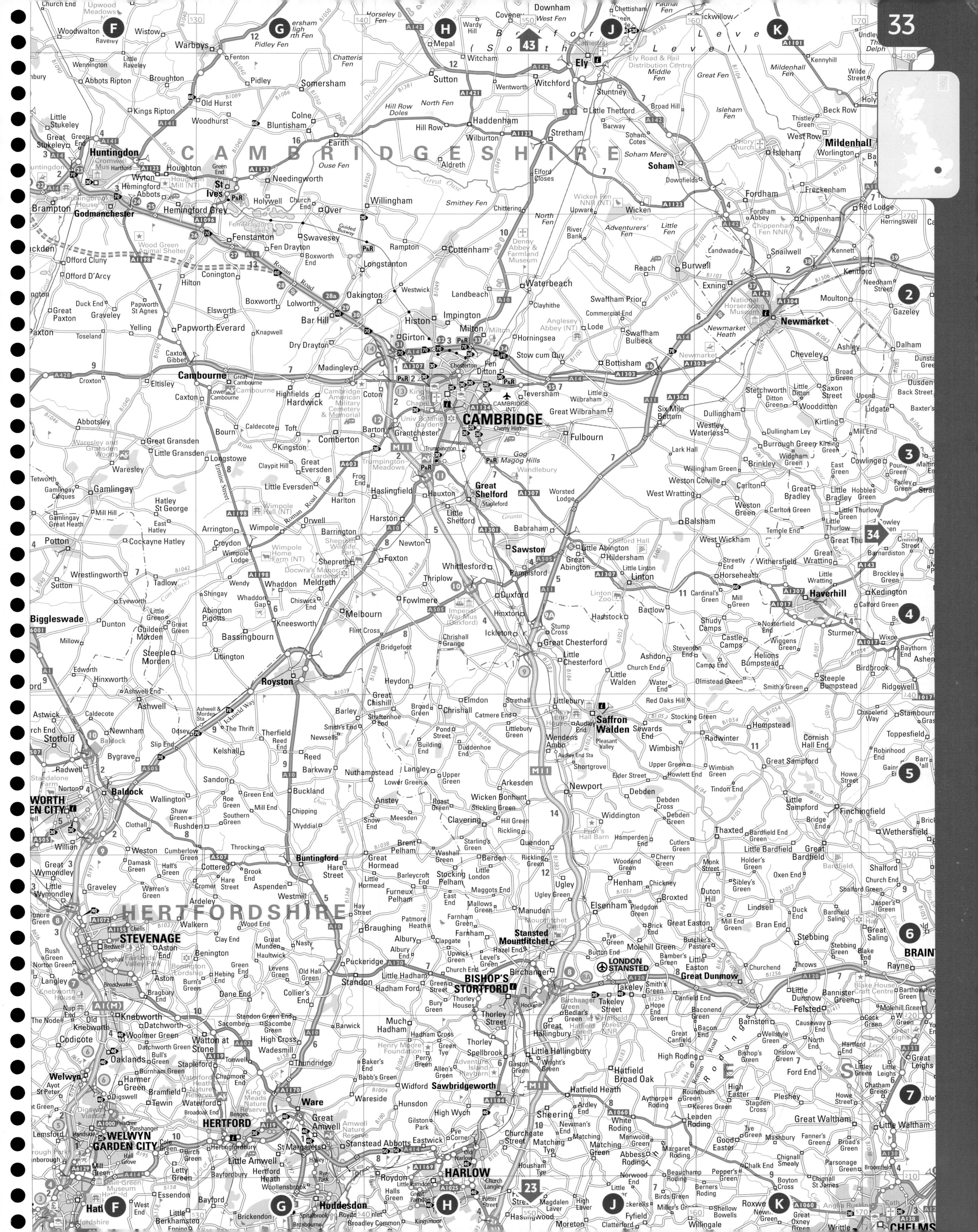

CAMBRIDGESHIRE
HERTFORDSHIRE
F
G
H
J
K
2
3
4
5
6
7
43
34
23
Huntingdon
Godmanchester
St Ives
Ely
Soham
Mildenhall
Newmarket
Cambourne
CAMBRIDGE
Haverhill
Royston
Saffron Walden
Baldock
STEVENAGE
BISHOP'S STORTFORD
LONDON STANSTED
Great Dunmow
HERTFORD
WELWYN GARDEN CITY
HARLOW
Ware
Sawbridgeworth
Hoddesdon
Biggleswade
Stotfold
Thaxted
Stansted Mountfitchet
Chatteris Fen
Great Ouse

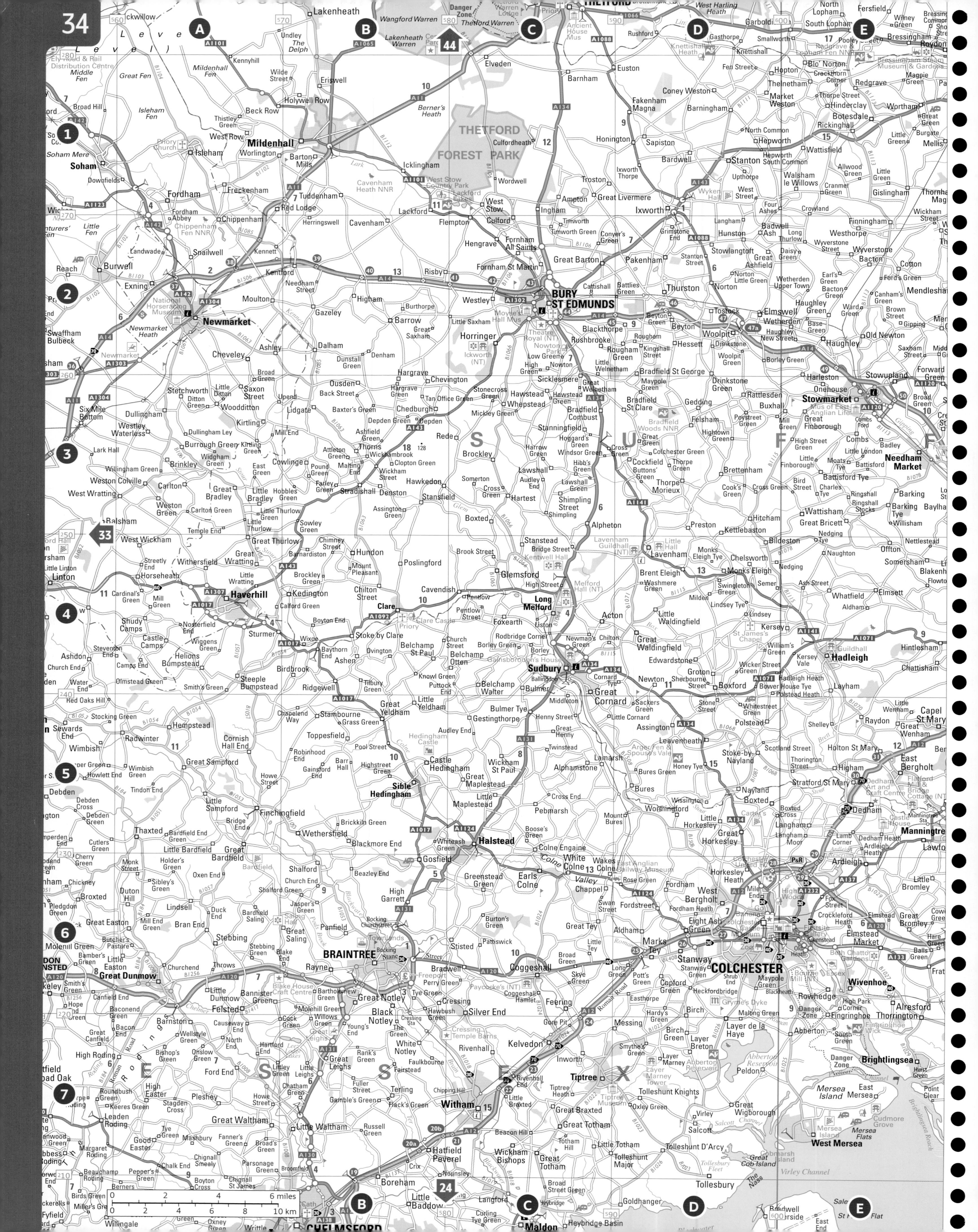
Lakenheath
Wangford Warren
Danger Zone
Thetford Warren
THETFORD
West Harling Heath
Lakenheath Warren
44
Undley
The Delph
Kennyhill
Mildenhall Fen
Wilde Street
Eriswell
Elveden
Euston
Rushford
Barnham
Great Fen
Isleham Fen
Beck Row
Holywell Row
Berner's Heath
THETFORD FOREST PARK
Culfordheath
Honington
Sapiston
Bardwell
Fakenham Magna
Coney Weston
Barningham
Hepworth
Stanton
Wattisfield
Rickinghall
Botesdale
Hinderclay
Thelnetham
Market Weston
Redgrave
Wortham
Mildenhall
West Row
Thistley Green
Isleham
Worlington
Barton Mills
Icklingham
Soham
Soham Mere
Fordham
Freckenham
Tuddenham
Red Lodge
Cavenham Heath NNR
West Stow Country Park
Lackford
Wordwell
Troston
Ixworth Thorpe
Ixworth
Great Livermere
Ampton
Ingham
West Stow
Culford
Flempton
Cavenham
Herringswell
Chippenham
Fordham Abbey
Little Fen
Landwade
Snailwell
Kennett
Kentford
Burwell
Reach
Exning
Hengrave
Fornham All Saints
Fornham St Martin
Great Barton
Pakenham
Risby
Thurston
Norton
BURY ST EDMUNDS
Westley
Moulton
Needham Street
Higham
Gazeley
Barrow
Newmarket
Newmarket Heath
National Horseracing Museum
Swaffham Bulbeck
Cheveley
Ashley
Dalham
Denham
Great Saxham
Little Saxham
Horringer
Ickworth (NT)
Theatre Royal (NT)
Nowton
Rushbrooke
Blackthorpe
Beyton
Woolpit
Elmswell
Wetherden
Haughley
Old Newton
Rougham
Rougham Green
Hessett
Drinkstone
Stowupland
Stowmarket
Onehouse
Harleston
Hargrave
Chevington
Ousden
Stetchworth
Dullingham
Six Mile Bottom
Westley Waterless
Kirtling
Lidgate
Chedburgh
Depden
Hawstead
Whepstead
Bradfield Combust
Bradfield St George
Bradfield St Clare
Gedding
Felsham
Buxhall
Great Finborough
Combs
Needham Market
Badley
Lark Hall
Burrough Green
Brinkley
Cowlinge
Stradishall
Denston
Rede
Brockley
Lawshall
Hartest
Shimpling
Cockfield
Thorpe Morieux
Brettenham
Hitcham
Wattisham
Barking
Weston Colville
West Wratting
Carlton
Great Bradley
Little Bradley
Little Thurlow
Great Thurlow
Wickhambrook
Hawkedon
Stansfield
Boxted
Somerton
Alpheton
Preston
Kettlebaston
Bildeston
Balsham
33
West Wickham
Withersfield
Great Wratting
Hundon
Poslingford
Glemsford
Lavenham
Brent Eleigh
Monks Eleigh
Chelsworth
Nedging
Naughton
Offton
Somersham
Linton
Horseheath
Haverhill
Kedington
Cavendish
Clare
Chilton Street
Long Melford
Melford Hall (NT)
Acton
Little Waldingfield
Great Waldingfield
Kersey
Lindsey
Whatfield
Elmsett
Aldham
Hintlesham
Hadleigh
Chattisham
Shudy Camps
Castle Camps
Sturmer
Stoke by Clare
Wixoe
Baythorne End
Ashen
Ovington
Belchamp St Paul
Belchamp Otten
Foxearth
Borley
Sudbury
Gainsborough's House
Edwardstone
Groton
Boxford
Newton
Layham
Ashdon
Helions Bumpstead
Steeple Bumpstead
Birdbrook
Ridgewell
Tilbury Green
Belchamp Walter
Bulmer
Great Cornard
Little Cornard
Assington
Polstead
Raydon
Capel St Mary
Great Wenham
Hempstead
Radwinter
Great Yeldham
Little Yeldham
Toppesfield
Gestingthorpe
Henny Street
Leavenheath
Stoke-by-Nayland
Holton St Mary
East Bergholt
Wimbish
Great Sampford
Castle Hedingham
Sible Hedingham
Wickham St Paul
Alphamstone
Lamarsh
Bures
Nayland
Boxted
Stratford St Mary
Dedham
Debden
Little Sampford
Finchingfield
Great Maplestead
Little Maplestead
Pebmarsh
Wormingford
Great Horkesley
Manningtree
Thaxted
Wethersfield
Blackmore End
Halstead
Colne Engaine
White Colne
Wakes Colne
Ardleigh
Great Bardfield
Shalford
Gosfield
Greenstead Green
Earls Colne
Chappel
Horkesley Heath
West Bergholt
Lindsell
High Garrett
Fordstreet
Great Easton
Stebbing
Great Saling
Panfield
Aldham
Marks Tey
Stanway
COLCHESTER
Wivenhoe
Elmstead Market
Great Bromley
Great Dunmow
BRAINTREE
Rayne
Bradwell
Coggeshall
Copford
Rowhedge
Alresford
Thorrington
Felsted
Great Notley
Black Notley
Cressing
Silver End
Feering
Messing
Birch
Layer de la Haye
Abberton
Fingringhoe
Barnston
High Roding
White Notley
Rivenhall
Kelvedon
Inworth
Tiptree
Layer Marney
Peldon
Brightlingsea
Pleshey
Great Leighs
Little Leighs
Terling
Witham
Tolleshunt Knights
Mersea Island
East Mersea
West Mersea
Great Waltham
Little Waltham
Great Braxted
Great Totham
Great Wigborough
Salcott
Hatfield Peverel
Wickham Bishops
Tolleshunt D'Arcy
Tolleshunt Major
Boreham
Little Baddow
24
Langford
Goldhanger
Tollesbury
Heybridge Basin
Maldon
CHELMSFORD
0 2 4 6 miles
0 2 4 6 8 10 km
A
B
C
D
E
1
2
3
4
5
6
7
S
U
F
E
S
S
E
X

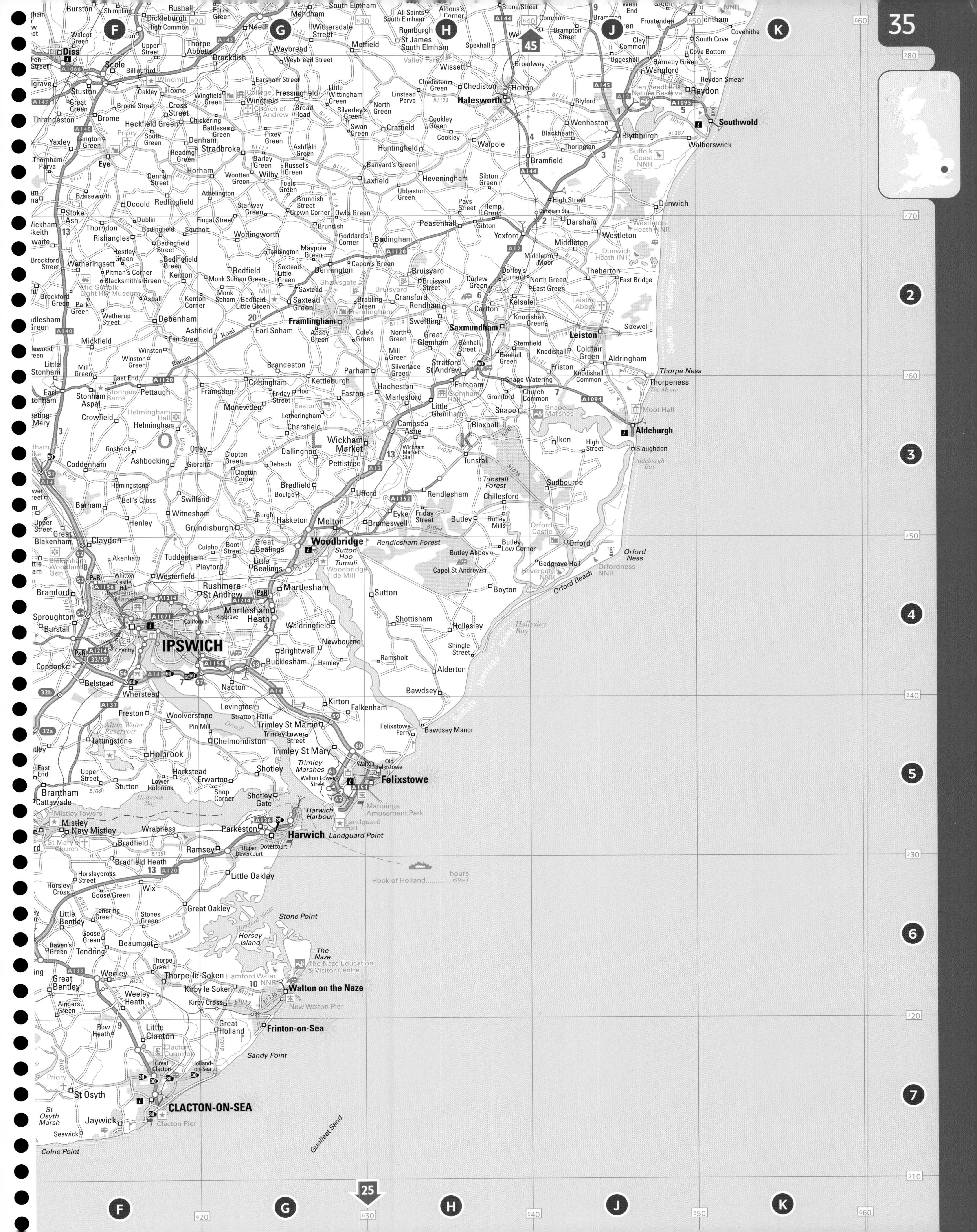

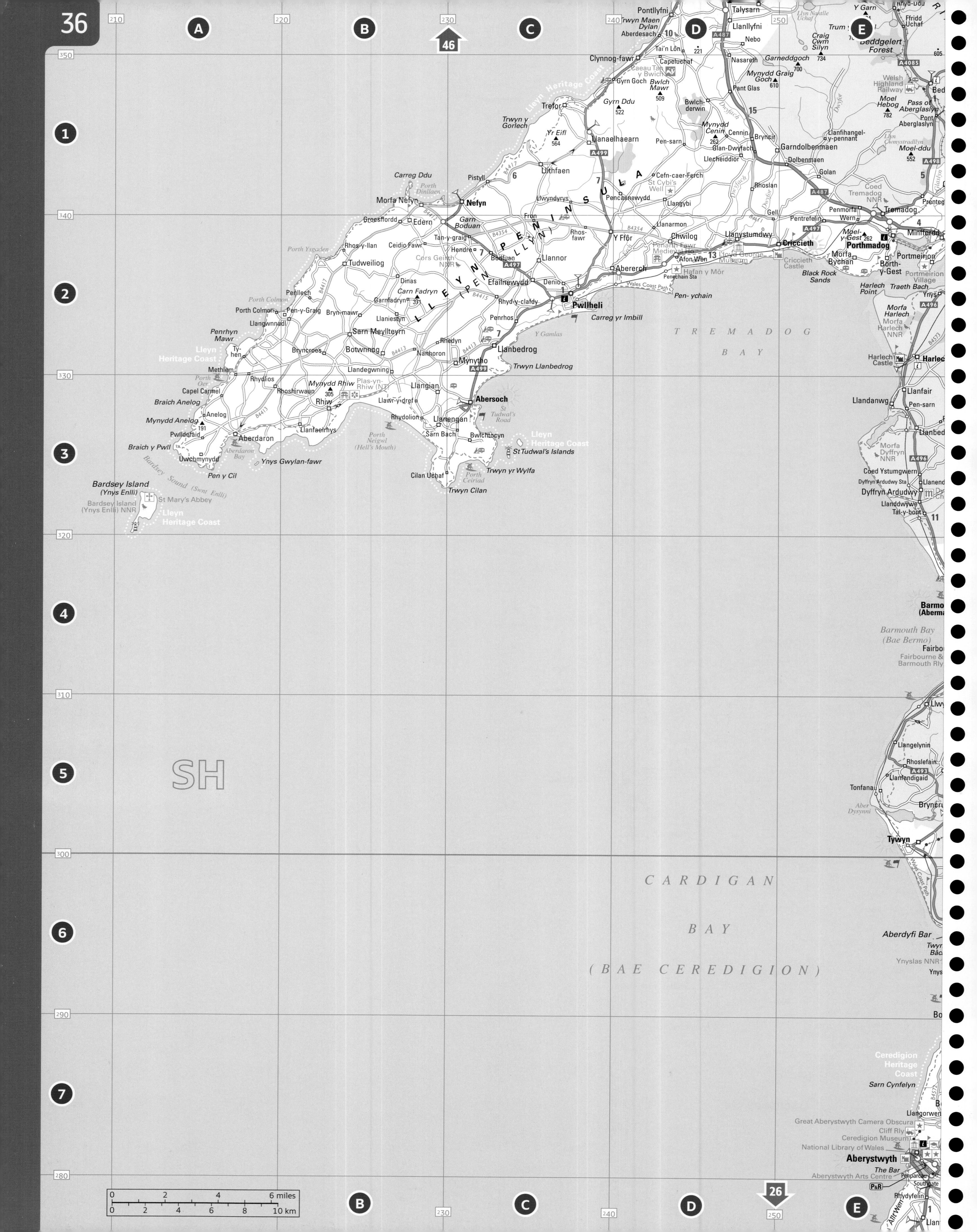

46
26
SH
Lleyn Peninsula (Penllyn)
Tremadog Bay
Cardigan Bay (Bae Ceredigion)
Bardsey Island (Ynys Enlli)
St Mary's Abbey
Bardsey Island (Ynys Enlli) NNR
Lleyn Heritage Coast
Bardsey Sound (Swnt Enlli)
Aberdaron
Pen y Cil
Ynys Gwylan-fawr
Uwchmynydd
Braich y Pwll
Mynydd Anelog
Anelog
Braich Anelog
Capel Carmel
Methlem
Penrhyn Mawr
Porth Colmon
Penllech
Tudweiliog
Porth Ysgaden
Rhos-y-llan
Groesffordd
Edern
Morfa Nefyn
Nefyn
Carreg Ddu
Porth Dinllaen
Pistyll
Llithfaen
Yr Eifl
Trwyn y Gorlech
Trefor
Llanaelhaearn
Clynnog-fawr
Gyrn Goch
Gyrn Ddu
Pontllyfni
Talysarn
Llanllyfni
Aberdesach
Dylan
Nebo
Nasareth
Garneddgoch
Mynydd Graig Goch
Pant Glas
Bwlch Mawr
Bwlch-derwin
Mynydd Cenin
Cennin
Bryncir
Garndolbenmaen
Dolbenmaen
Golan
Pen-sarn
Llecheiddior
Cefn-caer-Ferch
St Cybi's Well
Llangybi
Pencaenewydd
Llwyndyrys
Fron
Rhos-fawr
Y Ffôr
Llanarmon
Chwilog
Llanystumdwy
Criccieth
Criccieth Castle
Afon Wen
Penarth Fawr Medieval Ho
Lloyd George Museum
Hafan y Môr
Penychain Sta
Abererch
Pen-ychain
Wales Coast Path
Pwllheli
Carreg yr Imbill
Denio
Efailnewydd
Rhyd-y-clafdy
Llannor
Bodfuan
Garn Boduan
Tan-y-graig
Hendre
Cors Geirch NNR
Ceidio Fawr
Dinas
Carn Fadryn
Garnfadryn
Llaniestyn
Sarn Meyllteyrn
Bryn-mawr
Pen-y-Graig
Porth Colmon
Llangwnnadl
Tŷ-hen
Bryncroes
Botwnnog
Rhedyn
Nanhoron
Mynytho
Llandegwning
Penrhos
Llanbedrog
Trwyn Llanbedrog
Y Gamlas
Rhydlios
Rhoshirwaun
Mynydd Rhiw
Plas-yn-Rhiw (NT)
Rhiw
Llanfaelrhys
Llangian
Llawr-y-dref
Rhydolion
Llanengan
Sarn Bach
Abersoch
St Tudwal's Road
Bwlchtocyn
St Tudwal's Islands
Trwyn yr Wylfa
Porth Neigwl (Hell's Mouth)
Cilan Uchaf
Porth Ceiriad
Trwyn Cilan
Porth Oer
Pwlldefaid
Aberdaron Bay
Y Garn
Ffridd Uchaf
Trum
Craig Cwm Silyn
Beddgelert Forest
Welsh Highland Railway
Moel Hebog
Pass of Aberglaslyn
Aberglaslyn
Llanfihangel-y-pennant
Moel-ddu
Coed Tremadog NNR
Tremadog
Penmorfa
Wern
Pentrefelin
Gell
Rhoslan
Minffordd
Moel-y-Gest
Porthmadog
Portmeirion
Morfa Bychan
Borth-y-Gest
Portmeirion Village
Black Rock Sands
Harlech Point
Traeth Bach
Morfa Harlech
Morfa Harlech NNR
Harlech Castle
Llanfair
Llandanwg
Pen-sarn
Llanbed
Morfa Dyffryn NNR
Coed Ystumgwern
Dyffryn Ardudwy Sta
Dyffryn Ardudwy
Llanddwywe
Tal-y-bont
Barmouth Bay (Bae Bermo)
Fairbourne & Barmouth Rly
Llangelynin
Rhoslefain
Llanfendigaid
Tonfanau
Aber Dysynni
Tywyn
Aberdyfi Bar
Ynyslas NNR
Ceredigion Heritage Coast
Sarn Cynfelyn
Llangorwen
Great Aberystwyth Camera Obscura
Cliff Rly
Ceredigion Museum
National Library of Wales
Aberystwyth
The Bar
Aberystwyth Arts Centre
Penparcau
Southgate
Rhydyfelin
0 2 4 6 miles
0 2 4 6 8 10 km

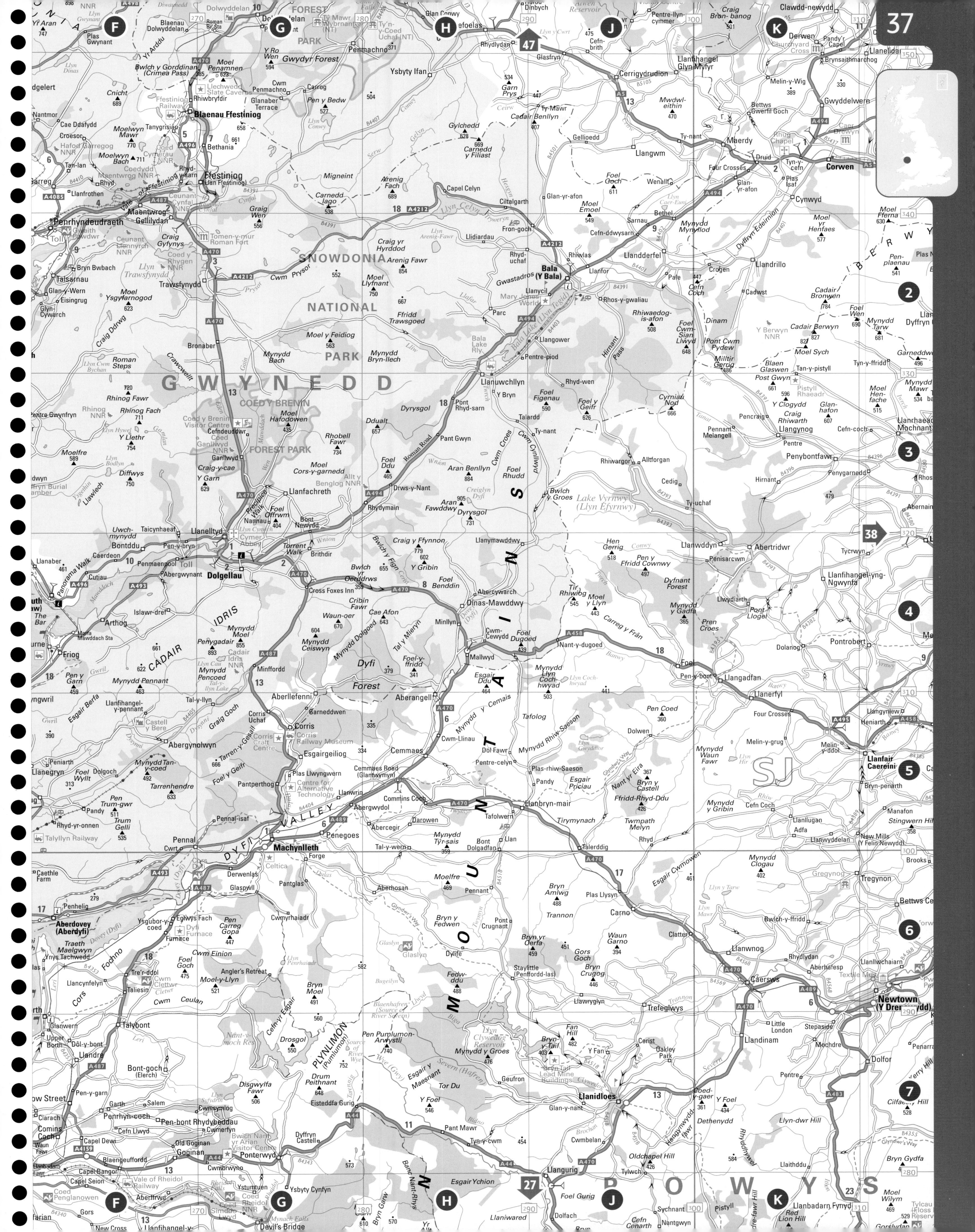

37
F
G
H
J
K
47
38
27
2
3
4
5
6
7
Dolwyddelan
Blaenau Dolwyddelan
Pentrefoelas
Rhydlydan
Glasfryn
Cerrigydrudion
Llanfihangel Glyn Myfyr
Clawdd-newydd
Derwen
Bryn Saith Marchog
Pentre-llyn-cymmer
Craig Bron-banog
Cefn-brith
Penmachno
Ysbyty Ifan
Gwydyr Forest
Moel Penamnen
Bwlch y Gorddinan (Crimea Pass)
Llechwedd Slate Caverns
Blaenau Ffestiniog
Ffestiniog (Llan Ffestiniog)
Cnicht
Moelwyn Mawr
Moelwyn Bach
Tanygrisiau
Bethania
Glanaber Terrace
Pen y Bedw
Garn Prys
Ty-Mawr
Cadair Benllyn
Gylchedd
Carnedd y Filiast
Mwdwl-eithin
Melin-y-Wig
Bettws Gwerfil Goch
Gwyddelwern
Llangwm
Ty-nant
Maerdy
Druid
Corwen
Four Crosses
Glan-yr-afon
Gellioedd
Foel Goch
Wenallt
Migneint
Arenig Fach
Capel Celyn
Ciltalgarth
Carnedd Iago
Llyn Celyn
A4212
Maentwrog
Gellilydan
Penrhyndeudraeth
Tomen-y-mur Roman Fort
Craig Gyfynys
Llyn Trawsfynydd
Trawsfynydd
Talsarnau
Glan-y-Wern
Eisingrug
Cwm Prysor
Moel Ysgyfarnogod
Craig yr Hyrddod
Arenig Fawr
Moel Llyfnant
Fron-goch
Llidiardau
Rhyd-uchaf
Bala (Y Bala)
Llanycil
Gwastadros
Llanfor
Rhos-y-gwaliau
Sarnau
Bethel
Cefn-ddwysarn
Llandderfel
Moel Emoel
Mynydd Mynyllod
Rhiwlas
Crogen
Llandrillo
Cynwyd
Dyffryn Edeirnion
Cefn Coch
Cadwst
Moel Fferna
Moel y Henfaes
Pen-plaenau
Cadair Bronwen
Foel Wen
Mynydd Tarw
Cadair Berwyn
Y Berwyn NNR
Moel Sych
Tan-y-pistyll
Tyn-y-ffridd
Garneddwen
SNOWDONIA
NATIONAL
PARK
GWYNEDD
Bronaber
Roman Steps
Moel y Feidiog
Mynydd Bach
Mynydd Bryn-llech
Ffridd Trawsgoed
Parc
Mary Jones World
Bala Lake (Llyn Tegid)
Bala Lake Rly
Llangower
Pentre-piod
Rhiwaedog-is-afon
Foel Cwm-Sian Llŵyd
Dinam
Pont Cwm Pydew
Milltir Gerrig
Hirnant Pass
Blaen Glaswen
Post Gwyn
Pistyll Rhaeadr
Y Clogydd
Craig Rhiwarth
Glan-hafon
Llangynog
Moel Hen-fache
Mynydd Mawr
Llanrhaeadr-ym-Mochnant
Cefn-coch
Pentre
Penybontfawr
Penygarnedd
Rhinog Fawr
Rhinog Fach
Coed y Brenin Visitor Centre
COED Y BRENIN FOREST PARK
Moel Hafodowen
Cefndeuddwr
Ganllwyd
Y Llethr
Moelfre
Diffwys
Craig-y-cae
Y Garn
Llanuwchllyn
Y Bryn
Pont Rhyd-sarn
Dyrysgol
Dduallt
Rhobell Fawr
Pant Gwyn
Roman Road
Foel Ddu
Moel Cors-y-garnedd
Aran Benllyn
Foel Rhudd
Cwm Croes
Cwm Cynllwyd
Talardd
Ty-nant
Foel Figenau
Foel y Geifr
Rhyd-wen
Cyrniau Nod
Pennant Melangell
Pencraig
Rhiwargor
Alltforgan
Cedig
Hirnant
Lake Vyrnwy (Llyn Efyrnwy)
Bwlch y Groes
Ty-uchaf
Abernant
Llanfachreth
Drws-y-Nant
Rhydymain
Creiglyn Dyfi
Aran Fawddwy
Dyrysgol
Foel Offrwm
Nannau
Bont Newydd
Llanelltyd
Cymer Abbey
Uwch-mynydd
Taicynhaeaf
Bontddu
Pen-y-bryn
Penmaenpool
Dolgellau
Torrent Walk
Brithdir
Craig y Ffynnon
Bwlch y Fign
Y Gribin
Bwlch yr Oerddrws
Foel Benddin
Llanymawddwy
Hen Gerrig
Pen y Ffridd Cownwy
Dyfnant Forest
Llanwddyn
Abertridwr
Penisarcwm
Tycrwyn
Llanfihangel-yng-Ngwynfa
Llanaber
Panorama Walk
Caerdeon
Abergwynant
Cutiau
Cross Foxes Inn
A470
Cribin Fawr
Dinas-Mawddwy
Aberceywarch
Tir Rhiwiog
Moel y Llyn
Carreg y Frân
Mynydd y Gadfa
Llwydiarth
Pont Llogel
Pren Croes
Arthog
Islawr-dref
Mawddach Sta
Friog
The Bar
CADAIR IDRIS
Mynydd Moel
Penygadair
Cadair Idris NNR
Waun-oer
Cae Afon
Mynydd Ceiswyn
Mynydd Dolgoed
Tal y Mieryn
Dyfi Forest
Foel-y-ffridd
Minllyn
Cwm-Cewydd
Foel Dugoed
Mallwyd
Nant-y-dugoed
Foel
Pen-y-bont
Llangadfan
Dolanog
Pontrobert
Pen y Garn
Mynydd Pennant
Llyn Cau
Mynydd Pencoed
Minffordd
Aberllefenni
Aberangell
Esgair Ddu
Mynydd y Cemais
Mynydd y Glyn Coch-hwyad
Llanerfyl
Four Crosses
Esgair Berfa
Llanfihangel-y-pennant
Castell y Bere
Tal-y-llyn
Graig Goch
Corris Uchaf
Corris
Garneddwen
Corris Craft Centre
Corris Railway Museum
Cwm-Llinau
Tafolog
Pen Coed
Dolwen
Melin-y-grug
Melin-y-ddôl
Heniarth
Llangyniew
Llanfair Caereinion
Abergynolwyn
Tarren-y-Gesail
Esgairgeiliog
Cemmaes
Cemmaes Road (Glantwymyn)
Dol Fawr
Pentre-celyn
Mynydd Rhiw-Saeson
Plas-rhiw-Saeson
Pandy
Esgair Priciau
Nant yr Eira
Bryn y Castell
Ffridd-Rhyd-Ddu
Mynydd Waun Fawr
SJ
Bryn-penarth
Peniarth
Llanegryn
Foel Wyllt
Dolgoch
Mynydd Tan-y-coed
Tarrenhendre
Foel y Geifr
Pantperthog
Plas Llwyngwern
Centre for Alternative Technology
Llanwrin
Commins Coch
Abergwydol
Darowen
Llanbryn-mair
Tafolwern
Tirymynach
Twmpath Melyn
Mynydd y Gribin
Cefn Coch
Llanllugan
Adfa
Manafon
Stingwern Hill
Pen Trum-gwr
Pandy
Trum Gelli
Rhyd-yr-onnen
Talyllyn Railway
Pennal-isaf
Pennal
Cwrt
DYFI VALLEY
Machynlleth
Penegoes
Abercegir
Tal-y-wern
Mynydd Tyr-sais
Bont Dolgadfan
Llan
Talerddig
Rhyd
New Mills (Y Felin Newydd)
Llanwyddelan
Caethle Farm
Forge
Derwenlas
Celtica
Glaspwll
Pantglas
Aberhosan
Moelfre
Pennant
Bryn Amlwg
Plas Llysyn
Carno
Trannon
Esgair Cwmowen
Mynydd Clogau
Gregynog
Tregynon
Brooks
Bettws Cedewain
Penhelig
Aberdovey (Aberdyfi)
Eglwys Fach
Ysgubor-y-coed
Dyfi Furnace
Furnace
Pen Carreg Gopa
Cwmyrhaiadr
Bryn y Fedwen
Pont Crugnant
Bryn yr Oerfa
Waun Garno
Gors Goch
Clatter
Lanwnog
Bwlch-y-ffridd
Traeth Maelgwyn
Ynys Tachwedd
Fochno
Cwm Einion
Glaslyn
Dylife
Staylittle (Penffordd-las)
Bryn Crugog
Rhydlydan
Aberhafesp
Llanllwchaiarn
Textile Museum
Tre'r-ddôl
Llancynfelyn
Taliesin
Cwm Cletwr
Foel Goch
Moel-y-Llyn
Angler's Retreat
Cwm
Ceulan
Cors Fochno
Bryn Moel
Fedw-ddu
Caersws
Llawryglyn
Trefeglwys
Newtown (Y Drenewydd)
Talybont
Glanwern
Upper Borth
Dôl-y-bont
Llandre
Cefn-yr-Esgair
Drosgol
Nant-y-moch Res
Blaenhafren (Source of River Severn)
PLYNLIMON (Pumlumon)
Pen Pumlumon-Arwystli
Llyn Clywedog Reservoir
Mynydd y Groes
Bryn-y-Tail
Bryn Tail Lead Mine Buildings
Fan Hill
Y Fan
Cerist
Oakley Park
Little London
Llandinam
Stepaside
Mochdre
Dolfor
Kerry Hill
Penarra
Bont-goch (Elerch)
Disgwylfa Fawr
Drum Peithnant
Eisteddfa Gurig
Esgair y Maesnant
Tor Du
Geufron
Llanidloes
Glan-y-nant
Coed-y-gaer
Y Foel
Dethenydd
Rhydyronen
Llyn-dwr Hill
Pentre
Cilfaesty Hill
Bow Street
Pen-y-garn
Garth
Salem
Penrhyn-coch
Pen-bont Rhydybeddau
Cwmsymlog
Cefn Llwyd
Cwmerfyn
Clarach
Comins Coch
Waun Fawr
Capel Dewi
Capel Bangor
Capel Seion
Dyffryn Castell
Bwlch Nant yr Arian Visitor Centre
Old Goginan
Goginan
Blaengeuffordd
Ponterwyd
Cwmbrwyno
Pant Mawr
Tyn-y-cwm
Cwmbelan
Oldchapel Hill
Tylwch
Llangurig
Foel Gurig
Llaithddu
Bryn Gydfa
Glyndŵr's Way
Vale of Rheidol Railway
Coed Penglanowen
Aberffrwd
Ystumtuen
Coed Rheidol NNR
Ysbyty Cynfyn
Mynach Falls
Devil's Bridge
Rheidol Falls
Gors
New Cross
Llanfihangel-y-
Simdde Lwyd
Bryn Garw
Banc Nant-Rhys
Esgair Ychion
Llaniwared
Dolfach
Sychnant
Nantgwyn
Cefn Cenarth
Pistyll
Llanbadarn Fynydd
Red Lion Hill
Moel Wylm
Tylcau
POWYS
MOUNTAINS
BERWYN
A5
A494
A487
A493
A496
A470
A458
A495
A489
A483
A44
A4159
B4391
B4393
B4396

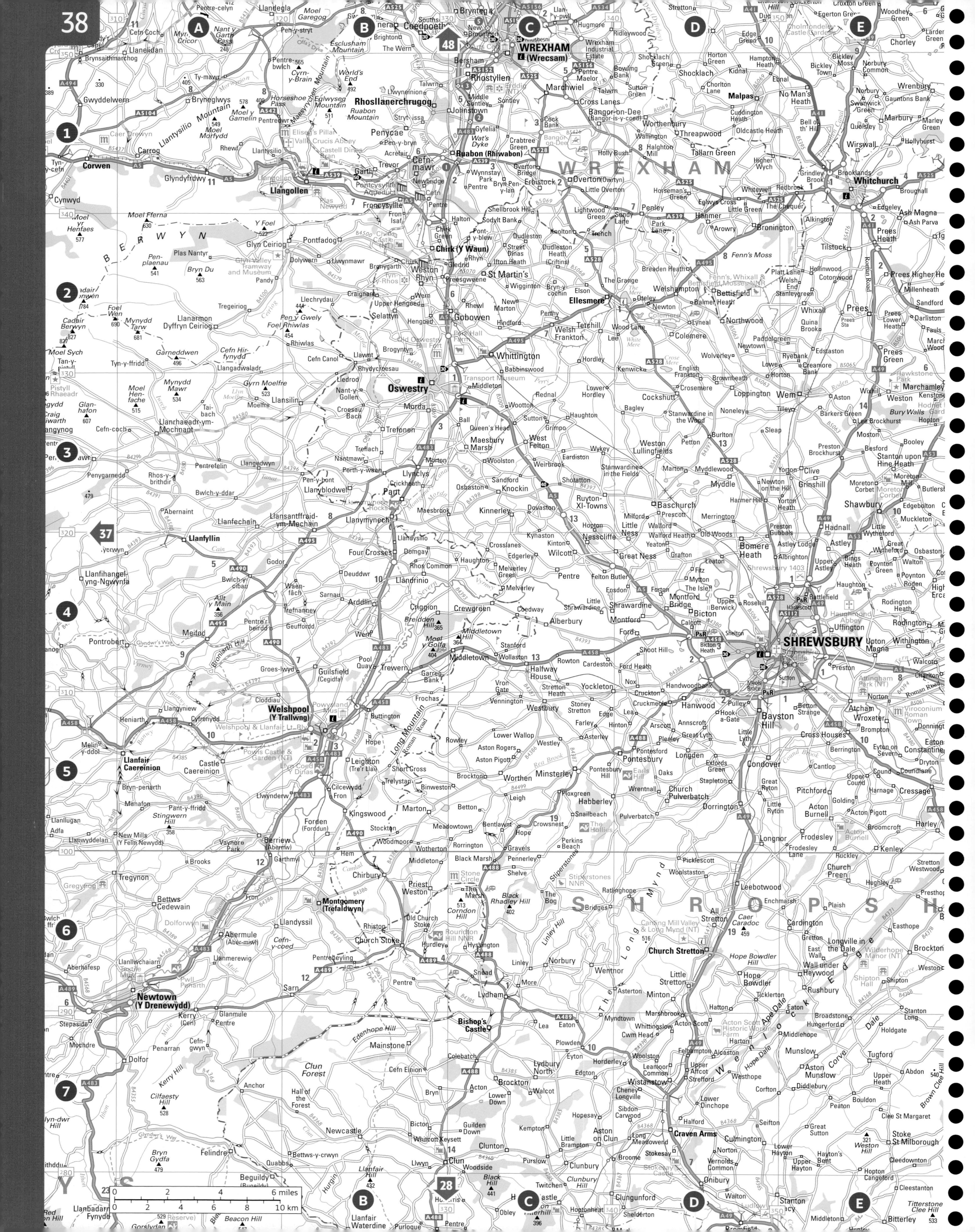
A
B
C
D
E
1
2
3
4
5
6
7
48
37
28
WREXHAM
(Wrecsam)
WREXHAM
SHROPSHIRE
BERWYN
Corwen
Llangollen
Ruabon (Rhiwabon)
Rhosllanerchrugog
Chirk (Y Waun)
Oswestry
Whittington
Ellesmere
Overton
Bangor-on-Dee
Malpas
Whitchurch
Wem
Prees
Hanmer
St Martin's
Gobowen
Llanfyllin
Welshpool
(Y Trallwng)
Llanfair Caereinion
Newtown
(Y Drenewydd)
Montgomery
(Trefaldwyn)
Church Stoke
Bishop's Castle
Clun
Craven Arms
Church Stretton
Minsterley
Pontesbury
SHREWSBURY
Bayston Hill
Condover
Nesscliffe
Baschurch
Llanymynech
Meifod
Berriew
Chirbury
Lydbury North
Wistanstow
Long Mynd
Wenlock Edge
Long Mountain
Stiperstones
Clun Forest
Kerry Hill
Offa's Dyke
0 2 4 6 miles
0 2 4 6 8 10 km

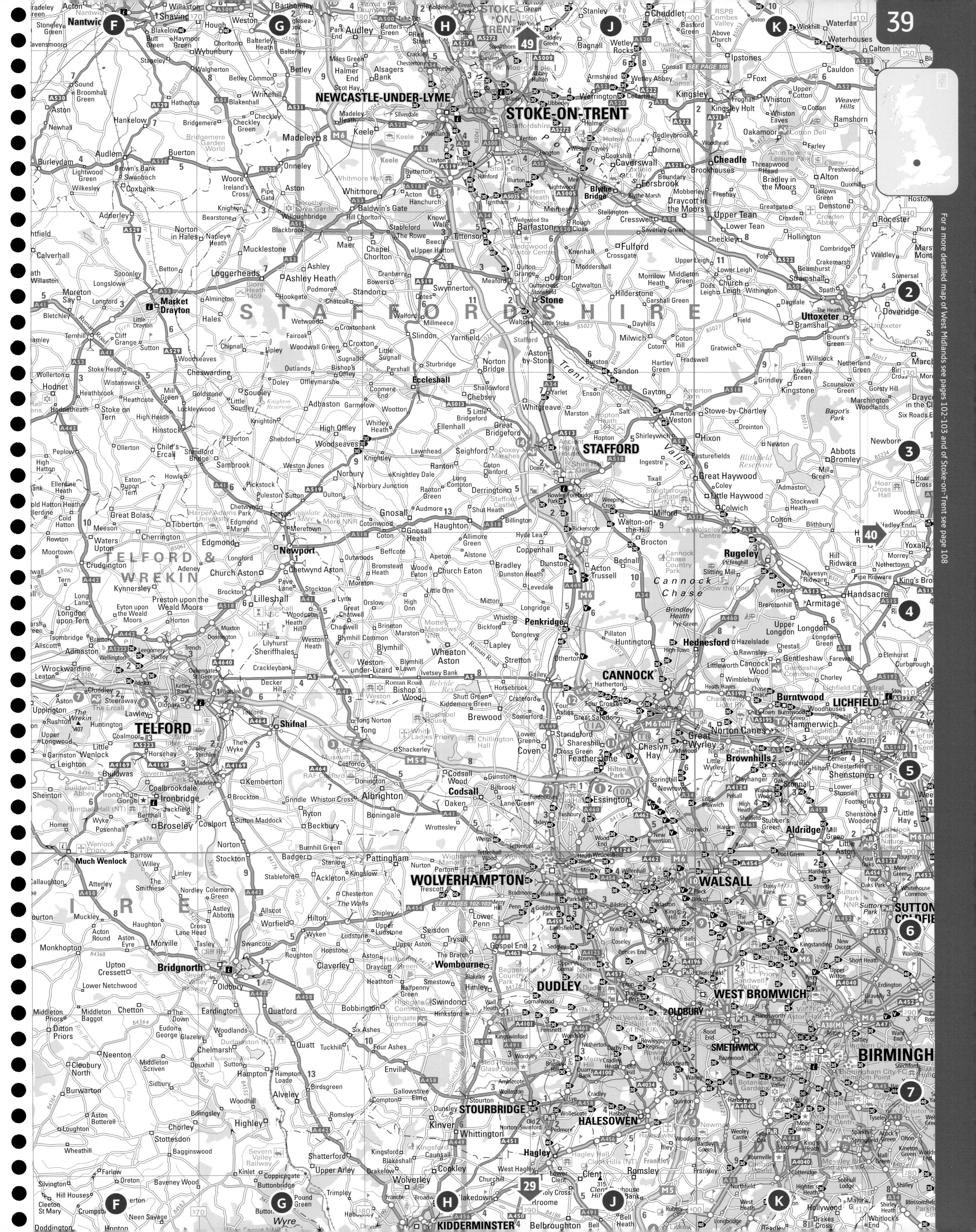

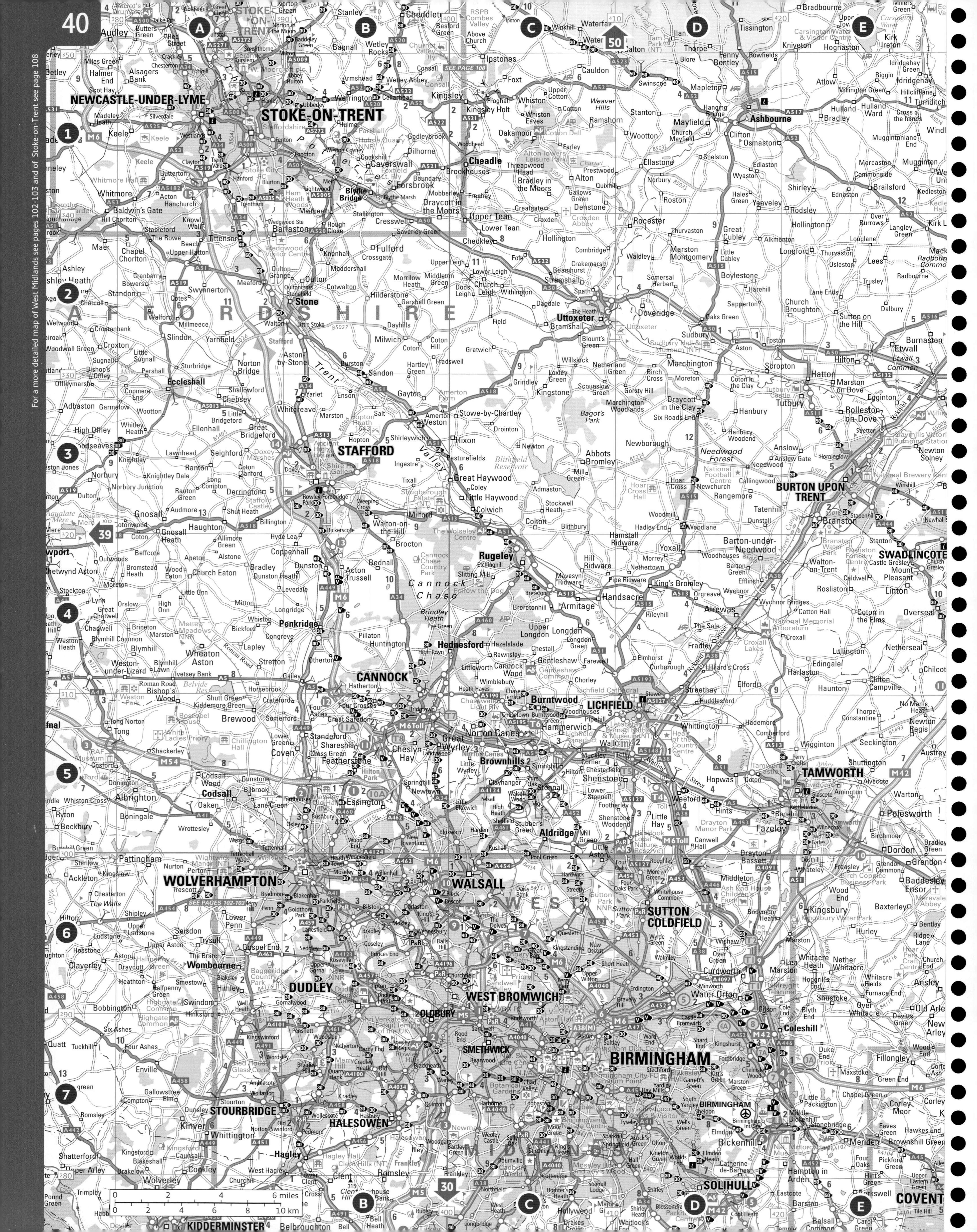

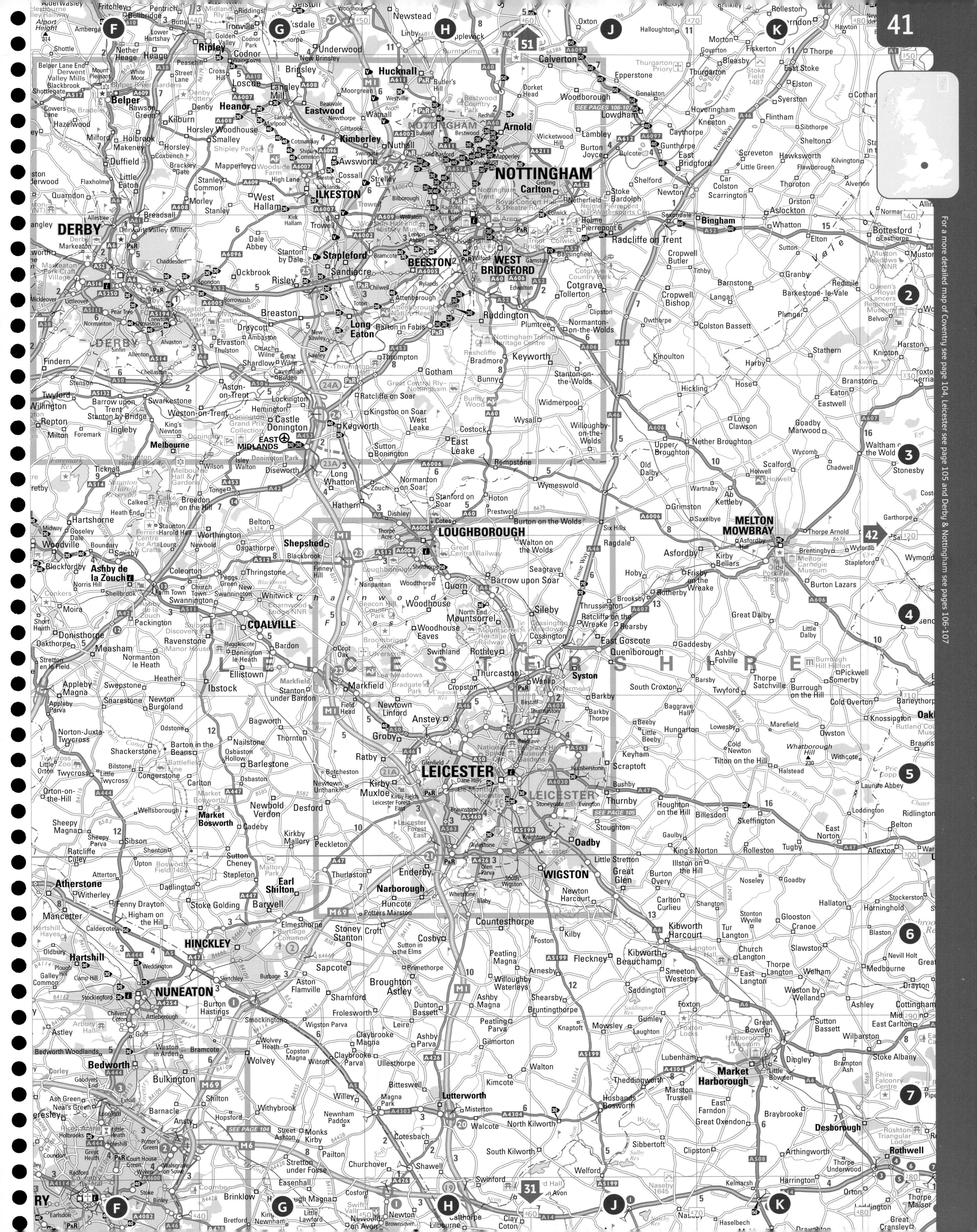
DERBY
NOTTINGHAM
LEICESTER
LOUGHBOROUGH
COALVILLE
ILKESTON
BEESTON
WEST BRIDGFORD
MELTON MOWBRAY
WIGSTON
HINCKLEY
NUNEATON
Market Harborough
Lutterworth
Bedworth
Belper
Ripley
Heanor
Eastwood
Hucknall
Arnold
Carlton
Bingham
Radcliffe on Trent
Stapleford
Long Eaton
Castle Donington
East Midlands
Kegworth
Shepshed
Ashby de la Zouch
Melbourne
Market Bosworth
Earl Shilton
Atherstone
Oadby
Syston
Narborough
Desborough
Rothwell
L E I C E S T E R S H I R E
SEE PAGES 106-107
SEE PAGE 105
SEE PAGE 104
For a more detailed map of Coventry see page 104, Leicester see page 105 and Derby & Nottingham see pages 106-107

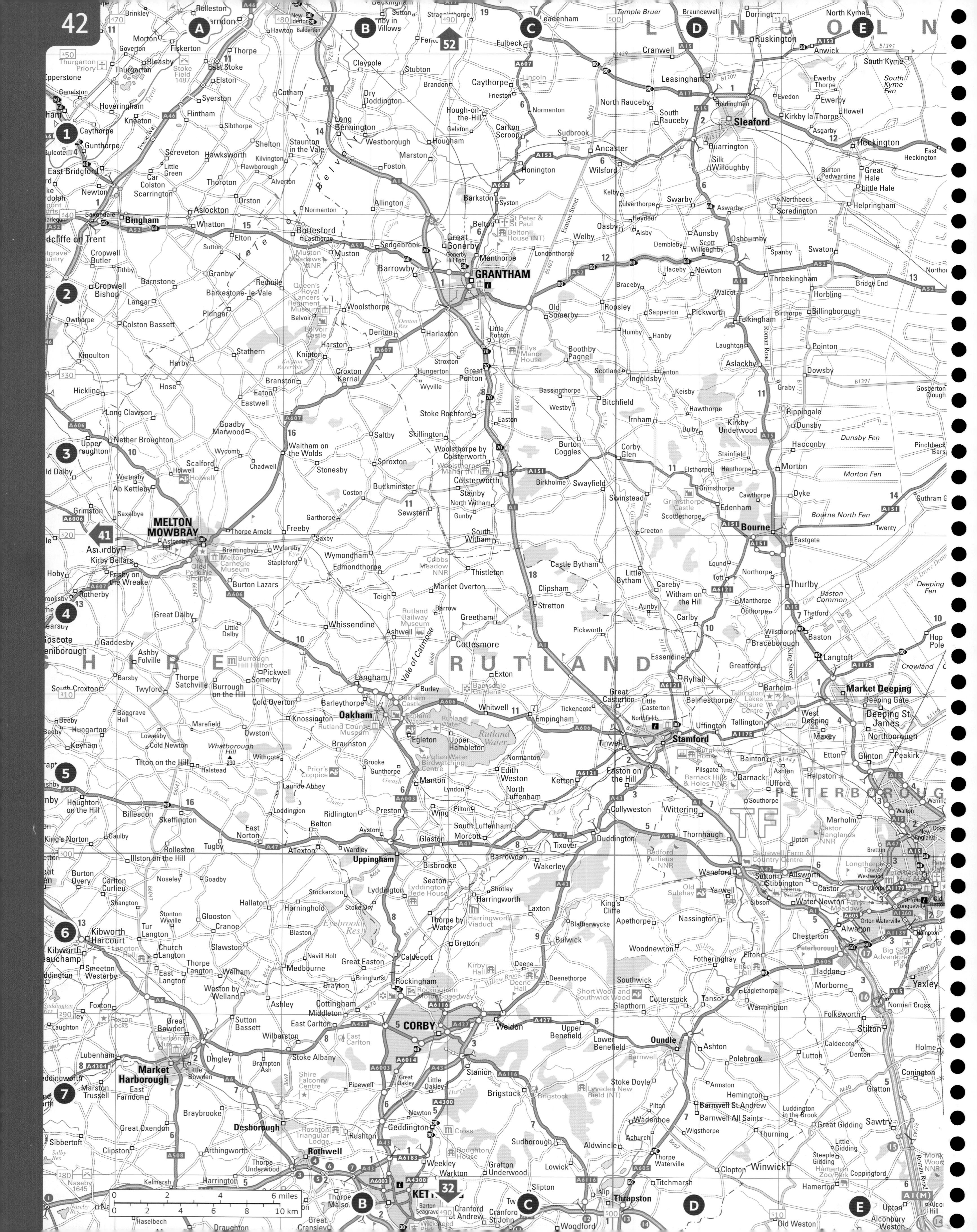
A
B
C
D
E
1
2
3
4
5
6
7
LINCOLN
SHIRE
RUTLAND
PETERBOROUGH
TF
52
41
32
GRANTHAM
MELTON MOWBRAY
Sleaford
Bourne
Stamford
Oakham
Uppingham
CORBY
Market Harborough
Desborough
Rothwell
Oundle
Thrapston
Market Deeping
Deeping St James
Bingham
Radcliffe on Trent
East Bridgford
Bottesford
Barrowby
Great Gonerby
Belton
Belton House (NT)
Manthorpe
Harlaxton
Denton
Woolsthorpe
Belvoir
Belvoir Castle
Knipton
Croxton Kerrial
Great Ponton
Little Ponton
Ellys Manor House
Colsterworth
Woolsthorpe by Colsterworth
Skillington
Stoke Rochford
Corby Glen
Swinstead
Grimsthorpe Castle
Edenham
Castle Bytham
Little Bytham
Clipsham
Stretton
Greetham
Cottesmore
Exton
Barnsdale Gardens
Rutland Water
Empingham
Whitwell
Edith Weston
Ketton
Easton on the Hill
Collyweston
Duddington
Wittering
Barnack Hills & Holes NNR
Burghley House
Helpston
Glinton
Peakirk
Castor
Water Newton
Wansford
Yarwell
Nassington
Apethorpe
King's Cliffe
Fotheringhay
Elton
Elton Hall
Warmington
Polebrook
Barnwell
Lyveden New Bield (NT)
Brigstock
Stanion
Weldon
Rockingham
Rockingham Castle
Rockingham Motor Speedway
Cottingham
Middleton
East Carlton
Wilbarston
Stoke Albany
Dingley
Braybrooke
Great Oxendon
Clipston
Arthingworth
Harrington
Kelmarsh
Naseby
Sibbertoft
Lubenham
Marston Trussell
East Farndon
Kibworth Harcourt
Kibworth Beauchamp
Smeeton Westerby
Great Bowden
Welham
Slawston
Medbourne
Hallaton
Horninghold
Blaston
Great Easton
Caldecott
Gretton
Harringworth
Laxton
Bulwick
Deene
Deene Hall
Kirby Hall
Lyddington
Lyddington Bede House
Seaton
Bisbrooke
Glaston
Morcott
South Luffenham
North Luffenham
Wing
Preston
Ridlington
Ayston
Belton
Launde Abbey
Loddington
Tugby
East Norton
Skeffington
Billesdon
Rolleston
Illston on the Hill
Noseley
Goadby
Carlton Curlieu
Houghton on the Hill
Tilton on the Hill
Halstead
Whatborough Hill
Owston
Withcote
Knossington
Cold Overton
Somerby
Pickwell
Burrough on the Hill
Burrough Hill Hillfort
Thorpe Satchville
Twyford
Ashby Folville
Gaddesby
Great Dalby
Little Dalby
Burton Lazars
Asfordby
Kirby Bellars
Frisby on the Wreake
Rotherby
Hoby
Brentingby
Wyfordby
Stapleford
Freeby
Saxby
Wymondham
Edmondthorpe
Teigh
Whissendine
Ashwell
Langham
Barleythorpe
Burley
Egleton
Upper Hambleton
Braunston
Brooke
Manton
Lyndon
Pilton
Barrowden
Tixover
Thornhaugh
Sacrewell Farm & Country Centre
Sutton
Stibbington
Ailsworth
Marholm
Castor Hanglands NNR
Ufford
Bainton
Barnack
Pilsgate
Southorpe
Uffington
Tallington
Tallington Lakes Leisure Centre
West Deeping
Maxey
Etton
Northborough
Deeping Gate
Barholm
Greatford
Braceborough
Wilsthorpe
Baston
Langtoft
Thurlby
Northorpe
Toft
Lound
Manthorpe
Obthorpe
Carlby
Witham on the Hill
Careby
Aunby
Pickworth
Essendine
Ryhall
Belmesthorpe
Great Casterton
Little Casterton
Tickencote
Tinwell
Thistleton
South Witham
North Witham
Gunby
Sewstern
Buckminster
Stainby
Coston
Garthorpe
Sproxton
Stonesby
Saltby
Waltham on the Wolds
Wycomb
Chadwell
Scalford
Holwell
Wartnaby
Ab Kettleby
Nether Broughton
Upper Broughton
Long Clawson
Hose
Harby
Eastwell
Eaton
Branston
Stathern
Plungar
Redmile
Barkestone-le-Vale
Langar
Granby
Barnstone
Cropwell Bishop
Cropwell Butler
Colston Bassett
Kinoulton
Owthorpe
Hickling
Elton
Orston
Whatton
Aslockton
Scarrington
Car Colston
Screveton
Hawksworth
Thoroton
Flintham
Sibthorpe
Syerston
Elston
East Stoke
Thorpe
Fiskerton
Rolleston
Farndon
Morton
Thurgarton
Bleasby
Hoveringham
Gonalston
Caythorpe
Gunthorpe
Newton
Kneeton
Cotham
Shelton
Staunton in the Vale
Long Bennington
Dry Doddington
Westborough
Claypole
Stubton
Brandon
Hougham
Marston
Foston
Allington
Sedgebrook
Muston
Normanton
Barkston
Syston
Honington
Carlton Scroop
Hough-on-the-Hill
Gelston
Caythorpe
Frieston
Fulbeck
Leadenham
Normanton
Ancaster
Wilsford
Sudbrook
Kelby
Oasby
Welby
Heydour
Aisby
Culverthorpe
Londonthorpe
Old Somerby
Ropsley
Braceby
Sapperton
Pickworth
Walcot
Humby
Hanby
Folkingham
Laughton
Aslackby
Scotland
Lenton
Ingoldsby
Keisby
Bitchfield
Boothby Pagnell
Bassingthorpe
Westby
Irnham
Hawthorpe
Bulby
Kirkby Underwood
Rippingale
Dunsby
Hacconby
Morton
Dyke
Cawthorpe
Stainfield
Hanthorpe
Elsthorpe
Grimsthorpe
Scottlethorpe
Creeton
Swayfield
Burton Coggles
Birkholme
Cranwell
Leasingham
North Rauceby
South Rauceby
Quarrington
Silk Willoughby
Aswarby
Swarby
Aunsby
Dembleby
Scott Willoughby
Osbournby
Newton
Haceby
Spanby
Swaton
Threekingham
Horbling
Billingborough
Birthorpe
Pointon
Dowsby
Graby
Ruskington
Anwick
South Kyme
Evedon
Ewerby
Ewerby Thorpe
Howell
Kirkby la Thorpe
Asgarby
Heckington
East Heckington
Burton Pedwardine
Great Hale
Little Hale
Helpringham
Scredington
Northbeck
Dorrington
North Kyme
Brauncewell
Temple Bruer
Holdingham
Wigsthorpe
Titchmarsh
Islip
Thorpe Waterville
Achurch
Wadenhoe
Aldwincle
Lowick
Sudborough
Slipton
Grafton Underwood
Cranford St Andrew
Cranford St John
Warkton
Weekley
Geddington
Newton
Boughton House
Great Oakley
Little Oakley
Pipewell
Rushton
Rushton Triangular Lodge
Thorpe Malsor
Thorpe Underwood
Barnwell St Andrew
Barnwell All Saints
Luddington in the Brook
Hemington
Armston
Thurning
Clopton
Winwick
Great Gidding
Little Gidding
Steeple Gidding
Hamerton
Hamerton Zoo Park
Coppingford
Sawtry
Glatton
Conington
Holme
Stilton
Caldecote
Denton
Folksworth
Norman Cross
Yaxley
Morborne
Haddon
Chesterton
Alwalton
Lutton
Southwick
Woodnewton
Cotterstock
Glapthorn
Tansor
Fotheringhay
Ashton
Stoke Doyle
Upper Benefield
Lower Benefield
Blatherwycke
Bedford Purlieus NNR
Old Sulehay
Ferry Meadows
Nene Valley Rly
Orton Waterville
Longthorpe
Longthorpe Tower
Westwood
Brampton Ash
Shire Falconry Centre
Sutton Bassett
Weston by Welland
Ashley
Bringhurst
Drayton
Thorpe Langton
Church Langton
East Langton
Tur Langton
Glooston
Cranoe
Stonton Wyville
Shangton
Burton Overy
Evebrook Res
Thorpe by Water
Harringworth Viaduct
Stockerston
Nevill Holt
Fleckney
Saddington
Foxton
Foxton Locks
Laughton
Gumley
Theddingworth
Husbands Bosworth
Old Weston
Alconbury Weston
Upton
Woodford
Great Cransley
Barton Seagrave
Haselbech
Draughton
KETTERING
0 2 4 6 miles
0 2 4 6 8 10 km
A1
A1(M)
A6
A15
A17
A43
A46
A47
A52
A153
A151
A427
A508
A605
A606
A607
A1175
A4304
A6003
A6116
A6121
A6183
A14
B1174
B6403
B1176
B1177
Ermine Street
Roman Road
Fosse Way
Vale of Belvoir
Vale of Catmose
King Street
Witham
Welland
Nene
Eye Brook
Willow Brook
Chater
Gwash
Glen
Slea
Wreake
Sence
Eye
Devon
Trent
Dunsby Fen
Morton Fen
Bourne North Fen
Baston Common
Deeping Fen
Crowland
Twenty
Guthram Gowt
Pinchbeck Bars
Bridge End
Hop Pole
Thetford
Eastgate
Rutland County Museum
Rutland Railway Museum
Anglian Water Birdwatching Centre
Prior's Coppice
Melton Carnegie Museum
Old Pork Pie Shoppe
Queen's Royal Lancers Regiment Museum
Cribbs Meadow NNR
Muston Meadows NNR
Knipton Reservoir
Denton Res
St Peter & St Paul
Oakham Castle
Rutland Water
Whitwell
Stoke Field 1487
Naseby 1645
Big Sky Adventure Play
Peterborough
Thurgarton Priory
Elton
Epperstone
Flawborough
Kilvington
Alverton
Little Green
Cold Newton
Lowesby
Baggrave Hall
Hungarton
Keyham
Beeby
South Croxton
Barsby
Marefield
Nether Broughton
Grimston
Saxelbye
Old Dalby
Thorpe Arnold
Asfordby Hill
Kirby Bellars
Brooksby
Hoby
Rearsby
Goscote
Queniborough
Gaulby
King's Norton
Houghton on the Hill
Humberstone

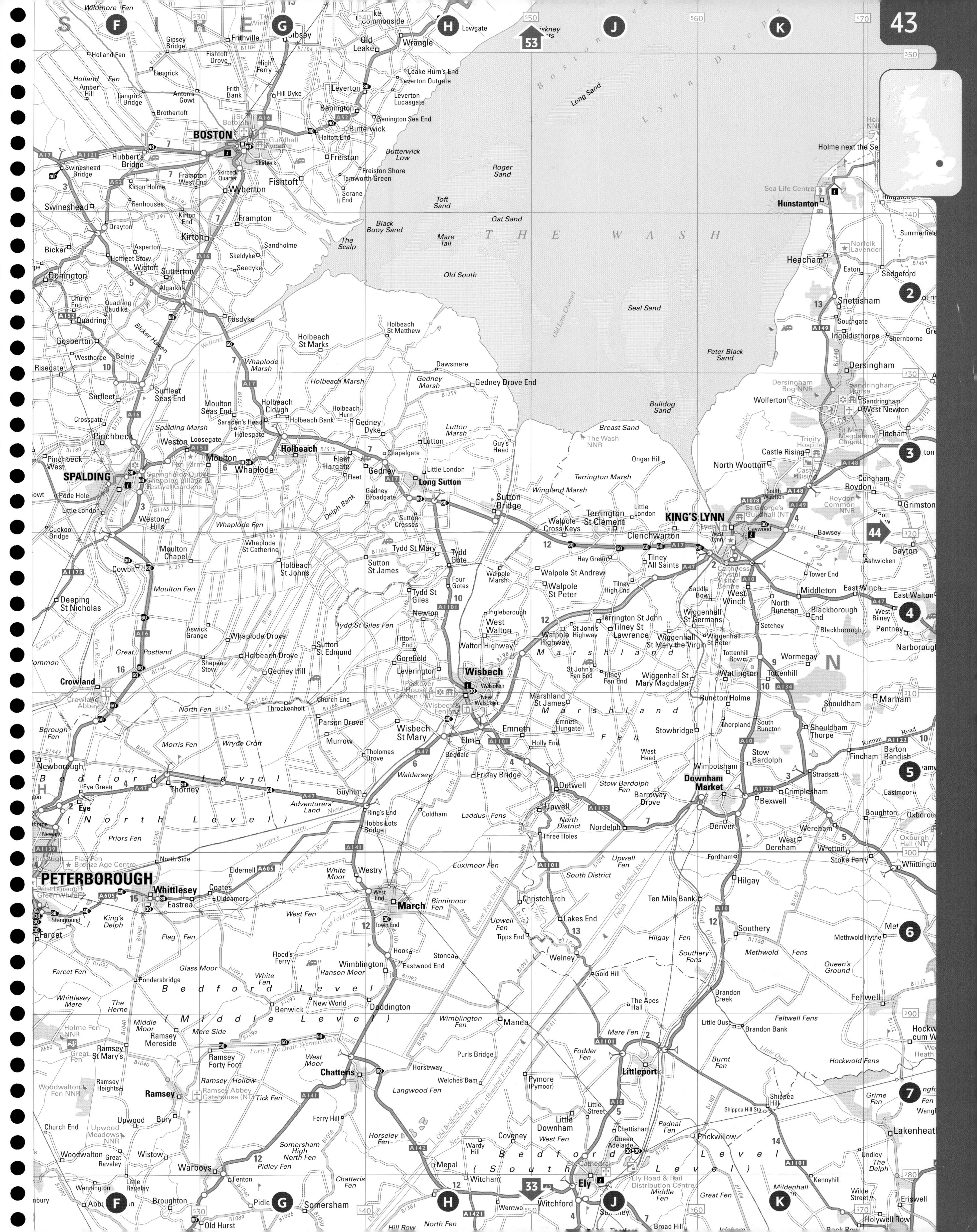

F
G
H
J
K
53
44
33
2
3
4
5
6
7
BOSTON
Wyberton
Fishtoft
Freiston
Butterwick
Benington
Leverton
Wrangle
Old Leake
Sibsey
Frithville
Kirton
Frampton
Swineshead
Donington
Sutterton
Gosberton
Quadring
Surfleet
Pinchbeck
SPALDING
Moulton
Whaplode
Holbeach
Fleet Hargate
Gedney
Long Sutton
Sutton Bridge
Crowland
Deeping St Nicholas
Cowbit
Moulton Chapel
Holbeach St Johns
Sutton St James
Tydd St Mary
Tydd Gote
Gorefield
Leverington
Wisbech
Walsoken
Wisbech St Mary
Parson Drove
Murrow
Thorney
Eye
Eye Green
Newborough
PETERBOROUGH
Whittlesey
Eastrea
Coates
Eldernell
March
Westry
Guyhirn
Elm
Emneth
Outwell
Upwell
Christchurch
Welney
Manea
Chatteris
Doddington
Wimblington
Benwick
Ramsey
Warboys
Somersham
Pidley
Bury
Upwood
Wistow
Littleport
Ely
Little Downham
Coveney
Witchford
Witcham
Mepal
Sutton
Downham Market
Denver
Hilgay
Southery
Stow Bardolph
Wimbotsham
Watlington
Tottenhill
KING'S LYNN
Clenchwarton
Terrington St Clement
Walpole St Andrew
Walpole St Peter
Tilney All Saints
West Winch
Middleton
Grimston
Gayton
Roydon
Congham
Castle Rising
North Wootton
Wolferton
Dersingham
Snettisham
Heacham
Hunstanton
Sedgeford
Ingoldisthorpe
Flitcham
Sandringham
West Newton
Marham
Shouldham
Fincham
Barton Bendish
Crimplesham
Stradsett
Boughton
Wereham
Stoke Ferry
Whittington
Methwold Hythe
Feltwell
Lakenheath
Eriswell
Prickwillow
Shippea Hill Sta.
Brandon Creek
Ten Mile Bank
Fordham
Nordelph
Three Holes
Barroway Drove
Stowbridge
Runcton Holme
Wormegay
Narborough
Pentney
East Winch
East Walton
Setchey
Blackborough End
North Runcton
Saddle Bow
Wiggenhall St Germans
Wiggenhall St Mary the Virgin
Wiggenhall St Mary Magdalen
Terrington St John
Tilney St Lawrence
Walpole Highway
Marshland St James
Emneth Hungate
Friday Bridge
THE WASH
Boston Deeps
Lynn Deeps
Long Sand
Roger Sand
Gat Sand
Toft Sand
Old South
Seal Sand
Peter Black Sand
Bulldog Sand
Breast Sand
Black Buoy Sand
The Scalp
Mare Tail
Bedford Level (North Level)
Bedford Level (Middle Level)
Bedford Level (South Level)
Marshland Fen
Holme next the Sea
Sea Life Centre
Norfolk Lavender
Springfields Outlet Shopping Village & Festival Gardens
Peckover House & Garden (NT)
Wisbech & Fenland Museum
Ely Road & Rail Distribution Centre
Ramsey Abbey Gatehouse (NT)
Flag Fen Bronze Age Centre
Oxburgh Hall (NT)
St George's Guildhall (NT)
A16
A17
A47
A10
A141
A142
A1101
A1122
A134
A149
A148
A605
A1121
A52
A151
A1175
A1139
A1421

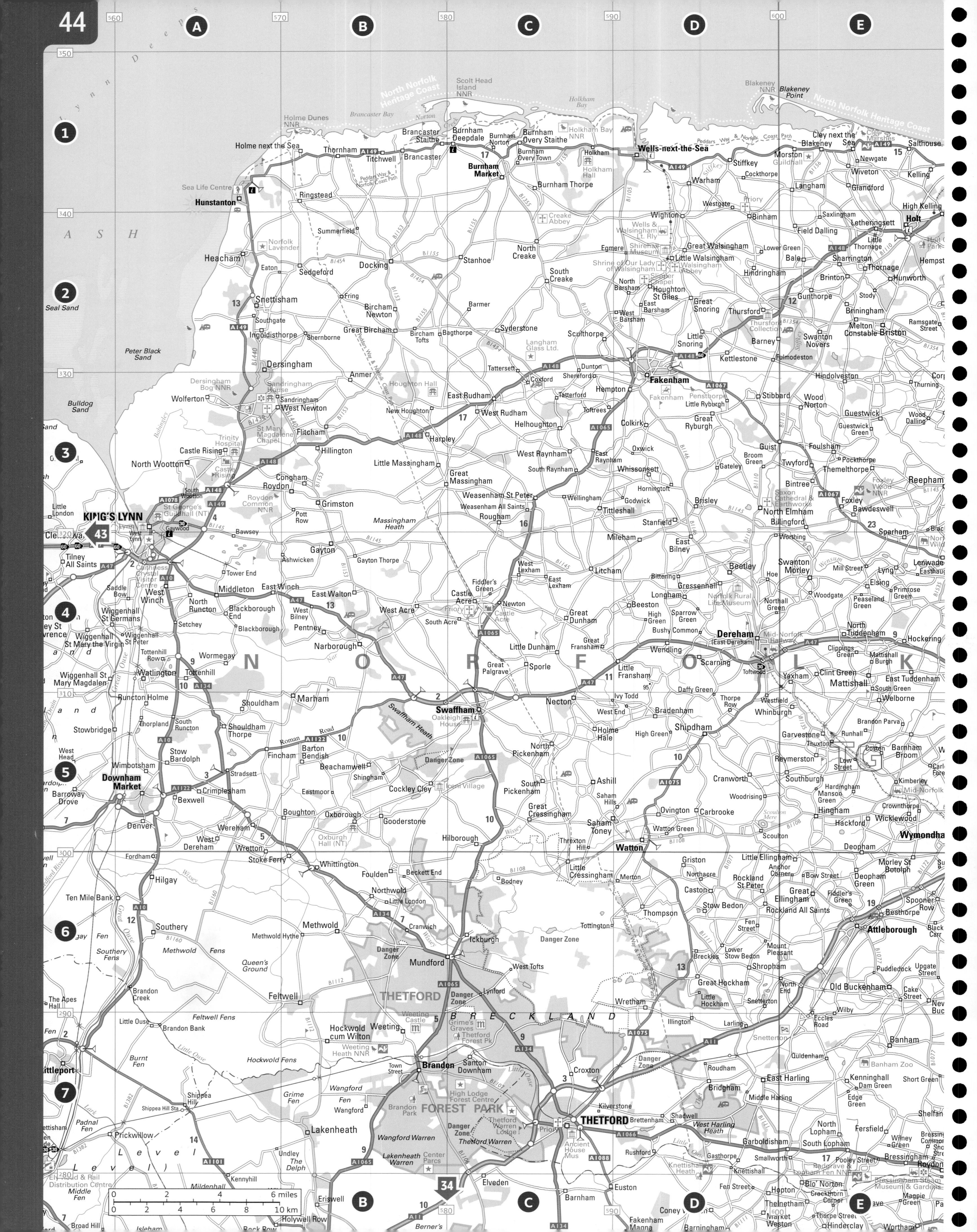
A
B
C
D
E
1
2
3
4
5
6
7
North Norfolk Heritage Coast
Scolt Head Island NNR
Holkham Bay
Blakeney NNR
Blakeney Point
Holme Dunes NNR
Brancaster Bay
Holme next the Sea
Thornham
Titchwell
Brancaster
Brancaster Staithe
Burnham Deepdale
Burnham Norton
Burnham Overy Staithe
Burnham Overy Town
Holkham
Holkham Hall
Wells-next-the-Sea
Stiffkey
Morston
Blakeney
Cley next the Sea
Salthouse
Burnham Market
Sea Life Centre
Hunstanton
Ringstead
Burnham Thorpe
Warham
Cockthorpe
Langham
Wiveton
Glandford
Kelling
High Kelling
Holt
Summerfield
Creake Abbey
Wighton
Binham
Saxlingham
Letheringsett
Field Dalling
Norfolk Lavender
Heacham
Docking
Stanhoe
North Creake
Egmere
Great Walsingham
Little Walsingham
Wells & Walsingham Lt. Rly
Shirehall Museum
Shrine of Our Lady of Walsingham
Walsingham Abbey
Lower Green
Bale
Sharrington
Thornage
Hunworth
Brinton
Eaton
Sedgeford
South Creake
North Barsham
Slipper Chapel
Houghton St Giles
East Barsham
West Barsham
Hindringham
Great Snoring
Thursford
Thursford Collection
Gunthorpe
Stody
Briningham
Melton Constable
Briston
Ramsgate Street
Snettisham
Fring
Bircham Newton
Barmer
Southgate
Ingoldisthorpe
Shernborne
Great Bircham
Bircham Tofts
Bagthorpe
Syderstone
Sculthorpe
Little Snoring
Barney
Swanton Novers
Kettlestone
Fulmodeston
Peter Black Sand
Seal Sand
Dersingham
Anmer
Langham Glass Ltd.
Tattersett
Coxford
Dunton
Shereford
Fakenham
Hindolveston
Dersingham Bog NNR
Sandringham House
Sandringham
Houghton Hall
Wolferton
West Newton
East Rudham
New Houghton
West Rudham
Tatterford
Hempton
Pensthorpe
Little Ryburgh
Stibbard
Wood Norton
Guestwick
Bulldog Sand
St Mary Magdalene Chapel
Flitcham
Harpley
Helhoughton
Toftrees
Colkirk
Great Ryburgh
Guestwick Green
Wood Dalling
Trinity Hospital
Castle Rising
Hillington
Little Massingham
West Raynham
East Raynham
Oxwick
Guist
Foulsham
North Wootton
Great Massingham
South Raynham
Whissonsett
Broom Green
Gateley
Twyford
Pockthorpe
Themelthorpe
Reepham
Congham
Roydon
South Wootton
Roydon Common NNR
Weasenham St Peter
Weasenham All Saints
Wellingham
Horningtoft
Godwick
Tittleshall
Bintree
Brisley
North Elmham
Foxley
Bawdeswell
Little London
KING'S LYNN
43
Grimston
Pott Row
Massingham Heath
Rougham
Stanfield
Billingford
Bawsey
St George's Guildhall (NT)
Gaywood
West Lynn
Mileham
East Bilney
Worthing
Sparham
Tilney All Saints
Gayton
Gayton Thorpe
Ashwicken
Tower End
West Lexham
East Lexham
Litcham
Beetley
Swanton Morley
Mill Street
Lyng
Lenwade
Elsing
Middleton
East Winch
West Winch
North Runcton
East Walton
Castle Acre
Fiddler's Green
Priory
Newton
Bittering
Gressenhall
Norfolk Rural Life Museum
Longham
Hoe
Woodgate
Peaseland Green
Northall Green
Saddle Bow
Wiggenhall St Germans
Setchey
Blackborough End
Blackborough
West Bilney
West Acre
South Acre
Beeston
High Green
Sparrow Green
Great Dunham
Bushy Common
Dereham
North Tuddenham
Hockering
Wiggenhall St Mary the Virgin
Wiggenhall St Peter
Pentney
Narborough
Little Dunham
Great Fransham
Wendling
Mid-Norfolk Railway
Tottenhill Row
Wormegay
Tottenhill
Watlington
Wiggenhall St Mary Magdalen
N
O
R
F
O
L
K
Great Palgrave
Sporle
Little Fransham
Scarning
Toftwood
Clippings Green
Mattishall Burgh
Yaxham
Clint Green
Mattishall
East Tuddenham
Runcton Holme
Shouldham
Marham
Swaffham
Oakleigh House
Necton
Ivy Todd
Daffy Green
Thorpe Row
Westfield
Welborne
South Green
West End
Bradenham
Whinburgh
Stowbridge
Thorpland
South Runcton
Shouldham Thorpe
Swaffham Heath
Roman Road
Holme Hale
High Green
Shipdham
Brandon Parva
Garvestone
Runhall
Thuxton
Coston
Barnham Broom
Low Street
West Head
Stow Bardolph
Barton Bendish
Fincham
Danger Zone
North Pickenham
Reymerston
Wimbotsham
Downham Market
Stradsett
Beachamwell
Shingham
South Pickenham
Ashill
Cranworth
Southburgh
Kimberley
Hardingham
Mid-Norfolk
Crimplesham
Bexwell
Eastmoor
Cockley Cley
Iceni Village
Saham Hills
Woodrising
Mansion Green
Crownthorpe
Barroway Drove
Denver
Boughton
Oxborough
Gooderstone
Great Cressingham
Ovington
Carbrooke
Hingham
Wicklewood
Hackford
Wereham
West Dereham
Wretton
Oxburgh Hall (NT)
Hilborough
Saham Toney
Watton Green
Scoulton Mere
Scoulton
Wymondham
Stoke Ferry
Threxton Hill
Watton
Deopham
Fordham
Whittington
Foulden
Beckett End
Bodney
Little Cressingham
Merton
Griston
Northacre
Little Ellingham
Anchor Street
Morley St Botolph
Hilgay
Northwold
Caston
Rockland St Peter
Bow Street
Deopham Green
Ten Mile Bank
Little London
Great Ellingham
Fiddler's Green
Stow Bedon
Rockland All Saints
Spooner Row
Southery
Methwold
Methwold Hythe
Cranwich
Ickburgh
Danger Zone
Tottington
Thompson
Fen Street
Besthorpe
Attleborough
Southery Fens
Methwold Fens
Queen's Ground
Mundford
West Tofts
Breckles
Lower Stow Bedon
Mount Pleasant
Shropham
Puddledock
Upgate Street
Brandon Creek
Feltwell
THETFORD
Danger Zone
Lynford
Great Hockham
Little Hockham
North End
Old Buckenham
Cake Street
The Apes Hall
Little Ouse
Brandon Bank
Feltwell Fens
Weeting Castle
B R E C K L A N D
Wretham
Illington
Larling
Eccles Road
Wilby
Hockwold cum Wilton
Weeting
Grime's Graves
Thetford Forest Pk
Snetterton
Banham
Weeting Heath NNR
Burnt Fen
Hockwold Fens
Town Street
Brandon
Santon Downham
Croxton
Danger Zone
Roudham
Quidenham
Banham Zoo
Littleport
Grime Fen
Wangford Fen
Wangford
High Lodge Forest Centre
Brandon Park
FOREST PARK
Kilverstone
Bridgham
East Harling
Middle Harling
Kenninghall
Dam Green
Short Green
Edge Green
Shippea Hill
Shippea Hill Sta
Padnal Fen
Prickwillow
Lakenheath
Wangford Warren
Lakenheath Warren
Danger Zone
Thetford Warren
Thetford Warren Lodge
Priory
THETFORD
Brettenham
Shadwell
West Harling Heath
North Lopham
South Lopham
Fersfield
Garboldisham
Ancient House Mus
Rushford
Little Ouse
Gasthorpe
Knettishall Heath
Knettishall
Smallworth
Pooley Street
Bressingham
Roydon
Level
Undley
The Delph
Center Parcs
Elveden
Euston
Fen Street
Hopton
Blo' Norton
Bressingham Steam Museum & Gardens
Mildenhall
Kennyhill
Ely Road & Rail Distribution Centre
Middle Fen
Eriswell
34
Barnham
Coney Weston
Thelnetham
Market Weston
Thorpe Street
Hinderclay
Magpie Green
Wortham
Broad Hill
Isleham
Rock Row
Holywell Row
Berner's
Fakenham Magna
Barningham
0 2 4 6 miles
0 2 4 6 8 10 km
560
570
580
590
600
350
340
330
320
310
300
290
280

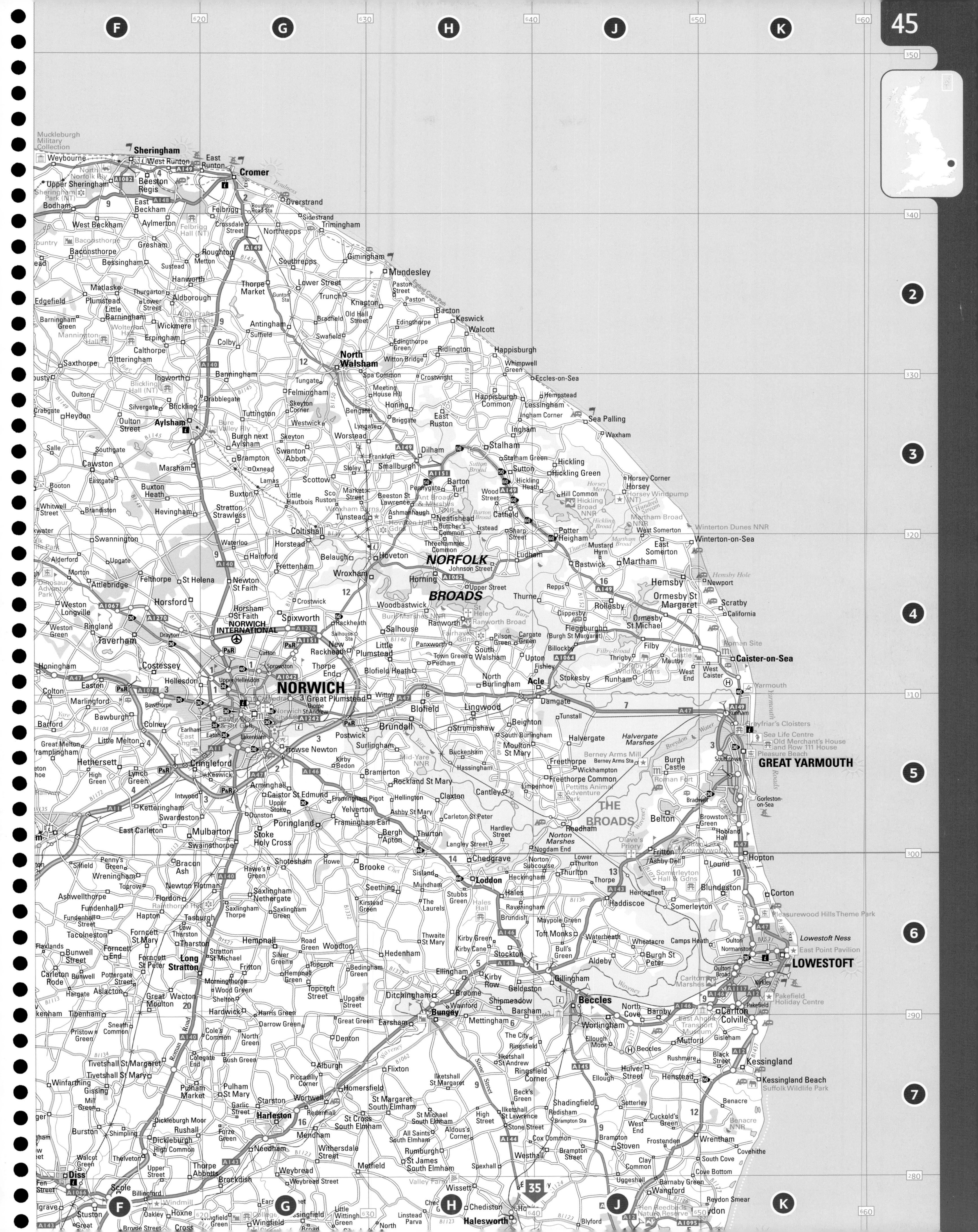

F
G
H
J
K
620
630
640
650
660
350
340
330
320
310
300
290
280
2
3
4
5
6
7
Muckleburgh Military Collection
Weybourne
Sheringham
West Runton
East Runton
Cromer
Upper Sheringham
Beeston Regis
Sheringham Park (NT)
Bodham
East Beckham
Felbrigg
Overstrand
Sidestrand
Trimingham
West Beckham
Aylmerton
Felbrigg Hall (NT)
Crossdale Street
Northrepps
Baconsthorpe
Gresham
Roughton
Bessingham
Metton
Southrepps
Gimingham
Mundesley
Sustead
Hanworth
Thorpe Market
Lower Street
Paston
Matlaske
Edgefield
Plumstead
Thurgarton
Aldborough
Little Barningham
Trunch
Knapton
Bacton
Keswick
Walcott
Barningham Green
Mannington Hall
Wolterton Hall
Wickmere
Erpingham
Antingham
Suffield
Bradfield
Old Hall Street
Swafield
Edingthorpe
Edingthorpe Green
Colby
Calthorpe
Itteringham
Saxthorpe
North Walsham
Ridlington
Happisburgh
Whimpwell Green
Eccles-on-Sea
Witton Bridge
Crostwight
Ingworth
Banningham
Tungate
Spa Common
Blickling Hall (NT)
Oulton
Felmingham
Meeting House Hill
Happisburgh Common
Hempstead
Lessingham
Heydon
Silvergate
Blickling
Brabblegate
Tuttington
Skeyton Corner
Honing
Briggate
East Ruston
Ingham Corner
Sea Palling
Waxham
Ingham
Oulton Street
Aylsham
Bure Valley Rly
Burgh next Aylsham
Westwick
Bengate
Lyngate
Worstead
Dilham
Stalham
Salle
Cawston
Southgate
Skeyton
Swanton Abbott
Brampton
Oxnead
Frankfort
Sloley
Smallburgh
Stalham Green
Sutton
Hickling
Hickling Green
Horsey Corner
Horsey
Horsey Windpump (NT)
Marsham
Lamas
Scottow
Barton Turf
Pennygate
Hickling Heath
Booton
Eastgate
Buxton Heath
Buxton
Little Hautbois
Scottow
Market Street
Beeston St Lawrence
Wood Street
Hill Common
Hickling Broad NNR
Whitwell Street
Brandiston
Hevingham
Stratton Strawless
Wroxham Barns
Tunstead
Ant Broads & Marshes NNR
Ashmanhaugh
Neatishead
Catfield
Martham Broad NNR
Winterton Dunes NNR
Hoveton Hall Gdns
Butcher's Common
Coltishall
Irstead
Sharp Street
Potter Heigham
West Somerton
Winterton-on-Sea
Swannington
Waterloo
Horstead
Threehammer Common
Mustard Hyrn
East Somerton
Alderford
Upgate
Hainford
Belaugh
Hoveton
NORFOLK BROADS
Ludham
Martham
Bastwick
Morton
Felthorpe
St Helena
Frettenham
Wroxham
Johnson Street
Attlebridge
Newton St Faith
Horning
Upper Street
Repps
Hemsby
Hemsby Hole
Newport
Weston Longville
Horsford
Crostwick
Woodbastwick
Thurne
Rollesby
Ormesby St Margaret
Scratby
California
Horsham St Faith
NORWICH INTERNATIONAL
Spixworth
Bure Marshes NNR
St Helen
Ranworth Broad
Ranworth
Clippesby
Ormesby St Michael
Weston Green
Ringland
Rackheath
Salhouse
Fleggburgh (Burgh St Margaret)
Taverham
Drayton
Little Plumstead
Panxworth
Pilson Green
Cargate Green
Billockby
Filby
Caister Castle
Roman Site
Caister-on-Sea
Honingham
Costessey
Catton
Sprowston
Thorpe End
Blofield Heath
Town Green
Pedham
South Walsham
Upton
Fishley
Thrigby
Thrigby Hall Wildlife Gdns
Mautby
West Caister
Colton
Easton
Hellesdon
NORWICH
Great Plumstead
Witton
North Burlingham
Acle
Stokesby
Runham
West End
Yarmouth
Marlingford
Bowthorpe
Thorpe St Andrew
Blofield
Lingwood
Damgate
Bawburgh
Norwich City
Castle Mus & Art Gall
Tunstall
Barford
Colney
Earlham
Eaton
Postwick
Brundall
Strumpshaw
Beighton
South Burlingham
Halvergate
Halvergate Marshes
Greyfriar's Cloisters
Sea Life Centre
Old Merchant's House and Row 111 House
Pleasure Beach
Little Melton
Great Melton
East Angle
Lakenham
Trowse Newton
Surlingham
Buckenham
Moulton St Mary
Berney Arms Mill
Berney Arms Sta
Breydon Water
GREAT YARMOUTH
Wramplingham
Hethersett
Cringleford
Keswick
Kirby Bedon
Mid-Yare NNR
Hassingham
Freethorpe
Wickhampton
Burgh Castle
Roman Fort
High Green
Lynch Green
Bramerton
Rockland St Mary
Freethorpe Common
Limpenhoe
Pettitts Animal Adventure Park
Arminghall
Caistor St Edmund
Framingham Pigot
Hellington
Claxton
Cantley
Gorleston-on-Sea
Bradwell
Intwood
Upper Stoke
Ketteringham
Swardeston
Dunston
Poringland
Yelverton
Ashby St Mary
Carleton St Peter
Framingham Earl
THE BROADS
Belton
Browston Green
Hobland Hall
East Carleton
Mulbarton
Swainsthorpe
Stoke Holy Cross
Bergh Apton
Thurton
Hardley Street
Reedham
Norton Marshes
Langley Street
St Olave's Priory
Nogdam End
Fritton
Fritton Lake Countryworld
Hopton
Siffield
Penny's Green
Bracon Ash
Shotesham
Howe
Brooke
Chedgrave
Norton Subcourse
Lower Thurlton
Thurlton
Ashby Dell
Lound
Wreningham
Hawe's Green
Sisland
Loddon
Heckingham
Somerleyton Hall & Gdns
Toprow
Newton Flotman
Mundham
Thorpe
Ashwellthorpe
Saxlingham Nethergate
Seething
Stubbs Green
Hales
Haddiscoe
Herringfleet
Blundeston
Corton
Fundenhall
Rainthorpe Hall
Kirstead Green
The Laurels
Hales Hall
Raveningham
Somerleyton
Fundenhall Street
Hapton
Saxlingham Thorpe
Saxlingham Green
Brundish
Maypole Green
Pleasurewood Hills Theme Park
Tacolneston
Forncett St Mary
Low Tharston
Tasburgh
Tharston
Hempnall
Thwaite St Mary
Kirby Green
Toft Monks
Waterheath
Wheatacre
Camps Heath
Oulton
Lowestoft Ness
Flaxlands
Bunwell Street
Forncett End
Forncett St Peter
Long Stratton
Stratton St Michael
Road Green
Woodton
Hedenham
Kirby Cane
Stockton
Bull's Green
Aldeby
Burgh St Peter
Normanston
East Point Pavilion
LOWESTOFT
Silver Green
Hempnall Green
Topcroft
Bedingham Green
Carleton Rode
Bunwell
Pottergate Street
Fritton
Morningthorpe
Ellingham
Kirby Row
Geldeston
Gillingham
Oulton Broad
Carlton Marshes
Kirkley
Aslacton
Wacton
Great Moulton
Wood Green
Topcroft Street
Ditchingham
Broome
Shipmeadow
Beccles
North Cove
Barnby
Pakefield
Pakefield Holiday Centre
Hargate
Shelton
Upgate Street
Wainford
Bungay
Barsham
Carlton Colville
Tibenham
Hardwick
Harris Green
Darrow Green
Great Green
Earsham
Mettingham
East Anglia Transport Museum
Sneath Common
Pristow Green
Cole's Common
North Green
Denton
The City
Worlingham
Ellough Moor
Beccles
Mutford
Gisleham
Ringsfield
Ilketshall St Andrew
Black Street
Tivetshall St Margaret
Colegate End
Bush Green
Alburgh
Flixton
Ringsfield Corner
Ellough
Hulver Street
Rushmere
Henstead
Kessingland
Winfarthing
Tivetshall St Mary
Pulham Market
Pulham St Mary
Piccadilly Corner
Homersfield
Ilketshall St Margaret
Beck's Green
Kessingland Beach
Suffolk Wildlife Park
Gissing
Mill Green
Garlic Street
Starston
Wortwell
St Margaret South Elmham
Shadingfield
Sotterley
Benacre
Harleston
Redenhall
St Michael South Elmham
High Street
Ilketshall St Lawrence
Redisham
Brampton Sta
Cuckold's Green
Burston
Shimpling
Dickleburgh Moor
Rushall
Forze Green
Mendham
St Cross South Elmham
All Saints South Elmham
Aldous's Corner
Stone Street
West End
Brampton
Benacre NNR
Dickleburgh
High Common
Needham
Withersdale Street
Rumburgh
Cox Common
Stoven
Frostenden
Wrentham
Walcot Green
Thelveton
Upper Street
St James South Elmham
Westhall
Brampton Street
Covehithe
Diss
Thorpe Abbotts
Brockdish
Weybread
Metfield
Spexhall
Clay Common
South Cove
Weybread Street
Valley Farm
Uggeshall
Cove Bottom
Scole
Billingford
Wissett
Barnaby Green
Wangford
Reydon Smear
Stuston
Oakley
Hoxne
Syleham
Wingfield
Linstead Parva
Chediston
Halesworth
Blyford
Hen Reedbeds Nature Reserve
Reydon
35
A140
A149
A148
A1082
A1067
A1270
A1151
A1062
A1064
A47
A1074
A1042
A11
A146
A1242
A143
A144
A145
A12
A1117
A1066
A1095
B1145
B1149
B1159
B1150
B1354
B1140
B1152
B1108
B1172
B1113
B1134
B1135
B1527
B1332
B1136
B1062
B1123
B1127
B1124
B1074
B1375

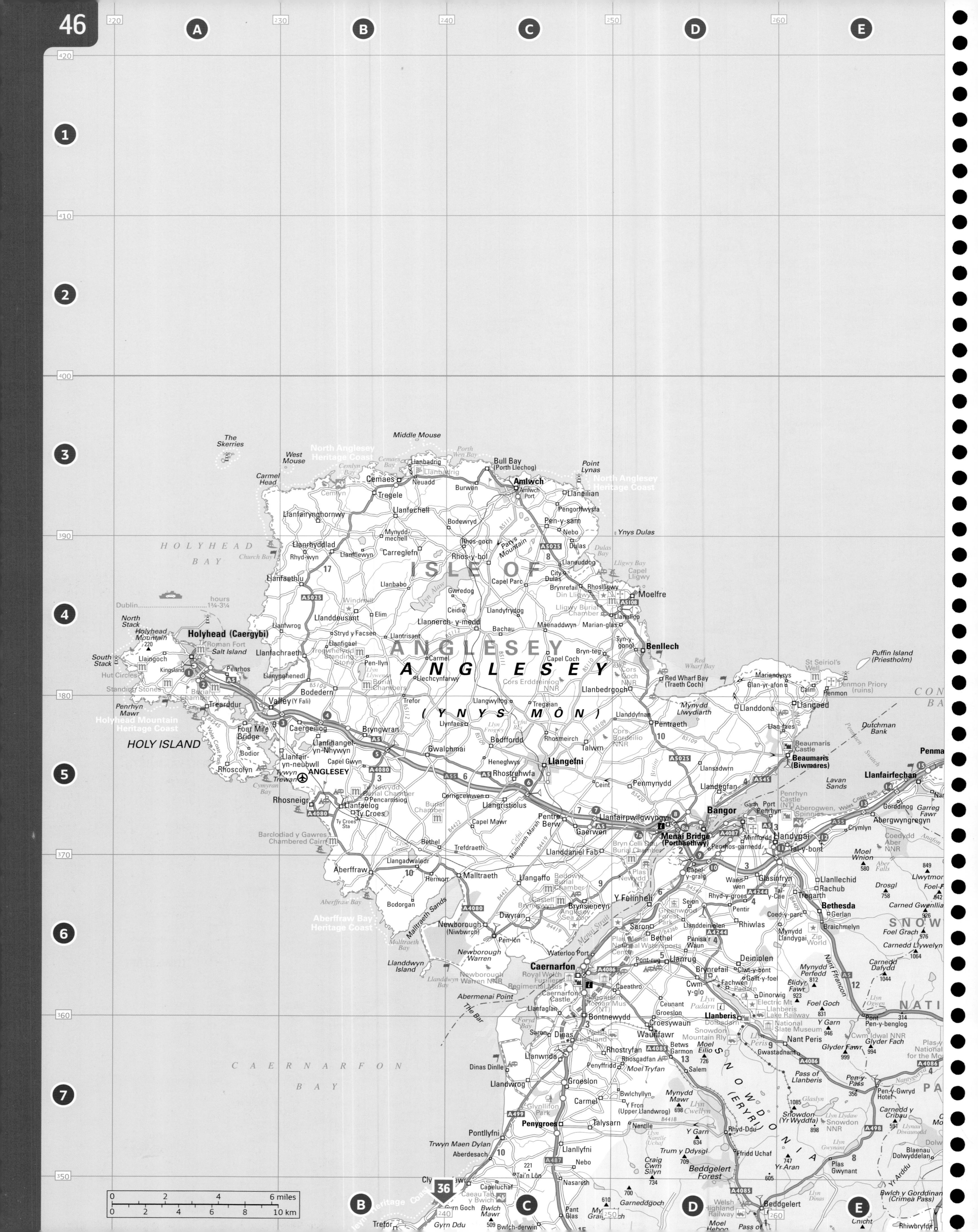

A
B
C
D
E
1
2
3
4
5
6
7
ISLE OF ANGLESEY
ANGLESEY
(YNYS MÔN)
HOLY ISLAND
HOLYHEAD BAY
CAERNARFON BAY
The Skerries
West Mouse
Middle Mouse
North Anglesey Heritage Coast
Carmel Head
Cemaes Bay
Cemaes
Llanbadrig
Bull Bay (Porth Llechog)
Amlwch
Amlwch Port
Point Lynas
Llaneilian
Tregele
Llanfechell
Llanfairynghornwy
Burwen
Bodewryd
Penysarn
Nebo
Ynys Dulas
Llanrhyddlad
Rhyd-wyn
Llanfflewyn
Carreglefn
Mynydd-mechell
Rhos-goch
Rhos-y-bol
Parys Mountain
Dulas
Dulas Bay
Church Bay
Llanfaethlu
Llanbabo
Llyn Alaw
Capel Parc
City Dulas
Llanallgo
Lligwy Bay
Capel Lligwy
Moelfre
Brynrefail
Rhosllugwy
Din Lligwy
Gwredog
Ceidio
Llandyfrydog
Lligwy Burial Chamber
Llanbedrgoch
Marian-glas
Llannerch-y-medd
Bachau
Maenaddwyn
Dublin
hours 1¼-3¼
North Stack
Holyhead Mountain
Holyhead (Caergybi)
Roman Fort
Salt Island
South Stack
Llaingoch
Kingsland
Penrhos
Hut Circles
Standing Stones
Burial Chamber
Trearddur
Penrhyn Mawr
Holyhead Mountain Heritage Coast
Four Mile Bridge
Bodior
Rhoscolyn
Llanddeusant
Elim
Llanfwrog
Llanfachraeth
Llanynghenedl
Stryd y Facsen
Llantrisant
Carmel
Pen-llyn
Llechcynfarwy
Bodedern
Valley (Y Fali)
Trefor
Caergeiliog
Llanfihangel-yn-Nhywyn
Bryngwran
Llanfair-yn-neubwll
Tywyn Trewan
ANGLESEY
Capel Gwyn
Gwalchmai
Rhosneigr
Llanfaelog
Ty Croes
Pencarnisiog
Barclodiad y Gawres Chambered Cairn
Aberffraw
Aberffraw Bay
Aberffraw Bay Heritage Coast
Llangadwaladr
Bethel
Trefdraeth
Hermon
Malltraeth
Malltraeth Sands
Bodorgan
Llanddwyn Island
Newborough (Niwbwrch)
Newborough Warren
Pen-lôn
Dwyran
Llangaffo
Brynsiencyn
Llangristiolus
Capel Mawr
Corrigceinwen
Heneglwys
Rhostrehwfa
Llangefni
Bodffordd
Llynfaes
Llangwyllog
Tregaian
Rhosmeirch
Talwrn
Cors Erddreiniog NNR
Capel Coch
Bryn-teg
Tyn-y-gongl
Benllech
Red Wharf Bay (Traeth Coch)
Mynydd Llwydiarth
Llanddyfnan
Pentraeth
Cors Bodeilio NNR
Llansadwrn
Penmynydd
Ceint
Llandegfan
Pentre Berw
Gaerwen
Llanfairpwllgwyngyll
Llanddaniel Fab
Bryn Celli Ddu Burial Chamber
Plas Newydd (NT)
Menai Bridge (Porthaethwy)
Bangor
Y Felinheli
Penmon Priory (ruins)
Puffin Island (Priestholm)
St Seiriol's Well
Mariandyrys
Glan-yr-afon
Penmon
Llangoed
Llanddona
Beaumaris Castle
Beaumaris (Biwmares)
Dutchman Bank
Lavan Sands
Llanfairfechan
Aber
Abergwyngregyn
Llandygai
Tal-y-bont
Penrhos-garnedd
Glasinfryn
Llanllechid
Rachub
Bethesda
Gerlan
Tregarth
Coed-y-parc
Pentir
Rhiwlas
Seion
Saron
Bethel
Caernarfon
Royal Welch Fusiliers Regimental Mus
Caernarfon Castle
Waterloo Port
Pont-rug
Llanrug
Deiniolen
Brynrefail
Cwm-y-glo
Llanberis
Llyn Padarn
Dinorwig
Electric Mt
National Slate Museum
Nant Peris
Caeathro
Bontnewydd
Llanfaglan
Abermenai Point
The Bar
Foryd Bay
Dinas
Saron
Llanwnda
Rhostryfan
Rhosgadfan
Waunfawr
Betws Garmon
Moel Eilio
Groeslon
Ceunant
Croesywaun
Dinas Dinlle
Llandwrog
Groeslon
Carmel
Penyffridd
Moel Tryfan
Salem
Y Fron (Upper Llandwrog)
Bwlchyllyn
Mynydd Mawr
Llyn Cwellyn
Penygroes
Talysarn
Nantlle
Pontllyfni
Trwyn Maen Dylan
Aberdesach
Llanllyfni
Nebo
Tai'n Lôn
Nasareth
Y Garn
Trum y Ddysgl
Craig Cwm Silyn
Rhyd-Ddu
Ffridd Uchaf
Beddgelert Forest
Snowdon (Yr Wyddfa)
Snowdon NNR
Yr Aran
Beddgelert
Plas Gwynant
Pass of Llanberis
Pen-y-Pass
Glyder Fawr
Glyder Fach
Pont Pen-y-benglog
Cwm Idwal NNR
Carnedd Dafydd
Carnedd Llywelyn
Yr Elen
Foel Grach
Mynydd Perfedd
Elidyr Fawr
Foel Goch
Nant Ffrancon
SNOWDONIA (ERYRI)
Welsh Highland Railway
Capeluchaf
Bwlch Mawr
Garnedd-goch
Pant Glas
Bwlch-derwin
Gyrn Goch
Gyrn Ddu
Trefor
Bwlch y Gorddinan (Crimea Pass)
Blaenau Dolwyddelan
Cnicht
Rhiwbryfdir
0 2 4 6 miles
0 2 4 6 8 10 km
36

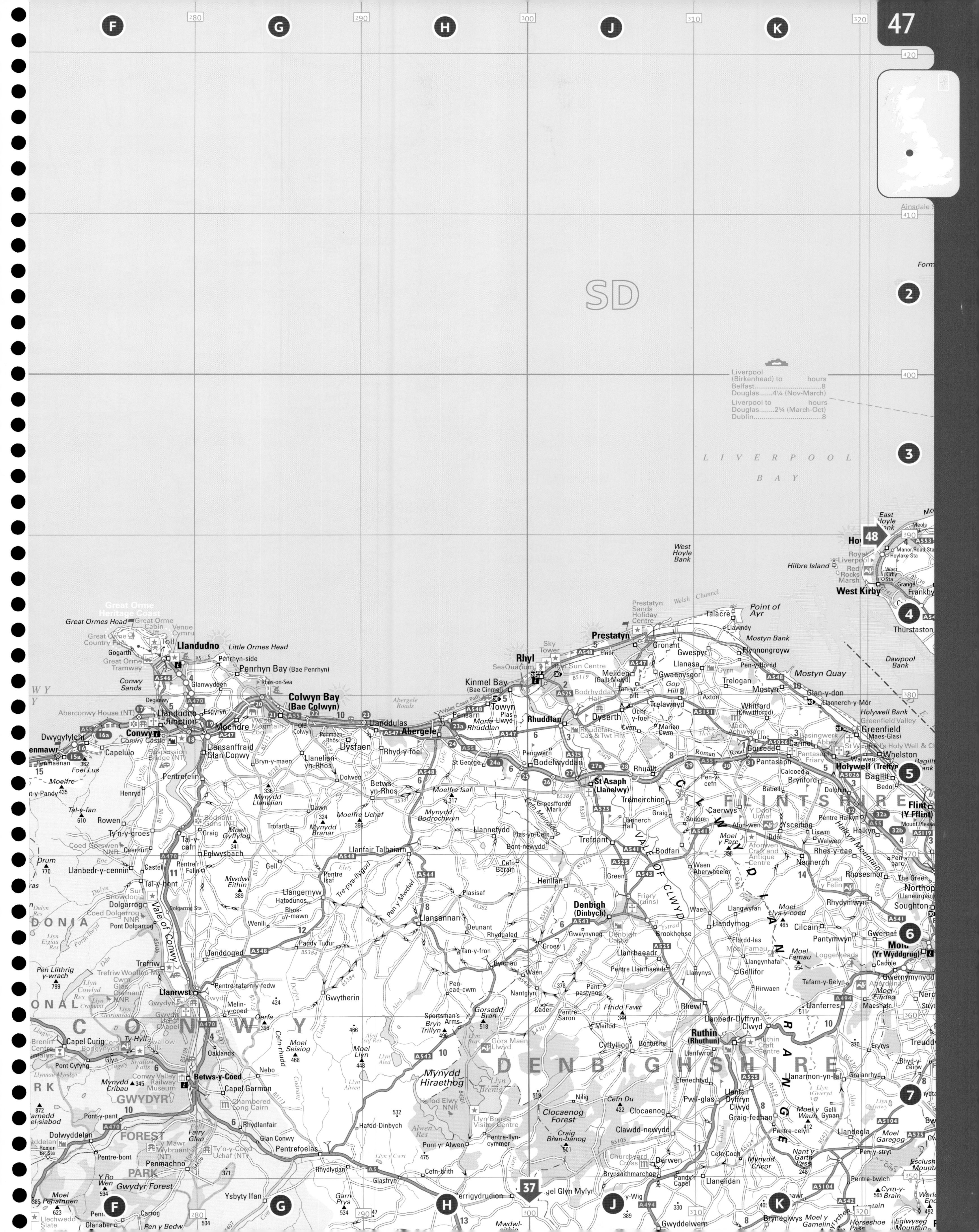

F
G
H
J
K
SD
Liverpool (Birkenhead) to hours
Belfast....8
Douglas......4¼ (Nov-March)
Liverpool to hours
Douglas......2¾ (March-Oct)
Dublin....8
L I V E R P O O L
B A Y
West Hoyle Bank
Hilbre Island
West Kirby
Thurstaston
Dawpool Bank
Mostyn Quay
Mostyn Bank
Point of Ayr
Talacre
Welsh Channel
Prestatyn
Rhyl
Kinmel Bay (Bae Cinmel)
Towyn
Abergele
Llanddulas
Colwyn Bay (Bae Colwyn)
Penrhyn Bay (Bae Penrhyn)
Little Ormes Head
Llandudno
Great Ormes Head
Great Orme Heritage Coast
Conwy Sands
Llandudno Junction
Conwy
Deganwy
Mochdre
Dwygyfylchi
Penmaenmawr
Capelulo
Foel Lus
Aberconwy House (NT)
Conwy Castle
Glan Conwy
Llansanffraid Glan Conwy
Rhuddlan
Dyserth
Meliden (Gallt Melyd)
Gronant
Gwespyr
Llanasa
Trelogan
Mostyn
Whitford (Chwitffordd)
Greenfield
Holywell (Treffynnon)
Bagillt
Flint (Y Fflint)
Halkyn Mountain
Halkyn
Northop
Soughton
Mold (Yr Wyddgrug)
Cilcain
Bodelwyddan
St Asaph (Llanelwy)
Rhuallt
Trefnant
Tremeirchion
Caerwys
Ysceifiog
Denbigh (Dinbych)
Henllan
Llansannan
Llanfair Talhaiarn
Llangernyw
Betws-yn-Rhos
Llysfaen
Llanelian-yn-Rhos
Eglwysbach
Llanrwst
Trefriw
Betws-y-Coed
Capel Curig
Capel Garmon
Pentrefoelas
Cerrigydrudion
Ruthin (Rhuthun)
Llanbedr-Dyffryn-Clwyd
Llanfair Dyffryn Clwyd
Clocaenog Forest
Derwen
Llandegla
Llanarmon-yn-Ial
Mynydd Hiraethog
Llyn Brenig
Alwen Res
FLINTSHIRE
DENBIGHSHIRE
CONWY
VALE OF CLWYD
CLWYDIAN RANGE
Vale of Conwy
Gwydyr Forest
Penmachno
Ysbyty Ifan
Dolwyddelan
Rhewl
Llanferres
Moel Famau
Gwernymynydd
Brynford
Pantasaph
Pentre Halkyn
Nannerch
Rhosesmor
Bodfari
Dolgarrog
Tal-y-bont
Rowen
Llanbedr-y-cennin
Henryd
Glanwydden
Llandudno
Rhos-on-Sea
Penmaenan
Pen Llithrig y-wrach
Moel Siabod
Gwytherin
Llanddoged
Pandy Tudur
Nantglyn
Cyffylliog
Graig-fechan
Clawdd-newydd
Craig Bron-banog
Llanelidan
Gwyddelwern
Bryneglwys
Horseshoe Pass
Eglwyseg Mountain
48
37

For a more detailed map of Merseyside see pages 110-111

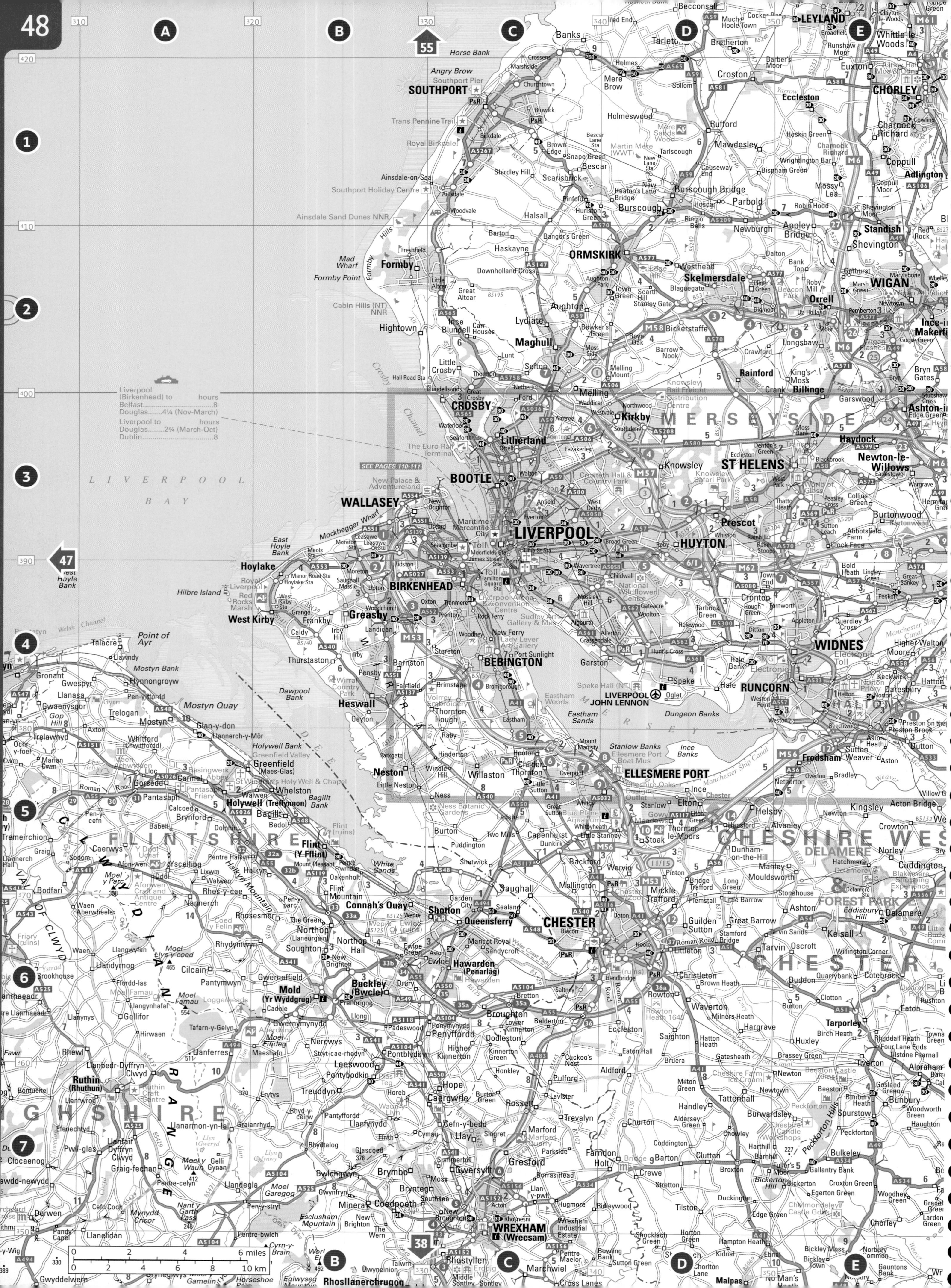

55

SOUTHPORT
Horse Bank
Banks
LEYLAND
CHORLEY
Southport Pier
Croston
Eccleston
Whittle-le-Woods
Euxton
Royal Birkdale
Rufford
Mawdesley
Standish
WIGAN
Ormskirk
Skelmersdale
Orrell
Formby
Ince-in-Makerfield
Ainsdale-on-Sea
Ainsdale Sand Dunes NNR

Liverpool (Birkenhead) to
Belfast.............8 hours
Douglas.......4¼ (Nov-March)
Liverpool to   hours
Douglas.........2¾ (March-Oct)
Dublin.............8

Maghull
Rainford
ST HELENS
Newton-le-Willows
Haydock

SEE PAGES 110-111

LIVERPOOL BAY

CROSBY
KIRKBY
MERSEYSIDE
Ashton-in-Makerfield
BOOTLE
Knowsley
PRESCOT
WALLASEY
LIVERPOOL
HUYTON
47
Hoylake
BIRKENHEAD
Greasby
WEST KIRBY
BEBINGTON
WIDNES
RUNCORN
Point of Ayr
Heswall
LIVERPOOL JOHN LENNON
Frodsham

FLINTSHIRE
Greenfield
Neston
ELLESMERE PORT
CHESHIRE WEST
Holywell (Treffynnon)
Flint (Y Fflint)
Connah's Quay
Shotton
Queensferry
CHESTER
Mold (Yr Wyddgrug)
Buckley (Bwcle)
Hawarden (Penarlâg)
Ruthin (Rhuthun)
Hope
Caergwrle
38
WREXHAM (Wrecsam)

0   2   4   6 miles
0   2   4   6   8   10 km

A   B   C   D   E

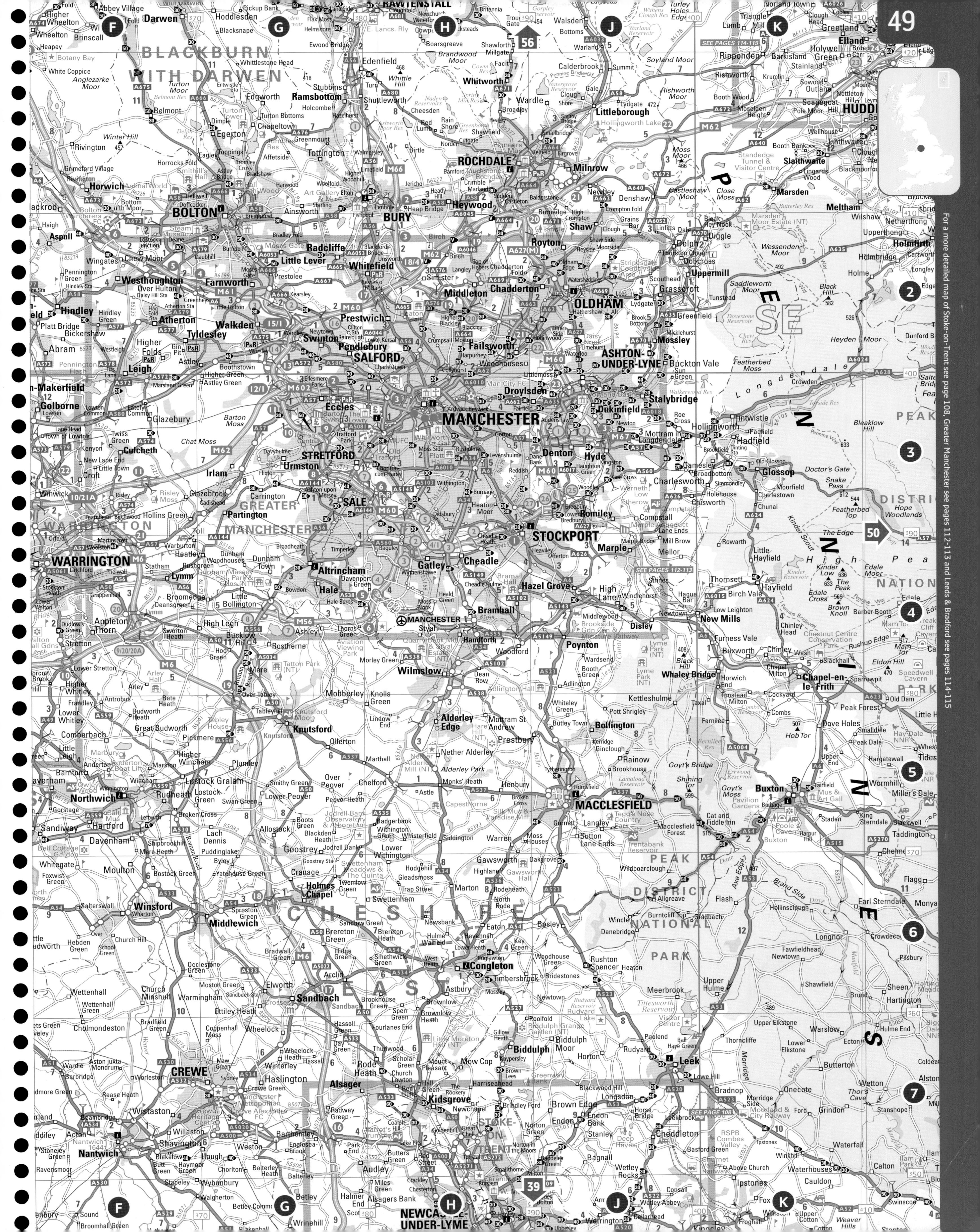

For a more detailed map of Stoke-on-Trent see page 108, Greater Manchester see pages 112-113 and Leeds & Bradford see pages 114-115

For a more detailed map of Stoke-on-Trent see page 108, Greater Manchester see pages 112-113 and Leeds & Bradford see pages 114-115

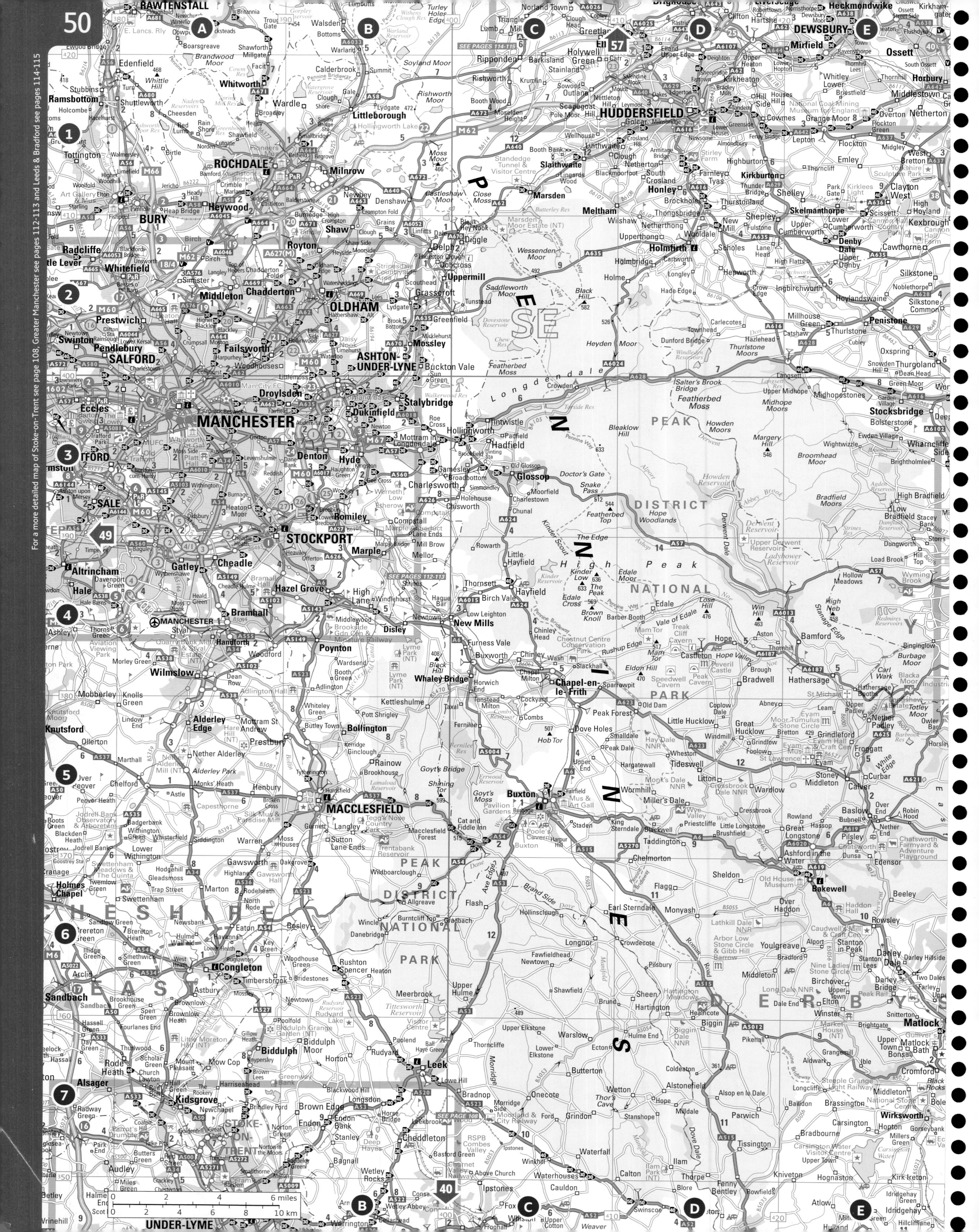

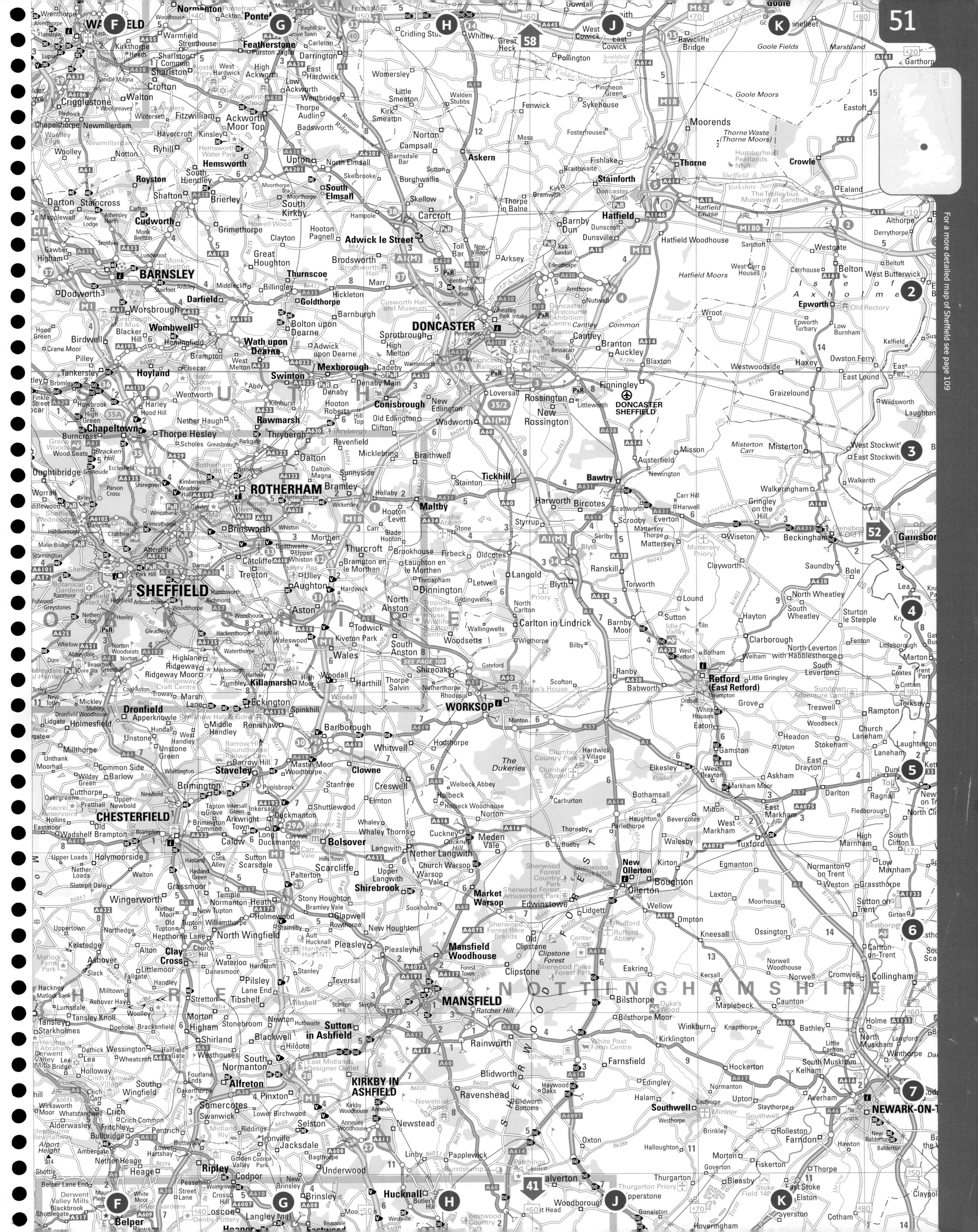

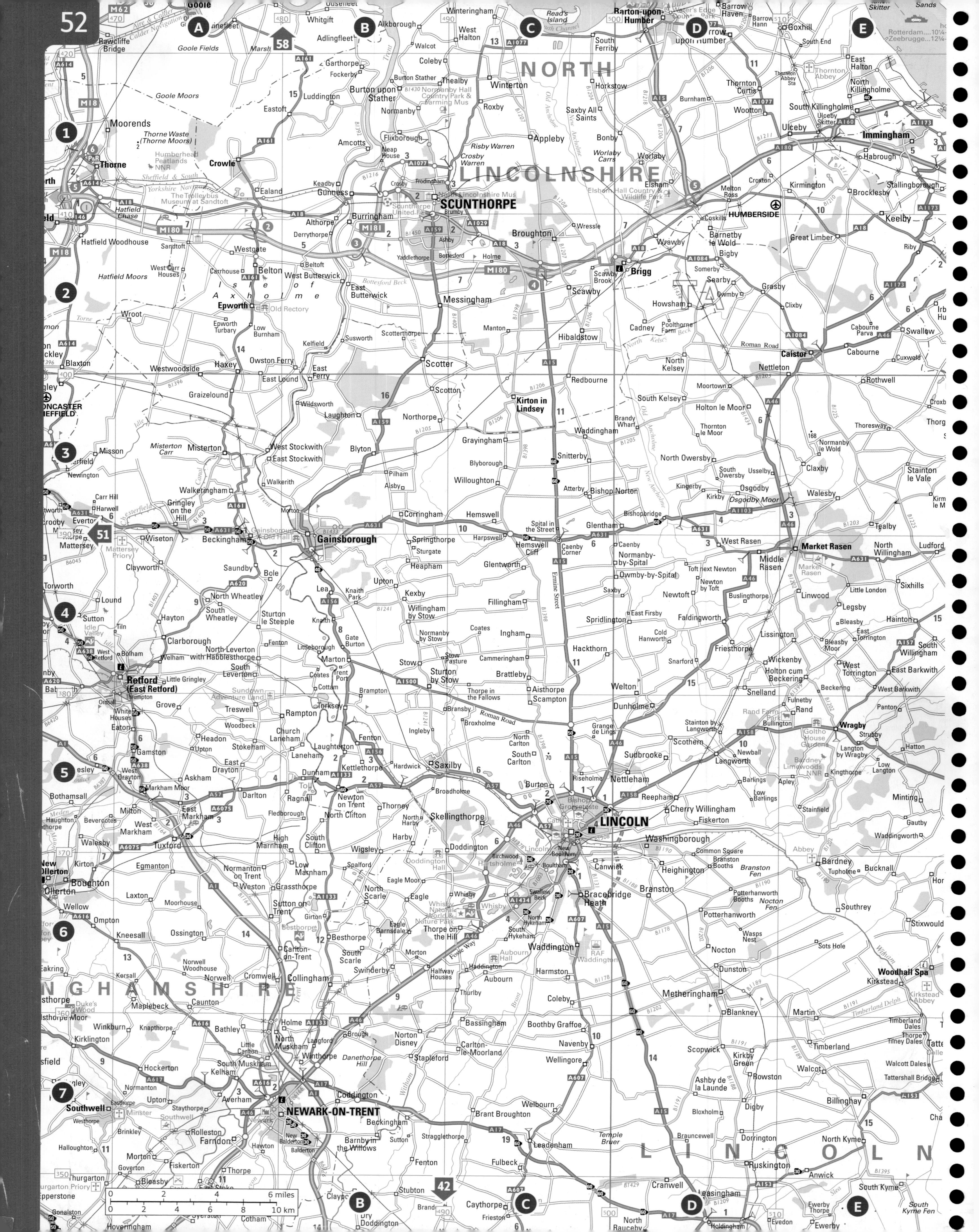
NORTH
LINCOLNSHIRE
LINCOLN
NGHAMSHIRE
Goole
Barton-upon-Humber
Immingham
SCUNTHORPE
HUMBERSIDE
Brigg
Thorne
Crowle
Epworth
Gainsborough
Caistor
Market Rasen
Kirton in Lindsey
Retford (East Retford)
Saxilby
Wragby
Woodhall Spa
NEWARK-ON-TRENT
Southwell
Tuxford
Bracebridge Heath
Waddington
Metheringham
Ermine Street
Isle of Axholme
Rotterdam....10¼
Zeebrugge...12¼
58
51
42
0 2 4 6 miles
0 2 4 6 8 10 km

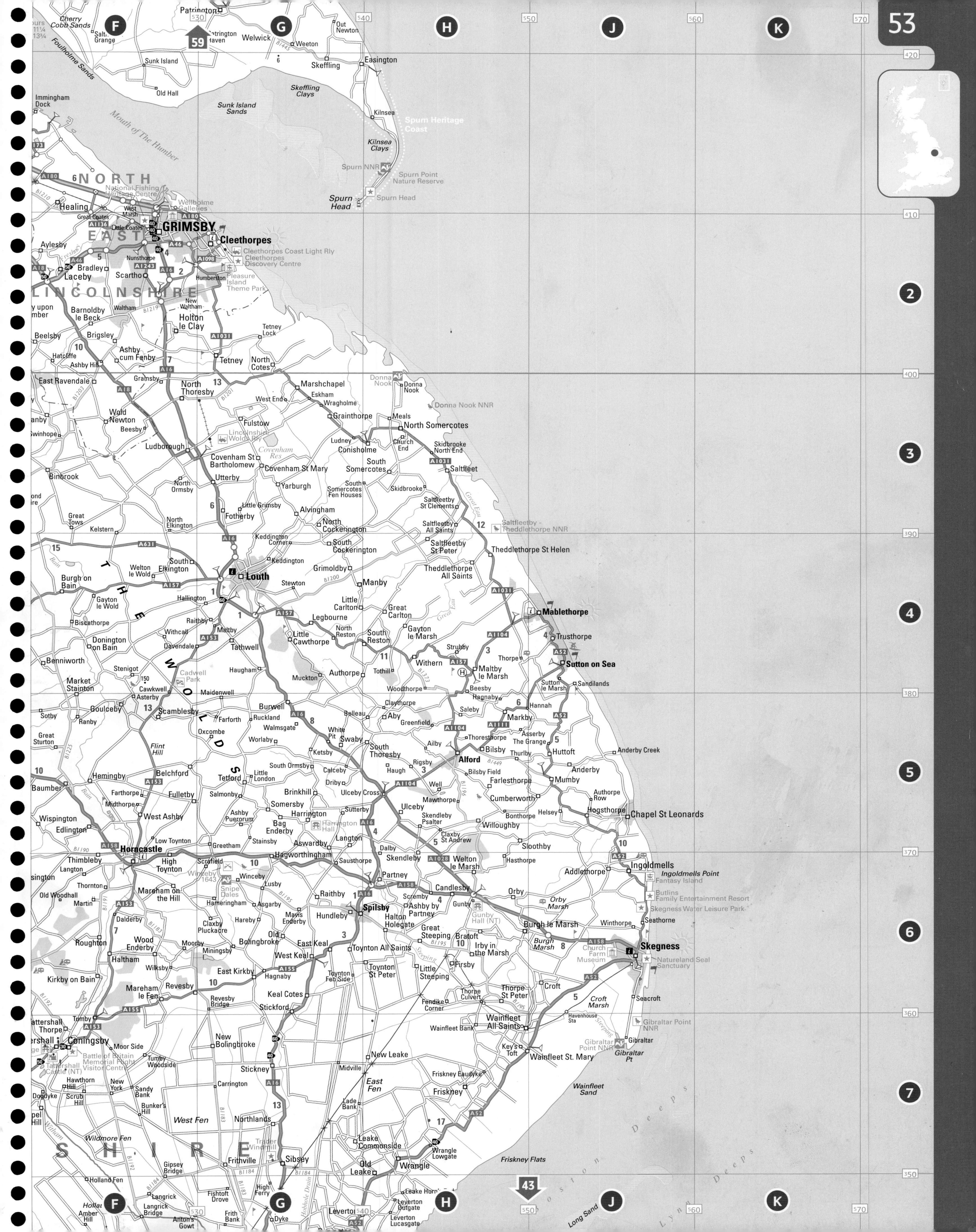

Patrington
Cherry Cobb Sands
Welwick
Weeton
Out Newton
Sunk Island
Old Hall
Skeffling
Easington
Skeffling Clays
Sunk Island Sands
Immingham Dock
Mouth of The Humber
Kilnsea
Spurn Heritage Coast
Kilnsea Clays
Spurn NNR
Spurn Point Nature Reserve
Spurn Head
NORTH
EAST
LINCOLNSHIRE
National Fishing Heritage Centre
Wellholme Galleries
Healing
Great Coates
Little Coates
GRIMSBY
Cleethorpes
Cleethorpes Coast Light Rly
Cleethorpes Discovery Centre
Pleasure Island Theme Park
Aylesby
Bradley
Laceby
Scartho
Nunsthorpe
Humberston
New Waltham
Waltham
Barnoldby le Beck
Holton le Clay
Beelsby
Brigsley
Tetney Lock
Ashby cum Fenby
Hatcliffe
Ashby Hill
Tetney
North Cotes
East Ravendale
Gransby
North Thoresby
Marshchapel
Donna Nook
Donna Nook NNR
West End
Eskham
Wragholme
Grainthorpe
Wold Newton
Beesby
Fulstow
Meals
North Somercotes
Lincolnshire Wolds Rly
Ludborough
Covenham Res
Ludney
Conisholme
Church End
Skidbrooke North End
Covenham St Bartholomew
Covenham St Mary
South Somercotes
Saltfleet
Binbrook
Utterby
North Ormsby
Yarburgh
South Somercotes Fen Houses
Skidbrooke
Saltfleetby St Clements
Little Grimsby
Fotherby
Alvingham
North Cockerington
Saltfleetby All Saints
Saltfleetby - Theddlethorpe NNR
Great Tows
Kelstern
North Elkington
Keddington Corner
South Cockerington
Saltfleetby St Peter
Theddlethorpe St Helen
Keddington
Welton le Wold
South Elkington
Grimoldby
Theddlethorpe All Saints
Burgh on Bain
Louth
Stewton
Manby
Gayton le Wold
Hallington
Little Carlton
Great Carlton
Mablethorpe
Biscathorpe
Raithby
Legbourne
Trusthorpe
Donington on Bain
Withcall
Maltby
Little Cawthorpe
North Reston
South Reston
Gayton le Marsh
Strubby
Thorpe
Davendale
Tathwell
Withern
Sutton on Sea
Benniworth
Stenigot
Haugham
Maltby le Marsh
Market Stainton
Cadwell Park
Muckton
Authorpe
Tothill
Sutton le Marsh
Sandilands
Cawkwell
Asterby
Maidenwell
Woodthorpe
Beesby
Hagnaby
Hannah
Goulceby
Scamblesby
Burwell
Claythorpe
Saleby
Sotby
Ranby
Farforth
Ruckland
Belleau
Aby
Greenfield
Markby
Walmsgate
White Pit
Swaby
Asserby
Great Sturton
Oxcombe
Worlaby
Ailby
Thoresthorpe
The Grange
Flint Hill
Ketsby
South Thoresby
Bilsby
Thurlby
Huttoft
Anderby Creek
Alford
Calceby
Rigsby
Bilsby Field
Anderby
Hemingby
Belchford
Tetford
Little London
South Ormsby
Driby
Haugh
Farlesthorpe
Mumby
Baumber
Farthorpe
Fulletby
Salmonby
Brinkhill
Ulceby Cross
Well
Cumberworth
Authorpe Row
Midthorpe
Somersby
Mawthorpe
Hogsthorpe
Chapel St Leonards
Wispington
West Ashby
Ashby Puerorum
Harrington
Sutterby
Ulceby
Bonthorpe
Helsey
Edlington
Bag Enderby
Harrington Hall
Skendleby Psalter
Willoughby
Low Toynton
Stainsby
Langton
Claxby St Andrew
Horncastle
Greetham
Aswardby
Dalby
Sloothby
Thimbleby
High Toynton
Scrafield
Hagworthingham
Sausthorpe
Skendleby
Welton le Marsh
Hasthorpe
Ingoldmells
Langton
Winceby 1643
Winceby
Partney
Addlethorpe
Ingoldmells Point
Fantasy Island
Thornton
Snipe Dales
Lusby
Candlesby
Butlins Family Entertainment Resort
Old Woodhall
Mareham on the Hill
Raithby
Scremby
Orby
Martin
Hameringham
Asgarby
Spilsby
Ashby by Partney
Gunby
Orby Marsh
Skegness Water Leisure Park
Dalderby
Claxby Pluckacre
Hareby
Mavis Enderby
Hundleby
Halton Holegate
Gunby Hall (NT)
Burgh le Marsh
Winthorpe
Seathorne
Roughton
Wood Enderby
Moorby
Old Bolingbroke
East Keal
Great Steeping
Bratoft
Irby in the Marsh
Burgh Marsh
Skegness
Haltham
Miningsby
West Keal
Toynton All Saints
Church Farm Museum
Natureland Seal Sanctuary
Wilksby
East Kirkby
Toynton St Peter
Little Steeping
Firsby
Kirkby on Bain
Hagnaby
Toynton Fen Side
Mareham le Fen
Revesby
Keal Cotes
Thorpe Culvert
Thorpe St Peter
Croft
Revesby Bridge
Stickford
Fendike Corner
Croft Marsh
Seacroft
Tattershall Thorpe
Tumby
Wainfleet All Saints
Havenhouse Sta
Gibraltar Point NNR
Coningsby
New Bolingbroke
Wainfleet Bank
Gibraltar
Moor Side
Key's Toft
Gibraltar Point NNR
Gibraltar Pt
Battle of Britain Memorial Flight Visitor Centre
Tumby Woodside
New Leake
Wainfleet St. Mary
Tattershall Castle (NT)
Stickney
Midville
Friskney Eaudyke
Hawthorn Hill
New York
Sandy Bank
Carrington
East Fen
Wainfleet Sand
Dogdyke
Scrub Hill
Friskney
Bunker's Hill
Lade Bank
West Fen
Northlands
Boston Deeps
Wildmore Fen
SHIRE
Trader Windmill
Leake Commonside
Wrangle Lowgate
Gipsey Bridge
Frithville
Sibsey
Wrangle
Friskney Flats
Holland Fen
Old Leake
Langrick
Fishtoft Drove
High Ferry
Leake Hurn's End
Langrick Bridge
Frith Bank
Leverton
Leverton Outgate
Amber Hill
Anton's Gowt
Dyke
Leverton Lucasgate
Long Sand
Lynn Deeps
59
43

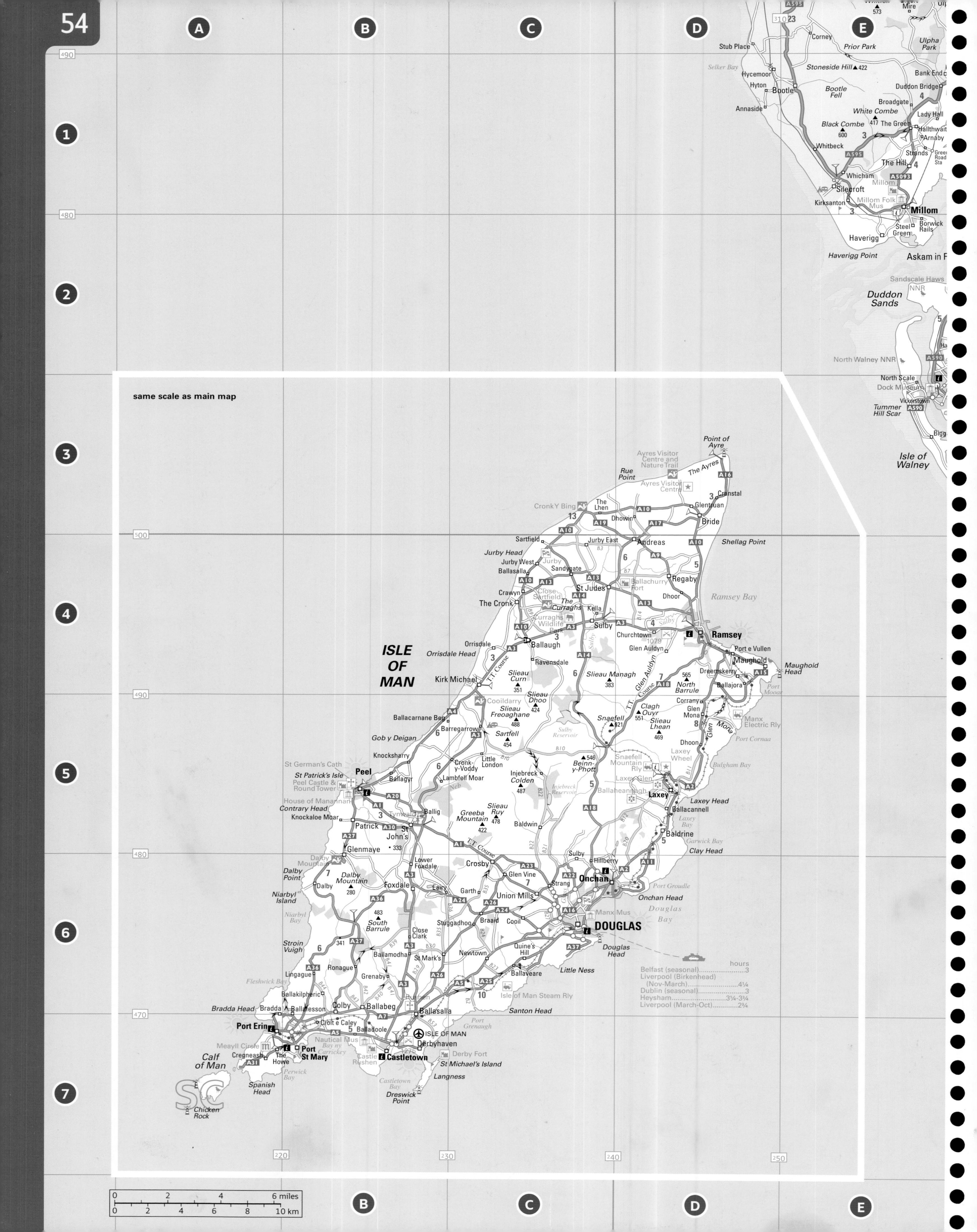

same scale as main map

ISLE
OF
MAN

Point of
Ayre

Ayres Visitor
Centre and
Nature Trail

Rue
Point

The Ayres

A16

Ayres Visitor
Centre

3 Cranstal

Glentuan

Bride

CronkY Bing

The
Lhen

A9

13 A19

Dhowin

A17

A10

Shellag Point

Sartfield

Jurby East

Andreas

A10

Jurby Head

A13

6

A9

5

Jurby West

Jurby

Sandygate

Ballachurry
Fort

Regaby

Ramsey Bay

Ballasalla

A13

St Judes

Dhoor

Crawyn

A10

A13

Close
Sartfield

Keila

A13

The Cronk

The
Curraghs

Sulby

Churchtown

Ramsey

Curraghs
Wildlife

A3

Glen Auldyn

Port e Vullen

Orrisdale

A10

A3

4 Sulby

1079

Maughold

Orrisdale Head

Ballaugh

Glen Auldyn

Maughold
Head

A3

Ravensdale

Slieau Managh

383

7

Dreemskerry

A15

Ballajora

Kirk Michael

TT Course

Slieau
Curn
351

6

565

North
Barrule

Corrany

Cooildarry

Slieau
Dhoo
424

Clagh
Ouyr
551

Glen Mona

A18

Slieau
Freoaghane
488

Sulby
Reservoir

Snaefell
621

Slieau
Lhean
469

8

Manx
Electric Rly

Ballacarnane Beg

A4

Dhoon

Barregarrow

A1

6

Sartfell
454

B10

Snaefell
Mountain
Rly

Port Cornaa

Gob y Deigan

Laxey
Wheel

Bulgham Bay

Knocksharry

Cronk-
y-Voddy

Little
London

546

Beinn-
y-Phott

5

St German's Cath

6

Injebreck
Colden
487

Injebreck
Reservoir

Laxey
Glen

Laxey Head

St Patrick's Isle
Peel Castle &
Round Tower

Peel

Ballagyr

Lambfell Moar

Neb

Ballaheannagh

Laxey

Ballacannell

House of Manannan

A20

Slieau
Ruy
478

Laxey
Bay

Contrary Head

5

Tynwald

Ballig

Greeba
Mountain
422

Baldwin

A18

Baldrine

Gurwick Bay

Knockaloe Moar

3

St
John's

A1

TT Course

B22

5

Clay Head

Patrick

A30

Sulby

Hillberry

A11

Glenmaye

333

Lower
Foxdale

Crosby

A23

A2

Port Groudle

Dalby
Mountain

A27

Glen Vine

A22

Strang

Onchan

Onchan Head

7

Dalby
Point

Dalby
Mountain
280

Foxdale

Eairy

Garth

A24

Union Mills

A16

Manx Mus

Douglas
Bay

Niarbyl
Island

A36

A24

A26

7

DOUGLAS

Niarbyl
Bay

483

South
Barrule

Close
Clark

Braaid

Cooil

Douglas
Head

Stroin
Vuigh

341

A27

6

Ballamodha

B30

Newtown

Quine's
Hill

A37

hours

Little Ness

Belfast (seasonal).........................3
Liverpool (Birkenhead)
(Nov-March)...........................4¼
Dublin (seasonal).........................3
Heysham.............................3¼-3¾
Liverpool (March-Oct)............2¾

Lingague

Ronague

St Mark's

A26

A5

Ballaveare

Fleshwick Bay

Grenaby

A25

Ballakilpheric

Colby

Ballabeg

10

Isle of Man Steam Rly

Bradda Head

Bradda

Ballasalla

Santon Head

Port
Grenaugh

Port Erin

Ballachesson

A7

Croit e Caley

Ballasalla

Calf
of Man

Meayll Circle

Port
St Mary

Balladoole

Nautical Mus

Castle
Rushen

Castletown

ISLE OF MAN

Derbyhaven

Derby Fort

Cregneash

The
Howe

A31

A36

St Michael's Island

Perwick
Bay

Castletown
Bay

Langness

SC

Spanish
Head

Dreswick
Point

Chicken
Rock

scale: 0 2 4 6 miles
0 2 4 6 8 10 km

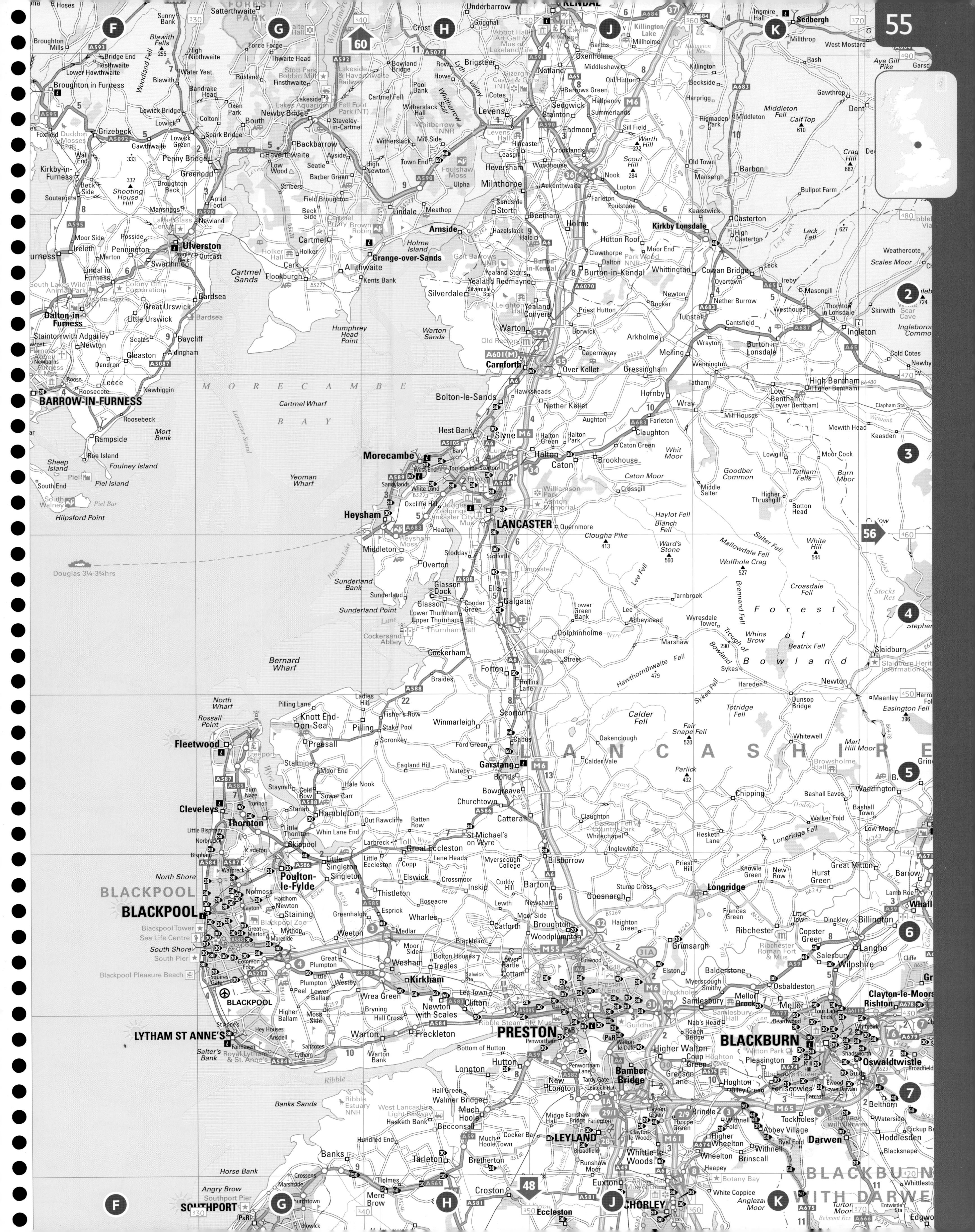

F
G
H
J
K
60
56
48
Kendal
Sedbergh
Killington Lake
Oxenholme
Natland
Sedgwick
Levens
Milnthorpe
Heversham
Arnside
Beetham
Holme
Kirkby Lonsdale
Casterton
Barbon
Dent
Broughton in Furness
Grizebeck
Kirkby-in-Furness
Ulverston
Newby Bridge
Haverthwaite
Backbarrow
Lindale
Cartmel
Grange-over-Sands
Allithwaite
Flookburgh
Cartmel Sands
Dalton-in-Furness
Barrow-in-Furness
Roosebeck
Rampside
Roa Island
Piel Island
Walney
Hilpsford Point
Morecambe Bay
Silverdale
Yealand Conyers
Warton
Carnforth
Over Kellet
Burton-in-Kendal
Whittington
Cowan Bridge
Tunstall
Burton in Lonsdale
Ingleton
High Bentham
Low Bentham
Wray
Hornby
Caton
Halton
Slyne
Bolton-le-Sands
Hest Bank
Morecambe
Heysham
Lancaster
Quernmore
Galgate
Glasson
Cockerham
Forton
Garstang
Scorton
Calder Fell
Forest of Bowland
Slaidburn
Newton
Waddington
Chipping
Longridge
Ribchester
Wilpshire
Langho
Fleetwood
Knott End-on-Sea
Preesall
Pilling
Cleveleys
Thornton
Poulton-le-Fylde
Blackpool
Blackpool Tower
Sea Life Centre
South Shore
South Pier
Blackpool Pleasure Beach
Kirkham
Wesham
Freckleton
Warton
Lytham St Anne's
Preston
Broughton
Woodplumpton
Grimsargh
Samlesbury
Blackburn
Clayton-le-Moors
Rishton
Oswaldtwistle
Darwen
Leyland
Bamber Bridge
Chorley
Eccleston
Croston
Tarleton
Banks
Southport
Douglas 3¼-3¾hrs
Lancashire
Blackburn with Darwen
M6
M55
M65
M61

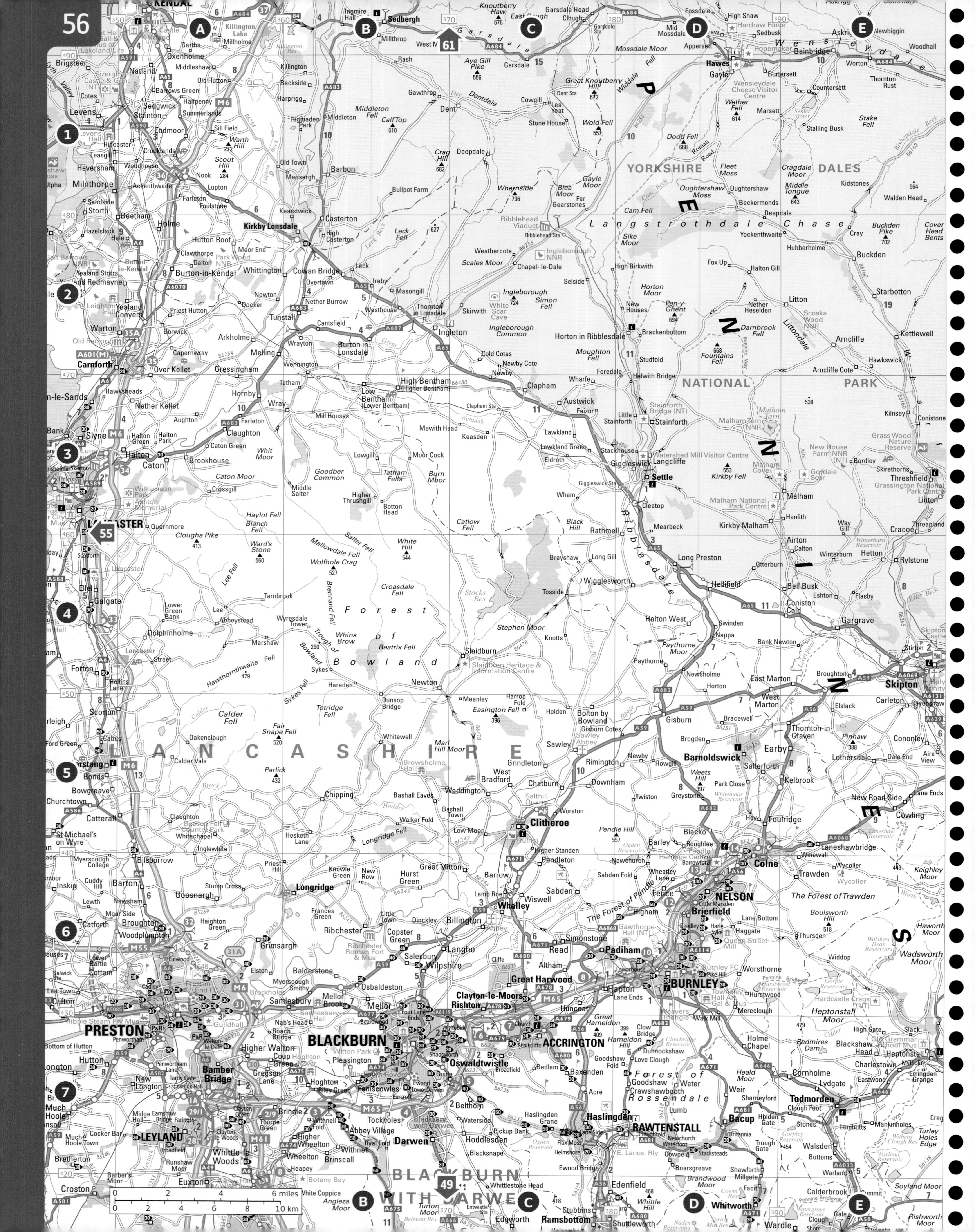

Kendal
Killington Lake
Sedbergh
Garsdale Head
Garsdale
Dentdale
Dent
Hawes
Gayle
Wensleydale
Bainbridge
Askrigg
Newbiggin
Woodhall
Worton
Aysgarth
Thornton Rust
Countersett
Burtersett
Wensleydale Cheese Visitor Centre
Hardraw Force
Sedbusk
Appersett
Mossdale Moor
Great Knoutberry Hill
Wold Fell
Dodd Fell
Wether Fell
Marsett
Semer Water
Stalling Busk
Stake Fell
YORKSHIRE DALES
NATIONAL PARK
PENNINES
Langstrothdale Chase
Oughtershaw
Beckermonds
Deepdale
Yockenthwaite
Hubberholme
Cray
Buckden
Buckden Pike
Starbotton
Kettlewell
Litton
Littondale
Arncliffe
Arncliffe Cote
Hawkswick
Kilnsey
Conistone
Grass Wood Nature Reserve
Threshfield
Grassington National Park Centre
Linton
Cracoe
Threapland
Rylstone
Hetton
Winterburn
Airton
Calton
Malham
Malham Tarn
Malham Cove
Gordale Scar
Malham National Park Centre
Kirkby Malham
Hanlith
Kirkby Fell
Watershed Mill Visitor Centre
Langcliffe
Settle
Giggleswick
Stainforth
Little Stainforth
Stainforth Bridge (NT)
Helwith Bridge
Studfold
Horton in Ribblesdale
Brackenbottom
Pen-y-Ghent
Fountains Fell
Darnbrook Fell
Moughton Fell
Ingleborough
Ingleborough Common
Simon Fell
White Scar Cave
Ingleton
Thornton in Lonsdale
Chapel-le-Dale
Ribblehead Viaduct
Ribblehead Sta
Ingleborough NNR
Whernside
Blea Moor
Gearstones
Cam Fell
Selside
High Birkwith
Scales Moor
Weathercote
Clapham
Austwick
Feizor
Wharfe
Lawkland
Lawkland Green
Eldroth
Stackhouse
Cleatop
Mearbeck
Rathmell
Wigglesworth
Long Preston
Ribblesdale
Hellifield
Otterburn
Bell Busk
Coniston Cold
Eshton
Flasby
Gargrave
Swinden
Nappa
Halton West
Paythorne
Paythorne Moor
Newsholme
Horton
Gisburn
East Marton
West Marton
Broughton
Skipton
Carleton
Elslack
Thornton-in-Craven
Earby
Pinhaw
Lothersdale
Cononley
Barnoldswick
Salterforth
Kelbrook
Bracewell
Brogden
Weets Hill
Greystone
Park Close
Foulridge
New Road Side
Cowling
Laneshawbridge
Colne
Trawden
Wycoller
The Forest of Trawden
Boulsworth Hill
Nelson
Brierfield
Barrowford
Fence
Higham
Sabden
Pendle Hill
Barley
Newchurch
Roughlee
Blacko
Downham
Twiston
Rimington
Sawley
Bolton by Bowland
Grindleton
Chatburn
West Bradford
Waddington
Clitheroe
Worston
Pendleton
Whalley
Wiswell
Barrow
Billington
Langho
Read
Simonstone
Padiham
Burnley
Worsthorne
Hurstwood
Haggate
Thursden
Widdop
Hebden Bridge
Heptonstall
Heptonstall Moor
Blackshaw Head
Charlestown
Todmorden
Cornholme
Lydgate
Walsden
Bacup
Whitworth
Rawtenstall
Haslingden
Edenfield
Ramsbottom
Stubbins
Shuttleworth
Edgworth
Accrington
Clayton-le-Moors
Rishton
Great Harwood
Oswaldtwistle
Blackburn
Darwen
Turton Moor
Whittlestone Head
Blackburn with Darwen
Mellor
Samlesbury
Salesbury
Wilpshire
Ribchester
Copster Green
Longridge
Goosnargh
Grimsargh
Chipping
Bashall Eaves
Slaidburn
Newton
Dunsop Bridge
Whitewell
Forest of Bowland
Trough of Bowland
Lancashire
Calder Fell
Totridge Fell
Fair Snape Fell
Parlick
Calder Vale
Oakenclough
Scorton
Garstang
Bowgreave
Churchtown
Catterall
St Michael's on Wyre
Bilsborrow
Barton
Broughton
Woodplumpton
Preston
Penwortham
Walton-le-Dale
Bamber Bridge
Leyland
Lostock Hall
Hutton
Longton
Much Hoole
Croston
Eccleston
Euxton
Chorley
Whittle-le-Woods
Brindle
Hoghton
Feniscowles
Lancaster
Galgate
Ellel
Dolphinholme
Abbeystead
Quernmore
Caton
Halton
Brookhouse
Caton Moor
Claughton
Hornby
Wray
Farleton
Gressingham
Melling
Wennington
Tatham
Low Bentham
High Bentham
Burton in Lonsdale
Cantsfield
Tunstall
Kirkby Lonsdale
Casterton
Cowan Bridge
Leck
Barbon
Middleton
Nether Kellet
Over Kellet
Carnforth
Warton
Yealand Conyers
Yealand Redmayne
Silverdale
Arnside
Milnthorpe
Heversham
Levens
Sedgwick
Natland
Oxenholme
Burton-in-Kendal
Holme
Hutton Roof
Whittington
Arkholme
Slyne
Bolton-le-Sands
Morecambe

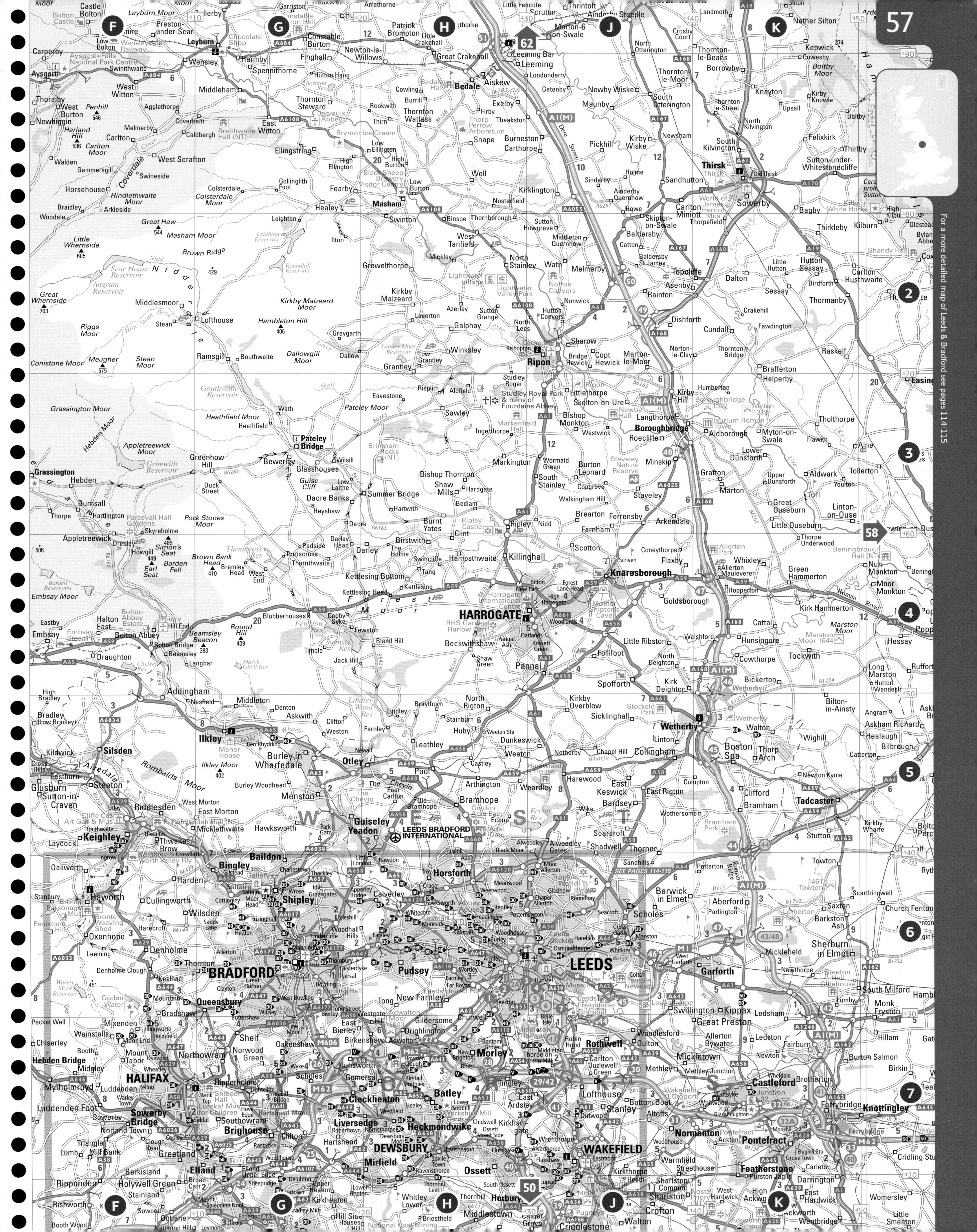
For a more detailed map of Leeds & Bradford see pages 114-115
SEE PAGES 114-115
62
58
50
F
G
H
J
K
2
3
4
5
6
7
Leyburn
Wensley
Middleham
Bedale
Leeming
Aiskew
Masham
Thirsk
Sowerby
Ripon
Boroughbridge
Pateley Bridge
Grassington
Knaresborough
HARROGATE
Wetherby
Ilkley
Otley
Keighley
Silsden
Bingley
Baildon
Shipley
Haworth
BRADFORD
LEEDS
Pudsey
Horsforth
Guiseley
Yeadon
LEEDS BRADFORD INTERNATIONAL
Queensbury
HALIFAX
Sowerby Bridge
Brighouse
Elland
Cleckheaton
Heckmondwike
Batley
DEWSBURY
Mirfield
Morley
Rothwell
WAKEFIELD
Ossett
Horbury
Normanton
Castleford
Pontefract
Featherstone
Knottingley
Garforth
Tadcaster
Boston Spa
Kirkby Hill
Masham Moor
Nidderdale
Grassington Moor
Rombalds Moor
Ilkley Moor
Great Whernside
Little Whernside
Fountains Abbey
Studley Royal Park & ruins of Fountains Abbey
Bolton Abbey
Embsay Moor
Harewood
Kirkby Malzeard
Pannal
Spofforth
Thorner
Scholes
Sherburn in Elmet
Ferrybridge
Wetherby
A1(M)
A61
A59
A65
A658
A660
M1
M62
M606
A168
A6055
A684
A6108
A1

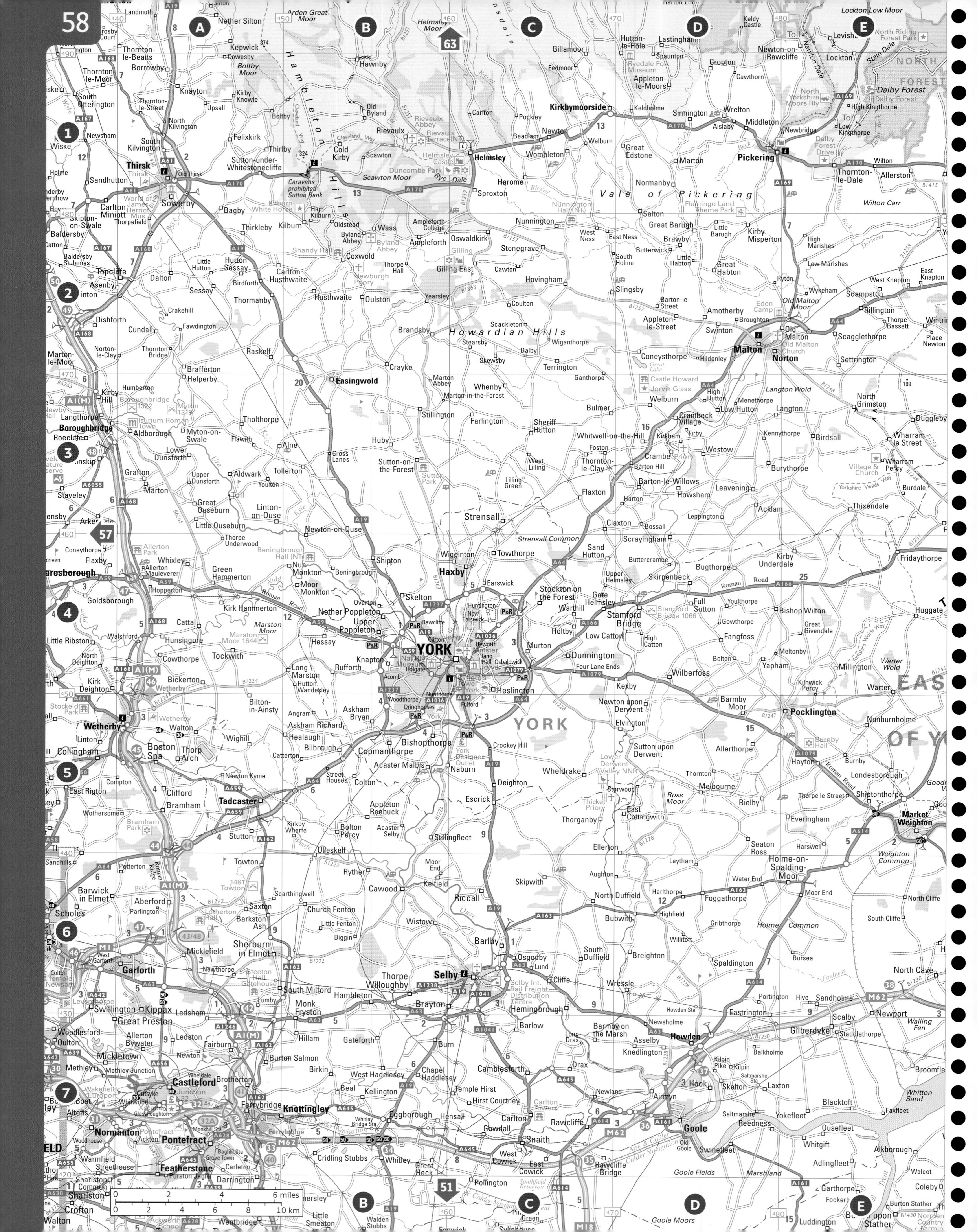
A
B
C
D
E
1
2
3
4
5
6
7
63
57
51
Thirsk
Sowerby
Helmsley
Kirkbymoorside
Pickering
Malton
Norton
Easingwold
Boroughbridge
Knaresborough
Wetherby
Boston Spa
Tadcaster
Haxby
YORK
Stamford Bridge
Pocklington
Market Weighton
Selby
Howden
Goole
Garforth
Castleford
Pontefract
Knottingley
Normanton
Featherstone
Snaith
Gilberdyke
Hambleton Hills
Howardian Hills
Vale of Pickering
Scawton Moor
Strensall Common
Dalby Forest
North Riding Forest Park
Sutton Bank
Marston Moor
Bishopthorpe
Copmanthorpe
Acaster Malbis
Naburn
Escrick
Riccall
Barlby
Cawood
Sherburn in Elmet
Church Fenton
Heslington
Dunnington
Elvington
Wheldrake
Sutton upon Derwent
Holme-on-Spalding-Moor
Bubwith
North Duffield
Hemingbrough
Camblesforth
Drax
Rawcliffe
Airmyn
Swinefleet
Whitley
Hensall
Carlton
Great Heck
Eggborough
Brayton
Thorpe Willoughby
Hambleton
Monk Fryston
South Milford
Kippax
Swillington
Great Preston
Allerton Bywater
Micklefield
Selby Int. Rail Freight Distribution Centre
0 2 4 6 miles
0 2 4 6 8 10 km

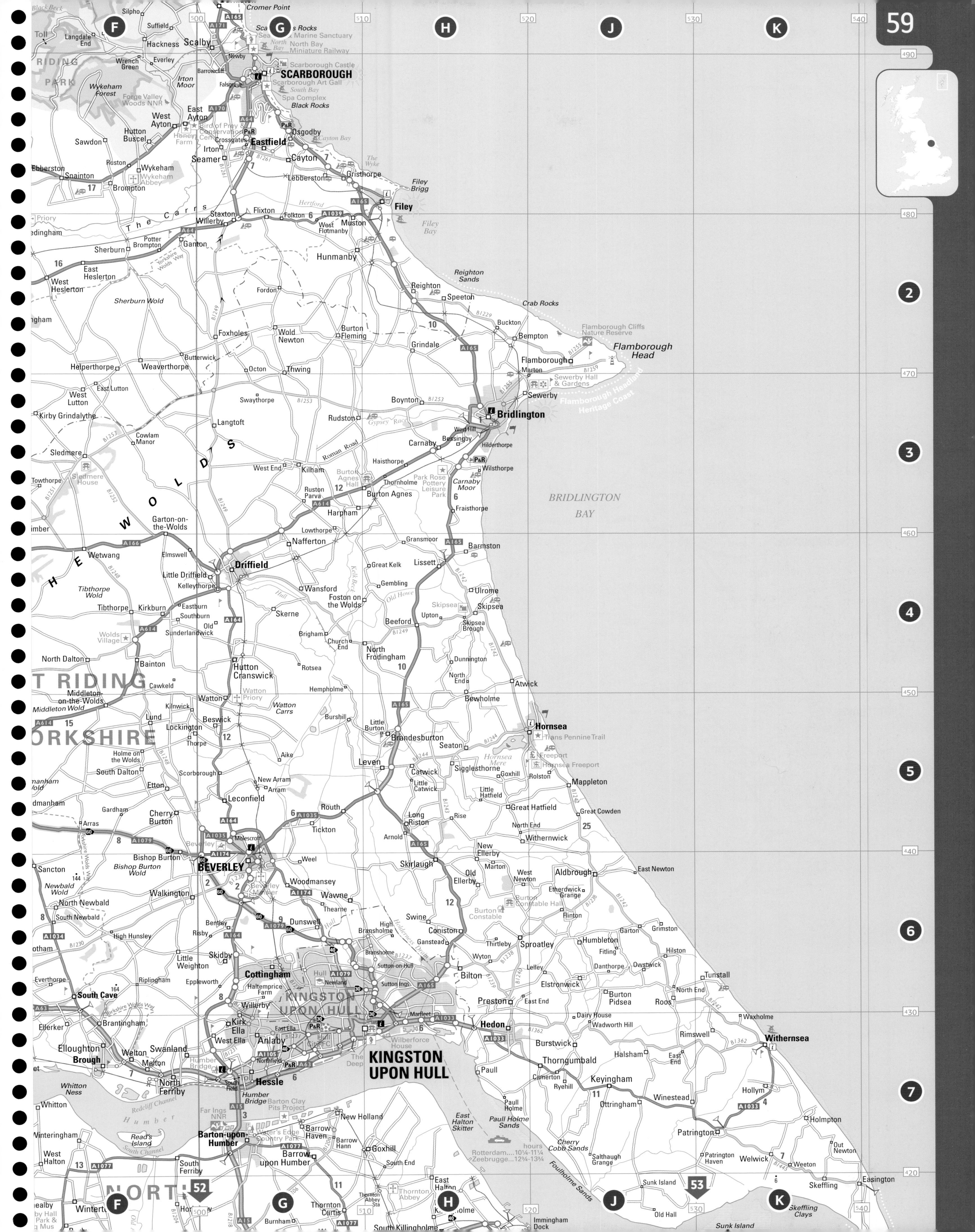
F
G
H
J
K
Cromer Point
Scalby
Hackness
Suffield
Silpho
Langdale End
Everley
Wrench Green
Irton Moor
SCARBOROUGH
Scarborough Castle
Scarborough Art Gall
South Bay
Spa Complex
Black Rocks
North Bay
North Bay Miniature Railway
Sea Life & Marine Sanctuary
Newby
Barrowcliff
Falsgrave
Wykeham Forest
Forge Valley Woods NNR
East Ayton
West Ayton
Hutton Buscel
Bird of Prey & Conservation Cen
Honey Farm
Crossgates
Eastfield
Osgodby
Cayton Bay
Irton
Seamer
Cayton
The Wyke
Gristhorpe
Lebberston
Filey Brigg
Filey
Filey Bay
Sawdon
Ruston
Wykeham
Wykeham Abbey
Ebberston
Snainton
Brompton
The Carrs
Staxton
Willerby
Flixton
Folkton
Hertford
West Flotmanby
Muston
Priory
Potter Brompton
Ganton
Sherburn
East Heslerton
West Heslerton
Yorkshire Wolds Way
Hunmanby
Reighton Sands
Reighton
Speeton
Crab Rocks
Buckton
Bempton
Flamborough Cliffs Nature Reserve
Flamborough Head
Flamborough
Marton
Sewerby Hall & Gardens
Sewerby
Flamborough Headland Heritage Coast
Sherburn Wold
Fordon
Foxholes
Wold Newton
Burton Fleming
Grindale
Helperthorpe
Weaverthorpe
Butterwick
Octon
Thwing
West Lutton
East Lutton
Kirby Grindalythe
Swaythorpe
Boynton
Bridlington
Rudston
Gypsey Race
Langtoft
Cowlam Manor
Sledmere
Sledmere House
Towthorpe
THE WOLDS
West End
Kilham
Roman Road
Haisthorpe
Carnaby
West Hill
Bessingby
Hilderthorpe
Wilsthorpe
Burton Agnes Hall
Thornholme
Park Rose Pottery Leisure Park
Carnaby Moor
Ruston Parva
Burton Agnes
Harpham
Fraisthorpe
BRIDLINGTON BAY
Garton-on-the-Wolds
Lowthorpe
Nafferton
Gransmoor
Barmston
Wetwang
Elmswell
Driffield
Little Driffield
Kelleythorpe
Great Kelk
Lissett
Gembling
Ulrome
Skipsea
Tibthorpe Wold
Tibthorpe
Kirkburn
Eastburn
Southburn
Wansford
Foston on the Wolds
Old Howe
Upton
Skipsea Brough
Skerne
Beeford
Wolds Village
Sunderlandwick
Brigham
Church End
North Dalton
Bainton
North Frodingham
Dunnington
Hutton Cranswick
Rotsea
North End
Atwick
EAST RIDING OF YORKSHIRE
Cawkeld
Middleton-on-the-Wolds
Middleton Wold
Watton
Watton Priory
Watton Carrs
Hempholme
Bewholme
Kilnwick
Lund
Beswick
Lockington
Burshill
Little Burton
Brandesburton
Hornsea
Trans Pennine Trail
Thorpe
Seaton
Hornsea Mere
Hornsea Freeport
Holme on the Wolds
Aike
Leven
Catwick
Sigglesthorne
Goxhill
Rolston
South Dalton
Scorborough
New Arram
Arram
Little Catwick
Mappleton
Etton
Little Hatfield
Leconfield
Great Hatfield
Great Cowden
Gardham
Cherry Burton
Routh
Long Riston
Rise
North End
Arras
Tickton
Arnold
Withernwick
Beverley
Molescroft
New Ellerby
Bishop Burton
Bishop Burton Wold
Weel
Skirlaugh
Marton
Aldbrough
East Newton
Sancton
BEVERLEY
Old Ellerby
West Newton
Newbald Wold
Beverley Minster
Woodmansey
Etherdwick Grange
Walkington
Wawne
North Newbald
Thearne
South Newbald
Burton Constable
Burton Constable Hall
Flinton
Bentley
Dunswell
Swine
Risby
High Bransholme
Coniston
Garton
Grimston
High Hunsley
Holderness Drain
Ganstead
Thirtleby
Sproatley
Humbleton
Skidby
Bransholme
Fitling
Hilston
Little Weighton
Wyton
Sutton-on-Hull
Bilton
Lelley
Danthorpe
Owstwick
Everthorpe
Cottingham
Hull
Newland
Sutton Ings
Elstronwick
Tunstall
Riplingham
Eppleworth
Haltemprice Farm
North End
South Cave
KINGSTON UPON HULL
Preston
East End
Burton Pidsea
Roos
Willerby
Marfleet
Dairy House
Waxholme
Brantingham
Kirk Ella
Hedon
Wadworth Hill
Ellerker
West Ella
Anlaby
East Ella
Rimswell
Withernsea
Elloughton
Welton
Swanland
Northfield
Wilberforce House
Burstwick
Halsham
Brough
Melton
Humber Bridge
The Deep
Paull
Thorngumbald
East End
Hessle
North Ferriby
Camerton
Ryehill
Keyingham
Hollym
Whitton Ness
Redcliff Channel
Humber Bridge
Barton Clay Pits Project
Paull Holme
Ottringham
Winestead
Whitton
Humber
Far Ings NNR
New Holland
East Halton Skitter
Paull Holme Sands
Patrington
Holmpton
Winteringham
Read's Island
Barton-upon-Humber
Water's Edge Country Park
Barrow Haven
Barrow Hann
Goxhill
Rotterdam....10¼-11¼ hours
Zeebrugge....12¼-13¼
Cherry Cobb Sands
Salthaugh Grange
Patrington Haven
Out Newton
Welwick
Weeton
West Halton
South Ferriby
Barrow upon Humber
South End
Foulholme Sands
Sunk Island
Easington
Skeffling
NORTH
Winterton
Thornton Abbey
East Halton
Old Hall
Skeffling Clays
Burnham
Thornton Curtis
Killingholme
South Killingholme
Immingham Dock
Sunk Island
52
53

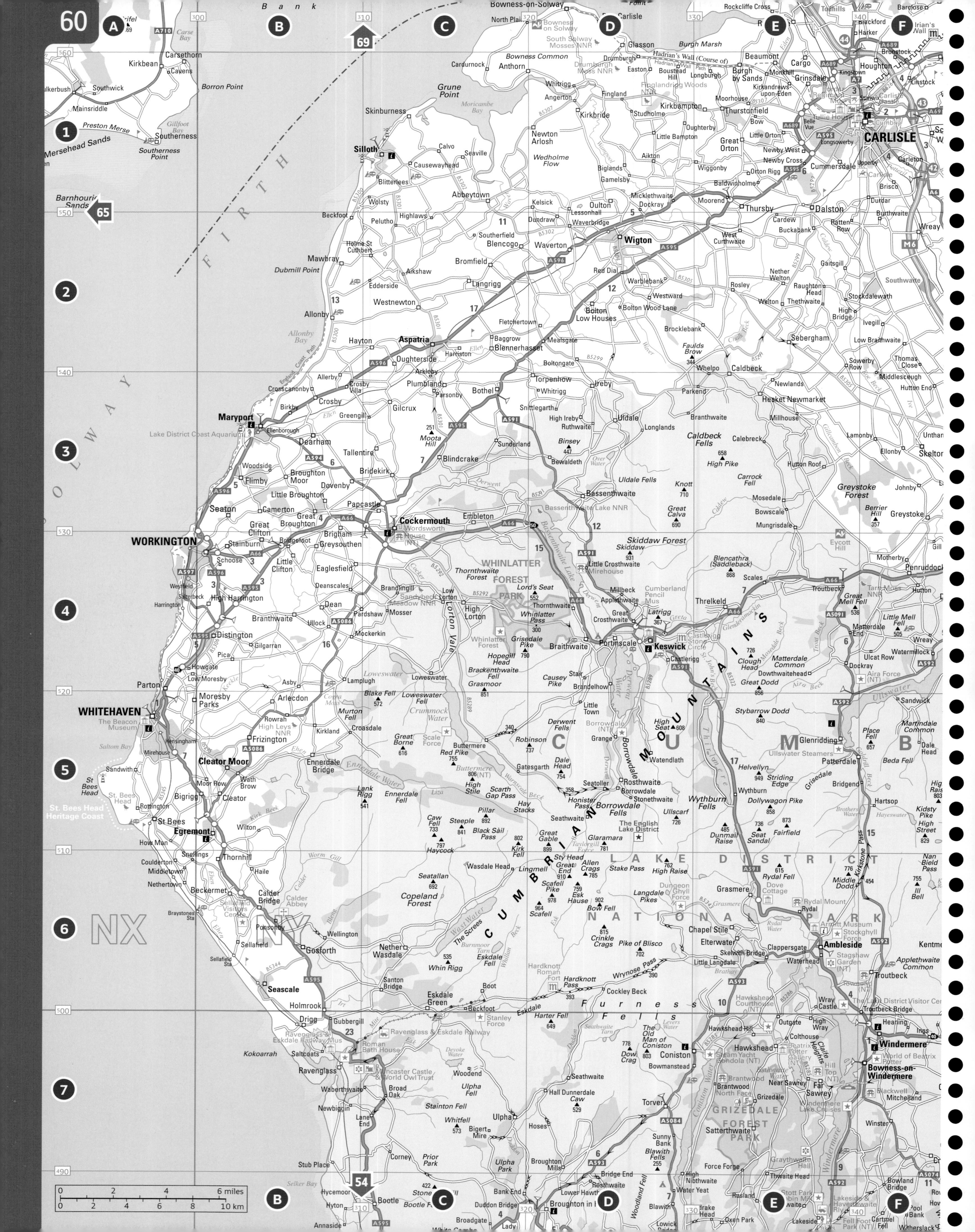

SOLWAY FIRTH
CARLISLE
WORKINGTON
WHITEHAVEN
Maryport
Cockermouth
Keswick
Wigton
Silloth
Aspatria
Cleator Moor
Egremont
Seascale
Gosforth
Ravenglass
Ambleside
Windermere
Bowness-on-Windermere
Coniston
Hawkshead
Grasmere
Patterdale
Penruddock
Dalston
Thursby
Abbeytown
Bowness-on-Solway
Burgh by Sands
Sebergham
Caldbeck
Hesket Newmarket
Greystoke
Threlkeld
Glenridding
Troutbeck
St Bees
St. Bees Head Heritage Coast
CUMBRIAN MOUNTAINS
LAKE DISTRICT NATIONAL PARK
CUMBRIA
WHINLATTER FOREST PARK
GRIZEDALE FOREST PARK
Furness Fells
Skiddaw Forest
Skiddaw 931
Blencathra (Saddleback) 868
Helvellyn 949
Scafell 964
Great Gable 899
Bow Fell 902
Coniston
Wastwater
Ullswater
Bassenthwaite Lake NNR
Hardknott Pass
Wrynose Pass
Honister Pass
Kirkstone Pass
Ravenglass & Eskdale Railway
Lake District Coast Aquarium
NX
NY
0 2 4 6 miles
0 2 4 6 8 10 km

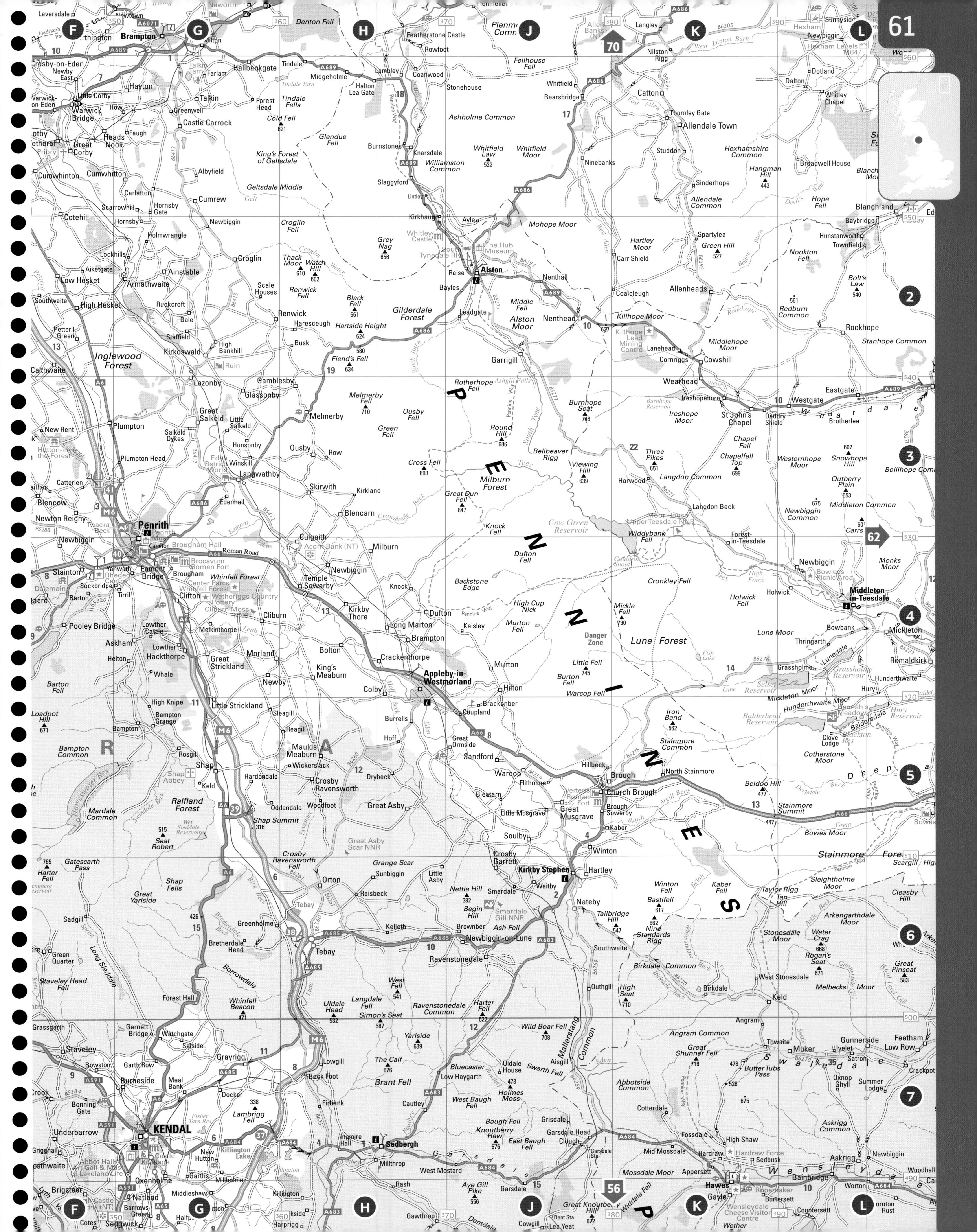

61
F
G
H
J
K
L
Laversdale
Newtown
Brampton
A6071
Denton Fell
Featherstone Castle
Rowfoot
Plenmeller Common
Allen Banks (NT)
70
Langley
Nilston Rigg
West Dipton Burn
Hexham
Sunnyside
Newbiggin
Hexham Levels
Crosby-on-Eden
Newby East
Hallbankgate
Tindale
Midgeholme
Lambley
Coanwood
Fellhouse Fell
Whitfield
Dotland
Dalton
Hayton
Farlam
Talkin
Tindale Tarn
Tindale Fells
Forest Head
Halton Lea Gate
Stonehouse
Bearsbridge
Catton
Thornley Gate
Allendale Town
Whitley Chapel
Warwick-on-Eden
Little Corby
Warwick Bridge
How
Greenwell
Castle Carrock
Cold Fell
621
Ashholme Common
East Allen
Great Corby
Heads Nook
Faugh
Glendue Fell
Burnstones
Knarsdale
Williamston Common
Whitfield Law
522
Whitfield Moor
Ninebanks
Studdon
Hexhamshire Common
Hangman Hill
443
Broadwell House
Cumwhinton
Cumwhitton
King's Forest of Geltsdale
Albyfield
Slaggyford
Sinderhope
Allendale Common
Geltsdale Middle
Carlatton
Cumrew
Hornsby Gate
Scarrowhill
Hornsby
Lintley
Kirkhaugh
Ayle
Hope Fell
Devil's Water
Blanchland
Baybridge
Cotehill
Newbiggin
Croglin Fell
Mohope Moor
Holmwrangle
Whitley Castle
Grey Nag
656
The Hub Museum
South Tynedale Rly
Hartley Moor
Spartylea
Green Hill
527
Hunstanworth
Townfield
Nookton Fell
Lockhills
Croglin
Thack Moor
610
Watch Hill
602
Alston
Carr Shield
Aiketgate
Ainstable
Low Hesket
Armathwaite
Scale Houses
Raise
Bayles
Nenthall
Renwick Fell
Black Fell
661
Middle Fell
Coalcleugh
Allenheads
Bolt's Law
540
Southwaite
High Hesket
Ruckcroft
Dale
Renwick
Gilderdale Forest
Leadgate
Alston Moor
Nenthead
Killhope Moor
561
Redburn Common
Harescough
Hartside Height
624
Killhope Lead Mining Centre
Rookhope
Petteril Green
Staffield
Busk
Stanhope Common
Inglewood Forest
Kirkoswald
High Bankhill
580
Fiend's Fell
634
Garrigill
Lanehead
Cornriggs
Middlehope Moor
Cowshill
Calthwaite
Ruin
Lazonby
Gamblesby
Glassonby
Rotherhope Fell
Ashgill Falls
Wearhead
Ireshopeburn
Eastgate
Westgate
Weardale
Melmerby Fell
710
Ousby Fell
Burnhope Seat
746
Ireshope Moor
St John's Chapel
Daddry Shield
Brotherlee
New Rent
Plumpton
Great Salkeld
Little Salkeld
Melmerby
Green Fell
Round Hill
686
Hunsonby
Salkeld Dykes
Chapel Fell
Hutton-in-the-Forest
Ousby
Row
Plumpton Head
Winskill
Eden Ostrich World
Langwathby
Bellbeaver Rigg
Three Pikes
651
Chapelfell Top
699
Westernhope Moor
607
Snowhope Hill
Bollihope Common
Cross Fell
893
Milburn Forest
Tees
Viewing Hill
639
Harwood
Langdon Common
Outberry Plain
653
Catterlen
Blencow
Skirwith
Kirkland
Edenhall
Great Dun Fell
847
Langdon Beck
Middleton Common
675
Newbiggin Common
Newton Reigny
Penrith
Penrith Museum
Blencarn
Crowdundle Beck
Knock Fell
Cow Green Reservoir
Moor House Upper Teesdale NNR
601
Carrs
62
Newbiggin
Brougham Hall
Culgaith
Acorn Bank (NT)
Milburn
Widdybank Fell
Forest-in-Teesdale
Carleton
Roman Road
Dufton Fell
Cauldron Snout
Monks Moor
Stainton
Yanwath
Eamont Bridge
Brocavum Roman Fort
Brougham
Newbiggin
Rheged Centre
Whinfell Forest
Temple Sowerby
Newbiggin
Bowlees Picnic Area
Dalemain
Sockbridge
Tirril
Center Parcs Whinfell Forest
Knock
Backstone Edge
Cronkley Fell
High Force
Holwick
Middleton-in-Teesdale
Barton
Clifton
Wetheriggs Country Pottery
High Cup Nick
Holwick Fell
Pooley Bridge
Cliburn Moss NNR
Cliburn
Kirkby Thore
Dufton
Mickle Fell
790
Lowther Castle
Melkinthorpe
Long Marton
Keisley
Murton Fell
Danger Zone
Lune Forest
Lune Moor
Bowbank
Mickleton
Askham
Lowther
Brampton
Thringarth
Helton
Hackthorpe
Morland
Bolton
Crackenthorpe
Fish Lake
Lunedale
Romaldkirk
Whale
Great Strickland
King's Meaburn
Appleby-in-Westmorland
Murton
Little Fell
745
Grassholme
Grassholme Reservoir
Hunderthwaite
Barton Fell
Newby
Burton Fell
Warcop Fell
Selset Reservoir
Mickleton Moor
Hury
Colby
Hilton
Hunderthwaite Moor
High Knipe
Little Strickland
Brackenber
Coupland
Hannah's Meadow
Hury Reservoir
Loadpot Hill
671
Bampton Grange
Sleagill
Iron Band
562
Balderhead Reservoir
Baldersdale
Bampton
Reagill
Burrells
Blackton Res
Hoff
Great Ormside
Stainmore Common
Clove Lodge
Pennine Way
Bampton Common
Rosgill
Maulds Meaburn
Cotherstone Moor
Shap
Shap Abbey
Wickerslack
Drybeck
Sandford
Hillbeck
Brough
North Stainmore
Haweswater Res
Keld
Hardendale
Crosby Ravensworth
Warcop
Flitholme
Deepdale
Ralfland Forest
Bleatarn
Verteris Roman Fort
Church Brough
Beldoo Hill
477
Mardale Common
Oddendale
Woodfoot
Great Asby
Little Musgrave
Great Musgrave
Brough Sowerby
Argill Beck
Stainmore Summit
447
Shap Summit
316
Kaber
Bowes Moor
515
Seat Robert
Wet Sleddale Reservoir
Great Asby Scar NNR
Soulby
Winton
Stainmore Forest
765
Harter Fell
Gatescarth Pass
Crosby Ravensworth Fell
Grange Scar
Crosby Garrett
Kirkby Stephen
Hartley
Scargill
Shap Fells
Orton
Sunbiggin
Little Asby
Waitby
Winton Fell
Kaber Fell
Taylor Rigg
Sleightholme Moor
Great Yarlside
Raisbeck
Nettle Hill
382
Smardale
Bastifell
617
Tan Hill
Cleasby Hill
Sadgill
Tebay
Begin Hill
Smardale Gill NNR
Nateby
Tailbridge Hill
547
Arkengarthdale Moor
426
Greenholme
Kelleth
Brownber
Ash Fell
Nine Standards Rigg
662
Stonesdale Moor
Water Crag
668
Bretherdale Head
Newbiggin-on-Lune
Rogan's Seat
671
Green Quarter
Long Sleddale
Tebay
Ravenstonedale
Southwaite
Birkdale Common
Great Pinseat
583
Borrowdale
West Stonesdale
Staveley Head Fell
West Fell
541
Outhgill
High Seat
710
Birkdale
Melbecks Moor
Forest Hall
Whinfell Beacon
471
Uldale Head
532
Langdale Fell
Simon's Seat
587
Ravenstonedale Common
Harter Fell
522
Keld
Grassgarth
Garnett Bridge
Watchgate
Selside
Yarlside
639
Wild Boar Fell
708
Mallerstang Common
Angram
Angram Common
Staveley
Garth Row
Grayrigg
Lowgill
The Calf
676
Aisgill
Great Shunner Fell
716
Thwaite
Muker
Ivelet
Gunnerside
Feetham
Low Row
Bowston
Burneside
Meal Bank
Beck Foot
Bluecaster
Low Haygarth
Uldale House
Swarth Fell
478
Butter Tubs Pass
Satron
Swaledale
Crackpot
Brant Fell
473
Holmes Moss
Abbotside Common
526
Oxnop Ghyll
Summer Lodge
Docker
338
Lambrigg Fell
Firbank
Cautley
West Baugh Fell
675
Bonning Gate
Fisher Tarn Res
Cotterdale
Askrigg Common
Underbarrow
KENDAL
Baugh Fell
Knoutberry Haw
676
East Baugh Fell
Grisdale
Garsdale Head
Clough
Fossdale
High Shaw
Grigghall
Killington Lake
Ingmire Hall
Sedbergh
Garsdale Sta
Mid Mossdale
Hardraw Force
Hardraw
Sedbusk
Askrigg
Newbiggin
Abbot Hall Art Gall & Mus of Lakeland Life
New Hutton
Millthrop
Garsdale
Mossdale Moor
Appersett
Hawes
Bainbridge
Woodhall
Wensleydale
Oxenholme
Garths
Millholme
West Mostard
Worton
Natland
Middleshaw
Killington
Rash
Aye Gill Pike
556
Garsdale
56
Widdale Fell
Gayle
Burtersett
Brigsteer
Levens
Sizergh Castle
Barrows Green
Cotes
Sedgwick
Halfpenny
Gawthrop
Dentdale
Great Knoutberry Hill
672
Dent Sta
Lea Yeat
Cowgill
Wensleydale Cheese Visitor Centre
Countersett
Wether Fell
Thornton Rust
Harprigg
Oxenholme
Kendal
M6
A6
A66
A686
A689
A683
A684
A685
A591

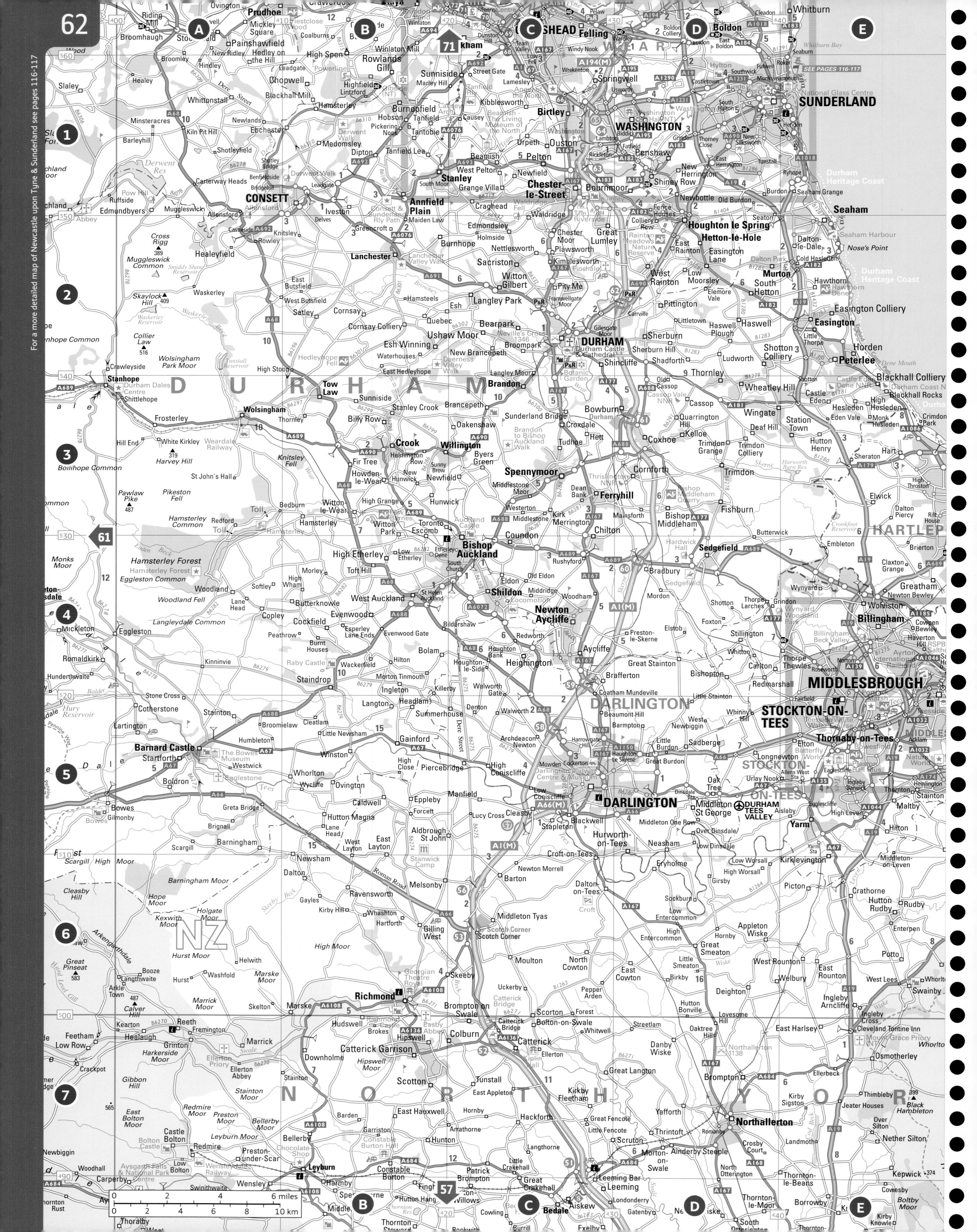

62
For a more detailed map of Newcastle upon Tyne & Sunderland see pages 116-117
SEE PAGES 116-117
A
B
C
D
E
1
2
3
4
5
6
7
71
61
57
DURHAM
NORTH YORKSHIRE
HARTLEPOOL
NZ
SUNDERLAND
WASHINGTON
Whitburn
Boldon
Felling
Seaham
CONSETT
Stanley
Chester-le-Street
Houghton le Spring
Hetton-le-Hole
Murton
Peterlee
Easington Colliery
DURHAM
Durham Castle & Cathedral
Spennymoor
Ferryhill
Crook
Willington
Bishop Auckland
Shildon
Newton Aycliffe
Sedgefield
Billingham
MIDDLESBROUGH
STOCKTON-ON-TEES
Thornaby-on-Tees
Yarm
DARLINGTON
DURHAM TEES VALLEY
Barnard Castle
Richmond
Catterick Garrison
Catterick
Northallerton
Leyburn
Bedale
Stanhope
Wolsingham
Hamsterley Forest
Durham Heritage Coast
Hartlepool
0 2 4 6 miles
0 2 4 6 8 10 km

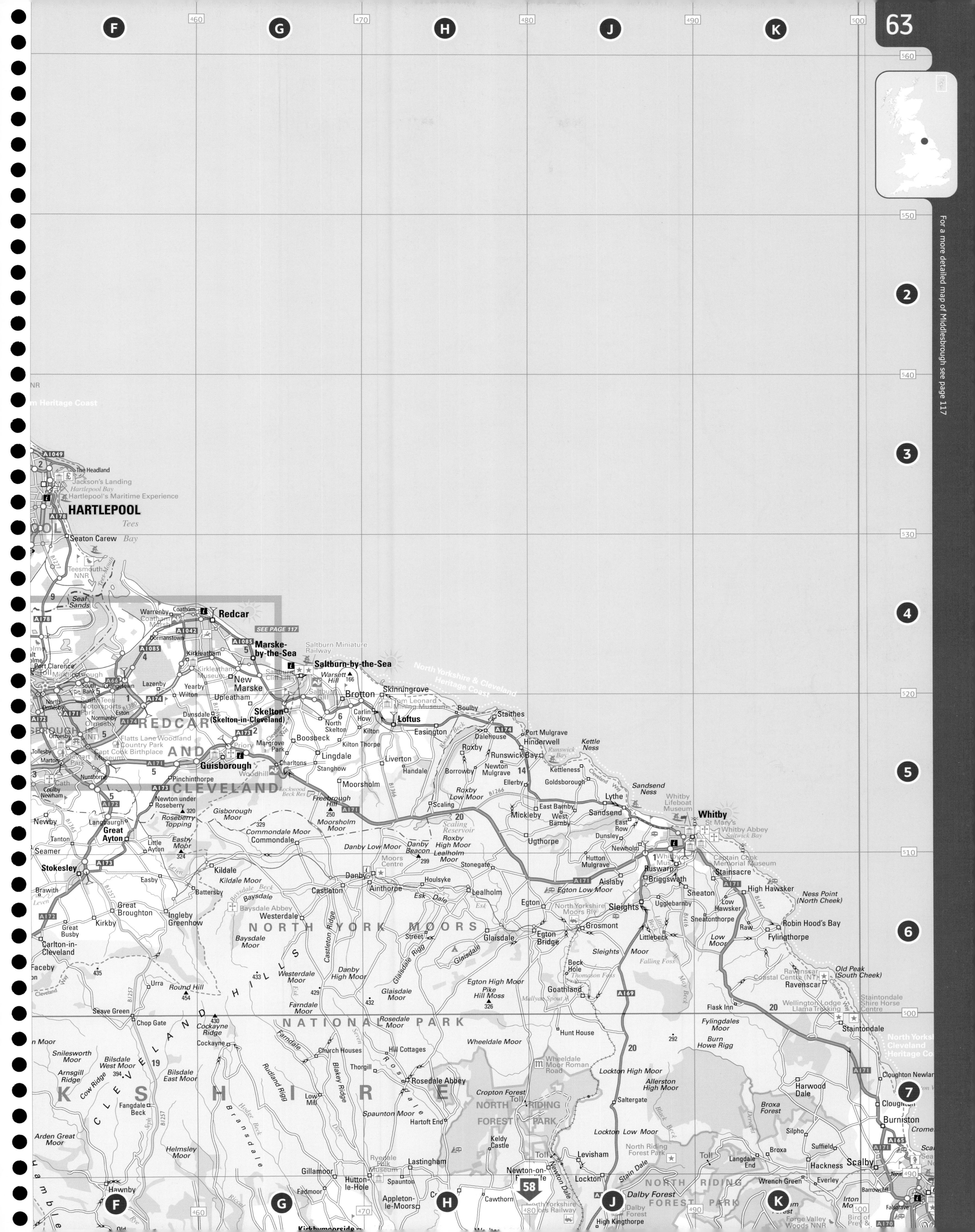

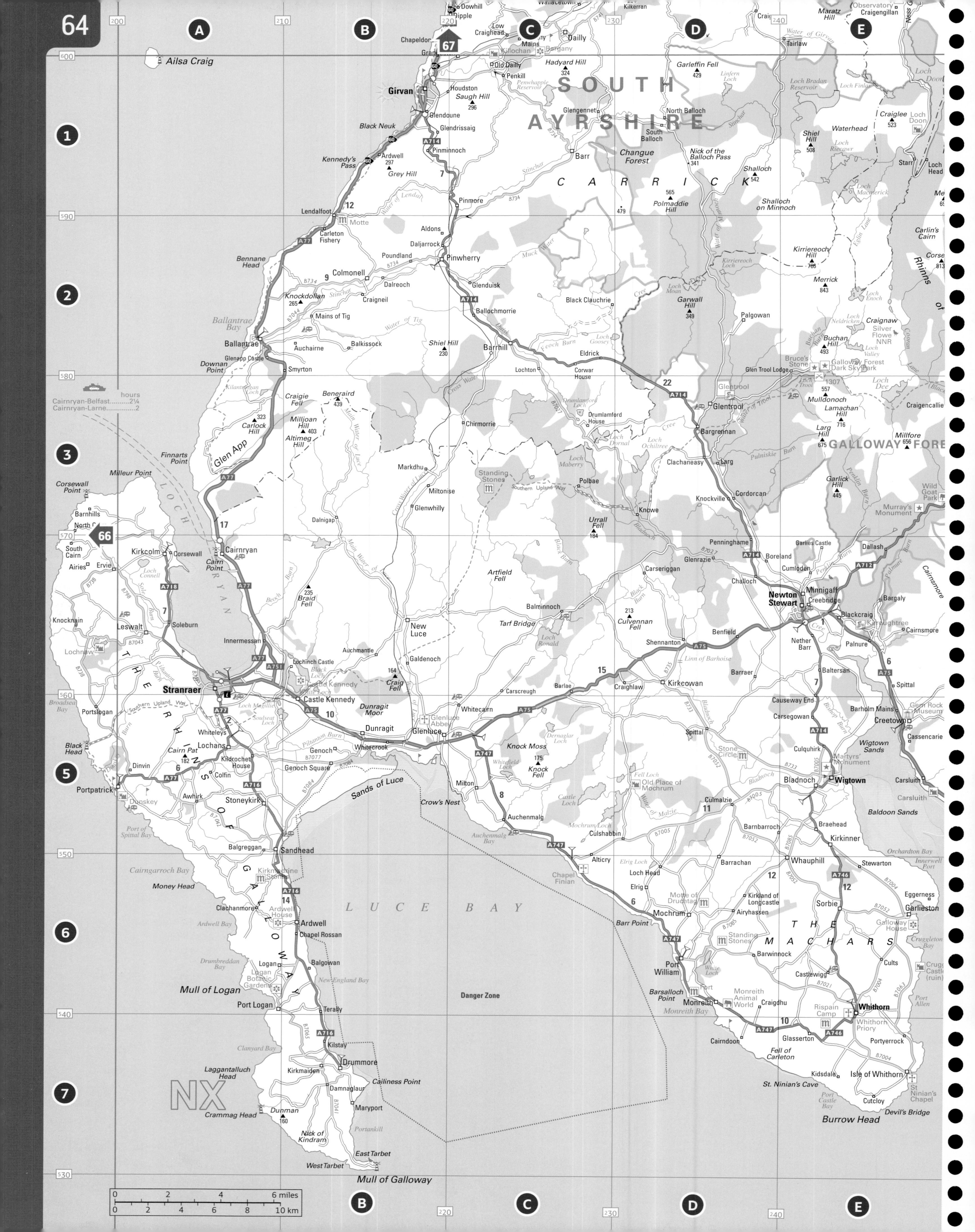

64
A
B
C
D
E
1
2
3
4
5
6
7
67
66
NX
Ailsa Craig
SOUTH AYRSHIRE
CARRICK
Girvan
Dailly
Old Dailly
Penkill
Killochan
Low Craighead
Chapeldonan
Dowhill
Dipple
Houston
Saugh Hill
Hadyard Hill
Glendoune
Glendrissaig
Black Neuk
Pinminnoch
Kennedy's Pass
Ardwell
Grey Hill
Lendalfoot
Motte
Carleton Fishery
Bennane Head
Pinmore
Aldons
Daljarrock
Poundland
Pinwherry
Colmonell
Dalreoch
Glenduisk
Knockdollian
Craigneil
Mains of Tig
Ballantrae Bay
Ballantrae
Glenapp Castle
Downan Point
Auchairne
Balkissock
Shiel Hill
Barrhill
Smyrton
Eldrick
Corwar House
Lochton
Ballochmorrie
Black Clauchrie
Glengennet
North Balloch
South Balloch
Barr
Changue Forest
Nick of the Balloch Pass
Garleffin Fell
Loch Bradan Reservoir
Shalloch
Shalloch on Minnoch
Polmaddie Hill
Tairlaw
Maratz Hill
Observatory
Craigengillan
Craiglee
Loch Doon
Waterhead
Shiel Hill
Starr
Loch Head
Loch Macaterick
Carlin's Cairn
Corserine
Rhinns of Kells
Kirriereoch Hill
Merrick
Garwall Hill
Palgowan
Craignaw
Silver Flowe NNR
Buchan Hill
Bruce's Stone
Galloway Forest Dark Sky Park
Glen Trool Lodge
Glentrool
Mulldonoch
Lamachan Hill
Larg Hill
Millfore
Craigencallie
GALLOWAY FOREST
Bargrennan
Clachaneasy
Larg
Cordorcan
Knockville
Garlick Hill
Wild Goat Park
Murray's Monument
Cairnryan-Belfast hours 2¼
Cairnryan-Larne 2
Craigie Fell
Beneraird
Carlock Hill
Milljoan Hill
Altimeg Hill
Glen App
Finnarts Point
Milleur Point
Corsewall Point
Chirmorrie
Drumlamford House
Loch Dornal
Loch Ochiltree
Markdhu
Miltonise
Standing Stones
Southern Upland Way
Polbae
Loch Maberry
Glenwhilly
Knowe
Urrall Fell
Dalnigap
Barnhills
North Cairn
South Cairn
Airies
Ervie
Kirkcolm
Corsewall
Cairnryan
Cairn Point
Loch Ryan
Loch Connell
Braid Fell
Artfield Fell
Penninghame
Glenrazie
Carseriggan
Challoch
Boreland
Cumloden
Garlies Castle
Dallash
Cairnsmore
Minnigaff
Newton Stewart
Creebridge
Blackcraig
Kirroughtree
Bargaly
Cairnsmore
Palnure
Knocknain
Leswalt
Soleburn
Lochnaw
Innermessan
Balminnoch
Culvennan Fell
Tarf Bridge
Loch Ronald
New Luce
Shennanton
Benfield
Nether Barr
Auchmantle
Galdenoch
Lochinch Castle
Castle Kennedy
Craig Fell
Carscreugh
Barlae
Craighlaw
Kirkcowan
Barraer
Baltersan
Spittal
Stranraer
Broadsea Bay
Portslogan
Dunragit Moor
Dunragit
Glenwhan
Glenluce Abbey
Glenluce
Whitecairn
Causeway End
Carsegowan
Barholm Mains
Creetown
Gem Rock Museum
Cassencarie
Whiteleys
Lochans
Cairn Pat
Black Head
Dinvin
Portpatrick
Kildrochet House
Colfin
Genoch
Genoch Square
Whitecrook
Knock Moss
Knock Fell
Dernaglar Loch
Stone Circle
Culquhirk
Martyrs' Monument
Wigtown Sands
Wigtown
Bladnoch
Carsluith
Awhirk
Stoneykirk
Dunskey
Sands of Luce
Crow's Nest
Milton
Old Place of Mochrum
Castle Loch
Culmalzie
Baldoon Sands
Port of Spittal Bay
Auchenmalg
Auchenmalg Bay
Mochrum Loch
Culshabbin
Barnbarroch
Braehead
Kirkinner
Balgreggan
Sandhead
Alticry
Elrig Loch
Loch Head
Elrig
Chapel Finian
Barrachan
Whauphill
Stewarton
Orchardton Bay
Innerwell Port
Cairngarroch Bay
Money Head
Kirkmadrine Stones
Motte of Druchtag
Kirkland of Longcastle
Airyhassen
Eggerness
Garlieston
Sorbie
Galloway House
Clachanmore
Ardwell House
Ardwell
Ardwell Bay
Chapel Rossan
LUCE BAY
Barr Point
Mochrum
Standing Stones
THE MACHARS
Cruggleton Bay
Cruggleton Castle (ruin)
Barwinnock
Drumbreddan Bay
Logan
Logan Botanic Gardens
Balgowan
New England Bay
Mull of Logan
Port Logan
Port William
White Loch
Castlewigg
Cults
Danger Zone
Barsalloch Point
Fort
Monreith
Monreith Bay
Monreith Animal World
Craigdhu
Rispain Camp
Whithorn
Port Allen
Whithorn Priory
Terally
Kilstay
Clanyard Bay
Drummore
Kirkmaiden
Cailiness Point
Damnaglaur
Maryport
Portankill
Cairndoon
Glasserton
Portyerrock
Fell of Carleton
Kidsdale
Isle of Whithorn
St. Ninian's Cave
St Ninian's Chapel
Port Castle Bay
Cutcloy
Devil's Bridge
Burrow Head
Laggantalluch Head
Crammag Head
Dunman
Nick of Kindram
East Tarbet
West Tarbet
Mull of Galloway
THE RHINNS OF GALLOWAY
A77
A714
A718
A751
A75
A747
A716
A712
A746
B7065
B7041
B7084
B7077
B7005
B7052
B7085
B7004
B7021
B7063
B734
B741
B7043
B738
B798
B7027
B733
B735
B7079
B7007
0 2 4 6 miles
0 2 4 6 8 10 km

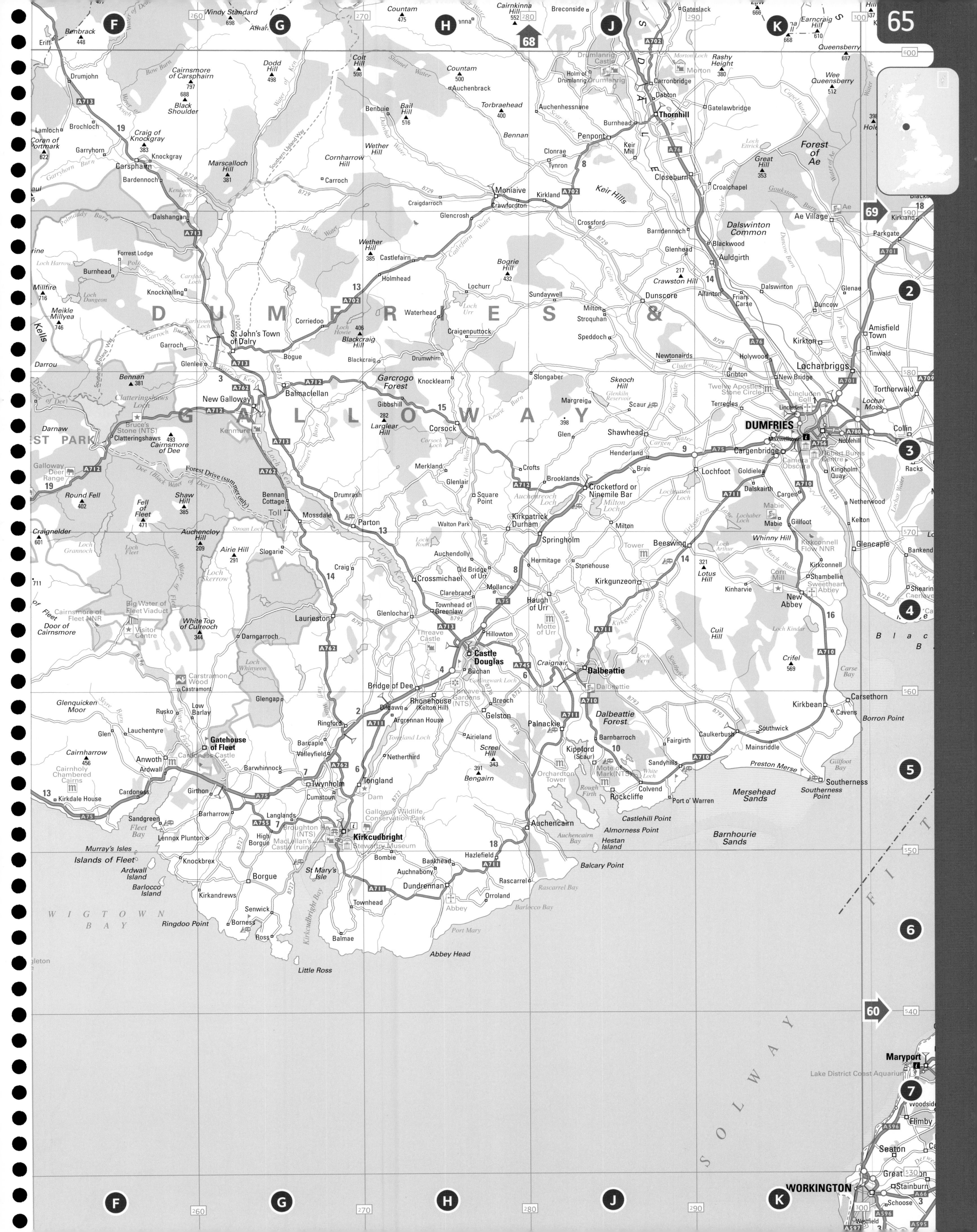
F
G
H
J
K
Windy Standard
Countam
Cairnkinna Hill
Breconside
Gateslack
Queensberry
68
Bennbrack
Eriff
Drumjohn
Cairnsmore of Carsphairn
Dodd Hill
Colt Hill
Countam
Auchenbrack
Drumlanrig Castle
Morton
Rashy Height
Wee Queensberry
Carronbridge
Dabton
Thornhill
Gatelawbridge
Black Shoulder
Benbuie
Bail Hill
Torbraehead
Auchenhessnane
Burnhead
Lamloch
Brochloch
Craig of Knockgray
Coran of Portmark
Garryhorn
Knockgray
Carsphairn
Bardennoch
Marscalloch Hill
Wether Hill
Cornharrow Hill
Carroch
Bennan
Penpont
Keir Mill
Clonrae
Tynron
Closeburn
Forest of Ae
Great Hill
Croalchapel
Moniaive
Crawfordton
Kirkland
Keir Hills
Craigdarroch
Glencrosh
Dalshangan
Crossford
Barndennoch
Dalswinton Common
Ae Village
Kirkland
Parkgate
69
Forrest Lodge
Burnhead
Knocknalling
Millfire
Meikle Millyea
Wether Hill
Castlefairn
Holmhead
Bogrie Hill
Lochurr
Glenhead
Blackwood
Auldgirth
Crawston Hill
Allanton
Friars Carse
Dalswinton
Glenae
Duncow
Sundaywell
Dunscore
Milton
Stroquhan
D U M F R I E S &
Waterhead
Corriedoo
Blackcraig Hill
Craigenputtock
Speddoch
St John's Town of Dalry
Garroch
Glenlee
Bogue
Blackcraig
Drumwhirn
Newtonairds
Holywood
Kirkton
Amisfield Town
Tinwald
Locharbriggs
Bennan
New Galloway
Balmaclellan
Garcrogo Forest
Knocklearn
Slongaber
Skeoch Hill
Gribton
New Bridge
Twelve Apostles Stone Circle
Lincluden Coll
Terregles
Torthorwald
Lochar Moss
G A L L O W A Y
Gibbshill
Larglear Hill
Corsock
Margreig
Scaur
Glen
Shawhead
DUMFRIES
Collin
Darnaw
Bruce's Stone (NTS)
Clatteringshaws
Cairnsmore of Dee
Kenmure
Maxwelltown
Cargenbridge
Henderland
Galloway Deer Range
Forest Drive (summer only)
Merkland
Glenlair
Crofts
Brooklands
Brae
Lochfoot
Goldielea
Dalskairth
Robert Burns Centre
Kingholm Quay
Racks
Round Fell
Shaw Hill
Fell of Fleet
Bennan Cottage
Drumrash
Square Point
Crocketford or Ninemile Bar
Auchenreoch Loch
Milton Loch
Cargen
Netherwood
Mossdale
Parton
Walton Park
Kirkpatrick Durham
Milton
Mabie
Gillfoot
Kelton
Craignelder
Auchenoloy Hill
Airie Hill
Slogarie
Springholm
Beeswing
Whinny Hill
Kirkconnell Flow NNR
Glencaple
Bankend
Auchendolly
Craig
Crossmichael
Old Bridge of Urr
Hermitage
Stonehouse
Lotus Hill
Kirkgunzeon
Kirkconnell
Shambellie
Kinharvie
Sweetheart Abbey
Shearington
Caerlaverock
Big Water of Fleet Viaduct
Cairnsmore of Fleet NNR
Door of Cairnsmore
Visitor Centre
White Top of Culreoch
Darngarroch
Laurieston
Glenlochar
Clarebrand
Townhead of Greenlaw
Mollance
Haugh of Urr
Motte of Urr
Hillowton
New Abbey
Cuil Hill
Threave Castle
Castle Douglas
Buchan
Craignair
Dalbeattie
Crifel
Carse Bay
Carstramon Wood
Castramont
Glengap
Bridge of Dee
Rhonehouse (Kelton Hill)
Threave Gardens (NTS)
Breoch
Gelston
Dalbeattie Forest
Carsethorn
Kirkbean
Cavens
Borron Point
Glenquicken Moor
Rusko
Low Barlay
Glen
Lauchentyre
Gatehouse of Fleet
Cardoness Castle
Ringford
Barcaple
Valleyfield
Argrennan House
Netherthird
Airieland
Screel Hill
Bengairn
Palnackie
Barnbarroch
Kippford (Scaur)
Fairgirth
Caulkerbush
Southwick
Mainsriddle
Cairnharrow
Cairnholy Chambered Cairns
Anwoth
Ardwall
Barwhinnock
Twynholm
Tongland
Cumstoun
Orchardton Tower
Mote of Mark (NTS)
Rockcliffe
Colvend
Sandyhills
Port o' Warren
Preston Merse
Southerness
Southerness Point
Mersehead Sands
Kirkdale House
Cardoness
Girthon
Sandgreen
Barharrow
Langlands
Fleet Bay
Lennox Plunton
High Borgue
Broughton Ho
MacLellan's Castle (ruin)
Kirkcudbright
Stewartry Museum
Galloway Wildlife Conservation Park
Auchencairn
Castlehill Point
Almorness Point
Hestan Island
Barnhourie Sands
Murray's Isles
Islands of Fleet
Knockbrex
Ardwall Island
Barlocco Island
Borgue
St Mary's Isle
Bombie
Bankhead
Auchnabony
Hazlefield
Balcary Point
Rascarrel
Dundrennan
Abbey
Orroland
Rascarrel Bay
Barlocco Bay
Kirkandrews
Townhead
Senwick
W I G T O W N B A Y
Ringdoo Point
Borness
Ross
Balmae
Port Mary
Abbey Head
Little Ross
S O L W A Y F I R T H
60
Maryport
Lake District Coast Aquarium
Flimby
Seaton
Great Clifton
WORKINGTON
Stainburn
Schoose
Westfield
A713
A702
A76
A701
A762
A712
A75
A780
A756
A711
A710
A745
A755
A596
A66
A595
A597
B729
B7000
B794
B795
B736
B727
B793
B725
1
2
3
4
5
6
7

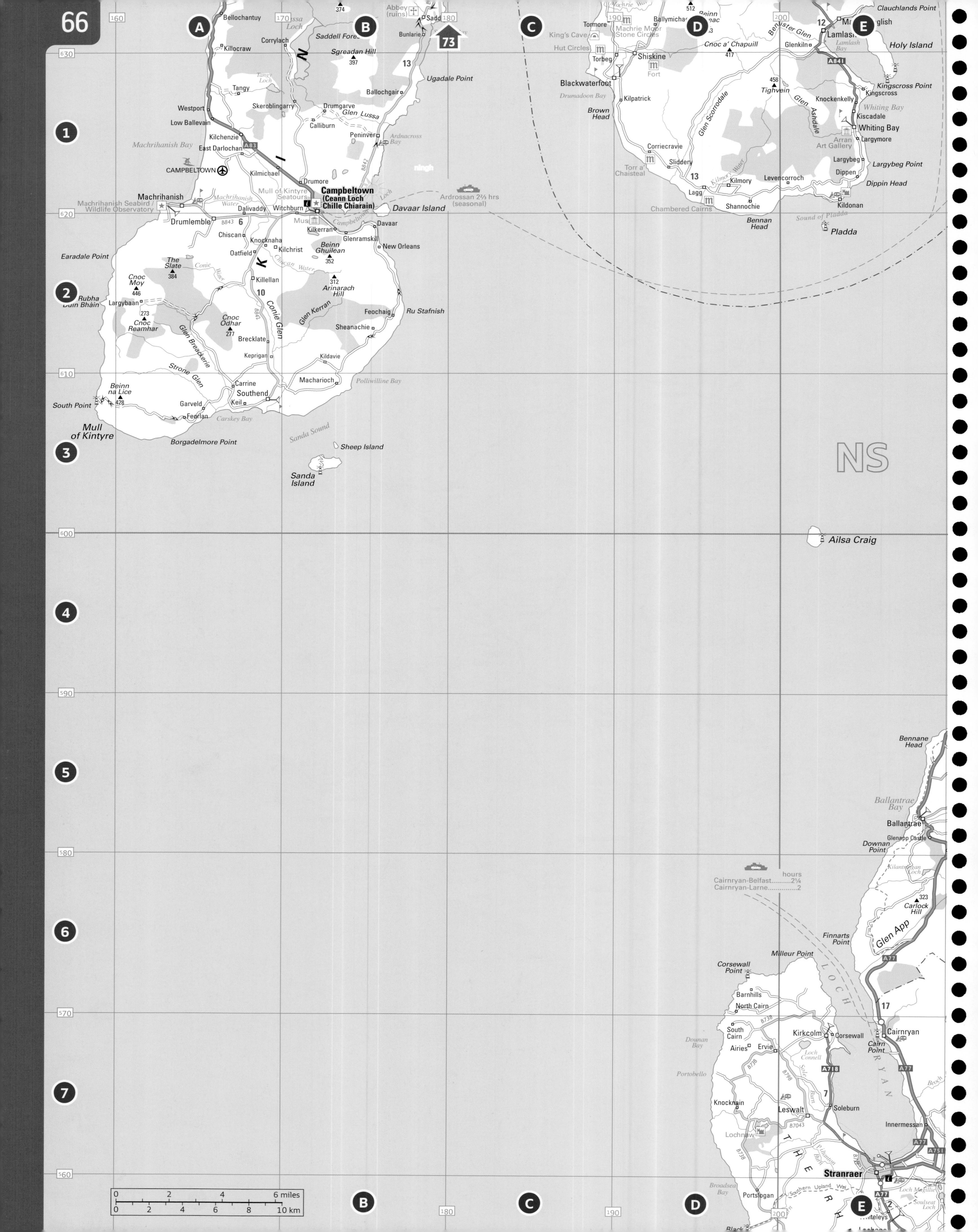
A
B
C
D
E
1
2
3
4
5
6
7
73
NS
Bellochantuy
Killocraw
Corrylach
Saddell Forest
Sgreadan Hill
397
Bunlarie
Abbey (ruins)
13
Ugadale Point
Ballochgair
Tangy
Westport
Low Ballevain
Skeroblingarry
Drumgarve
Glen Lussa
Calliburn
Peninver
Ardnacross Bay
Machrihanish Bay
Kilchenzie
East Darlochan
A83
CAMPBELTOWN
Kilmichael
Drumore
B842
Machrihanish
Machrihanish Seabird / Wildlife Observatory
Mull of Kintyre Seatours
Campbeltown (Ceann Loch Chille Chiarain)
Davaar Island
Ardrossan 2¾ hrs (seasonal)
Dalivaddy
Witchburn
Drumlemble
B843
6
Mus
Kilkerran
Davaar
Glenramskill
Chiscan
Knocknaha
Kilchrist
Oatfield
Beinn Ghuilean
352
New Orleans
Earadale Point
The Slate
384
Conie Water
Chiscan Water
Cnoc Moy
446
Killellan
312
Arinarach Hill
10
Rubha Duin Bhàin
Largybaan
273
Cnoc Reamhar
Conie Glen
Glen Kerran
Feochaig
Ru Stafnish
Cnoc Odhar
277
B842
Glen Breackerie
Brecklate
Sheanachie
Kepригan
Kildavie
Strone Glen
Carrine
Macharioch
Polliwilline Bay
Beinn na Lice
428
Southend
South Point
Garveld
Keil
Feorlan
Carskey Bay
Mull of Kintyre
Borgadelmore Point
Sanda Sound
Sheep Island
Sanda Island
K
I
N
Tormore
King's Cave
Hut Circles
Machrie Moor Stone Circles
Ballymichael
512
Beinn
Cnoc a' Chapuill
417
Glen
Benlister Glen
12
Lamlash
Lamlash Bay
Clauchlands Point
Holy Island
Glenkiln
A841
Torbeg
Shiskine
Fort
Blackwaterfoot
Drumadoon Bay
Kilpatrick
Brown Head
458
Tighvein
Glen Scorrodale
Glen Ashdale
Kingscross Point
Kingscross
Knockenkelly
Whiting Bay
Kiscadale
Whiting Bay
Arran Art Gallery
Largymore
Corriecravie
Sliddery
Torr a' Chaisteal
Kilmory Water
Kilmory
Largybeg
Largybeg Point
Dippen
Dippen Head
Levencorroch
Lagg
Chambered Cairns
Shannochie
Kildonan
Bennan Head
Sound of Pladda
Pladda
Ailsa Craig
Bennane Head
Ballantrae Bay
Ballantrae
Glenapp Castle
Downan Point
Kilantringan Loch
Cairnryan-Belfast..........2¼ hours
Cairnryan-Larne...............2
323
Carlock Hill
Finnarts Point
Glen App
Milleur Point
Corsewall Point
A77
Barnhills
North Cairn
B738
LOCH RYAN
17
South Cairn
Kirkcolm
Corsewall
Cairnryan
Dounan Bay
Airies
Ervie
Loch Connell
Cairn Point
A718
A77
Portobello
B798
B738
7
Knocknain
Leswalt
Soleburn
Innermessan
B7043
Lochnaw
THE RHINS
A77
A751
Stranraer
Broadsea Bay
Portslogan
Southern Upland Way
Soulseat Loch
Loch Magillie
2
Black
160
170
180
190
200
630
620
610
600
590
580
570
560
0
2
4
6 miles
0
2
4
6
8
10 km

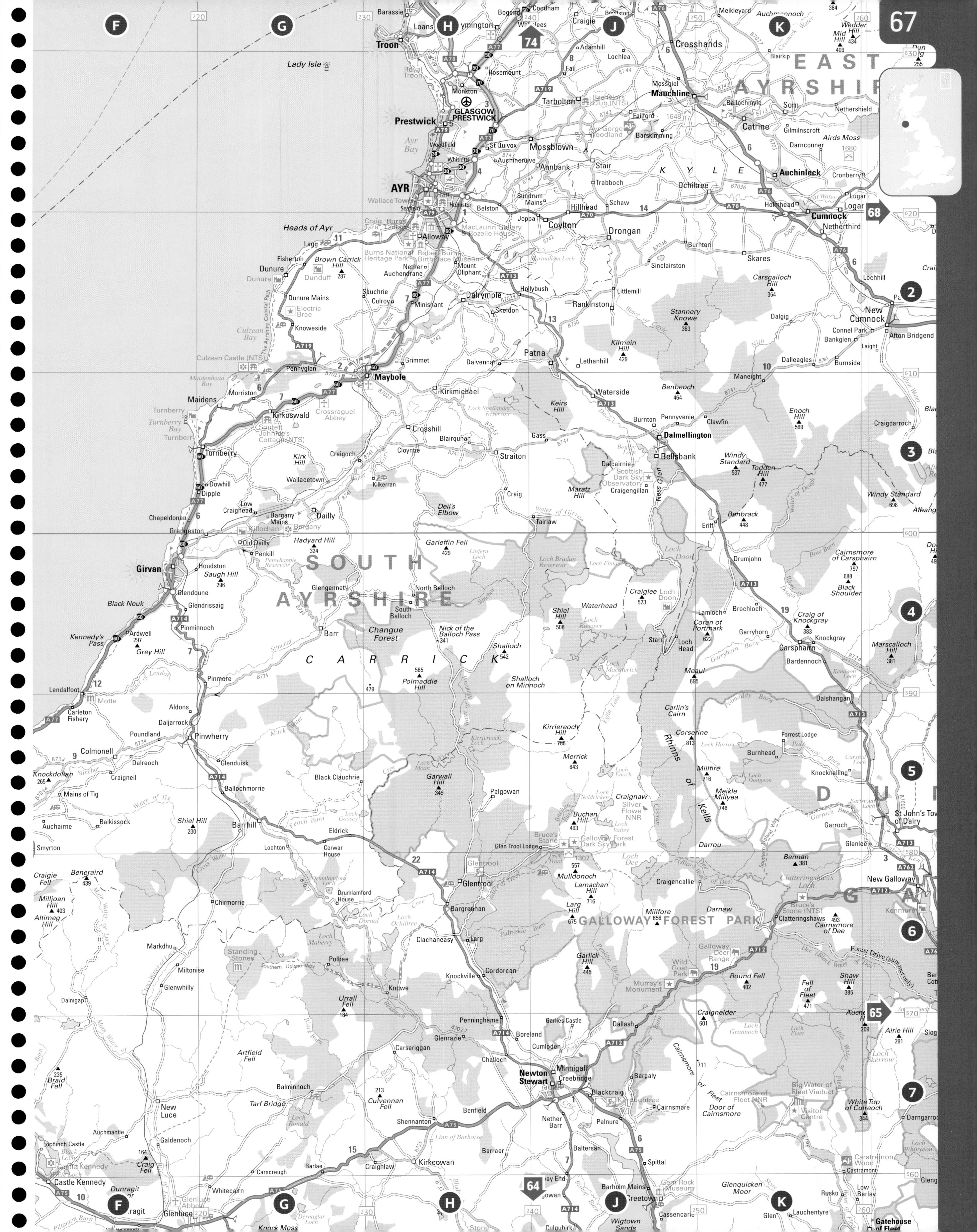

EAST AYRSHIRE
SOUTH AYRSHIRE
C A R R I C K
K Y L E
GALLOWAY FOREST PARK
74
68
64
65
Troon
Lady Isle
Prestwick
GLASGOW PRESTWICK
Ayr Bay
AYR
Heads of Ayr
Alloway
Monkton
Mossblown
Tarbolton
Mauchline
Catrine
Sorn
Auchinleck
Cumnock
New Cumnock
Crosshands
Ochiltree
Coylton
Drongan
Dalrymple
Patna
Dalmellington
Bellsbank
Maybole
Kirkmichael
Crosshill
Straiton
Dunure
Culzean Castle (NTS)
Culzean Bay
Maidens
Turnberry
Kirkoswald
Dailly
Girvan
Barr
Pinwherry
Colmonell
Barrhill
Glentrool
Bargrennan
Newton Stewart
Minnigaff
Creebridge
Kirkcowan
Glenluce
Castle Kennedy
Dunragit
New Luce
Lendalfoot
Carsphairn
Dalry
New Galloway
Creetown
Gatehouse of Fleet
Loch Doon
Rhinns of Kells
Merrick 843
Corserine 813
Shalloch on Minnoch
Nick of the Balloch Pass
Changue Forest
Black Neuk
Kennedy's Pass
Heads of Ayr
Glen Trool Lodge
Galloway Forest Dark Sky Park
Murray's Monument
Carstramon Wood
Big Water of Fleet Viaduct
Cairnsmore of Fleet NNR

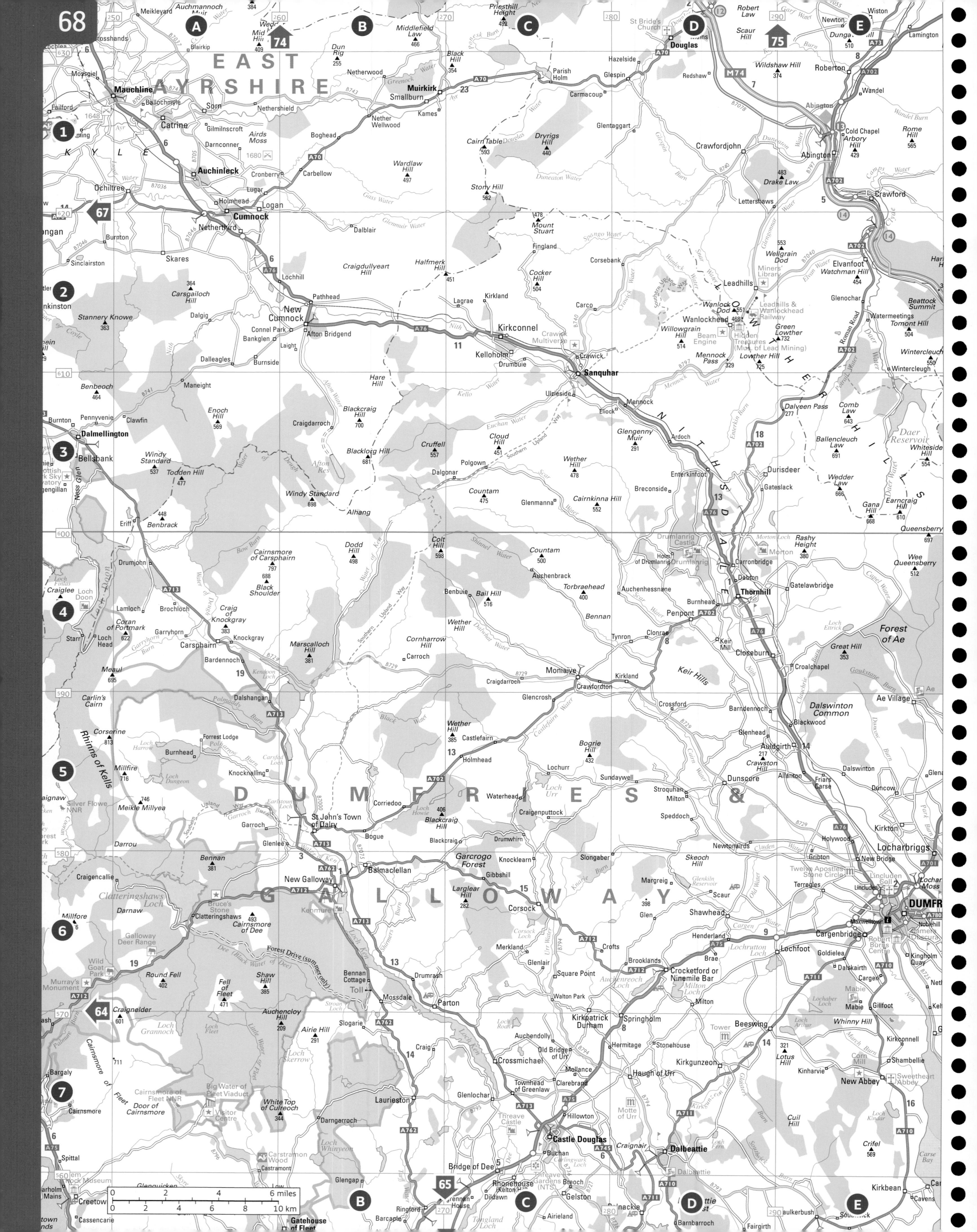
EAST AYRSHIRE
KYLE
DUMFRIES & GALLOWAY
NITHSDALE
LOWTHER HILLS
Rhinns of Kells
74
75
67
64
65
A
B
C
D
E
1
2
3
4
5
6
7
Auchmannoch Muir
Meikleyard
Blairkip
Mid Wedder Hill
Priesthill Height
St Bride's Church
Douglas
Robert Law
Scaur Hill
Wiston
Newton
Lamington
Middlefield Law
Dun Rig
Black Hill
Netherwood
Hazelside
Glespin
Parish Holm
Redshaw
Wildshaw Hill
Roberton
Wandel
Mauchline
Mossgiel
Ballochmyle
Sorn
Failford
Catrine
Nethershield
Muirkirk
Smallburn
Kames
Nether Wellwood
Carmacoup
Glentaggart
Abington
Cold Chapel
Arbory Hill
Rome Hill
Gilmilnscroft
Airds Moss
Darnconner
Boghead
Cairn Table
Dryrigs Hill
Crawfordjohn
Auchinleck
Cronberry
Carbellow
Wardlaw Hill
Duneaton Water
Drake Law
Ochiltree
Lugar
Logan
Holmhead
Cumnock
Netherthird
Stony Hill
Lettershaws
Crawford
Burnton
Dalblair
Mount Stuart
Skares
Sinclairston
Fingland
Corsebank
Wellgrain Dod
Miners' Library
Elvanfoot
Watchman Hill
Craigdullyeart Hill
Halfmerk Hill
Cocker Hill
Leadhills
Lochhill
Glenochar
Carsgailoch Hill
Pathhead
Lagrae
Kirkland
Carco
Wanlock Dod
Leadhills & Wanlockhead Railway
Beattock Summit
Stannery Knowe
New Cumnock
Connel Park
Afton Bridgend
Bankglen
Laight
Dalgig
Kirkconnel
Crawick Multiverse
Wanlockhead
Willowgrain Hill
Beam Engine
Green Lowther
Museum of Lead Mining
Watermeetings
Tomont Hill
Kelloholm
Drumbuie
Crawick
Dalleagles
Burnside
Mennock Pass
Lowther Hill
Wintercleuch
Sanquhar
Hare Hill
Benbeoch
Maneight
Ulzieside
Eliock
Mennock
Dalveen Pass
Comb Law
Enoch Hill
Burnton
Pennyvenie
Clawfin
Craigdarroch
Blackcraig Hill
Dalmellington
Bellsbank
Cloud Hill
Glengenny Muir
Ardoch
Ballencleuch Law
Daer Reservoir
Windy Standard
Todden Hill
Blacklorg Hill
Cruffell
Polgown
Wether Hill
Durisdeer
Gateslack
Whiteside Hill
Dalgonar
Enterkinfoot
Afton Res
Countam
Glenmanna
Cairnkinna Hill
Breconside
Wedder Law
Windy Standard
Alhang
Gana Hill
Earncraig Hill
Ness Glen
Eriff
Benbrack
Queensberry
Drumjohn
Cairnsmore of Carsphairn
Dodd Hill
Colt Hill
Countam
Drumlanrig Castle
Morton Loch
Morton
Rashy Height
Black Shoulder
Auchenbrack
Torbraehead
Holm of Drumlanrig
Carronbridge
Dabton
Wee Queensberry
Loch Doon
Craiglee
Lamloch
Brochloch
Craig of Knockgray
Benbuie
Ball Hill
Auchenhessnane
Thornhill
Gatelawbridge
Burnhead
Coran of Portmark
Garryhorn
Knockgray
Wether Hill
Bennan
Penpont
Loch Head
Carsphairn
Cornharrow Hill
Tynron
Clonrae
Keir Mill
Loch Etrick
Forest of Ae
Marscalloch Hill
Carroch
Closeburn
Great Hill
Bardennoch
Craigdarroch
Moniaive
Crawfordton
Kirkland
Keir Hills
Croalchapel
Meaul
Carlin's Cairn
Dalshangan
Glencrosh
Crossford
Ae Village
Dalswinton Common
Barndennoch
Blackwood
Corserine
Forrest Lodge
Burnhead
Wether Hill
Castlefairn
Holmhead
Bogrie Hill
Glenhead
Auldgirth
Crawston Hill
Dalswinton
Millfire
Knocknalling
Loch Dungeon
Lochurr
Sundaywell
Dunscore
Allanton
Friars Carse
Duncow
Silver Flowe NNR
Meikle Millyea
Corriedoo
Waterhead
Loch Urr
Stroquhan
Milton
Craigenputtock
Speddoch
St John's Town of Dalry
Blackcraig Hill
Kirkton
Garroch
Bogue
Blackcraig
Drumwhirn
Holywood
Glenlee
Newtonairds
Locharbriggs
Darrou
Bennan
Garcrogo Forest
Knocklearn
Slongaber
Skeoch Hill
Gribton
New Bridge
Craigencallie
New Galloway
Balmaclellan
Gibbshill
Margreig
Glenkiln Reservoir
Twelve Apostles Stone Circle
Terregles
Lincluden
Clatteringshaws Loch
Bruce's Stone
Kenmure
Larglear Hill
Corsock
Scaur
Shawhead
Glen
Dumfries
Maxwelltown
Camera Obscura
Millfore
Darnaw
Clatteringshaws
Cairnsmore of Dee
Corsock Loch
Henderland
Cargenbridge
Robert Burns Centre
Galloway Deer Range
Merkland
Crofts
Lochfoot
Goldielea
Kingholm Quay
Forest Drive (summer only)
Glenlair
Brooklands
Brae
Crocketford or Ninemile Bar
Dalskairth
Wild Goat Park
Round Fell
Shaw Hill
Square Point
Auchenreoch Loch
Milton Loch
Cargen
Murray's Monument
Fell of Fleet
Bennan Cottage
Toll
Drumrash
Mossdale
Parton
Walton Park
Milton
Mabie
Gillfoot
Craignelder
Loch Grannoch
Loch Fleet
Auchencloy Hill
Stroan Loch
Slogarie
Kirkpatrick Durham
Springholm
Beeswing
Whinny Hill
Cairnsmore of Fleet
Airie Hill
Loch Skerrow
Auchendolly
Tower
Lotus Hill
Craig
Old Bridge of Urr
Hermitage
Stonehouse
Kirkconnell
Corn Mill
Crossmichael
Mollance
Kirkgunzeon
Haugh of Urr
Kinharvie
Shambellie
Bargaly
Big Water of Fleet Viaduct
Cairnsmore of Fleet NNR
White Top of Culreoch
Laurieston
Glenlochar
Townhead of Greenlaw
Clarebrand
Hillowton
Motte of Urr
New Abbey
Sweetheart Abbey
Cairnsmore
Visitor Centre
Door of Cairnsmore
Darngarroch
Threave Castle
Castle Douglas
Craignair
Dalbeattie
Cuil Hill
Crifel
Spittal
Carstramon Wood
Castramont
Loch Whinyeon
Bridge of Dee
Buchan
Carlingwark Loch
Threave Gardens (NTS)
Breoch
Dalbeattie
Carse Bay
Glenquicken
Creetown
Glengap
Rhonehouse
Kelton Hill
Dildawn
Gelston
Kirkbean
Cassencarie
Gatehouse of Fleet
Ringford
Barcaple
Tongland Loch
Airieland
Palnackie
Barnbarroch
Aulkerbush
Fairgirth
Cavens
Kirkconnell
0 2 4 6 miles
0 2 4 6 8 10 km

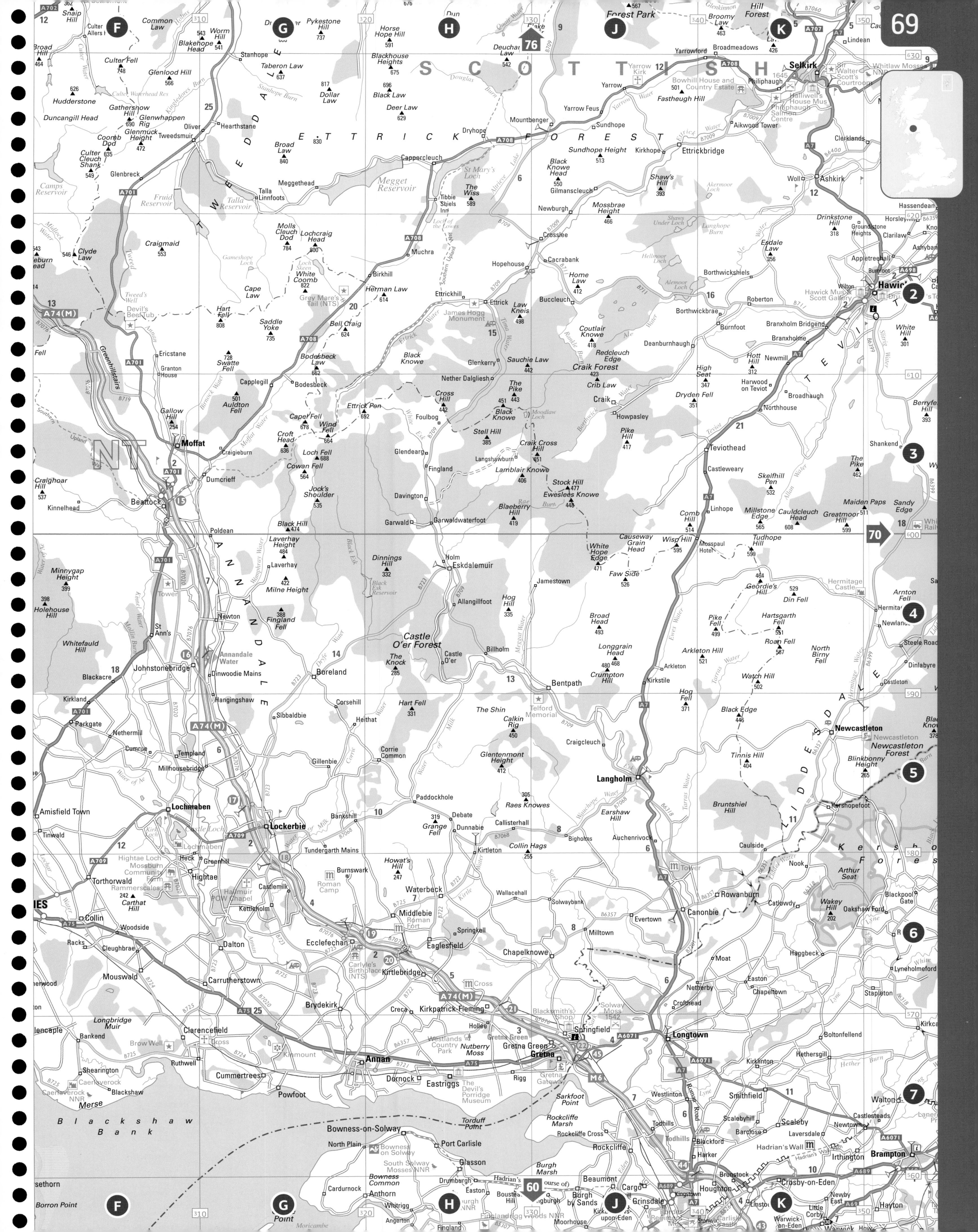
F
G
H
J
K
76
70
60
SCOTTISH
ETTRICK FOREST
TWEEDDALE
ANNANDALE
TEVIOTDALE
LIDDESDALE
Kershope Forest
Blackshaw Bank
Selkirk
Hawick
Moffat
Beattock
Lockerbie
Lochmaben
Annan
Gretna
Gretna Green
Langholm
Longtown
Newcastleton
Eskdalemuir
Ettrickbridge
Ashkirk
Teviothead
Canonbie
Ecclefechan
Kirtlebridge
Johnstonebridge
Boreland
Bentpath
Craik
Craik Forest
Castle O'er Forest
Ettrick
Tweedsmuir
Yarrow
Yarrow Feus
Mountbenger
Cappercleuch
Megget Reservoir
St Mary's Loch
Talla Reservoir
Fruid Reservoir
Camps Reservoir
Tibbie Shiels Inn
Hart Fell
White Coomb
Grey Mare's Tail (NTS)
Devil's Beef Tub
Ettrick Pen
Bowhill House and Country Estate
Halliwell's House Mus
Philiphaugh Salmon Centre
Sir Walter Scott's Courtroom
Hawick Mus & Scott Gallery
James Hogg Monument
Telford Memorial
Carlyle's Birthplace (NTS)
Hallmuir POW Chapel
Bowness-on-Solway
Port Carlisle
Glasson
Drumburgh
Burgh by Sands
Beaumont
Rockcliffe
Brampton
Irthington
Crosby-on-Eden
Houghton
Harker
Smithfield
Westlinton
Springfield
Eastriggs
Dornock
Powfoot
Cummertrees
Ruthwell
Clarencefield
Carrutherstown
Dalton
Collin
Torthorwald
Amisfield Town
Tinwald
Mouswald
Caerlaverock
Merse
Brydekirk
Middlebie
Waterbeck
Eaglesfield
Chapelknowe
Kirkpatrick-Fleming
Roman Camp
Roman Fort
A74(M)
A701
A708
A7
A709
A75
A6071
A698
A689
M6
B7009
B709
B723
B7068
B6357
B6318
South Solway Mosses NNR
Solway Moss 1542
Hadrian's Wall
Hadrian's Wall Path
Teviot
Annan Water
Moffat Water
White Esk
Black Esk Reservoir
Tudhope Hill
Cauldcleuch Head
Greatmoor Hill
Maiden Paps
Hermitage Castle
Arkleton Hill
Burnswark
Castlemilk
Broadmeadows
Philiphaugh
Aikwood Tower
Sundhope Height
Kirkhope
Eskdalemuir
Jamestown
Billholm
Craigcleuch
Bigholms
Auchenrivock
Rowanburn
Catlowdy
Kershopefoot
Blackpool Gate
Oakshaw Ford
Stapleton
Bewcastle

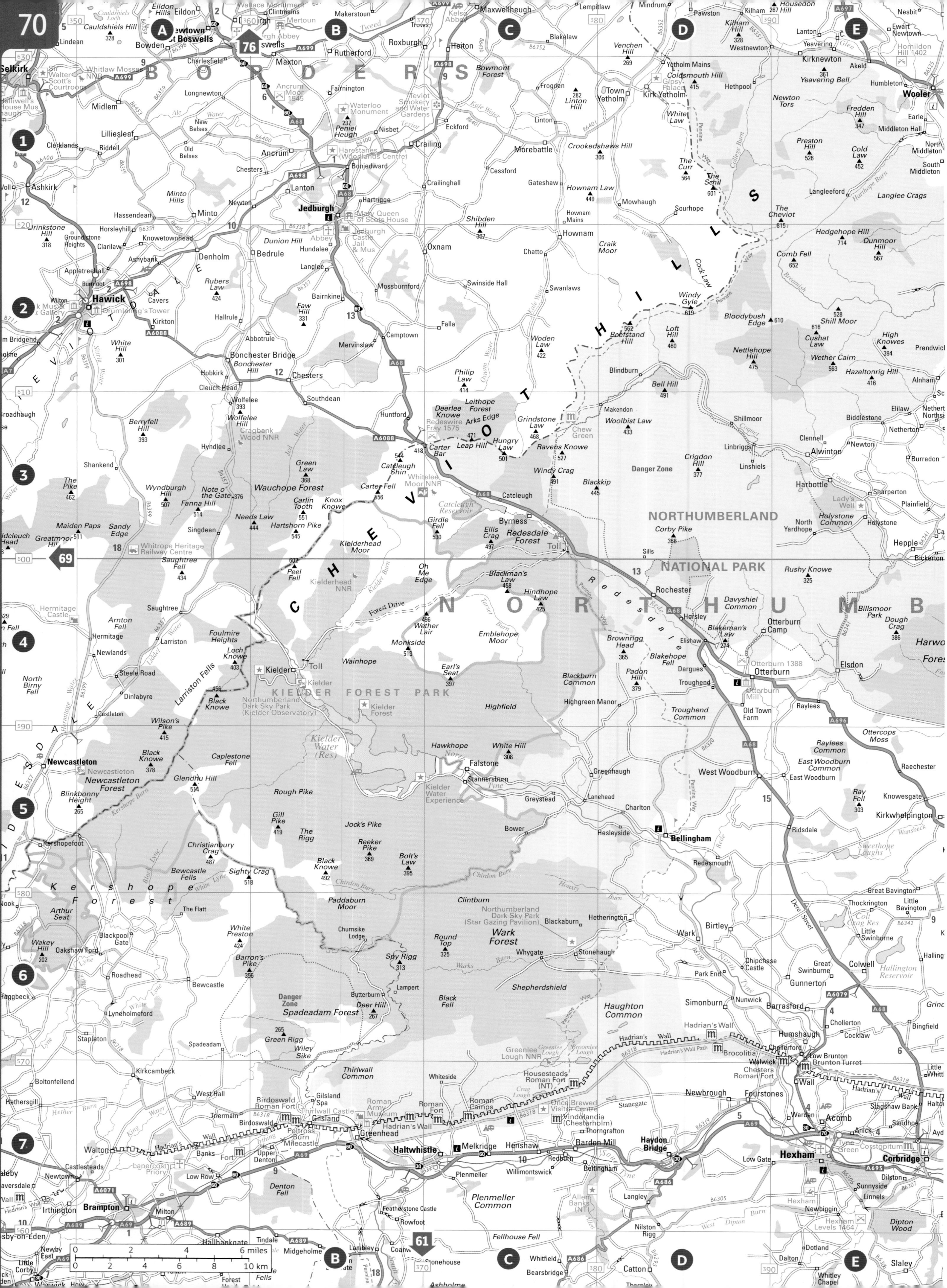

BORDERS

NORTHUMBERLAND

NATIONAL PARK

NORTHUMB

CHEVIOT HILLS

REDESDALE

KIELDER FOREST PARK

Selkirk
Newtown St Boswells
Bowden
Eildon Hills
Eildon
Wallace Monument
Clintmains
Makerstoun
Maxwellheugh
Kelso
Kilham
Housedon
Pawston
Mindrum
Nesbit
Ewart Newtown
Homildon Hill 1402

Lindean
Cauldshiels Hill
Dryburgh Abbey
Mertoun
Trows
Roxburgh
Heiton
Blakelaw
Venchen Hill
Kirknewton
Akeld
Wooler

Whitlaw Moss NNR
Midlem
Charlesfield
Maxton
Rutherford
Fairnington
Bowmont Forest
Frogden
Town Yetholm
Kirk Yetholm
Yeavering Bell
Humbleton

Clerklands
Lilliesleaf
New Belses
Ancrum
Waterloo Monument
Nisbet
Crailing
Eckford
Linton
Linton Hill
Hethpool
Newton Tors
Fredden Hill
Earle
Middleton Hall

Ashkirk
Riddell
Old Belses
Chesters
Crailinghall
Cessford
Gateshaw
Hownam Law
Mowhaugh
Sourhope
The Curr
The Schil
Langleeford
Langlee Crags

Minto Hills
Minto
Newton
Lanton
Jedburgh
Mary Queen of Scots House
Morebattle
Crookedshaws Hill
Hownam Mains
The Cheviot
Hedgehope Hill
Dunmoor Hill
Comb Fell

Hassendean
Knowetownhead
Clarilaw
Denholm
Bedrule
Hundalee
Dunion Hill
Oxnam
Chatto
Craik Moor
Hownam
Windy Gyle
Bloodybush Edge
Shill Moor
High Knowes
Prendwick

Hawick
Cavers
Kirkton
Rubers Law
Langlee
Mossburnford
Swinside Hall
Swanlaws
Woden Law
Bell Hill
Nettlehope Hill
Wether Cairn
Hazeltonrig Hill
Alnham

White Hill
Bonchester Bridge
Bonchester Hill
Chesters
Faw Hill
Camptown
Falla
Philip Law
Grindstone Law
Shillmoor
Eliaw
Biddlestone
Netherton

Hobkirk
Cleuch Head
Southdean
Huntford
Deerlee Knowe
Leithope Forest
Arks Edge
Leap Hill
Hungry Law
Chew Green
Makendon
Woolbist Law
Ravens Knowe
Linbriggs
Clennell
Newton
Burradon

Berryfell Hill
Wolfelee
Wolfelee Hill
Cragbank Wood NNR
Redeswire Fray 1575
Carter Bar
Catcleugh Shin
Windy Crag
Blackkip
Danger Zone
Crigdon Hill
Linshiels
Harbottle
Sharperton
Plainfield

The Pike
Wyndburgh Hill
Note o' the Gate
Wauchope Forest
Green Law
Whitelee Moor NNR
Carter Fell
Catcleugh Reservoir
Girdle Fell
Byrness
Catcleugh
NORTHUMBERLAND
Corby Pike
North Yardhope
Holystone Common
Holystone

Maiden Paps
Sandy Edge
Fanna Hill
Needs Law
Carlin Tooth
Knox Knowe
Hartshorn Pike
Singdean
Kielderhead Moor
Ellis Crag
Redesdale Forest
Toll
Sills
Rushy Knowe
Hepple
Bickerton

Greatmoor Hill
Whitrope Heritage Railway Centre
Saughtree
Peel Fell
Kielderhead NNR
Oh Me Edge
Blackman's Law
Hindhope Law
Rochester
Davyshiel Common
NATIONAL PARK

Hermitage Castle
Arnton Fell
Hermitage
Larriston
Foulmire Heights
Loch Knowe
Forest Drive
Wether Lair
Monkside
Emblehope Moor
Earl's Seat
Brownrigg Head
Blakehope Fell
Dargues
Horsley
Blakeman's Law
Otterburn Camp
Billsmoor Park
Dough Crag
Harwood Forest

North Birny Fell
Newlands
Steele Road
Kielder
Toll
Kielder
Northumberland Dark Sky Park (Kielder Observatory)
Kielder Forest
Highfield
Blackburn Common
Highgreen Manor
Padon Hill
Troughend
Otterburn 1388
Otterburn
Elsdon

Larriston Fells
Black Knowe
KIELDER FOREST PARK
Kielder Water (Res)
Troughend Common
Old Town Farm
Raylees
Rochester

Newcastleton
Newcastleton Forest
Wilson's Pike
Black Knowe
Caplestone Fell
Kielder Water Experience
Hawkhope
White Hill
Falstone
Greenhaugh
West Woodburn
Raylees Common
East Woodburn Common
East Woodburn
Raechester

Blinkbonny Height
Glendhu Hill
Rough Pike
Stannersburn
Greystead
Lanehead
Charlton
Hesleyside
Bellingham
Ridsdale
Ray Fell
Knowesgate
Kirkwhelpington

Kershopefoot
Christianbury Crag
Bewcastle Fells
Gill Pike
Jock's Pike
Reeker Pike
Bolt's Law
Bower
Redesmouth
Great Bavington
Little Bavington

Arthur Seat
Sighty Crag
The Flatt
The Rigg
Black Knowe
Chirdon Burn
Wark
Birtley
Thockrington
Colt Crag Res

KERSHOPE FOREST
Paddaburn Moor
Clintburn
Northumberland Dark Sky Park (Star Gazing Pavilion)
Blackaburn
Hetherington
Wark
Colwell
Hallington Reservoir

Wakey Hill
Oakshaw Ford
Blackpool Gate
White Preston
Churnsike Lodge
Round Top
Wark Forest
Whygate
Stonehaugh
Chipchase Castle
Park End
Great Swinburne
Gunnerton
Bingfield

Roadhead
Barron's Pike
Spy Rigg
Warks Burn
Shepherdshield
Simonburn
Nunwick
Barrasford
Chollerton
Cocklaw

Bewcastle
Danger Zone
Spadeadam Forest
Butterburn
Deer Hill
Lampert
Black Fell
Haughton Common
Humshaugh
Chollerford
Walwick
Wall

Haggbeck
Lyneholmeford
Green Rigg
Wiley Sike
Thirlwall Common
Whiteside
Housesteads Roman Fort (NT)
Once Brewed Visitor Centre
Vindolanda (Chesterholm)
Stanegate
Newbrough
Brocolitia
Chesters Roman Fort
Brunton Turret
Acomb

Stapleton
Spadeadam
Greenlee Lough NNR
Broomlee Lough
Hadrian's Wall
Hadrian's Wall Path
Hadrian's Wall

Boltonfellend
West Hall
Birdoswald Roman Fort
Gilsland Spa
Roman Army Museum
Roman Camps
Haydon Bridge
Hexham

Hethersgill
Kirkcambeck
Triermain
Birdoswald
Gilsland
Thirlwall Castle
Greenhead
Haltwhistle
Melkridge
Henshaw
Bardon Mill
Beltingham
Low Gate
Warden
Corbridge

Walton
Banks
Lanercost Priory
Low Row
Upper Denton
Milecastle
Featherstone Castle
Plenmeller
Williamsrigg
Redburn
Langley
Newbiggin
Dilston

Brampton
Milton
Denton Fell
Rowfoot
Plenmeller Common
Nilston Rigg
Allen Banks (NT)
Hexham Levels 1464
Dipton Wood

Irthington
Tindale
Midgeholme
Fellhouse Fell
Whitfield
Bearsbridge
Catton
Slaley

Newby-on-Eden
Hallbankgate
Lambley
Coanwood
Stonehouse
Ashholme
Dotland
Whitley Chapel

Little Corby
Newby East
Hallhankgate
Forest

0 2 4 6 miles
0 2 4 6 8 10 km

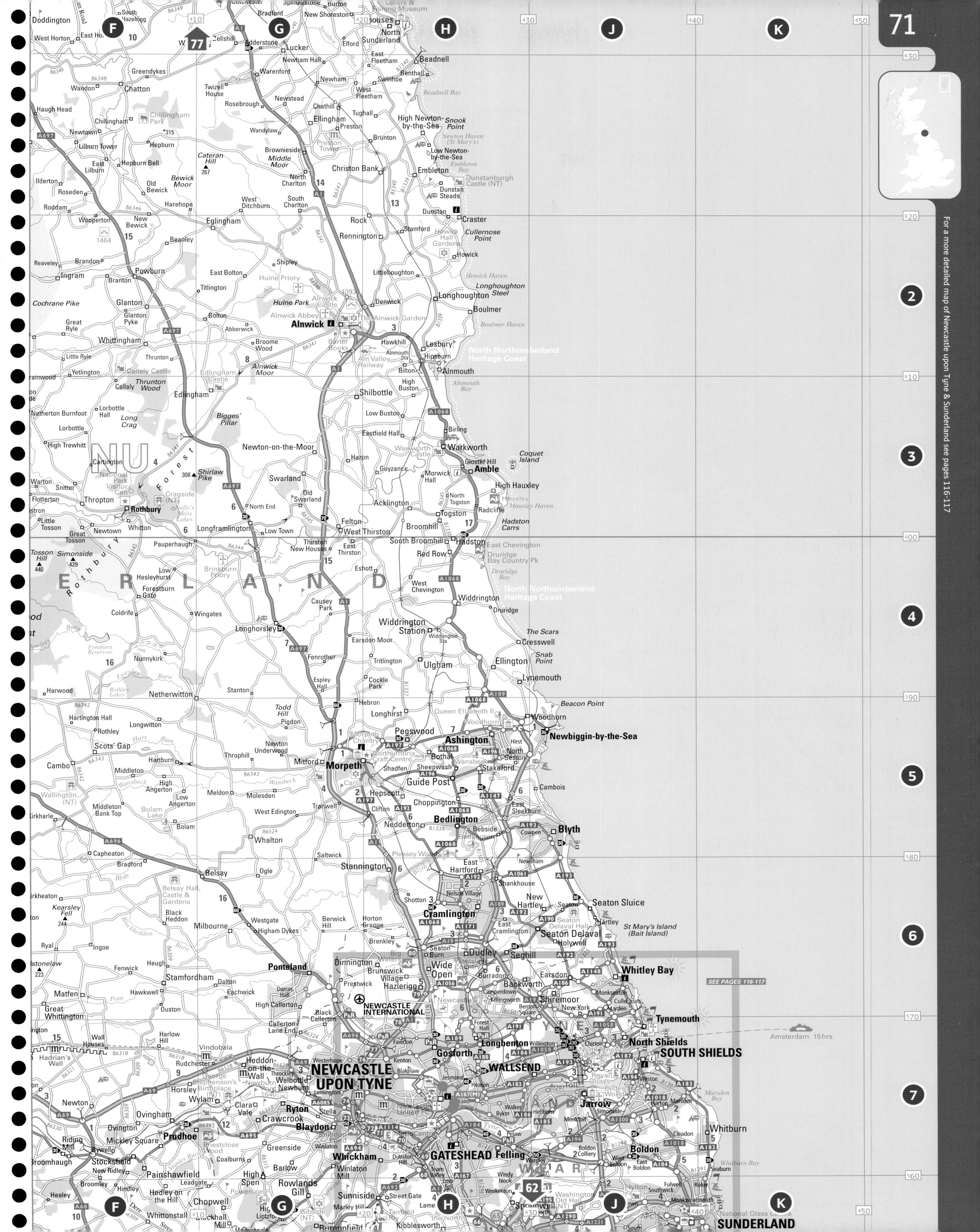

For a more detailed map of Newcastle upon Tyne & Sunderland see pages 116-117

SEE PAGES 116-117

Amsterdam 15hrs

NEWCASTLE UPON TYNE

GATESHEAD

SUNDERLAND

SOUTH SHIELDS

Alnwick

Morpeth

Ashington

Newbiggin-by-the-Sea

Blyth

Cramlington

Bedlington

Whitley Bay

Tynemouth

North Shields

Wallsend

Jarrow

Boldon

Amble

Warkworth

Rothbury

Ponteland

Gosforth

Longbenton

Prudhoe

Blaydon

Whickham

Felling

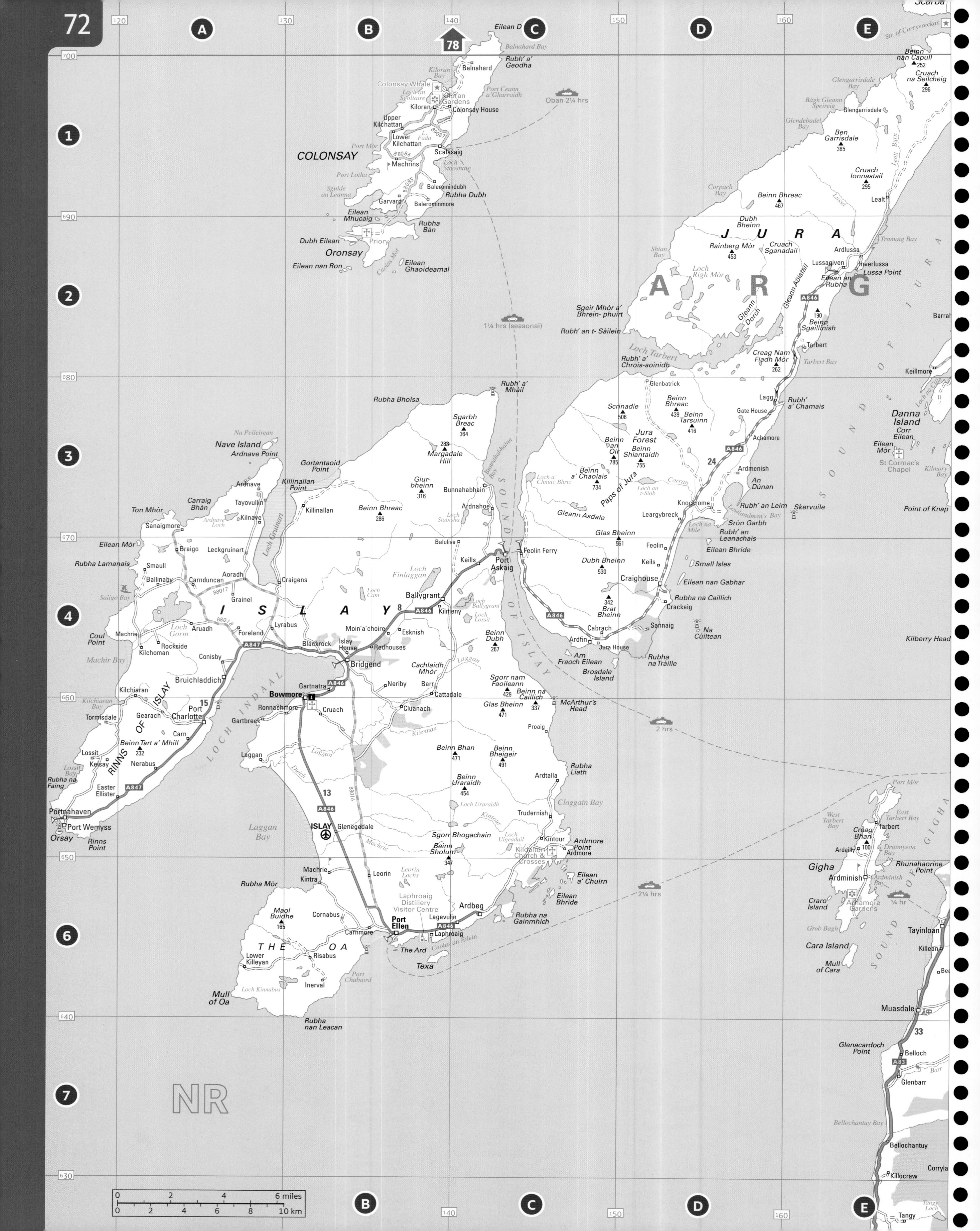
78
Eilean D
Balnahard Bay
Balnahard
Rubh' a' Geodha
Kiloran Bay
Colonsay Whale
Kiloran Gardens
Kiloran
Colonsay House
Port Ceann a' Gharraidh
Oban 2¼ hrs
Upper Kilchattan
Lower Kilchattan
Port Mòr
COLONSAY
Scalasaig
Machrins
Loch Staosnaig
Port Lotha
Balerominduhb
Rubha Dubh
Garvard
Baleromimmore
Eilean Mhucaig
Rubha Bàn
Priory
Dubh Eilean
Oronsay
Eilean nan Ron
Eilean Ghaoideamal
1¼ hrs (seasonal)
Sgeir Mhòr a' Bhrein- phuirt
Rubh' an t- Sàilein
JURA
ARGYLL
Beinn Bhreac
Dubh Bheinn
Rainberg Mòr
Cruach Sganadail
Loch Righ Mòr
Gleann Dorch
Gleann Aoistail
Ardlussa
Lussagiven
Inverlussa
Lussa Point
Eilean an Rubha
Tramaig Bay
Lealt
Cruach Ionnastail
Ben Garrisdale
Glengarrisdale
Glengarrisdale Bay
Beinn nan Capull
Cruach na Seilcheig
Str. of Corryvreckan
Scarba
Beinn Sgaillinish
Tarbert
Tarbert Bay
Creag Nam Fiadh Mòr
Loch Tarbert
Rubh' a' Chrois-aoinidh
Glenbatrick
Scrinadle
Beinn Bhreac
Beinn Tarsuinn
Lagg
Rubh' a' Chamais
Gate House
Achamore
Jura Forest
Beinn an Oir
Beinn Shiantaidh
Beinn a' Chaolais
Paps of Jura
Ardmenish
An Dùnan
Knockrome
Rubh' an Leim
Skervuile
Leargybreck
Gleann Asdale
Glas Bheinn
Sròn Garbh
Rubh' an Leanachais
Feolin
Feolin Ferry
Eilean Bhride
Dubh Bheinn
Keils
Small Isles
Craighouse
Eilean nan Gabhar
Rubha na Caillich
Crackaig
Brat Bheinn
Cabrach
Sannaig
Na Cùiltean
Ardfin
Jura House
Am Fraoch Eilean
Brosdale Island
Rubha na Tràille
SOUND OF JURA
Danna Island
Corr Eilean
Eilean Mòr
St Cormac's Chapel
Keillmore
Point of Knap
Kilberry Head
Barrahormid
Rubh' a' Mhàil
Rubha Bholsa
Sgarbh Breac
Margadale Hill
Giur-bheinn
Bunnahabhain
Ardnahoe
Beinn Bhreac
Loch Staoisha
Balulive
Keills
Port Askaig
SOUND OF ISLAY
Loch Finlaggan
Ballygrant
Kilmeny
Loch Ballygrant
Loch Lossit
Nave Island
Ardnave Point
Na Peileirean
Gortantaoid Point
Killinallan Point
Killinallan
Ardnave
Tayovullin
Kilnave
Carraig Bhàn
Ton Mhòr
Sanaigmore
Ardnave Loch
Loch Gruinart
Eilean Mòr
Braigo
Leckgruinart
Rubha Lamanais
Smaull
Ballinaby
Carnduncan
Aoradh
Craigens
Saligo Bay
Grainel
ISLAY
Loch Gorm
Aruadh
Foreland
Lyrabus
Moin'a'choire
Esknish
Coul Point
Machrie
Rockside
Kilchoman
Blackrock
Islay House
Redhouses
Bridgend
Beinn Dubh
Machir Bay
Conisby
Bruichladdich
Cachlaidh Mhòr
Gartnatra
Neriby
Barr
Sgorr nam Faoileann
Beinn na Caillich
McArthur's Head
Kilchiaran
Kilchiaran Bay
Bowmore
Cattadale
Glas Bheinn
Port Charlotte
Ronnachmore
Cruach
Cluanach
Tormisdale
Gearach
Gartbreck
Proaig
2 hrs
RINNS OF ISLAY
Carn
LOCH INDAAL
Beinn Tart a' Mhill
Beinn Bhan
Beinn Bheigeir
Lossit
Kelsay
Laggan
Rubha Liath
Nerabus
Rubha na Faing
Easter Ellister
Beinn Uraraidh
Ardtalla
Claggain Bay
Portnahaven
Port Wemyss
Orsay
Rinns Point
Laggan Bay
ISLAY
Glenegedale
Trudernish
Sgorr Bhogachain
Beinn Sholum
Kintour
Ardmore Point
Ardmore
Kildalton Church & Crosses
Machrie
Kintra
Leorin
Rubha Mòr
Eilean a' Chùirn
Eilean Bhride
2¼ hrs
Laphroaig Distillery Visitor Centre
Ardbeg
Rubha na Gainmhich
Maol Buidhe
Cornabus
Port Ellen
Lagavulin
Laphroaig
Carnmore
THE OA
The Ard
Texa
Lower Killeyan
Risabus
Inerval
Port Chubaird
Loch Kinnabus
Mull of Oa
Rubha nan Leacan
Port Mòr
Creag Bhan
Tarbert
Ardailly
Rhunahaorine Point
Gigha
Ardminish
Achamore Gardens
¼ hr
Craro Island
Cara Island
Mull of Cara
SOUND OF GIGHA
Tayinloan
Killean
Muasdale
Glenacardoch Point
Belloch
Glenbarr
Bellochantuy Bay
Bellochantuy
Killocraw
Tangy
NR
6 miles
10 km

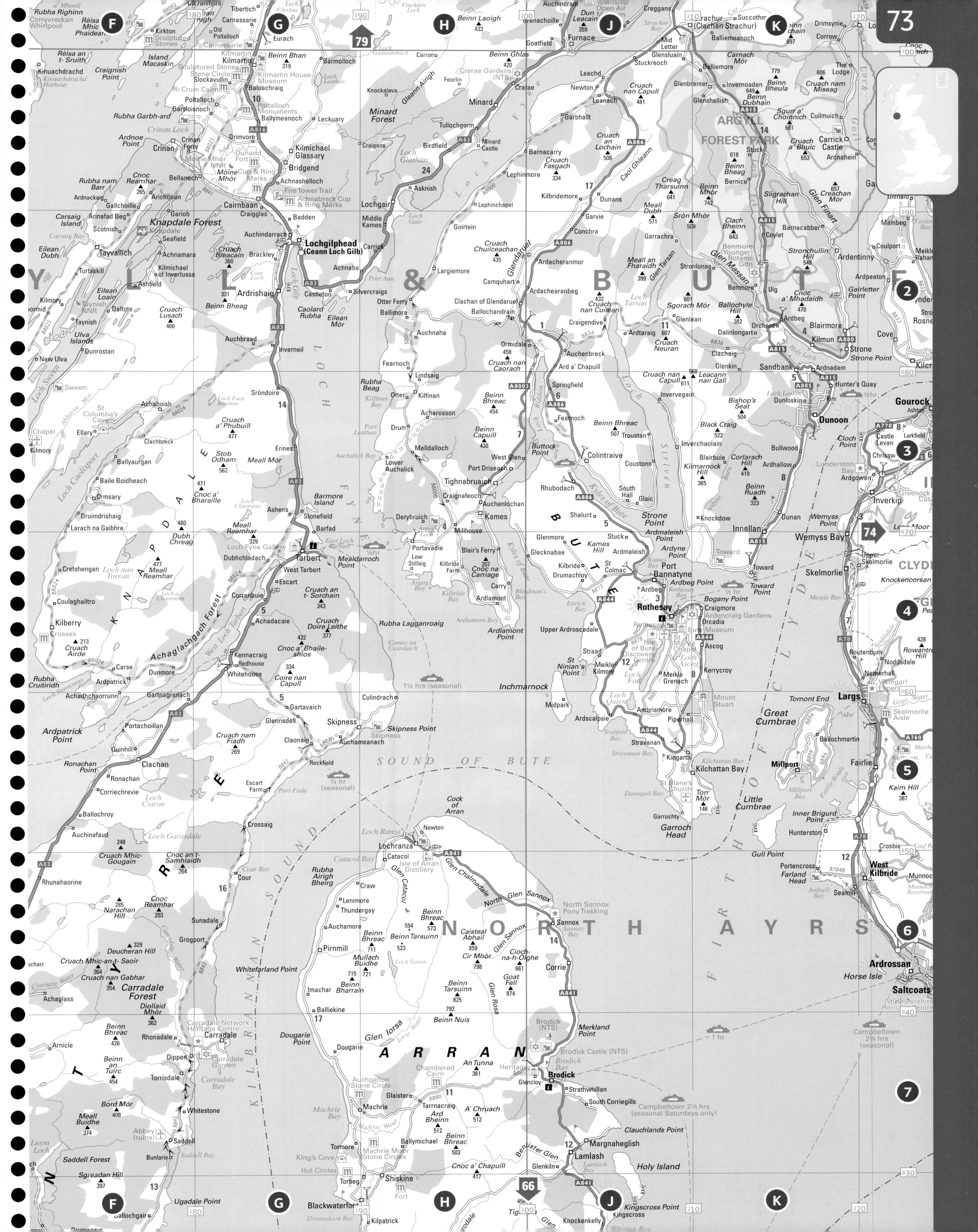

79
Rubha Righinn
Corryvreckan Whirlpool
Rèisa Mhic Phaidean
Kirkton
Sculptured Stones
Carnassarie
Tibertich
Eurach
Kilmartin
Beinn Bhan 319
Kilmartin House Museum
Rèisa an t- Sruith
Island Macaskin
Kinuachdrachd
Craignish Point
Kinuachdrachd Harbour
Sculptured Stones
Stone Circle
Ri Cruin Cairn
Slockavullin
Baluachraig
Poltalloch
Poltalloch Monuments
Ballymeanoch
Barsloisnoch
Rubha Garbh-ard
Crinan Loch
Ardnoe Point
Crinan
Crinan Ferry
Drimvore
Dunadd Fort
Moine Mhòr NNR
Kilmichael Glassary
Bridgend
Cup & Ring Marks
Achnashelloch
Fire Tower Trail
Achnabreck Cup & Ring Marks
Bellanoch
Cnoc Reamhar 265
Rubha nam Barr
Ardnackaig
Gallchoille
Arichonan
Gariob
Cairnbaan
Craiglass
Badden
Knapdale Forest
Carsaig Island
Arinafad Beg
Scotnish
Seafield
Auchindarrach
Lochgilphead (Ceann Loch Gilb)
Eilean Dubh
Taynvallich
Achnamara
Cruach Breacain 360
Brackley
Carrick
Achnaba
Turbiskill
Kilmichael of Inverlussa
Ashfield
Eilean Loain
Ardrishaig
Castleton
Silvercraigs
Kilmory
Taynish NNR
Daltote
Beinn Bheag
Taynish
Cruach Lusach 466
Caolard Rubha
Eilean Mòr
Ulva Islands
Dunrostan
Auchbraad
Inverneil
New Ulva
Loch Sween
Sween
Sròndoire
Achahoish
Cruach a' Phubuill 477
St Columba's Cave
Chapel
Ellary
Kilmory
Clachbreck
Ballyaurgan
Stob Odham 562
Meall Mòr
Erines
Loch Caolisport
Baile Boidheach
Cnoc a' Bharaille 471
Ormsary
Druimdrishaig
Dubh Chreag 480
Larach na Gaibhre
Meall Reamhar 329
Loch Fyne Gallery
Dubhchladach
Barmore Island
Ashens
Stonefield
Barfad
Tarbert
West Tarbert
Mealdarroch Point
Cretshengan
Loch nan Torran
Meall Reamhar 477
Escart
Cruach an t- Sorchain 343
Coulaghailtro
Corranbuie
Achadacaie
Cruach Doire Leithe 377
Kilberry
Crosses
Cruach Airde 213
Achaglachgach Forest
Cnoc a' Bhaile-shios 422
Kennacraig
Redhouse
Whitehouse
Coire nan Capull 334
Carse
Dunmore
Rubha Cruitiridh
Ardpatrick
Achadhchaorrunn
Gartnagrenach
Gartavaich
Glenrisdell
Skipness
Skipness Point
Claonaig
Auchameanach
Rockfield
Culindrach
Ardpatrick Point
Portachoillan
Quinhill
Cruach nam Fiadh 269
Ronachan Point
Clachan
Ronachan
Corriechrevie
Loch Ciaran
Escart Farm
Port Fada
Ballochroy
Crossaig
Auchinafaud
Loch Garasdale
Crauch Mhic-Gougain 248
Cnoc an t-Samhlaidh 264
Rhunahaorine
Cour
Cour Bay
Narachan Hill 285
Cnoc Reamhar 203
Sunadale
Grogport
Deucheran Hill 329
Cruach Mhic-an-t-Saoir 364
Cruach nan Gabhar 354
Carradale Forest
Achaglass
Diollaid Mhòr 362
Carradale Network Heritage Centre
Carradale
Rhonadale
Beinn Bhreac 426
Arnicle
Dippen
Carradale Garden
Beinn an Tuirc 454
Torrisdale
Carradale Bay
Bord Mòr 408
Whitestone
Meall Buidhe 374
Abbey (ruins)
Saddell
Bunlarie
Saddell Bay
Saddell Forest
Sgreadan Hill 397
Ugadale Point
Ballochgair
Kilbrannan Sound
Minard Forest
Gleann Airigh
Knockalava
Leckuary
Craigans
Birdfield
Loch Glashan
Asknish
Lochgair
Middle Kames
Carron
Beinn Ghlas 420
Crarae Gardens (NTS)
Crarae
Feorlin
Minard
Tullochgorm
Minard Castle
Goatfield
Furnace
Auchindrain
Dun Leacainn 359
Beinn Laoigh 433
Loch Fyne
Lephinmore
Lephinchapel
Goirtein
Largiemore
Otter Ferry
Ballimore
Auchnaha
Fearnoch
Lindsaig
Kilfinan
Kilfinan Bay
Rubha Beag
Otter
Acharosson
Drum
Port Leathan
Melldalloch
Lower Auchalick
Auchalick Bay
Tighnabruaich
Craignafeoch
Auchenlochan
Kames
Derybruich
Millhouse
Portavadie
Low Stillaig
Kilbride Farm
Blair's Ferry
Cnoc na Carraige 207
Carry
Ardlamont
Ardlamont Bay
Ardlamont Point
Rubha Lagganroaig
Camas na Ceardaich
1½ hrs (seasonal)
Inchmarnock
Midpark
Sound of Bute
½ hr (seasonal)
Cock of Arran
Newton
Loch Ranza
Lochranza
Catacol
Catacol Bay
Isle of Arran Distillery
Glen Catacol
Glen Chalmadale
Rubha Airigh Bheirg
Craw
Lenimore
Thundergay
Auchamore
Pirnmill
Beinn Bhreac 711
Beinn Tarsuinn
Beinn Bhreac 573
Caisteal Abhail 859
Cir Mhòr 798
North Glen Sannox
North Sannox Pony Trekking
Sannox
Cioch-na-h-Oighe 661
Mullach Buidhe 721
Beinn Bharrain
Whitefarland Point
Imachar
Goat Fell 874
Corrie
Glen Rosa
Beinn Tarsuinn 825
Beinn Nuis 792
Balliekine
Dougarie Point
Glen Iorsa
Dougarie
Brodick (NTS)
Merkland Point
Brodick Castle (NTS)
Brodick Bay
A R R A N
An Tunna 361
Chambered Cairn
Auchgallon Stone Circle
Heritage Museum
Glencloy
Brodick
Strathwhillan
South Corriegills
Glaister
Machrie
Machrie Bay
Tarrnacraig
Ard Bheinn 512
A' Chruach 512
Beinn Bhreac 503
Ballymichael
Tormore
Machrie Moor Stone Circles
King's Cave
Hut Circles
Torbeg
Shiskine
Fort
Cnoc a' Chapuill 417
Blackwaterfoot
Kilpatrick
Benlister Glen
Glenkiln
Lamlash
Lamlash Bay
Margnaheglish
Clauchlands Point
Holy Island
Kingscross Point
Kingscross
Knockenkelly
Whiting Bay
Campbeltown 2½ hrs (seasonal Saturdays only)
66
Strachur (Clachan Strachur)
Creggans
Succoth
Drimsynie
Corrow
Mid Letter
Balliemeanoch
Glensluain
Stuckreoch
Carnach Mòr
Balliemore
Leachd
Newton
Leanach
Cruach nan Capull 481
Glenbranter
Invernoaden
Beinn Dubhain 649
Beinn Bheula 779
The Lodge
Cruach nam Miseag 606
Glenshellish
Garbhallt
Argyll Forest Park
Sgurr a' Choinnich 661
Cuilmuich
Cruach an Lochain 508
Barnacarry
Cruach Fasgach 334
Caol Ghleann
Stuck
Beinn Bheag 618
Bernice
Cruach a' Bhuic 653
Carrick Castle
Ardnahein
Creag Tharsuinn 641
Beinn Mhòr 742
Creachan Mòr 657
Sligrachan Hill
Glen Finart
Kilbridemore
Dunans
Meall Dubh 571
Sròn Mhòr 509
Garvie
Conchra
Clach Bheinn 643
Coylet
Barnacabber
Mambeg
Garrachra
Benmore (Younger) Botanic Gdn
Stronchullin Hill 548
Ardentinny
Coulport
Cruach Chuilceachan 435
Glendaruel
Ardacheranmor
Meall an Fharaidh 399
Glen Tarsan
Stronlonag
Glen Massan
Ardpeaton
Camquhart
Ardacheranbeg
A R G Y L L & B U T E
Cruach nan Cuilean 432
Loch Tarsan
Benmore
Uig
Sgorach Mòr 601
Ballochyle Hill 382
Cnoc a' Mhadaidh 470
Gairletter Point
Clachan of Glendaruel
Ballochandrain
Craigendive
Glenlean
Ardbeg
Blairmore
Ardtaraig
Cruach Neuran 607
Dalinlongart
Orchard
Kilmun
Cove
Strone
Strone Point
Ormidale
Cruach nan Caorach 458
Auchenbreck
Ard a' Chapuill
Clachaig
Glenkin
Sandbank
Ardnadam
Hunter's Quay
Cruach nan Capull 611
Leacann nan Gall 568
Invervegain
Kirn
½ hr
Gourock
Ashton
Dunloskin
Dunoon
Springfield
Beinn Bhreac 454
Bishop's Seat 504
Black Craig 522
Feamoch
Beinn Bhreac 507
Troustan
Beinn Capuill 438
Buttock Point
Colintraive
Inverchaolain
Cloch Point
Castle Levan
Larkfield
West Glen
Port Driseach
Coustonn
Blairbuie
Kilmarnock Hill 365
Corlarach Hill 419
Bullwood
Ardhallow
Lunderston Bay
Ardgowan
Inverkip
Rhubodach
Kyles of Bute
South Hall
Glaic
Beinn Ruadh 323
Shalunt
Strone Point
Knockdow
Dunan
Wemyss Point
Innellan
Wemyss Bay
74
Lunderston Bay
Glenmore
Stuck
Kames Hill
Ardmaleish Point
Ardyne Point
Ardmaleish
Toward
Glecknabae
Kilbride
Drumachloy
St Colmac
Port Bannatyne
Ardbeg Point
Rothesay Bay
Toward Point
Skelmorlie
Upper Skelmorlie
Knockencorsan
Ettrick Bay
Ardbeg
Rothesay
Bogany Point
Craigmore
Ardencraig Gardens
Meigle Bay
Upper Ardroscadale
Bute Museum
Ascog
Isle of Bute Discovery Centre
St Mary's Chapel
Rowantree Hill 428
Straad
St Ninian's Point
Meikle Kilmory
Loch Fad
Loch Ascog
Kerrycroy
Routenburn
Noddsdale
Meikle Grenach
Netherhall
Vikingar! Experience
Loch Quien
Mount Stuart
Tomont End
Largs
Ardscalpsie
Ambrisbeg
Ambrismore
Great Cumbrae
Skelmorlie Aisle
Piperhall
Scalpsie Bay
Stravanan
Stravanan Bay
Ballochmartin
Kingarth
Kilchattan Bay
Millport
Fairlie
St Blane's Church
Dunagoil Bay
Torr Mòr 146
Millport Bay
Kaim Hill 387
Garrochty
Little Cumbrae
Inner Brigurd Point
Garroch Head
Firth of Clyde
Hunterston
Crosbie
Gull Point
Portencross
Farland Head
West Kilbride
Munnoch Reservoir
Seamill
Ardneil Bay
N O R T H A Y R S
Ardrossan
Horse Isle
Saltcoats
North Ayrshire Heritage Centre
1 hr
Campbeltown 2½ hrs (seasonal)
Clyde

For a more detailed map of Glasgow see pages 118-119

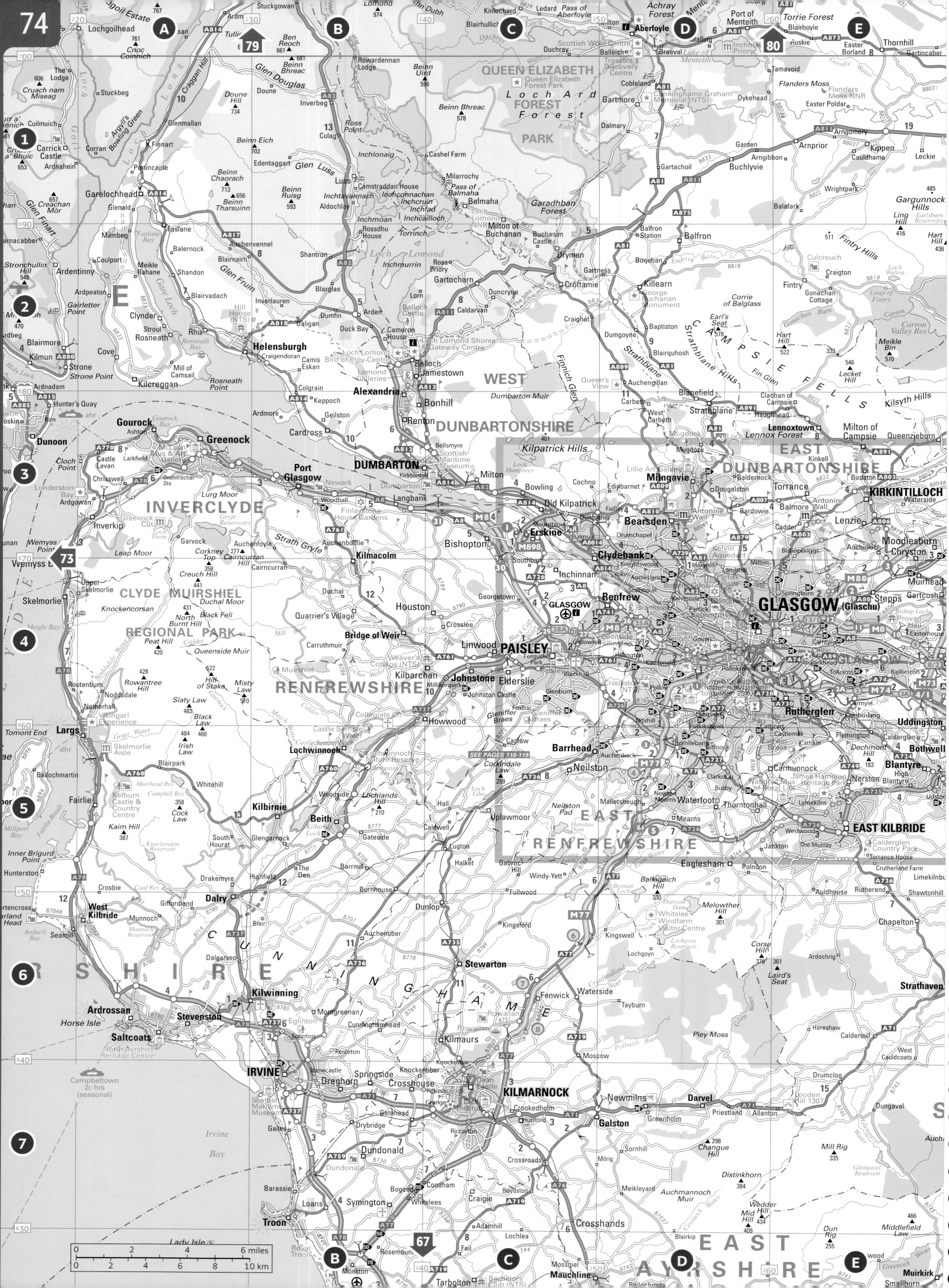

Map of the Glasgow and Clyde region

Grid references: A B C D E (top and bottom); 1-7 (left side)

Major places and features (reading roughly across the map):

Lochgoilhead, Stuckgowan, Lochgoil Estate, Ben Reoch, Aberfoyle, Achray Forest, Port of Menteith, Torrie Forest, Blairhoyle, Easter Borland, Thornhill, Gartincaber

Carrick Castle, Ardnahein, Portincaple, Garelochhead, Glen Douglas, Rowardennan Lodge, QUEEN ELIZABETH FOREST PARK, Queen Elizabeth Forest Park, Trossachs Discovery Centre, Braeval, Ruskie, Easter Poldar, Arngomery, Kippen, Leckie, Cauldhame

Cuilmuich, Argyll's Bowling Green, Cruach a' Bhuic, Glenmallan, Doune Hill, Inverbeg, Loch Ard, Gartmore, Flanders Moss, Flanders Moss NNR, Wrightpark, Gargunnock Hills, Earlsburn Reservoirs

Stronchullin Hill, Ardentinny, Coulport, Meikle Rahane, Shandon, Blairmairn, Shantron, Inchlonaig, Cashel Farm, Dalmary, Balfron Station, Balfron, Craigton, Culcreuch, Fintry, Fintry Hills, Hart Hill, Meikle Bin, Lecket Hill, Kilsyth Hills

Blairmore, Cove, Kilmun, Strone, Ardnadam, Helensburgh, Craigendoran, Rhu, Faslane, Shandon, Luss, Camstraddan House, Inchtavannach, Inchconnachan, Inchcruin, Inchmoan, Inchfad, Inchcailloch, Balmaha, Milton of Buchanan, Buchanan Castle, Drymen, Gartness, Killearn, George Buchanan Monument, Strathblane, Blairhosh, Blanefield, Clachan of Campsie, CAMPSIE FELLS, Fin Glen, Carron Valley Res

Dunoon, Hunter's Quay, Gourock, Gourock Bay, Ashton, Greenock, Port Glasgow, Newark, Bowling, Old Kilpatrick, Kilpatrick Hills, Milngavie, Baldernock, Torrance, Lennoxtown, Milton of Campsie, Queenzieburn, EAST DUNBARTONSHIRE, KIRKINTILLOCH, Lenzie, Moodiesburn, Chryston

Cloch Point, Inverkip, INVERCLYDE, Ardgowan, Houston, Georgetown, Bishopton, Erskine, Clydebank, Bearsden, Bardowie, Antonine Wall, Bishopbriggs, Muirhead, Gartcosh, Stepps, Easterhouse

Wemyss Bay, Skelmorlie, CLYDE MUIRSHIEL REGIONAL PARK, Kilmacolm, Bridge of Weir, Quarrier's Village, Linwood, PAISLEY, Renfrew, GLASGOW (Glaschu), Rutherglen, Uddingston, Bothwell

Largs, Fairlie, Kilbirnie, Beith, Lochwinnoch, Howwood, Johnstone, Elderslie, Barrhead, Neilston, EAST RENFREWSHIRE, Waterfoot, Thorntonhall, Newton Mearns, East Kilbride, The Murray, Jackton, Eaglesham, Calderglen Country Park

West Kilbride, Dalry, Dunlop, Stewarton, Fenwick, Waterside, Strathaven

Ardrossan, Saltcoats, Stevenston, Kilwinning, Montgreenan, Cunninghamhead, Kilmaurs, Moscow

Irvine, Dreghorn, Springside, Crosshouse, KILMARNOCK, Newmilns, Darvel, Galston, Mauchline

Troon, Loans, Symington, Barassie, Dundonald, Craigie, Crosshands, EAST AYRSHIRE

Scale: 0 2 4 6 miles / 0 2 4 6 8 10 km

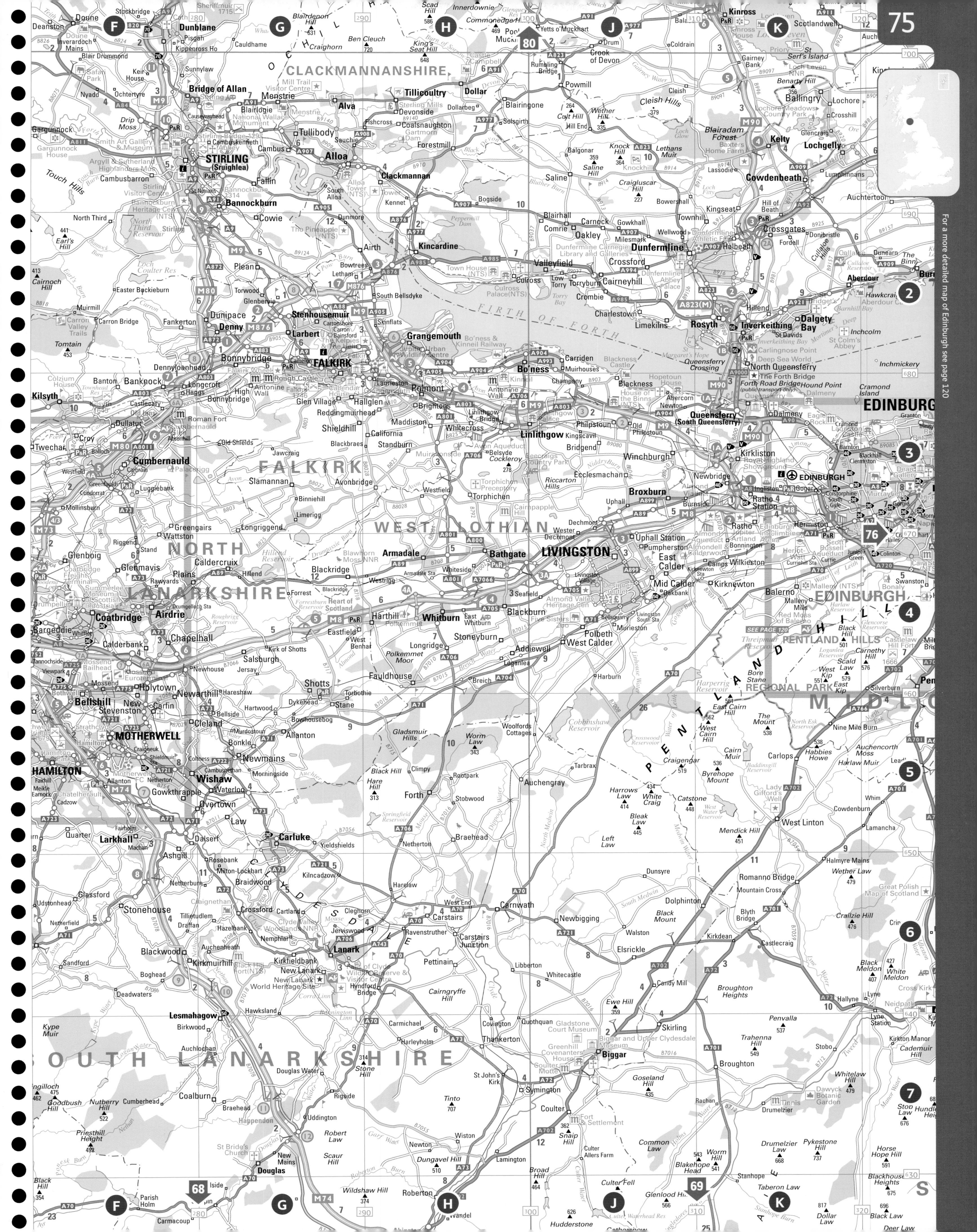
CLACKMANNANSHIRE
FALKIRK
WEST LOTHIAN
NORTH LANARKSHIRE
SOUTH LANARKSHIRE
PENTLAND HILLS REGIONAL PARK
FIRTH OF FORTH
EDINBURGH
LIVINGSTON
STIRLING
(Sruighlea)
MOTHERWELL
HAMILTON
FALKIRK
Dunblane
Bridge of Allan
Menstrie
Tillicoultry
Dollar
Alva
Tullibody
Alloa
Clackmannan
Bannockburn
Kincardine
Dunfermline
Rosyth
Inverkeithing
Dalgety Bay
Aberdour
Cowdenbeath
Kelty
Lochgelly
Kinross
Grangemouth
Stenhousemuir
Larbert
Denny
Bonnybridge
Bo'ness
Linlithgow
Polmont
Queensferry
(South Queensferry)
North Queensferry
Kirkliston
Broxburn
Winchburgh
Bathgate
Armadale
Whitburn
Blackburn
West Calder
Mid Calder
East Calder
Balerno
Cumbernauld
Kilsyth
Airdrie
Coatbridge
Bellshill
Wishaw
Shotts
Carluke
Larkhall
Lanark
Lesmahagow
Biggar
Carnwath
Douglas
Symington
West Linton
Carlops
Forth
Stonehouse
Blackwood
Strathaven
Newarthill
Harthill
Fauldhouse
Stoneyburn
Addiewell
Uphall Station
Ratho
Newbridge
Pumpherston
Kirknewton
Dechmont
Edinburgh Airport
The Forth Bridge
Queensferry Crossing
Cramond Island
Inchcolm
Inchmickery
Ochil Hills
Cleish Hills
Touch Hills
Black Hill
Tinto
Culter Fell
Dollar Law
Broughton
Skirling
Dolphinton
Elsrickle
Carstairs
Carstairs Junction
Newbigging
New Lanark
World Heritage Site
Falkirk Wheel
The Kelpies & The Helix
Rough Castle
Antonine Wall
Stirling Castle
National Wallace Monument
Culross Palace (NTS)
Hopetoun House
Blackness Castle
House of the Binns (NTS)
Linlithgow Palace
Almond Valley Heritage Cen
Five Sisters
Heart of Scotland
Polkemmet Country Park
Clyde Valley Woodlands NNR
Gladstone Court Museum
Biggar and Upper Clydesdale Museum
Greenhill Covenanters House
Dawyck Botanic Garden
Great Polish Map of Scotland
Cobbinshaw Reservoir
Harperrig Reservoir
Threipmuir Reservoir
Gladsmuir Hills
Worm Law
Harrows Law
White Craig
Craigengar
East Cairn Hill
West Cairn Hill
Carnethy Hill
Scald Law
The Mount
Mendick Hill
Broughton Heights
Cairngryffe Hill
SEE PAGE 120
80
76
68
69
F G H J K
2 3 4 5 6 7

For a more detailed map of Edinburgh see page 120

For a more detailed map of Edinburgh see page 120

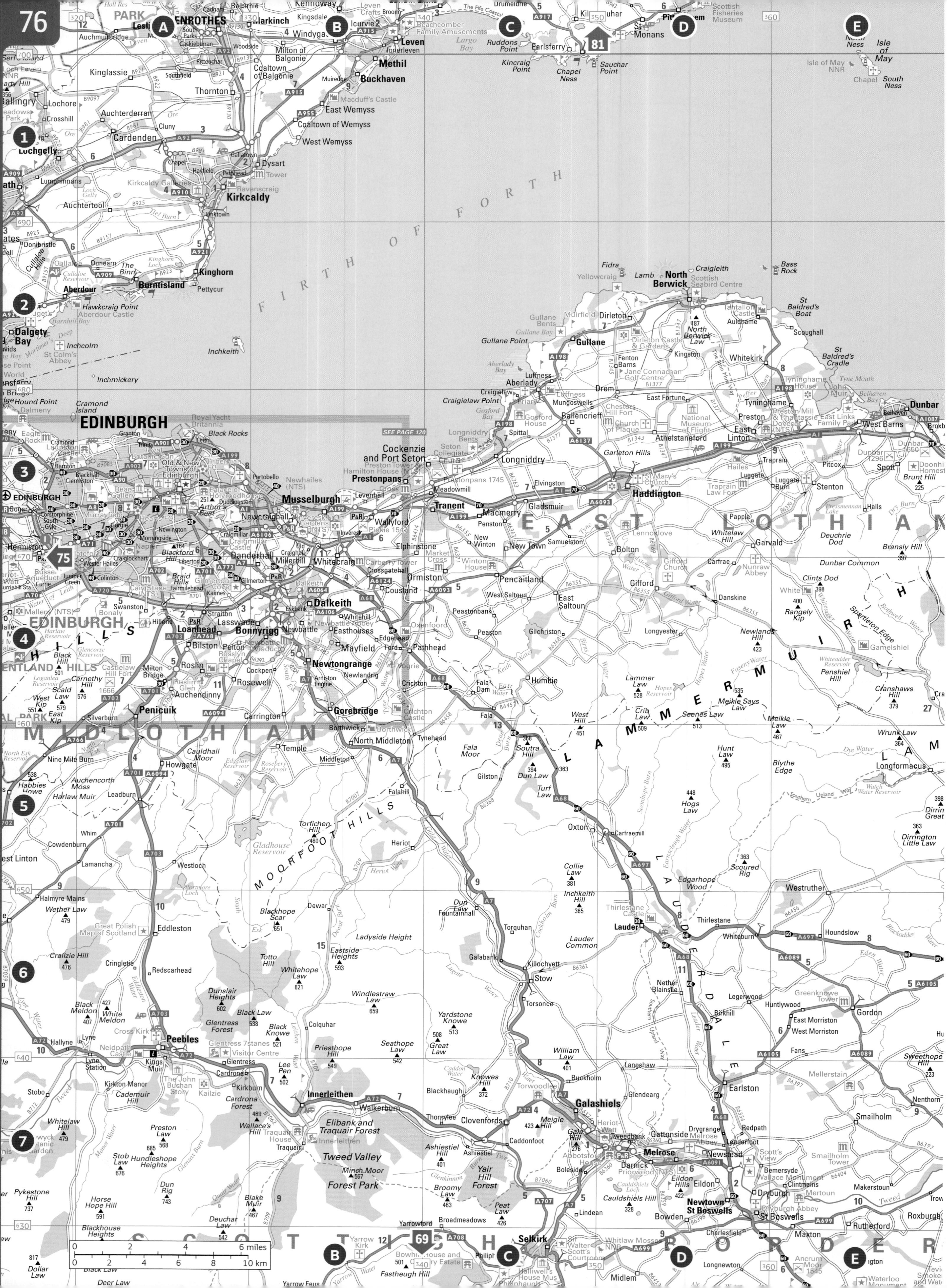

FIRTH OF FORTH

EAST LOTHIAN

MIDLOTHIAN

LAMMERMUIR HILLS

LAUDERDALE

SCOTTISH BORDER

PENTLAND HILLS

MOORFOOT HILLS

Grid references: A B C D E · 1 2 3 4 5 6 7

81 · 75 · 69 · SEE PAGE 120

Isle of May · North Ness · South Ness · Chapel Ness · Scottish Fisheries Museum

Pittenweem · St Monans · Earlsferry · Elie · Chapel Point · Kincraig Point · Ruddons Point · Largo Bay · Sauchar Point

GLENROTHES · Leslie · Markinch · Kennoway · Windygates · Leven · Methil · Buckhaven · East Wemyss · Coaltown of Wemyss · West Wemyss · Macduff's Castle

Auchmuirbridge · Kinglassie · Cardenden · Auchterderran · Cluny · Thornton · Coaltown of Balgonie · Milton of Balgonie · Muiredge

Lochgelly · Lumphinnans · Auchtertool · Kirkcaldy · Kirkcaldy Galleries · Linktown · Dysart · Ravenscraig Castle · Pathhead

Aberdour · Burntisland · Kinghorn · Pettycur · The Binn · Aberdour Castle · Hawkcraig Point

Dalgety Bay · Inchcolm · St Colm's Abbey · Inchmickery · Inchkeith · Cramond Island · Dalmeny · Hound Point

EDINBURGH · Granton · Leith · Portobello · Black Rocks · Royal Yacht Britannia · Musselburgh · Newcraighall · Newhailes (NTS)

Cockenzie and Port Seton · Prestonpans · Preston Tower · Hamilton House (NTS) · Prestonpans 1745 · Seton Collegiate Church · Longniddry · Aberlady · Gullane · Dirleton · North Berwick · North Berwick Law · Tantallon Castle · Bass Rock · Craigleith · Lamb · Fidra · Yellowcraig · Gullane Bents · Gullane Bay · Gullane Point · Aberlady Bay · Luffness · Craigielaw Point · Muirfield · Dirleton Castle & Gardens · Scottish Seabird Centre · Auldhame · Scoughall · St Baldred's Boat · St Baldred's Cradle · Whitekirk

Drem · East Fortune · National Museum of Flight · Athelstaneford · Haddington · East Linton · Preston Mill & Phantassie Doocot (NTS) · Tyninghame · Tyninghame House · Dunbar · West Barns · Belhaven · Belhaven Bay · Tyne Mouth · Broxburn

Tranent · Macmerry · Pencaitland · Ormiston · Gladsmuir · Samuelston · Bolton · Lennoxlove · Gifford · Gifford Water · Nunraw Abbey · Garvald · Whitelaw Hill · Deuchrie Dod · Bransly Hill · Dunbar Common · Clints Dod · Gamelshiel · Penshiel Hill · Whiteadder Reservoir

Dalkeith · Bonnyrigg · Lasswade · Loanhead · Bilston · Roslin · Rosslyn Chapel · Newbattle · Mayfield · Newtongrange · Rosewell · Gorebridge · Ford · Pathhead · Crichton Castle · Fala · Humbie · Gilchriston · East Saltoun · West Saltoun

Penicuik · Carrington · Temple · Middleton · North Middleton · Tynehead · Fala Moor · Soutra Hill · Oxton · Carfraemill · Fala Dam · Heriot · Stow · Lauder · Thirlestane Castle · Westruther · Houndslow

Peebles · Innerleithen · Walkerburn · Clovenfords · Galashiels · Stow · Torsonce · Killochyett · Fountainhall · Collie Law · Edgarhope Wood · Lauder Common · Legerwood · Birkhill · Gordon · Earlston · Greenknowe Tower

Eddleston · Cringletie · Redscarhead · Glentress Forest · Dunslair Heights · Black Law · Windlestraw Law · Priesthope Hill · Seathope Law · Yardstone Knowe · William Law · Langshaw · Buckholm · Torwoodlee · East Morriston · West Morriston · Fans · Mellerstain

Melrose · Galashiels · Tweedbank · Abbotsford · Darnick · Newstead · Gattonside · Bemersyde · Wallace Monument · Scott's View · Dryburgh Abbey · Eildon Hills · Newtown St Boswells · St Boswells · Maxton · Smailholm · Smailholm Tower · Redpath · Makerstoun · Roxburgh

Selkirk · Yarrow · Ettrick · Philiphaugh · Bowhill House · Ashiestiel · Yair Hill Forest · Cardrona Forest · Glentress Forest · Elibank and Traquair Forest · Traquair House · Tweed Valley Forest Park · Minch Moor · Broadmeadows · Yarrowford · Ancrum

Halmyre Mains · Wether Law · Lamancha · Westloch · West Linton · Cowdenburn · Whim · Leadburn · Auchencorth Moss · Harlaw Muir · Habbies Howe · Nine Mile Burn · Silverburn · Carlops · Howgate

Crailzie Hill · Black Meldon · White Meldon · Cross Kirk · Neidpath Castle · Kings Muir · Kirkton Manor · Cademuir Hill · The John Buchan Story · Kailzie · Lyne Station · Stobo · Hallyne · Lyne · Kirkburn

Stob Law · Hundleshope Heights · Dun Rig · Blake Muir · Deuchar Law · Blackhope Heights · Horse Hope Hill · Pykestone Hill · Dollar Law · Black Law · Deer Law

Scale: 0 2 4 6 miles / 0 2 4 6 8 10 km

Heights noted: 380 · 225 · 397 · 400 · 398 · 423 · 535 · 509 · 513 · 495 · 451 · 394 · 363 · 365 · 381 · 448 · 479 · 476 · 602 · 521 · 538 · 621 · 659 · 542 · 549 · 513 · 401 · 372 · 479 · 568 · 685 · 676 · 743 · 591 · 542 · 463 · 737 · 817 · 640 · 650 · 700 · 690 · 576 · 501

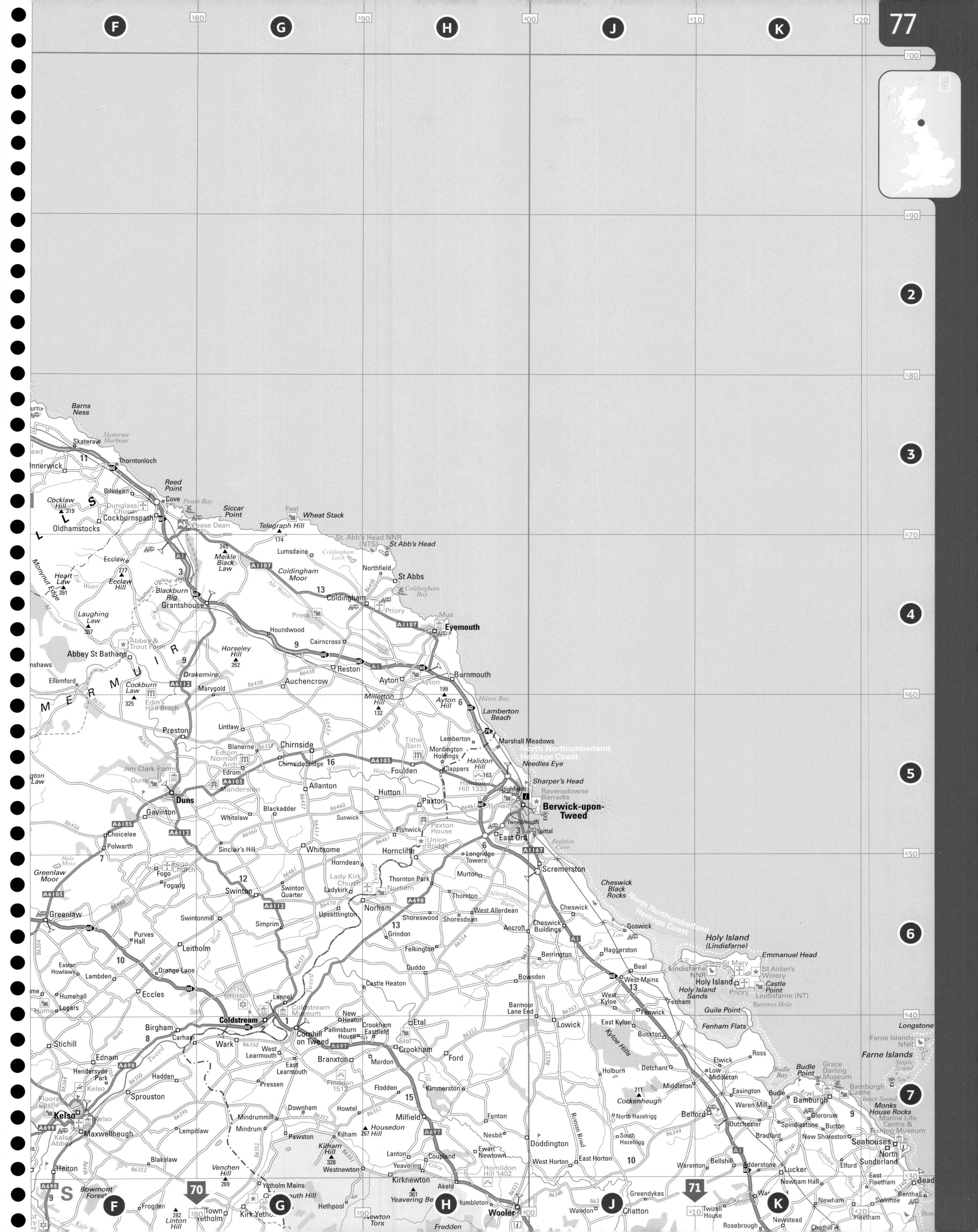

F
G
H
J
K
380
390
400
410
420
700
690
680
670
660
650
640
630
2
3
4
5
6
7
Barns Ness
Skateraw
Skateraw Harbour
Innerwick
Thorntonloch
Reed Point
Bilsdean
Cove
Pease Bay
Siccar Point
Cocklaw Hill
319
Dunglass Church
Cockburnspath
Oldhamstocks
Pease Dean
Fast
Wheat Stack
Telegraph Hill
174
St. Abb's Head NNR (NTS)
St Abb's Head
Lumsdaine
Coldingham Loch
Meikle Black Law
245
A1107
Coldingham Moor
Northfield
St Abbs
Coldingham Bay
Ecclaw
277
Ecclaw Hill
Heart Law
391
Monynut Edge
Blackburn Rig
Grantshouse
Coldingham
13
Priory
Press
Mus
Eyemouth
Laughing Law
307
Houndwood
Abbey & Trout Farm
Abbey St Bathans
Horseley Hill
262
Cairncross
9
Reston
Burnmouth
Ayton
Drakemire
Auchencrow
Ellemford
Cockburn Law
325
Edin's Hall Broch
Marygold
Millerton Hill
132
199
Ayton Hill
Hilton Bay
Lamberton Beach
Lintlaw
Preston
Blanerne
Chirnside
Tithe Barn
Lamberton
Marshall Meadows
Mordington Holdings
North Northumberland Heritage Coast
Needles Eye
Edrom Norman Arch
Chirnsidebridge
16
A6105
Foulden
Halidon Hill
Clappers
163
Jim Clark Room
Duns
Edrom
Allanton
Halidon Hill 1333
Sharper's Head
Manderston
Hutton
Highfields
Ravensdowne Barracks
Berwick-upon-Tweed
Duns
Gavinton
Blackadder
Paxton
Berwick
Whitelaw
Sunwick
Tweedmouth
Spittal
Choicelee
Fishwick
Paxton House
East Ord
Polwarth
Sinclair's Hill
Whitsome
Union Bridge
Horncliffe
Redshin Cove
Longridge Towers
Hule Moss
Fogo Church
Horndean
Scremerston
Greenlaw Moor
Fogo
Fogorig
Lady Kirk Church
Ladykirk
Thornton Park
Murton
Swinton
Swinton Quarter
12
Norham
Cheswick Black Rocks
Thornton
West Allerdean
Cheswick
Greenlaw
Swintonmill
Upsettlington
Simprim
Shoreswood
Shoresdean
Ancroft
Cheswick Buildings
Goswick
Purves Hall
Grindon
Holy Island (Lindisfarne)
Leitholm
Felkington
Berrington
Haggerston
Emmanuel Head
Easter Howlaws
Orange Lane
Duddo
Lindisfarne NNR
St Mary
St Aidan's Winery
Lambden
10
Bowsden
Beal
Holy Island
Castle Point
Castle Heaton
West Mains
Humehall
Eccles
The Hirsel
Lindisfarne (NT)
Holy Island Sands
Priory
Legars
Lennel
Coldstream Museum
West Kyloe
Fenham
Burrows Hole
Hume
Barmoor Lane End
Fenwick
Guile Point
Coldstream
New Heaton
Etal
Lowick
Birgham
Crookham Eastfield
East Kyloe
Fenham Flats
Longstone
Carham
Pallinsburn House
Cornhill on Tweed
Buckton
Farne Islands NNR
Stichill
Wark
West Learmouth
Crookham
Kyloe Hills
Ross
Farne Islands
Ednam
Ford
Staple Sound
Hendersyde Park
East Learmouth
Branxton
Mardon
Holburn
Detchant
Elwick
Budle Bay
Budle Point
Grace Darling Museum
Hadden
Pressen
Flodden 1513
Kimmerston
Low Middleton
Kelso
Flodden
211
Middleton
Easington
Bamburgh Castle
Sprouston
15
Cockenheugh
Budle
Bamburgh
Inner Sound
Floors Castle
Kelso
Downham
Howtel
Milfield
Waren Mill
Monks House Rocks
Kelso
Mindrummill
Fenton
North Hazelrigg
Belford
Glororum
Marine Life Centre & Fishing Museum
Maxwellheugh
Mindrum
Housedon Hill
267
Nesbit
Outchester
Spindlestone
Burton
Kelso Abbey
Lempitlaw
Pawston
Kilham
Doddington
South Hazelrigg
Bradford
New Shoreston
Seahouses
Kilham Hill
328
Ewart Newtown
Roman Road
North Sunderland
Heiton
Blakelaw
Venchen Hill
269
Westnewton
Lanton
Coupland
West Horton
East Horton
Warenton
Bellshill
Adderstone
Lucker
Elford
Yeavering
Glen
Humildon Hill 1402
East Fleetham
Fleetham
Kirknewton
Newham Hall
Bowmont Forest
70
Yetholm Mains
Akeld
71
Greendykes
War
Newham
Swinhoe
Beadnell
Yeavering Bell
361
Humbleton
Chatton
Benthall
Frogden
Town Yetholm
Kirk Yetholm
Hethpool
Newton Tors
Wooler
Wandon
Twizell House
Newstead
Linton Hill
282
Fredden
Rosebrough
Chathill
Kale Water
A1
A1167
A697
A698
A6112
A6105
A699
B6438
B6437
B6460
B6461
B6470
B6350
B6352
B6353
B6354
B6525
B6349
B6348
B1342
B1341
B1340
B6364
B6456
B6355
B6365
B6396
B6397
B6351
B6436

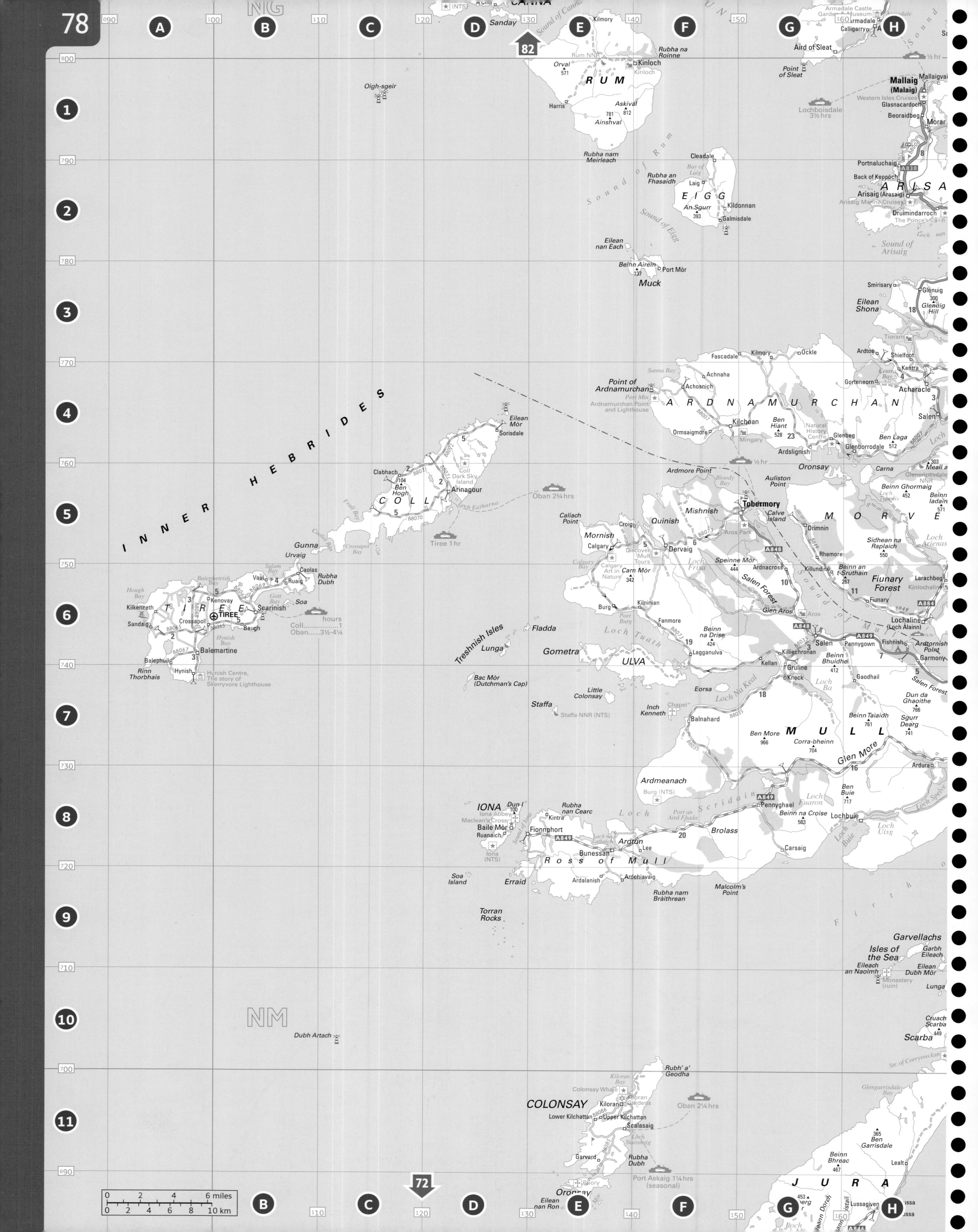

NG
NM
INNER HEBRIDES
82
72
RUM
Kinloch
Orval 571
Askival 812
Ainshval 781
Harris
Rubha na Roinne
Rubha nam Meirleach
Sound of Rum
Sound of Canna
Sanday
Kilmory
Oigh-sgeir
EIGG
Cleadale
Laig
An Sgurr 393
Kildonnan
Galmisdale
Rubha an Fhasaidh
Sound of Eigg
Eilean nan Each
Beinn Airein 137
Port Mòr
Muck
Aird of Sleat
Point of Sleat
Armadale
Calligarry
Mallaig (Malaig)
Mallaigvaig
Western Isles Cruises
Glasnacardoch
Beoraidbeg
Morar
Lochboisdale 3½ hrs
Portnaluchaig
Back of Keppoch
Arisaig (Arasaig)
Arisaig Marine Cruises
Druimindarroch
The Prince's Cairn
ARISAIG
Sound of Arisaig
Smirisary
Glenuig
Glenuig Hill 300
Eilean Shona
Tioram
Ardtoe
Shielfoot
Kentra
Gorteneorn
Acharacle
Salen
Fascadale
Kilmory
Ockle
Sanna Bay
Achnaha
Achosnich
Point of Ardnamurchan
Port Min
Ardnamurchan Point and Lighthouse
ARDNAMURCHAN
Kilchoan
Ormsaigmore
Mingary
Ben Hiant 528
Natural History Centre
Glenbeg
Glenborrodale
Ben Laga 512
Ardslignish
Oronsay
Carna
Ardmore Point
Bloody Bay
Auliston Point
Tobermory
Calve Island
Mishnish
Quinish
Drimnin
MORVERN
Beinn Ghormaig 452
Beinn Iadain 571
Sidhean na Raplaich 550
Rhemore
Killundine
Beinn an t-Sruthain 267
Funiary Forest
Larachbeg
Kinlochaline
Lochaline (Loch Alainn)
Ardtornish Point
Fishnish
Pennygown
Salen
Garmony
Coll
Eilean Mòr
Sorisdale
Clabhach
Ben Hogh 104
Coll Dark Sky Island
Arinagour
COLL
Loch Eatharna
Oban 2¼ hrs
Tiree 1 hr
Feall Bay
Crossapol Bay
Caolas Bàn
Gunna
Urvaig
Caolas
Rubha Dubh
Ruaig
Vaul
Salum Bay
Balephetrish Bay
Hough Bay
Kilkenneth
Kenovay
Gott Bay
Scarinish
Soa
TIREE
Sandaig
Crossapol
Baugh
Hynish Bay
Balemartine
Balephuil
Hynish
Rinn Thorbhais
Hynish Centre, The story of Skerryvore Lighthouse
Coll 1
Oban 3½-4¼
Caliach Point
Mornish
Croig
Calgary
Calgary Bay
Calgary Art in Nature
Discover Mull Tours
Dervaig
Carn Mòr 342
Aros Park
Speinne Mòr 444
Ardnacross
Salen Forest
Glen Aros
Aros
Burg
Kilninian
Port Burg
Fanmore
Beinn na Drise 424
Loch Tuath
Lagganulva
Killiechronan
Kellan
Gruline
Knock
Beinn Bhuidhe 412
Gaodhail
Treshnish Isles
Lunga
Fladda
Gometra
ULVA
Bac Mòr (Dutchman's Cap)
Little Colonsay
Staffa
Staffa NNR (NTS)
Inch Kenneth
Eorsa
Loch Na Keal
Balnahard
Ben More 966
MULL
Corra-bheinn 704
Beinn Talaidh 761
Sgurr Dearg 741
Dun da Ghaoithe 766
Glen More
Ardura
Ardmeanach
Burg (NTS)
Ben Buie 717
IONA
Dun-I 100
Iona Abbey
Maclean's Cross
Baile Mòr
Ruanaich
Iona (NTS)
Rubha nan Cearc
Kintra
Fionnphort
Sound of Iona
Loch Scridain
Port an Aird Fhada
Pennyghael
Beinn na Croise 503
Loch Fuaron
Lochbuie
Loch Uisg
Ardtun
Lee
Bunessan
Brolass
Carsaig
Ross of Mull
Ardalanish
Ardchiavaig
Erraid
Soa Island
Rubha nam Bràithrean
Malcolm's Point
Torran Rocks
Firth of Lorn
Garvellachs
Isles of the Sea
Garbh Eileach
Eileach an Naoimh
Eilean Dubh Mòr
Monastery (ruin)
Lunga
Cruach Scarba 449
Scarba
Dubh Artach
Rubh' a' Geodha
Kiloran Bay
Colonsay Whale
Kiloran Gardens
Kiloran
COLONSAY
Lower Kilchattan
Upper Kilchattan
Scalasaig
Oban 2¼ hrs
Garvard
Rubha Dubh
Port Askaig 1¼ hrs (seasonal)
Oronsay
Eilean nan Ron
Glengarrisdale Bay
Ben Garrisdale 365
Beinn Bhreac 467
Lealt
JURA
Lussagiven
0 2 4 6 miles
0 2 4 6 8 10 km

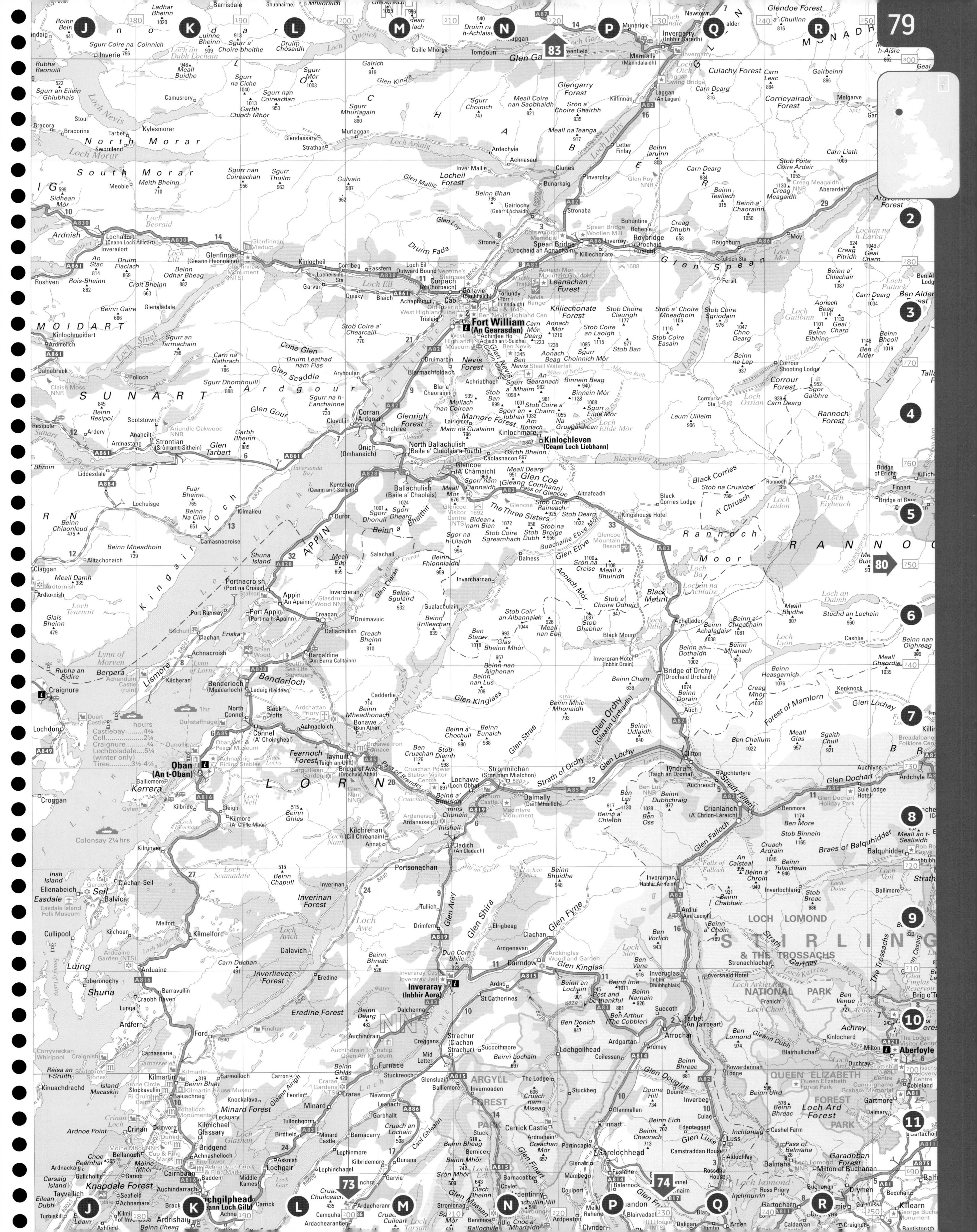
J
K
L
M
N
P
Q
R
North Morar
South Morar
MOIDART
SUNART
Ardgour
APPIN
LORN
Loch Nevis
Loch Morar
Loch Eil
Loch Linnhe
Loch Leven
Loch Lochy
Loch Etive
Loch Awe
Loch Fyne
Loch Lomond
Loch Long
Loch Eck
Loch Tulla
Loch Laidon
Loch Ossian
Loch Treig
Fort William
(An Gearasdan)
Oban
(An t-Oban)
Inveraray
(Inbhir Aora)
Kinlochleven
(Ceann Loch Liobhann)
Aberfoyle
Lochgilphead
Invergarry
Mallaig
Glenfinnan
Corpach
Banavie
Caol
Spean Bridge
Roybridge
Tulloch Sta
Onich
Ballachulish
Glencoe
Kingshouse Hotel
Rannoch Moor
Bridge of Orchy
Tyndrum
Crianlarich
Dalmally
Lochawe
Taynuilt
Connel
Benderloch
Appin
Port Appin
Lismore
Kerrera
Seil
Luing
Shuna
Strontian
Corran
Ben Nevis
1345
Nevis Forest
Mamore Forest
Glen Coe
Glen Etive
Black Mount
Glen Orchy
Glen Lochy
Strath Fillan
Glen Falloch
Glen Dochart
Glen Kinglas
Glen Shira
Glen Fyne
Glen Aray
Tarbet
Arrochar
Ardlui
Lochgoilhead
Strachur
St Catherines
Carrick Castle
Garelochhead
Kilmartin
Crinan
Cairnbaan
Ford
Kilmelford
Knapdale Forest
ARGYLL FOREST PARK
LOCH LOMOND & THE TROSSACHS NATIONAL PARK
QUEEN ELIZABETH FOREST PARK
STIRLING
Balquhidder
Braes of Balquhidder
Drymen
Milton of Buchanan
Balmaha
Loch Ard Forest
Colonsay 2¼ hrs
Castlebay 4¾
Coll 2¼
Craignure ¾
Lochboisdale 5¼
(winter only)
Tiree 3½-4¼
A82
A830
A861
A86
A87
A828
A85
A819
A83
A815
A816
A814
A884
A81
A811
A821
A84
83
80
73
74
2
3
4
5
6
7
8
9
10
11

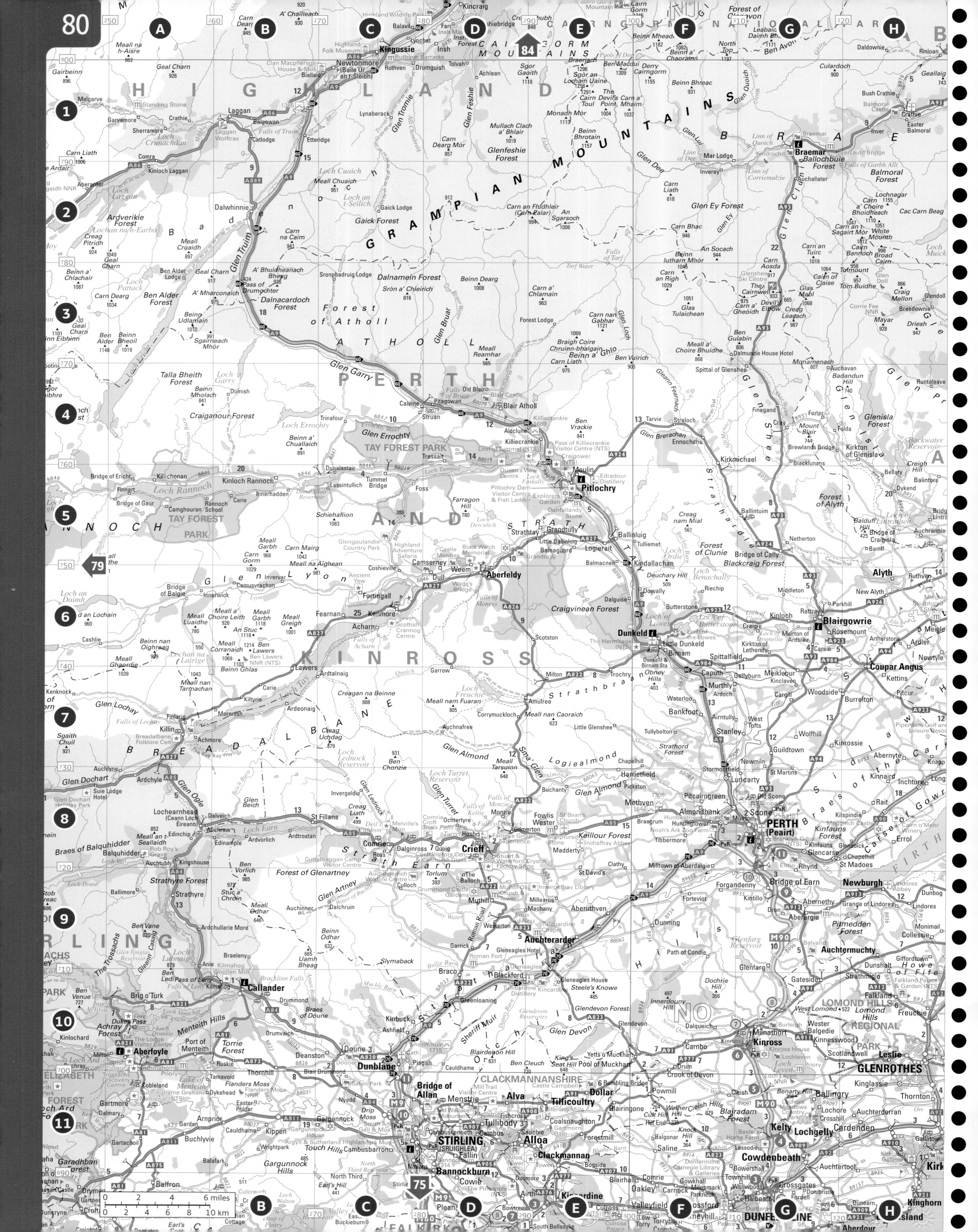
A
B
C
D
E
F
G
H
1
2
3
4
5
6
7
8
9
10
11
84
79
75
H I G H L A N D
G R A M P I A N   M O U N T A I N S
C A I R N G O R M   M O U N T A I N S
P E R T H   A N D   K I N R O S S
B R A E M A R
B R E A D A L B A N E
F o r e s t   o f   A t h o l l
A T H O L L
S t r a t h   E a r n
O c h i l   H i l l s
Kingussie
Newtonmore
Kincraig
Ruthven Barracks
Highland Folk Museum
Highland Wildlife Park
Insh
Drumguish
Tolvah
Achlean
Laggan
Kinloch Laggan
Dalwhinnie
Pass of Drumochter
Dalnacardoch Forest
Gaick Forest
Glen Truim
Glen Tromie
Glen Feshie
Glenfeshie Forest
Mar Lodge
Inverey
Braemar
Linn of Dee
Glen Dee
Glen Ey Forest
Glen Shee
Spittal of Glenshee
Dalmunzie House Hotel
Glenshee Ski Centre
Devil's Elbow
Balmoral Forest
Crathie
Easter Balmoral
Inver
Glen Clunie
Forest Lodge
Glen Tilt
Beinn a' Ghlo
Glen Bruar
Glen Garry
Calvine
Pitagowan
Struan
Trinafour
Blair Castle
Blair Atholl
Old Blair
Killiecrankie
Pass of Killiecrankie
Moulin
Pitlochry
Edradour Distillery
Glen Errochty
Tay Forest Park
Tummel Bridge
Dunalastair
Loch Tummel
Loch Rannoch
Kinloch Rannoch
Bridge of Ericht
Killichonan
Finnart
Bridge of Gaur
Rannoch School
Carie
Schiehallion
Foss
Strathtay
Grandtully
Logierait
Ballinluig
Aberfeldy
Weem
Castle Menzies
Dull
Camserney
Coshieville
Glen Lyon
Fortingall
Kenmore
Fearnan
Acharn
Loch Tay
Lawers
Ben Lawers
Ardtalnaig
Killin
Finlarig
Achmore
Glen Lochay
Glen Dochart
Crianlarich
Lochearnhead
Loch Earn
St Fillans
Comrie
Crieff
Muthill
Auchterarder
Gleneagles Hotel
Blackford
Braco
Greenloaning
Dunblane
Doune
Bridge of Allan
Stirling
Bannockburn
Cambusbarron
Callander
Kilmahog
Pass of Leny
Brig o'Turk
Aberfoyle
Port of Menteith
Menteith Hills
Kippen
Gargunnock
Buchlyvie
Balfron
Drymen
Gartmore
The Trossachs
Strathyre
Balquhidder
Kingshouse
Ardvorlich
Glen Artney
Glen Almond
Sma' Glen
Amulree
Strathbraan
Dunkeld
Birnam
Little Dunkeld
Caputh
Murthly
Stanley
Luncarty
Methven
Almondbank
Perth
Scone
Bridge of Earn
Abernethy
Newburgh
Auchtermuchty
Kinross
Milnathort
Loch Leven
Glenrothes
Leslie
Kelty
Lochgelly
Cowdenbeath
Dunfermline
Kincardine
Clackmannan
Alloa
Tillicoultry
Alva
Menstrie
Dollar
Castle Campbell
Clackmannanshire
Glen Devon
Glendevon
Blairgowrie
Rattray
Coupar Angus
Alyth
Kirkmichael
Bridge of Cally
Glenisla
Kirkton of Glenisla
Glenisla Forest
Forest of Alyth
Meikleour
Kinclaven
Burrelton
Balbeggie
Errol
Kinfauns
Glencarse
St Madoes
Braes of the Carse
Sidlaw Hills
Lomond Hills
Falkland
Freuchie
Strathmiglo
Kinnesswood
Scotlandwell
Powmill
Crook of Devon
Rumbling Bridge
Saline
Carnock
Oakley
Kinghorn
Crossgates
0 2 4 6 miles
0 2 4 6 8 10 km

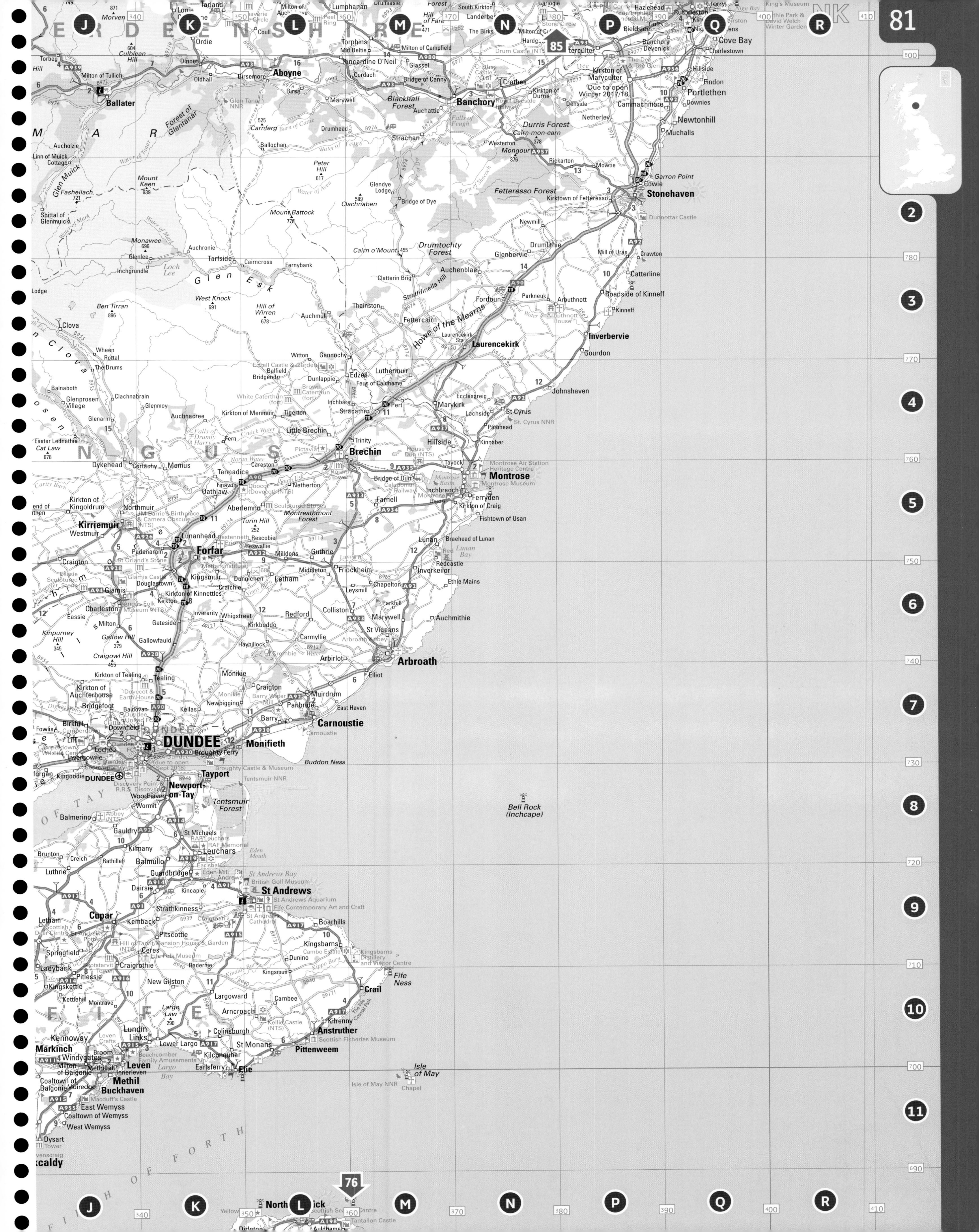
ABERDEENSHIRE
ANGUS
FIFE
MAR
J
K
L
M
N
P
Q
R
NK
85
76
2
3
4
5
6
7
8
9
10
11
Aboyne
Ballater
Banchory
Stonehaven
Laurencekirk
Inverbervie
Brechin
Montrose
Forfar
Kirriemuir
Arbroath
Carnoustie
Monifieth
DUNDEE
Tayport
Newport-on-Tay
Leuchars
St Andrews
Cupar
Crail
Anstruther
Pittenweem
Elie
Leven
Methil
Buckhaven
Markinch
Portlethen
Cove Bay
Newtonhill
Muchalls
Johnshaven
Edzell
Letham
Glamis
Tentsmuir Forest
Bell Rock (Inchcape)
Isle of May
Fife Ness
Buddon Ness
Howe of the Mearns
Fetteresso Forest
Durris Forest
Drumtochty Forest
Montreathmont Forest
Glen Esk
Glen Clova
Mount Keen
Mount Battock
Largo Bay
FIRTH OF FORTH
FIRTH OF TAY
Due to open Winter 2017/18

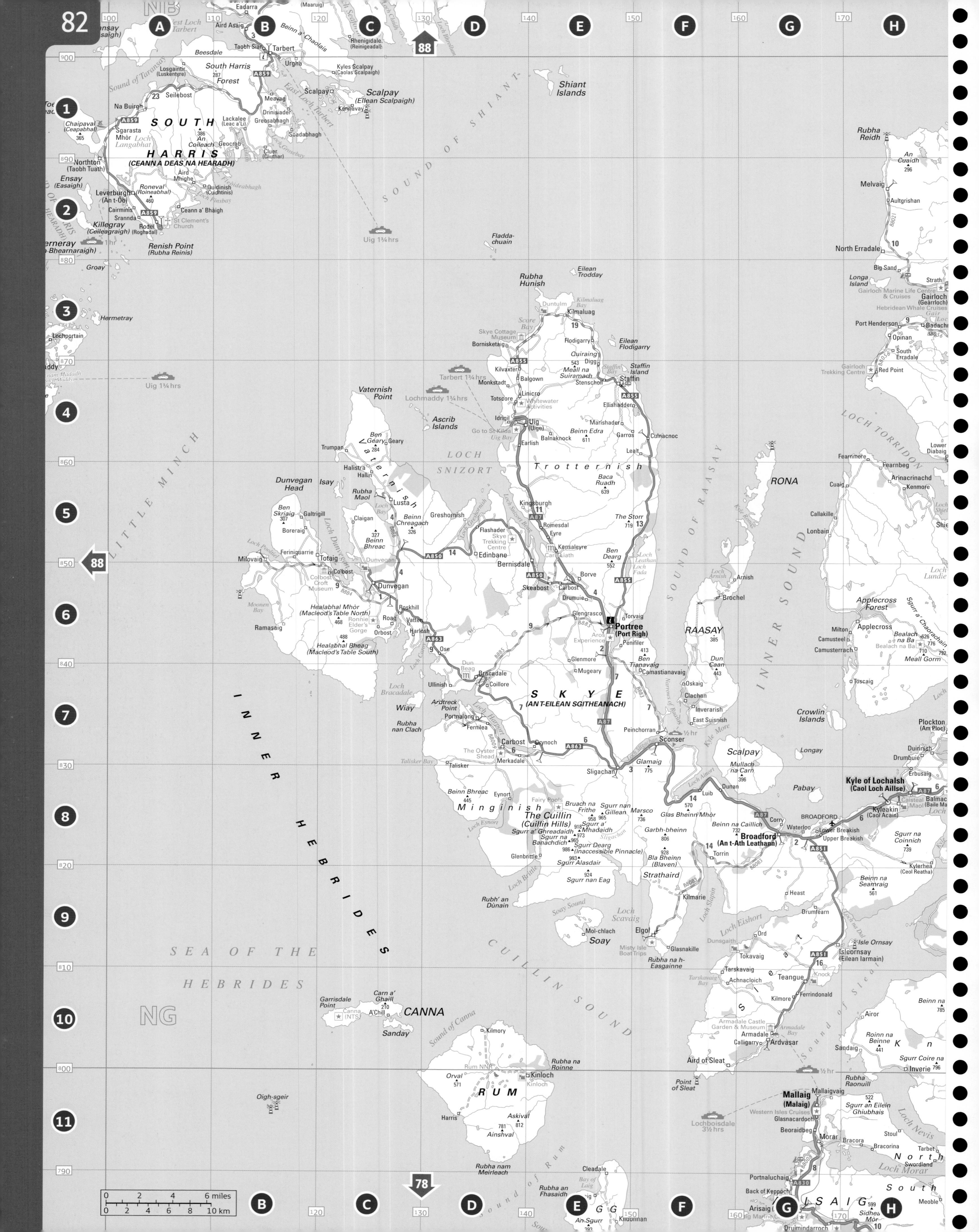
SOUTH HARRIS
(CEANN A DEAS NA HEARADH)
South Harris Forest
Tarbert
Scalpay
(Eilean Scalpaigh)
Leverburgh
(An t-Òb)
Rodel
(Roghadal)
Renish Point
(Rubha Reinis)
Shiant Islands
SOUND OF SHIANT
Uig 1¾ hrs
Tarbert 1¾ hrs
Lochmaddy 1¾ hrs
LITTLE MINCH
Trotternish
Vaternish
Duntulm
Kilmaluag
Staffin
Uig
(Uige)
Dunvegan
Portree
(Port Rìgh)
RAASAY
RONA
SOUND OF RAASAY
INNER SOUND
SKYE
(AN T-EILEAN SGITHEANACH)
Minginish
The Cuillin
(Cuillin Hills)
Broadford
(An t-Ath Leathann)
Kyle of Lochalsh
(Caol Loch Aillse)
Kyleakin
(Caol Acain)
Sconser
Sligachan
Elgol
Soay
CUILLIN SOUND
Sleat
Armadale
Ardvasar
Isleornsay
(Eilean Iarmain)
Mallaig
(Malaig)
CANNA
Sanday
RUM
Kinloch
INNER HEBRIDES
SEA OF THE HEBRIDES
LOCH TORRIDON
Gairloch
(Geàrrloch)
Applecross
Plockton
(Am Ploc)
Loch Nevis
North Morar
Loch Morar
EIGG
ARISAIG
NB
NG
88
78
0 2 4 6 miles
0 2 4 6 8 10 km

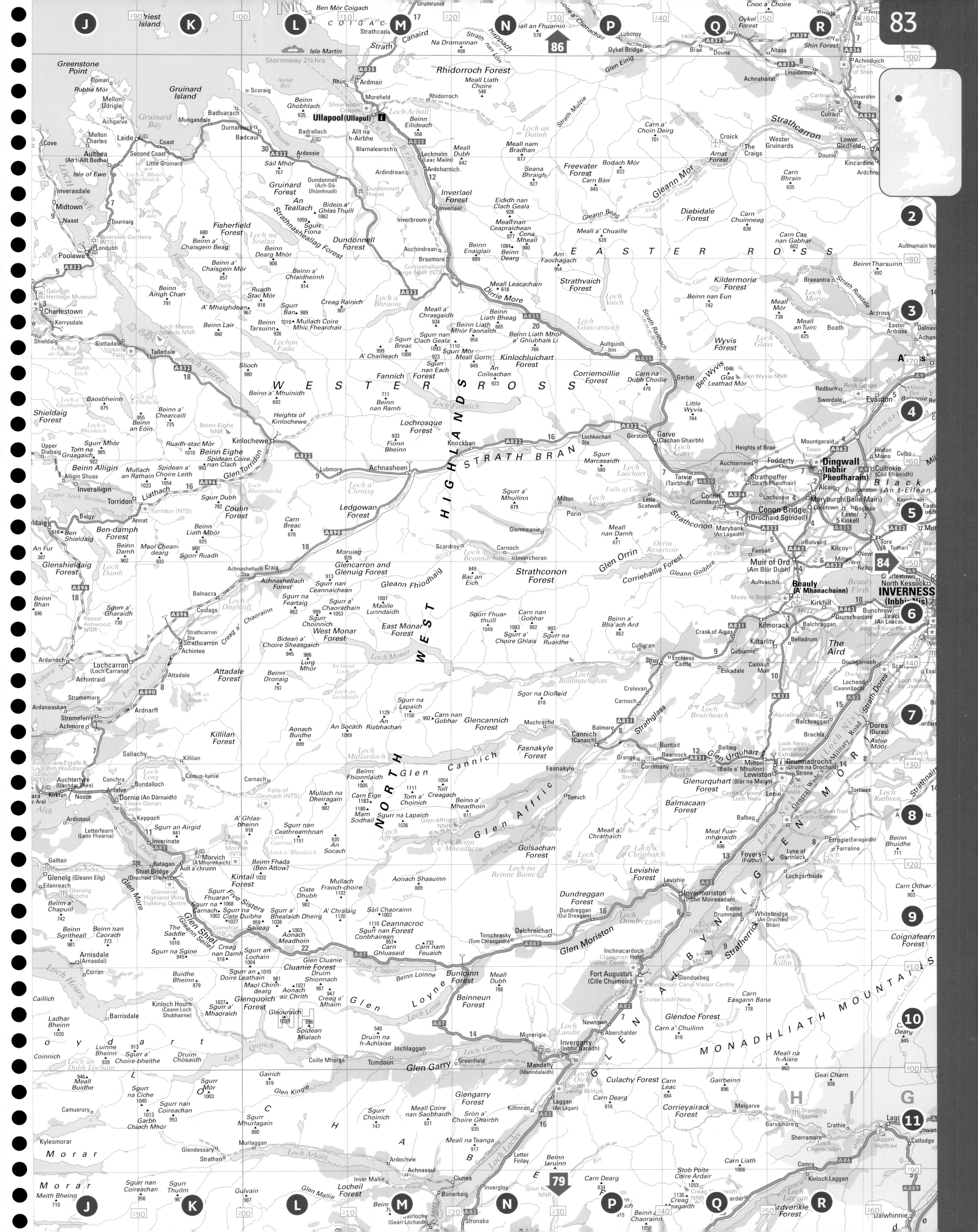

WESTER ROSS
EASTER ROSS
NORTH WEST HIGHLANDS
GLEN ALBYN (GLEN MÒR)
MONADHLIATH MOUNTAINS
STRATH BRAN
Ullapool (Ullapul)
Dingwall (Inbhir Pheofharain)
INVERNESS (Inbhir Nis)
Beauly (A' Mhanachainn)
Muir of Ord (Am Blàr Dubh)
Conon Bridge (Drochaid Sgùideil)
Maryburgh (Baile Màiri)
Strathpeffer (Srath Pheofhair)
Contin (Cunndainn)
Garve (Clachan Ghairbh)
Kinlochewe
Achnasheen
Torridon
Lochcarron (Loch Carrann)
Shieldaig
Gairloch
Poolewe
Aultbea (An t-Allt Beithe)
Dornie (An Dòrnaidh)
Kyle of Lochalsh
Shiel Bridge (Drochaid Sheile)
Glenelg (Gleann Eilg)
Arnisdale (Arnasdal)
Invermoriston (Inbhir Moireasdain)
Fort Augustus (Cille Chuimein)
Drumnadrochit (Druim na Drochaid)
Invergarry (Inbhir Garadh)
Cannich (Canaich)
Struy
Kiltarlity
Dores (Duras)
Foyers (Foithir)
Laggan (An Lagan)
Tomdoun
Glen Garry
Glen Moriston
Glen Affric
Glen Cannich
Glen Shiel
Glen Torridon
Glen Orrin
Glen Urquhart
Strathconon
Strathglass
Strathcarron
Loch Maree
Loch Broom
Loch Fannich
Loch Ness
Loch Monar
Loch Mullardoch
Loch Quoich
Loch Arkaig
Loch Lochy
Loch Garry
Loch Loyne
Loch Cluanie
Ben Wyvis
Beinn Eighe
Beinn Alligin
Slioch
Black Isle (An t-Eilean Dubh)
The Aird
Dirrie More
Five Sisters
Kintail Forest
Fisherfield Forest
Strathvaich Forest
Freevater Forest
Rhidorroch Forest
Inverewe Gardens (NTS)
Eilean Donan Castle
Corrieshalloch Gorge NNR (NTS)
Torridon (NTS)
Stornoway 2½ hrs
A835 A832 A890 A896 A87 A887 A82 A831 A833 A862 A9 A86 A889 A836 A837 A839
J K L M N P Q R
2 3 4 5 6 7 8 9 10 11
86 84 79

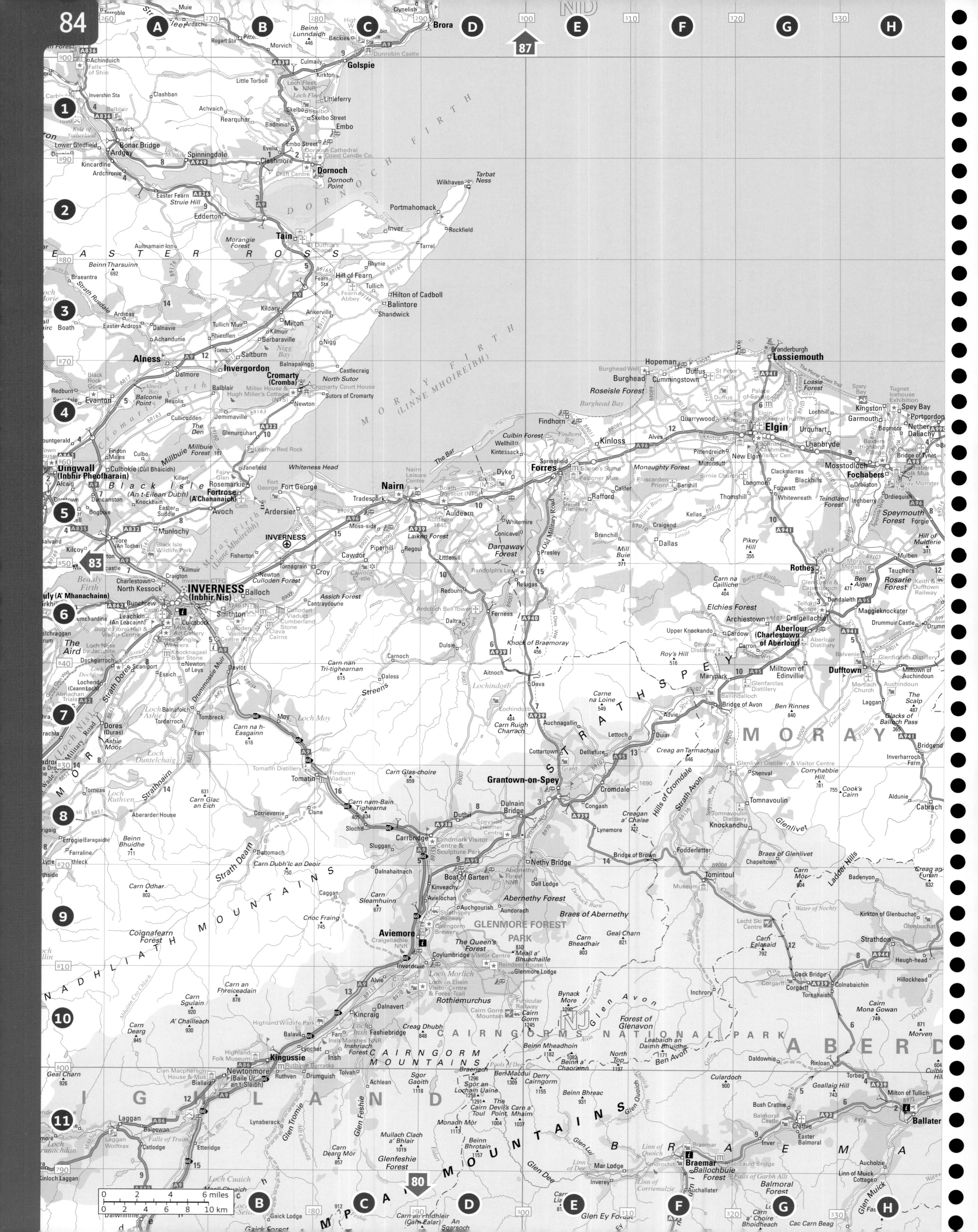
A
B
C
D
E
F
G
H
87
Brora
Golspie
Dunrobin Castle
Loch Fleet NNR
Littleferry
Skelbo
Embo
Bonar Bridge
Ardgay
Spinningdale
Clashmore
Dornoch
Dornoch Cathedral
Dornoch Point
Tarbat Ness
Wilkhaven
Portmahomack
Inver
Rockfield
DORNOCH FIRTH
Edderton
Tain
Morangie Forest
EASTER ROSS
Tarrel
Rhynie
Hill of Fearn
Fearn Abbey
Tullich
Hilton of Cadboll
Balintore
Shandwick
Beinn Tharsuinn
692
Strath Rusdale
Ardross
Easter Ardross
Dalnavie
Kildary
Milton
Kilmuir
Barbaraville
Nigg
Nigg Bay
Saltburn
Alness
Invergordon
Cromarty (Cromba)
North Sutor
Cromarty Court House
Sutors of Cromarty
Hugh Miller's Cottage
Newton
Balblair
Resolis
Evanton
Balconie Point
Cromarty Firth
MORAY FIRTH (LINNE MHOIREIBH)
Jemimaville
Glenurquhart
The Den
Millbuie Forest
Dingwall (Inbhir Pheofharain)
Culbokie (Cùil Bhàicidh)
Black Isle (An t-Eilean Dubh)
Rosemarkie
Fortrose (A'Chananaich)
Avoch
Munlochy
Fort George
Ardersier
Whiteness Head
The Bar
Culbin Forest
Findhorn
Findhorn Bay
Burghead Bay
Burghead
Hopeman
Cummingstown
Duffus
Roseisle Forest
Lossiemouth
Brandenburgh
Lossie Forest
Spey Bay
Kingston
Garmouth
Portgordon
Kinloss
Alves
Quarrywood
Elgin
Urquhart
Lhanbryde
New Elgin
Mosstodloch
Fochabers
Forres
Sueno's Stone
Dyke
Monaughty Forest
Pluscarden Priory
Nairn
Auldearn
Tradespark
Moss-side
Cawdor
Piperhill
Darnaway Forest
Rafford
Dallas
Kellas
Thornhill
Longmorn
Teindland Forest
Speymouth Forest
INVERNESS
Tore
Kilcoy
Charlestown
North Kessock
INVERNESS (Inbhir Nis)
Balloch
Smithton
Culloden Forest
Croy
Culloden
Clava Cairns
Dochgarroch
Scaniport
Essich
Dores (Duras)
Strath Dores
Loch Ashie
Loch Duntelchaig
Daviot
Moy
Loch Moy
Tomatin
Findhorn Viaduct
Carn Glas-choire
Streens
Lochindorb
Dava
Carne na Loine
Knock of Braemoray
Relugas
Ferness
Logie
Pikey Hill
Rothes
Ben Aigan
Rosarie Forest
Elchies Forest
Archiestown
Craigellachie
Aberlour (Charlestown of Aberlour)
Dufftown
Maggieknockater
Marypark
Bridge of Avon
Ben Rinnes
STRATHSPEY
MORAY
Advie
Cromdale
Grantown-on-Spey
Dulnain Bridge
Duthil
Carrbridge
Boat of Garten
Nethy Bridge
Abernethy Forest
Braes of Abernethy
Aviemore
GLENMORE FOREST PARK
Loch Morlich
Glenmore Lodge
Rothiemurchus
Kincraig
Kingussie
Newtonmore
Ruthven
MONADHLIATH MOUNTAINS
Tomintoul
Glenlivet
Knockandhu
Tomnavoulin
Hills of Cromdale
Cock Bridge
Corgarff
Strathdon
CAIRNGORMS NATIONAL PARK
CAIRNGORM MOUNTAINS
Cairn Gorm
Ben Macdui
Ben Avon
Beinn a' Bhuird
Braeriach
Glen Feshie
Glen Tromie
Laggan
Braemar
Ballochbuie Forest
Inverey
Mar Lodge
Linn of Dee
Crathie
Balmoral Castle
Ballater
ABERD
BRAEMAR
GRAMPIAN MOUNTAINS
HIGHLAND
NJ
80
83
0 2 4 6 miles
0 2 4 6 8 10 km

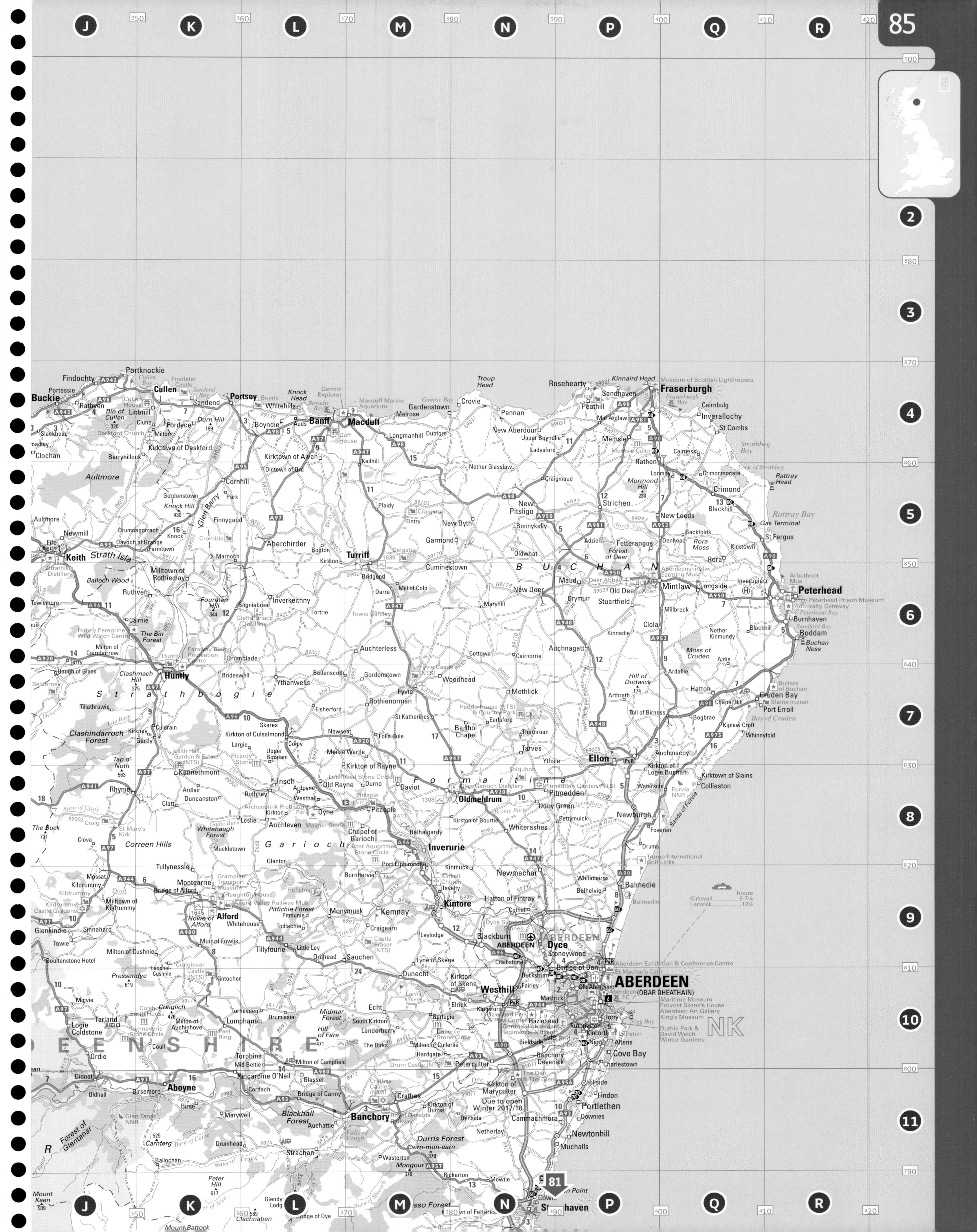
J
K
L
M
N
P
Q
R
2
3
4
5
6
7
8
9
10
11
NK
81
Findochty
Portknockie
Portessie
Buckie
Rathven
Cullen
Cullen Bay
Findlater Castle
Sandend
Sandend Bay
Portsoy
Boyne
Knock Head
Boyndie Bay
Gemini Explorer
Macduff Marine Aquarium
Gamrie Bay
Troup Head
Rosehearty
Kinnaird Head
Museum of Scottish Lighthouses
Fraserburgh
Fraserburgh Bay
Sandhaven
Pitsligo
Peathill
Cairnbulg
Inverallochy
St Combs
Whitehills
Banff
Macduff
Duff House
Gardenstown
Melrose
Crovie
Pennan
New Aberdour
Upper Boyndlie
Ladysford
Mid Ardlaw
Memsie
Memsie Cairn
Rathen
Cairness
Strathbeg Bay
Loch of Strathbeg
Lonmay
Crimonmogate
Rattray Head
Crimond
Blackhill
Rattray Bay
Gas Terminal
St Fergus
Bin of Cullen
Clune
Lintmill
Fordyce
Durn Hill
Boyndie
Slackhead
Deskford Church
Milton
Kirktown of Deskford
Clochan
Berryhillock
Aultmore
Longmanhill
Dubford
Kirktown of Alvah
Oldtown of Ord
Keilhill
Eden
Plaidy
Craigston
Fintry
New Byth
Nether Glasslaw
Craigmaud
New Pitsligo
Strichen
Mormond Hill
New Leeds
Backfolds
Cornhill
Gordonstown
Knock Hill
Park
Glen Barry
Finnygaud
Knock
Crombie
Newmill
Keith
Strath Isla
Drumnagorrach
Davoch of Grange
Farmtown
Marnoch
Aberchirder
Bogton
Turriff
Delgatie
Garmond
Kirkton
Cuminestown
Bonnykelly
Oldwhat
Adziel
Fetterangus
Forest of Deer
Denhead
Rora Moss
Kirktown
Rora
Strathisla Distillery
Balloch Wood
Milltown of Rothiemay
Ruthven
Bridgend
Mill of Colp
Darra
B U C H A N
Maud
New Deer
Deer Abbey
Aberdeenshire Farming Mus
Old Deer
Aden
Mintlaw
Longside
Inverugie
Arbuthnot Mus
Peterhead
Peterhead Prison Museum
Admiralty Gateway
Peterhead Bay
Towiemore
Fourman Hill
Bognie brae
Inverkeithny
Fortrie
Towie Barclay
Maryhill
Drymuir
Stuartfield
Millbreck
Burnhaven
Sandford Bay
Boddam
Buchan Ness
Cairnie
Glendronach Distillery
Huntly Peregrine Wild Watch Centre
The Bin Forest
Milton of Cairnborrow
Huntly
Faraway Riding Recreation Centre
Drumblade
Auchterless
Cottown
Cairnorrie
Auchnagatt
Kinnadie
Clola
Nether Kinmundy
Blackhill
Moss of Cruden
Aldie
Torry
Haugh of Glass
Clashmach Hill
Brideswell
Ythanwells
Badenscoth
Gordonstown
Fyvie Castle (NTS)
Woodhead
Fyvie
Methlick
Hill of Dudwick
Arthrath
Ardallie
Hatton
Bullers of Buchan
Cruden Bay
Slains (ruins)
Port Erroll
Bay of Cruden
Beldorney
Strathbogie
Tillathrowie
Clashindarroch Forest
Rothienorman
Fisherford
St Katherines
Haddo House (NTS) & Country Park
Toll of Birness
Bogbrae
Kiplaw Croft
Whinnyfold
Kirkney
Culdrain
Gartly
Skares
Kirkton of Culsalmond
Largie
Colpy
Newseat
Folla Rule
Barthol Chapel
Thornroan
Tarves
Ythsie
Auchmacoy
Ellon
Kirkton of Logie Buchan
Kirktown of Slains
Collieston
Forvie NNR
Sands of Forvie
Leith Hall, Garden & Estate (NTS)
Picardy Stone
Upper Boddam
Meikle Wartle
Kirkton of Rayne
Tap o' North
Kennethmont
Insch
Loanhead Stone Circle
Old Rayne
Durno
Daviot
Formartine
Tolquhon
Glen Garioch Distillery
Pitmedden Garden (NTS)
Pitmedden
Oldmeldrum
Waterside
Rhynie
Ardlair
Duncanston
Rothney
Ardoyne
Westhall
Pitcaple
Burn of Craig
Clatt
Archaeolink Prehistory Park
Kirkton
Oyne
Udny Green
Pettymuick
Newburgh
Foveran
The Buck
Craig
Leslie
Auchleven
Maiden Stone
Chapel of Garioch
Balhalgardy
Kirkton of Bourtie
Whiterashes
Drums
St Mary's Kirk
Clova
Correen Hills
Whitehaugh Forest
Muckletown
Garioch
Easter Aquorthies Stone Circle
Inverurie
Trump International Golf Links
Tullynessle
Glenton
Port Elphinstone
Kinmuck
Newmachar
Whitecairns
Mossat
Kildrummy
Montgarrie
Bridge of Alford
Grampian Transport Museum
Haughton House
Pitfichie
Burnhervie
Kirkell Church
Tavelty
Belhelvie
Balmedie
Alford Valley Railway Mus
Pitfichie Forest
Pitmunie
Kintore
Hatton of Fintray
Cothall
Kildrummy Castle Gardens
Milton of Kildrummy
Howe of Alford
Alford
Whitehouse
Monymusk
Kemnay
Kirkwall
Lerwick
hours
Todlachie
Craigearn
Dyce Symbol Stones
Glenkindie
Sinnahard
Castle Fraser (NTS)
Leylodge
Blackburn
Dyce
Aberdeen
Towie
Muir of Fowlis
Tillyfourie
Little Ley
Ordhead
Sauchen
Lyne of Skene
Stoneywood
Craibstone
Milton of Cushnie
Boultenstone Hotel
Leochel Cushnie
Craigievar Castle (NTS)
Kintocher
Dunecht
Bucksburn
Bridge of Don
Aberdeen Exhibition & Conference Centre
St Machar's Cath
Pressendye
Kirkton of Skene
Westhill
Fairley
Old Aberdeen
ABERDEEN
(OBAR DHEATHAIN)
Aberdeen FC
Maritime Museum
Provost Skene's House
Aberdeen Art Gallery
King's Museum
Migvie
Culsh Earth House
Craiglich
Tornaveen
Echt
Elrick
Mastrick
Kingsford
Hazlehead
Torry
Nigg Bay
Tarland
Logie Coldstone
Tomnaverie Stone Circle
Milton of Auchinhove
Lumphanan
Drumlasie
Midmar Forest
South Kirkton
Garlogie
Cullerlie Stone Circle
Hazlehead Park
Gordon Highlanders Regimental Museum
Peel Ring
Landerberry
Hill of Fare
Duthie Park & David Welch Winter Gardens
Cults
Bieldside
Ruthrieston
Kincorth
Loirston
Altens
Nigg
Cove Bay
Coull
Torphins
Mid Beltie
The Birks
Milton of Cullerlie
Hardgate
Banchory Devenick
Charlestown
ABERDEENSHIRE
Ordie
Milton of Campfield
Drum Castle (NTS)
Peterculter
Dinnet
Oldhall
Aboyne
Kincardine O'Neil
Glassel
Kirkton of Maryculter
The Deer & The Glen
Hillside
Birsemore
Cordach
Crathes Castle (NTS)
Crathes
Findon
Glen Tanar NNR
Birse
Marywell
Blackhall Forest
Bridge of Canny
Banchory
Royal Deeside Railway
Kirkton of Durris
Due to open Winter 2017/18
Portlethen
Downies
Forest of Glentanar
Auchattie
Falls of Feugh
Durris Forest
Cairn-mon-earn
Denside
Cammachmore
Netherley
Newtonhill
Carnferg
Burn of Cattie
Drumhead
Strachan
Muchalls
Ballochan
Water of Feugh
Westerton
Mongour
Rickarton
Mowtie
Peter Hill
Glendye Lodge
Bridge of Dye
Mount Keen
Clachnaben
Mount Battock
Cowie
Stonehaven
Fetteresso Forest
Garron Point

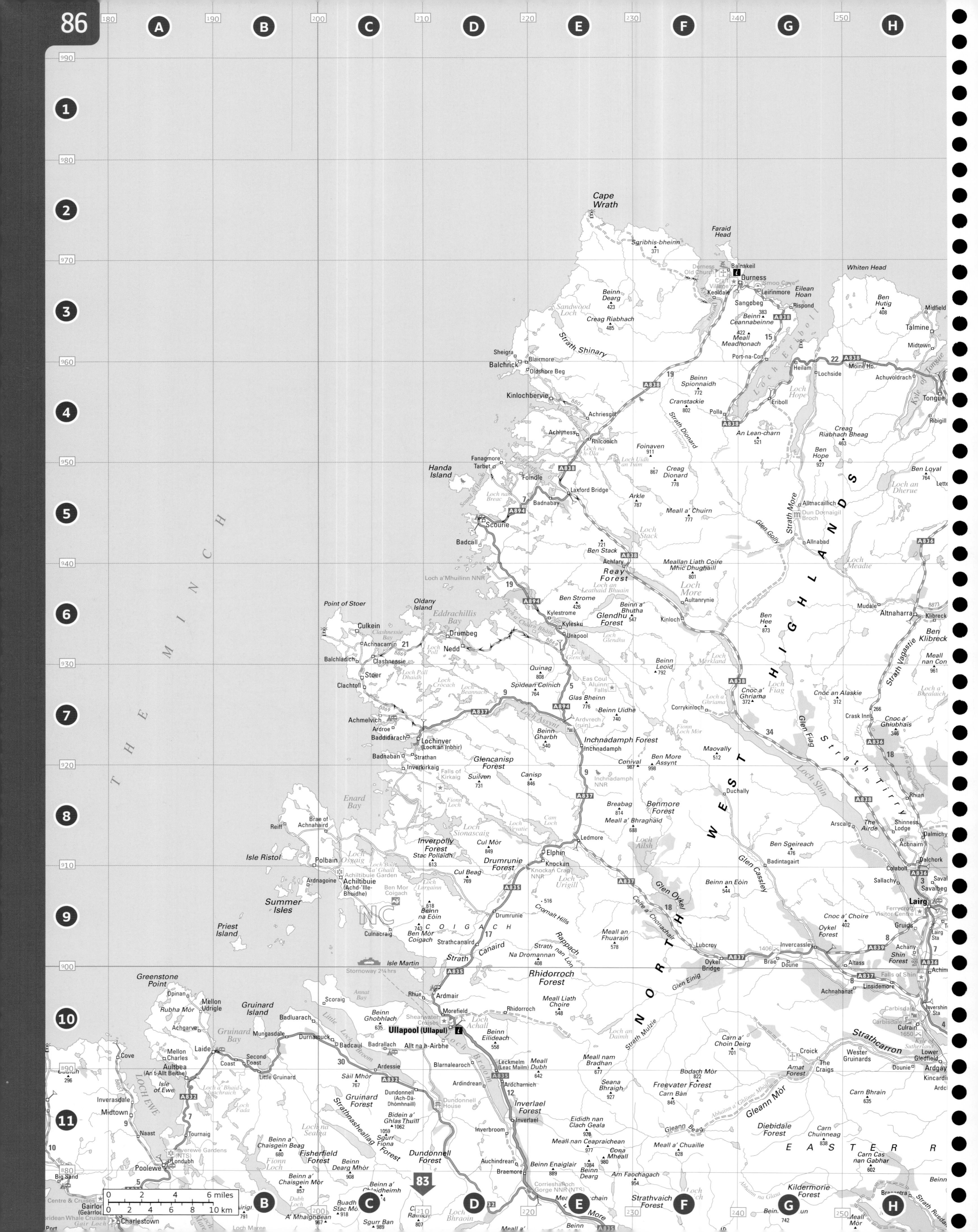
A
B
C
D
E
F
G
H
THE MINCH
Cape Wrath
Faraid Head
Sgribhis-bheinn
Balnakeil
Durness
Smoo Cave
Eilean Hoan
Whiten Head
Keoldale
Leirinmore
Rispond
Sangobeg
Sandwood Loch
Beinn Dearg
Creag Riabhach
Strath Shinary
Beinn Ceannabeinne
Meall Meadhonach
Port-na-Con
Loch Eriboll
Ben Hutig
Midfield
Talmine
Midtown
Heilam
Lochside
Moine Ho.
Achuvoldrach
Tongue
Kyle of Tongue
Sheigra
Balchrick
Blairmore
Oldshore Beg
Kinlochbervie
Achriesgill
Beinn Spionnaidh
Cranstackie
Strath Dionard
Polla
Eriboll
Loch Hope
Ribigill
Achlyness
Rhiconich
Foinaven
An Lean-charn
Creag Riabhach Bheag
Ben Hope
Ben Loyal
Loch an Dherue
Fanagmore
Tarbet
Handa Island
Foindle
Laxford Bridge
Creag Dionard
Arkle
Meall a' Chuirn
Badnabay
Scourie
Badcall
Alltnacaillich
Dun Dornaigil Broch
Strath More
Glen Golly
Allnabad
Loch Stack
Ben Stack
Achfary
Reay Forest
Meallan Liath Coire Mhic Dhughaill
Loch More
Aultanrynie
Loch Meadie
Loch a'Mhuilinn NNR
Ben Strome
Beinn a' Bhutha
Glendhu Forest
Kylestrome
Kylesku
Unapool
Kinloch
Ben Hee
Mudale
Altnaharra
Klibreck
Ben Klibreck
Point of Stoer
Oldany Island
Eddrachillis Bay
Culkein
Drumbeg
Nedd
Achnacarnin
Balchladich
Clashnessie
Stoer
Clachtoll
Quinag
Spidean Coinich
Glas Bheinn
Eas Coul Aulinn Falls
Beinn Leoid
Loch Merkland
Loch Fiag
Cnoc a' Ghriama
Cnoc an Alaskie
Crask Inn
Cnoc a' Ghiubhais
Strath Vagastie
Meall nan Con
Achmelvich
Ardroe
Baddidarach
Lochinver (Loch an Inbhir)
Strathan
Loch Assynt
Beinn Uidhe
Ardvreck (ruin)
Corrykinloch
Beinn Gharbh
Inchnadamph Forest
Inchnadamph
Maovally
Glen Fiag
Strath Tirry
Badnaban
Inverkirkaig
Falls of Kirkaig
Glencanisp Forest
Suilven
Canisp
Conival
Ben More Assynt
Inchnadamph NNR
Duchally
Loch Shin
Rhian
Enard Bay
Reiff
Brae of Achnahaird
Fionn Loch
Loch Sionascaig
Breabag
Benmore Forest
Meall a' Bhraghaid
Arscaig
The Airde
Shinness Lodge
Achnairn
Dalmichy
Inverpolly Forest
Stac Pollaidh
Cul Mòr
Drumrunie Forest
Elphin
Knockan
Knockan Crag NNR
Ledmore
Loch Urigill
Loch Ailsh
Ben Sgeireach
Badintagairt
Colaboll
Dalchork
Isle Ristol
Polbain
Achiltibuie Garden
Achiltibuie (Achd-'Ille-Bhuidhe)
Ardnagoine
Cul Beag
Ben Mor Coigach
Glen Oykel
Beinn an Eòin
Glen Cassley
Sallachy
Lairg
Summer Isles
Priest Island
NC
COIGACH
Culnacraig
Strathcanaird
Drumrunie
Cromalt Hills
Rappach
Strath nan Lon
Meall an Fhuarain
Coire a Chonachair
NORTH WEST HIGHLANDS
Lubcroy
Oykel Bridge
Invercassley
Brae
Doune
Cnoc a' Choire
Oykel Forest
Achany
Shin Forest
Altass
Isle Martin
Stornoway 2¾ hrs
Strath Canaird
Na Dromannan
Rhidorroch Forest
Glen Einig
Falls of Shin
Linsidemore
Achnahanat
Greenstone Point
Opinan
Mellon Udrigle
Gruinard Island
Rubha Mòr
Achgarve
Gruinard Bay
Mungasdale
Scoraig
Badluarach
Rhue
Ardmair
Morefield
Beinn Ghobhlach
Ullapool (Ullapul)
Rhidorroch
Meall Liath Choire
Shearwater Cruises
Loch Achall
Beinn Eilideach
Loch an Daimh
Strath Mulzie
Carbisdale
Culrain
Strathcarron
Carn a' Choin Deirg
Laide
Second Coast
Durnamuck
Badcaul
Badrallach
Allt na h-Airbhe
Little Loch Broom
Loch Broom
Leckmelm (Leac Mailm)
Meall Dubh
Meall nam Bradhan
Croick
Western Gruinards
Lower Gledfield
Ardgay
Kincardine
Mellon Charles
Aultbea (An t-Allt Beithe)
Isle of Ewe
Cove
Coast
Little Gruinard
Ardessie
Sàil Mhòr
Blarnalearoch
Ardindrean
Ardcharnich
Seana Bhraigh
Freevater Forest
Bodach Mòr
Carn Bàn
Amat Forest
The Craigs
Dounie
Carn Bhrain
Gleann Mòr
Diebidale Forest
Inverasdale
Midtown
Loch Ewe
Gruinard Forest
Strathnasheallag Forest
Dundonnell (Ach-Dà-Dhòmhnaill)
Dundonnell House
Bidein a' Ghlas Thuill
Sgurr Fiona
Inverlael Forest
Inverlael
Eididh nan Clach Geala
Gleann Beag
Carn Chuinneag
EASTER R
Naast
Tournaig
Inverewe Gardens (NTS)
Londubh
Poolewe
Beinn a' Chaisgein Beag
Fisherfield Forest
Dundonnell Forest
Inverbroom
Meall nan Ceapraichean
Cona Mheall
Meall a' Chuaille
Carn Cas nan Gabhar
Beinn Dearg Mhòr
Beinn a' Chaisgein Mòr
Auchindrean
Braemore
Beinn Enaiglair
Beinn Dearg
Am Faochagach
Big Sand
Gairloch
Charlestown
Ruadh Stac Mòr
A' Mhaighdean
Sgurr Ban
Beinn a' Chlaidheimh
Loch a' Bhraoin
Corrieshalloch Gorge NNR (NTS)
Strathvaich Forest
Kildermorie Forest
Loch Maree
0 2 4 6 miles
0 2 4 6 8 10 km
83

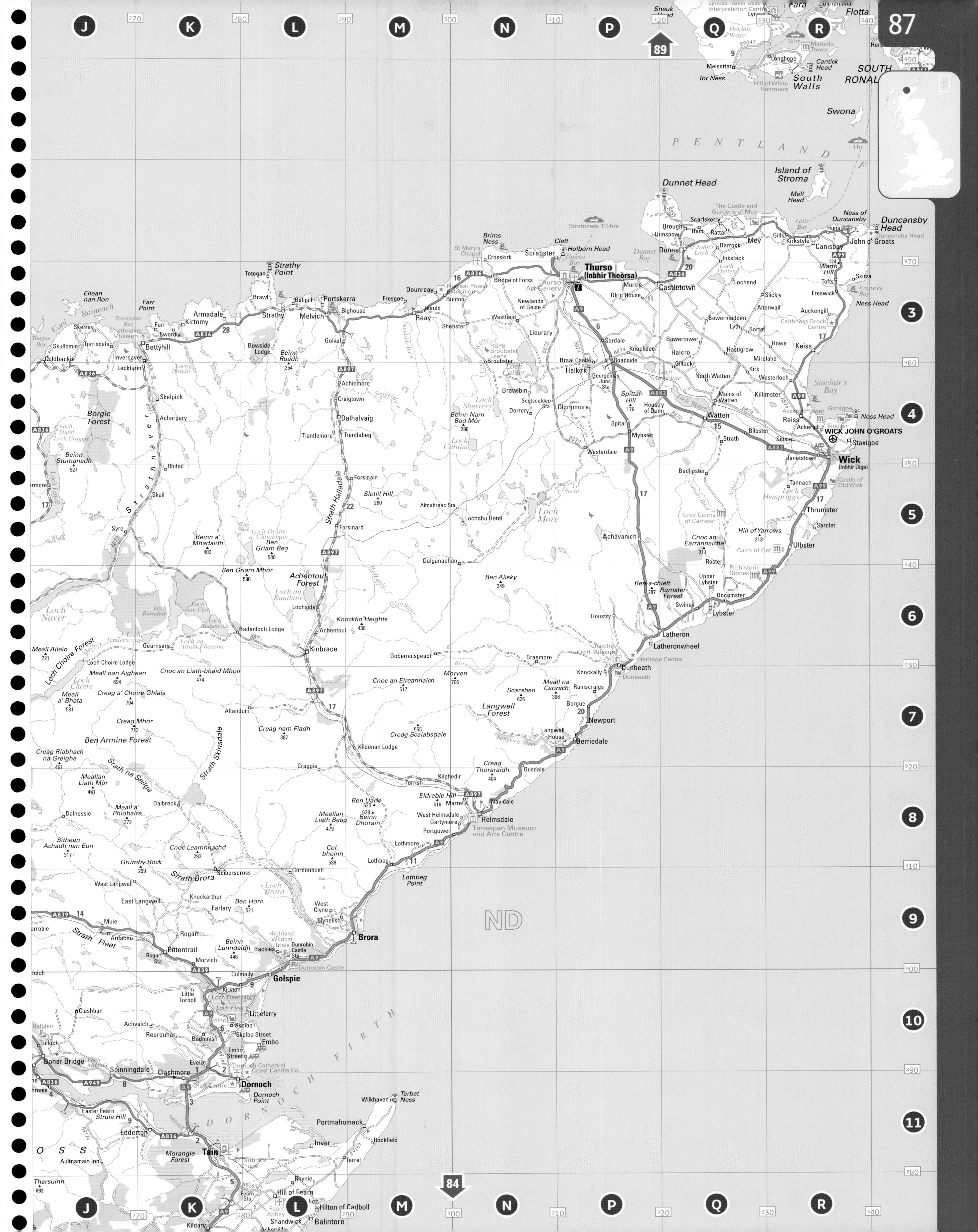
J
K
L
M
N
P
Q
R
89
84
ND
P E N T L A N D F
Flotta
Fara
Heldale Water
Melsetter
Longhope
Cantick Head
Tor Ness
South Walls
SOUTH RONALDSAY
Swona
Island of Stroma
Mell Head
Dunnet Head
The Castle and Gardens of Mey
Brough
Scarfskerry
Ham
Rattar
Mey
Gills Bay
Huna
Ness of Duncansby
Duncansby Head
John o' Groats
Canisbay
Kirkstyle
Dunnet
Dunnet Bay
Barrock
Inkstack
Castletown
Lochend
Warth Hill
Tofts
Skirza
Freswick
Ness Head
Slickly
Alterwall
Auckengill
Bowermadden
Lyth
Sortat
Stromness 1½ hrs
Brims Ness
Holborn Head
Scrabster
Crosskirk
Thurso (Inbhir Theòrsa)
Thurso Art Gallery
Murkle
Olrig House
Bridge of Forss
Dounreay
Buldoo
Isauld
Reay
Newlands of Geise
Westfield
Lieurary
Shebster
Strathy Point
Totegan
Brawl
Baligill
Portskerra
Melvich
Bighouse
Fresgoe
Strathy
Armadale
Kirtomy
Farr Point
Farr
Swordly
Bettyhill
Eilean nan Ron
Skerray
Skullomie
Torrisdale
Coldbackie
Invernaver
Leckfurin
Golval
Bowside Lodge
Beinn Ruadh 254
A836
A897
A9
A99
A882
Sordale
Knockdee
Roadside
Braal Castle
Halkirk
Georgemas Junc. Sta
Broubster
Loch Calder
Brawlbin
Scotscalder Sta
Dorrery
Olgrinmore
Spittal Hill 176
Spittal
Houstry of Dunn
Mybster
Westerdale
Bowertower
Halcro
Gillock
North Watten
Hastigrow
Howe
Keiss
Mireland
Kirk
Westerloch
Killimster
Sinclair's Bay
Mains of Watten
Watten
Reiss
Ackergill
Noss Head
Bilbster
Sibster
Strath
Staxigoe
WICK JOHN O'GROATS
Wick (Inbhir Ùige)
Janetstown
Castle of Old Wick
Badlipster
Tannach
Loch Hempriggs
Thrumster
Sarclet
Grey Cairns of Camster
Hill of Yarrows 213
Cnoc an Earrannaiche 211
Cairn of Get
Ulbster
Prehistoric Stones
Roster
Upper Lybster
Occumster
Lybster
Swiney
Rumster Forest
Ben-a-chielt 287
Houstry
Latheron
Latheronwheel
Dunbeath
Heritage Centre
Knockally
Ramscraigs
Borgue
Newport
Berriedale
Langwell House
Langwell Forest
Scaraben 626
Meall na Caorach 396
Braemore
Morven 706
Gobernuisgeach
Cnoc an Eireannaich 517
Creag Scalabsdale 555
Ousdale
Creag Thoraraidh 404
Kilphedir
Torrish
Eldrable Hill 416
Marrel
Navidale
Helmsdale
Timespan Museum and Arts Centre
West Helmsdale
Gartymore
Portgower
Lothmore
Lothbeg
Lothbeg Point
Ben Uarie 623
Beinn Dhorain 628
Meallan Liath Beag 479
Col-bheinn 538
Gordonbush
Loch Brora
West Clyne
Clynelish
Brora
Highland Wildcat Trails
Dunrobin Castle Sta
Dunrobin Castle
Golspie
Culmaily
Backies
Beinn Lunndaidh 446
Ben Horn 521
Farlary
Knockarthur
East Langwell
West Langwell
Sciberscross
Strath Brora
Cnoc Leamhnachd 293
Grumby Rock 299
Dalbreck
Kildonan Lodge
Craggie
Creag nam Fiadh 387
Altanduin
Kinbrace
Achentoul
Knockfin Heights 438
Lochside
Achentoul Forest
Loch an Ruathair
Ben Griam Mhòr 590
Ben Griam Beg 580
Loch Druim a' Chliabhain
Beinn a' Mhadaidh 403
Forsinard
Forsinain
Sletill Hill 280
Strath Halladale
Altnabreac Sta
Lochdhu Hotel
Loch More
Achavanich
Dalganachan
Ben Alisky 349
Beinn Nam Bad Mòr 290
Loch Shurrery
Loch Caluim
Trantlemore
Trantlebeg
Dalhalvaig
Craigtown
Achiemore
Skelpick
Achargary
Rhifail
Skail
Syre
Strathnaver
Borgie Forest
Beinn Stumanadh 527
Loch Naver
Badanloch Lodge
Loch Rimsdale
Loch nan Clàr
Loch Badanloch
Gearnsary
Loch an Alltan Fhearna
Meall Ailein 721
Loch Choire Forest
Loch Choire Lodge
Meall nan Aighean 694
Cnoc an Liath-bhaid Mhòir 474
Creag a' Choire Ghlais 704
Meall a' Bhata 581
Creag Mhòr 713
Ben Armine Forest
Creag Riabhach na Greighe 461
Strath na Seilge
Strath Skinsdale
Meallan Liath Mòr 461
Meall a' Phiobaire 372
Dalnessie
Sithean Achadh nan Eun 317
Muie
Ardachu
Strath Fleet
Rogart
Rogart Sta
Pittentrail
Morvich
Little Torboll
Kirkton
Loch Fleet NNR
Littleferry
Skelbo
Skelbo Street
Embo
Embo Street
Badninish
Rearquhar
Achvaich
Clashban
Clashmore
Evelix
Dornoch
Dornoch Cathedral
Coast Candle Co.
Craft Centre
Dornoch Point
DORNOCH FIRTH
Spinningdale
Bonar Bridge
Tulloch
Easter Fearn
Struie Hill
Edderton
Tain
Morangie Forest
Aultnamain Inn
Tharsuinn 692
Wilkhaven
Tarbat Ness
Portmahomack
Rockfield
Inver
Tarrel
Rhynie
Hill of Fearn
Fearn Abbey
Hilton of Cadboll
Balintore
Shandwick
Kildary
Tongue Bay
Caol Raineach
A839
A949
A882
B874
B870
B876
B871
B873

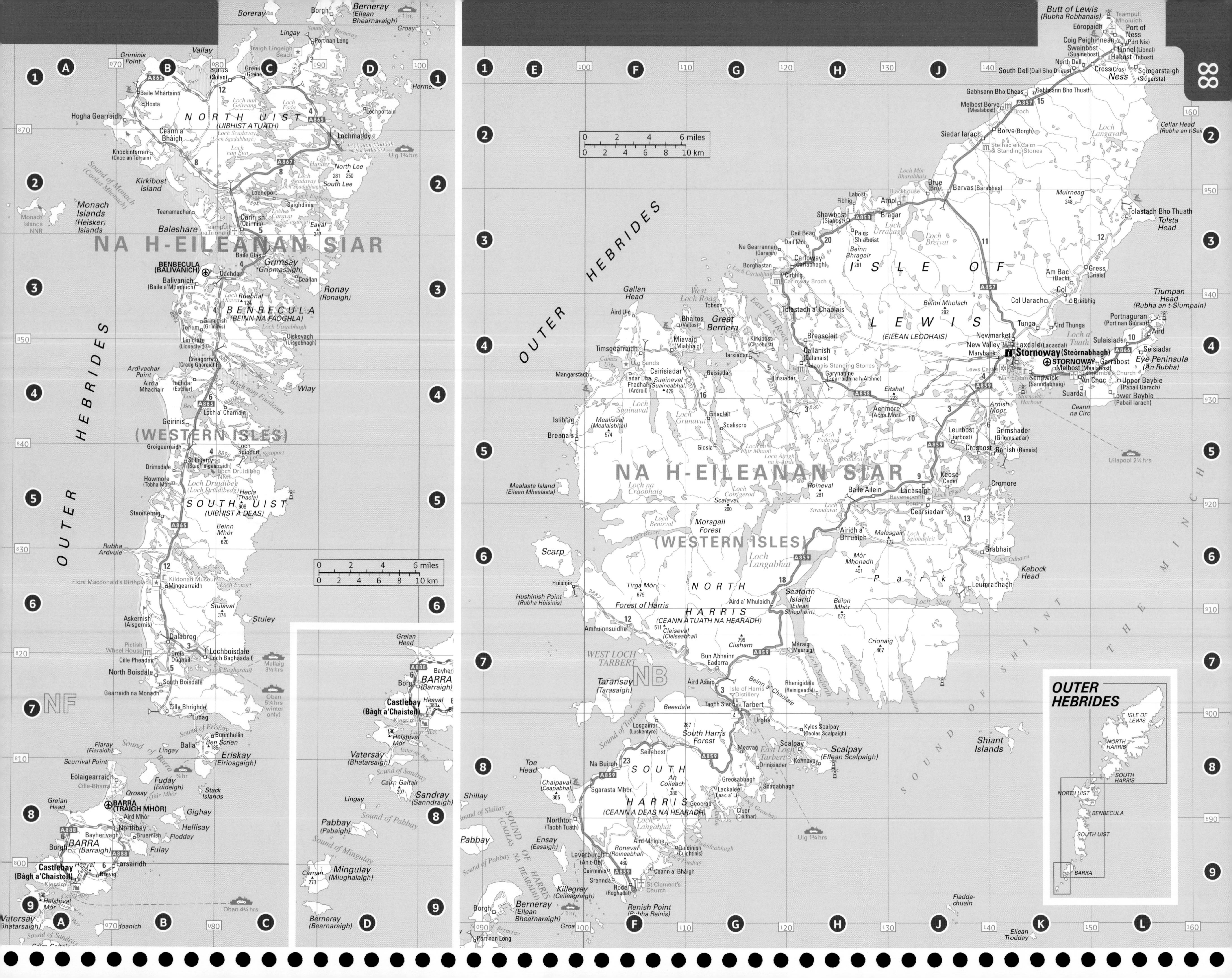
OUTER HEBRIDES
NA H-EILEANAN SIAR
(WESTERN ISLES)
NORTH UIST
(UIBHIST A TUATH)
BENBECULA
(BEINN NA FAOGHLA)
SOUTH UIST
(UIBHIST A DEAS)
BARRA
(Barraigh)
ISLE OF LEWIS
(EILEAN LEODHAIS)
NORTH HARRIS
(CEANN A TUATH NA HEARADH)
SOUTH HARRIS
(CEANN A DEAS NA HEARADH)
Park
THE MINCH
SOUND OF SHIANT
SOUND OF HARRIS
(CAOLAS NA HEARADH)
Butt of Lewis
(Rubha Robhanais)
Port of Ness
Ness
Cellar Head
(Rubha an t-Seil)
Tolstadh Bho Thuath
Tolsta Head
Tiumpan Head
(Rubha an t-Siumpain)
Eye Peninsula
(An Rubha)
Stornoway (Steòrnabhagh)
STORNOWAY
Ullapool 2½ hrs
Uig 1¼ hrs
Kebock Head
Shiant Islands
Scalpay
(Eilean Scalpaigh)
Tarbert
Leverburgh
(An t-Ob)
Renish Point
(Rubha Reinis)
Berneray
(Eilean Bhearnaraigh)
Pabbay
Shillay
Taransay
(Tarasaigh)
Scarp
Hushinish Point
(Rubha Huisinis)
Mealasta Island
(Eilean Mhealasta)
Gallan Head
Great Bernera
Callanish
(Calanais)
Carloway
(Carlabhagh)
Barvas (Barabhas)
Lochmaddy
Monach Islands
(Heisker)
Islands
Kirkibost Island
Baleshare
BENBECULA
(BALIVANICH)
Grimsay
(Griomasaigh)
Ronay
(Ronaigh)
Wiay
Lochboisdale
(Loch Baghasdail)
Mallaig 3½ hrs
Oban 5¼ hrs (winter only)
Eriskay
(Eiriosgaigh)
Stuley
Hecla
(Thacla)
Beinn Mhòr
Flora Macdonald's Birthplace
Kildonan Museum
BARRA
(TRAIGH MHÒR)
Castlebay
(Bàgh a'Chaisteil)
Vatersay
(Bhatarsaigh)
Sandray
(Sanndraigh)
Pabbay
(Pabaigh)
Mingulay
(Miughalaigh)
Berneray
(Bearnaraigh)
Oban 4¾ hrs
NB
NF
0 2 4 6 miles
0 2 4 6 8 10 km
OUTER HEBRIDES
ISLE OF LEWIS
NORTH HARRIS
SOUTH HARRIS
NORTH UIST
BENBECULA
SOUTH UIST
BARRA

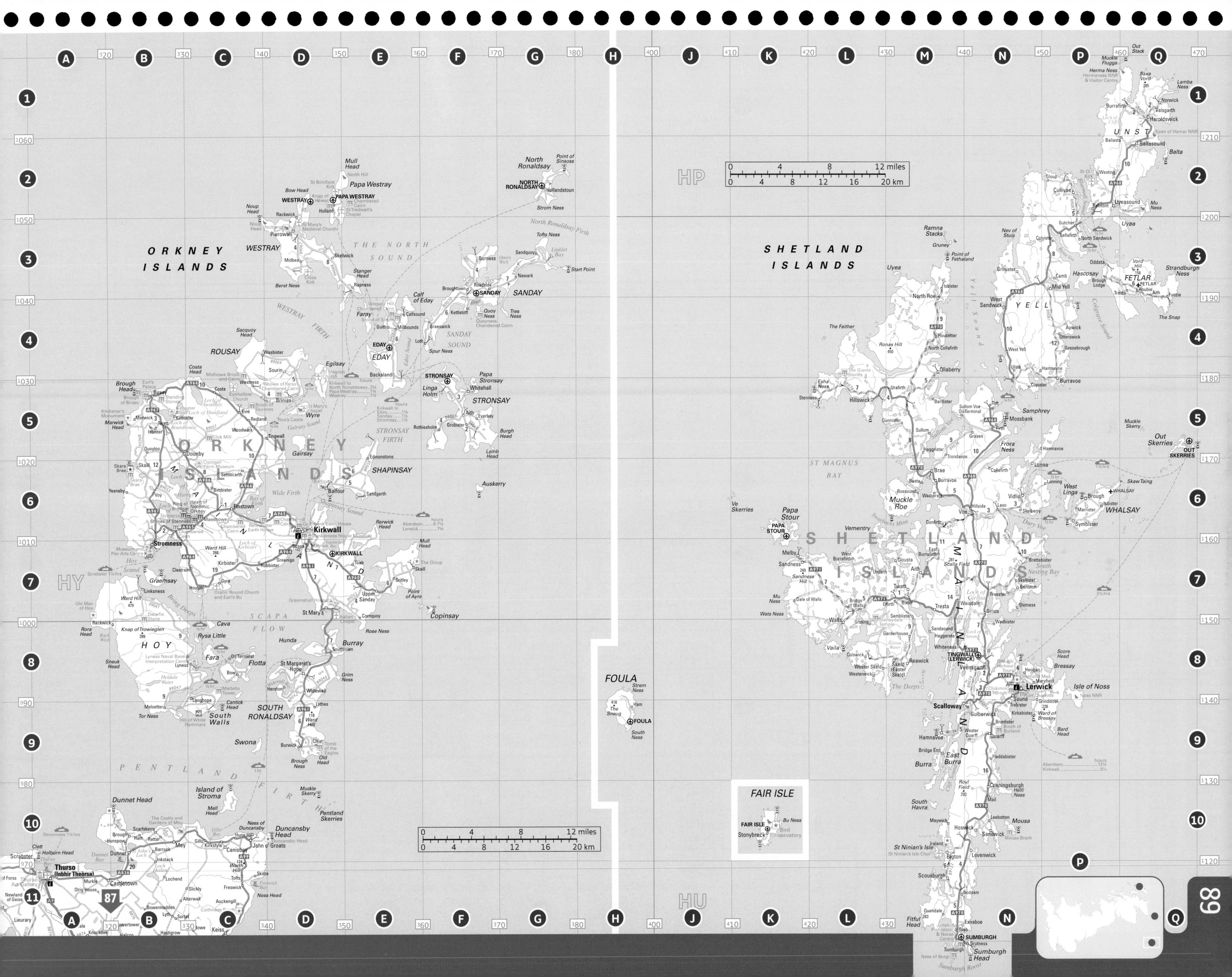
ORKNEY ISLANDS
SHETLAND ISLANDS
FAIR ISLE
FOULA
HY
HP
HU
0 4 8 12 miles
0 4 8 12 16 20 km
Kirkwall
Stromness
Lerwick
Scalloway
Thurso (Inbhir Theòrsa)
Castletown
John o' Groats
Dunnet Head
Duncansby Head
Pentland Skerries
Island of Stroma
Swona
South Ronaldsay
Burray
Hoy
Flotta
Fara
Cava
Rysa Little
Graemsay
Scapa Flow
Pentland Firth
Westray
Papa Westray
Rousay
Egilsay
Wyre
Gairsay
Shapinsay
Eday
Stronsay
Sanday
North Ronaldsay
Auskerry
Copinsay
Westray Firth
The North Sound
Sanday Sound
Stronsay Firth
Wide Firth
Mainland
Unst
Yell
Fetlar
Whalsay
Out Skerries
Papa Stour
Muckle Roe
Vementry
Vaila
Bressay
Isle of Noss
Mousa
Burra
East Burra
St Magnus Bay
Yell Sound
Colgrave Sound
Sumburgh Head
Fitful Head
Sumburgh
St Ninian's Isle
Ronas Hill
Esha Ness
The Faither
Ramna Stacks
Muckle Flugga
Out Stack
87

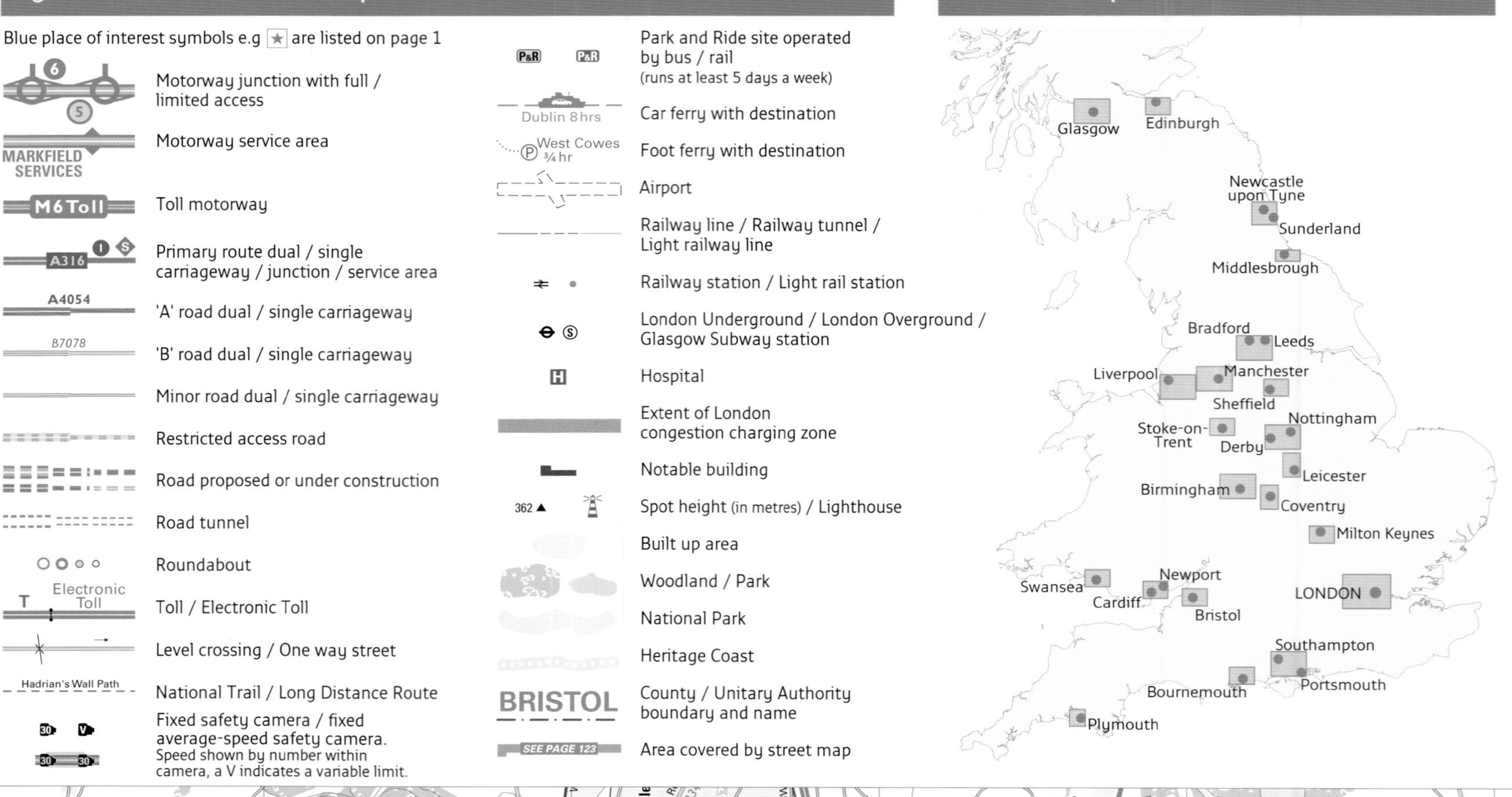

COVENTRY

Symbols used on the map

Blue place of interest symbols e.g ★ are listed on page 1

Motorway junction with full / limited access

MARKFIELD SERVICES — Motorway service area

M6 Toll — Toll motorway

A316 — Primary route dual / single carriageway / junction / service area

A4054 — 'A' road dual / single carriageway

B7078 — 'B' road dual / single carriageway

Minor road dual / single carriageway

Restricted access road

Road proposed or under construction

Road tunnel

Roundabout

Toll / Electronic Toll

Level crossing / One way street

Hadrian's Wall Path — National Trail / Long Distance Route

Fixed safety camera / fixed average-speed safety camera. Speed shown by number within camera, a V indicates a variable limit.

P&R — Park and Ride site operated by bus / rail (runs at least 5 days a week)

Dublin 8 hrs — Car ferry with destination

West Cowes ¾ hr — Foot ferry with destination

Airport

Railway line / Railway tunnel / Light railway line

Railway station / Light rail station

London Underground / London Overground / Glasgow Subway station

Hospital

Extent of London congestion charging zone

Notable building

362 ▲ — Spot height (in metres) / Lighthouse

Built up area

Woodland / Park

National Park

Heritage Coast

BRISTOL — County / Unitary Authority boundary and name

SEE PAGE 123 — Area covered by street map

Locator map

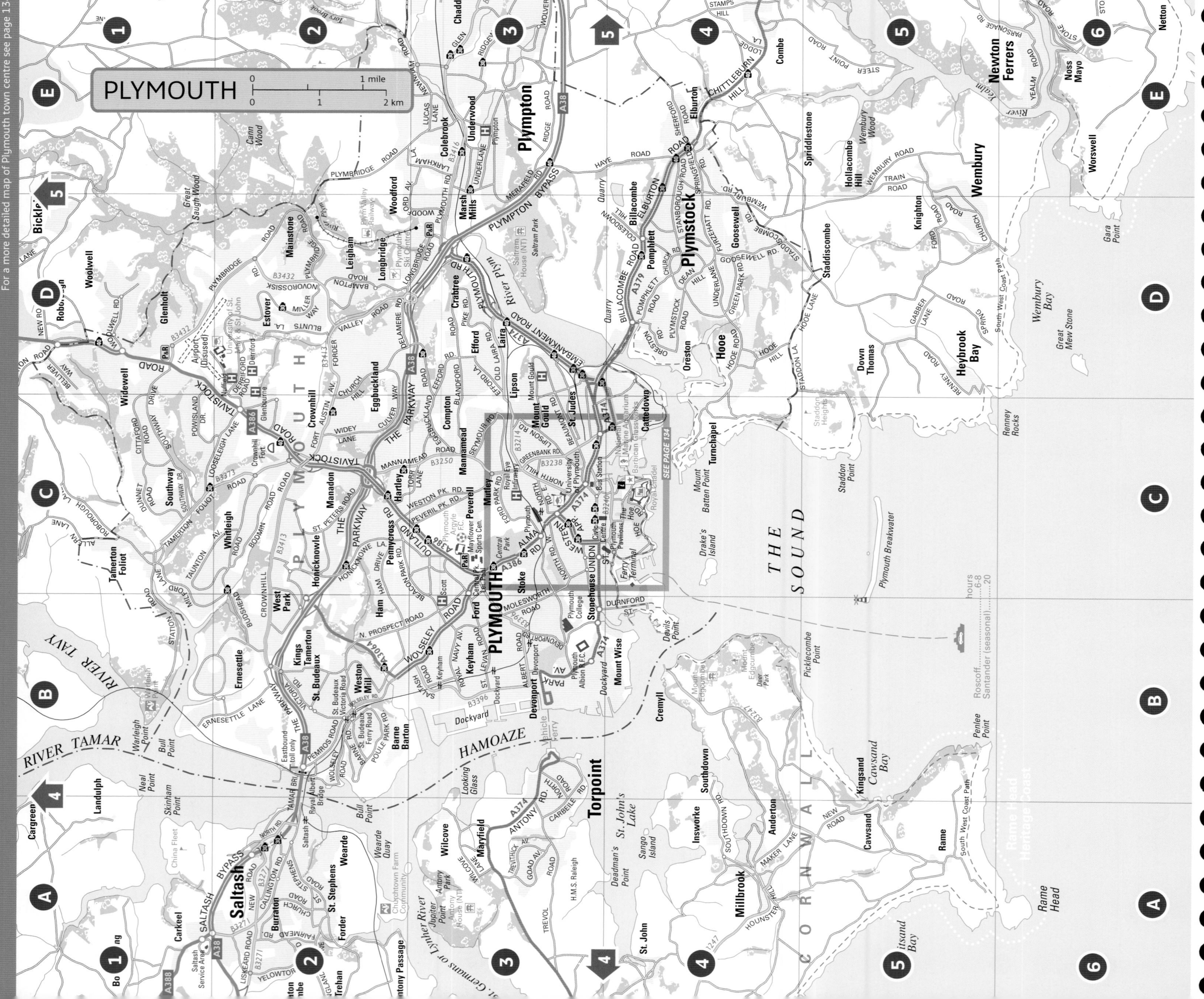

PLYMOUTH

0 — 1 mile
0 — 1 — 2 km

For a more detailed map of Plymouth town centre see page 134

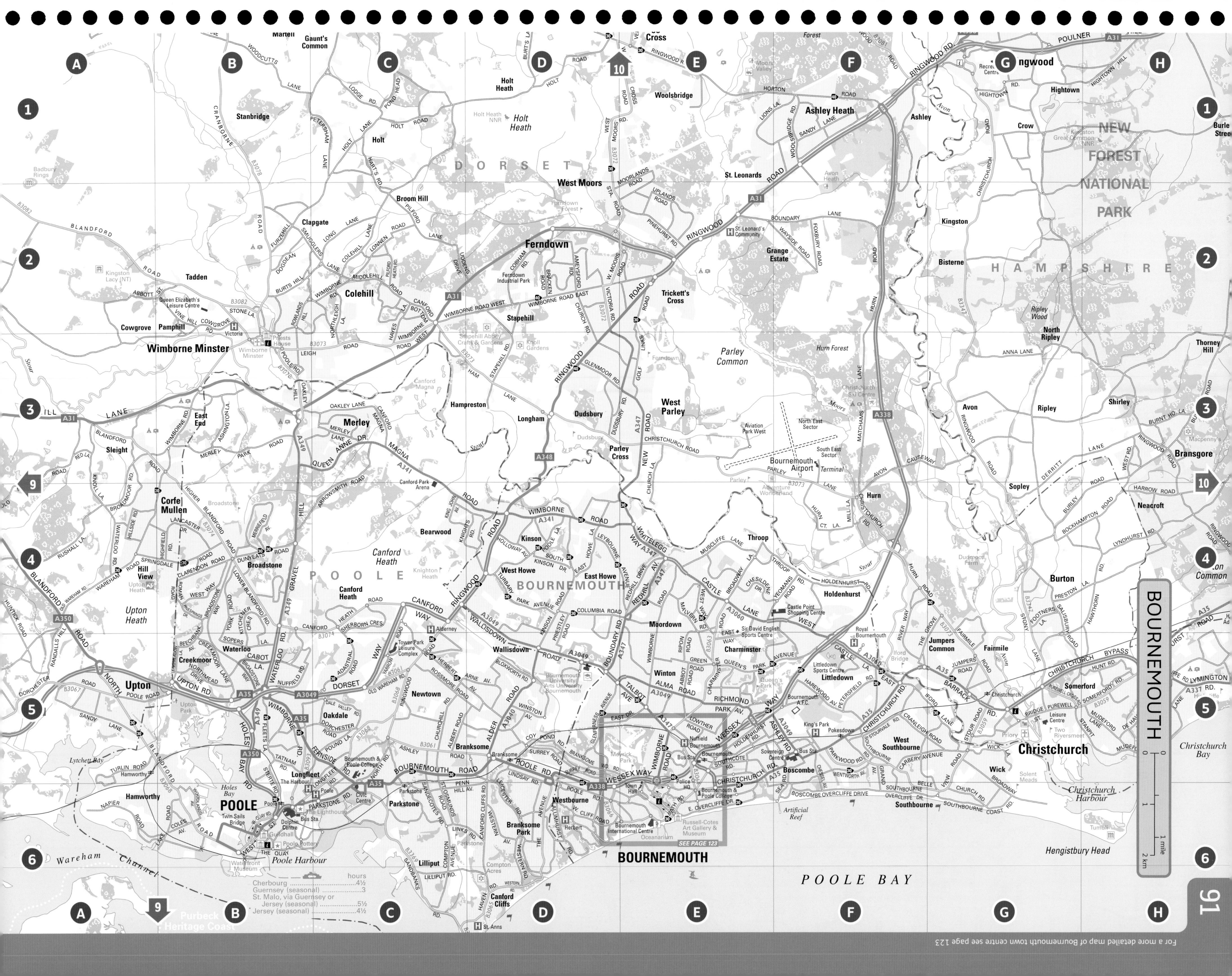

BOURNEMOUTH

For a more detailed map of Bournemouth town centre see page 123

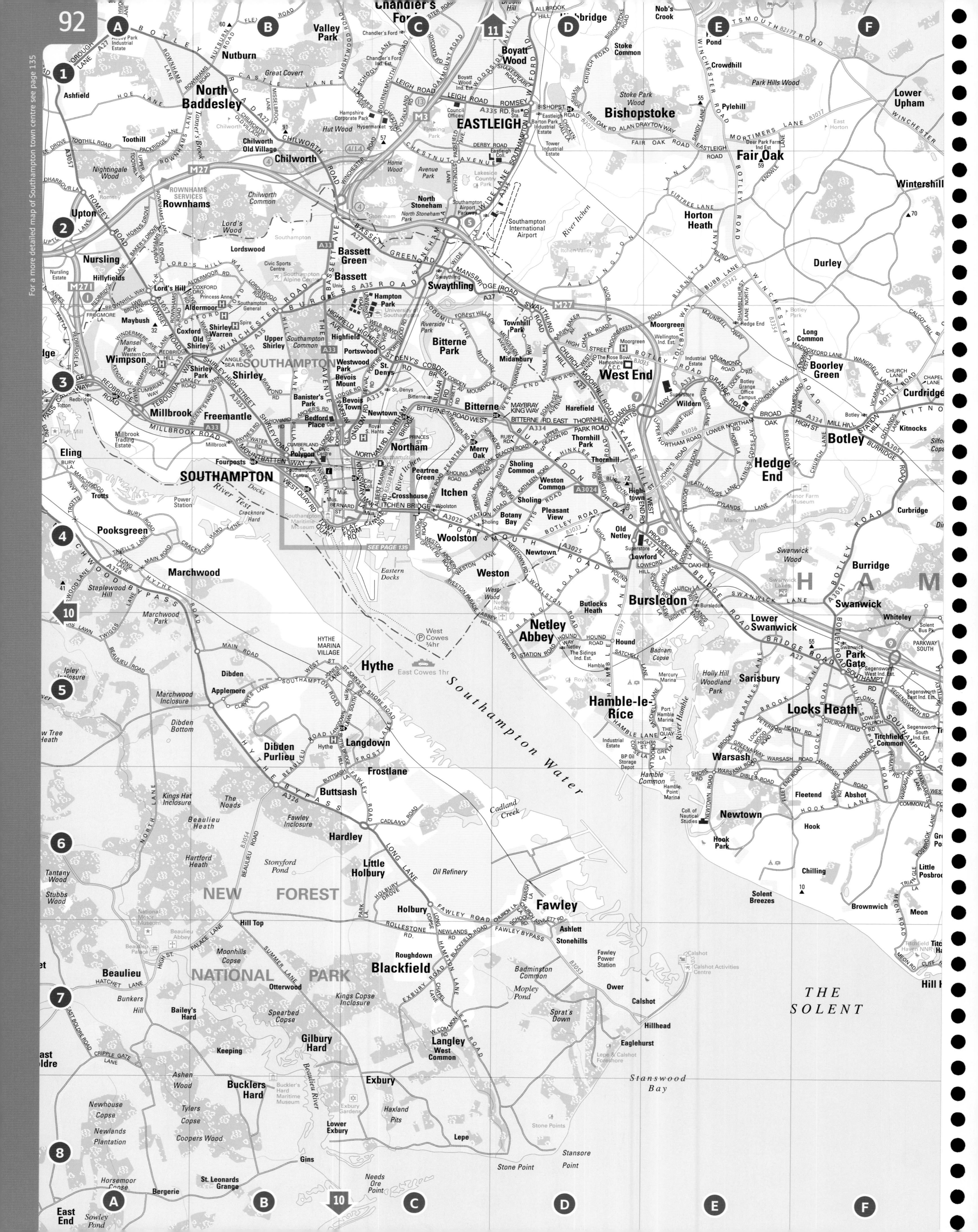

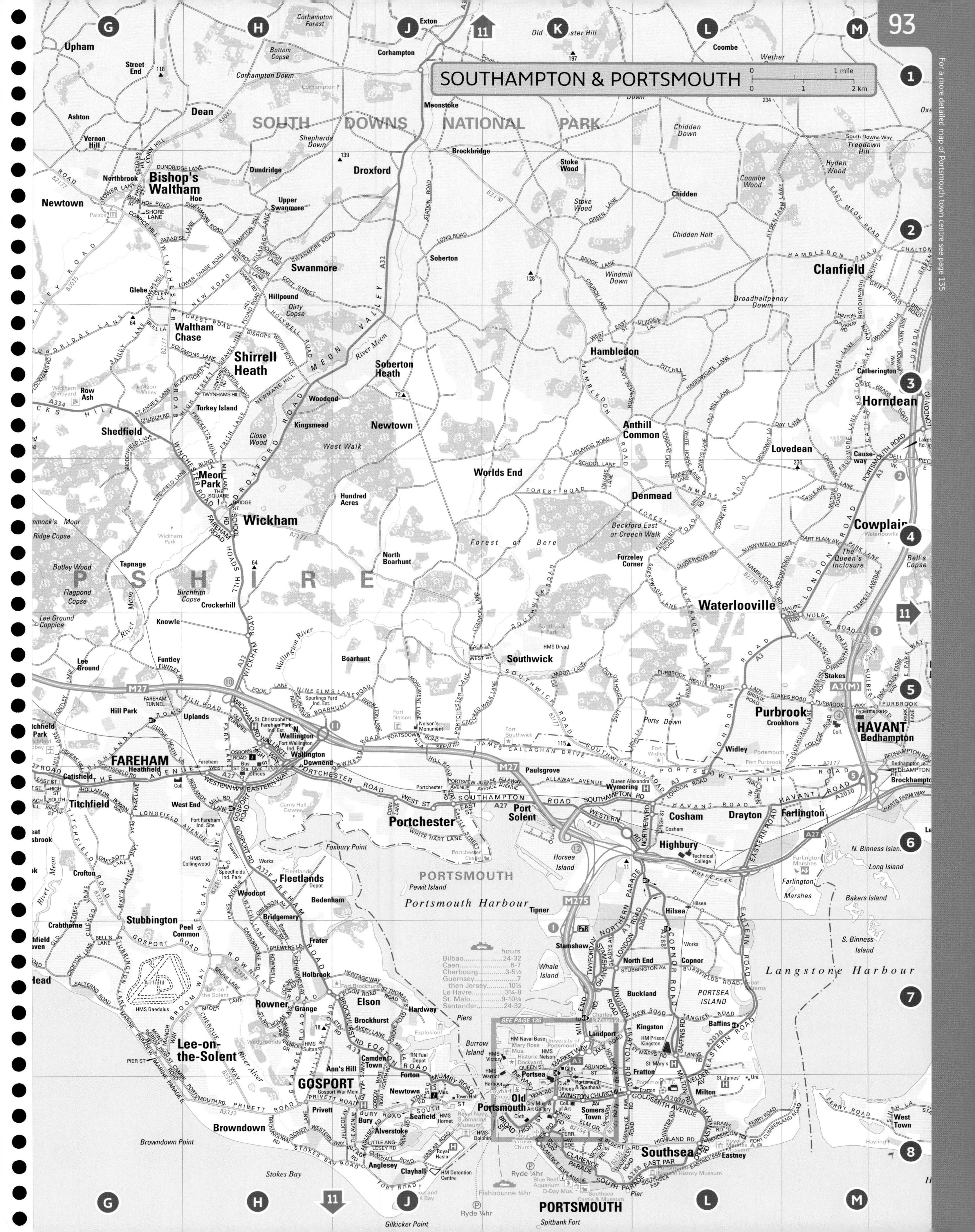
SOUTHAMPTON & PORTSMOUTH
0 1 mile
0 1 2 km
For a more detailed map of Portsmouth town centre see page 135
SOUTH DOWNS NATIONAL PARK
H A M P S H I R E
Upham
Corhampton
Exton
Meonstoke
Droxford
Dean
Bishop's Waltham
Newtown
Swanmore
Shirrell Heath
Waltham Chase
Soberton
Soberton Heath
Hambledon
Clanfield
Horndean
Catherington
Lovedean
Cowplain
Denmead
Anthill Common
Worlds End
Wickham
Shedfield
Meon Park
Waterlooville
Southwick
Purbrook
HAVANT
Bedhampton
Widley
Cosham
Drayton
Farlington
Highbury
Paulsgrove
Portchester
Port Solent
FAREHAM
Titchfield
Stubbington
Lee-on-the-Solent
Fleetlands
Bridgemary
Rowner
Elson
GOSPORT
Alverstoke
Browndown
Stokes Bay
Portsmouth Harbour
Whale Island
Stamshaw
North End
Copnor
Buckland
Kingston
Baffins
Fratton
Milton
Eastney
Southsea
Old Portsmouth
Portsea
Landport
Hilsea
PORTSEA ISLAND
Langstone Harbour
West Town
PORTSMOUTH
Spitbank Fort
Gilkicker Point
Forest of Bere
SEE PAGE 135
hours
Bilbao 24-32
Caen 6-7
Cherbourg 3-5½
Guernsey 7
then Jersey 10½
Le Havre 3¼-8
St. Malo 9-10½
Santander 24-32
Ryde ¼hr
Fishbourne ¾hr
M27
M275
A3(M)

SWANSEA

For a more detailed map of Swansea town centre see page 136

0 ___ 1 mile
0 ___ 1 ___ 2 km

For a more detailed map of Swansea town centre see page 136

SEE PAGE 136

SWANSEA (ABERTAWE)

Neath (Castell-nedd)

Port Talbot

The Mumbles

NEATH PORT TALBOT

SWANSEA BAY

Margam
Margam Sands
Margam Road
Commercial Road
Goytre
Pwll-y-glaw
Pontrhydyfen
Cwmafan
Cimla
Baglan
Afan
Aberavon
Aberavon Sands
Briton Ferry (Llansawel)
Skewen
Llandarcy
Jersey Marine
Birchgrove
Llansamlet
Peniel Green
Trallwyn
Pentre-Dwr
Winsh-wen
Bon-y-maen
Pentre-chwyth
Port Tennant
Clydach
Ynystawe
Morriston
Trewyddfa
Landore
Treboeth
Llangyfelach
Cwmdu
Townhill
Sketty
Killay
Dunvant
Upper Killay
Mayals
Norton
West Cross
Black Pill
Bishopston
Newton
Pennard
Southgate
Kittle
Ilston
Three Crosses
Gowerton
Waunarlwydd
Penllergaer
Garden Village
Grovesend
Gorseinon
Loughor
Penclawdd
Llangennech
Felindre
Tircoed Forest Village
Swansea Services
Penllergaer Forest
Cadle
Portmead
Fforest-fach
Cockett
Caswell Bay
Langland Bay
Mumbles Head
Bracelet Bay
Clyne Common
Clyne Wood
Singleton Park
Swansea Airport

M4
A48
A483
A4067
A4118
A4216
A4241
A474
A465
A4107
A483
B4290
B4291
B4489
B4436

Swansea Bay
River Neath (Afon Nedd)
River Tawe (Afon Tawe)
River Loughor
Clyne River
Mynydd Drumau
Mynydd Gelliwastad
Tennant Canal
Neath Canal
Wales Coast Path

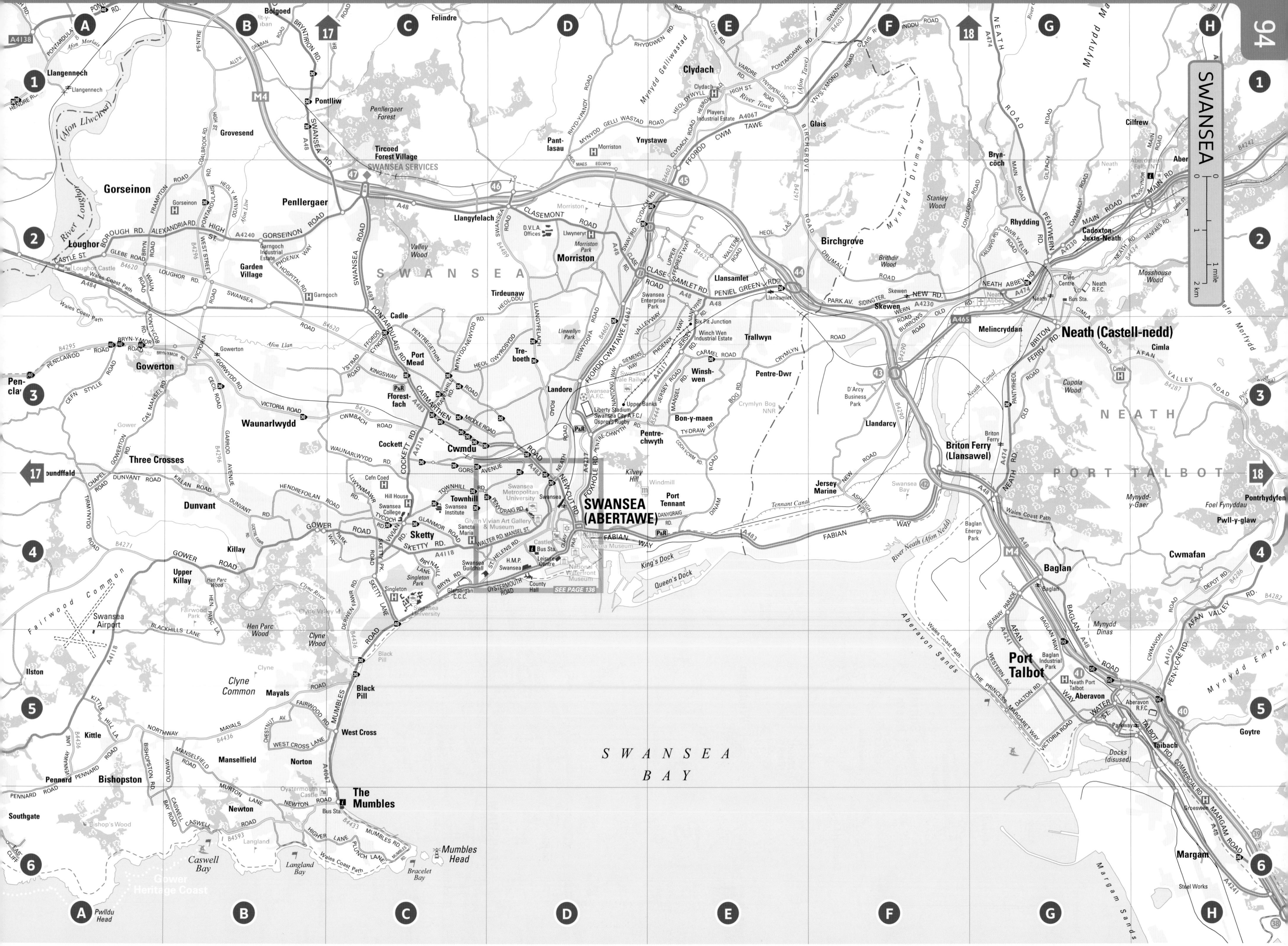

CARDIFF & NEWPORT

0   1 mile
0   1   2 km

For a more detailed map of Cardiff town centre see page 124

BRISTOL CHANNEL

NEWPORT

CARDIFF

CAERPHILLY

RHONDDA CYNON TAFF

VALE OF GLAMORGAN

St Julians, Christchurch, Lawrence Hill, Bishpool, Ringland, Alway, Somerton, Liswerry, Lliswerry, Pye Corner, Broadstreet Common, Nash, Uskmouth, Christchurch, Summerhill, Maindee, Beechwood, Barnardtown, Crindau, Brynglas, Malpas, Barrack Hill, Allt-yr-yn, Pillgwenlly, Stelvio, Mendalgief, Maes-glas, Duffryn, Gaer, Glasllwch, Ridgeway, Cefn, High Cross, Rogerstone, Bassaleg, Pye Corner, Garth, Pentre-poeth, Rhiwderin, Michaelston-y-Fedw, Castleton, Marshfield, Blacktown, St Brides Wentlooge, Peterstone Wentlooge, Coedkernew, Bryn Newydd

Bettws, Bassaleg, Machen, Chatham, Draethen, Lower Machen, Ochrwyth, Rudry, Waterloo, Nant-y-ceisiad, Craig-y-Rhacca, Trethomas, Bedwas, Greenacre, Pwllypant, Morningtown Meadows, Lansbury Park, Van, Castle Park, Caerphilly, Energlyn, Trecenydd, Penyheol, Caledfryn, Highfields, Hendredenny Park, Senghenydd, Abertridwr, Llanbradach, Nantgarw, Groes-wen, Taff's Well, Glan-y-llyn, Ty Rhiw, Morganstown, Coryton, Radyr, Pentrebane, St Fagans, Michaelston-super-Ely, Llanmaes, Caerau, Cyntwell, Leckwith, Tongwynlais, Pantmawr, Whitchurch, Llandaff North, Fairwater, Gabalfa, Llandaff, Birchgrove, Rhiwbina, Mynachdy, Pontcanna, Canton, Riverside, Blackweir, Maendy, Cathays, Heath, Llanishen, Thornhill, Lisvane, Cyncoed, Roath Park, Pen-y-lan, Roath, Tremorfa, Splott, Butetown, Grangetown, Pentwyn, Llanedeyrn, Llanrumney, Rumney, Pontprennau, St Mellons, Trowbridge, Newton, Wentloog

M4, A48, A48(M), A470, A469, A467, A468, A4232, A4161, A4119, A4232, A4050, A48, A4055

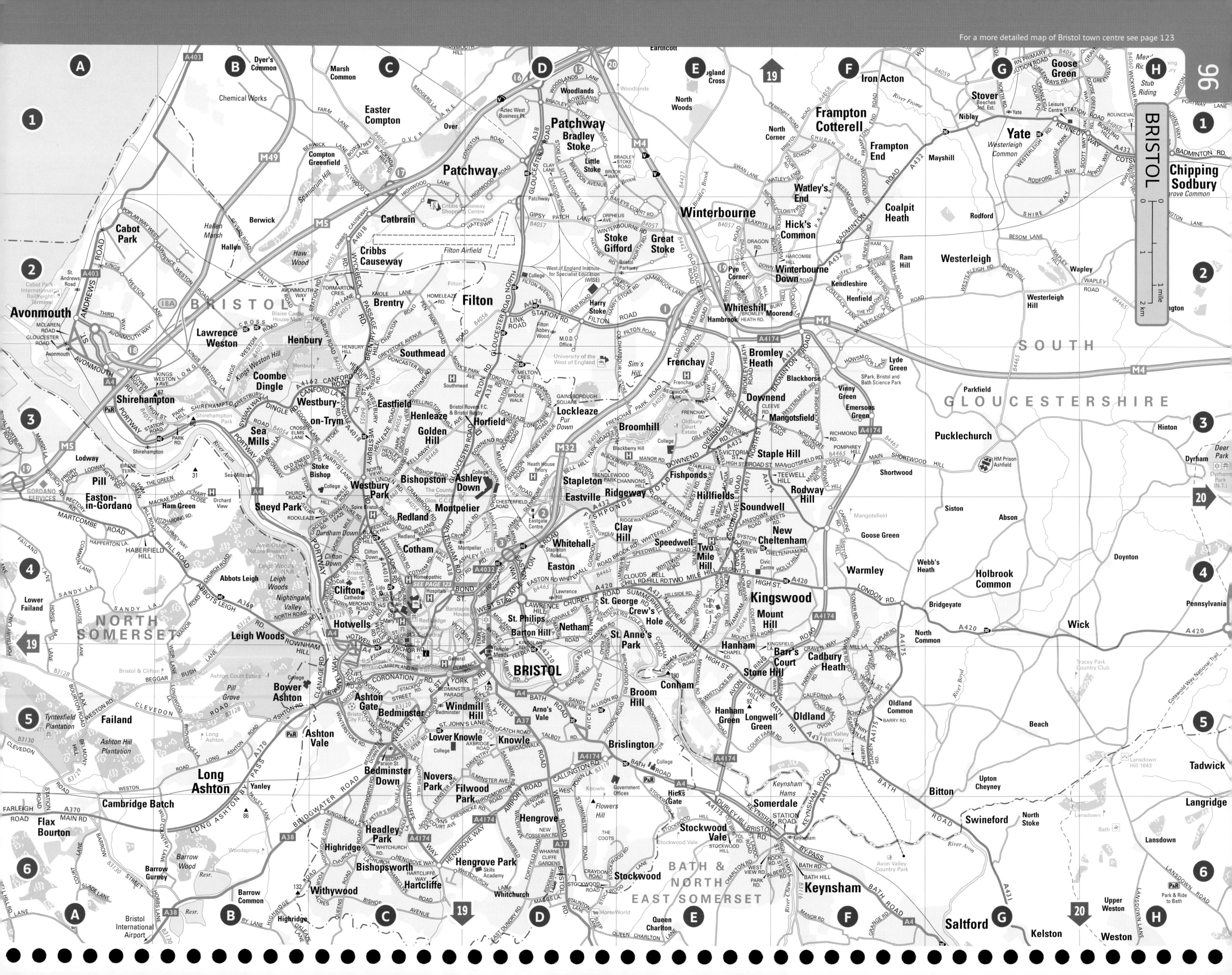

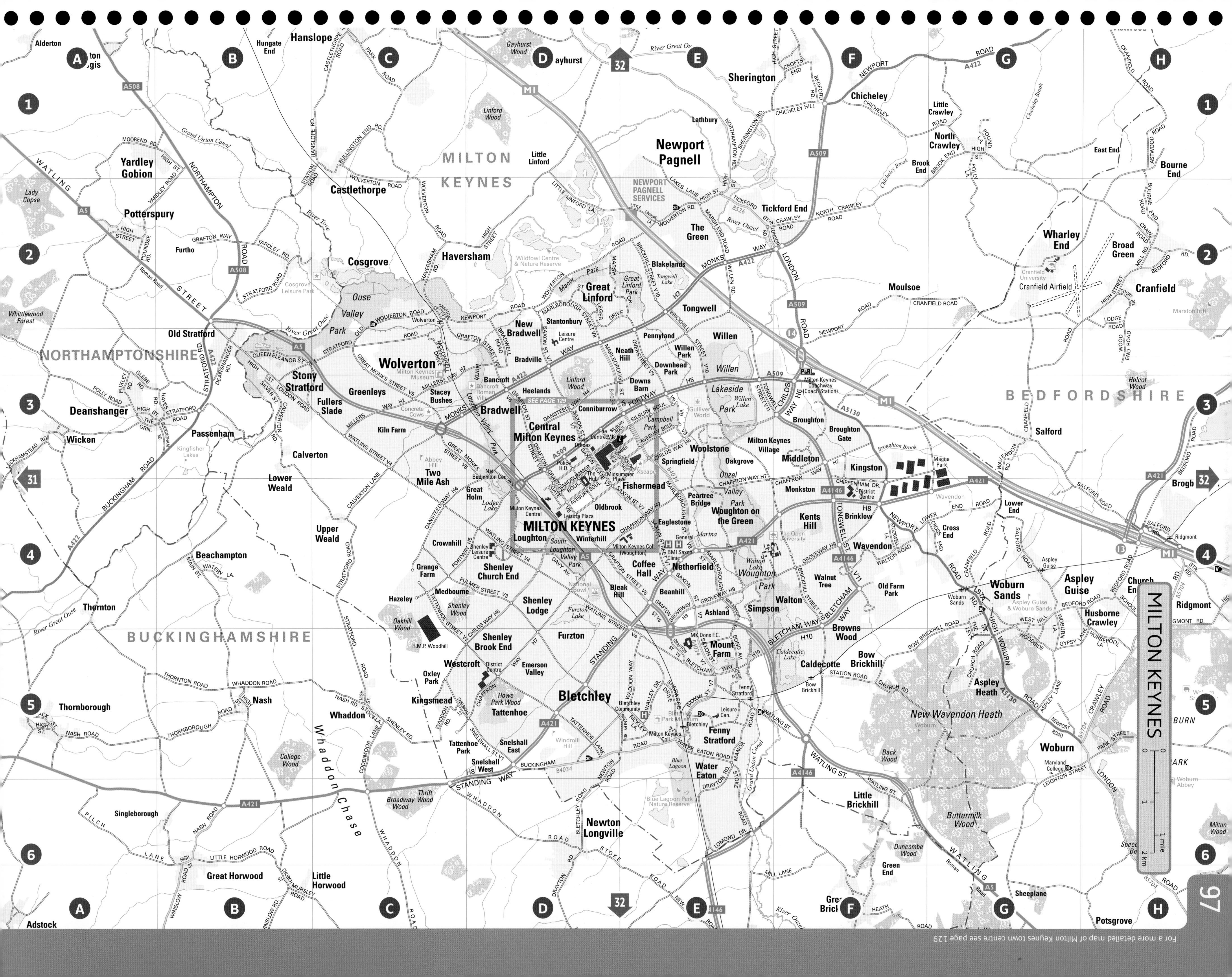

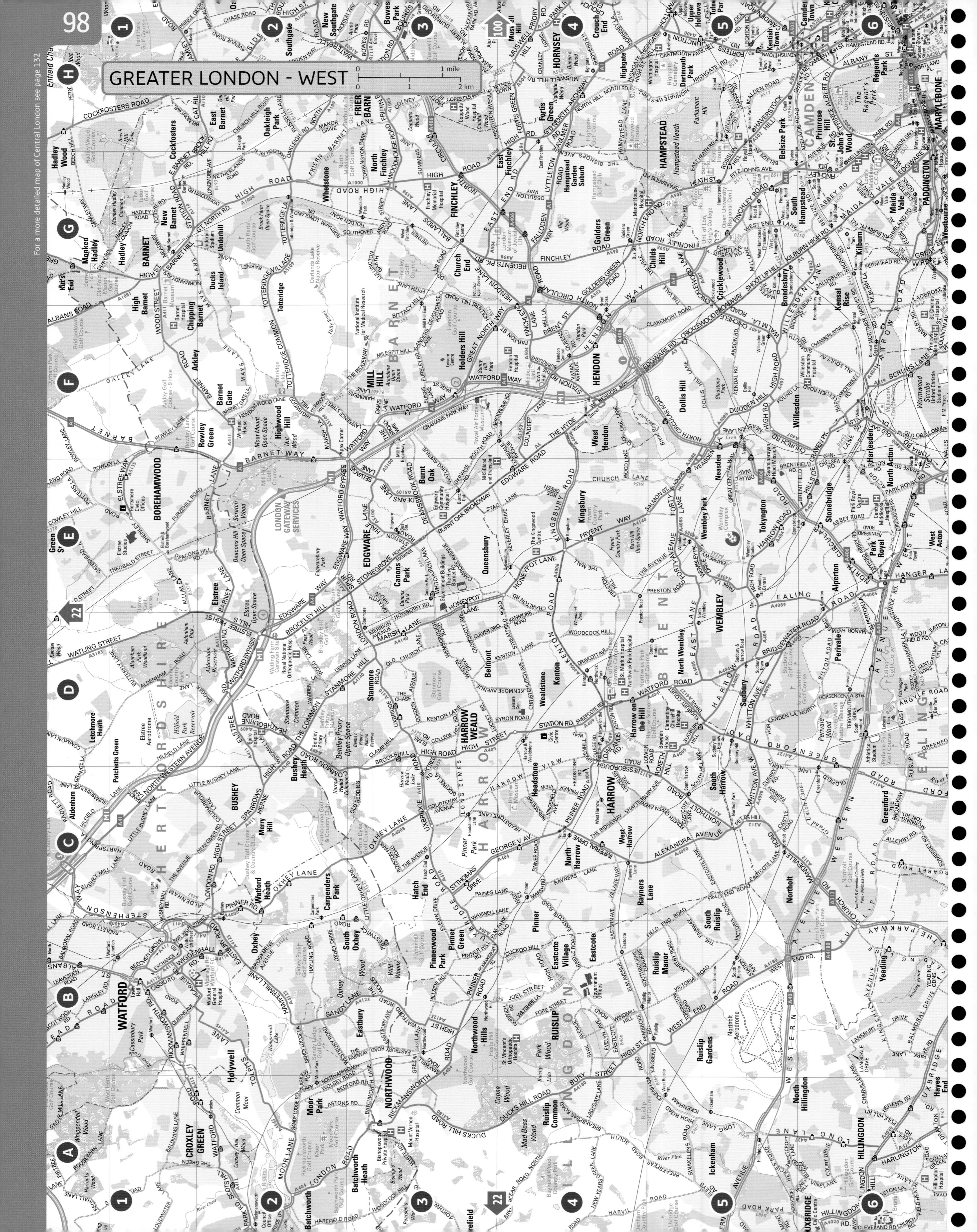

For a more detailed map of Central London see page 132

GREATER LONDON - EAST

For a more detailed map of Central London see page 132

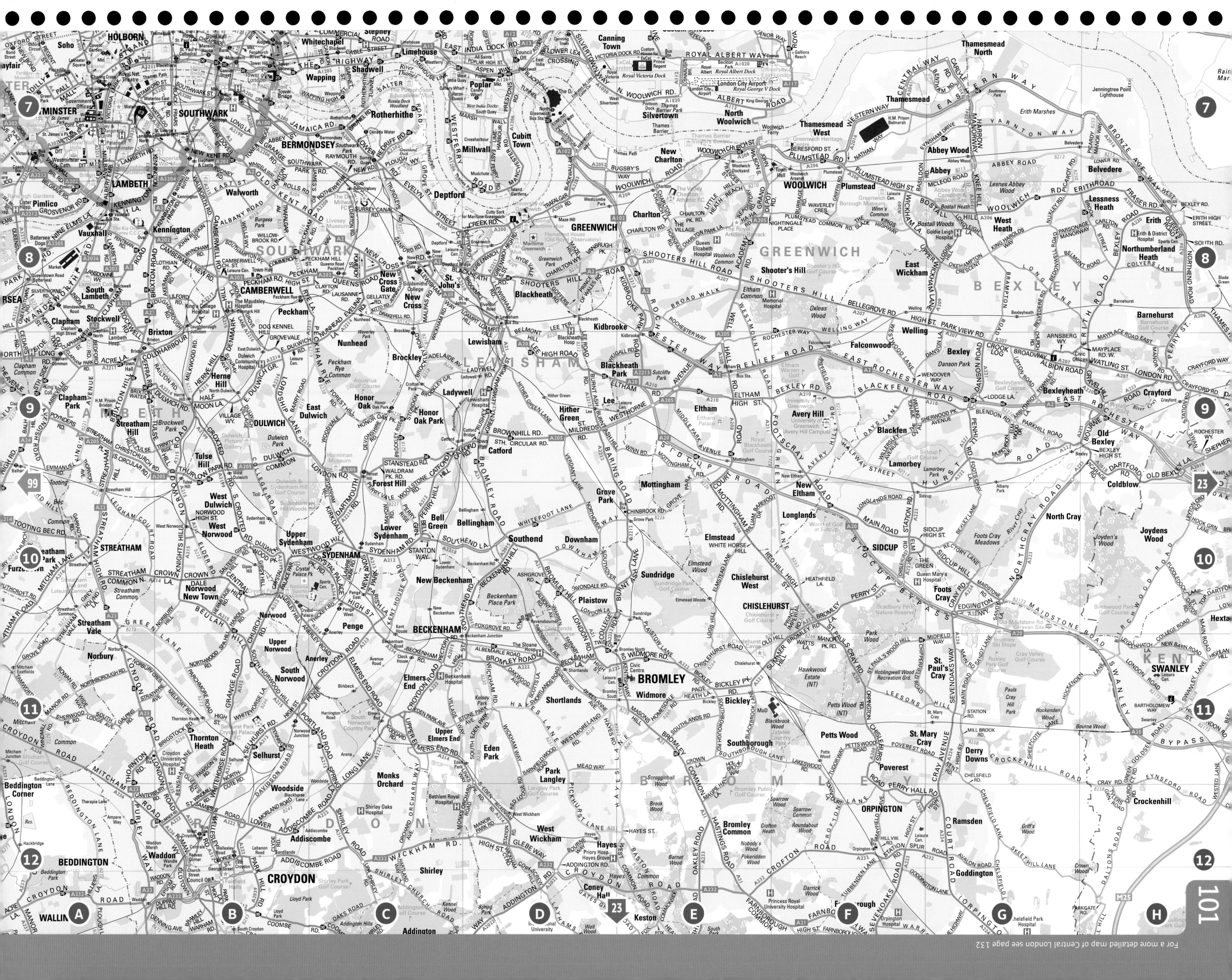

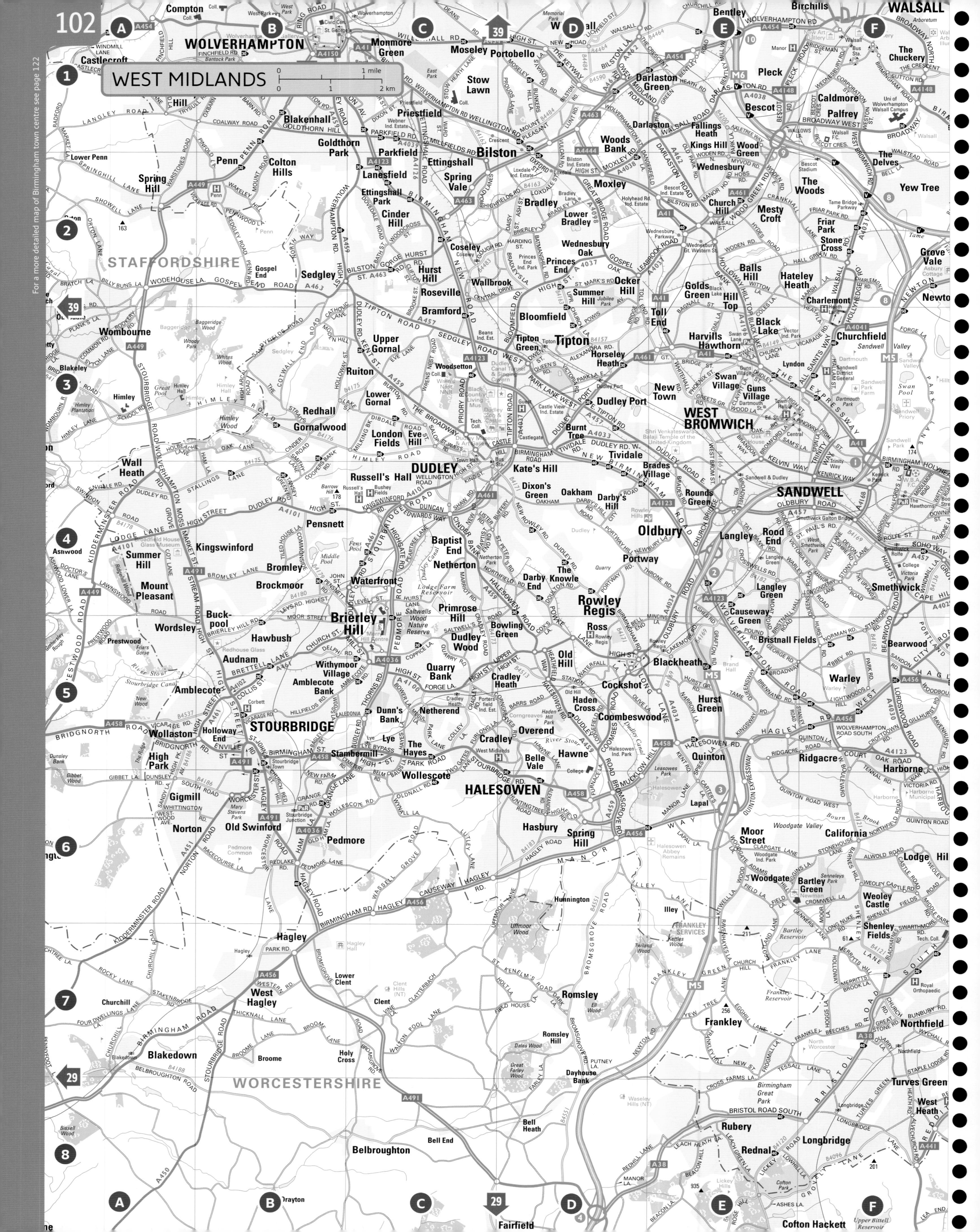

WEST MIDLANDS
1 mile
2 km
For a more detailed map of Birmingham town centre see page 122
A B C D E F
1 2 3 4 5 6 7 8
39
29
WOLVERHAMPTON
WALSALL
Compton
Castlecroft
Monmore Green
Moseley
Portobello
Bentley
Birchills
Pleck
The Chuckery
Darlaston Green
Stow Lawn
Priestfield
Blakenhall
Goldthorn Park
Parkfield
Bilston
Darlaston
Fallings Heath
Kings Hill
Wednesbury
Woods Bank
Wood Green
Bescot
Caldmore
Palfrey
Lower Penn
Penn
Colton Hills
Spring Hill
Lanesfield
Ettingshall
Ettingshall Park
Spring Vale
Moxley
The Delves
Yew Tree
The Woods
Bradley
Lower Bradley
Church Hill
Mesty Croft
Friar Park
Stone Cross
Cinder Hill
Coseley
Wednesbury Oak
STAFFORDSHIRE
Gospel End
Sedgley
Hurst Hill
Princes End
Ocker Hill
Balls Hill
Hateley Heath
Grove Vale
Wallbrook
Roseville
Summer Hill
Golds Green
Hill Top
Charlemont
Newton
Bramford
Bloomfield
Toll End
Wombourne
Upper Gornal
Tipton Green
Tipton
Black Lake
Harvills Hawthorn
Churchfield
Blakeley
Ruiton
Woodsetton
Horseley Heath
Himley
Lower Gornal
Swan Village
Guns Village
New Town
Dudley Port
Redhall
Gornalwood
WEST BROMWICH
London Fields
Eve Hill
Burnt Tree
Tividale
Wall Heath
DUDLEY
Kate's Hill
Brades Village
Russell's Hall
Dixon's Green
Oakham
Darby's Hill
SANDWELL
Rounds Green
Pensnett
Oldbury
Kingswinford
Summer Hill
Baptist End
Netherton
Langley
Rood End
Smethwick
Bromley
Portway
Brockmoor
Mount Pleasant
Waterfront
The Knowle
Darby End
Langley Green
Rowley Regis
Buckpool
Brierley Hill
Primrose Hill
Ross
Causeway Green
Wordsley
Hawbush
Bowling Green
Dudley Wood
Bristnall Fields
Bearwood
Audnam
Withymoor Village
Quarry Bank
Old Hill
Blackheath
Amblecote
Amblecote Bank
Cradley Heath
Cockshot
Warley
Hurst Green
Netherend
Haden Cross
Dunn's Bank
Coombeswood
Wollaston
STOURBRIDGE
Holloway End
High Park
Lye
Stambermill
The Hayes
Cradley
Overend
Hawne
Quinton
Ridgacre
Harborne
Belle Vale
Wollescote
HALESOWEN
Gigmill
Norton
Old Swinford
Pedmore
Hasbury
Spring Hill
Lapal
Moor Street
California
Lodge Hill
Woodgate
Bartley Green
Weoley Castle
Shenley Fields
Hunnington
Illey
FRANKLEY SERVICES
Hagley
Lower Clent
Clent
Churchill
West Hagley
Romsley
Frankley
Northfield
Blakedown
Broome
Holy Cross
Romsley Hill
Dayhouse Bank
WORCESTERSHIRE
Turves Green
West Heath
Bell Heath
Rubery
Rednal
Longbridge
Belbroughton
Bell End
Drayton
Fairfield
Cofton Hackett
M5
M6

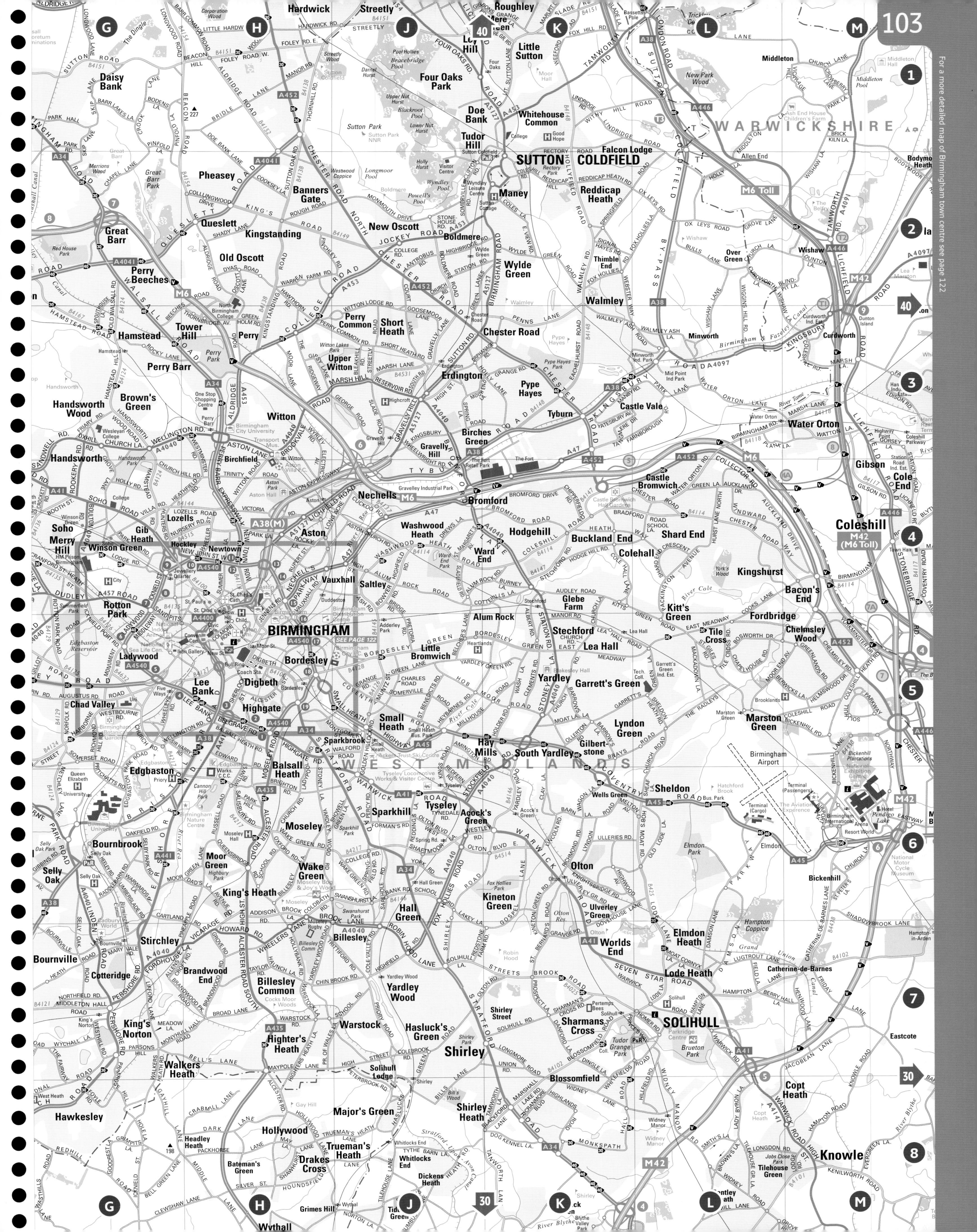

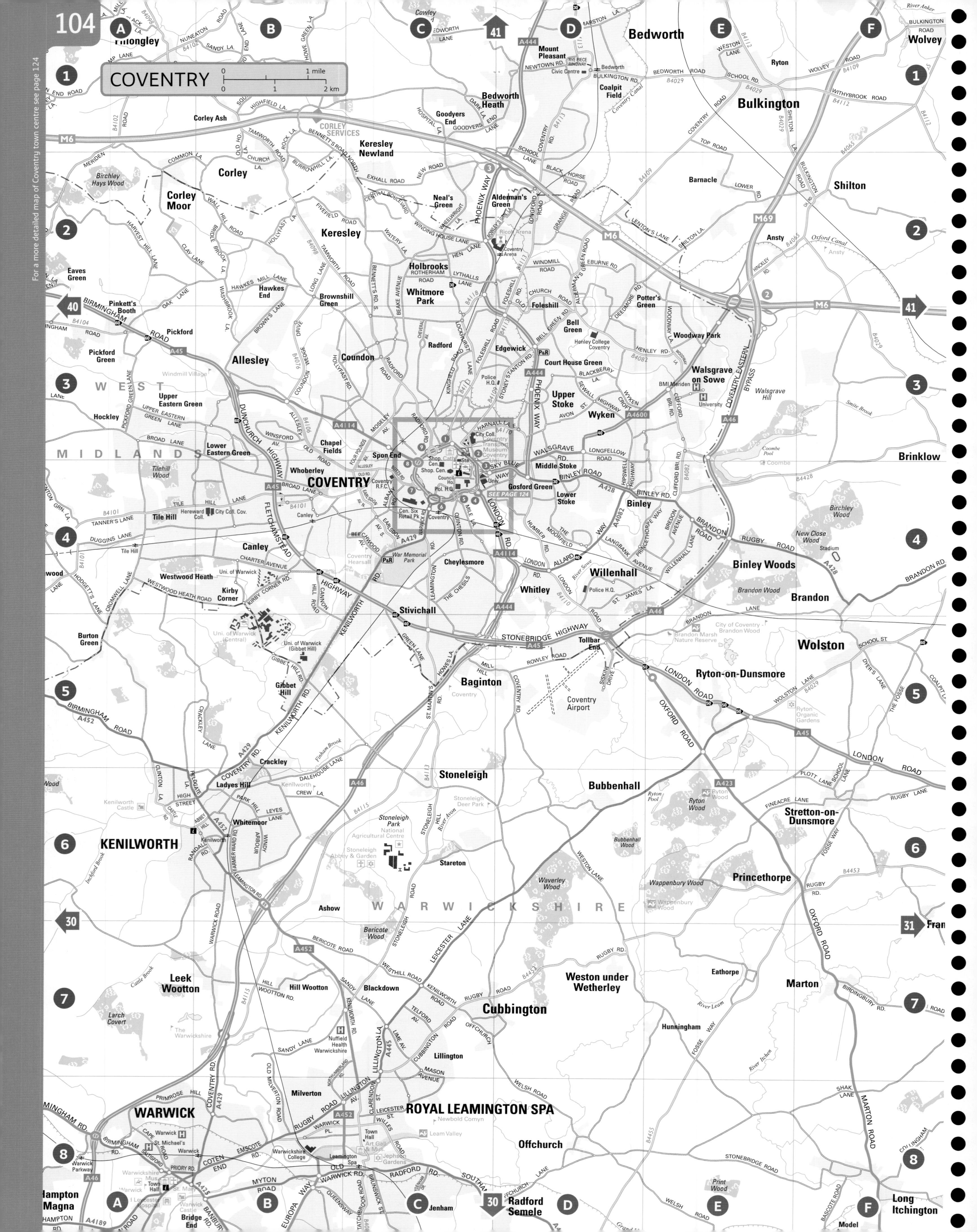

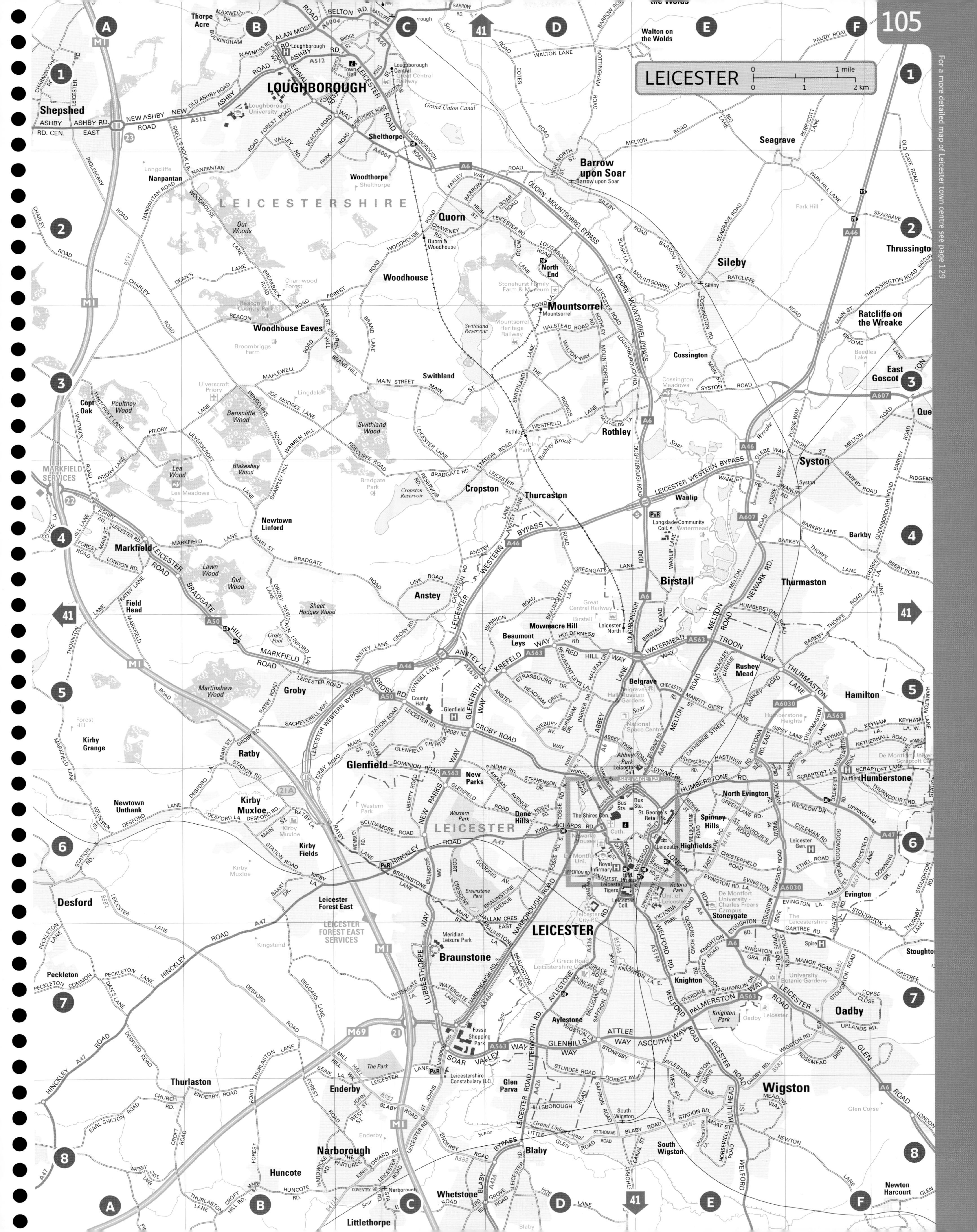

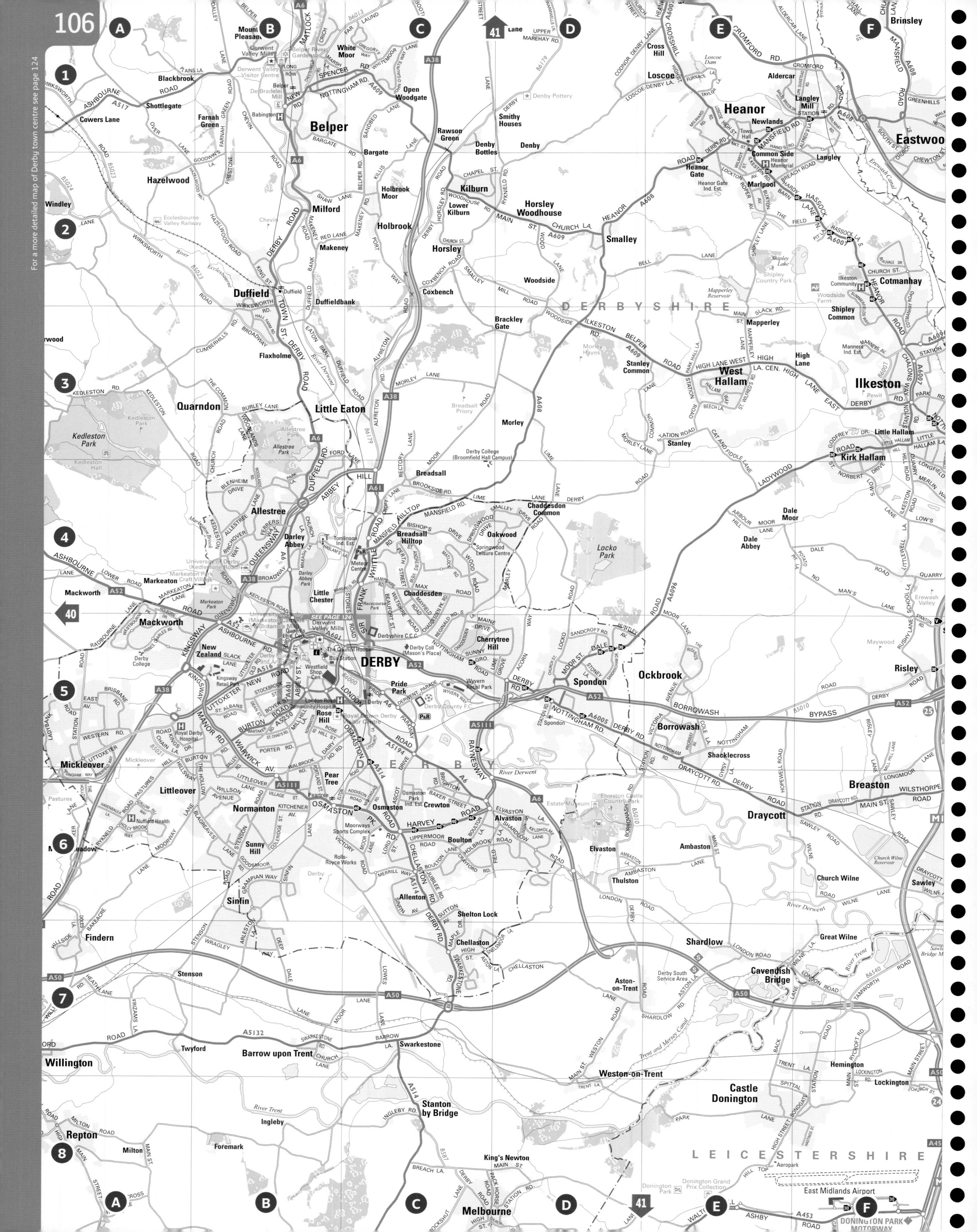

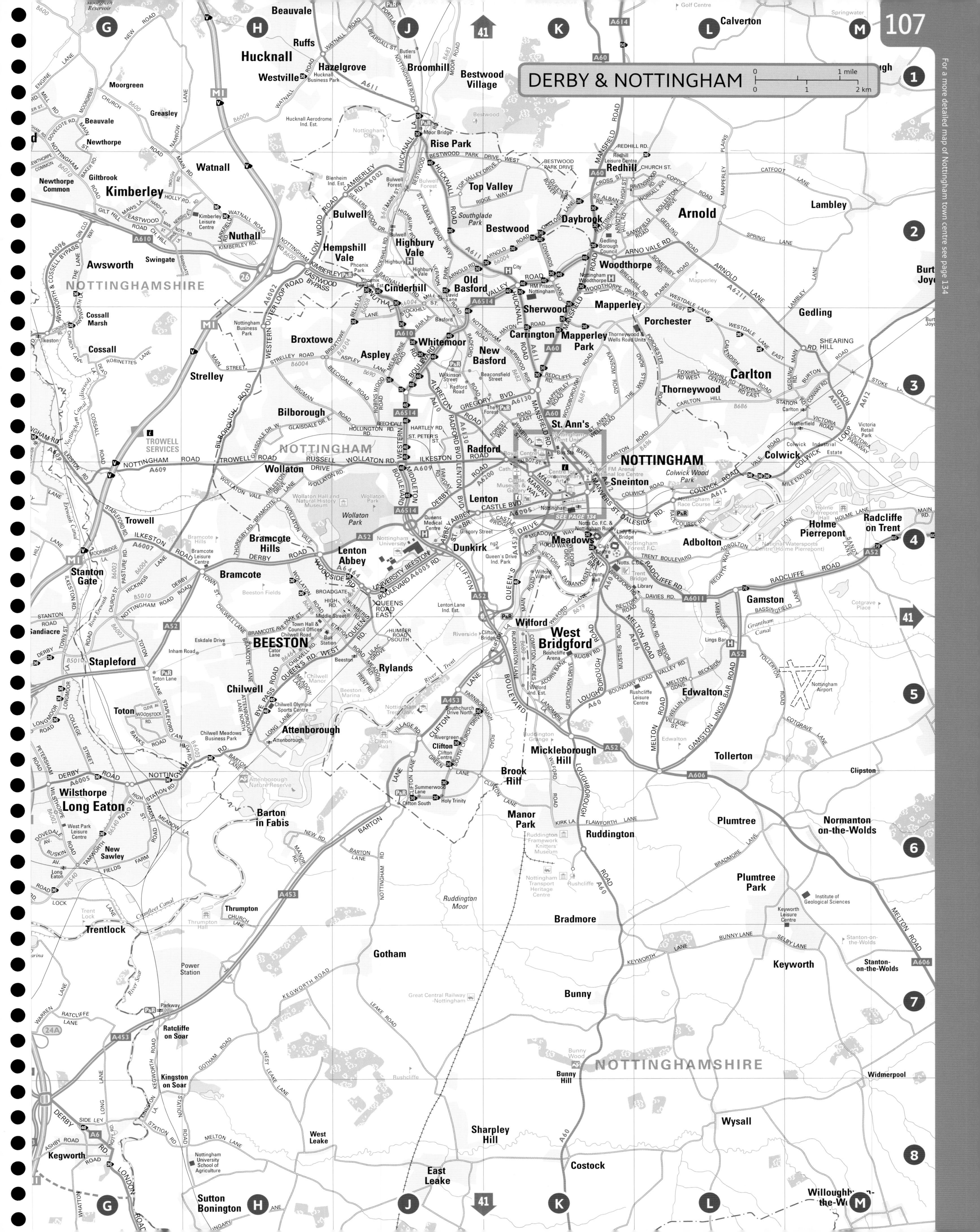
DERBY & NOTTINGHAM
0 1 mile
0 1 2 km
For a more detailed map of Nottingham town centre see page 134
NOTTINGHAMSHIRE
NOTTINGHAM
Beauvale
Ruffs
Hucknall
Hazelgrove
Westville
Broomhill
Bestwood Village
Calverton
Moorgreen
Greasley
Newthorpe
Watnall
Giltbrook
Newthorpe Common
Kimberley
Rise Park
Redhill
Top Valley
Bulwell
Nuthall
Daybrook
Arnold
Lambley
Bestwood
Hempshill Vale
Highbury Vale
Awsworth
Swingate
Woodthorpe
Old Basford
Cinderhill
Sherwood
Mapperley
Gedling
Cossall Marsh
Broxtowe
Whitemoor
Carrington
Porchester
Cossall
Aspley
New Basford
Mapperley Park
Strelley
Carlton
Thorneywood
Bilborough
TROWELL SERVICES
St. Ann's
Radford
Colwick
Wollaton
Sneinton
Colwick Wood Park
Lenton
Trowell
Bramcote Hills
Meadows
Radcliffe on Trent
Holme Pierrepont
Adbolton
Lenton Abbey
Dunkirk
Stanton Gate
Bramcote
BEESTON
Gamston
Wilford
West Bridgford
Stapleford
Rylands
Chilwell
Edwalton
Toton
Attenborough
Clifton
Mickleborough Hill
Tollerton
Brook Hill
Clipston
Wilsthorpe
Long Eaton
Barton in Fabis
Manor Park
Ruddington
Plumtree
Normanton on-the-Wolds
New Sawley
Plumtree Park
Nottingham Airport
Thrumpton
Ruddington Moor
Trentlock
Bradmore
Gotham
Keyworth
Stanton-on-the-Wolds
Bunny
Ratcliffe on Soar
Kingston on Soar
Bunny Hill
Widmerpool
Kegworth
West Leake
Sharpley Hill
Wysall
Costock
East Leake
Sutton Bonington
Willoughby-on-the-Wolds
SEE PAGE 134
41
G
H
J
K
L
M
1
2
3
4
5
6
7
8

STOKE-ON-TRENT

For a more detailed map of Stoke-on-Trent town centre see page 136

0   1 mile
0   1   2 km

STAFFORDSHIRE
CHESHIRE EAST

Major places:

- Birchall
- Cheddleton
- Cheadle
- Draycott in the Moors
- Cresswell
- Saverley Green
- Stallington
- Cellarhead
- Wetley Rocks
- Werrington
- Weston Coyney
- Caverswall
- Cookshill
- Blythe Bridge
- Meir
- Meir Heath
- Rough Close
- Bagnall
- Milton
- Abbey Hulton
- Bucknall
- Ubberley
- STOKE-ON-TRENT
- Fenton
- Longton
- Dresden
- Blurton
- Blythe Marsh
- Forsbrook
- Endon
- Brown Edge
- Norton-in-the-Moors
- Sneyd Green
- Northwood
- Hanley (City Centre)
- Etruria
- Stoke
- Hanford
- Newstead
- Wedgwood
- Barlaston
- Newchapel
- Kidsgrove
- Tunstall
- Burslem
- Cobridge
- Waterloo
- Port Hill
- Wolstanton
- Springfields
- Hartshill
- Penkhull
- Trent Vale
- Hanford
- Oak Hill
- Trentham
- Hanchurch
- Church Lawton
- Red Street
- Talke
- Talke Pits
- Red Bull
- Harding's Wood
- Acres Nook
- High Carr
- Bradwell
- Chesterton
- May Bank
- Cross Heath
- Knutton
- Newcastle-under-Lyme
- Clayton
- Seabridge
- Westbury Park
- Westlands
- Whitmore
- Baldwin's Gate
- Acton
- Butterton
- Radway Green
- Barthomley
- Alsager's Bank
- Audley
- Wood Lane
- Boon Hill
- Halmer End
- Miles Green
- Butters Green
- Scott Hay
- Silverdale
- Keele
- Keele Services
- Madeley
- Madeley Heath
- Mid Madeley
- Shraleybrook
- Whitmore
- Snapehall
- Blackbrook
- Maer Hills
- Hill Chorlton

Roads: A53, A522, A52, A521, A50, A520, A500, A34, A527, A519, A51, M6, B5051

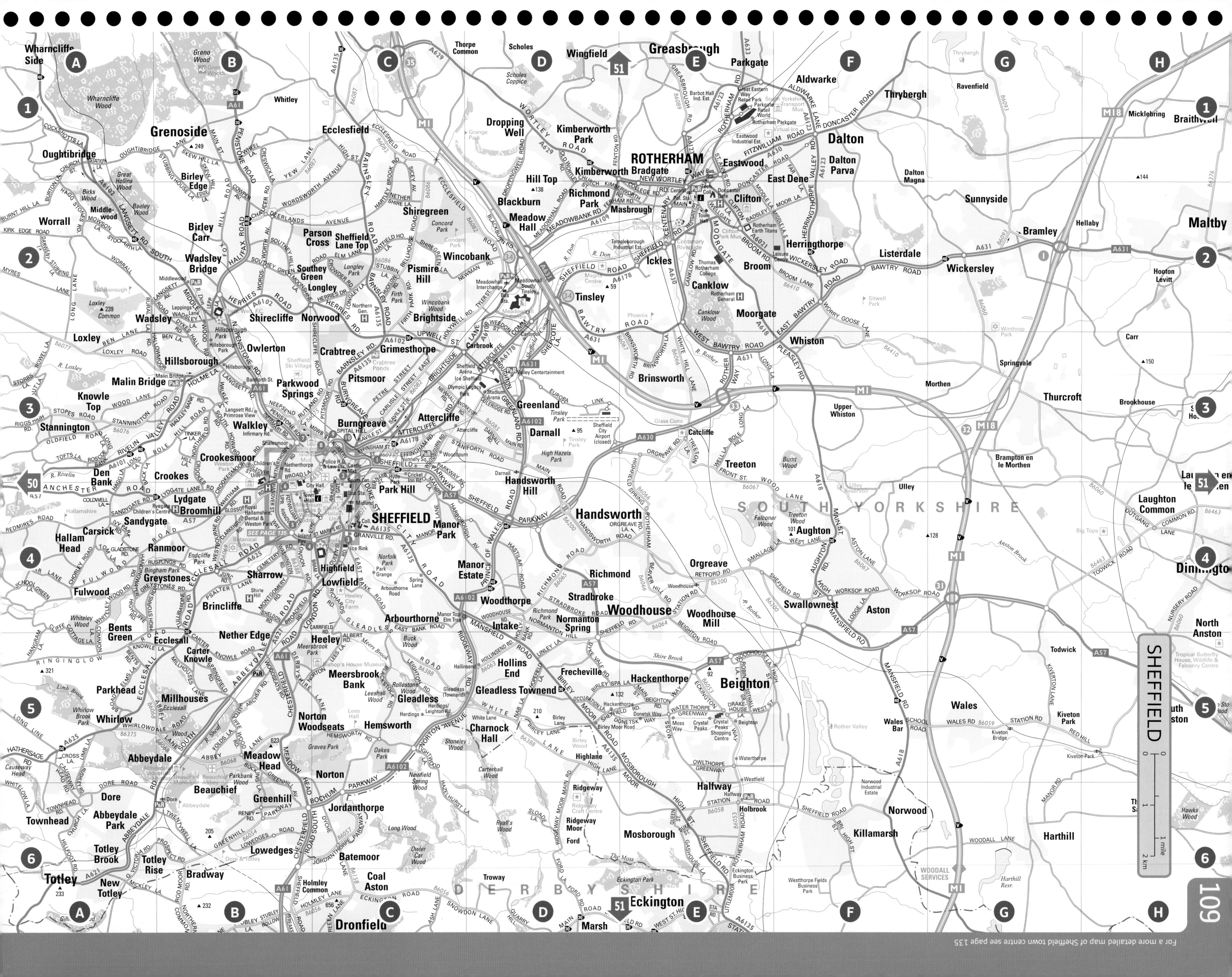

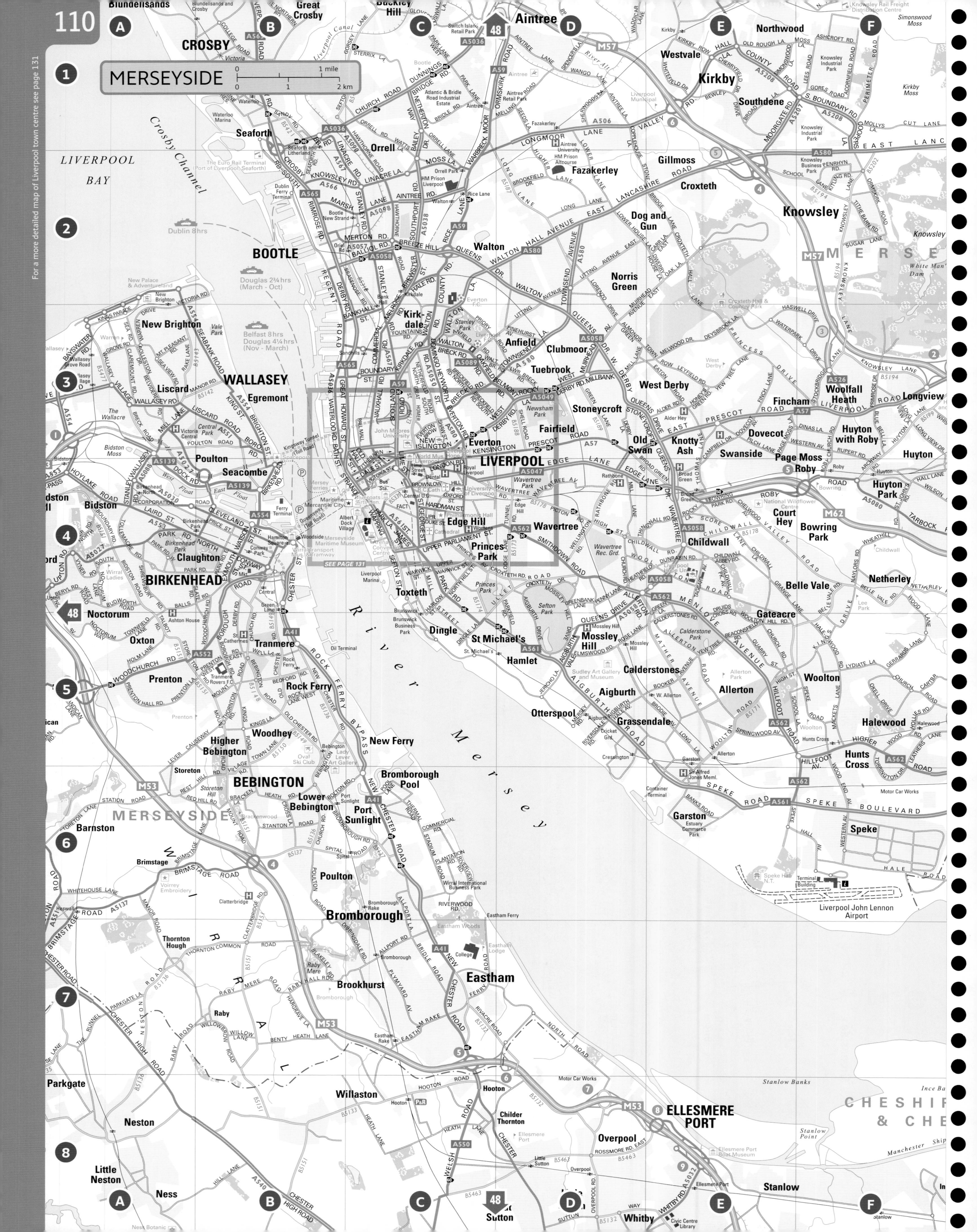

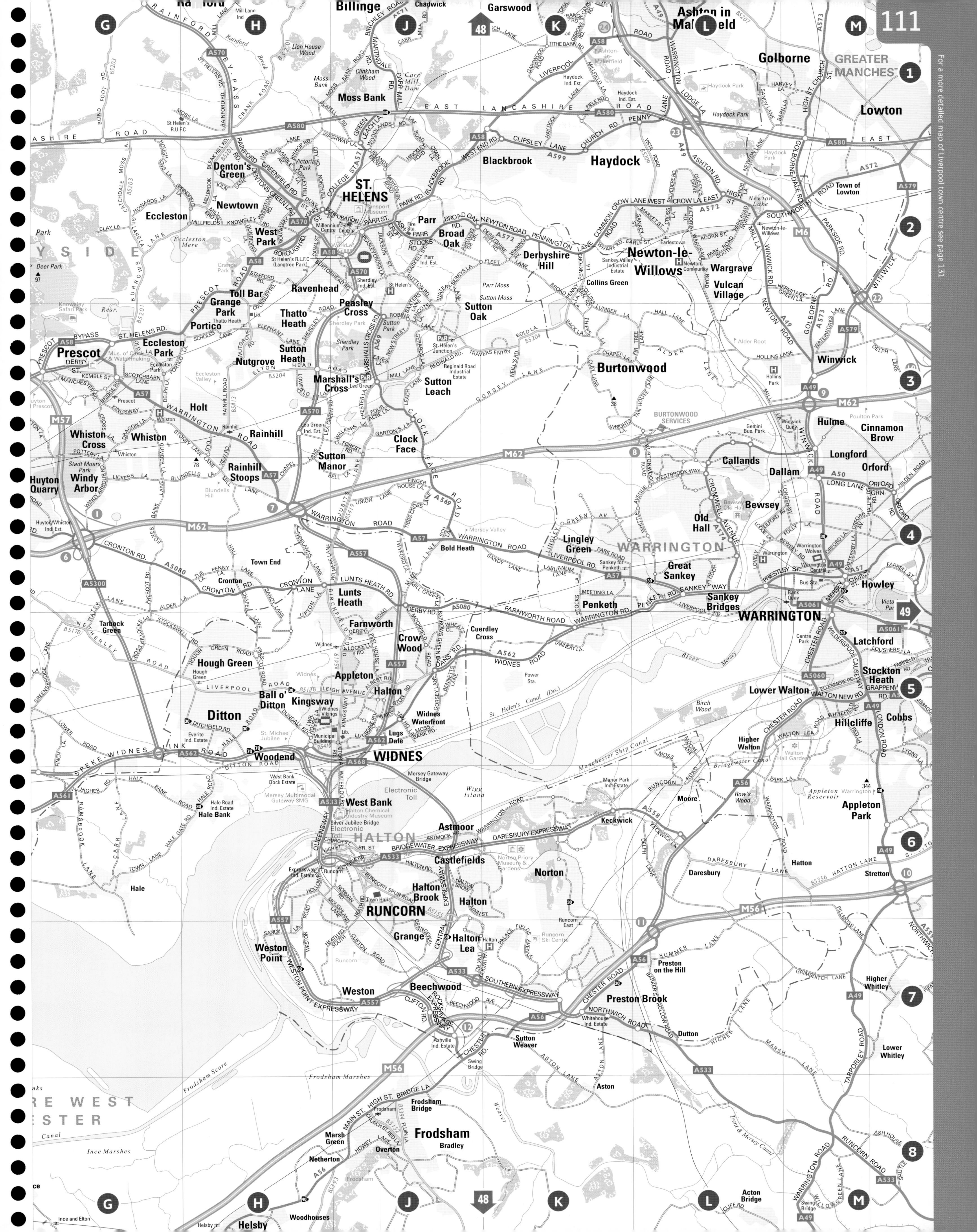

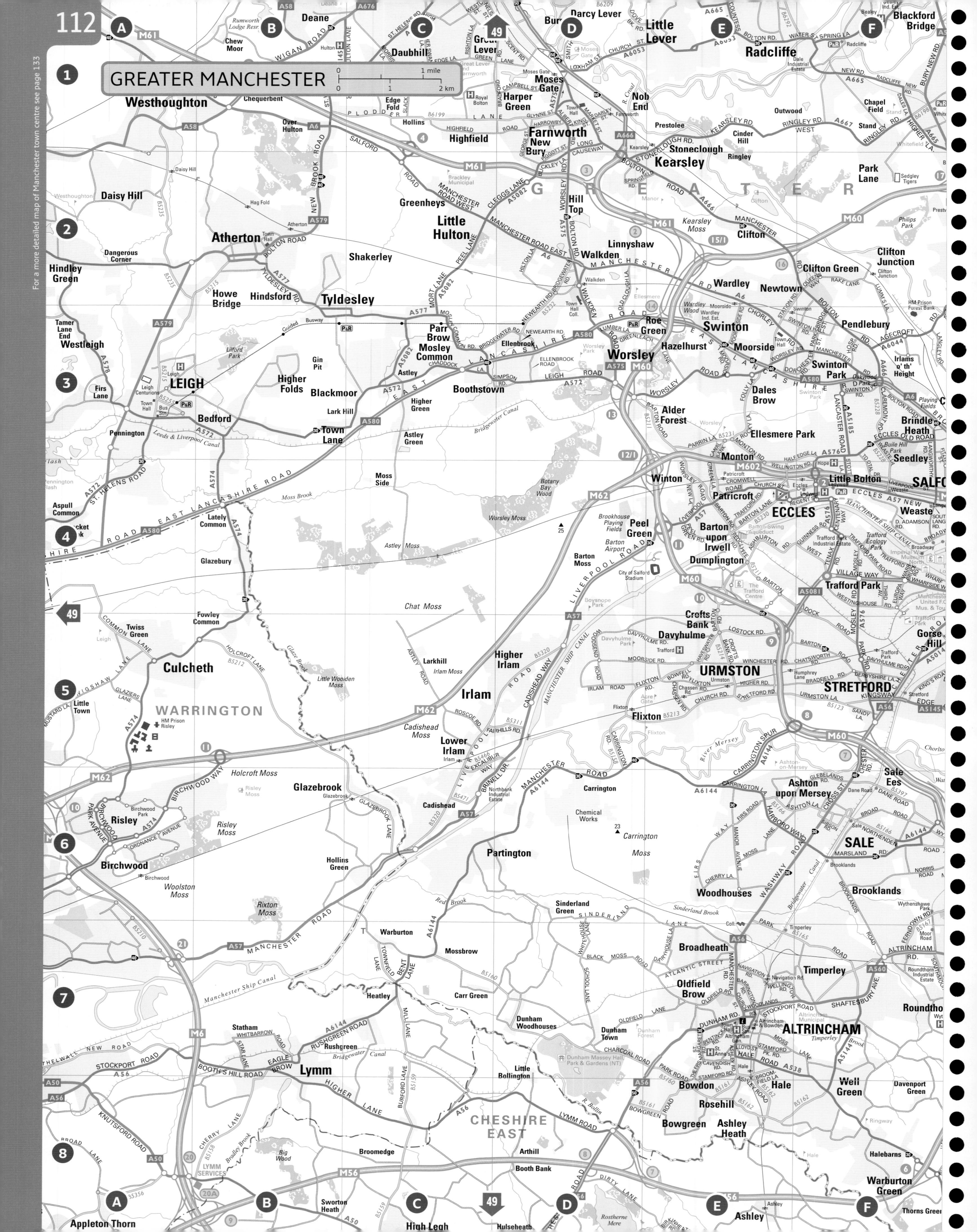
GREATER MANCHESTER
0 1 mile
0 1 2 km
For a more detailed map of Manchester town centre see page 133
Westhoughton
Daisy Hill
Atherton
Hindley Green
Westleigh
LEIGH
Tyldesley
Little Hulton
Farnworth
Kearsley
Radcliffe
Walkden
Swinton
Worsley
Pendlebury
ECCLES
URMSTON
STRETFORD
Irlam
Culcheth
WARRINGTON
Birchwood
Partington
SALE
ALTRINCHAM
Lymm
Timperley
Hale
Bowdon
CHESHIRE EAST
Chat Moss
Carrington Moss
Manchester Ship Canal
Bridgewater Canal
Leeds & Liverpool Canal
River Mersey
Glazebrook
Cadishead
Flixton
Davyhulme
Ashton upon Mersey
Broadheath
Dunham Massey Hall, Park & Gardens (NT)
49

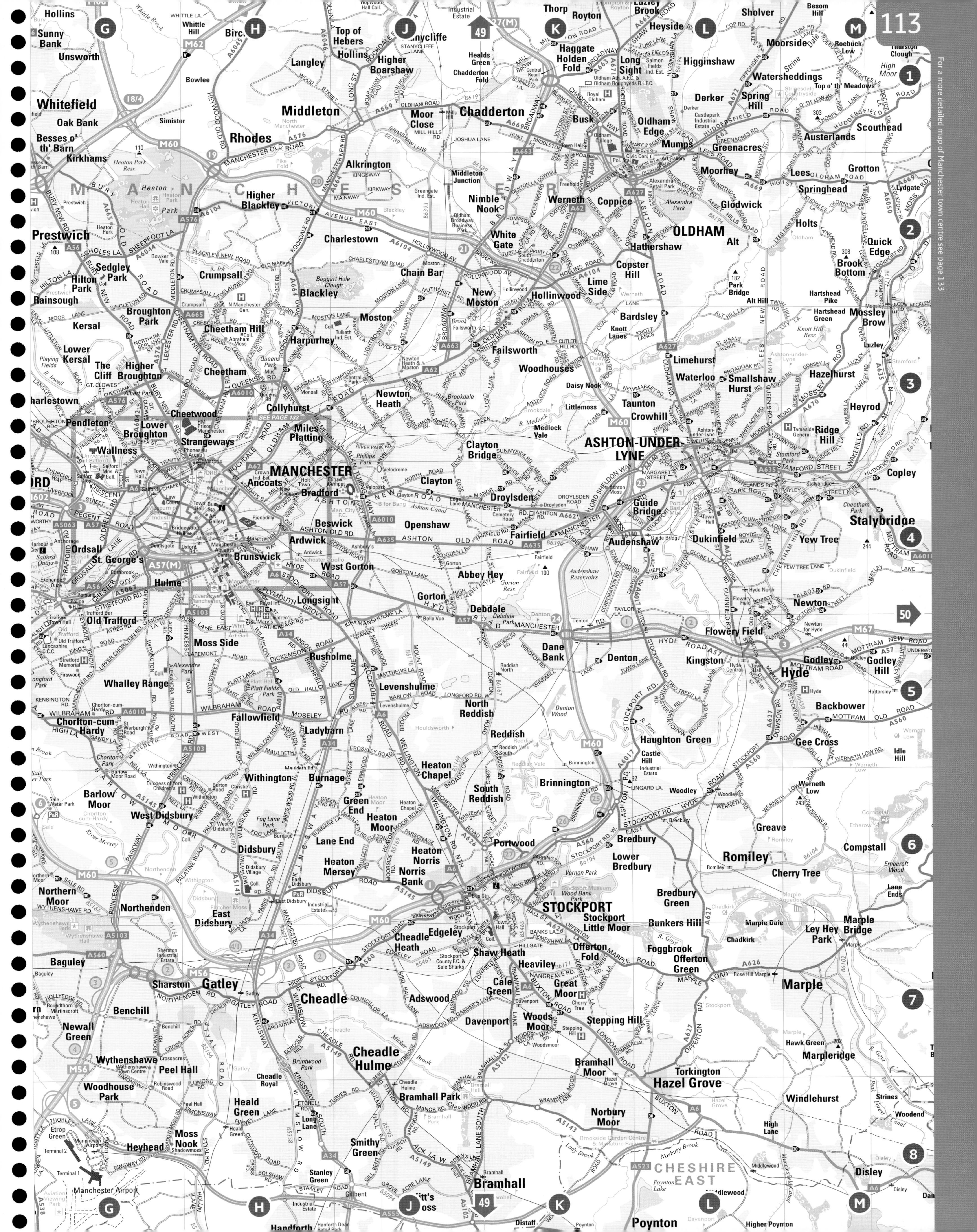

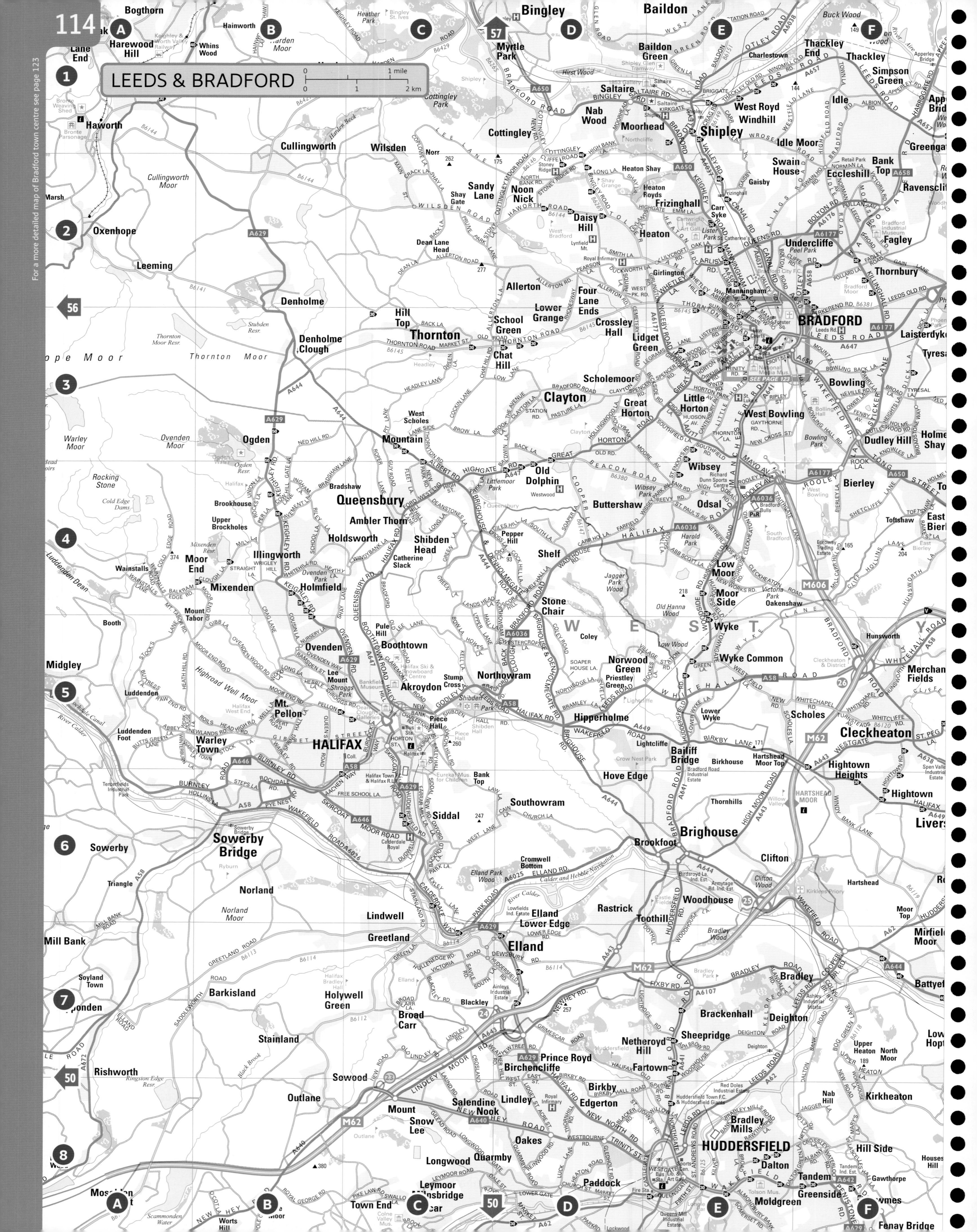
LEEDS & BRADFORD
0 1 mile
0 1 2 km
For a more detailed map of Bradford town centre see page 123
A B C D E F
1 2 3 4 5 6 7 8
57
56
50
SEE PAGE 123
Bogthorn
Harewood Hill
Hainworth
Harden Moor
Whins Wood
Bingley
Baildon
Buck Wood
Thackley End
Thackley
Baildon Green
Charlestown
Simpson Green
Myrtle Park
Hirst Wood
Saltaire
Nab Wood
Moorhead
Shipley
West Royd
Windhill
Idle
Idle Moor
Greengates
Haworth
Cottingley
Cullingworth
Wilsden
Cullingworth Moor
Sandy Lane
Shay Gate
Noon Nick
Heaton Shay
Heaton Royds
Frizinghall
Swain House
Gaisby
Eccleshill
Bank Top
Ravenscliffe
Marsh
Oxenhope
Daisy Hill
Heaton
Carr Syke
Undercliffe
Fagley
Leeming
Dean Lane Head
Allerton
Girlington
Manningham
Thornbury
Denholme
Hill Top
Thornton
School Green
Lower Grange
Four Lane Ends
Crossley Hall
Lidget Green
BRADFORD
Laisterdyke
Tyersal
Denholme Clough
Chat Hill
Thornton Moor
Scholemoor
Clayton
Great Horton
Little Horton
Bowling
West Bowling
West Scholes
Mountain
Ogden
Warley Moor
Ovenden Moor
Rocking Stone
Old Dolphin
Wibsey
Dudley Hill
Holme Shay
Bierley
Bradshaw
Queensbury
Ambler Thorn
Buttershaw
Odsal
Brookhouse
Upper Brockholes
Holdsworth
Shibden Head
Catherine Slack
Pepper Hill
Shelf
Tofshaw
East Bierley
Moor End
Illingworth
Wainstalls
Mixenden
Holmfield
Low Moor
Moor Side
Oakenshaw
Booth
Mount Tabor
Stone Chair
Coley
Wyke
Hunsworth
Midgley
Ovenden
Boothtown
Norwood Green
Wyke Common
Lee Mount
Akroydon
Stump Cross
Northowram
Priestley Green
Merchant Fields
Luddenden
Mt. Pellon
Hipperholme
Lower Wyke
Scholes
Cleckheaton
Luddenden Foot
Warley Town
HALIFAX
Piece Hall
Lightcliffe
Bailiff Bridge
Birkhouse
Hartshead Moor Top
Hightown Heights
Hightown
Bank Top
Hove Edge
Southowram
Siddal
Thornhills
Sowerby
Sowerby Bridge
Brookfoot
Brighouse
Clifton
Triangle
Norland
Cromwell Bottom
Elland Park Wood
Hartshead
Norland Moor
Lindwell
Rastrick
Woodhouse
Elland
Lower Edge
Toothill
Moor Top
Mill Bank
Greetland
Mirfield Moor
Soyland Town
Barkisland
Holywell Green
Bradley
Battyeford
Ripponden
Broad Carr
Blackley
Brackenhall
Deighton
Stainland
Netheroyd Hill
Sheepridge
Upper Heaton
North Moor
Rishworth
Prince Royd
Birchencliffe
Fartown
Sowood
Birkby
Edgerton
Kirkheaton
Outlane
Lindley
Salendine Nook
Nab Hill
Mount
Snow Lee
Bradley Mills
Oakes
HUDDERSFIELD
Hill Side
Longwood
Quarmby
Dalton
Paddock
Tandem
Gawthorpe
Leymoor
Greenside
Moldgreen
Town End
Worts Hill
Fenay Bridge
WEST YORKSHIRE
M62
M606
A629
A644
A58
A646
A641
A643
A62
A640
A650
A6036
A6177
A647
A658

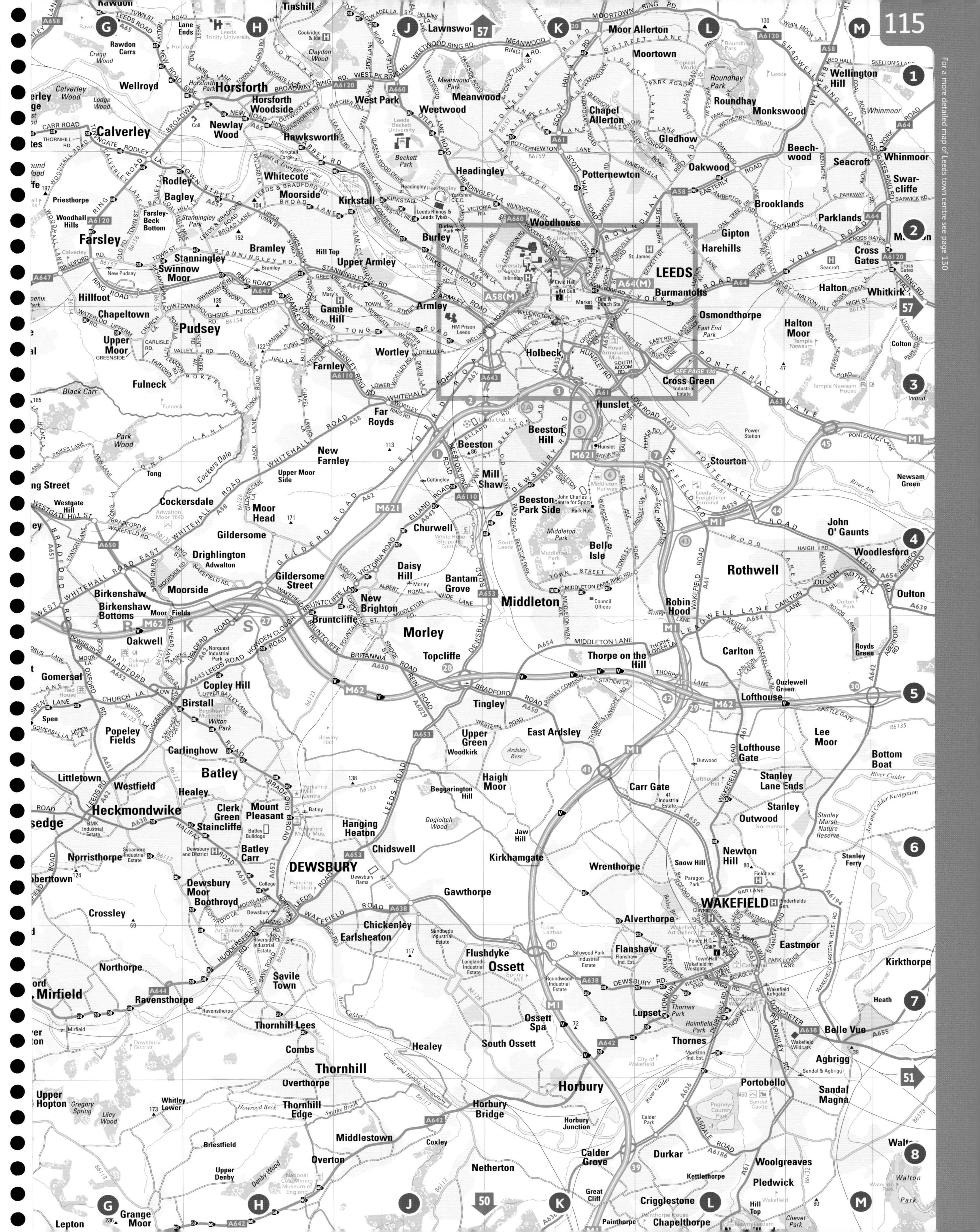

NEWCASTLE UPON TYNE & SUNDERLAND

For a more detailed map of Newcastle upon Tyne town centre see page 129 and of Sunderland town centre see page 136

Scale: 1 mile / 2 km

Major places:

Whitley Bay, Monkseaton, Earsdon, Shiremoor, Backworth, Holystone, Killingworth, Camperdown, Burradon, Dudley, West Moor, Forest Hall, Benton, Longbenton, Palmersville, Gosforth, South Gosforth, East Brunton, Brunswick Village, Hazlerigg, Seaton Burn, Wide Open, Coxlodge, Fawdon, Kenton Bar, Kenton, Kingston Park, Fawdon, Blakelaw, Woolsington, Newcastle International Airport, Dinnington, Prestwick, Mason, North Walbottle, West Denton, East Denton, Newbiggin Hall Estate, Denton Burn, Blakelaw

Tynemouth, North Shields, Preston Grange, Preston, Marden, Chirton, New York, Murton, Billy Mill, West Chirton, West Allotment, Howdon, Willington Quay, Howdon Pans, Percy Main, Rosehill, Battle Hill, Wallsend, Walker, Byker, Heaton, High Heaton, Jesmond, South Jesmond, Sandyford, Shieldfield

NEWCASTLE UPON TYNE, Elswick, Benwell, Old Benwell, Scotswood, Fenham, Cowgate, Ponteland, Kingston Park

South Shields, Marsden, Cleadon Park, Harton, Westoe, West Park, West Harton, Simonside, Whiteleas, Cleadon, Boldon, East Boldon, West Boldon, Boldon Colliery, Witherwack, Hylton Red House, Southwick, Hylton Castle, North Hylton, South Hylton, Pallion, Millfield, SUNDERLAND, Roker, Fulwell, Seaburn, Monkwearmouth, Carley Hill

Jarrow, Hebburn, Monkton, Primrose, Hedworth, Felling, High Felling, Wardley, Pelaw, Bill Quay, Heworth, Windy Nook, Wrekenton, Eighton Banks, Springwell, Usworth, Concord, Washington, Donwell, Springwell

Gateshead, Team Valley, Teams, Bensham, Saltwell, Mount Pleasant, Carr Hill, Low Fell, Chowdene, Deckham, Sheriff Hill, Dunston, Dunston Hill, Lobley Hill, Whickham, Swalwell, Winlaton, Blaydon, Sunniside, Marley Hill, Street Gate, Lamesley, Kibblesworth

Rivers: River Tyne, River Wear, River Team, River Don, River Derwent

TYNE & WEAR, NORTHUMBERLAND

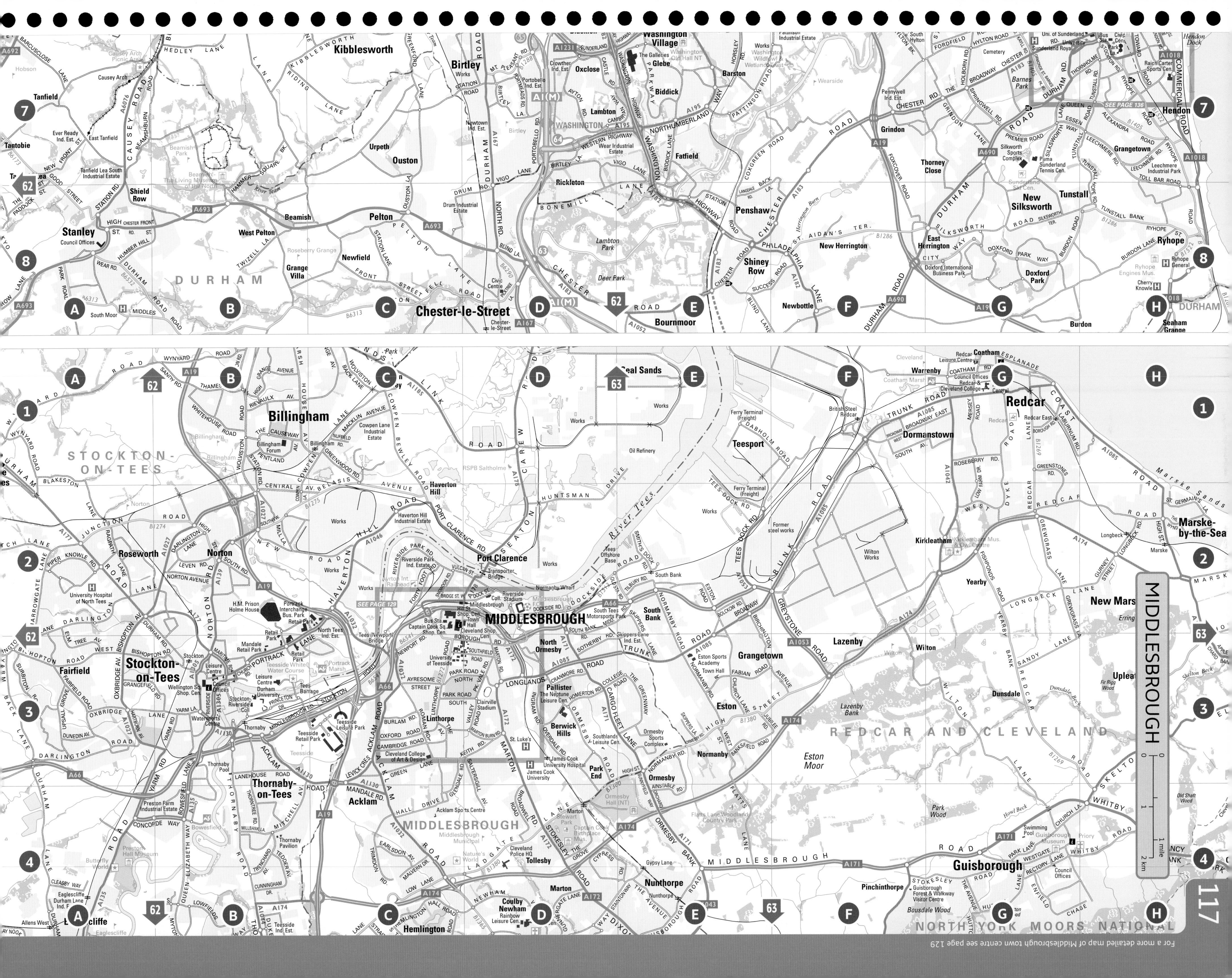

For a more detailed map of Middlesbrough town centre see page 129

MIDDLESBROUGH

0   1 mile
0   1   2 km

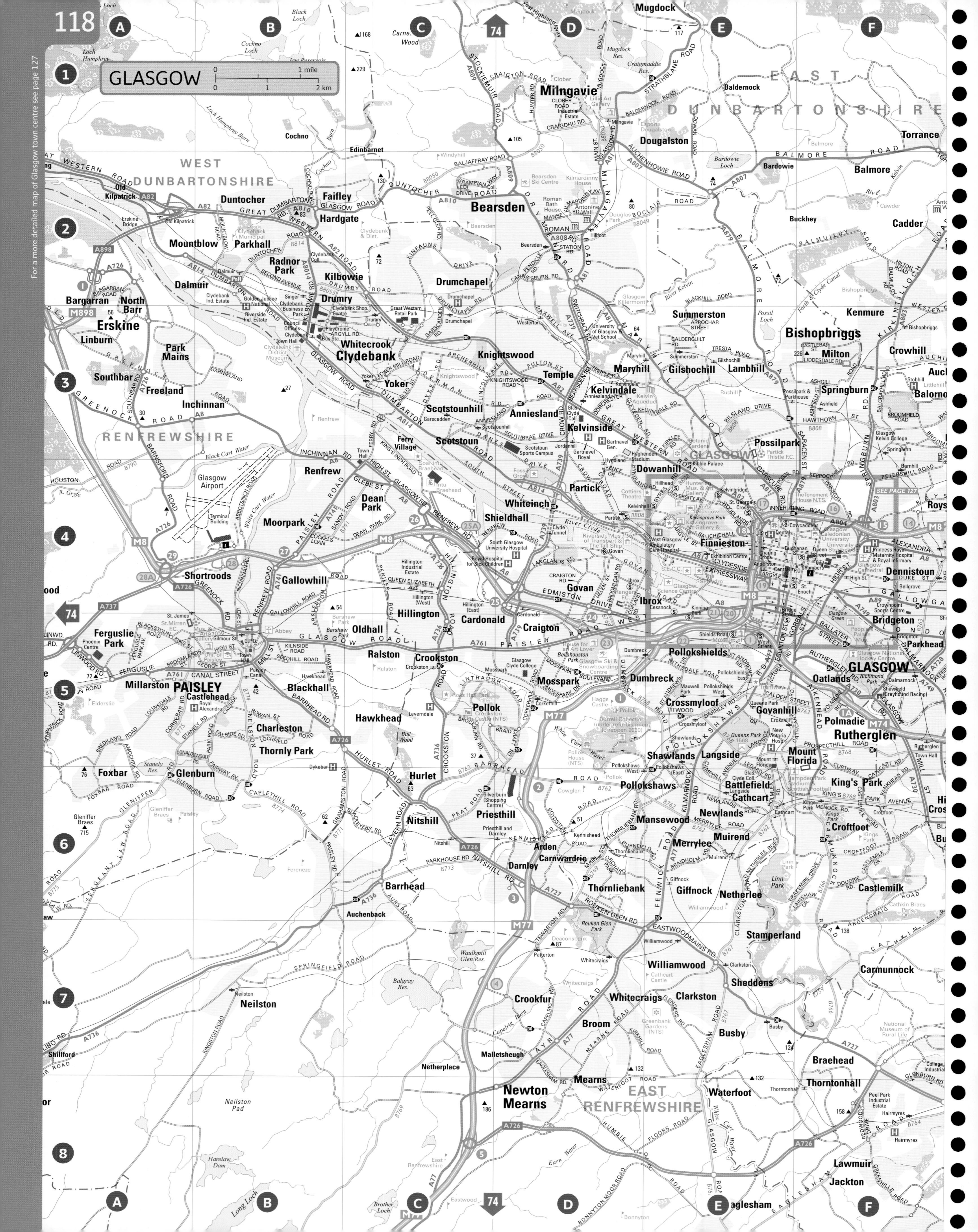

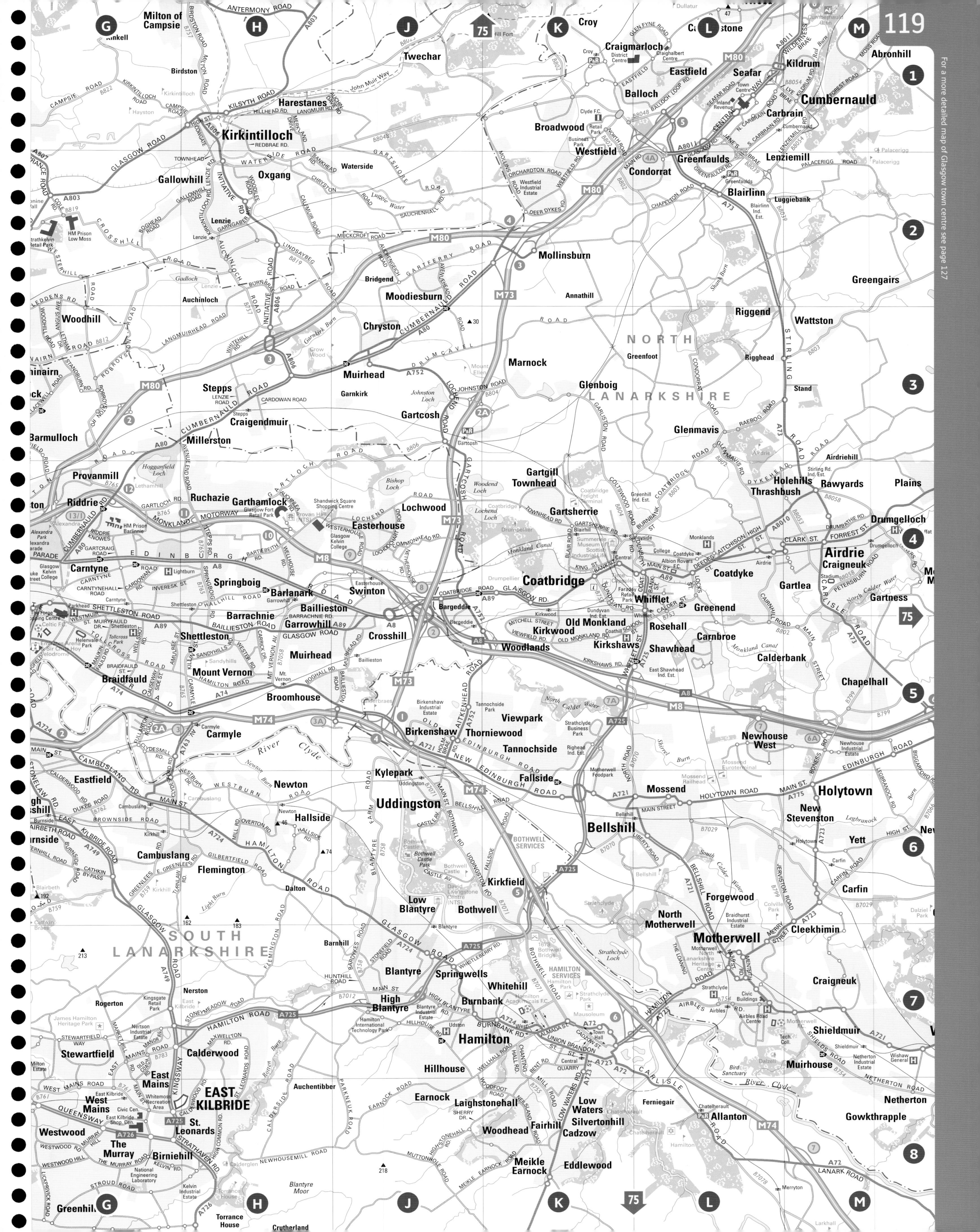

EDINBURGH

For a more detailed map of Edinburgh town centre see page 126

FIRTH OF FORTH

EAST LOTHIAN

MIDLOTHIAN

PENTLAND HILLS REGIONAL PARK

Musselburgh · Inveresk · Monktonhall · Whitecraig · Dalkeith · Dewartown · Chesterhill · Whitehill · Mayfield · Edgehead · Cousland · Southfield · Cranstoun Riddel

Newhailes · Joppa · Portobello · Newcraighall · Newton · Danderhall · Newbattle · Easthouses · Newtongrange · Bonnyrigg · Lasswade · Arniston · Corebridge

Leven · Dolphingstone · Preston Grange

Newhaven · Leith · Trinity · Granton · Pilton · West Pilton · Restalrig · Niddrie · Duddingston · Bingham · Craigmillar · Edmonstone · Gilmerton · Moredun · Gracemount · Liberton · Inch · Loanhead · Roslin · Rosewell · Thorton

EDINBURGH · Old Town · New Town · Comely Bank · Warriston · Inverleith · Merchiston · Morningside · Newington · Prestonfield · Kaimes · Straiton · Bilston · Damhead · Easter Bush · Milton Bridge · Auchendinny

Murrayfield · Dalry · Gorgie · Craiglockhart · Stenhouse · Saughton · Longstone · Colinton Mains · Colinton · Fairmilehead · Swanston · Penicuik

Craigleith · Ravelston · Blackhall · Davidson's Mains · Silverknowes · Muirhouse · Cramond · Barnton · Clermiston · Corstorphine · North Gyle · South Gyle · Sighthill · Wester Hailes · Juniper Green · Currie · Balerno

Craigiehall · Cramond Bridge · Turnhouse · Edinburgh Airport · Hermiston · Riccarton · East Rigg · Wester Bavelaw · Easter Bavelaw

Scale: 0 — 1 mile / 0 — 1 — 2 km

Street maps

Symbols used on the map

M8	Motorway	
A4 ①	Primary route dual / single carriageway / Junction	
A40	'A' road dual / single carriageway	
B507	'B' road dual / single carriageway	
Toll	Other road dual / single carriageway / Toll	
	One way street / Orbital route	
	Access restriction	
	Pedestrian street / Street market	
	Minor road / Track	
FB	Footpath / Footbridge	
	Road under construction	
	Extent of London congestion charging zone	

Main / other National Rail station	
London Underground / Overground station	
Light Rail / Station	
Bus / Coach station	
P&R Park and Ride site - rail operated (runs at least 5 days a week)	
Dublin 8hrs Vehicle / Pedestrian ferry	
P **P** Car park	
Theatre	
Major hotel	
Public House	
Police station	
Library	
Post Office	

i i Visitor information centre (open all year / seasonally)	
Toilet	
JAPAN Embassy	
Cinema	
Cathedral / Church	
Mosque / Synagogue / Other place of worship	
Park / Garden / Sports ground	
Cemetery	

Leisure & tourism	
Shopping	
Administration & law	
Health & welfare	
Education	
Industry / Office	
Other notable building	

Locator map

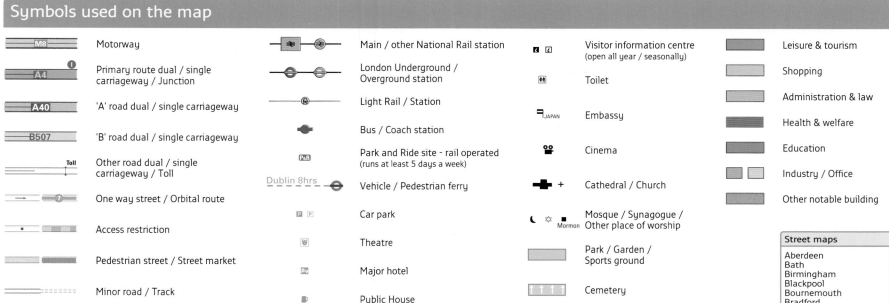

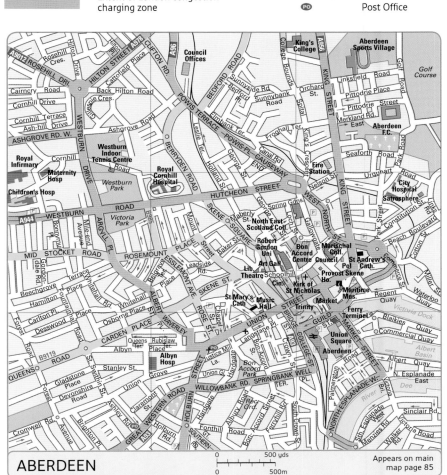

ABERDEEN

Appears on main map page 85

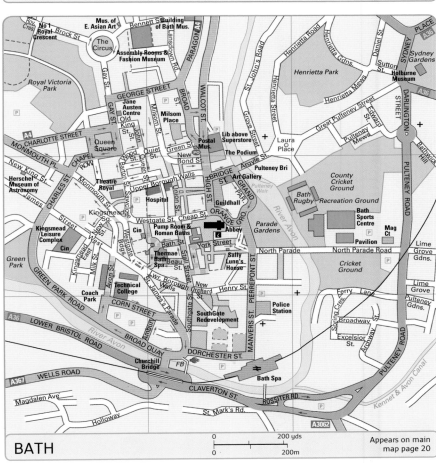

BATH

Appears on main map page 20

BLACKPOOL

Appears on main map page 55

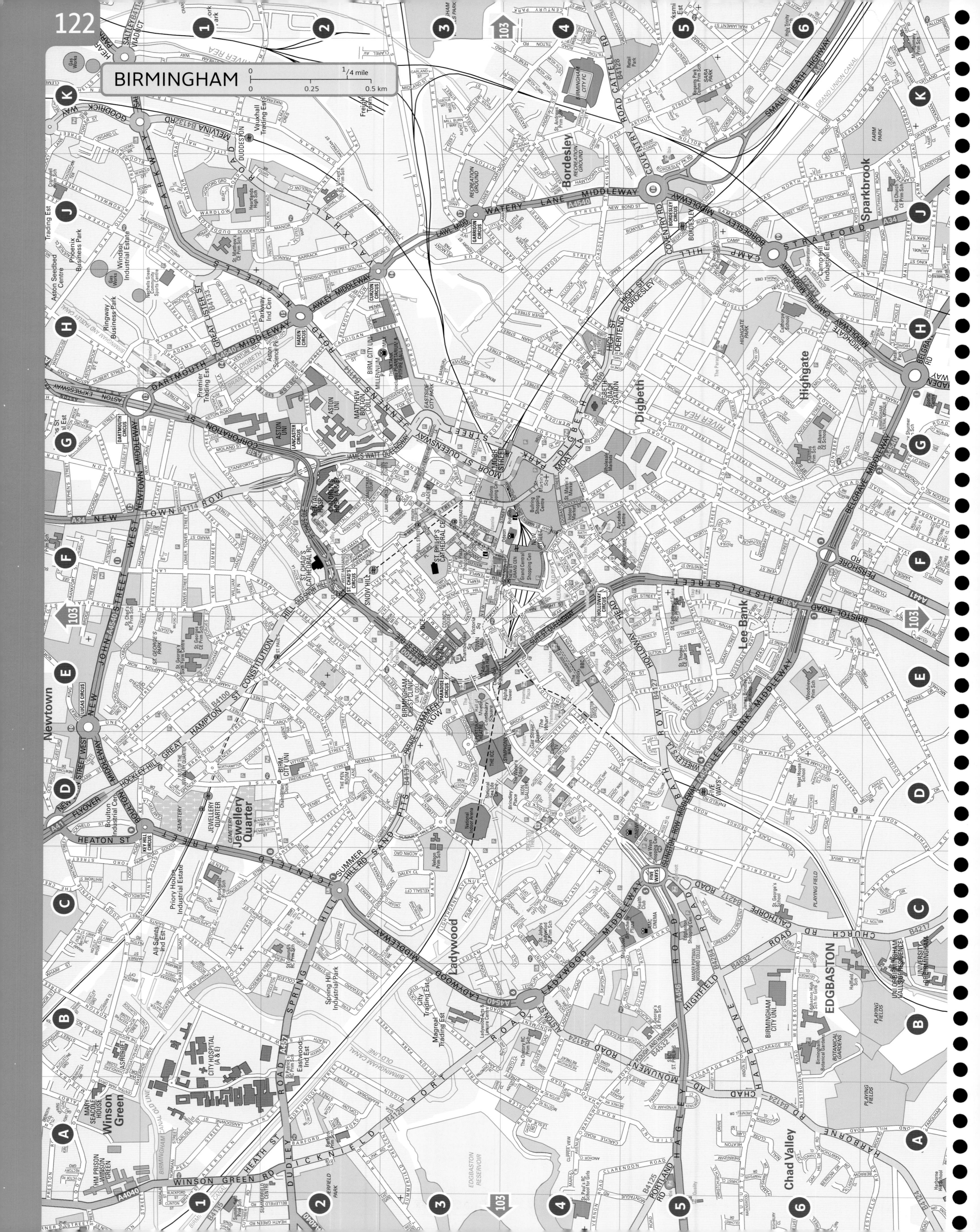

BIRMINGHAM
1/4 mile
0
0.25
0.5 km
Bordesley
Digbeth
Highgate
Sparkbrook
Lee Bank
Edgbaston
Chad Valley
Ladywood
Jewellery Quarter
Newtown
Winson Green
Birmingham City FC
Aston Uni
Birm. City Uni
St Chad's Cathedral RC
St Philip's Cathedral CE
Bullring Shopping Centre
Grand Central Shopping Cen
Five Ways
City Hospital (A & E)
HM Prison Winson Green
Edgbaston Reservoir
Botanical Gardens
University of Birmingham
Birmingham City Uni
National Indoor Arena
Symphony Hall
The ICC
Brindley Place
BBC
Digbeth Coach Station
Priory House Industrial Estate
All Saints Ind Est
Spring Hill Industrial Park
City Trading Est
Magreal Trading Est
Eastwood Ind Est
Ringway Business Park
Aston Seedbed Centre
Phoenix Business Park
Windsor Industrial Estate
Parkway Ind Cen
Premier Trading Est
Aston Science Park
Camp Hill Industrial Est
Boulton Industrial Cen
Vauxhall Trading Est
Heartlands High Sch
Matthew Bolton Coll
Eastside City Park
Highgate Park
River Rea
Birmingham Canal Old Line
Grand Union Canal
Dartmouth Circus
Lancaster Circus
Haden Circus
Curzon Circus
Garrison Circus
Bordesley Circus
St Chad's Circus
Paradise Circus
Holloway Circus
Lucas Circus
Key Hill Circus
A34
A38
A41
A441
A456
A457
A4040
A4540
B4100
B4114
B4124
B4126
B4127
B4132
B4135
B4217
B4284
B4532
103
1
2
3
4
5
6
A
B
C
D
E
F
G
H
J
K

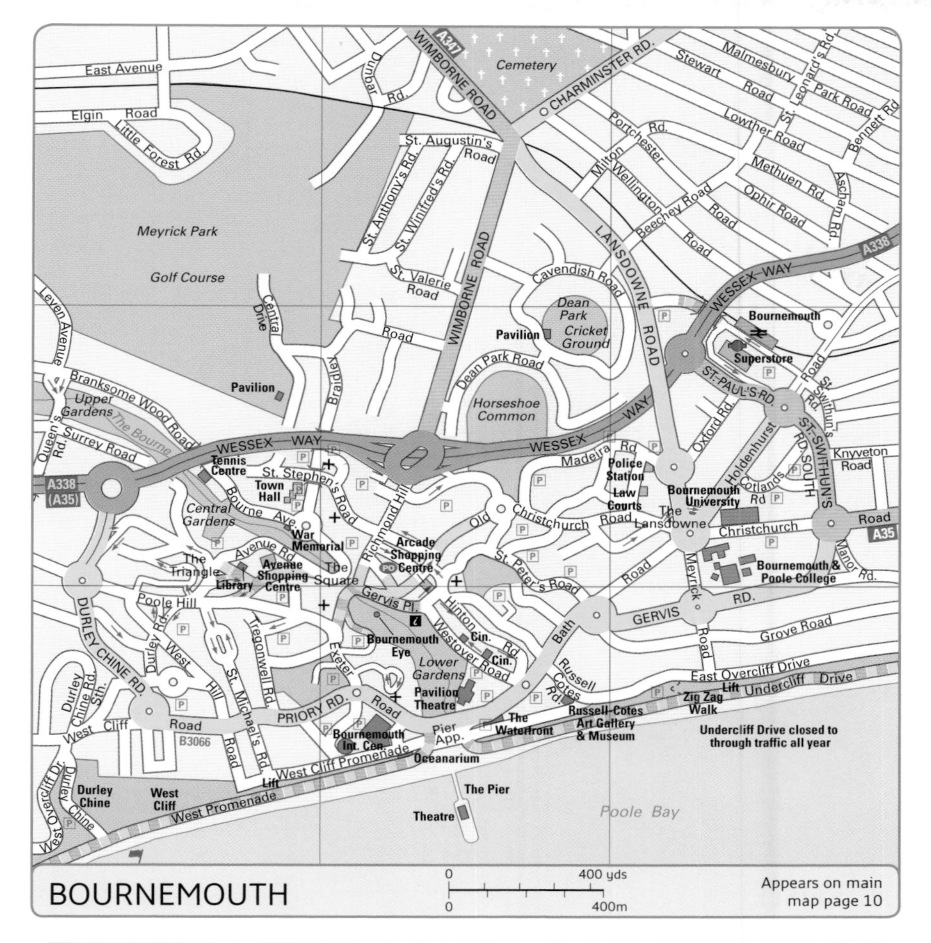

BOURNEMOUTH

| 0 | 400 yds |
| 0 | 400m |

Appears on main
map page 10

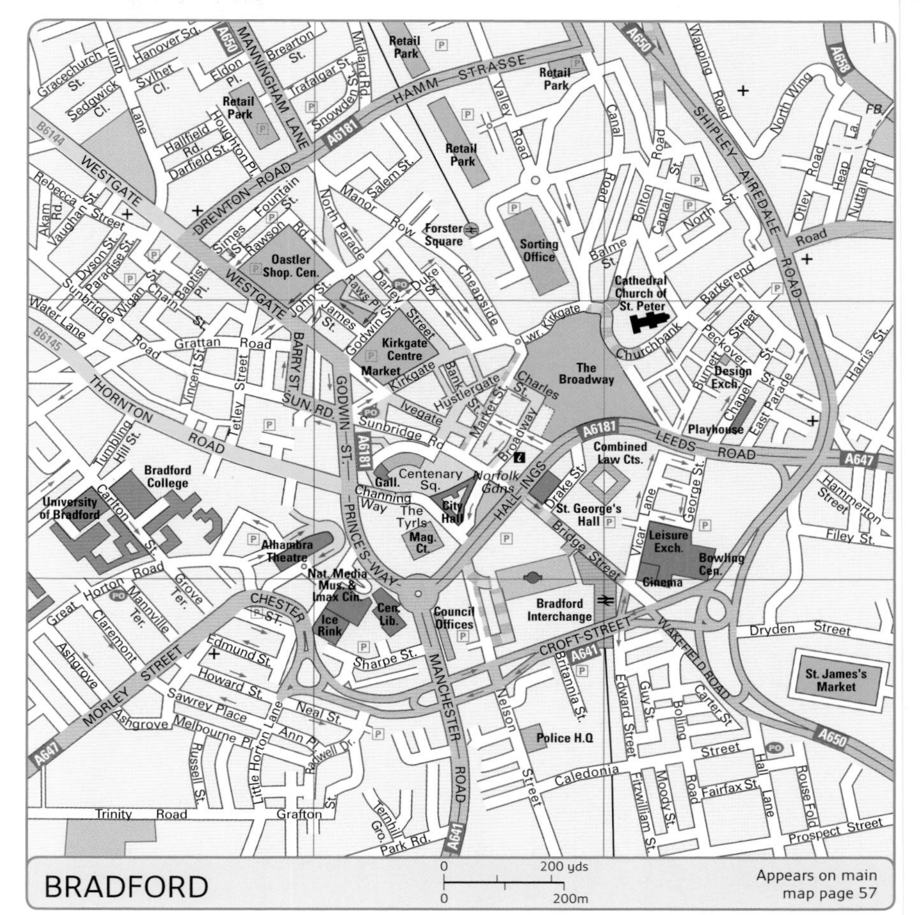

BRADFORD

| 0 | 200 yds |
| 0 | 200m |

Appears on main
map page 57

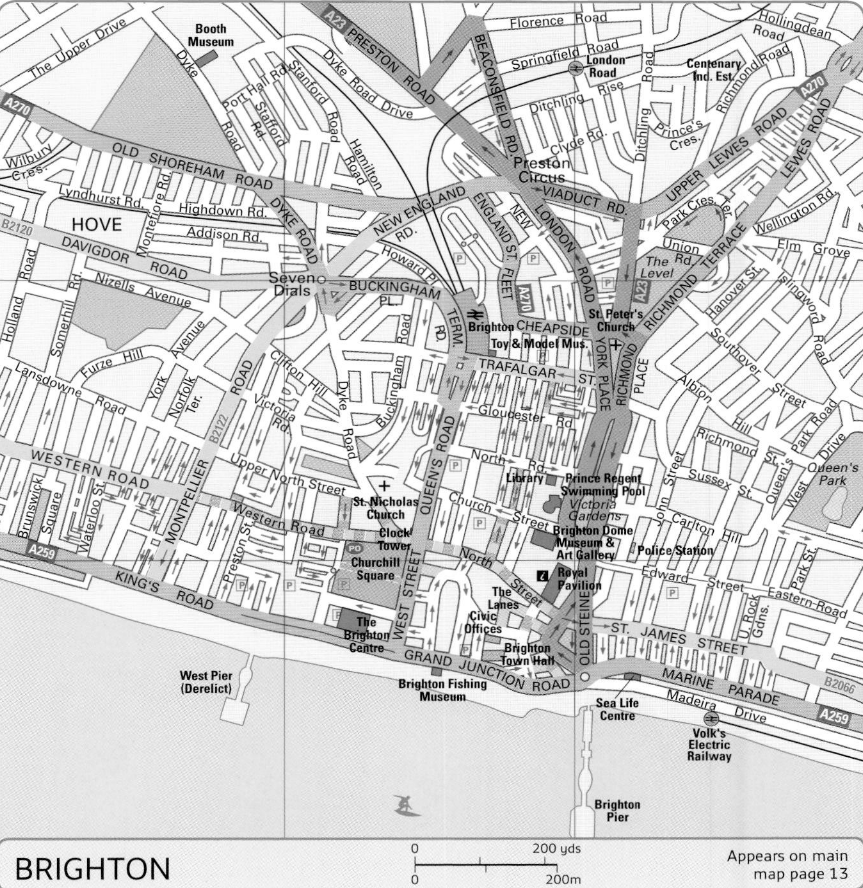

BRIGHTON

| 0 | 200 yds |
| 0 | 200m |

Appears on main
map page 13

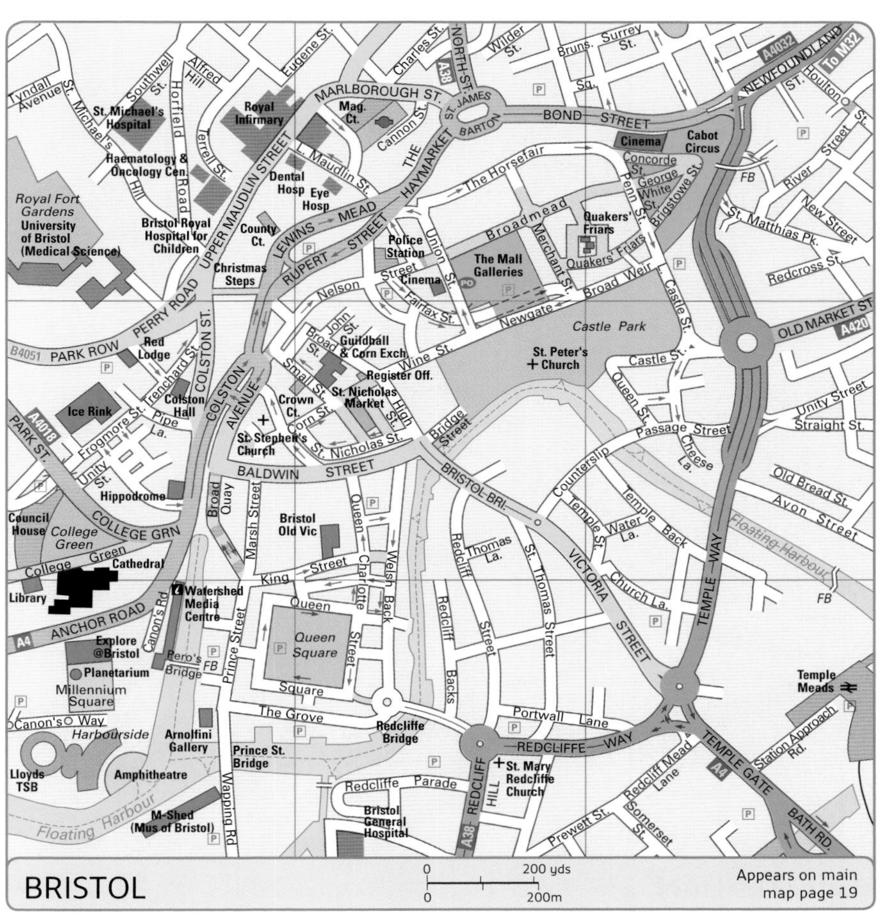

BRISTOL

| 0 | 200 yds |
| 0 | 200m |

Appears on main
map page 19

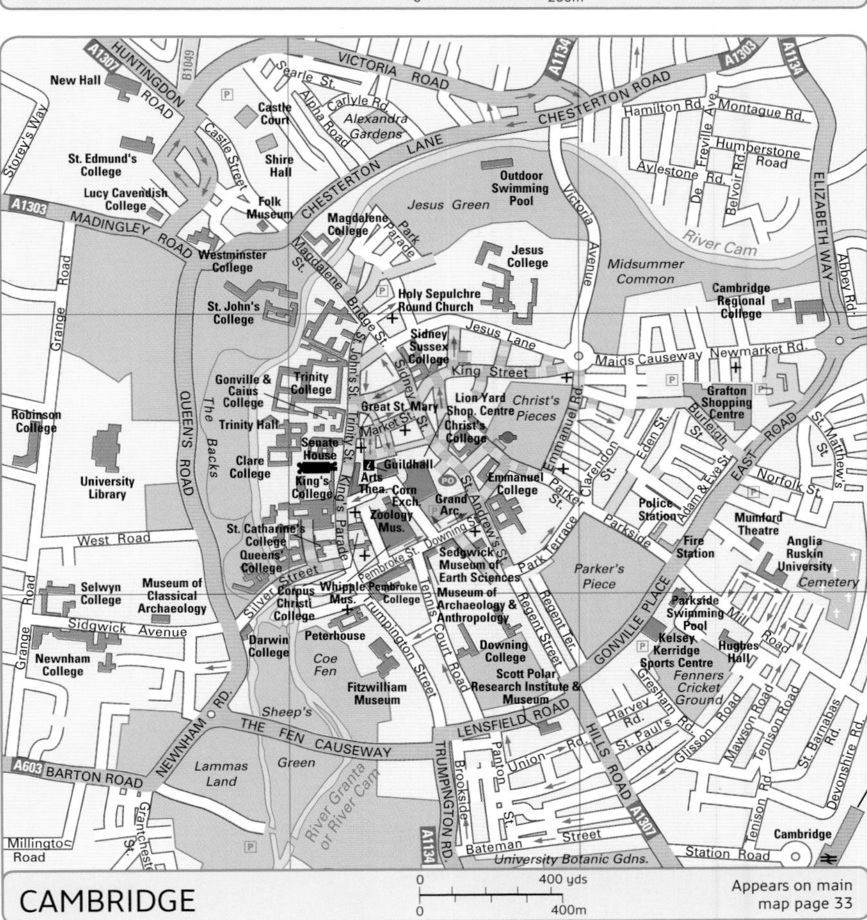

CAMBRIDGE

| 0 | 400 yds |
| 0 | 400m |

Appears on main
map page 33

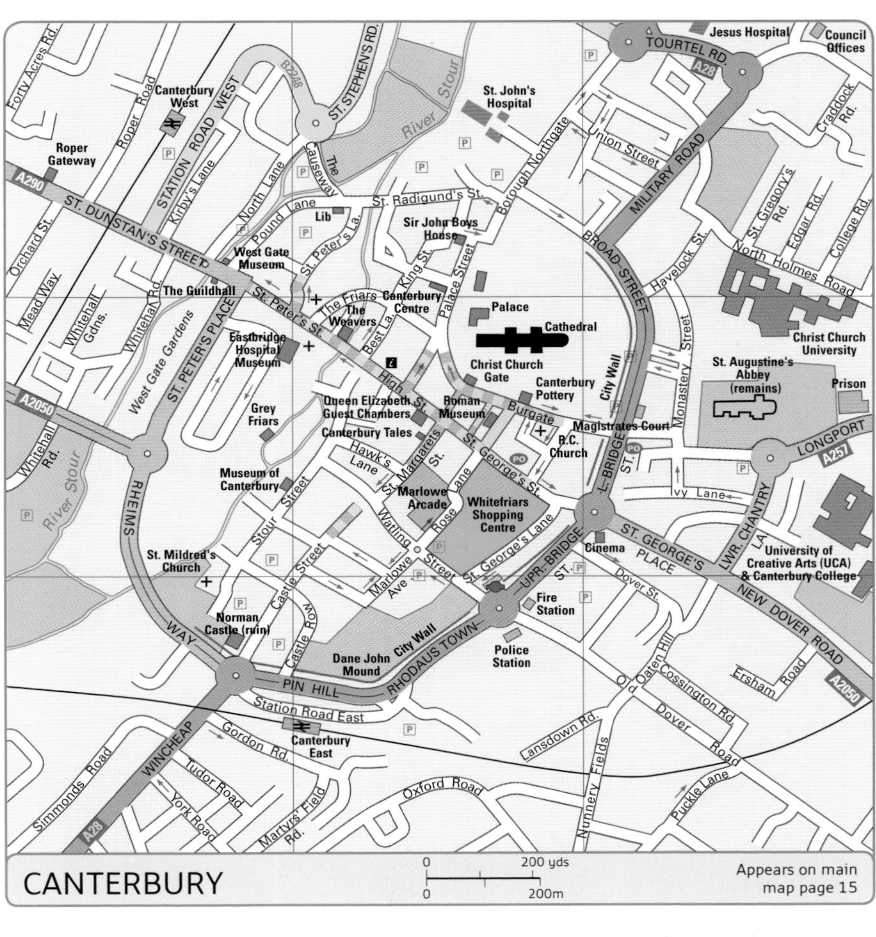

CANTERBURY

| 0 | 200 yds |
| 0 | 200m |

Appears on main
map page 15

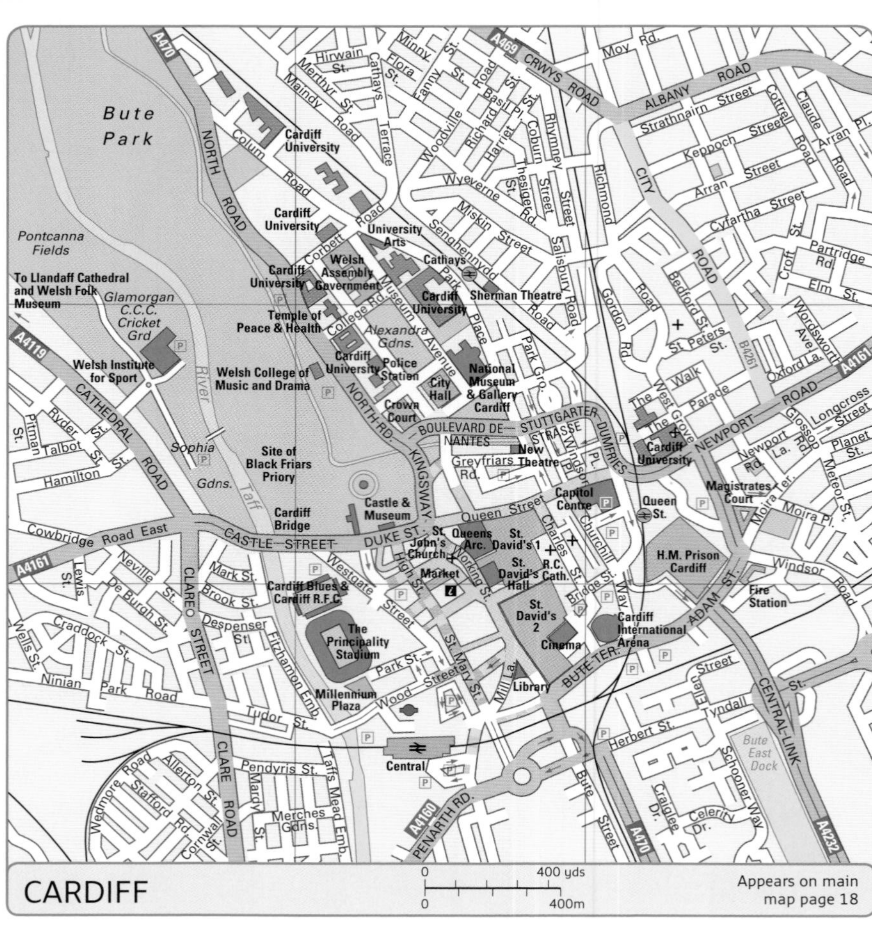

CARDIFF

Appears on main map page 18

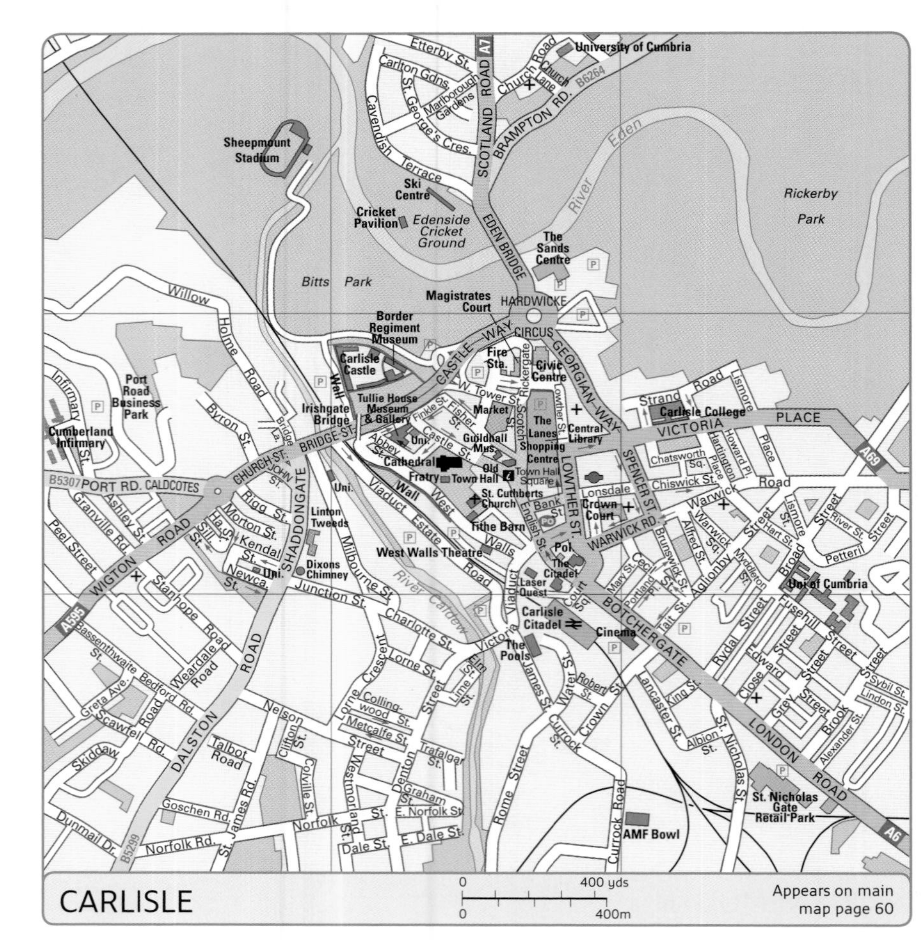

CARLISLE

Appears on main map page 60

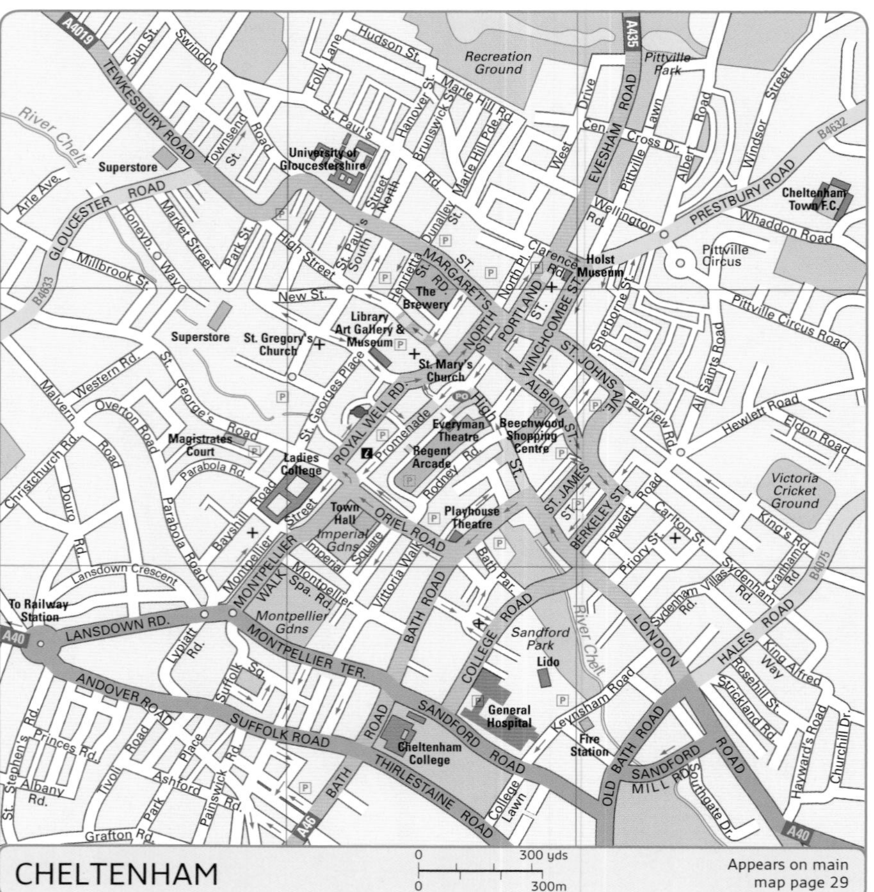

CHELTENHAM

Appears on main map page 29

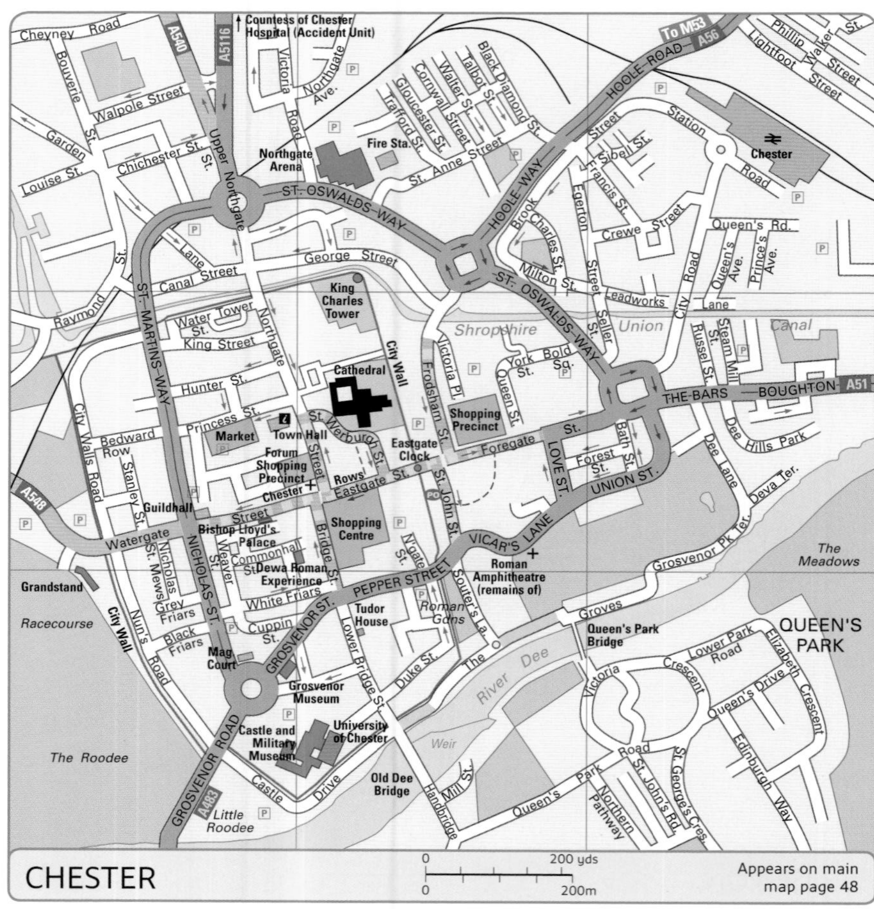

CHESTER

Appears on main map page 48

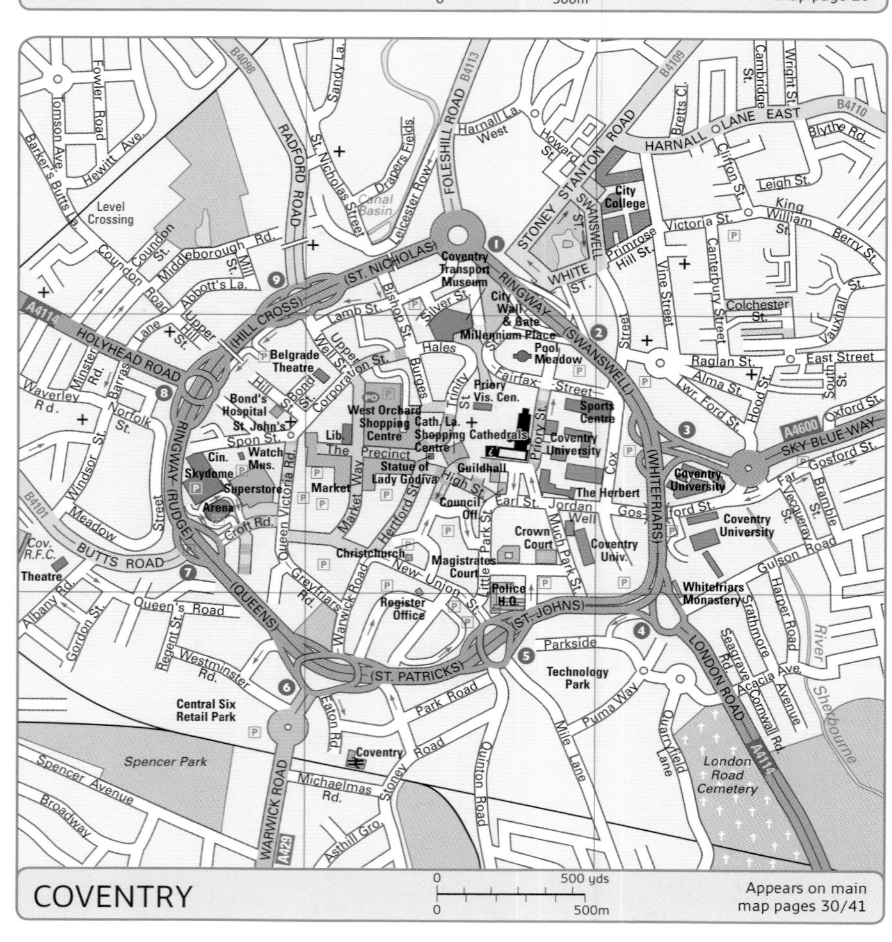

COVENTRY

Appears on main map pages 30/41

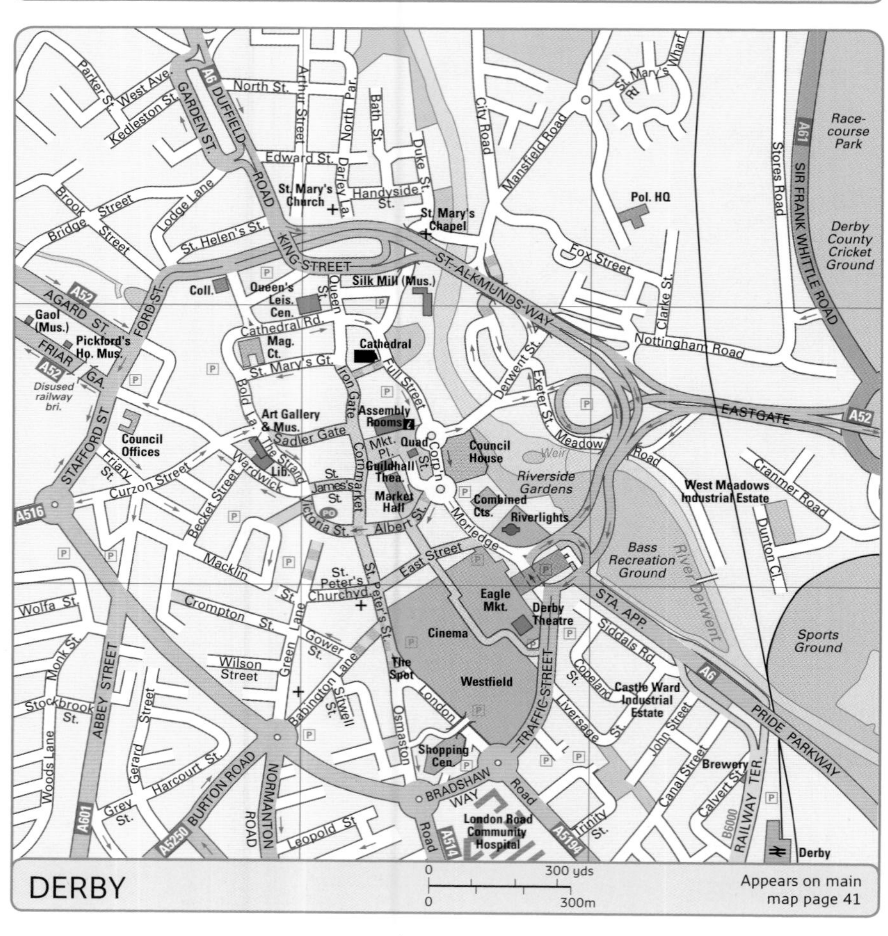

DERBY

Appears on main map page 41

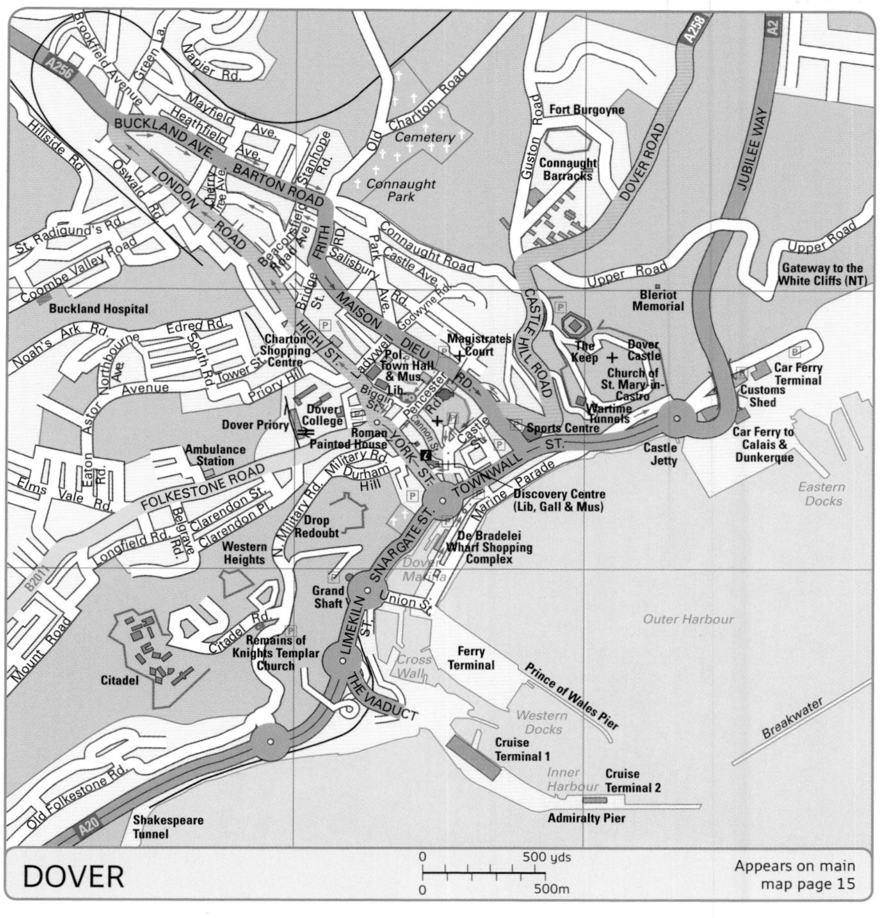

DOVER

500 yds
500m

Appears on main
map page 15

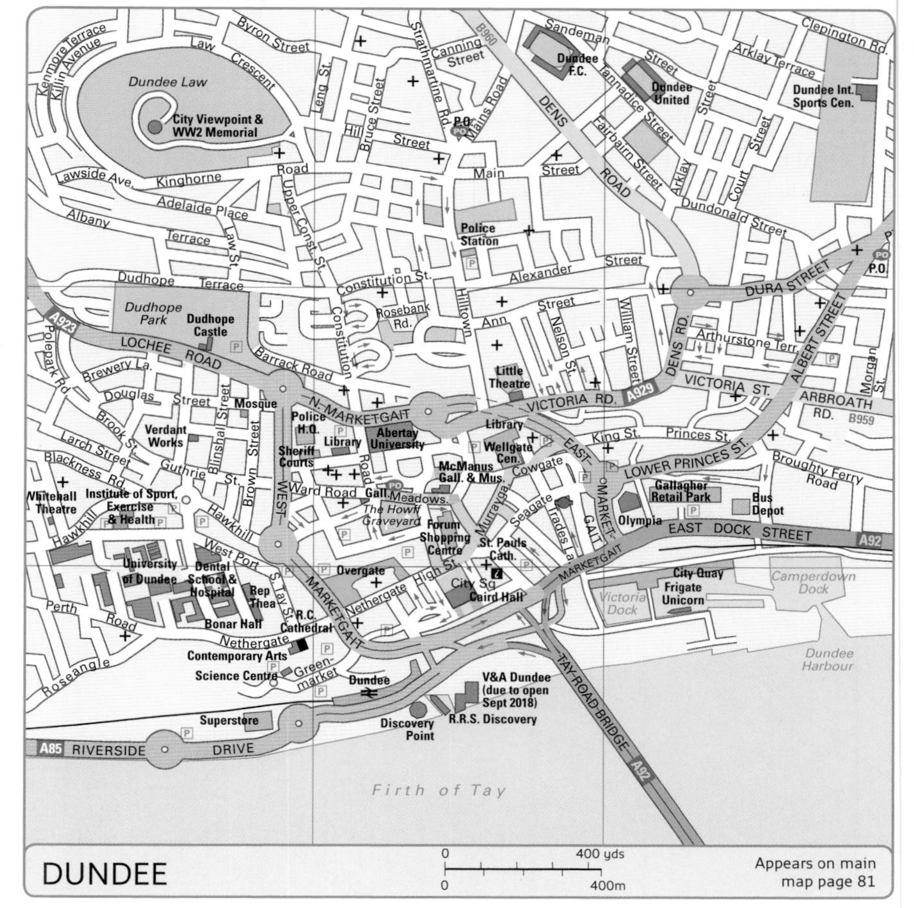

DUNDEE

400 yds
400m

Appears on main
map page 81

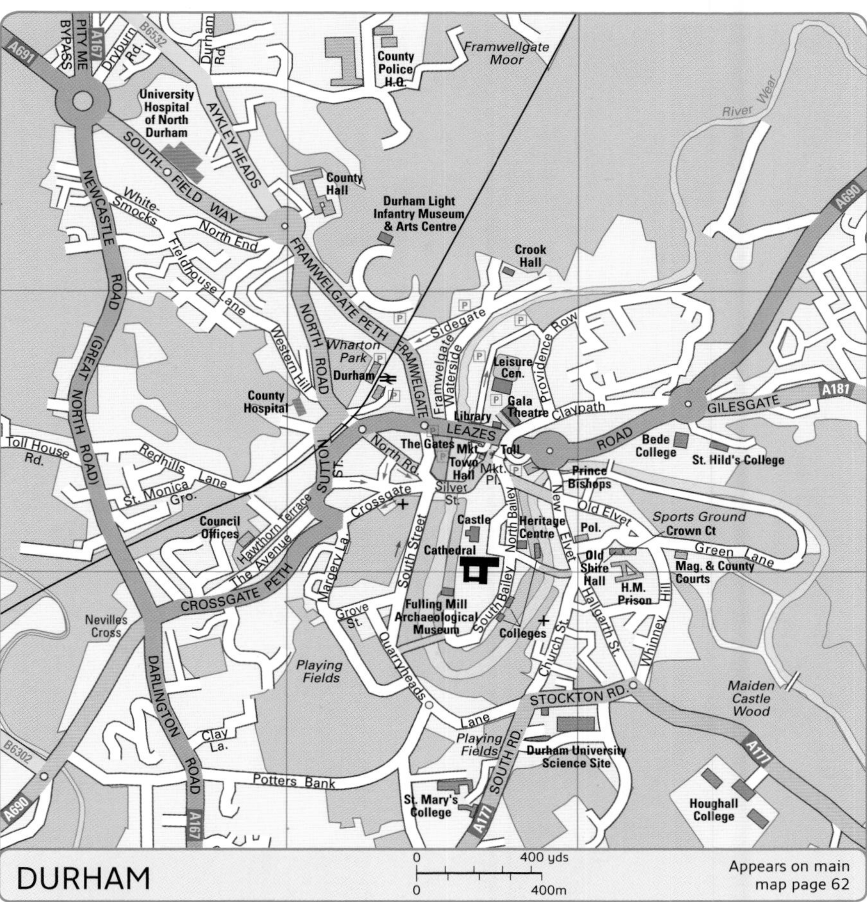

DURHAM

400 yds
400m

Appears on main
map page 62

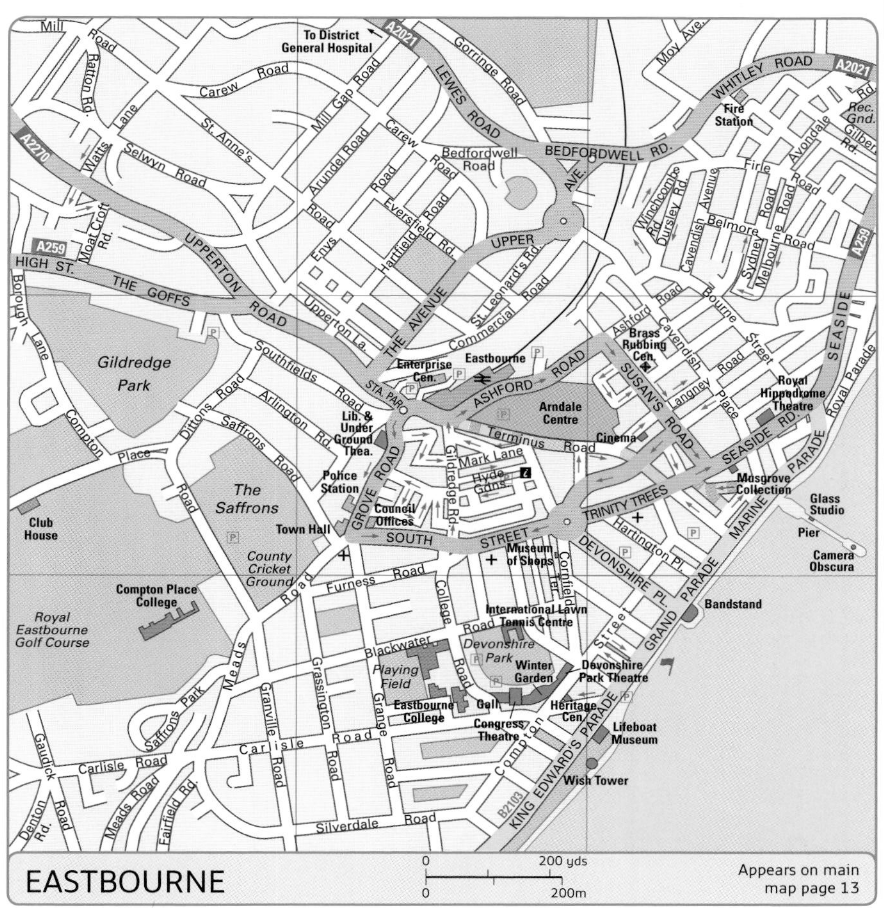

EASTBOURNE

200 yds
200m

Appears on main
map page 13

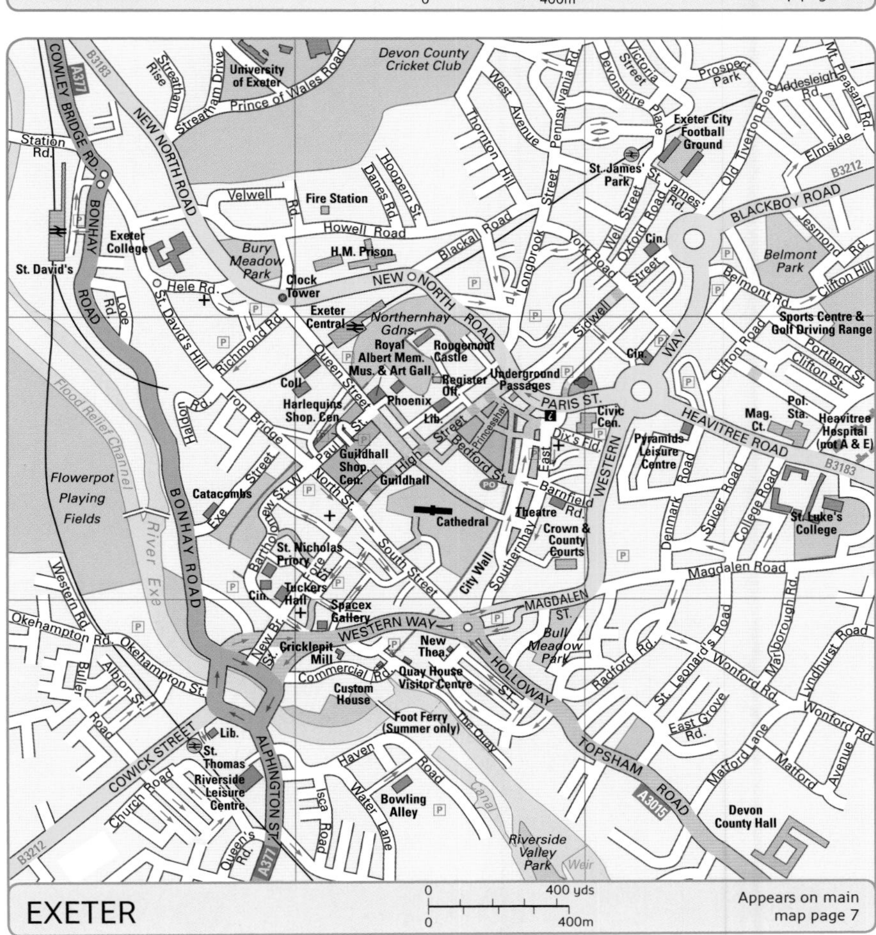

EXETER

400 yds
400m

Appears on main
map page 7

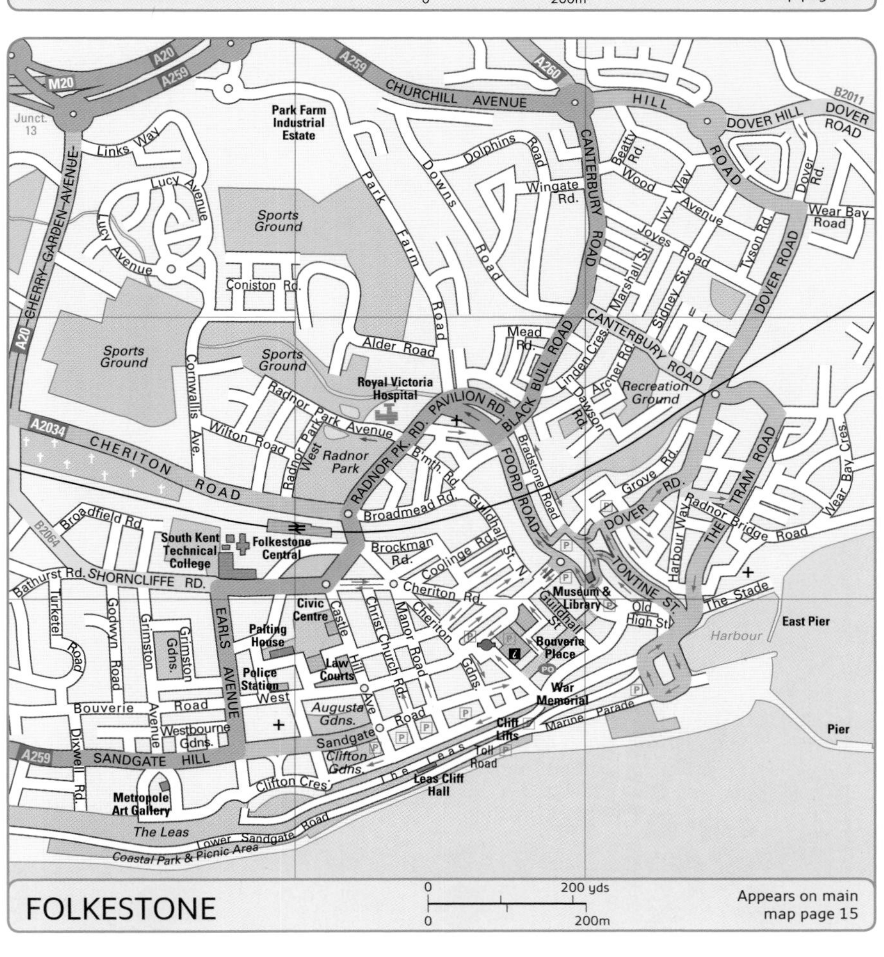

FOLKESTONE

200 yds
200m

Appears on main
map page 15

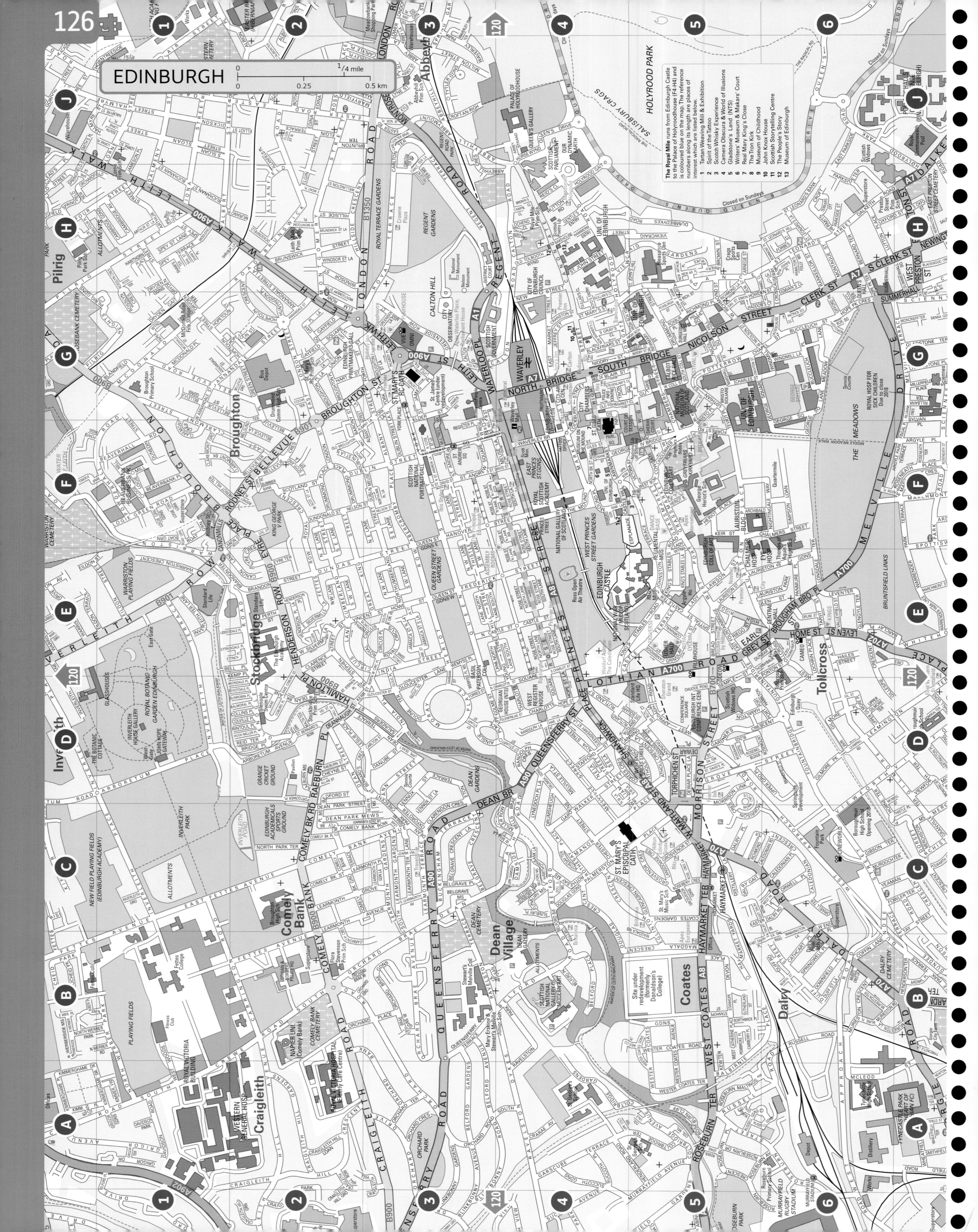

EDINBURGH

1/4 mile
0   0.25   0.5 km

The Royal Mile runs from Edinburgh Castle to the Palace of Holyroodhouse (F4-H4) and is coloured blue on the map. The reference numbers along its length are places of interest which are listed below.

1   Tartan Weaving Mill & Exhibition
2   Spirit of the Tattoo
3   Scotch Whisky Experience
4   Camera Obscura & World of Illusions
5   Gladstone's Land (NTS)
6   Writers' Museum & Makars' Court
7   Real Mary King's Close
8   Museum of Childhood
9   John Knox House
10 Scottish Storytelling Centre
11 The People's Story
12 The Museum of Edinburgh
13 The Museum of Edinburgh

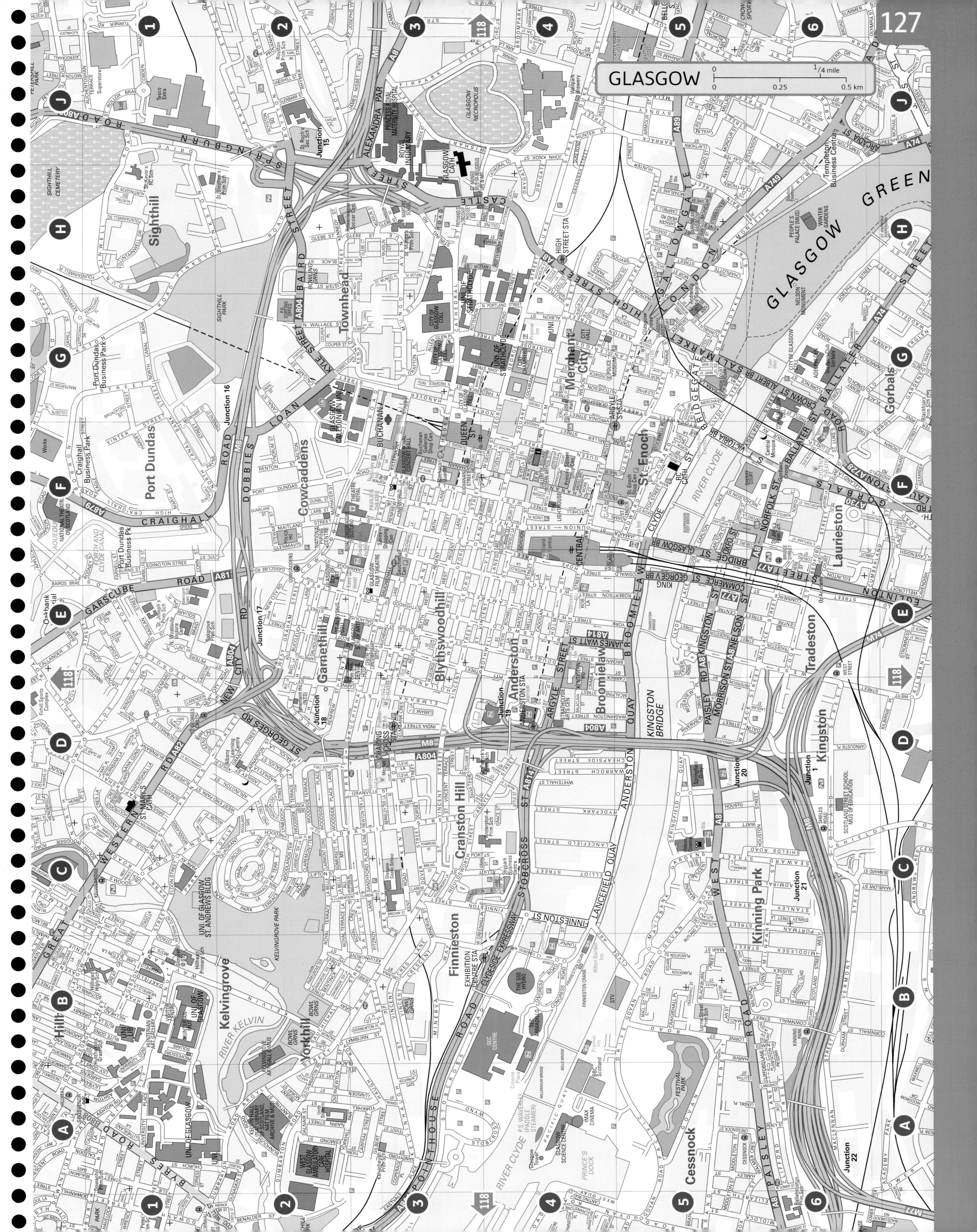
GLASGOW
1/4 mile
0.25
0.5 km
Sighthill
Sighthill Cemetery
Sighthill Park
Port Dundas
Port Dundas Business Park
Craighall Business Park
Townhead
Cowcaddens
Garnethill
Blythswoodhill
Anderston
Broomielaw
Merchant City
St. Enoch
Gorbals
Laurieston
Tradeston
Kingston
Kinning Park
Cessnock
Cranston Hill
Finnieston
Yorkhill
Kelvingrove
Hill
Glasgow Necropolis
Glasgow Cath
Royal Infirmary
Princess Royal Maternity Hosp
Wellpark Brewery
Glasgow Green
People's Palace (Mus)
Winter Gardens
Nelson Monument
Templeton Business Centre
City of Glasgow Coll
Univ. of Strathclyde
Glasgow Caledonian Univ.
Buchanan Galleries Shop Cen
Glasgow Royal Concert Hall
Theatre Royal
Central Mosque
Citizens Theatre
Sheriff Court
St. Enoch Shop Cen
RC Cath
River Clyde
Kingston Bridge
The SSE Hydro
SEC Centre
SEC Armadillo
Finnieston Crane
Exhibition Centre Sta
Clydeside Expressway
Pointhouse Road
Glasgow Science Centre
Glasgow Tower
IMAX Cinema
Festival Park
Prince's Dock
Pacific Quay
BBC Scotland
STV
Kelvingrove Art Gall & Mus
River Kelvin
Kelvingrove Park
Univ. of Glasgow
Hunterian Art Gallery
Univ. of Glasgow St. Andrews Bldg
St. Mary's Cath
West Glasgow Ambulatory Care Hospital
Great Western Rd
St. Georges Rd
New City Rd
Garscube Road
Craighall Road
Springburn Road
Alexandra Par
Castle Street
Baird Street
Dobbies Loan
Sauchiehall St
High Street
Gallowgate
London Rd
Saltmarket
Crown St
Ballater St
Gorbals St
Eglinton St
Kingston St
Paisley Rd
Broomielaw
Anderston Quay
Lancefield Quay
Stobcross St
Argyle Street
Finnieston St
Glasgow Br
Clyde St
Norfolk St
High Street Sta
Central
Queen St
Argyle St Sta
Charing Cross
Junction 15
Junction 16
Junction 17
Junction 18
Junction 19
Junction 20
Junction 21
Junction 22
M8
M74
M77
A8
A74
A77
A728
A730
A749
A803
A804
A814
A879
A81
A82
A89
118
J H G F E D C B A
1 2 3 4 5 6

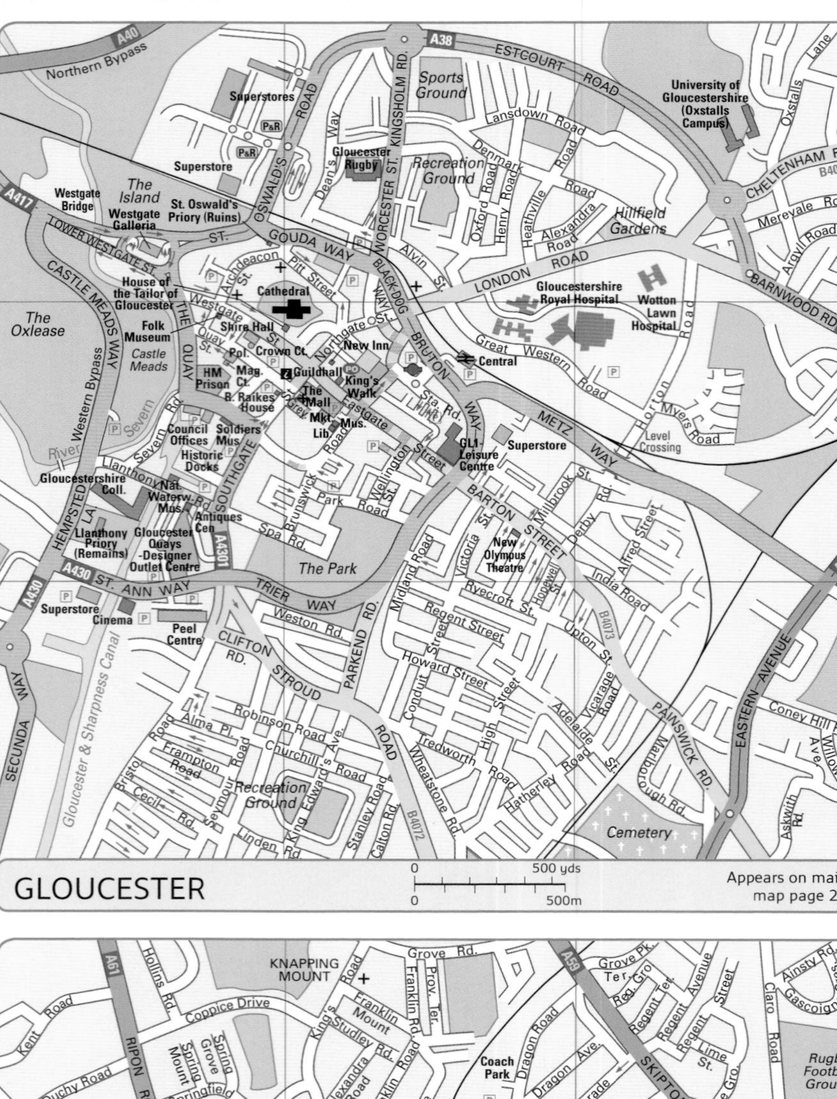

GLOUCESTER

Appears on main
map page 29

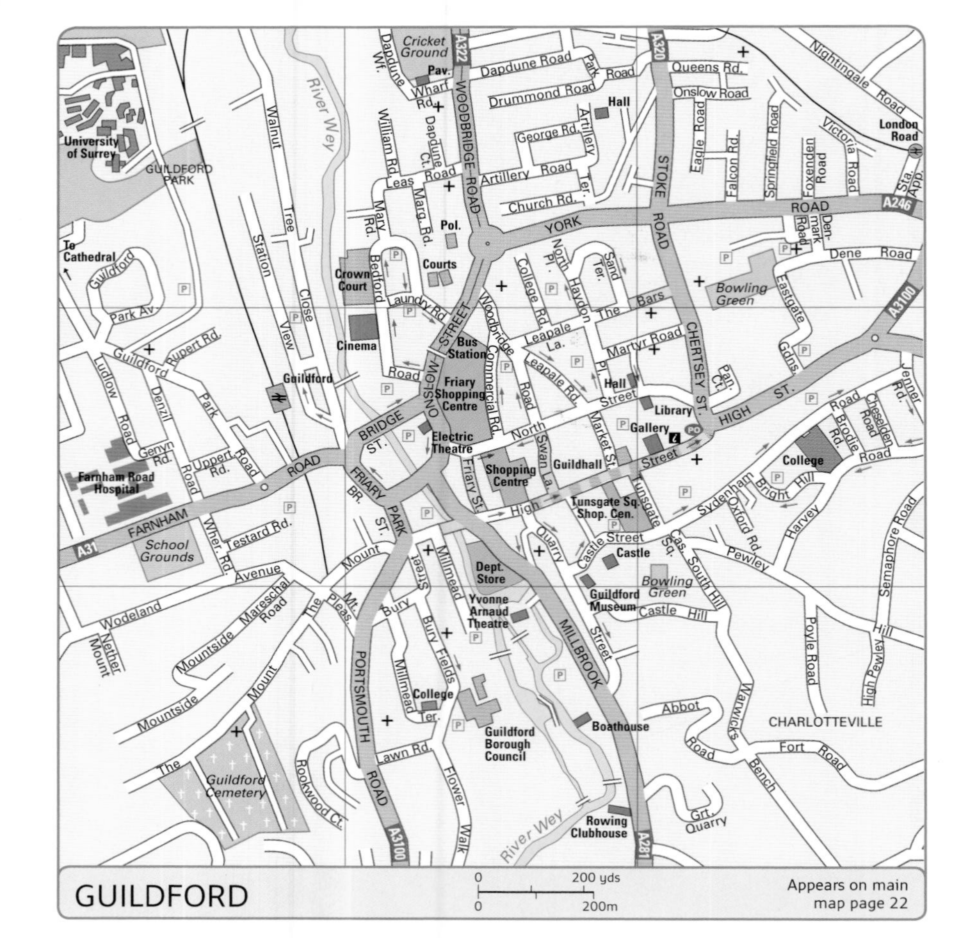

GUILDFORD

Appears on main
map page 22

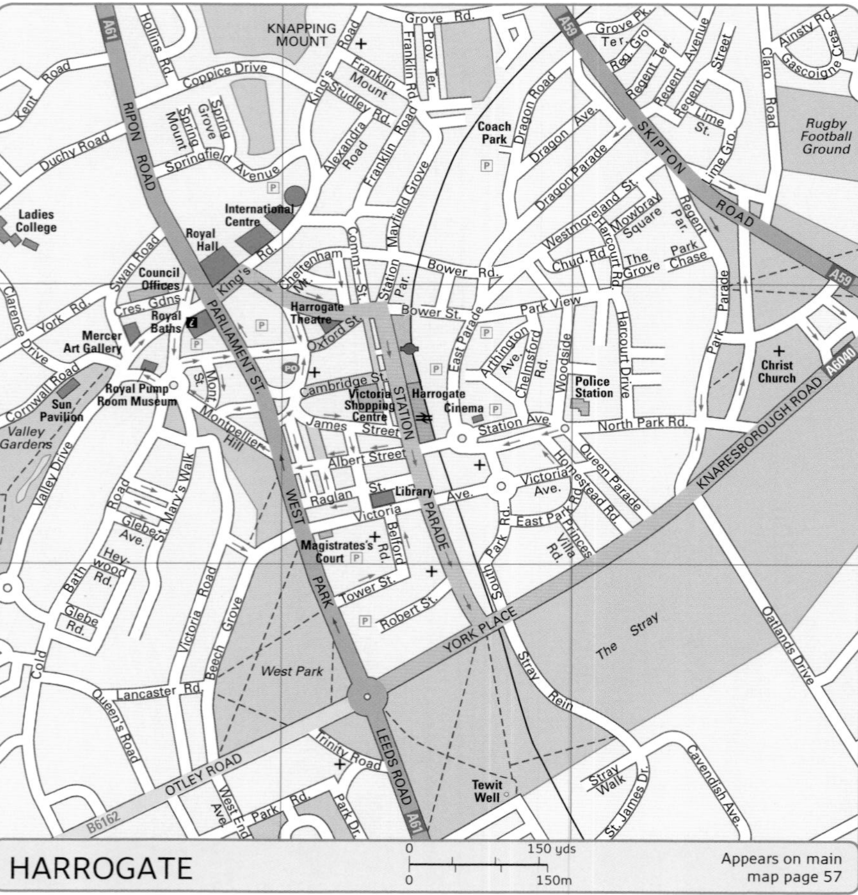

HARROGATE

Appears on main
map page 57

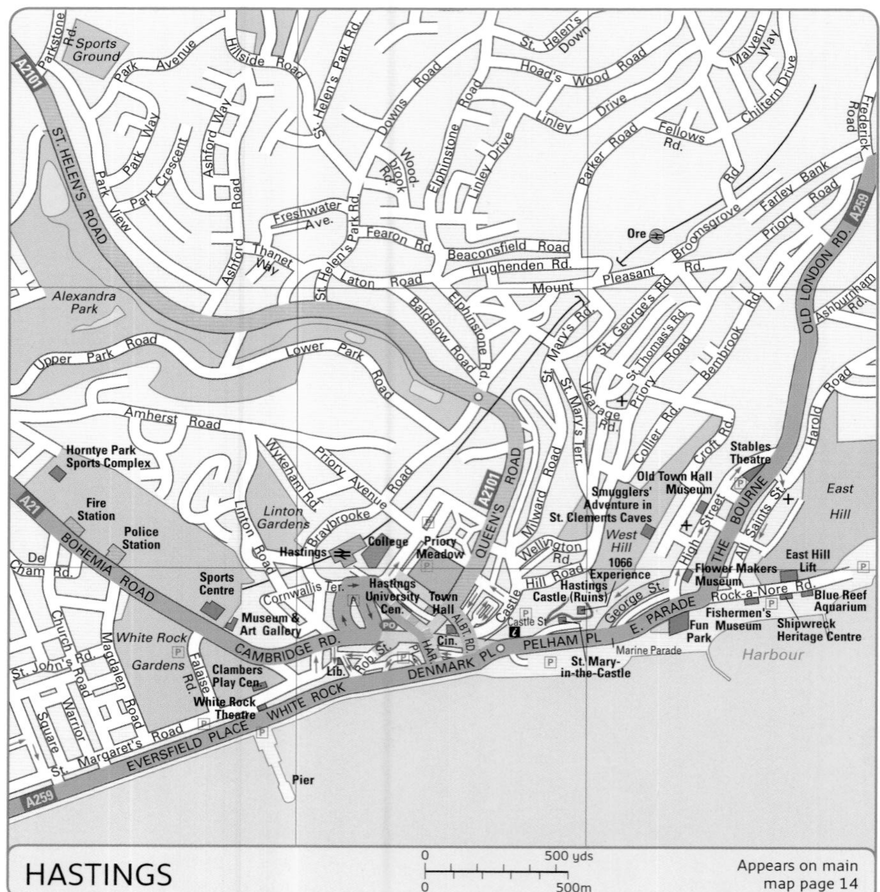

HASTINGS

Appears on main
map page 14

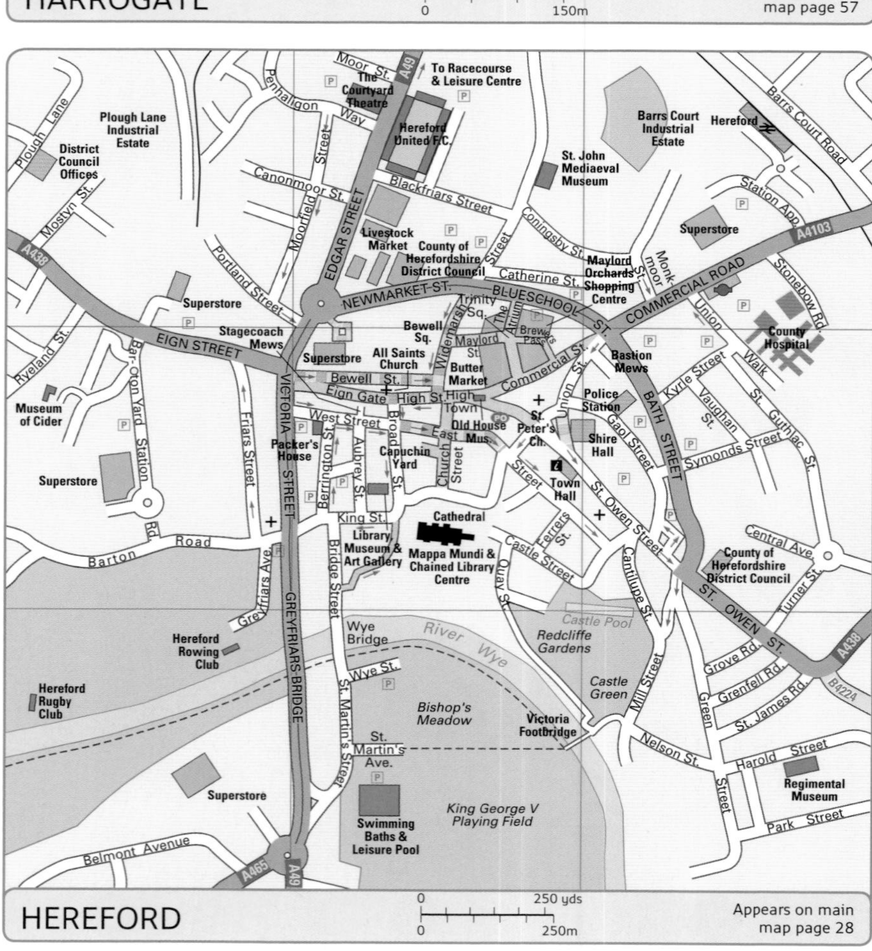

HEREFORD

Appears on main
map page 28

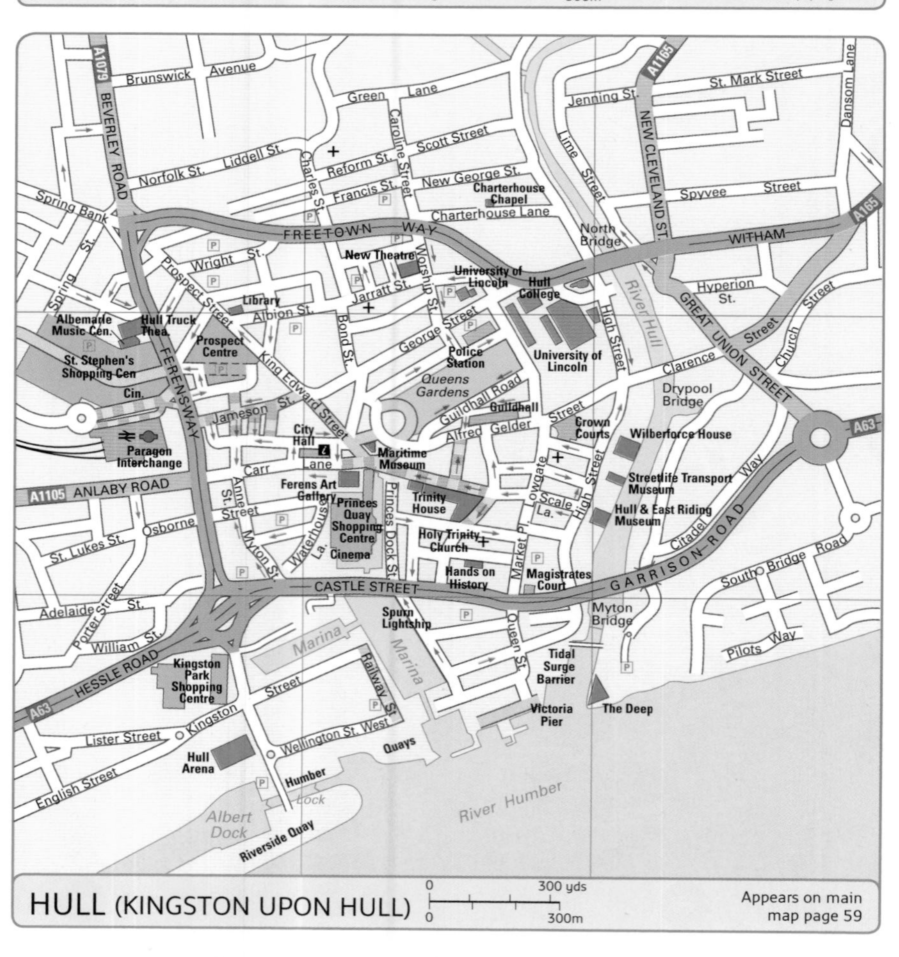

HULL (KINGSTON UPON HULL)

Appears on main
map page 59

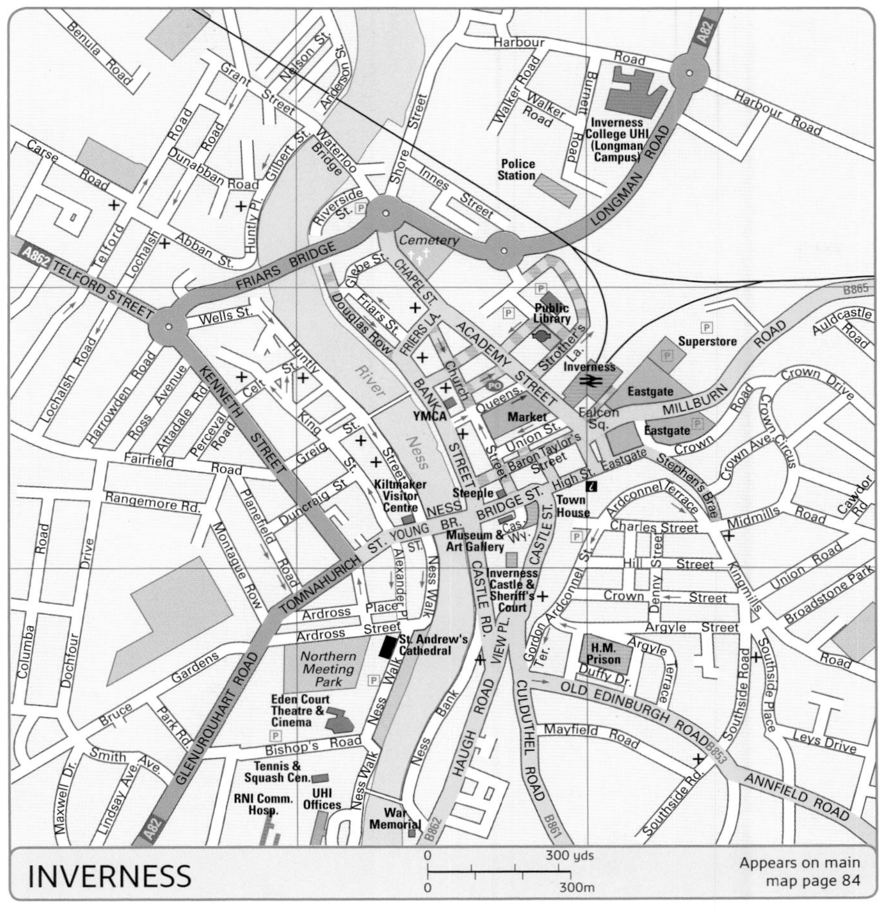

INVERNESS

Appears on main map page 84

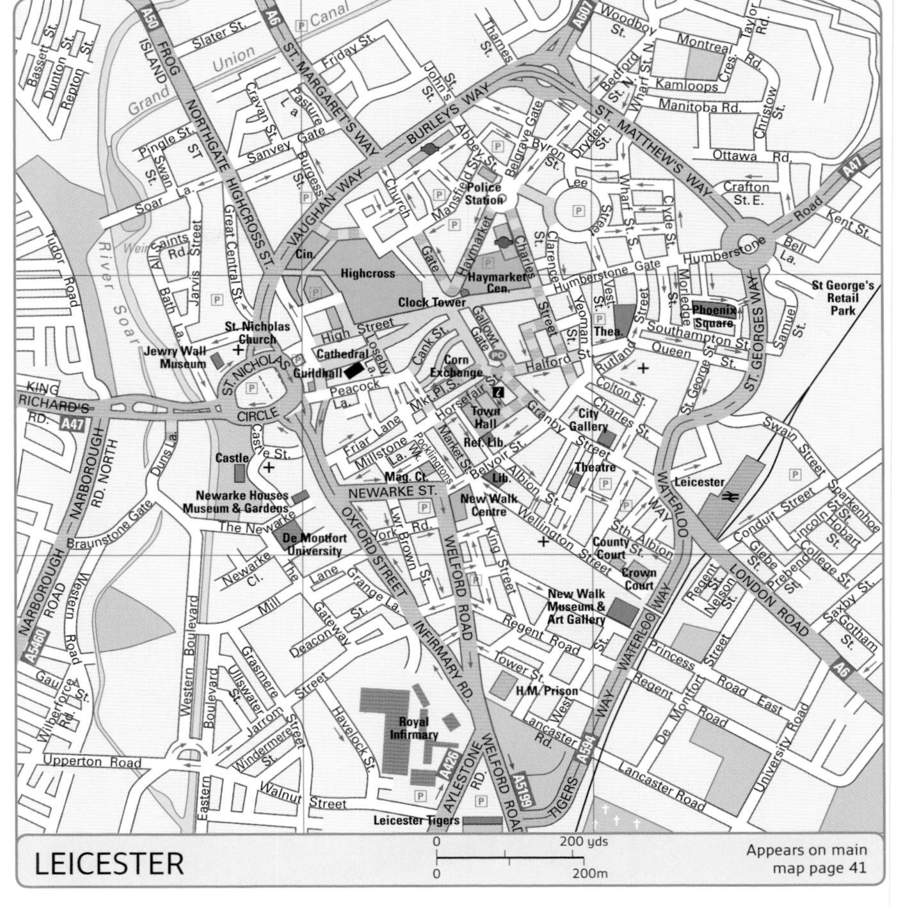

LEICESTER

Appears on main map page 41

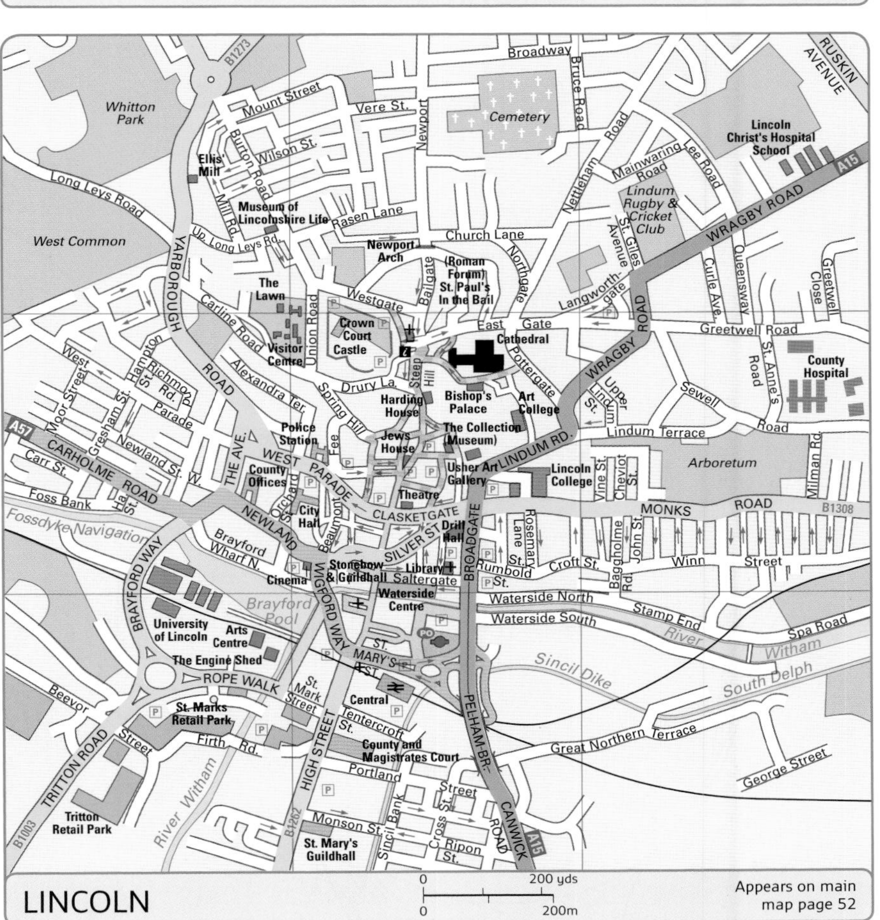

LINCOLN

Appears on main map page 52

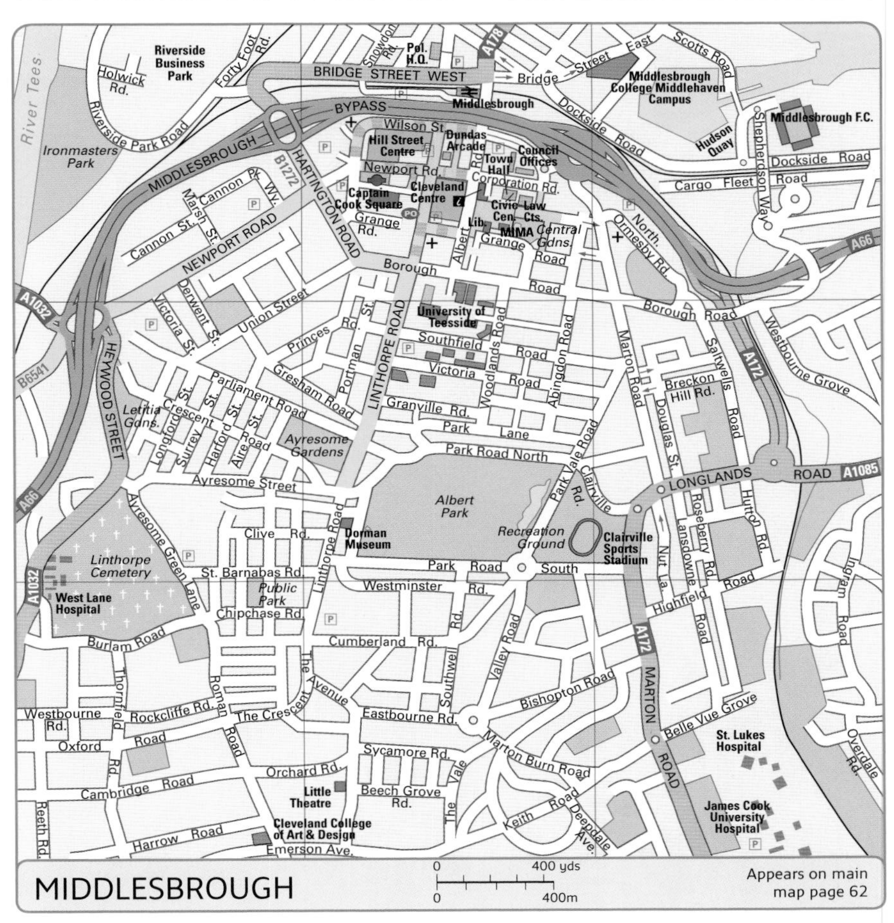

MIDDLESBROUGH

Appears on main map page 62

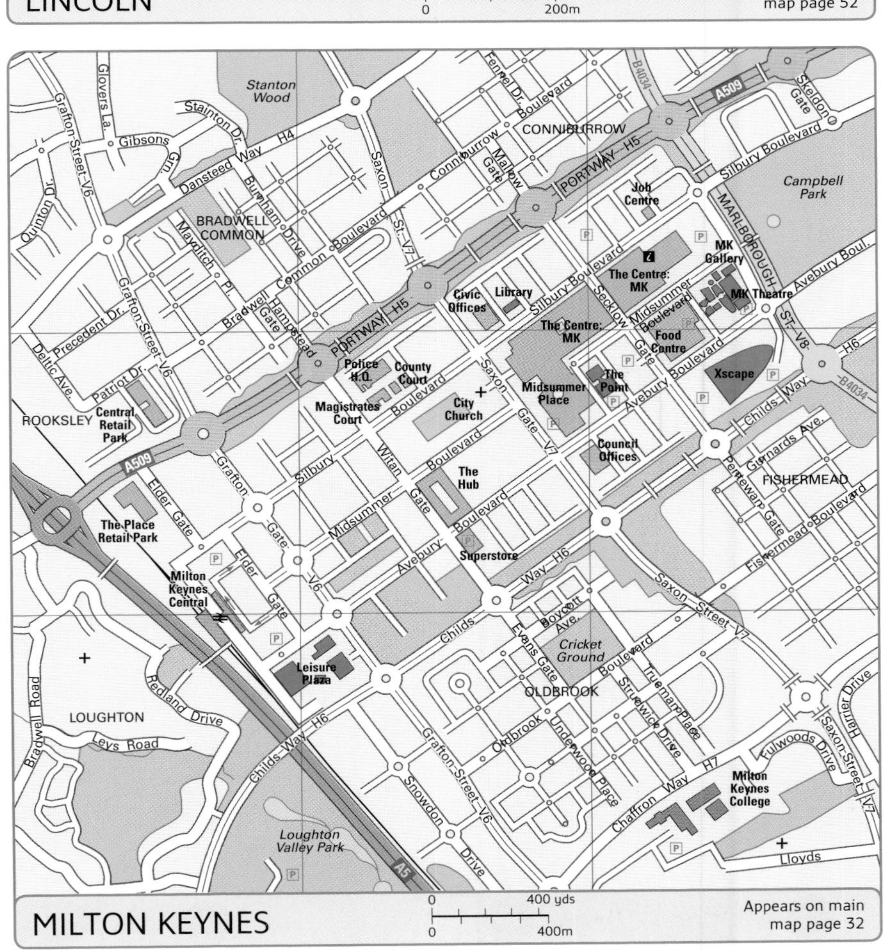

MILTON KEYNES

Appears on main map page 32

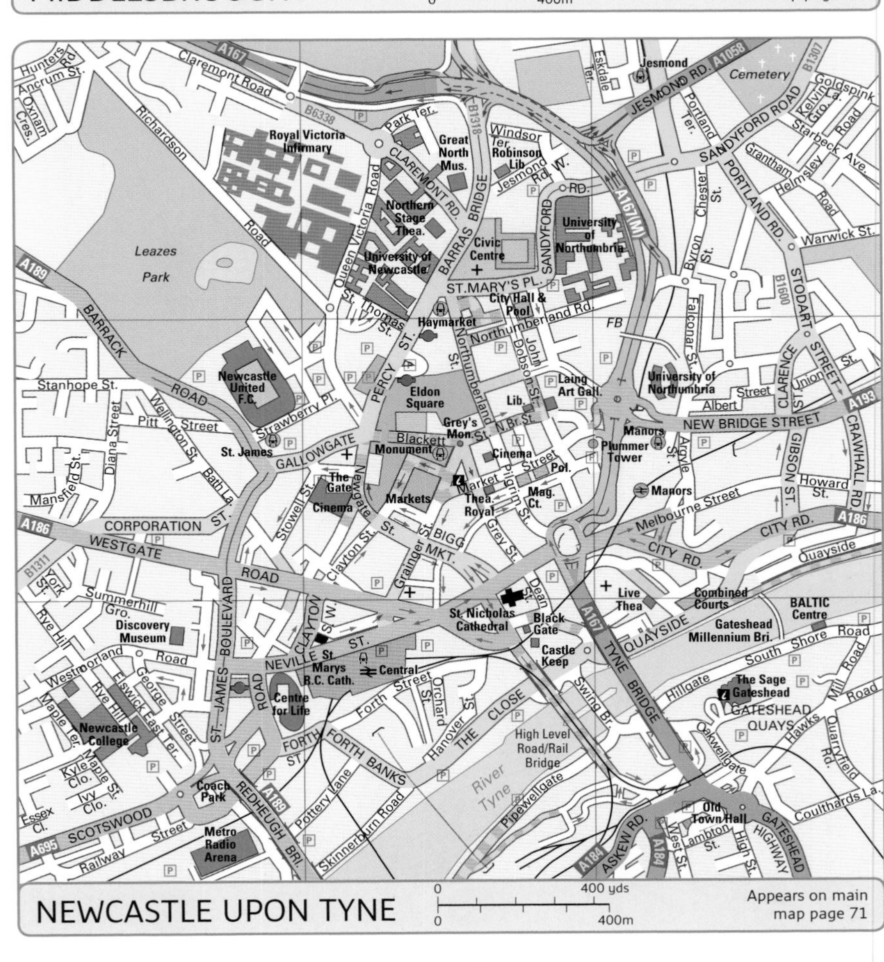

NEWCASTLE UPON TYNE

Appears on main map page 71

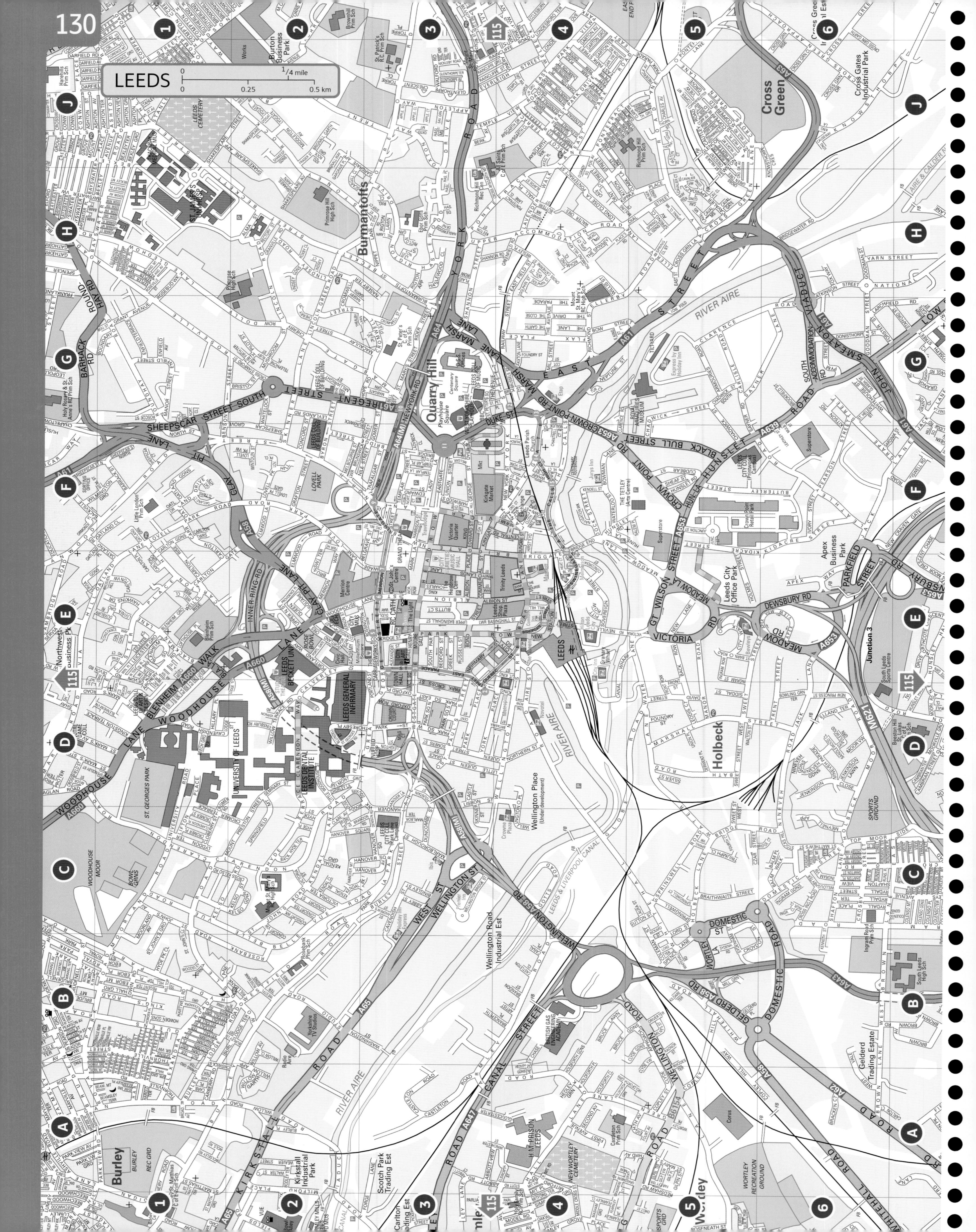
LEEDS
1/4 mile
0
0.25
0.5 km
Cross Green
Cross Gates Industrial Park
Burmantofts
Quarry Hill
Playhouse Square
Kirkgate Market
Victoria Quarter
Trinity Leeds
Merrion Centre
LEEDS GENERAL INFIRMARY
UNIVERSITY OF LEEDS
LEEDS BECKETT UNIV
ST. GEORGES PARK
WOODHOUSE MOOR
St. James's University Hospital
LEEDS CEMETERY
Burton Business Park
Leeds City Office Park
Apex Business Park
Holbeck
Wellington Place (Under development)
Wellington Road Industrial Est
RIVER AIRE
LEEDS & LIVERPOOL CANAL
LEEDS
Granary Wharf
Junction 3
South Leeds Sports Centre
SPORTS GROUND
Gelderd Trading Estate
WORTLEY RECREATION GROUND
H.M.PRISON LEEDS
NEW WORTLEY CEMETERY
BRITISH GAS ENGINEERING ACADEMY
Yorkshire TV Studios
Kirkstall Industrial Park
Scotch Park Trading Est
Burley
Northwest Business Pk
Crown Point Retail Park
THE TETLEY (Arts Centre)
ROYAL ARMOURIES MUSEUM
Superstore
Richmond Hill Rec Cen
YORK ROAD
REGENT STREET
SHEEPSCAR STREET SOUTH
CLAY PIT LANE
INNER RING RD
WOODHOUSE LANE
BLENHEIM WALK
KIRKSTALL ROAD
WELLINGTON ROAD
CANAL STREET
GELDERD RD
DOMESTIC ROAD
WHITEHALL ROAD
MEADOW ROAD
DEWSBURY RD
GT WILSON STREET
VICTORIA RD
HUNSLET RD
BLACK BULL STREET
CROWN POINT RD
EAST STREET
MARSH LANE
SMEATON VIADUCT
BARRACK ROAD
A58(M)
A64(M)
A61
A63
A65
A643
A653
A639
A660
A647
A58
A62
M621
115

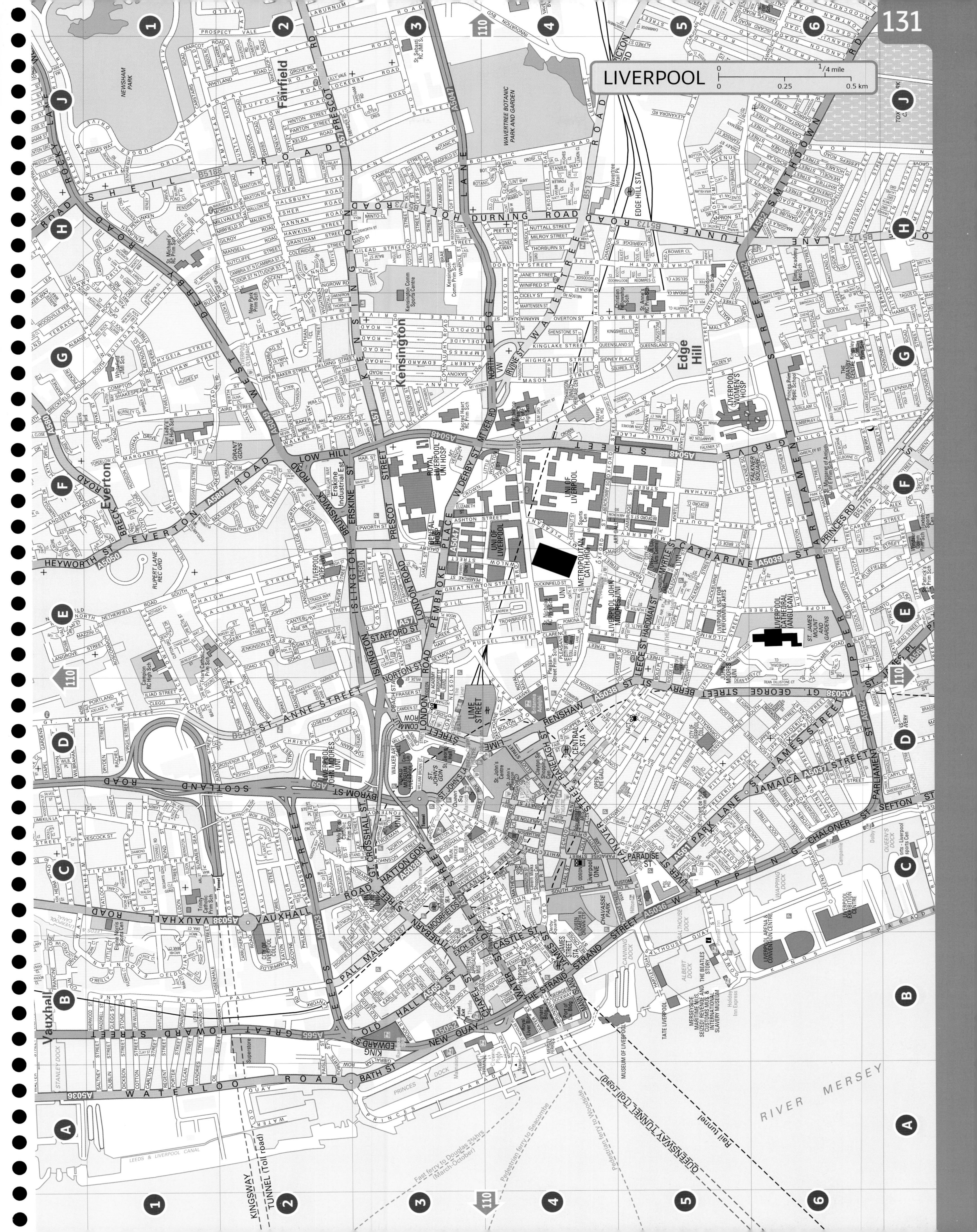

LIVERPOOL
0
1/4 mile
0
0.25
0.5 km
Fairfield
Kensington
Edge Hill
Everton
Vauxhall
NEWSHAM PARK
WAVERTREE BOTANIC PARK AND GARDEN
RUPERT LANE REC GRD
SHEIL ROAD
DERBY LANE
WEST DERBY ROAD
PRESCOT ROAD
KENSINGTON
EDGE LANE
DURNING ROAD
WAVERTREE ROAD
TUNNEL ROAD
SMITHDOWN ROAD
UPPER PARLIAMENT STREET
PRINCES RD
CATHARINE ST
HOPE ST
GROVE STREET
MULGRAVE STREET
CROWN STREET
PEMBROKE PLACE
LONDON ROAD
BRUNSWICK ST
ERSKINE ST
ISLINGTON
LOW HILL
EVERTON ROAD
HEYWORTH ST
NETHERFIELD ROAD NORTH
SCOTLAND ROAD
BYROM ST
ST ANNE STREET
LIME STREET
RENSHAW ST
BERRY ST
GT GEORGE STREET
ST JAMES STREET
PARK LANE
JAMAICA STREET
CHALONER ST
SEFTON ST
WAPPING
STRAND STREET
THE STRAND
NEW QUAY
OLD HALL ST
DALE ST
VAUXHALL ROAD
GREAT HOWARD STREET
WATERLOO ROAD
BATH ST
LEEDS STREET
PALL MALL
HANOVER STREET
PARADISE ST
ROYAL LIVERPOOL UNI HOSP
UNI OF LIVERPOOL
METROPOLITAN CATH (RC)
LIVERPOOL JOHN MOORES UNI
LIVERPOOL CATHEDRAL (ANGLICAN)
ST. JAMES MOUNT AND GARDENS
LIVERPOOL WOMEN'S HOSP
EDGE HILL STA
LIME STREET
CENTRAL STA
MOORFIELDS
ST. JOHN'S GDN
WALKER ART GALLERY
CHAVASSE PARK
Liverpool ONE
ALBERT DOCK
CANNING DOCK
WAPPING DOCK
QUEEN'S DOCK
PRINCES DOCK
STANLEY DOCK
LIVERPOOL ARENA & CONVENTION CENTRE
LIVERPOOL EXHIBITION CENTRE
TATE LIVERPOOL
THE BEATLES STORY
MERSEYSIDE MARITIME MUS., SEIZED! REVENUE AND CUSTOMS MUS. & INTERNATIONAL SLAVERY MUSEUM
MUSEUM OF LIVERPOOL
Royal Liver Bldg
Port of Lpl Bldg
RIVER MERSEY
QUEENSWAY TUNNEL (Toll road)
KINGSWAY TUNNEL (Toll road)
Rail tunnel
Fast ferry to Douglas 2¾hrs (March-October)
Pedestrian ferry to Seacombe
Pedestrian ferry to Woodside
LEEDS & LIVERPOOL CANAL
A5036
A565
A5052
A5038
A5053
A59
A57
A580
A5049
A5047
A5048
A5039
A562
A561
A5050
B5173
B5178
B5175
B5188
110
1
2
3
4
5
6
A
B
C
D
E
F
G
H
J

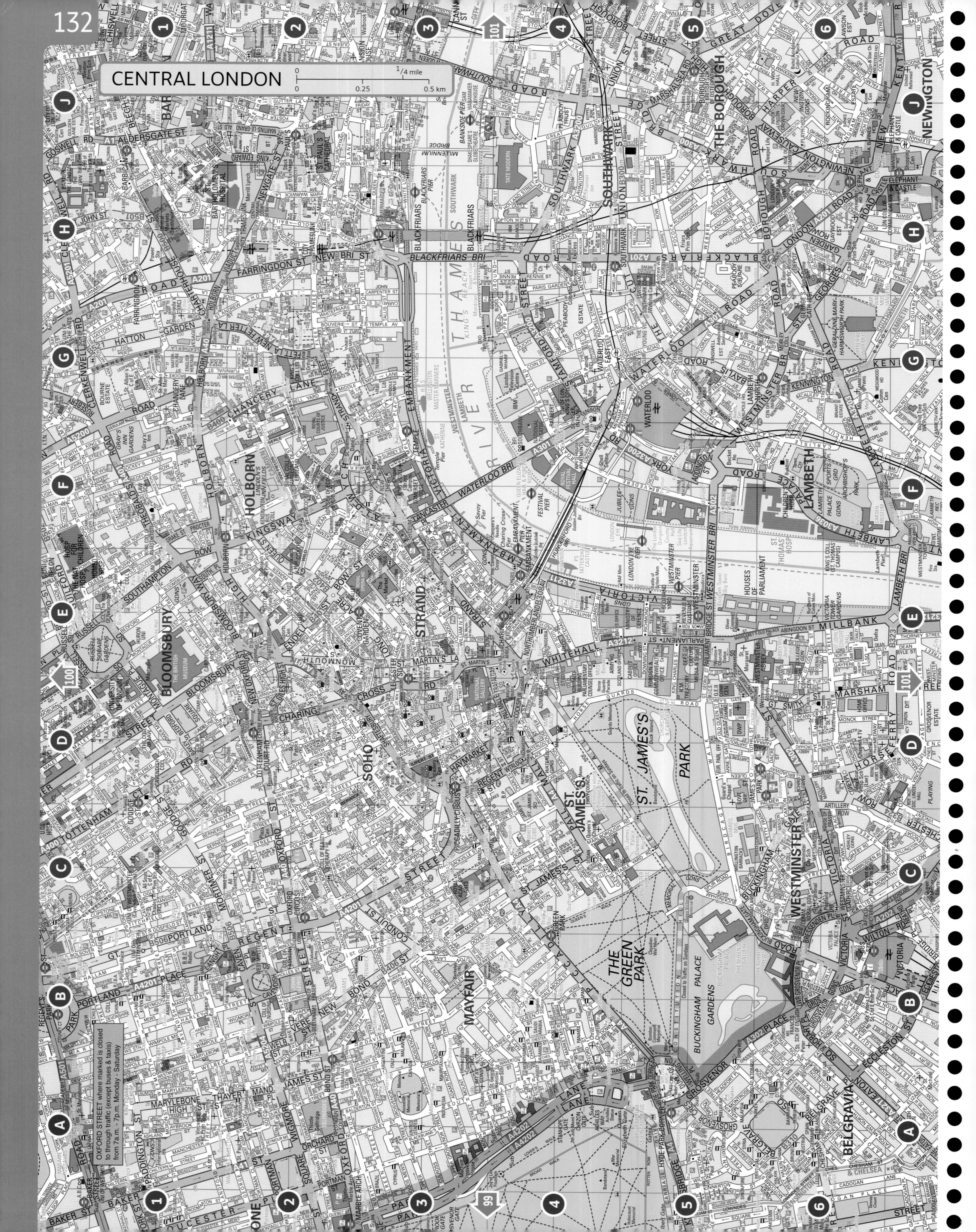

CENTRAL LONDON
1/4 mile
0
0.25
0.5 km
OXFORD STREET where marked is closed to through traffic (except buses & taxis) from 7a.m. - 7p.m. Monday - Saturday
BLOOMSBURY
HOLBORN
SOHO
STRAND
MAYFAIR
ST. JAMES'S
ST. JAMES'S PARK
THE GREEN PARK
BUCKINGHAM PALACE GARDENS
WESTMINSTER
BELGRAVIA
LAMBETH
SOUTHWARK
THE BOROUGH
NEWINGTON
RIVER THAMES
HOUSES OF PARLIAMENT
WATERLOO
TATE MODERN
ST. PAUL'S CATHEDRAL
THE BRITISH MUSEUM
NATIONAL GALLERY
BLACKFRIARS BRI
WATERLOO BRI
WESTMINSTER BRI
LAMBETH BRI
MILLENNIUM BRIDGE
SOUTHWARK BRIDGE
HUNGERFORD BRI
VICTORIA EMBANKMENT
ELEPHANT & CASTLE
VICTORIA
100
101
99

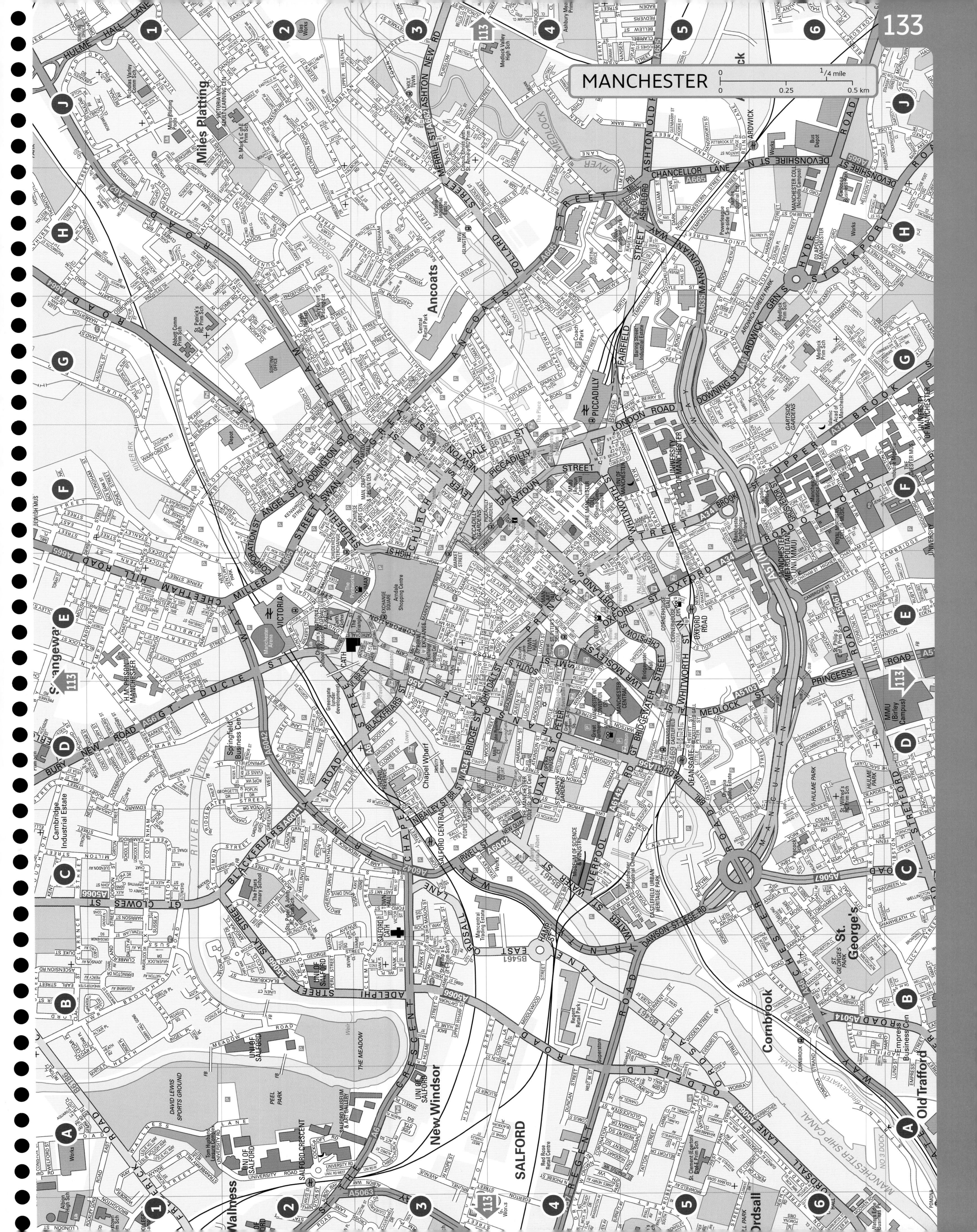
MANCHESTER
0
1/4 mile
0
0.25
0.5 km
Miles Platting
Ancoats
ARDWICK
PICCADILLY
VICTORIA
Manchester Arena
Arndale Shopping Centre
EXCHANGE SQUARE
Chapel Wharf
UNIVERSITY OF MANCHESTER
MANCHESTER METROPOLITAN UNI (MMU)
MMU (Birley Campus)
HULME PARK
St. George's
Cornbrook
Old Trafford
SALFORD
New Windsor
Ordsall
Walness
Strangeways
H.M. PRISON MANCHESTER
Cambridge Industrial Estate
Springfield Business Cen
The Friars Primary School
UNI OF SALFORD
PEEL PARK
DAVID LEWIS SPORTS GROUND
THE MEADOW
SALFORD MUSEUM & ART GALLERY
SALFORD CRESCENT
Regent Retail Park
Red Rose Retail Centre
Mancentral Trading Estate
CASTLEFIELD URBAN HERITAGE PARK
MUSEUM OF SCIENCE & INDUSTRY
MANCHESTER SHIP CANAL
BRIDGEWATER CANAL
RIVER IRWELL
RIVER MEDLOCK
ROCHDALE CANAL
ASHTON CANAL
GARTSIDE GARDENS
Central Retail Park
SORTING OFFICE
Bus Depot
Medlock Valley High Sch
Empress Business Cen
Chancellor Lane
Devonshire St
London Road
Oxford Road
Deansgate
Great Ancoats Street
Rochdale Road
Cheetham Hill Road
Bury New Road
Chapel Street
Trinity Way
Mancunian Way
Stockport Road
Hyde Road
Stretford Road
Chester Road
Regent Road
Ordsall Lane
Water St
Liverpool Rd
Quay St
Peter St
Princess St
Portland St
Piccadilly
Market St
Corporation St
Ducie St
Oldham Road
Ashton Old Road
Downing St
Upper Brook St
Oxford St
Whitworth St
Medlock St
Great Bridgewater St
Blackfriars Rd
Cross Lane
Eccles New Road
Gt Clowes St
Trafford Rd
A665
A6
A34
A56
A57(M)
A635
A6042
A5103
A5066
A5063
A5014
A6067
A6143
A6041
B5461
B6469
A662
113
1
2
3
4
5
6
A
B
C
D
E
F
G
H
J

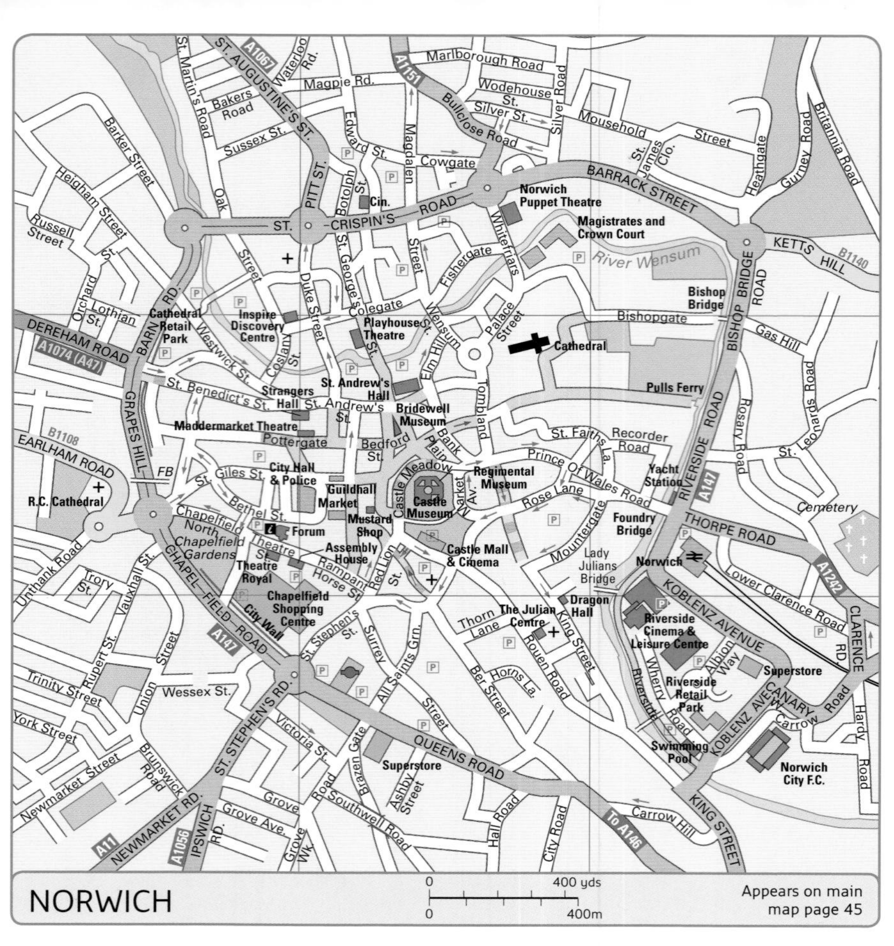

NORWICH

0 400 yds
0 400m

Appears on main
map page 45

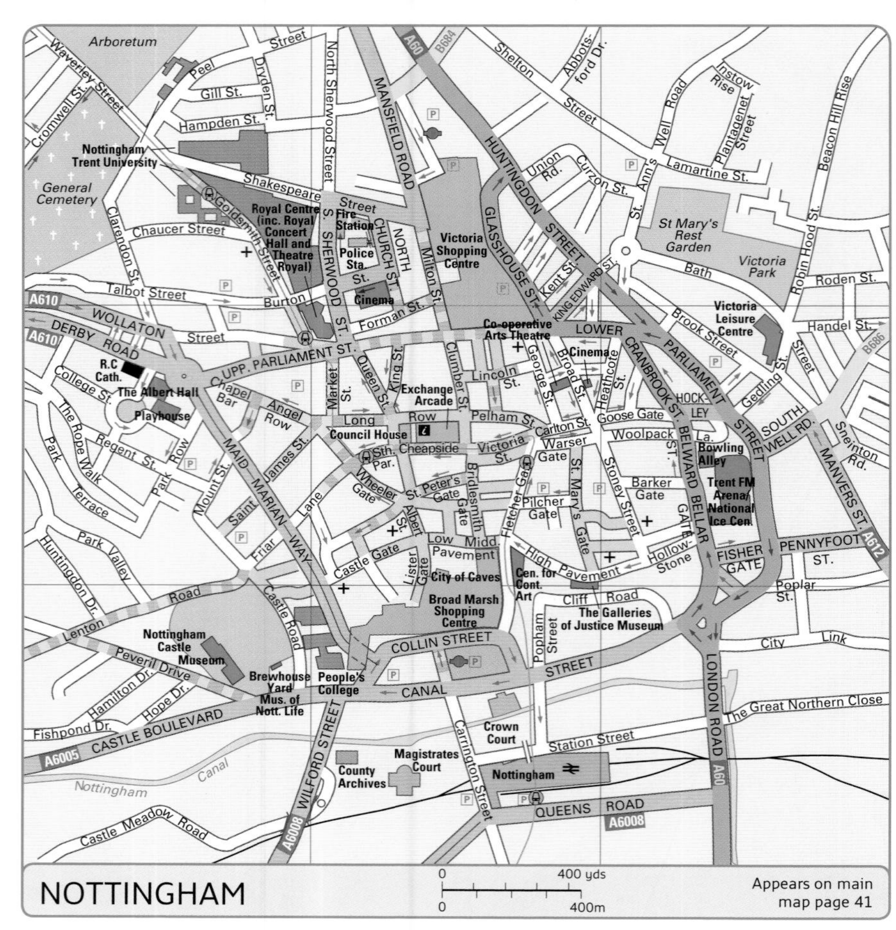

NOTTINGHAM

0 400 yds
0 400m

Appears on main
map page 41

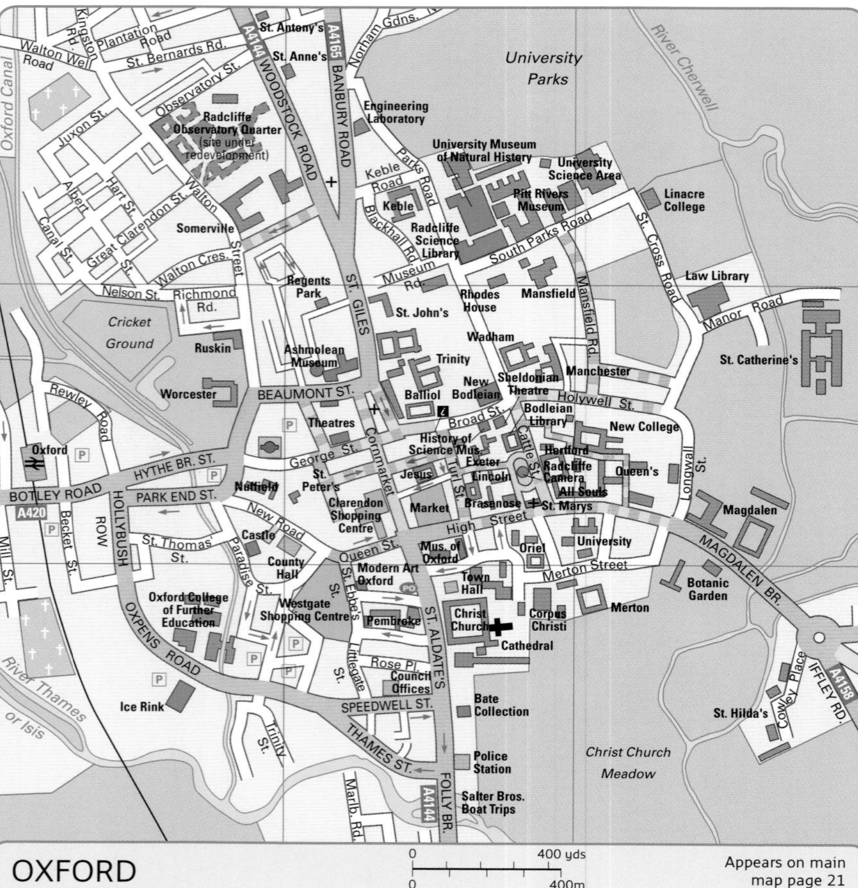

OXFORD

0 400 yds
0 400m

Appears on main
map page 21

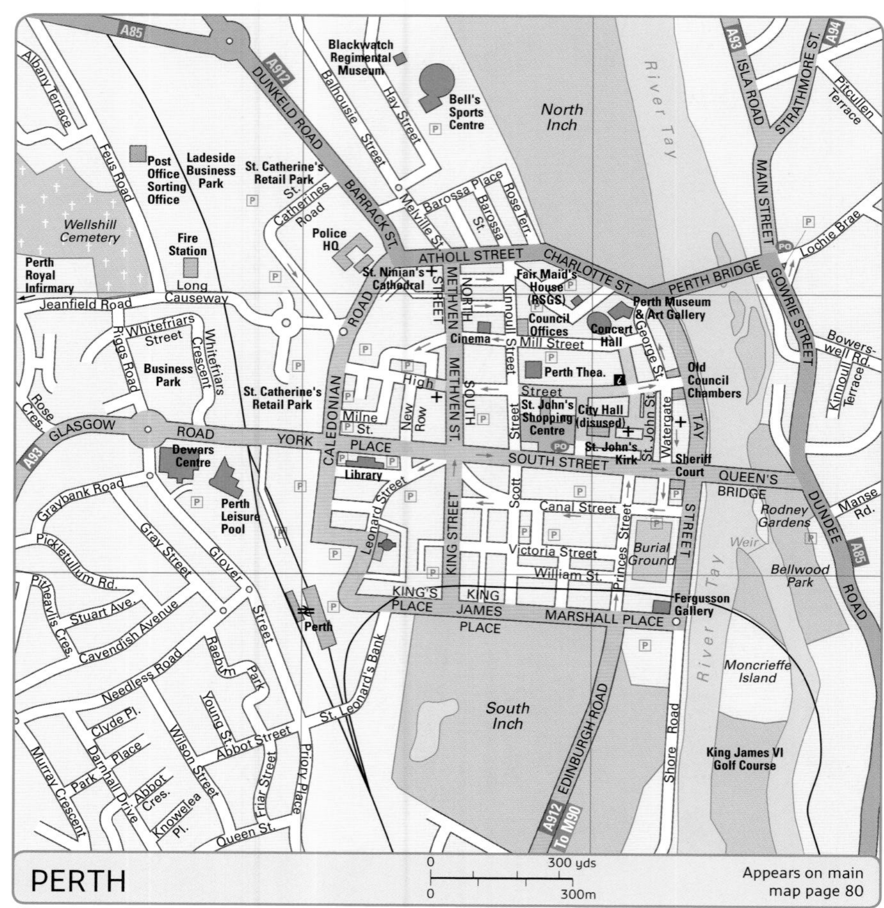

PERTH

0 300 yds
0 300m

Appears on main
map page 80

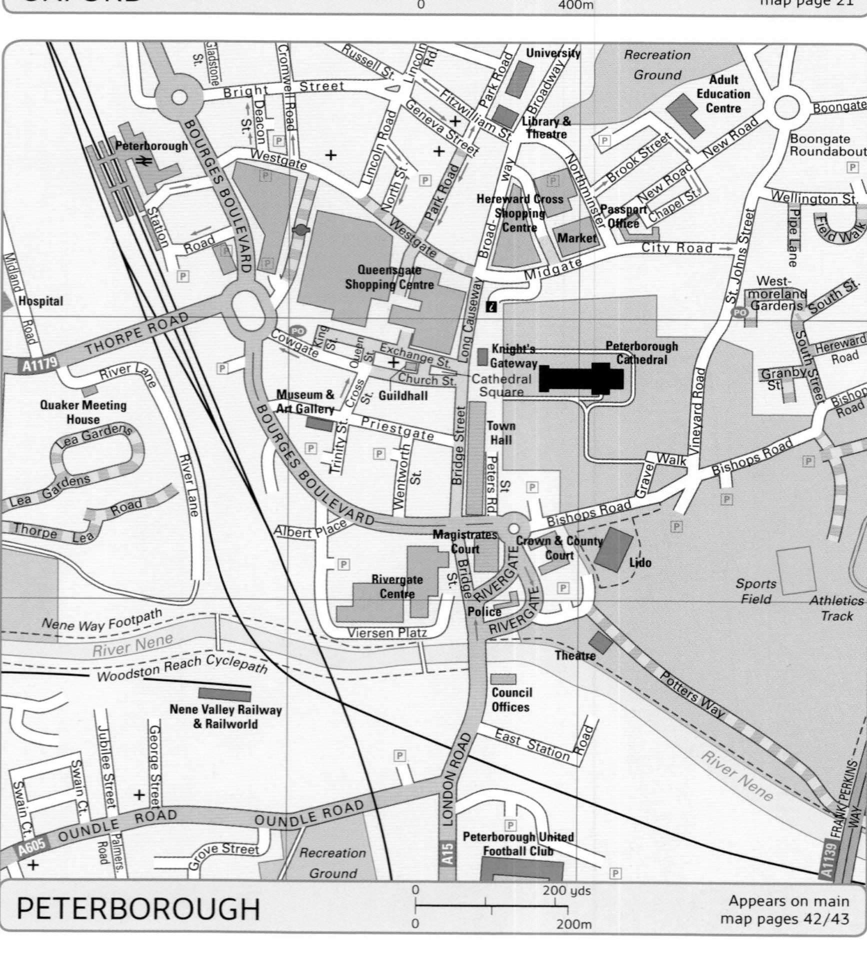

PETERBOROUGH

0 200 yds
0 200m

Appears on main
map pages 42/43

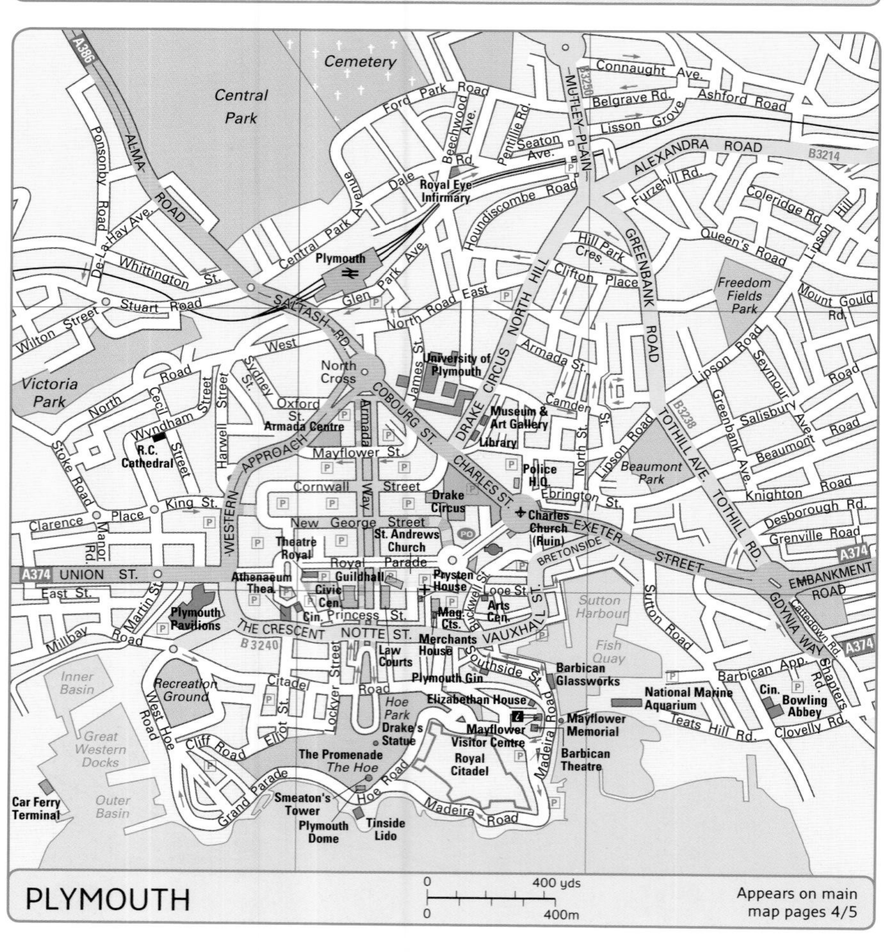

PLYMOUTH

0 400 yds
0 400m

Appears on main
map pages 4/5

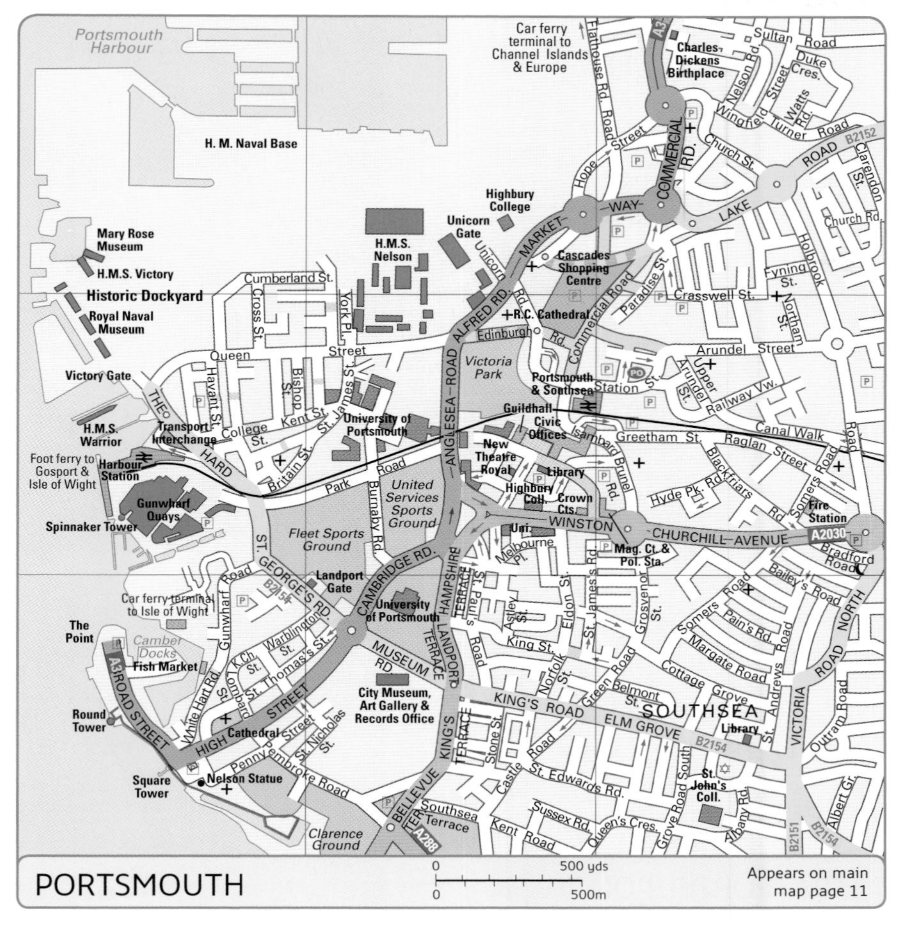

PORTSMOUTH
Appears on main map page 11

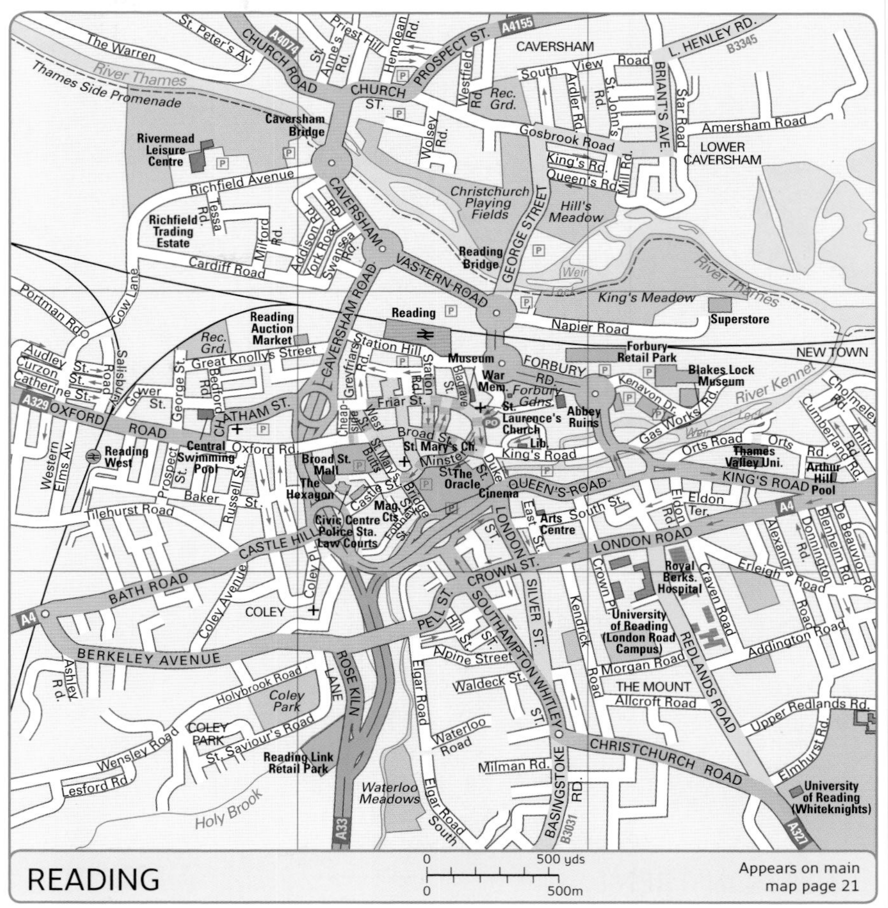

READING
Appears on main map page 21

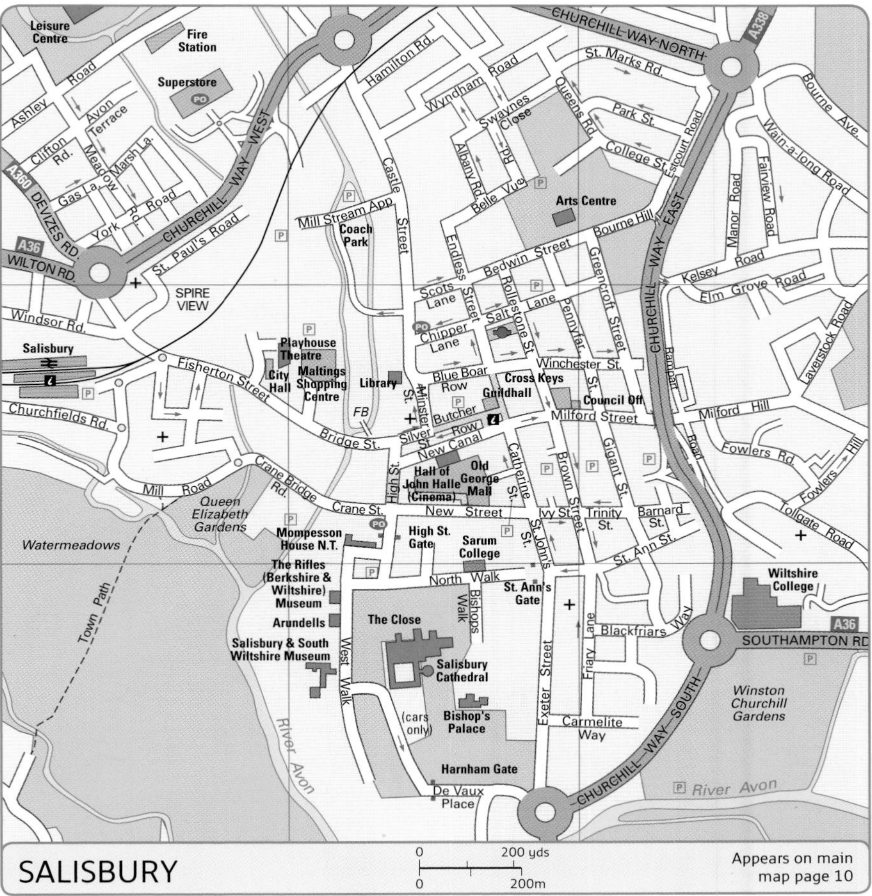

SALISBURY
Appears on main map page 10

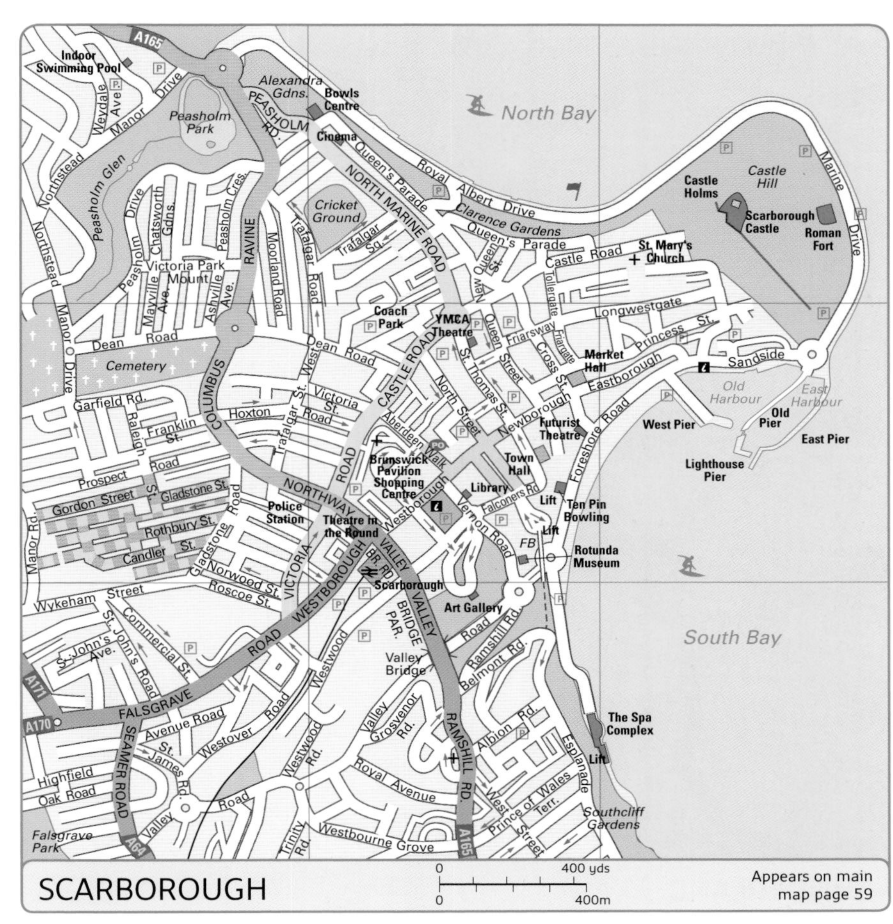

SCARBOROUGH
Appears on main map page 59

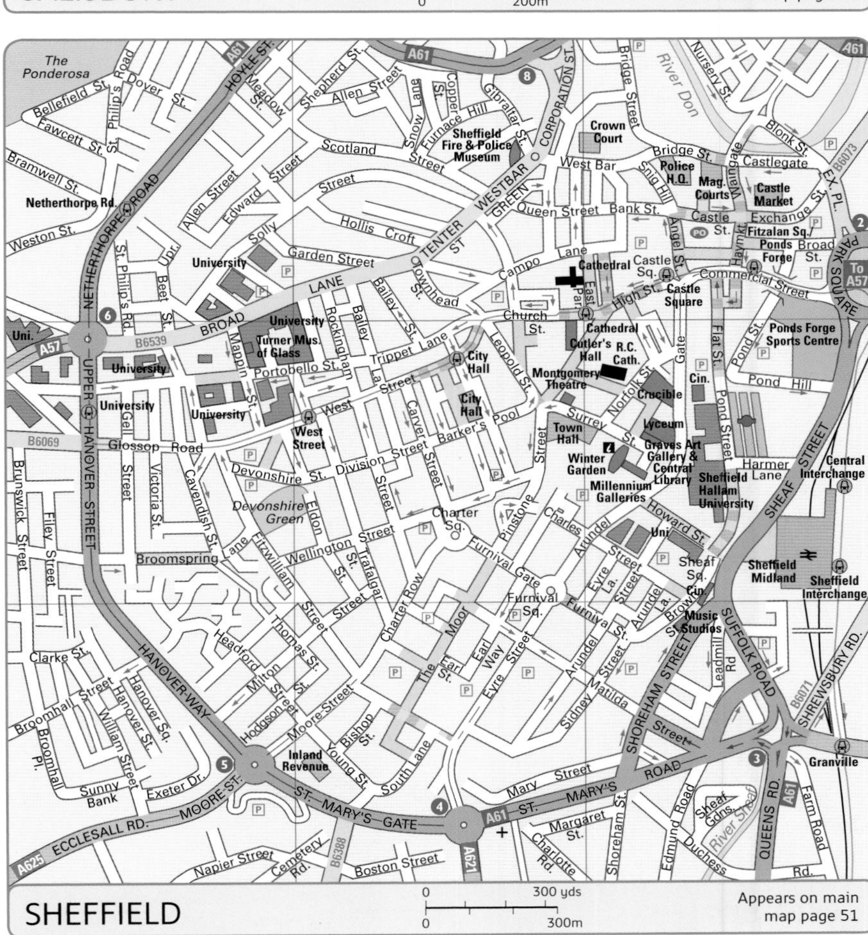

SHEFFIELD
Appears on main map page 51

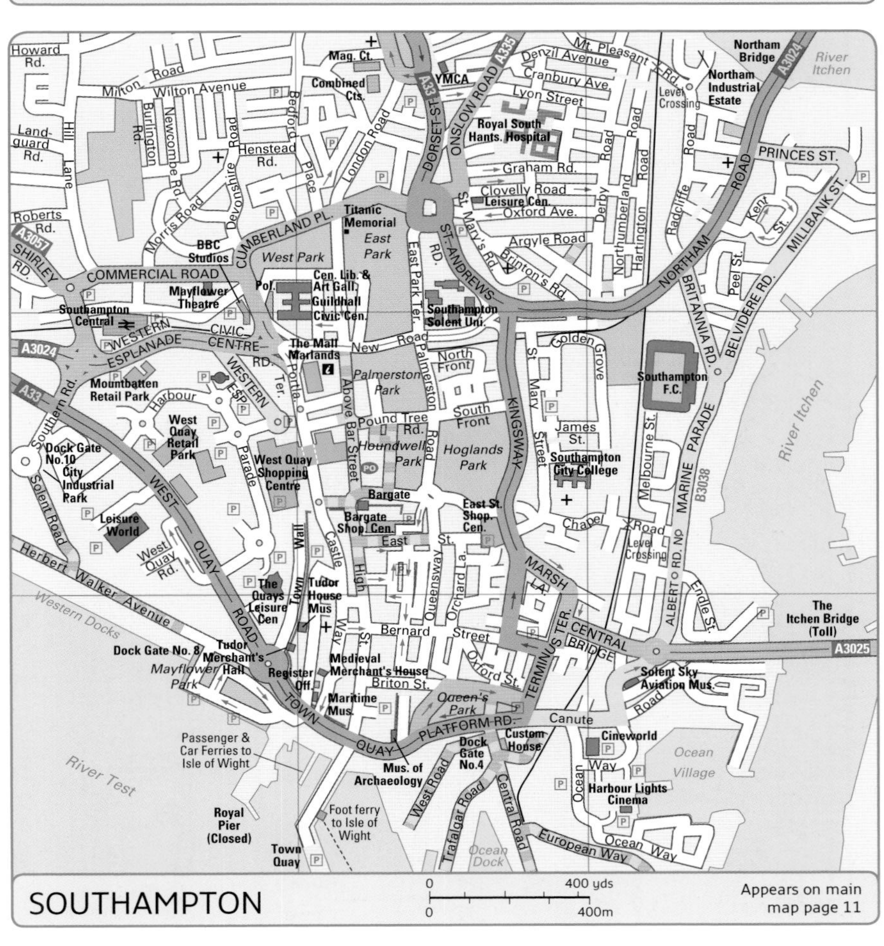

SOUTHAMPTON
Appears on main map page 11

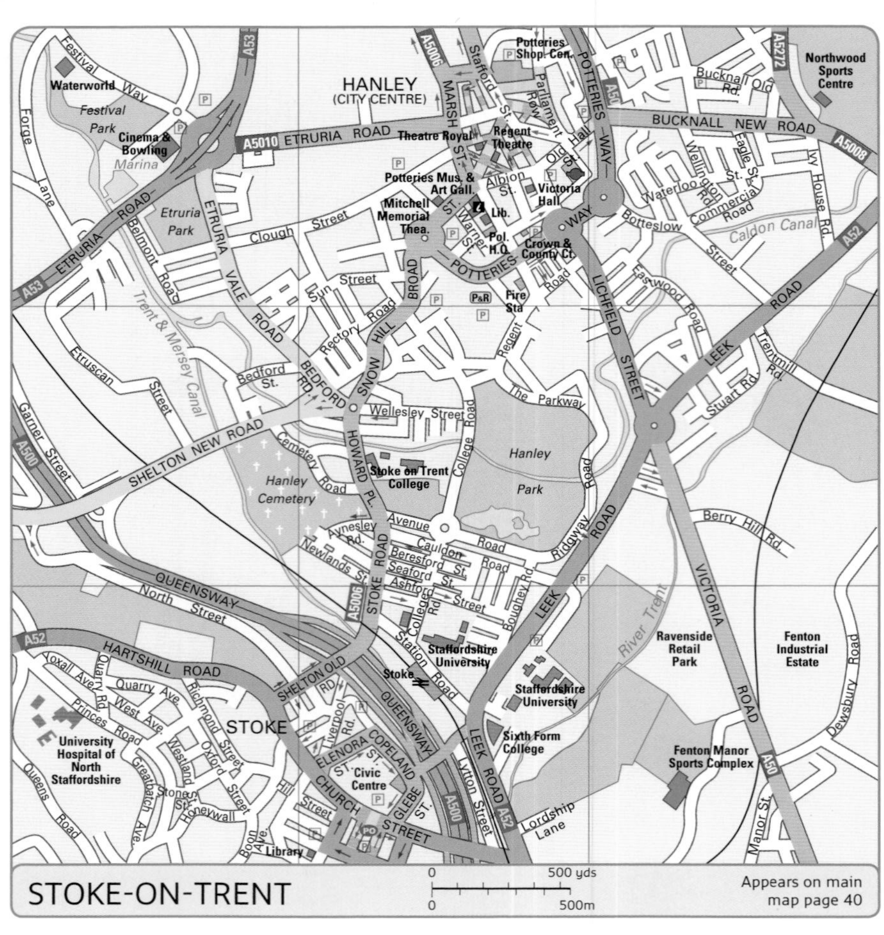

STOKE-ON-TRENT

0   500 yds
0   500m

Appears on main
map page 40

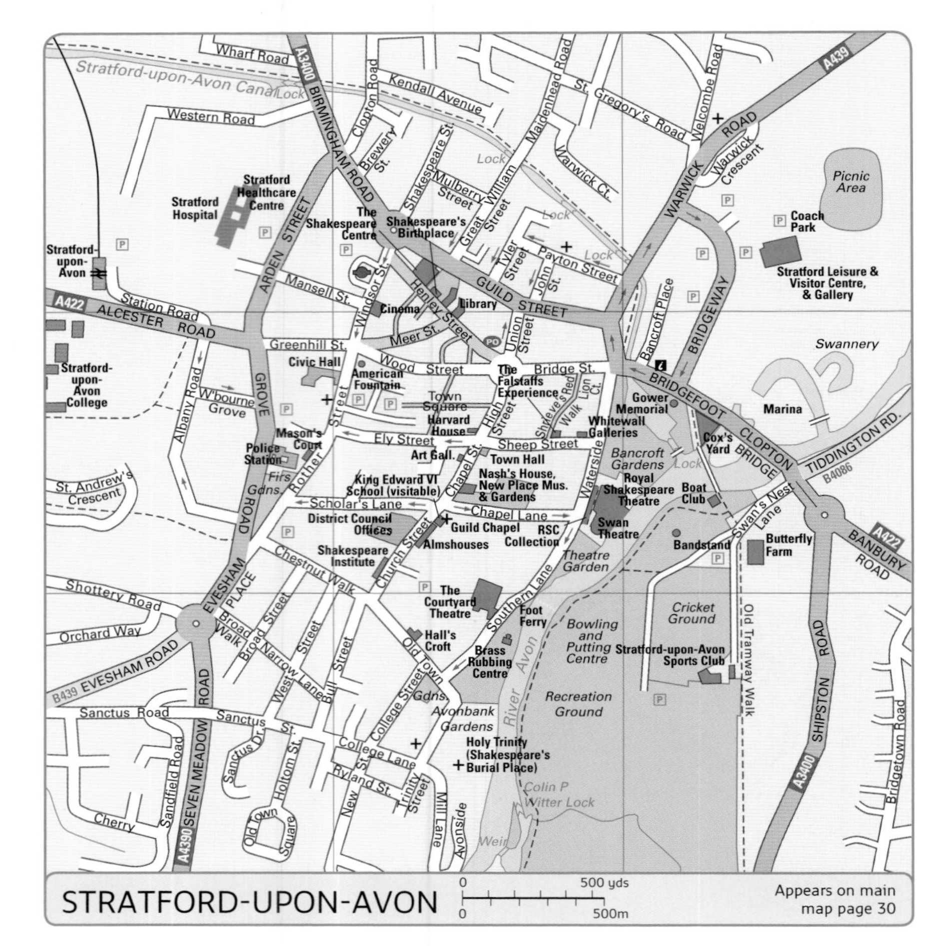

STRATFORD-UPON-AVON

0   500 yds
0   500m

Appears on main
map page 30

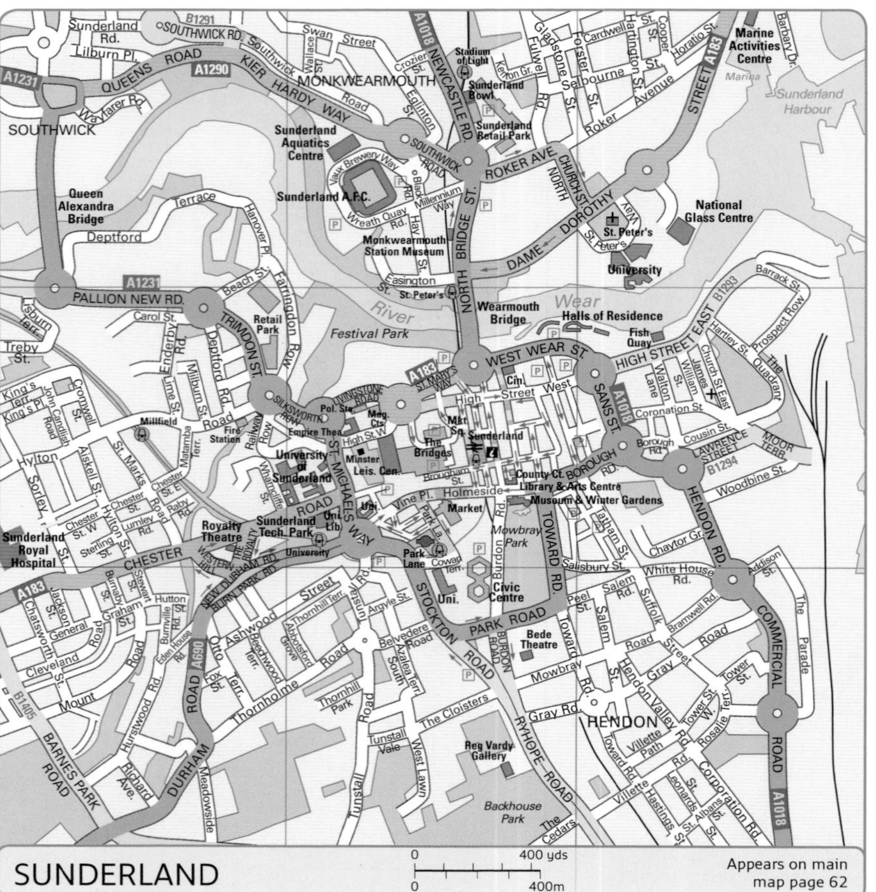

SUNDERLAND

0   400 yds
0   400m

Appears on main
map page 62

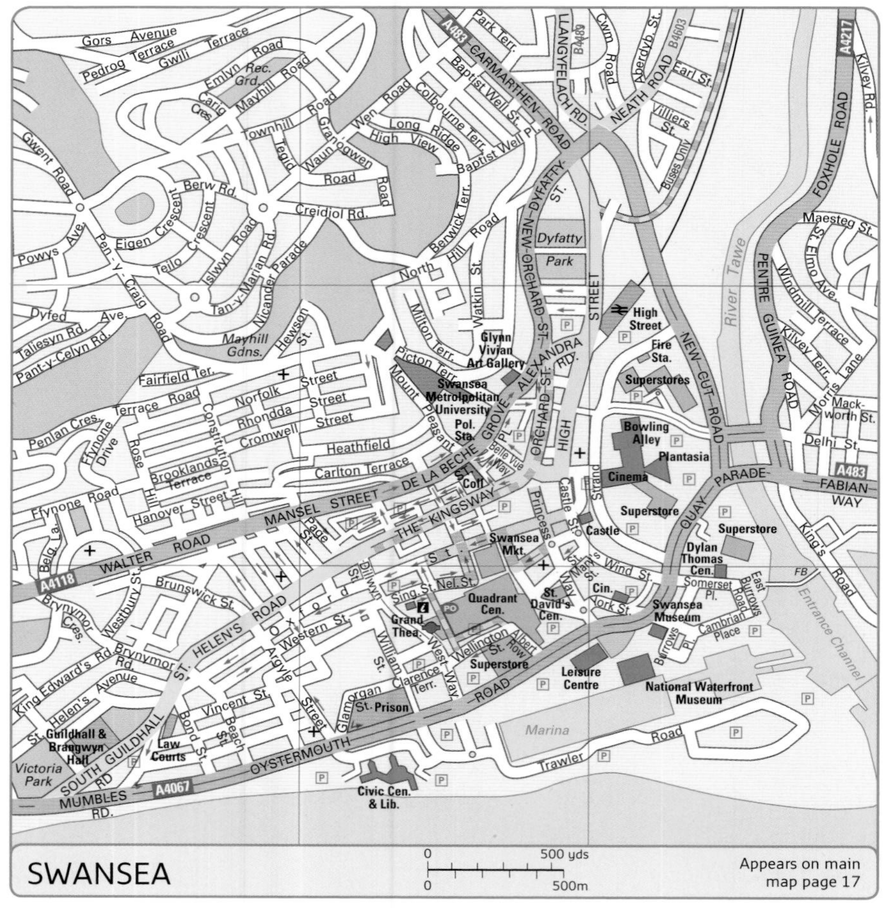

SWANSEA

0   500 yds
0   500m

Appears on main
map page 17

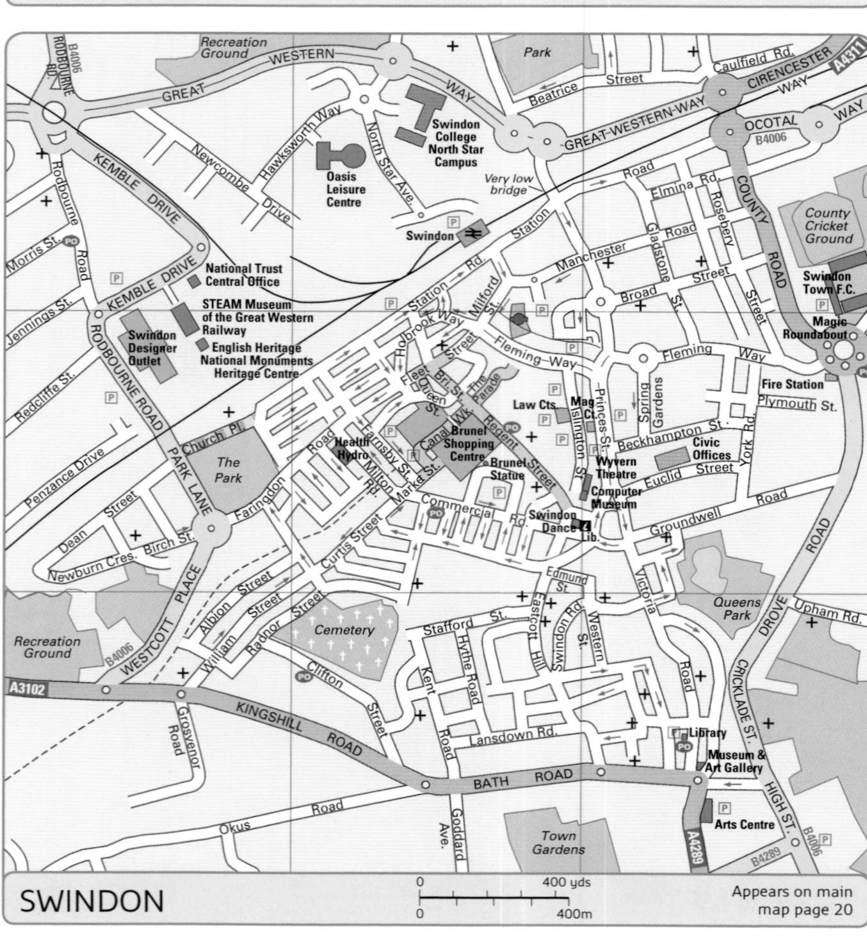

SWINDON

0   400 yds
0   400m

Appears on main
map page 20

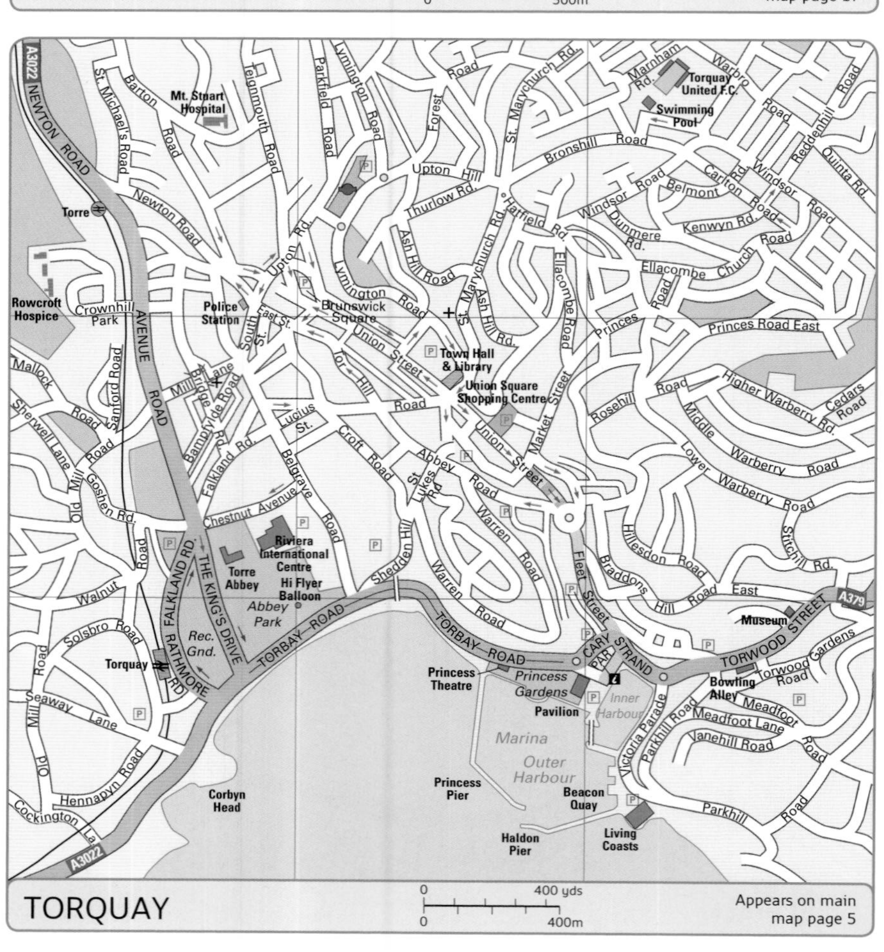

TORQUAY

0   400 yds
0   400m

Appears on main
map page 5

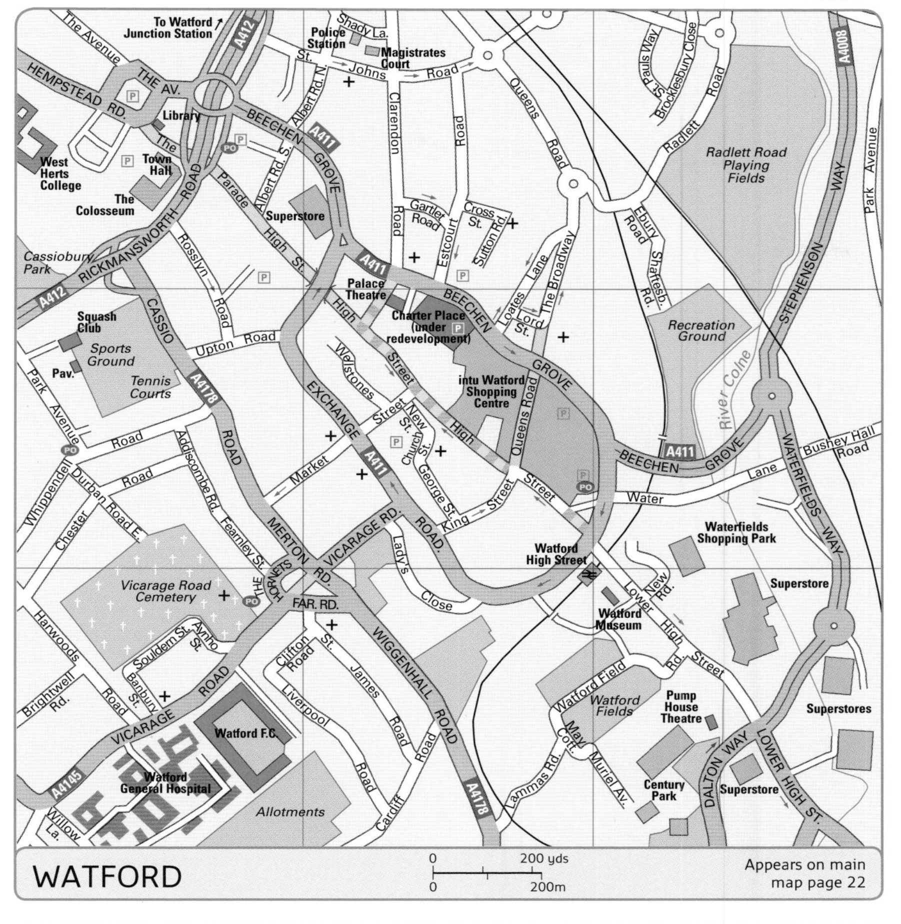

WATFORD

0   200 yds

0   200m

Appears on main
map page 22

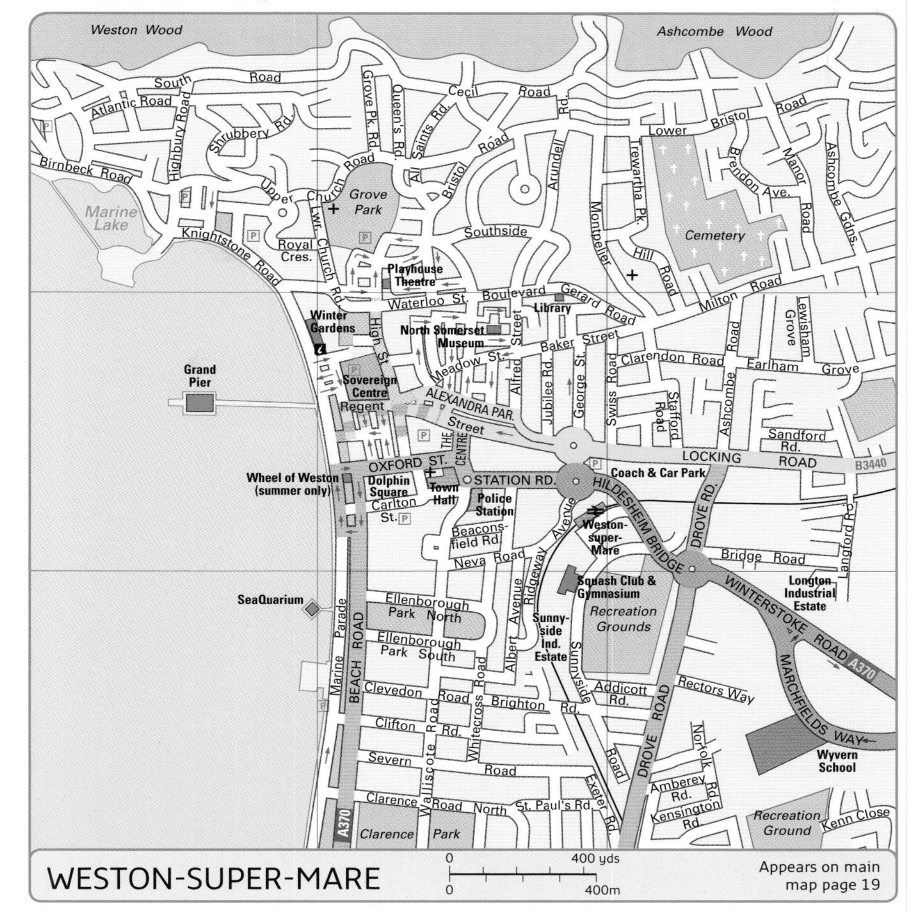

WESTON-SUPER-MARE

0   400 yds

0   400m

Appears on main
map page 19

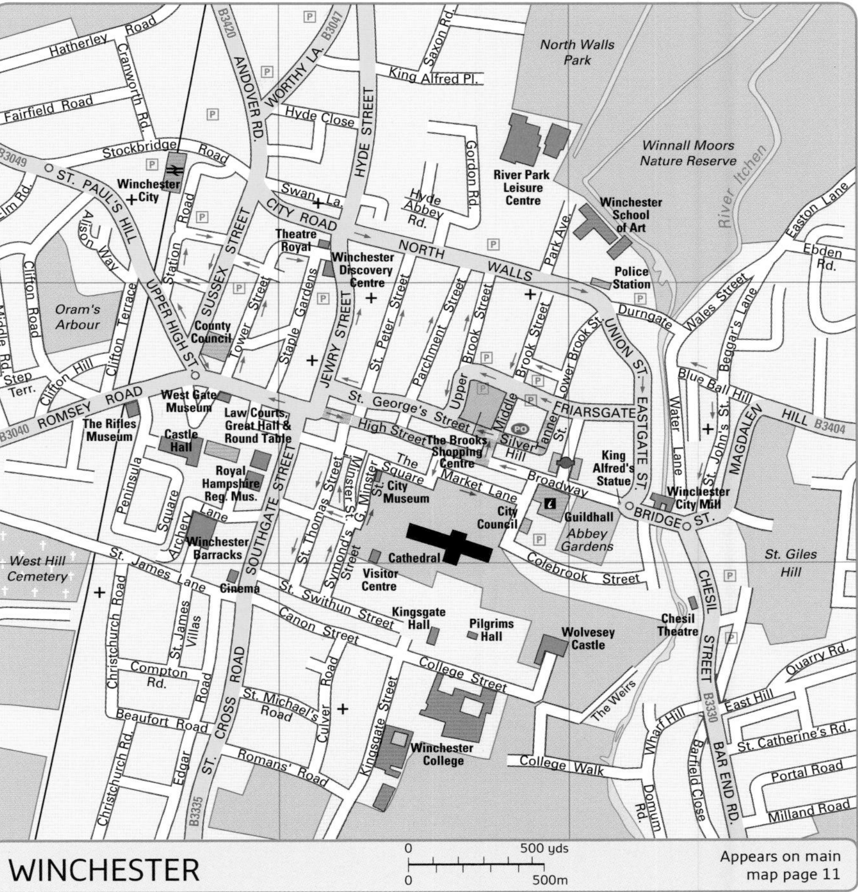

WINCHESTER

0   500 yds

0   500m

Appears on main
map page 11

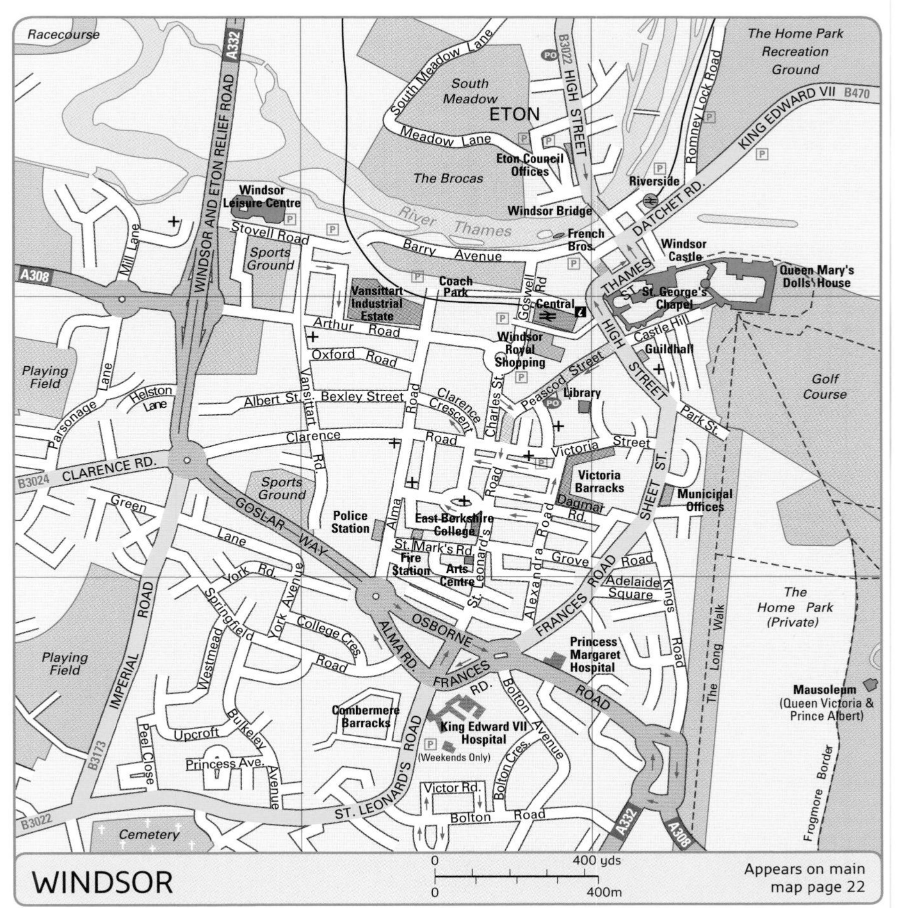

WINDSOR

0   400 yds

0   400m

Appears on main
map page 22

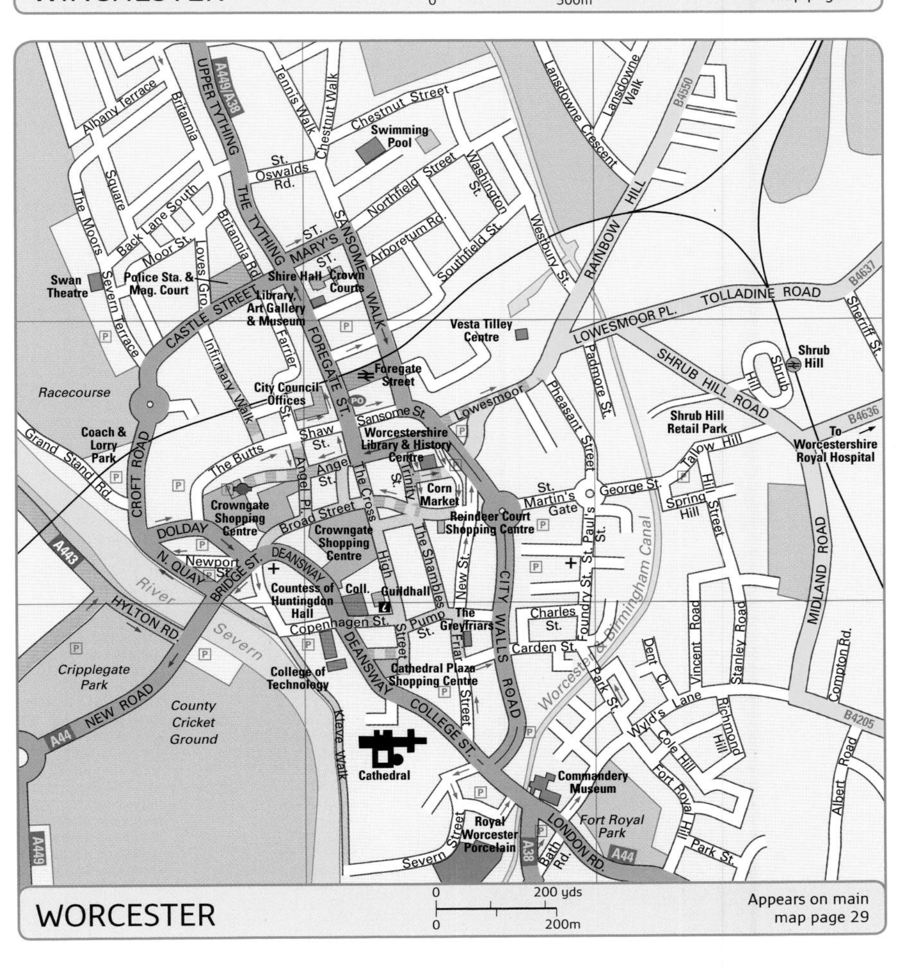

WORCESTER

0   200 yds

0   200m

Appears on main
map page 29

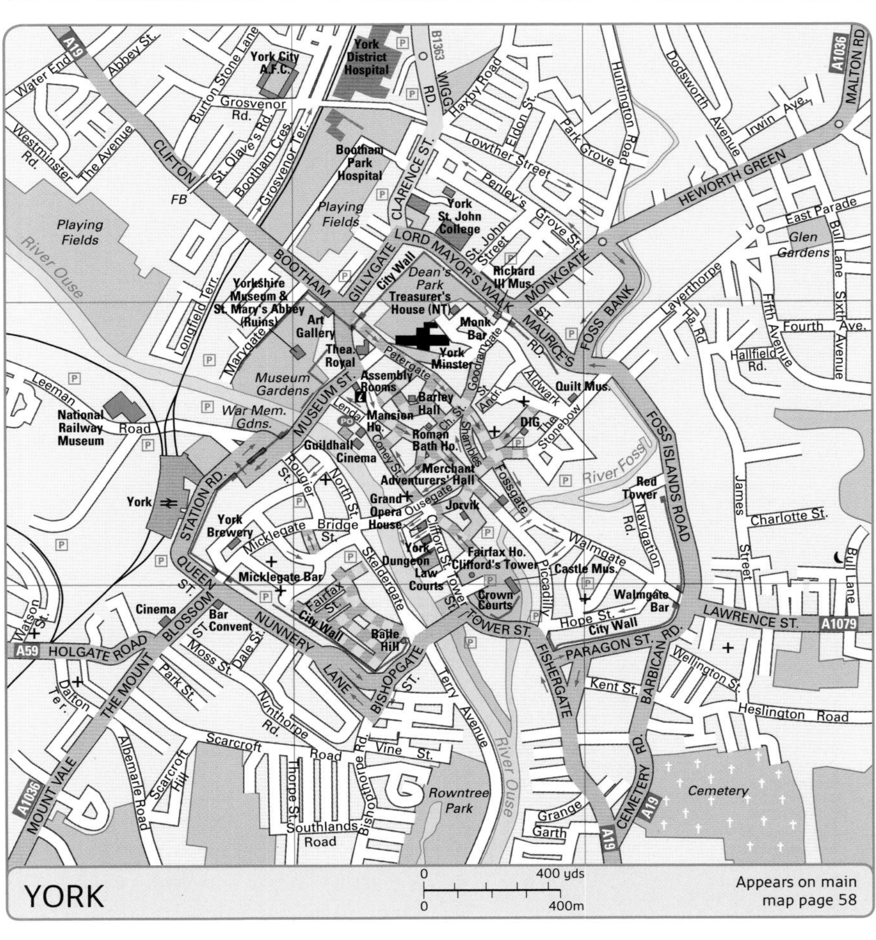

YORK

0   400 yds

0   400m

Appears on main
map page 58

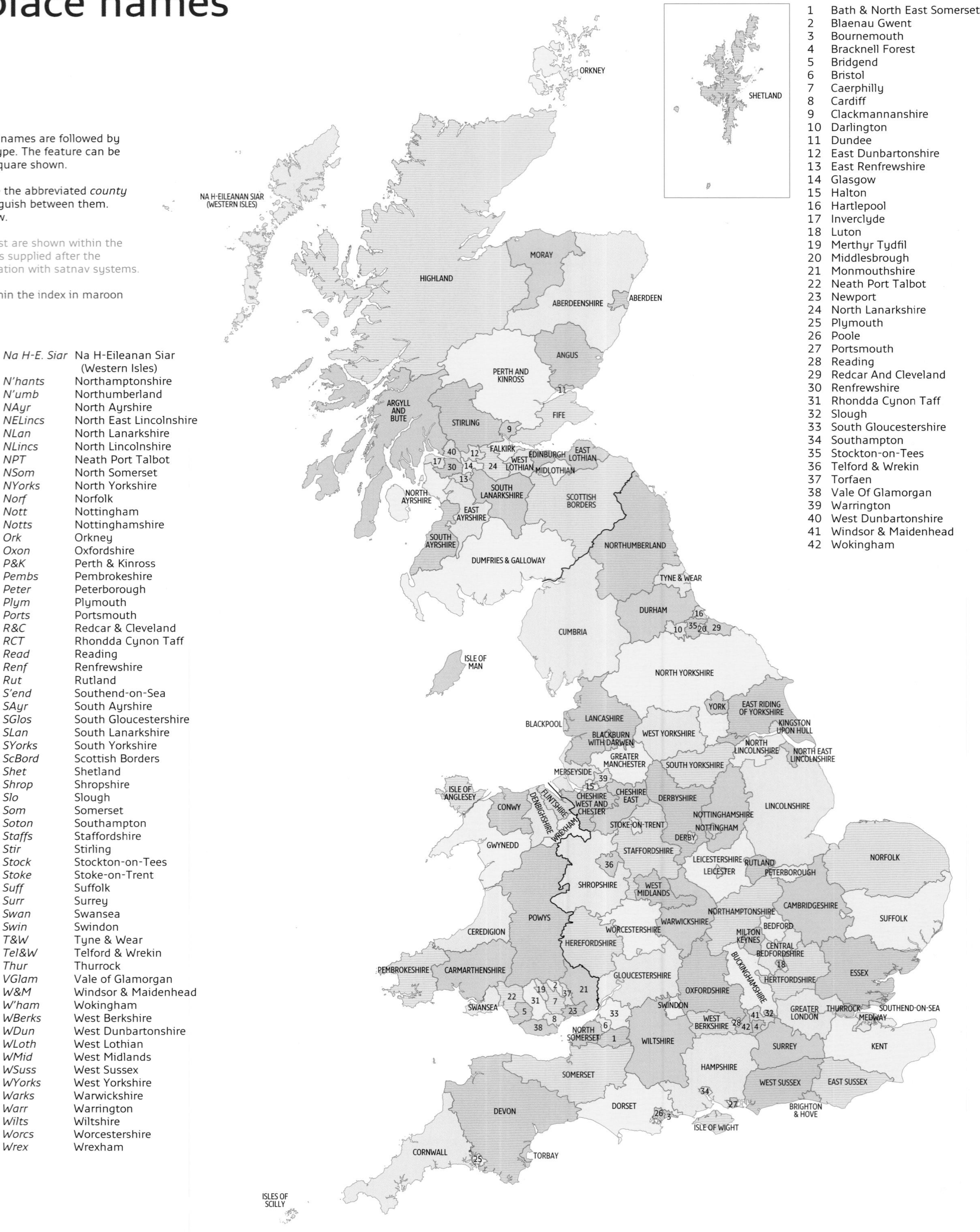

Using the index

Place, place of interest and World Heritage Site names are followed by a **page number** and a grid reference in black type. The feature can be found on the map somewhere within the grid square shown.

Where two or more places have the same name the abbreviated *county* or *unitary authority* names are shown to distinguish between them. A list of these abbreviated names appears below.

A selection of the most popular places of interest are shown within the index in blue type. Their postcode information is supplied after the county / unitary authority names to aid integration with satnav systems.

Sites with World Heritage Status are shown within the index in maroon type.

A&B	Argyll & Bute	*Na H-E. Siar*	Na H-Eileanan Siar (Western Isles)
Aber	Aberdeenshire		
B&H	Brighton & Hove	*N'hants*	Northamptonshire
B&NESom	Bath & North East Somerset	*N'umb*	Northumberland
B'burn	Blackburn with Darwen	*NAyr*	North Ayrshire
B'pool	Blackpool	*NELincs*	North East Lincolnshire
BGwent	Blaenau Gwent	*NLan*	North Lanarkshire
Bed	Bedford	*NLincs*	North Lincolnshire
Bourne	Bournemouth	*NPT*	Neath Port Talbot
BrackF	Bracknell Forest	*NSom*	North Somerset
Bucks	Buckinghamshire	*NYorks*	North Yorkshire
Caerp	Caerphilly	*Norf*	Norfolk
Cambs	Cambridgeshire	*Nott*	Nottingham
Carmar	Carmarthenshire	*Notts*	Nottinghamshire
CenBeds	Central Bedfordshire	*Ork*	Orkney
Cere	Ceredigion	*Oxon*	Oxfordshire
Chanl	Channel Islands	*P&K*	Perth & Kinross
ChesE	Cheshire East	*Pembs*	Pembrokeshire
ChesW&C	Cheshire West & Chester	*Peter*	Peterborough
Corn	Cornwall	*Plym*	Plymouth
Cumb	Cumbria	*Ports*	Portsmouth
D&G	Dumfries & Galloway	*R&C*	Redcar & Cleveland
Darl	Darlington	*RCT*	Rhondda Cynon Taff
Denb	Denbighshire	*Read*	Reading
Derbys	Derbyshire	*Renf*	Renfrewshire
Dur	Durham	*Rut*	Rutland
EAyr	East Ayrshire	*S'end*	Southend-on-Sea
EDun	East Dunbartonshire	*SAyr*	South Ayrshire
ELoth	East Lothian	*SGlos*	South Gloucestershire
ERenf	East Renfrewshire	*SLan*	South Lanarkshire
ERid	East Riding of Yorkshire	*SYorks*	South Yorkshire
ESuss	East Sussex	*ScBord*	Scottish Borders
Edin	Edinburgh	*Shet*	Shetland
Falk	Falkirk	*Shrop*	Shropshire
Flints	Flintshire	*Slo*	Slough
Glas	Glasgow	*Som*	Somerset
Glos	Gloucestershire	*Soton*	Southampton
GtLon	Greater London	*Staffs*	Staffordshire
GtMan	Greater Manchester	*Stir*	Stirling
Gwyn	Gwynedd	*Stock*	Stockton-on-Tees
Hants	Hampshire	*Stoke*	Stoke-on-Trent
Hart	Hartlepool	*Suff*	Suffolk
Here	Herefordshire	*Surr*	Surrey
Herts	Hertfordshire	*Swan*	Swansea
High	Highland	*Swin*	Swindon
Hull	Kingston upon Hull	*T&W*	Tyne & Wear
Invcly	Inverclyde	*Tel&W*	Telford & Wrekin
IoA	Isle of Anglesey	*Thur*	Thurrock
IoM	Isle of Man	*VGlam*	Vale of Glamorgan
IoS	Isles of Scilly	*W&M*	Windsor & Maidenhead
IoW	Isle of Wight	*W'ham*	Wokingham
Lancs	Lancashire	*WBerks*	West Berkshire
Leic	Leicester	*WDun*	West Dunbartonshire
Leics	Leicestershire	*WLoth*	West Lothian
Lincs	Lincolnshire	*WMid*	West Midlands
MK	Milton Keynes	*WSuss*	West Sussex
MTyd	Merthyr Tydfil	*WYorks*	West Yorkshire
Med	Medway	*Warks*	Warwickshire
Mersey	Merseyside	*Warr*	Warrington
Middl	Middlesbrough	*Wilts*	Wiltshire
Midlo	Midlothian	*Worcs*	Worcestershire
Mon	Monmouthshire	*Wrex*	Wrexham

#	
1	Bath & North East Somerset
2	Blaenau Gwent
3	Bournemouth
4	Bracknell Forest
5	Bridgend
6	Bristol
7	Caerphilly
8	Cardiff
9	Clackmannanshire
10	Darlington
11	Dundee
12	East Dunbartonshire
13	East Renfrewshire
14	Glasgow
15	Halton
16	Hartlepool
17	Inverclyde
18	Luton
19	Merthyr Tydfil
20	Middlesbrough
21	Monmouthshire
22	Neath Port Talbot
23	Newport
24	North Lanarkshire
25	Plymouth
26	Poole
27	Portsmouth
28	Reading
29	Redcar And Cleveland
30	Renfrewshire
31	Rhondda Cynon Taff
32	Slough
33	South Gloucestershire
34	Southampton
35	Stockton-on-Tees
36	Telford & Wrekin
37	Torfaen
38	Vale Of Glamorgan
39	Warrington
40	West Dunbartonshire
41	Windsor & Maidenhead
42	Wokingham

Barrowford 56 D6
Barrow-in-Furness 55 F3
Barrows Green 55 J1
Barry Angus 81 L7
Barry VGlam 18 E5
Barry Island Pleasure Park VGlam CF62 5TR 18 E5
Barsby 41 J4
Barsham 45 H7
Barskimming 67 J1
Barsloisnoch 73 G1
Barston 30 D1
Barter Books, Alnwick N'umb NE66 2NP 71 G2
Bartestree 28 E4
Barthol Chapel 85 N7
Bartholomew Green 34 B6
Barthomley 49 G7
Bartley 10 E3
Bartley Green 40 C7
Bartlow 33 J4
Barton Cambs 33 H3
Barton ChesW&C 48 D7
Barton Cumb 61 F4
Barton Glos 30 B6
Barton Lancs 48 C2
Barton Lancs 55 J6
Barton NYorks 62 C6
Barton Oxon 21 J1
Barton Torbay 5 K4
Barton Warks 30 C3
Barton Bendish 44 B5
Barton End 20 B2
Barton Green 40 D4
Barton Hartshorn 31 H5
Barton Hill 58 D3
Barton in Fabis 41 H2
Barton in the Beans 41 F5
Barton Mills 34 B1
Barton on Sea 10 D5
Barton Seagrave 32 B1
Barton St. David 8 E1
Barton Stacey 21 H7
Barton Town 6 E1
Barton Turf 45 H3
Bartongate 31 F6
Barton-le-Clay 32 D5
Barton-le-Street 58 D2
Barton-le-Willows 58 D3
Barton-on-the-Heath 30 D5
Barton-under-Needwood 40 D4
Barton-upon-Humber 59 G7
Barvas (Barabhas) 88 J3
Barway 33 J1
Barwell 41 G6
Barwhinnock 65 G5
Barwick Herts 33 G7
Barwick Som 8 E3
Barwick in Elmet 57 J6
Barwinnock 64 D6
Baschurch 38 D3
Bascote 31 F2
Base Green 34 E2
Basford Green 49 J7
Bashall Eaves 56 B5
Bashall Town 56 C5
Bashley 10 D5
Basildon Essex 24 D3
Basildon WBerks 21 K4
Basingstoke 21 K6
Baslow 50 E5
Bason Bridge 19 G7
Bassaleg 19 F3
Bassenthwaite 60 D3
Basset's Cross 6 D5
Bassett 11 F3
Bassingbourn 33 G4
Bassingfield 41 J2
Bassingham 52 C7
Bassingthorpe 42 C3
Baston 42 E4
Bastonford 29 H3
Bastwick 45 J4
Batch 19 G6
Batchley 30 B2
Batchworth 22 D2
Batchworth Heath 22 D2
Batcombe Dorset 9 F4
Batcombe Som 9 F1
Bate Heath 49 F5
Bath 20 A5
Bath Abbey B&NESom BA1 1LT 121 Bath
Bathampton 20 A5
Bathealton 7 J3
Batheaston 20 A5
Bathford 20 A5
Bathgate 75 H4
Bathley 51 K7
Bathpool Corn 4 C3
Bathpool Som 8 B2
Bathway 19 J6
Batley 57 H7
Batsford 30 C5
Batson 5 H7
Battersby 63 F6
Battersea 23 F4
Battersea Cats & Dogs Home GtLon SW8 4AA 99 H8
Battersea Park Children's Zoo GtLon SW11 4NJ 99 H8
Battisborough Cross 5 G6
Battisford 34 E3
Battisford Tye 34 E3
Battle ESuss 14 C6
Battle Powys 27 K5
Battle Abbey ESuss TN33 0AD 14 C6
Battledown 29 J6
Battlefield 38 E4
Battlesbridge 24 D2
Battlesden 32 C6
Battlesea Green 35 G1
Battleton 7 H3
Battlies Green 34 D2
Battramsley 10 E5
Batt's Corner 22 B7
Baugh 78 B6
Baughton 29 H4
Baughurst 21 J5
Baulking 21 G2
Baumber 53 F5
Baunton 20 D1
Baveney Wood 29 F1
Baverstock 10 B1
Bawburgh 45 F5
Bawdeswell 44 E3
Bawdrip 8 C1
Bawdsey 35 H4
Bawdsey Manor 35 H5
Bawsey 44 A4
Bawtry 51 J3
Baxenden 56 C7
Baxterley 40 E6
Baxter's Green 34 B3
Baxters Highland Village Moray IV32 7LD 84 H4
Baxters Home Farm Fife KY4 0JR 75 K1
Bay 9 H2
Baybridge 61 L2
Baycliff 55 F2
Baydon 21 F4
Bayford Herts 23 G1
Bayford Som 9 G2
Bayfordbury 33 G7
Bayham Abbey 13 K3
Bayles 61 J2
Baylham 35 F3
Baynards Green 31 G6
Baysham 28 E6
Bayston Hill 38 D5
Bayswater 23 F3
Baythorn End 34 B4
Bayton 29 F1
Bayworth 21 J1
Beach 20 A4
Beachampton 31 J5
Beachamwell 44 B5
Beacharr 72 E6
Beachley 19 J2
Beacon Devon 7 K5
Beacon Devon 8 B4
Beacon Fell Country Park Lancs PR3 2NL 55 J5
Beacon Hill Dorset 9 J5
Beacon Hill Essex 34 C7
Beacon Hill Surr 12 B3
Beacon Hill Country Park Leics LE12 8SR 105 B3
Beacon Park, Up Holland Lancs WN8 7RU 48 E2
Beacon's Bottom 22 A2
Beaconsfield 22 C2
Beadlam 58 C1
Beadlow 32 E5
Beadnell 71 H1
Beaford 6 D4
Beal NYorks 58 B7
Beal N'umb 77 J6
Beale Park WBerks RG8 9NH 21 K4
Bealsmill 4 D3
Beamhurst 40 C2
Beaminster 8 D4
Beamish 62 C1
Beamish - The Living Museum of the North Dur DH9 0RG 117 B7
Beamsley 57 F4
Bean 23 J4
Beanacre 20 C5
Beanley 71 F2
Beardon 6 D7
Beardwood 56 B7
Beare 7 H5
Beare Green 22 E7
Bearley 30 C2
Bearnock 83 Q7
Bearpark 62 C2
Bearsbridge 61 J1
Bearsden 74 D3
Bearsted 14 C2
Bearstone 39 G2
Bearwood Poole 10 B5
Bearwood WMid 40 C7
Beatles Story Mersey L3 4AD 131 B5
Beattock 69 F3
Beauchamp Roding 33 J7
Beauchief 51 F4
Beaudesert 30 C2
Beaufort 28 A7
Beaulieu 10 E4
Beaulieu: National Motor Museum, Abbey & Palace House Hants SO42 7ZN 92 A7
Beauly (A' Mhanachainn) 83 R6
Beaumaris (Biwmares) 46 E5
Beaumaris Castle (Castles & Town Walls of King Edward in Gwynedd) Gwyn LL58 8AP 46 E5
Beaumont Chanl 3 J7
Beaumont Cumb 60 E1
Beaumont Essex 35 F6
Beaumont Hill 62 C5
Beaumont Leys 41 H5
Beausale 30 D1
Beauvale 41 G1
Beauworth 11 G2
Beaver Green 14 E3
Beaworthy 6 C6
Beazley End 34 B6
Bebington 48 C4
Bebside 71 H5
Beccles 45 J6
Beccles Heliport 45 J7
Becconsall 55 H7
Beck Foot 61 H7
Beck Hole 63 J6
Beck Row 33 K1
Beck Side Cumb 55 F1
Beck Side Cumb 55 G1
Beckbury 39 G5
Beckenham 23 G5
Beckering 52 E4
Beckermet 60 B6
Beckermonds 56 D1
Beckett End 44 B6
Beckfoot Cumb 60 B2
Beckfoot Cumb 60 C6
Beckford 29 J5
Beckhampton 20 D5
Beckingham Lincs 52 B7
Beckingham Notts 51 K4
Beckington 20 B6
Beckley ESuss 14 D5
Beckley Oxon 31 G7
Beck's Green 45 H7
Beckside 56 B1
Beckton 23 H3
Beckwithshaw 57 H4
Becontree 23 H3
Bedale 57 H1
Bedburn 62 A3
Bedchester 9 H3
Beddau 18 D3
Beddgelert 36 E1
Beddingham 13 H6
Beddington 23 F5
Beddington Corner 23 F5
Bedfield 35 G2
Bedfield Little Green 35 G2
Bedford 32 D4
Bedgebury Cross 14 C4
Bedgrove 32 B7
Bedham 12 D4
Bedhampton 11 J4
Bedingfield 35 F2
Bedingfield Green 35 F2
Bedingfield Street 35 F2
Bedingham Green 45 G6
Bedlam Lancs 56 C7
Bedlam NYorks 57 H3
Bedlar's Green 33 J6
Bedlington 71 H5
Bedlinog 18 D1
Bedminster 19 J4
Bedmond 22 D1
Bednall 40 B4
Bedol 48 B5
Bedrule 70 B2
Bedstone 28 C1
Bedwas 18 E3
Bedwell 33 F6
Bedwellty 18 E1
Bedworth 41 F7
Bedworth Woodlands 41 F7
Beeby 41 J5
Beech Hants 11 H1
Beech Staffs 40 A2
Beech Hill 21 K5
Beechingstoke 20 D6
Beechwood 48 E4
Beecraigs Country Park WLoth EH49 6PL 75 H3
Beedon 21 H4
Beeford 59 H4
Beeley 50 E6
Beelsby 53 F2
Beenham 21 J5
Beeny 4 B1
Beer 8 B6
Beer Hackett 9 F3
Beercrocombe 8 C2
Beesands 5 J6
Beesby Lincs 53 H4
Beesby NELincs 53 F3
Beeson 5 J6
Beeston CenBeds 32 E4
Beeston ChesW&C 48 E7
Beeston Norf 44 D4
Beeston Notts 41 H2
Beeston WYorks 57 H6
Beeston Regis 45 F1
Beeston St. Lawrence 45 H3
Beeswing 65 J4
Beetham Cumb 55 H2
Beetham Som 8 B3
Beetley 44 D4
Beffcote 40 A4
Began 19 F3
Begbroke 31 F7
Begdale 43 H5
Begelly 16 E5
Beggar's Bush 28 B2
Beggearn Huish 7 J2
Beguildy (Bugeildy) 28 A1
Beighton Norf 45 H5
Beighton SYorks 51 G4
Beili-glas 19 G1
Beith 74 B5
Bekesbourne 15 G2
Bekonscot Model Village Bucks HP9 2PL 22 C2
Belaugh 45 G4
Belbroughton 29 J1
Belchalwell 9 G4
Belchalwell Street 9 G4
Belchamp Otten 34 C4
Belchamp St. Paul 34 B4
Belchamp Walter 34 C4
Belchford 53 F5
Belfield 49 J1
Belford 77 K7
Belgrave 41 H5
Belhaven 76 E3
Belhelvie 85 P9
Bell Bar 23 F1
Bell Busk 56 E4
Bell End 29 J1
Bell Heath 29 J1
Bell Hill 11 J2
Bell o' th' Hill 38 E1
Belladrum 83 R6
Bellanoch 73 G1
Bellaty 80 H5
Belle Isle 57 J7
Belle Vue 60 E1
Belleau 53 H5
Bellerby 62 B7
Bellever 5 G3
Bellfields 22 C6
Bellingdon 22 C1
Bellingham GtLon 23 G4
Bellingham N'umb 70 D5
Belloch 72 E7
Bellochantuy 72 E7
Bell's Cross 35 F3
Bells Yew Green 13 K3
Bellsbank 67 J3
Bellshill NLan 75 F4
Bellshill N'umb 77 K7
Bellside 75 G5
Bellsmyre 74 C3
Bellsquarry 75 J4
Belluton 19 K5
Belmesthorpe 42 D4
Belmont B'burn 49 F1
Belmont GtLon 22 E2
Belmont GtLon 23 F5
Belmont Shet 89 P2
Belnie 43 F2
Belowda 3 G2
Belper 41 F1
Belper Lane End 41 F1
Belsay 71 G6
Belsford 5 H5
Belsize 22 D1
Belstead 35 F4
Belston 67 H1
Belstone 6 E6
Belstone Corner 6 E6
Belsyde 75 H3
Belthorn 56 C7
Beltinge 25 H5
Beltingham 70 C7
Beltoft 52 B2
Belton Leics 41 G3
Belton Lincs 42 C2
Belton NLincs 51 K2
Belton Norf 45 J5
Belton Rut 42 B5
Belton House Lincs NG32 2LS 42 C2
Beltring 23 K7
Belvedere 23 H4
Belvoir 42 B2
Bembridge 11 H6
Bemersyde 76 D7
Bemerton 10 C1
Bempton 59 H2
Ben Alder Lodge 80 A3
Ben Rhydding 57 G5
Benacre 45 K7
Benbecula (Balivanich Airport) 88 B3
Benbecula (Beinn na Faoghla) 88 C3
Benbuie 68 C4
Benderloch (Meadarloch) 79 L7
Bendish 32 E6
Benenden 14 D4
Benfield 64 D4
Benfieldside 62 A1
Bengate 45 H3
Bengeo 33 G7
Bengeworth 30 B4
Benhall 29 J6
Benhall Green 35 H2
Benhall Street 35 H2
Beningbrough 58 B4
Benington Herts 33 F6
Benington Lincs 43 G1
Benington Sea End 43 H1
Benllech 46 D4
Benmore A&B 73 K2
Benmore Stir 79 R8
Bennacott 4 C1
Bennan Cottage 65 G3
Bennett End 22 A2
Bennetts End 22 D1
Benniworth 53 F4
Benover 14 C3
Benson 21 K2
Benston 89 N7
Benthall N'umb 71 H1
Benthall Shrop 39 F5
Bentham 29 J7
Bentlawnt 38 C5
Bentley ERid 59 G6
Bentley Essex 23 J2
Bentley Hants 22 A7
Bentley Suff 35 F5
Bentley SYorks 51 H2
Bentley Warks 40 E6
Bentley WMid 40 B6
Bentley WYorks 57 H6
Bentley Heath Herts 23 F2
Bentley Heath WMid 30 C1
Bentley Rise 51 H2
Benton 6 E2
Benton Square 71 J6
Bentpath 69 J4
Bentworth 21 K7
Benville Lane 8 E4
Benwell 71 H7
Benwell Vallum Crossing & Temple (Frontiers of the Roman Empire) T&W NE15 6QP 116 B4
Benwick 43 G6
Beoley 30 B2
Beoraidbeg 82 G11
Bepton 12 B5
Berden 33 H6
Bere Alston 4 E4
Bere Ferrers 4 E4
Bere Regis 9 H5
Berea 16 A2
Berepper 2 D6
Bergh Apton 45 H5
Berinsfield 21 J2
Berkeley 19 K2
Berkhamsted 22 C1
Berkley 20 B7
Berkswell 30 D1
Bermondsey 23 G4
Bernera (Eilean Bhearnaraigh) 88 E9
Berners Roding 24 C1
Bernice 73 K1
Bernisdale 82 E5
Berrick Prior 21 K2
Berrick Salome 21 K2
Berriedale 87 P7
Berriew (Aberriw) 38 A5
Berrington N'umb 77 J6
Berrington Shrop 38 E5
Berrington Worcs 28 E2
Berrington Green 28 E2
Berriowbridge 4 C3
Berrow Som 19 F6
Berrow Worcs 29 G5
Berrow Green 29 G3
Berry Cross 6 C4
Berry Down Cross 6 D1
Berry Hill Glos 28 E7
Berry Hill Pembs 16 D1
Berry Pomeroy 5 J4
Berryhillock 85 K4
Berrynarbor 6 D1
Berry's Green 23 H6
Bersham 38 C1
Berthlwyd 17 J6
Berwick 13 J6
Berwick Bassett 20 E4
Berwick Hill 71 G6
Berwick St. James 10 B1
Berwick St. John 9 J2
Berwick St. Leonard 9 J1
Berwick-upon-Tweed 77 H5
Bescar 48 C1
Bescot 40 C6
Besford Shrop 38 E3
Besford Worcs 29 J4
Bessacarr 51 J2
Bessels Leigh 21 H1
Besses o' th' Barn 49 H2
Bessingby 59 H3
Bessingham 45 F2
Best Beech Hill 13 K3
Besthorpe Norf 44 E6
Besthorpe Notts 52 B6
Bestwood 41 H1
Bestwood Country Park Notts NG6 8UF 107 J1
Beswick ERid 59 G5
Beswick GtMan 49 H3
Betchworth 23 F6
Bethania Cere 26 E2
Bethania Gwyn 37 G1
Bethel Gwyn 37 J2
Bethel Gwyn 46 D6
Bethel IoA 46 B5
Bethersden 14 E4
Bethesda Gwyn 46 E6
Bethesda Pembs 16 D4
Bethlehem 17 K3
Bethnal Green 23 G3
Betley 39 G1
Betley Common 39 G1
Betsham 24 C4
Betteshanger 15 J2
Bettiscombe 8 C5
Bettisfield 38 D2
Betton Shrop 38 C5
Betton Shrop 39 F2
Betton Strange 38 E5
Bettws Bridgend 18 C3
Bettws Newport 19 F2
Bettws Cedewain 38 A6
Bettws Gwerfil Goch 37 K1
Bettws Newydd 19 G1
Bettws-y-crwyn 38 B7
Bettyhill 87 K3
Betws 17 K4
Betws Disserth 28 A3
Betws Garmon 46 D7
Betws Ifan 17 G1
Betws-y-Coed 47 F7
Betws-yn-Rhos 47 H5
Beulah Cere 17 F1
Beulah Powys 27 J3
Bevendean 13 G6
Bevercotes 51 J5
Beverley 59 G6
Beverston 20 B2
Bevington 19 K2
Bewaldeth 60 D3
Bewcastle 70 A6
Bewdley 29 G1
Bewerley 57 G3
Bewholme 59 H4
Bewl Water ESuss TN3 8JH 13 K3
Bewley Common 20 C5
Bexhill 14 C7
Bexley 23 H4
Bexleyheath 23 H4
Bexwell 44 A5
Beyton 34 D2
Beyton Green 34 D2
Bhaltos (Valtos) 88 F4
Biallaid 84 B11
Bibury 20 E1
Bicester 31 G6
Bickenhall 8 B3
Bickenhill 40 D7
Bicker 43 F2
Bickershaw 49 F2
Bickerstaffe 48 D2
Bickerton ChesW&C 48 E7
Bickerton Devon 5 J7
Bickerton NYorks 57 K4
Bickerton N'umb 70 E3
Bickford 40 A4
Bickham 7 H1
Bickham Bridge 5 H5
Bickham House 7 H7
Bickington Devon 5 H3
Bickington Devon 6 D2
Bickleigh Devon 5 F4
Bickleigh Devon 7 H5
Bickleton 6 D2
Bickley 23 H5
Bickley Moss 38 E1
Bickley Town 38 E1
Bicknacre 24 D1
Bicknoller 7 K2
Bicknor 14 D2
Bickton 10 C3
Bicton Here 28 D2
Bicton Shrop 38 B7
Bicton Shrop 38 D4
Bicton Heath 38 D4
Bicton Park Gardens Devon EX9 7BJ 7 J7
Bidborough 23 J7
Biddenden 14 D4
Biddenden Green 14 D3
Biddenham 32 D4
Biddestone 20 B4
Biddick 62 D1
Biddisham 19 G6
Biddlesden 31 H4
Biddlestone 70 E3
Biddulph 49 H7
Biddulph Moor 49 J7
Bideford 6 C3
Bidford-on-Avon 30 C3
Bidlake 6 C7
Bidston 48 B3
Bidwell 32 D6
Bielby 58 D5
Bieldside 85 N10
Bierley IoW 11 G7
Bierley WYorks 57 G6
Bierton 32 B7
Big Pit National Coal Museum (Blaenavon Industrial Landscape) Torfaen NP4 9XP 19 F1
Big Sand 82 H3
Big Sheep, The Devon EX39 5AP 6 C3
Big Sky Adventure Play Peter PE2 7BU 42 E6
Bigbury 5 G6
Bigbury-on-Sea 5 G6
Bigby 52 D2
Bigert Mire 60 C7
Biggar Cumb 54 E3
Biggar SLan 75 J7
Biggin Derbys 40 E1
Biggin Derbys 50 D7
Biggin NYorks 58 B6
Biggin Hill 23 H6
Biggin Hill Airport 23 H5
Biggleswade 32 E4
Bigholms 69 J5
Bighouse 87 L3
Bighton 11 H1
Biglands 60 D1
Bignor 12 C5
Bigrigg 60 B5
Bigton 89 M10
Bilberry 4 A5
Bilborough 41 H1
Bilbrook Som 7 J1
Bilbrook Staffs 40 A5
Bilbrough 58 B5
Bilbster 87 Q4
Bilby 51 J4
Bildershaw 62 B4
Bildeston 34 D4
Billericay 24 C2
Billesdon 42 A5
Billesley 30 C3
Billholm 69 H4
Billingborough 42 E2
Billinge 48 E3
Billingford Norf 35 F1
Billingford Norf 44 E3
Billingham 62 E4
Billinghay 52 E7
Billingley 51 G2
Billingshurst 12 D4
Billingsley 39 G7
Billington CenBeds 32 C6
Billington Lancs 56 C6
Billington Staffs 40 A3
Billockby 45 J4
Billy Row 62 B3
Bilsborrow 55 J6
Bilsby 53 H5
Bilsby Field 53 H5
Bilsdean 77 F3
Bilsham 12 C6
Bilsington 15 F4
Bilson Green 29 F7
Bilsthorpe 51 J6
Bilsthorpe Moor 51 J7
Bilston Midlo 76 A4
Bilston WMid 40 B6
Bilstone 41 F5
Bilting 15 F3
Bilton ERid 59 H6
Bilton NYorks 57 J4
Bilton N'umb 71 H2
Bilton Warks 31 F1
Bilton-in-Ainsty 57 K5
Bimbister 89 C6
Binbrook 53 F3
Bincombe 9 F6
Bindon 7 K3
Binegar 19 K7
Bines Green 12 E5
Binfield 22 B4
Binfield Heath 22 A4
Bingfield 70 E6
Bingham 42 A2
Bingham's Melcombe 9 G4
Bingley 57 G6
Bings Heath 38 E4
Binham 44 D2
Binley Hants 21 H6
Binley WMid 30 E1
Binniehill 75 G3
Binsoe 57 H2
Binstead 11 G5
Binsted Hants 22 A7
Binsted WSuss 12 C6
Binton 30 C3
Bintree 44 E3
Binweston 38 C5
Birch Essex 34 D6
Birch GtMan 49 H2
Birch Cross 40 D2
Birch Green Essex 34 D7
Birch Green Herts 33 F7
Birch Grove 13 H4
Birch Heath 48 E6
Birch Vale 50 C4
Birch Wood 8 B3
Bircham Newton 44 B2
Bircham Tofts 44 B2
Birchanger 33 J6
Bircher 28 D2
Bircher Common 28 D2
Birchgrove Cardiff 18 E3
Birchgrove Swan 18 A2
Birchington 25 K5
Birchmoor 40 E5
Birchover 50 E6
Birchwood Lincs 52 C6
Birchwood Warr 49 F3
Bircotes 51 J3
Bird Street 34 E3
Birdbrook 34 B4
Birdbush 9 J2
Birdfield 73 H1
Birdforth 57 K2
Birdham 12 B6
Birdingbury 31 F2
Birdland, Bourton-on-the-Water Glos GL54 2BN 30 C6
Birdlip 29 J7
Birdoswald 70 B7
Birdoswald Roman Fort (Frontiers of the Roman Empire) Cumb CA8 7DD 70 B7
Birds Green 23 J1
Birdsall 58 E3
Birdsgreen 39 G7
Birdsmoor Gate 8 C4
Birdston 74 E3
Birdwell 51 F2
Birdwood 29 G7
Birdworld, Farnham Hants GU10 4LD 22 B7
Birgham 77 F7
Birkby Cumb 60 B3
Birkby NYorks 62 D6
Birkdale Mersey 48 C1
Birkdale NYorks 61 K6
Birkenhead 48 C4
Birkenhills 85 M6
Birkenshaw 57 H7
Birkhill Angus 81 J7
Birkhill ScBord 69 H2
Birkhill ScBord 76 D6
Birkholme 42 C3
Birkin 58 B7
Birks 57 H7
Birkwood 75 G7
Birley 28 D3
Birley Carr 51 F3
Birling Kent 24 C5
Birling N'umb 71 H3
Birling Gap 13 J7
Birlingham 29 J4
Birmingham 40 C7
Birmingham Botanical Gardens WMid B15 3TR 122 B6
Birmingham City Museum & Art Gallery WMid B3 3DH 122 E3
Birmingham Airport 40 D7
Birnam 80 F6
Birsay 89 B5
Birse 85 K11
Birsemore 85 K11
Birstall 41 H5
Birstall Smithies 57 H7
Birstwith 57 H4
Birthorpe 42 E2
Birtle 49 H1
Birtley Here 28 C2
Birtley N'umb 70 D6
Birtley T&W 62 C1
Birts Street 29 G5
Birtsmorton 29 H5
Bisbrooke 42 B6
Biscathorpe 53 F4
Bish Mill 7 F3
Bisham 22 B3
Bishampton 29 J3
Bishop Auckland 62 C4
Bishop Burton 59 F5
Bishop Middleham 62 D3
Bishop Monkton 57 J3
Bishop Norton 52 C3
Bishop Sutton 19 J6
Bishop Thornton 57 H3
Bishop Wilton 58 D4
Bishopbridge 52 D3
Bishopbriggs 74 E3
Bishops Cannings 20 D5
Bishop's Castle 38 C7
Bishop's Caundle 9 F3
Bishop's Cleeve 29 J6
Bishop's Frome 29 F4
Bishops Gate 22 C4
Bishop's Green Essex 33 K7
Bishop's Green Hants 21 J5
Bishop's Hull 8 B2
Bishop's Itchington 30 E3
Bishops Lydeard 7 K3
Bishop's Norton 29 H6
Bishops Nympton 7 F3
Bishop's Offley 39 G3
Bishop's Stortford 33 H6
Bishop's Sutton 11 H1
Bishop's Tachbrook 30 E2
Bishop's Tawton 6 D2
Bishop's Waltham 11 G3
Bishop's Wood 40 A5
Bishopsbourne 15 G2
Bishopsteignton 5 K3
Bishopstoke 11 F3
Bishopston Bristol 19 J4
Bishopston Swan 17 J7
Bishopstone Bucks 32 B7
Bishopstone ESuss 13 H6
Bishopstone Here 28 D4
Bishopstone Swin 21 F3
Bishopstone Wilts 10 B2
Bishopstrow 20 B7
Bishopswood 8 B3
Bishopsworth 19 J5
Bishopthorpe 58 B5
Bishopton Darl 62 D4
Bishopton NYorks 57 H2
Bishopton Renf 74 C3
Bishopton Warks 30 C3
Bishton 19 G3
Bisley Glos 20 C1
Bisley Surr 22 C6
Bispham 55 G6
Bispham Green 48 D1
Bissoe 2 E4
Bisterne 10 C4
Bisterne Close 10 D4
Bitchet Green 23 J6
Bitchfield 42 C3
Bittadon 6 D1
Bittaford 5 G5
Bittering 44 D4
Bitterley 28 E1
Bitterne 11 F3
Bitteswell 41 H7
Bitton 19 K5
Bix 22 A3
Bixter 89 M7
Blaby 41 H6
Black Bourton 21 F1
Black Bridge 16 C5
Black Callerton 71 G7
Black Carr 44 E6
Black Clauchrie 67 G5
Black Corries Lodge 79 P5
Black Country Living Museum WMid DY1 4SQ 102 C3
Black Crofts 79 L7
Black Cross 3 G2
Black Dog 7 G5
Black Heddon 71 F6
Black Hill 30 D3
Black Marsh 38 C6
Black Moor 57 H5
Black Mount 79 P6
Black Notley 34 B6
Black Park Country Park Bucks SL3 6DR 22 D3
Black Pill 17 K6
Black Sheep Brewery Visitor Centre NYorks HG4 4EN 57 H1
Black Street 45 K7
Black Swan Guild, Frome Som BA11 1BB 20 A7
Black Torrington 6 C5
Blackaburn 70 C6
Blackacre 69 F4
Blackadder 77 G5
Blackawton 5 J5
Blackborough Devon 7 J5
Blackborough Norf 44 A4
Blackborough End 44 A4
Blackboys 13 J4
Blackbraes 75 H3
Blackbrook Derbys 41 F1
Blackbrook Leics 41 G4
Blackbrook Mersey 48 E3
Blackbrook Staffs 39 G2
Blackburn Aber 85 N9
Blackburn B'burn 56 B7
Blackburn WLoth 75 H4
Blackbushe 22 A6
Blackcraig D&G 64 E4
Blackcraig D&G 68 C5
Blackden Heath 49 G5
Blackdog 85 P9
Blackdown Devon 5 F3
Blackdown Dorset 8 C4
Blackdown Warks 30 E2
Blacker Hill 51 F2
Blackfen 23 H4
Blackfield 11 F4
Blackford Cumb 69 J7
Blackford P&K 80 D10
Blackford Som 8 E2
Blackford Som 19 H7
Blackford Bridge 49 H2
Blackfordby 41 F4
Blackgang 11 F7
Blackgang Chine IoW PO38 2HN 11 F7
Blackhall Edin 76 A3
Blackhall Renf 74 C4
Blackhall Colliery 62 E3
Blackhall Mill 62 B1
Blackhall Rocks 62 E3
Blackham 13 J3
Blackhaugh 76 C7
Blackheath Essex 34 E6
Blackheath GtLon 23 G4
Blackheath Suff 35 J1
Blackheath Surr 22 D7
Blackheath WMid 40 B7
Blackhill Aber 85 Q5
Blackhill Aber 85 Q6
Blackhills 84 G5
Blackland 20 D5
Blacklands 7 G2
Blackleach 55 H6
Blackley 49 H2
Blacklunans 80 G4
Blackmill 18 C3
Blackmoor Hants 11 J1
Blackmoor Som 7 K4
Blackmoor Gate 6 E1
Blackmoorfoot 50 C1
Blackmore 24 C1
Blackmore End Essex 34 B5
Blackmore End Herts 32 E7
Blackness 75 J3
Blacknest 22 A7
Blackney 8 D5
Blacko 56 D5
Blackpole 29 H3
Blackpool B'pool 55 G6
Blackpool Devon 5 J6
Blackpool Airport 55 G6
Blackpool Bridge 16 D4
Blackpool Gate 70 A6
Blackpool Piers B'pool FY4 1BB 55 G6
Blackpool Pleasure Beach B'pool FY4 1PL 55 G6
Blackpool Tower FY1 4BJ 121 Blackpool
Blackpool Zoo B'pool FY3 8PP 55 G6
Blackridge 75 G4
Blackrock A&B 72 B4
Blackrock Mon 28 B7
Blackrod 49 F1
Blackshaw 69 F7
Blackshaw Head 56 E7
Blacksmith's Green 35 F2
Blacksnape 56 C7
Blackstone 13 F5
Blackthorn 31 H7
Blackthorpe 34 D2
Blacktoft 58 E7
Blacktown 19 F3
Blackwater Corn 2 E4
Blackwater Hants 22 B6
Blackwater IoW 11 G6
Blackwater Norf 44 E3
Blackwater Som 8 B3
Blackwaterfoot 66 D1
Blackwell Darl 62 C5
Blackwell Derbys 50 D5
Blackwell Derbys 51 G7
Blackwell Warks 30 D4
Blackwell Worcs 29 J1
Blackwell WSuss 13 G3
Blackwells End 29 G6
Blackwood (Coed-duon) Caerp 18 E2
Blackwood D&G 68 E5
Blackwood SLan 75 F6
Blackwood Hill 49 J7
Blacon 48 C6
Bladbean 15 G3
Bladnoch 64 E5
Bladon 31 F7
Blaen Clydach 18 C2
Blaenannerch 17 F1
Blaenau Dolwyddelan 46 E7
Blaenau Ffestiniog 37 F1
Blaenavon 19 F1
Blaenavon Industrial Landscape Torfaen 19 F1
Blaenavon Ironworks (Blaenavon Industrial Landscape) Torfaen NP4 9RQ 19 F1
Blaenawey 28 B7
Blaencelyn 26 C3
Blaencwm 18 C2
Blaendyryn 27 J5
Blaenffos 16 E2
Blaengarw 18 C2
Blaengeuffordd 37 F7
Blaengweche 17 K4
Blaengwrach 18 B1
Blaengwynfi 18 B2
Blaenllechau 18 D2
Blaenos 27 G5
Blaenpennal 27 F2
Blaenplwyf 26 E1
Blaenporth 17 F1
Blaenrhondda 18 C1
Blaenwaun 17 F3
Blaen-y-coed 17 G3
Blagdon NSom 19 H6
Blagdon Torbay 5 J4
Blagdon Hill 8 B3
Blaguegate 48 D2
Blaich 79 M3
Blaina 18 E1
Blair 74 B6
Blair Atholl 80 D4
Blair Castle P&K PH18 5TL 80 D4
Blair Drummond 75 F1
Blair Drummond Safari & Adventure Park Stir FK9 4UR 75 F1
Blairbuie 73 K3
Blairgowrie 80 G6
Blairhall 75 J2
Blairhullichan 79 R10
Blairingone 75 H1
Blairkip 74 D7
Blairlogie 75 G1
Blairmore A&B 73 K2
Blairmore High 86 D4
Blairnairn 74 A2
Blairpark 74 A5
Blairquhan 67 H3
Blairquhosh 74 D2
Blair's Ferry 73 H4
Blairvadach 74 A2
Blaisdon 29 G7
Blake End 34 B6
Blake House Craft Centre Essex CM77 6RA 34 B6
Blakebrook 29 H1
Blakedown 29 H1
Blakelaw ScBord 77 F7
Blakelaw T&W 71 H7
Blakeley 40 A6
Blakelow 49 F7
Blakemere 28 C4
Blakemere Shopping Experience ChesW&C CW8 2EB 48 E5
Blakeney Glos 19 K1
Blakeney Norf 44 E1
Blakenhall ChesE 39 G1
Blakenhall WMid 40 B6
Blakeshall 40 A7
Blakesley 31 H3
Blanchland 61 L1
Bland Hill 57 H4
Blandford Camp 9 J4
Blandford Forum 9 H4
Blandford St. Mary 9 H4
Blanefield 74 D3
Blanerne 77 G5
Blankney 52 D6
Blantyre 74 E5
Blar a' Chaorainn 79 N4
Blarglas 74 B2
Blarmachfoldach 79 M4
Blarnalearoch 83 M1
Blashford 10 C4
Blaston 42 B6
Blatherwycke 42 C6
Blawith 55 F1
Blaxhall 35 H3
Blaxton 51 J2
Blaydon 71 G7
Bleadney 19 H7
Bleadon 19 G6
Bleak Hey Nook 50 C2
Blean 25 H5
Bleasby Lincs 52 E4
Bleasby Notts 42 A1
Bleasby Moor 52 E4
Bleatarn 61 J5
Bleathwood Common 28 E2
Bleddfa 28 B2
Bledington 30 D6
Bledlow 22 A1
Bledlow Ridge 22 A2
Blencarn 61 H3
Blencogo 60 C2
Blencow 61 F3
Blendworth 11 J3
Blenheim Palace Oxon OX20 1PP 31 F7
Blennerhasset 60 C2
Bletchingdon 31 G7
Bletchingley 23 G6
Bletchley MK 32 B5
Bletchley Shrop 39 F2
Bletherston 16 D3
Bletsoe 32 D3
Blewbury 21 J3
Blickling 45 F3
Blickling Hall Norf NR11 6NF 45 F3
Blidworth 51 H7
Blidworth Bottoms 51 H7
Blindburn 70 D2
Blindcrake 60 C3
Blindley Heath 23 G7
Blisland 4 A3
Bliss Gate 29 G1
Blissford 10 C3
Blisworth 31 J3
Blithbury 40 C3
Blitterlees 60 C1
Blo' Norton 34 E1
Blockley 30 C5
Blofield 45 H5
Blofield Heath 45 H4
Blore 40 D1
Blossomfield 30 C1
Blount's Green 40 C2
Blowick 48 C1
Bloxham 31 F5
Bloxholm 52 D7
Bloxwich 40 B5
Bloxworth 9 H5
Blubberhouses 57 G4
Blue Anchor Corn 3 G3
Blue Anchor Som 7 J1
Blue Bell Hill 24 D5
Blue Planet Aquarium, Ellesmere Port ChesW&C CH65 9LF 48 D5
Blue Reef Aquarium ESuss TN34 3DW 128 Hastings
Blue Reef Aquarium, Newquay Corn TR7 1DU 3 F2
Blue Reef Aquarium, Southsea Ports PO5 3PB 93 K8
Blue Reef Aquarium, Tynemouth T&W NE30 4JF 116 G2
Bluebell Railway WSuss TN22 3QL 13 G4
Bluewater 23 J4
Blundellsands 48 C3
Blundeston 45 K6
Blunham 32 E3
Blunsdon St. Andrew 20 E3
Bluntington 29 H1
Bluntisham 33 G1
Blunts 4 D4
Blurton 40 A1
Blyborough 52 C3
Blyford 35 J1
Blymhill 40 A4
Blymhill Common 39 G4
Blymhill Lawn 40 A4
Blyth Notts 51 J4
Blyth N'umb 71 J5
Blyth Bridge 75 K6
Blyth End 40 E6
Blythburgh 35 J1
Blythe Bridge 40 B1
Blythe Marsh 40 B1
Blyton 52 B3
Boarhills 81 L9
Boarhunt 11 H4
Boars Hill 21 H1
Boarsgreave 56 D7
Boarshead 13 J3
Boarstall 31 H7
Boarzell 14 C5
Boasley Cross 6 D6
Boat of Garten 84 D9
Boath 83 R3
Bobbing 24 E5
Bobbington 40 A6
Bobbingworth 23 J1
Bocaddon 4 B5
Bocketts Farm Park Surr KT22 9BS 22 E6
Bockhampton 21 G4
Bocking 34 B6
Bocking Churchstreet 34 B6
Bockleton 28 E2
Boconnoc 4 B4
Boddam Aber 85 R6
Boddam Shet 89 M11
Bodden 19 K7
Boddington 29 H6
Bodedern 46 B4
Bodelwyddan 47 J5
Bodenham Here 28 E3
Bodenham Wilts 10 C2
Bodenham Moor 28 E3
Bodesbeck 69 G3
Bodewryd 46 C3
Bodfari 47 J5
Bodffordd 46 C5
Bodfuan 36 C2
Bodham 45 F1
Bodiam 14 C5
Bodiam Castle ESuss TN32 5UA 14 C5
Bodicote 31 F5
Bodieve 3 G1
Bodinnick 4 B5
Bodior 46 A5
Bodle Street Green 13 K5
Bodleian Library Oxon OX1 3BG 134 Oxford
Bodmin 4 A4
Bodnant Garden Conwy LL28 5RE 47 G5
Bodney 44 C6
Bodorgan 46 B6
Bodrane 4 C4
Bodsham 15 G3
Bodwen 4 A4
Bodymoor Heath 40 D6
Bogbrae 85 Q7
Bogbuie 83 R5
Bogend 74 B7
Boghead EAyr 68 B1
Boghead SLan 75 F6
Bogmoor 84 H4
Bogniebrae 85 K6
Bognor Regis 12 C7
Bogside 75 H1
Bogton 85 L5
Bogue 68 B5
Bohemia 10 D3
Bohenie 79 P2
Bohetherick 4 E4
Bohortha 3 F5
Bohuntine 79 P2
Bojewyan 2 A5
Bokiddick 4 A4
Bolam Dur 62 B4
Bolam N'umb 71 F5
Bolam Lake Country Park N'umb NE20 0HE 71 F5
Bolberry 5 G7
Bold Heath 48 E4
Bolderwood 10 D4
Boldon 71 J7
Boldon Colliery 71 J7
Boldre 10 E5
Boldron 62 A5
Bole 51 K4
Bolehill 50 E7
Boleigh 2 B6
Bolenowe 2 D5
Boleside 76 C7
Bolgoed 17 K5
Bolham Devon 7 H4
Bolham Notts 51 K4
Bolham Water 7 K4
Bolingey 2 E3
Bollington 49 J5
Bolney 13 F4
Bolnhurst 32 D3
Bolsover 51 G5
Bolsterstone 50 E3
Bolstone 28 E5
Boltby 57 K1
Bolter End 22 A2
Bolton Cumb 61 H4
Bolton ELoth 76 D3
Bolton ERid 58 D4
Bolton GtMan 49 G2
Bolton N'umb 71 G2
Bolton Abbey 57 F4
Bolton Abbey Estate NYorks BD23 6EX 57 F4
Bolton Bridge 57 F4
Bolton by Bowland 56 C5
Bolton Houses 55 H6
Bolton Low Houses 60 D2
Bolton Museum & Art Gallery GtMan BL1 1SE 49 G2
Bolton Percy 58 B5
Bolton Priory NYorks BD23 6EX 57 F4
Bolton upon Dearne 51 G2
Bolton Wood Lane 60 D2
Boltonfellend 69 K7
Boltongate 60 D2
Bolton-le-Sands 55 H3
Bolton-on-Swale 62 C7
Bolventor 4 B3
Bombie 65 H6
Bomere Heath 38 D4
Bonar Bridge 84 A1
Bonawe (Bun Atha) 79 M7
Bonby 52 D1
Boncath 17 F2
Bonchester Bridge 70 A2
Bonchurch 11 G7
Bondleigh 6 E5
Bonds 55 H5
Bonehill 40 D5
Bo'ness 75 H2
Bonhill 74 B3
Boningale 40 A5
Bonjedward 70 B1
Bonkle 75 G5
Bonning Gate 61 F7
Bonnington Edin 75 K4
Bonnington Kent 15 F4
Bonnybridge 75 G2
Bonnykelly 85 N5
Bonnyrigg 76 B4
Bonsall 50 E7
Bont 28 C7
Bont Dolgadfan 37 H5
Bont Newydd 37 G3
Bontddu 37 F4
Bont-goch (Elerch) 37 F7
Bonthorpe 53 H5
Bont-newydd Conwy 47 J5
Bontnewydd Gwyn 46 C6
Bontuchel 47 J7
Bonvilston 18 D4
Bon-y-maen 17 K6
Boode 6 D2
Boohay 5 K5
Booker 22 B2
Booley 38 E3
Boorley Green 11 G3
Boosbeck 63 G5
Boose's Green 34 C5
Boot 60 C6
Boot Street 35 G4
Booth 57 F7
Booth Bank 50 C1
Booth Green 49 J4
Booth Wood 50 C1
Boothby Graffoe 52 C7
Boothby Pagnell 42 C2
Boothstown 49 G2
Boothville 31 J2
Bootle Cumb 54 E1
Bootle Mersey 48 C3
Booton 45 F3
Boots Green 49 G5
Booze 62 A6
Boquhan 74 D2
Boraston 29 F1
Bordeaux 3 J5
Borden Kent 24 E5
Borden WSuss 12 B4
Bordley 56 E3
Bordon 11 J1
Boreham Essex 24 D1
Boreham Wilts 20 B7
Boreham Street 13 K5
Borehamwood 22 E2
Boreland D&G 64 D4
Boreland D&G 69 G4
Boreley 29 H2
Boreraig 82 B5
Borgh Na H-E. Siar 88 A8
Borgh Na H-E. Siar 88 E9
Borghastan 88 G3
Borgie 86 J4
Borgue D&G 65 G6
Borgue High 87 P7
Borley 34 C4
Borley Green Essex 34 C4
Borley Green Suff 34 D2
Borness 65 G6
Bornisketaig 82 D3
Borough Green 23 K6
Boroughbridge 57 J3
Borras Head 48 C7
Borrowash 41 G2
Borrowby NYorks 57 K1
Borrowby NYorks 63 H5
Borrowdale 60 D5
Borstal 24 D5
Borth 37 F7
Borthwick 76 B5
Borthwickbrae 69 K2
Borthwickshiels 69 K2
Borth-y-Gest 36 E2
Borve High 82 E6
Borve (Borgh) Na H-E. Siar 88 K2
Borwick 55 J2
Borwick Rails 54 E2
Bosavern 2 A5
Bosbury 29 F4
Boscarne 4 A4
Boscastle 4 A1
Boscastle Pottery Corn PL35 0HE 4 B1
Boscombe Bourne 10 C5
Boscombe Wilts 10 D1
Bosham 12 B6
Bosham Hoe 12 B6
Bosherston 16 C6
Bosley 49 J6
Bossall 58 D3
Bossiney 4 A2
Bossingham 15 G3
Bossington Hants 10 E1
Bossington Som 7 G1
Bostock Green 49 F6
Boston 43 G1
Boston Spa 57 K5
Boswarthan 2 B5
Boswinger 3 G4
Bosworth Battlefield Country Park Leics CV13 0AD 41 G5
Botallack 2 A5
Botanic Garden Oxon OX1 4AZ 134 Oxford
Botany Bay Lancs PR6 9AF 48 E1
Botany Bay 23 G2
Botcheston 41 G5
Botesdale 34 E1
Bothal 71 H5
Bothamsall 51 J5
Bothel 60 C3
Bothenhampton 8 D5
Bothwell 75 F5
Botley Bucks 22 C1
Botley Hants 11 G3
Botley Oxon 21 H1
Botloe's Green 29 G6
Botolph Claydon 31 J6
Botolphs 12 E6
Botolph's Bridge 15 G4
Bottesford Leics 42 B2
Bottesford NLincs 52 B2
Bottisham 33 J2
Bottlesford 20 E6
Bottom Boat 57 J7
Bottom of Hutton 55 H7
Bottom o'th'Moor 49 F1
Bottoms 56 E7
Botton Head 56 B3
Botusfleming 4 E4
Botwnnog 36 B2
Bough Beech 23 H7
Boughrood 28 A5
Boughspring 19 J2
Boughton Norf 44 B5
Boughton Notts 51 J6
Boughton N'hants 31 J2
Boughton Aluph 15 F3
Boughton Lees 15 F3
Boughton Malherbe 14 D3
Boughton Monchelsea 14 C2
Boughton Street 15 F2
Boulby 63 H5
Bouldnor 10 E6
Bouldon 38 E7
Boulge 35 G3
Boulmer 71 H2
Boulston 16 C4
Boultenstone Hotel 85 J9
Boultham 52 C6
Boundary Derbys 41 F4
Boundary Staffs 40 B1
Bourn 33 G3
Bourne 42 D3
Bourne End Bucks 22 B3
Bourne End CenBeds 32 C4
Bourne End Herts 22 D1
Bournebridge 23 J2
Bournemouth 10 B5
Bournemouth Airport 10 C5
Bournemouth International Centre BH2 5BH 123 Bournemouth
Bournheath 29 J1
Bournmoor 62 D1
Bournville 40 C7

Bourton Bucks 31 J5
Bourton Dorset 9 G1
Bourton NSom 19 G5
Bourton Oxon 21 F3
Bourton Shrop 38 E6
Bourton Wilts 20 D5
Bourton on Dunsmore 31 F1
Bourton-on-the-Hill 30 C5
Bourton-on-the-Water 30 C6
Boustead Hill 60 D1
Bouth 55 G1
Bouthwaite 57 G2
Boveney 22 C4
Boveridge 10 B3
Boverton 18 C5
Bovey Tracey 5 J3
Bovingdon 22 D1
Bovinger 23 J1
Bovington Camp 9 H6
Bow Cumb 60 E1
Bow Devon 5 J5
Bow Devon 7 F5
Bow Ork 89 C8
Bow Brickhill 32 C5
Bow Street Cere 37 F7
Bow Street Norf 44 E6
Bowbank 61 L4
Bowburn 62 D3
Bowcombe 11 F6
Bowd 7 K6
Bowden Devon 5 J6
Bowden ScBord 76 D7
Bowden Hill 20 C5
Bowdon 49 G4
Bower 70 C5
Bower Hinton 8 D3
Bower House Tye 34 D4
Bowerchalke 10 B2
Bowerhill 20 C5
Bowermadden 87 Q3
Bowers 40 A2
Bowers Gifford 24 D3
Bowershall 75 J1
Bowertower 87 Q3
Bowes 61 L5
Bowgreave 55 H5
Bowhousebog 75 G5
Bowithick 4 B2
Bowker's Green 48 D2
Bowland Bridge 55 H1
Bowley 28 E3
Bowley Town 28 E3
Bowlhead Green 12 C3
Bowling WDun 74 C3
Bowling WYorks 57 G6
Bowling Bank 38 C1
Bowlish 19 K7
Bowmanstead 60 E7
Bowmore 72 B4
Bowness-on-Solway 69 H7
Bowness-on-Windermere 60 F7
Bowood House & Gardens Wilts SN11 0LZ 20 C5
Bowscale 60 E3
Bowsden 77 H6
Bowside Lodge 87 L3
Bowston 61 F7
Bowthorpe 45 F5
Bowtrees 75 H2
Box Glos 20 B1
Box Wilts 20 B5
Box End 32 D4
Boxbush Glos 29 F6
Boxbush Glos 29 G7
Boxford Suff 34 D4
Boxford WBerks 21 H4
Boxgrove 12 C6
Boxley 14 C2
Boxmoor 22 D1
Box's Shop 6 A5
Boxted Essex 34 D5
Boxted Suff 34 C3
Boxted Cross 34 D5
Boxwell 20 B2
Boxworth 33 G2
Boxworth End 33 G2
Boyden Gate 25 J5
Boydston 74 C7
Boylestone 40 D2
Boyndie 85 L4
Boynton 59 H3
Boys Hill 9 F4
Boyton Corn 6 B6
Boyton Suff 35 H4
Boyton Wilts 9 J1
Boyton Cross 24 C1
Boyton End 34 B4
Bozeat 32 C3
Braaid 54 C6
Braal Castle 87 P3
Brabling Green 35 G2
Brabourne 15 F3
Brabourne Lees 15 F3
Bracadale 82 D7
Braceborough 42 D4
Bracebridge Heath 52 C6
Braceby 42 D2
Bracewell 56 D5
Brachla 83 R7
Brackenber 61 J5
Brackenbottom 56 D2
Brackenfield 51 F7
Bracklesham 12 B7
Brackley A&B 73 G2
Brackley N'hants 31 G5
Brackley Gate 41 F1
Brackley Hatch 31 H4
Bracknell 22 B5
Braco 80 D10
Bracon Ash 45 F6
Bracora 82 H11
Bracorina 82 H11
Bradbourne 50 E7
Bradbury 62 D4
Bradda 54 A6
Bradden 31 H4
Braddock 4 B4
Bradenham Bucks 22 B2
Bradenham Norf 44 D5
Bradenstoke 20 D4
Bradfield Devon 7 J5
Bradfield Essex 35 F5
Bradfield Norf 45 G2
Bradfield WBerks 21 K4
Bradfield Combust 34 C3
Bradfield Green 49 F7
Bradfield Heath 35 F6
Bradfield Southend (Southend) 21 J4
Bradfield St. Clare 34 D3
Bradfield St. George 34 D2
Bradford Corn 4 B3
Bradford Derbys 50 E6
Bradford Devon 6 C5
Bradford N'umb 71 F6
Bradford N'umb 77 K7
Bradford WYorks 57 G6
Bradford Abbas 8 E3
Bradford Cathedral Church of St. Peter WYorks BD1 4EH 123 Bradford
Bradford Leigh 20 B5
Bradford Peverell 9 F5
Bradford-on-Avon 20 B5
Bradford-on-Tone 7 K3
Bradgate Park Leics LE6 0HE 105 C4
Bradiford 6 D2
Brading 11 H6
Bradley ChesW&C 48 E5
Bradley Derbys 40 E1
Bradley Hants 21 K7
Bradley NELincs 53 F2
Bradley (Low Bradley) NYorks 57 F5
Bradley Staffs 40 A4
Bradley WMid 40 B6
Bradley Fold 49 G2
Bradley Green Warks 40 E5
Bradley Green Worcs 29 J2
Bradley in the Moors 40 C1
Bradley Mills 50 D1
Bradley Stoke 19 K3
Bradlingill 60 C4
Bradmore Notts 41 H2
Bradmore WMid 40 A6
Bradney 8 C1
Bradninch 7 J5
Bradnop 50 C7
Bradnor Green 28 B3
Bradpole 8 D5
Bradshaw GtMan 49 G1
Bradshaw WYorks 57 F6
Bradstone 6 B7
Bradwall Green 49 G6
Bradwell Derbys 50 D4
Bradwell Devon 6 C1
Bradwell Essex 34 C6
Bradwell MK 32 B5
Bradwell Norf 45 K5
Bradwell Grove 21 F1
Bradwell Waterside 25 F1
Bradwell-on-Sea 25 G1
Bradworthy 6 B4
Brae D&G 65 J3
Brae High 86 G9
Brae Shet 89 M6
Brae of Achnahaird 86 C8
Braeantra 83 R3
Braedownie 80 H3
Braegrum 80 F8
Braehead D&G 64 E5
Braehead Glas 74 D4
Braehead SLan 75 G7
Braehead SLan 75 H5
Braehead of Lunan 81 M5
Braeleny 80 B9
Braemar 84 F11
Braemore High 83 M3
Braemore High 87 N6
Braeswick 89 F4
Brafferton Darl 62 C4
Brafferton NYorks 57 K2
Brafield-on-the-Green 32 B3
Bragar 88 H3
Bragbury End 33 F6
Bragenham 32 C6
Braichmelyn 46 E6
Braides 55 H4
Braidley 57 F1
Braidwood 75 G6
Braigo 72 A4
Brailsford 40 E1
Brain's Green 19 K1
Braintree 34 B6
Braiseworth 35 F1
Braishfield 10 E2
Braithwaite Cumb 60 D4
Braithwaite SYorks 51 J1
Braithwaite WYorks 57 F5
Braithwell 51 H3
Bramber 12 E5
Brambletye 13 H3
Brambridge 11 F2
Bramcote Notts 41 H2
Bramcote Warks 41 G7
Bramdean 11 H2
Bramerton 45 G5
Bramfield Herts 33 F7
Bramfield Suff 35 H1
Bramford 35 F4
Bramhall 49 H4
Bramham 57 K5
Bramhope 57 H5
Bramley Hants 21 K6
Bramley Surr 22 D7
Bramley SYorks 51 G3
Bramley Corner 21 K6
Bramley Head 57 G4
Bramley Vale 51 G6
Bramling 15 H2
Brampford Speke 7 H6
Brampton Cambs 33 F1
Brampton Cumb 61 H4
Brampton Cumb 70 A7
Brampton Derbys 51 F5
Brampton Lincs 52 B5
Brampton Norf 45 G3
Brampton Suff 45 J7
Brampton SYorks 51 G2
Brampton Abbotts 29 F6
Brampton Ash 42 A7
Brampton Bryan 28 C1
Brampton en le Morthen 51 G4
Brampton Street 45 J7
Brampton Valley Way Country Park N'hants NN6 9DG 31 J1
Bramshall 40 C2
Bramshaw 10 D3
Bramshill 22 A5
Bramshott 12 B3
Bramwell 8 D2
Bran End 33 K6
Brancaster 44 B1
Brancaster Staithe 44 B1
Brancepeth 62 C3
Branchill 84 E5
Brand Green 29 G6
Brandelhow 60 D4
Branderburgh 84 G3
Brandesburton 59 H5
Brandeston 35 G2
Brandis Corner 6 C5
Brandiston 45 F3
Brandlingill 60 C4
Brandon Dur 62 C3
Brandon Lincs 42 C1
Brandon N'umb 71 F2
Brandon Suff 44 B7
Brandon Warks 31 F1
Brandon Bank 44 A7
Brandon Creek 44 A6
Brandon Park Suff IP27 0SU 44 B7
Brandon Parva 44 E5
Brandon to Bishop Auckland Walk Dur DH7 7RJ 62 C3
Brandsby 58 B2
Brandy Wharf 52 D3
Brane 2 B6
Branksome 10 B5
Branksome Park 10 B5
Bransbury 21 H7
Bransby 52 B5
Branscombe 7 K7
Bransford 29 G3
Bransford Bridge 29 H3
Bransgore 10 C5
Bransholme 59 H6
Branson's Cross 30 B1
Branston Leics 42 B3
Branston Lincs 52 D6
Branston Staffs 40 E3
Branston Booths 52 D6
Branston Water Park Staffs DE14 3EZ 40 E3
Brant Broughton 52 C7
Brantham 35 F5
Branthwaite Cumb 60 B4
Branthwaite Cumb 60 D3
Brantingham 59 F7
Branton N'umb 71 F2
Branton SYorks 51 J2
Brantwood 60 E7
Branxholm Bridgend 69 K2
Branxholme 69 K2
Branxton 77 G7
Brassey Green 48 E6
Brassington 50 E7
Brasted 23 H6
Brasted Chart 23 H6
Bratoft 53 H6
Brattleby 52 C4
Bratton Som 7 H1
Bratton Tel&W 39 F4
Bratton Wilts 20 C6
Bratton Clovelly 6 C6
Bratton Fleming 6 E2
Bratton Seymour 9 F2
Braughing 33 G6
Brauncewell 52 D7
Braunston N'hants 31 G2
Braunston Rut 42 B5
Braunstone 41 H5
Braunton 6 C2
Brawby 58 D2
Brawdy 16 B3
Brawith 63 F6
Brawl 87 L3
Brawlbin 87 N4
Bray 22 C4
Bray Shop 4 D3
Bray Wick 22 B4
Braybrooke 42 A7
Braydon Side 20 D3
Brayford 6 E2
Brayshaw 56 C4
Braythorn 57 H5
Brayton 58 C6
Braywoodside 22 B4
Brazacott 4 C1
Brea 2 D4
Breach Kent 15 G3
Breach Kent 24 E5
Breachwood Green 32 E6
Breaden Heath 38 D2
Breadsall 41 F2
Breadstone 20 A1
Breage 2 D6
Breakish 82 G8
Bream 19 K1
Breamore 10 C3
Brean 19 F6
Breanais 88 E5
Brearton 57 J3
Breascleit 88 H4
Breaston 41 G2
Brechfa 17 J2
Brechin 81 M4
Brecklate 66 A2
Breckles 44 D6
Brecon (Aberhonddu) 27 K6
Brecon Beacons Int. Dark Sky Reserve Carmar/Powys 27 H7
Brecon Beacons Visitor Centre Powys LD3 8ER 27 J6
Breconside 68 D3
Bredbury 49 J3
Brede 14 D6
Bredenbury 29 F3
Bredfield 35 G3
Bredgar 24 E5
Bredhurst 24 D5
Bredon 29 J5
Bredon's Hardwick 29 J5
Bredon's Norton 29 J5
Bredwardine 28 C4
Breedon on the Hill 41 G3
Breibhig 88 K4
Breich 75 H4
Breightmet 49 G2
Breighton 58 D6
Breinton 28 D5
Breinton Common 28 D5
Bremhill 20 C4
Bremhill Wick 20 C4
Brenchley 23 K7
Brendon Devon 6 B4
Brendon Devon 6 B5
Brendon Devon 7 F1
Brenkley 71 H6
Brent Eleigh 34 D4
Brent Knoll 19 G6
Brent Pelham 33 H5
Brentford 22 E4
Brentingby 42 A4
Brentwood 23 J2
Brenzett 15 F5
Brenzett Green 15 F5
Breoch 65 H5
Brereton 40 C4
Brereton Green 49 G6
Brereton Heath 49 H6
Breretonhill 40 C4
Bressay 89 P8
Bressingham 44 E7
Bressingham Common 44 E7
Bressingham Steam Museum & Gardens Norf IP22 2AA 44 E7
Bretby 40 E3
Bretford 31 F1
Bretforton 30 B4
Bretherdale Head 61 G6
Bretherton 55 H7
Brettabister 89 N7
Brettenham Norf 44 D7
Brettenham Suff 34 D3
Bretton Derbys 50 D5
Bretton Flints 48 C6
Bretton Peter 42 E5
Brevig 88 A9
Brewlands Bridge 80 G4
Brewood 40 A5
Briantspuddle 9 H5
Brick End 33 J6
Brickendon 23 G1
Bricket Wood 22 E1
Brickfields Horse Country IoW PO33 3TH 11 G5
Brickkiln Green 34 B5
Bricklehampton 29 J4
Bride 54 D3
Bridekirk 60 C3
Bridell 16 E1
Bridestones 49 J6
Bridestowe 6 D7
Brideswell 85 K7
Bridford 7 G7
Bridge Corn 2 D4
Bridge Kent 15 G2
Bridge End Cumb 55 F1
Bridge End Devon 5 G6
Bridge End Essex 33 K5
Bridge End Lincs 42 E2
Bridge End Shet 89 M9
Bridge Hewick 57 J2
Bridge of Alford 85 K9
Bridge of Allan 75 F1
Bridge of Avon 84 F7
Bridge of Awe (Drochaid Abha) 79 M7
Bridge of Balgie 80 A6
Bridge of Brown 84 F8
Bridge of Cally 80 G5
Bridge of Canny 85 L11
Bridge of Craigisla 80 H5
Bridge of Dee 65 H4
Bridge of Don 85 P10
Bridge of Dun 81 M5
Bridge of Dye 81 M2
Bridge of Earn 80 G9
Bridge of Ericht 80 A5
Bridge of Feugh 85 M11
Bridge of Forss 87 N3
Bridge of Gaur 80 A5
Bridge of Orchy (Drochaid Urchaidh) 79 P7
Bridge of Tynet 84 H4
Bridge of Walls 89 L7
Bridge of Weir 74 B4
Bridge Reeve 6 E4
Bridge Sollers 28 D4
Bridge Street 34 C4
Bridge Trafford 48 D5
Bridgefoot Angus 81 J7
Bridgefoot Cambs 33 H4
Bridgefoot Cumb 60 B4
Bridgehampton 8 E2
Bridgehill 62 A1
Bridgemary 11 G4
Bridgemere 39 G1
Bridgend A&B 72 B4
Bridgend A&B 73 G1
Bridgend Aber 85 M6
Bridgend Angus 81 L4
Bridgend (Pen-y-bont ar Ogwr) Bridgend 18 C4
Bridgend Corn 4 B5
Bridgend Cumb 60 F5
Bridgend Moray 84 H7
Bridgend WLoth 75 J3
Bridgend of Lintrathen 80 H5
Bridgerule 6 A5
Bridges 38 C6
Bridgeton 74 E4
Bridgetown Corn 6 B7
Bridgetown Som 7 H2
Bridgewater Hall, Manchester GtMan M2 3WS 133 E5
Bridgeyate 19 K4
Bridgham 44 D7
Bridgnorth 39 G6
Bridgnorth Cliff Railway Shrop WV16 4AH 39 G6
Bridgtown 40 B5
Bridgwater 8 B1
Bridlington 59 H3
Bridport 8 D5
Bridstow 28 E6
Brierfield 56 D6
Brierley Glos 29 F7
Brierley Here 28 D3
Brierley SYorks 51 G1
Brierley Hill 40 B7
Brierton 62 E4
Briestfield 50 E1
Brig o'Turk 80 A10
Brigg 52 D2
Briggate 45 H3
Briggswath 63 J6
Brigham Cumb 60 B3
Brigham ERid 59 G4
Brighouse 57 G7
Brighstone 11 F6
Brightgate 50 E7
Brighthampton 21 G1
Brightholmlee 50 E3
Brightling 13 K4
Brightlingsea 34 E7
Brighton B&H 13 G6
Brighton Corn 3 G3
Brighton Centre, The B&H BN1 2GR 123 Brighton
Brighton Museum & Art Gallery B&H BN1 1EE 123 Brighton
Brighton City (Shoreham) Airport 12 E6
Brighton Pier B&H BN2 1TW 123 Brighton
Brightons 75 H3
Brightwalton 21 H4
Brightwalton Green 21 H4
Brightwell 35 G4
Brightwell Baldwin 21 K2
Brightwell Upperton 21 K2
Brightwell-cum-Sotwell 21 J2
Brignall 62 A5
Brigsley 53 F2
Brigsteer 55 H1
Brigstock 42 C7
Brill Bucks 31 H7
Brill Corn 2 E6
Brilley 28 B4
Brilley Mountain 28 B3
Brimaston 16 C3
Brimfield 28 E2
Brimington 51 G5
Brimington Common 51 G5
Brimley 5 J3
Brimpsfield 29 J7
Brimpton 21 J5
Brimscombe 20 B1
Brimstage 48 C4
Brindham 19 J7
Brindister 89 N9
Brindle 55 J7
Brindley Ford 49 H7
Brineton 40 A4
Bringhurst 42 B6
Brington 32 D1
Brinian 89 D5
Briningham 44 E2
Brinkhill 53 G5
Brinkley Cambs 33 K3
Brinkley Notts 51 K7
Brinklow 31 F1
Brinkworth 20 D3
Brinscall 56 B7
Brinsea 19 H5
Brinsley 41 G1
Brinsop 28 D4
Brinsworth 51 G3
Brinton 44 E2
Brisco 60 F1
Brisley 44 D3
Brislington 19 K4
Brissenden Green 14 E4
Bristol 19 J4
Bristol Cathedral BS1 5TJ 123 Bristol
Bristol City Museum & Art Gallery Bristol BS8 1RL 96 C4
Bristol Filton Airport 19 J3
Bristol International Airport 19 J5
Bristol Zoo Bristol BS8 3HA 96 C4
Briston 44 E2
Britannia 56 D7
Britford 10 C2
Brithdir Caerp 18 E1
Brithdir Gwyn 37 G4
Brithem Bottom 7 J4
British Empire & Commonwealth Museum BS1 6QH 123 Bristol
British Library (St. Pancras) GtLon NW1 2DB 100 A6
British Museum GtLon WC1B 3DG 132 E1
Briton Ferry (Llansawel) 18 A2
Britwell 22 C3
Britwell Salome 21 K2
Brixham 5 K5
Brixton Devon 5 F5
Brixton GtLon 23 G4
Brixton Deverill 9 H1
Brixworth 31 J1
Brixworth Country Park N'hants NN6 9DG 31 J1
Brize Norton 21 F1
Broad Alley 29 H2
Broad Blunsdon 20 E2
Broad Campden 30 C5
Broad Carr 50 C1
Broad Chalke 10 B2
Broad Ford 14 C4
Broad Green Cambs 33 K3
Broad Green CenBeds 32 C4
Broad Green Essex 33 H5
Broad Green Essex 34 C6
Broad Green Mersey 48 D3
Broad Green Suff 34 E3
Broad Green Worcs 29 G3
Broad Haven 16 B4
Broad Hill 33 J1
Broad Hinton 20 E4
Broad Laying 21 H5
Broad Marston 30 C4
Broad Oak Carmar 17 J3
Broad Oak Cumb 60 C7
Broad Oak ESuss 13 K4
Broad Oak ESuss 14 D6
Broad Oak Here 28 D6
Broad Road 35 G1
Broad Street ESuss 14 D6
Broad Street Kent 14 D2
Broad Street Wilts 20 E6
Broad Street Green 24 E1
Broad Town 20 D4
Broadbottom 49 J3
Broadbridge 12 B6
Broadbridge Heath 12 E3
Broadclyst 7 H6
Broadfield Lancs 55 J7
Broadfield Lancs 56 C7
Broadford (An t-Ath Leathann) 82 G8
Broadford Airport 82 G8
Broadford Bridge 12 D4
Broadgate 54 E1
Broadhaugh 69 K3
Broadheath GtMan 49 G4
Broadheath Worcs 29 F2
Broadhembury 7 K5
Broadhempston 5 J4
Broadholme 52 B5
Broadland Row 14 D6
Broadlay 17 G5
Broadley Lancs 49 H1
Broadley Moray 84 H4
Broadley Common 23 H1
Broadmayne 9 G6
Broadmeadows 76 C7
Broadmere 21 K7
Broadmoor 16 D5
Broadnymett 7 F5
Broadoak Dorset 8 D5
Broadoak Glos 29 F7
Broadoak Kent 25 H5
Broadoak End 33 G7
Broad's Green 33 K7
Broadstairs 25 K5
Broadstone Poole 10 B5
Broadstone Shrop 38 E7
Broadstreet Common 19 G3
Broadwas 29 G3
Broadwater Herts 33 F6
Broadwater WSuss 12 E6
Broadwater Down 13 J3
Broadwaters 29 H1
Broadway Carmar 17 F4
Broadway Carmar 17 G5
Broadway Pembs 16 B4
Broadway Som 8 C3
Broadway Suff 35 H1
Broadway Worcs 30 C5
Broadwell Glos 30 D6
Broadwell Oxon 21 F1
Broadwell Warks 31 F2
Broadwell House 61 L1
Broadwey 9 F6
Broadwindsor 8 D4
Broadwood Kelly 6 E5
Broadwoodwidger 6 C7
Brobury 28 C4
Brocastle 18 C4
Brochel 82 F6
Brochloch 67 K4
Brockamin 29 G3
Brockbridge 11 H3
Brockdish 35 G1
Brockenhurst 10 D4
Brockford Green 35 F2
Brockford Street 35 F2
Brockhall 31 H2
Brockham 22 E7
Brockhampton Glos 29 J6
Brockhampton Glos 30 B6
Brockhampton Here 28 E5
Brockhampton Here 29 F3
Brockhampton Green 9 G4
Brockholes Lancs PR5 0AG 55 J6
Brockholes 50 D1
Brockhurst Hants 11 G4
Brockhurst WSuss 13 H3
Brocklebank 60 E2
Brocklesby 52 E1
Brockley NSom 19 H5
Brockley Suff 34 C3
Brockley Green 34 B4
Brock's Green 21 H5
Brockton Shrop 38 B5
Brockton Shrop 38 C7
Brockton Shrop 38 E6
Brockton Shrop 39 G5
Brockton Tel&W 39 G4
Brockweir 19 J1
Brockwood Park 11 H2
Brockworth 29 H7
Brocolitia (Frontiers of the Roman Empire) N'umb 70 D6
Brocton 40 B4
Brodick 73 J7
Brodsworth 51 H2
Brogborough 32 C5
Brogden 56 D5
Brogyntyn 38 B2
Broken Cross ChesE 49 H5
Broken Cross ChesW&C 49 F5
Brokenborough 20 C3
Brokes 62 B7
Bromborough 48 C4
Brome 35 F1
Brome Street 35 F1
Bromeswell 35 H3
Bromfield Cumb 60 C2
Bromfield Shrop 28 D1
Bromham Bed 32 D3
Bromham Wilts 20 C5
Bromley GtLon 23 H5
Bromley SYorks 51 F3
Bromley Cross 49 G1
Bromley Green 14 E4
Brompton Med 24 D5
Brompton NYorks 59 F1
Brompton NYorks 62 D7
Brompton Shrop 38 E5
Brompton on Swale 62 C7
Brompton Ralph 7 J2
Brompton Regis 7 H2
Bromsash 29 F6
Bromsberrow 29 G5
Bromsberrow Heath 29 G5
Bromsgrove 29 J1
Bromstead Heath 40 A4
Bromyard 29 F3
Bromyard Downs 29 F3
Bronaber 37 G2
Brondesbury 23 F3
Brongest 17 G1
Bronington 38 D2
Bronllys 28 A5
Bronnant 27 F2
Bronte Weaving Shed WYorks BD22 8EP 114 A1
Bronwydd Arms 17 H3
Bronydd 28 B4
Bron-y-gaer 17 G4
Bronygarth 38 B2
Brook Carmar 17 F5
Brook Hants 10 D3
Brook Hants 10 E2
Brook IoW 10 E6
Brook Kent 15 F3
Brook Surr 12 C3
Brook Surr 22 D7
Brook Bottom 49 J2
Brook End Bed 32 D2
Brook End Herts 33 G6
Brook End MK 32 C4
Brook End Worcs 29 H4
Brook Hill 10 D3
Brook Street Essex 23 J2
Brook Street Kent 14 E4
Brook Street Suff 34 C4
Brook Street WSuss 13 G4
Brooke Norf 45 G6
Brooke Rut 42 B5
Brookend Glos 19 J2
Brookend Glos 19 K1
Brookfield 50 C3
Brookhampton 21 K2
Brookhouse ChesE 49 J5
Brookhouse Denb 47 J6
Brookhouse Lancs 55 J3
Brookhouse SYorks 51 H4
Brookhouse Green 49 H6
Brookhouses 40 B1
Brookland 14 E5
Brooklands D&G 65 J3
Brooklands Shrop 38 E1
Brooklands Museum Surr KT13 0QN 22 D5
Brookmans Park 23 F1
Brooks 38 A6
Brooks Green 12 E4
Brooksby 41 J4
Brookside Miniature Railway ChesE SK12 1BZ 113 L8
Brookthorpe 29 H7
Brookwood 22 C6
Broom CenBeds 32 E4
Broom Fife 81 J10
Broom Warks 30 B3
Broom Green 44 D3
Broom Hill Dorset 10 B4
Broom Hill Worcs 29 J1
Broomcroft 38 E5
Broome Norf 45 H6
Broome Shrop 38 D7
Broome Worcs 29 J1
Broome Wood 71 G2
Broomedge 49 G4
Broomer's Corner 12 E4
Broomfield Essex 34 B7
Broomfield Kent 14 D2
Broomfield Kent 25 H5
Broomfield Som 8 B1
Broomfleet 58 E7
Broomhall Green 39 F1
Broomhaugh 71 F7
Broomhill Bristol 19 K4
Broomhill N'umb 71 H3
Broomielaw 62 A5
Broomley 71 F7
Broompark 62 C2
Broom's Green 29 G5
Brora 87 M9
Broseley 39 F5
Brotherlee 61 L3
Brothertoft 43 F1
Brotherton 57 K7
Brotton 63 G5
Broubster 87 N3
Brough Cumb 61 J5
Brough Derbys 50 D4
Brough ERid 59 F7
Brough High 87 Q2
Brough Notts 52 B7
Brough Shet 89 P6
Brough Lodge 89 P3
Brough Sowerby 61 J5
Broughall 38 E1
Brougham 61 G4
Brougham Hall Cumb CA10 2DE 61 G4
Broughton Bucks 32 B7
Broughton Cambs 33 F1
Broughton Flints 48 C6
Broughton Hants 10 E1
Broughton Lancs 55 J6
Broughton MK 32 B4
Broughton NLincs 52 C2
Broughton NYorks 56 E4
Broughton NYorks 58 D2
Broughton N'hants 32 B1
Broughton Oxon 31 F5
Broughton ScBord 75 K7
Broughton VGlam 18 C4
Broughton Astley 41 H6
Broughton Beck 55 F1
Broughton Gifford 20 B5
Broughton Green 29 J2
Broughton Hackett 29 J3
Broughton in Furness 55 F1
Broughton Mills 60 D7
Broughton Moor 60 B3
Broughton Poggs 21 F1
Broughtown 89 F3
Broughty Ferry 81 K7
Brown Candover 11 G1
Brown Edge Lancs 48 C1
Brown Edge Staffs 49 J7
Brown Heath 48 D6
Brown Lees 49 H7
Brown Street 34 E2
Brownber 61 J6
Browndown 11 G5
Brownheath 38 D3
Brownhills 40 C5
Brownieside 71 G1
Brownlow 49 H6
Brownlow Heath 49 H6
Brown's Bank 39 F1
Brownsea Island Dorset BH13 7EE 10 B6
Brownshill 20 B1
Brownshill Green 41 F7
Brownsover 31 G1
Brownston 5 G5
Brownston Green 45 J5
Broxa 63 K7
Broxbourne 23 G1
Broxburn ELoth 76 E3
Broxburn WLoth 75 J3
Broxholme 52 C5
Broxted 33 J6
Broxton 48 D7
Broxwood 28 C3
Broyle Side 13 H5
Brue (Brù) 88 J3
Bruera 48 D6
Bruern 30 D6
Bruernish 88 B8
Bruichladdich 72 A4
Bruisyard 35 H2
Bruisyard Street 35 H2
Brumby 52 C1
Brund 50 D6
Brundall 45 H5
Brundish Norf 45 H6
Brundish Suff 35 G2
Brundish Street 35 G1
Brunstock 60 F1
Brunswick Village 71 H6
Bruntingthorpe 41 J6
Brunton Fife 81 J8
Brunton N'umb 71 H1
Brunton Wilts 21 F6
Brunton Turret (Frontiers of the Roman Empire) N'umb NE46 4EJ 70 E7
Brushfield 50 D5
Brushford Devon 6 E5
Brushford Som 7 H3
Bruton 9 F1
Bryanston 9 H4
Bryant's Bottom 22 B2
Brydekirk 69 G6
Brymbo 48 B7
Brymor Ice Cream NYorks HG4 4PG 57 G1
Brympton 8 E3
Bryn Caerp 18 E2
Bryn Carmar 17 J5
Bryn ChesW&C 49 F5
Bryn GtMan 48 E2
Bryn NPT 18 B2
Bryn Shrop 38 B7
Bryn Bach Country Park BGwent NP22 3AY 18 E1
Bryn Bwbach 37 F2
Bryn Gates 48 E2
Bryn Pen-y-lan 38 C1
Brynamman 27 G7
Brynberian 16 E2
Bryncae 18 C3
Bryncethin 18 C3
Bryncir 36 D1
Bryncoch Bridgend 18 C3
Bryn-côch NPT 18 A2
Bryncroes 36 B2
Bryncrug 37 F5
Brynegiwys 38 A1
Brynford 47 K5
Bryngwran 46 B5
Bryngwyn Mon 19 G1
Bryngwyn Powys 28 A4
Bryn-henllan 16 D2
Brynhoffnant 26 C3
Bryning 55 H6
Brynithel 19 F1
Brynmawr BGwent 28 A7
Bryn-mawr Gwyn 36 B2
Brynmenyn 18 C3
Brynna 18 C3
Brynnau Gwynion 18 C3
Brynog 26 E3
Bryn-penarth 38 A5
Brynrefail Gwyn 46 D6
Brynrefail IoA 46 C4
Brynsadler 18 D3
Brynsaithmarchog 47 J7
Brynsiencyn 46 C6
Bryn-teg IoA 46 C4
Brynteg Wrex 48 C7
Bryn-y-cochin 38 C2
Brynygwenin 28 C7
Bryn-y-maen 47 G5
Bubbenhall 30 E1
Bubnell 50 E5
Bubwith 58 D6
Buccleuch 69 J2
Buchan 65 H4
Buchanan Castle 74 C2
Buchanty 80 E8
Buchlyvie 74 D1
Buckabank 60 E2
Buckby Wharf 31 H2
Buckden Cambs 32 E2
Buckden NYorks 56 E2
Buckenham 45 H5
Buckerell 7 K5
Buckfast 5 H4
Buckfast Abbey Devon TQ11 0EE 5 H4
Buckfastleigh 5 H4
Buckhaven 76 B1
Buckholm 76 C7
Buckholt 28 E7
Buckhorn Weston 9 G2
Buckhurst Hill 23 H2
Buckie 85 J4
Buckingham 31 H5
Buckingham Palace GtLon SW1A 1AA 132 B5
Buckland Bucks 32 B7
Buckland Devon 5 G6
Buckland Glos 30 B5
Buckland Hants 10 E5
Buckland Here 28 E3
Buckland Herts 33 G5
Buckland Kent 15 J3
Buckland Oxon 21 G2
Buckland Surr 23 F6
Buckland Brewer 6 C3
Buckland Common 22 C1
Buckland Dinham 20 A6
Buckland Filleigh 6 C5
Buckland in the Moor 5 H3
Buckland Monachorum 4 E4
Buckland Newton 9 F4
Buckland Ripers 9 F6
Buckland St. Mary 8 B3
Buckland-tout-Saints 5 H6
Bucklebury 21 J4
Bucklers Hard 11 F5
Bucklesham 35 G4
Buckley (Bwcle) 48 B6
Buckley Green 30 C2
Bucklow Hill 49 G4
Buckman Corner 12 E4
Buckminster 42 B3
Bucknall 52 E6
Bucknell Oxon 31 G6
Bucknell Shrop 28 C1
Buckridge 29 G1
Buck's Cross 6 B3
Bucks Green 12 D3
Bucks Hill 22 D1
Bucks Horn Oak 22 B7
Buck's Mills 6 B3
Bucksburn 85 N10
Buckspool 16 C6
Buckton ERid 59 H2
Buckton Here 28 C1
Buckton N'umb 77 J7
Buckton Vale 49 J2
Buckworth 32 E1
Budbrooke 30 D2
Budby 51 J5
Bude 6 A5
Budge's Shop 4 D5
Budlake 7 H5
Budle 77 K7
Budleigh Salterton 7 J7
Budock Water 2 E5
Budworth Heath 49 F5
Buerton 39 F1
Bugbrooke 31 H3
Buglawton 49 H6
Bugle 4 A5
Bugthorpe 58 D4
Building End 33 H5
Buildwas 39 F5
Builth Road 27 K3
Builth Wells (Llanfair-ym-Muallt) 27 K3
Bulby 42 D3
Bulcote 41 J1
Buldoo 87 M3
Bulford 20 E7
Bulford Camp 20 E7
Bulkeley 48 E7
Bulkington Warks 41 F7
Bulkington Wilts 20 C6
Bulkworthy 6 B4
Bull Bay (Porth Llechog) 46 C3
Bull Green 14 E4
Bullbridge 51 F7
Bullbrook 22 B5
Bullen's Green 23 F1
Bulley 29 G7
Bullington 52 D5
Bullpot Farm 56 B1
Bulls Cross 23 G2
Bull's Green Herts 33 F7
Bull's Green Norf 45 J6
Bullwood 73 K3
Bulmer Essex 34 C4
Bulmer NYorks 58 C3
Bulmer Tye 34 C5
Bulphan 24 C3
Bulstone 7 K7
Bulverhythe 14 C7
Bulwell 41 H1
Bulwick 42 C6
Bumble's Green 23 H1
Bun Abhainn Eadarra 88 G7
Bunarkaig 79 N2
Bunbury 48 E7
Bunbury Heath 48 E7
Bunchrew 84 A6
Bundalloch 83 J8
Bunessan 78 E8
Bungay 45 H7
Bunker's Hill 53 F7
Bunlarie 73 F7
Bunmhullin 88 B7
Bunnahabhain 72 C3
Bunny 41 H3
Buntait 83 P7
Buntingford 33 G6
Bunwell 45 F6
Bunwell Street 45 F6
Burbage Derbys 50 C5
Burbage Leics 41 G6
Burbage Wilts 21 F5
Burbage Common Leics LE10 3DD 41 G6
Burchett's Green 22 B3
Burcombe 10 B1
Burcot Oxon 21 J2
Burcot Worcs 29 J1
Burcott 32 B6
Burdale 58 E3
Burdocks 12 D4
Burdon 62 D1
Burdrop 30 E5
Bure Valley Railway Norf NR11 6BW 45 G3
Bures 34 D5
Bures Green 34 D5
Burfa 28 B2
Burford Oxon 30 D7
Burford Shrop 28 E2
Burg 78 E6
Burgate 34 E1
Burgates 11 J2
Burge End 32 E5
Burgess Hill 13 G5
Burgh 35 G3
Burgh by Sands 60 E1
Burgh Castle 45 J5
Burgh Heath 23 F6
Burgh le Marsh 53 H6
Burgh next Aylsham 45 G3
Burgh on Bain 53 F4
Burgh St. Margaret (Fleggburgh) 45 J4
Burgh St. Peter 45 J6
Burghclere 21 H5
Burghead 84 F4
Burghfield 21 K5
Burghfield Common 21 K5
Burghfield Hill 21 K5
Burghill 28 D4
Burghwallis 51 H1
Burham 24 D5
Buriton 11 J2
Burland 49 F7
Burlawn 3 G2
Burleigh 22 B5
Burlescombe 7 J4
Burleston 9 G5
Burley Hants 10 D4
Burley Rut 42 B4
Burley WYorks 57 H6
Burley Gate 28 E4
Burley in Wharfedale 57 G5
Burley Street 10 D4
Burley Woodhead 57 G5
Burleydam 39 F1
Burlingjobb 28 B3
Burlow 13 J5
Burlton 38 D3
Burmarsh 15 G4
Burmington 30 D5
Burn 58 B7
Burn Naze 55 G5
Burnage 49 H3
Burnaston 40 E2
Burnby 58 E5
Burncross 51 F3
Burndell 12 C6
Burnden 49 G2
Burnedge 49 J1
Burneside 61 G7
Burness 89 F3
Burneston 57 J1
Burnett 19 K5
Burnfoot ScBord 69 K2
Burnfoot ScBord 70 A2
Burnham Bucks 22 C3
Burnham NLincs 52 D1
Burnham Deepdale 44 C1
Burnham Green 33 F7
Burnham Market 44 C1
Burnham Norton 44 C1
Burnham Overy Staithe 44 C1
Burnham Overy Town 44 C1
Burnham Thorpe 44 C1
Burnham-on-Crouch 25 F2
Burnham-on-Sea 19 G7
Burnhaven 85 R6
Burnhead D&G 67 K5
Burnhead D&G 68 D4
Burnhervie 85 M9
Burnhill Green 39 G5
Burnhope 62 B2
Burnhouse 74 B5
Burniston 63 K7
Burnley 56 D6
Burnmouth 77 H4
Burnopfield 62 B1
Burn's Green 33 G6
Burns National Heritage Park SAyr KA7 4PQ 67 H2
Burnsall 57 F3
Burnside EAyr 67 K2
Burnside WLoth 75 J3
Burnstones 61 H1
Burnswark 69 G6
Burnt Hill 21 J4
Burnt Houses 62 B4
Burnt Oak 23 F2
Burnt Yates 57 H3
Burntcliff Top 49 J6
Burntisland 76 A2
Burnton EAyr 67 J2
Burnton EAyr 67 K3
Burntwood 40 C5
Burntwood Green 40 C5
Burnworthy 7 K4
Burpham Surr 22 D6
Burpham WSuss 12 D6
Burra 89 M9
Burradon N'umb 70 E3
Burradon T&W 71 H6
Burrafirth 89 Q1
Burras 2 D5
Burraton Corn 4 E4
Burraton Corn 4 E5
Burravoe Shet 89 M6
Burravoe Shet 89 P5
Burray 89 D8
Burrell Collection Glas G43 1AT 118 D5
Burrells 61 H5
Burrelton 80 H7
Burridge Devon 8 C4
Burridge Hants 11 G3
Burrill 57 H1
Burringham 52 B2
Burrington Devon 6 E4
Burrington Here 28 D1
Burrington NSom 19 H6
Burrough Green 33 K3
Burrough on the Hill 42 A4
Burrow Som 7 H1
Burrow Som 8 D3
Burrow Bridge 8 C1
Burrowhill 22 C5
Burrows Cross 22 D7
Burry 17 H6
Burry Green 17 H6
Burry Port 17 H5
Burscough 48 D1
Burscough Bridge 48 D1
Bursea 58 E6
Burshill 59 G5
Bursledon 11 F4
Burslem 40 A1
Burstall 35 F4
Burstock 8 D4
Burston Norf 45 F7
Burston Staffs 40 B2
Burstow 23 G7
Burstwick 59 J7
Burtersett 56 D1
Burthorpe 34 B2
Burthwaite 60 F2
Burtle 19 H7
Burtle Hill 19 G7
Burton ChesW&C 48 C5
Burton ChesW&C 48 E6
Burton Dorset 10 C5
Burton Lincs 52 C5
Burton N'umb 77 K7
Burton Pembs 16 C5
Burton Som 7 K1
Burton Wilts 9 H1
Burton Wilts 20 B4
Burton Agnes 59 H3
Burton Bradstock 8 D6
Burton Coggles 42 C3
Burton End 33 J6
Burton Ferry 16 C5
Burton Fleming 59 G2
Burton Green Warks 30 D1
Burton Green Wrex 48 C7
Burton Hastings 41 G7
Burton in Lonsdale 56 B2
Burton Joyce 41 J1
Burton Latimer 32 C1
Burton Lazars 42 A4
Burton Leonard 57 J3
Burton on the Wolds 41 H3
Burton Overy 41 J6
Burton Pedwardine 42 E1
Burton Pidsea 59 J6
Burton Salmon 57 K7
Burton Stather 52 B1
Burton upon Stather 52 B1
Burton upon Trent 40 E3
Burton-in-Kendal 55 J2
Burton's Green 34 C6
Burtonwood 48 E3
Burwardsley 48 E7
Burwarton 39 F7
Burwash 13 K4
Burwash Common 13 K4
Burwash Weald 13 K4
Burwell Cambs 33 J2
Burwell Lincs 53 G5
Burwen 46 C3
Burwick 89 D9
Bury Cambs 43 F7
Bury GtMan 49 H1
Bury Som 7 H3
Bury WSuss 12 D5
Bury End 30 B5
Bury Green 33 H6
Bury St. Edmunds 34 C2
Buryas Bridge 2 B6
Burythorpe 58 D3
Busbridge 22 C7
Busby 74 D5
Buscot 21 F2
Bush 6 A5
Bush Bank 28 D3
Bush Crathie 84 G11
Bush Green 45 G7
Bushbury 40 B5
Bushby 41 J5
Bushey 22 E2
Bushey Heath 22 E2
Bushley 29 H5
Bushley Green 29 H5
Bushton 20 D4
Bushy Common 44 D4
Business Design Centre, Islington GtLon N1 0QH 100 B6
Busk 61 H2
Buslingthorpe 52 D4
Bussage 20 B1
Busta 89 M6
Butcher's Common 45 H3
Butcher's Cross 13 J4
Butcher's Pasture 33 K6
Butcombe 19 J5
Bute 73 J4
Bute Town 18 E1
Butleigh 8 E1
Butleigh Wootton 8 E1
Butler's Cross 22 B1
Butler's Hill 41 H1
Butlers Marston 30 E4
Butlersbank 38 E3
Butley 35 H3
Butley Abbey 35 H4
Butley Low Corner 35 H4
Butley Mills 35 H3
Butley Town 49 J5
Butt Green 49 F7
Butt Lane 49 H7
Butterburn 70 B6
Buttercrambe 58 D4
Butterknowle 62 B4
Butterleigh 7 H5
Butterley 51 G7
Buttermere Cumb 60 C5
Buttermere Wilts 21 G5
Butters Green 49 H7
Buttershaw 57 G6
Butterstone 80 F6
Butterton Staffs 40 A1
Butterton Staffs 50 C7
Butterwick Dur 62 D4
Butterwick Lincs 43 G1
Butterwick NYorks 58 D2
Butterwick NYorks 59 F2
Buttington 38 B5
Buttonbridge 29 G1
Buttonoak 29 G1
Buttons' Green 34 D3
Butts 7 G7
Butt's Green Essex 24 D1
Butt's Green Hants 10 D2
Buttsash 11 F4
Buxhall 34 E3
Buxted 13 H4
Buxton Derbys 50 C5
Buxton Norf 45 G3
Buxton Heath 45 F3
Buxworth 50 C4
Bwlch 28 A6
Bwlch-clawdd 17 G2
Bwlch-derwin 36 D1
Bwlchgwyn 48 B7
Bwlch-llan 26 E3
Bwlchnewydd 17 G3
Bwlchtocyn 36 C3
Bwlch-y-cibau 38 A4
Bwlch-y-ddar 38 A3
Bwlchyfadfa 17 H1
Bwlch-y-ffridd 37 K6
Bwlch-y-groes 17 F2
Bwlchyllyn 46 D7
Bwlchymynydd 17 J6
Bwlch-y-sarnau 27 K1
Byers Green 62 C3
Byfield 31 G3
Byfleet 22 D5
Byford 28 C4
Bygrave 33 F5
Byker 71 H7
Byland Abbey 58 B2
Bylane End 4 C5
Bylchau 47 H6
Byley 49 G6
Bynea 17 J6
Byrness 70 C3
Bystock 7 J7
Bythorn 32 D1
Byton 28 C2
Bywell 71 F7
Byworth 12 C4

C

Cabourne 52 E2
Cabourne Parva 52 E2
Cabrach A&B 72 C4
Cabrach Moray 84 H8
Cabus 55 H5
Cackle Street ESuss 13 H4
Cackle Street ESuss 14 D6
Cacrabank 69 J2
Cadbury 7 H5
Cadbury Barton 6 E4
Cadbury Heath 19 K4
Cadbury World WMid B30 1JR 103 G7
Cadder 74 E3
Cadderlie 79 M7
Caddington 32 D7
Caddonfoot 76 C7
Cade Street 13 K4
Cadeby Leics 41 G5
Cadeby SYorks 51 H2
Cadeleigh 7 H5
Cader 47 J6
Cadgwith 2 E7
Cadishead 49 G3
Cadle 17 K6
Cadley Lancs 55 J6
Cadley Wilts 21 F5
Cadmore End 22 A2
Cadnam 10 D3
Cadney 52 D2
Cadole 48 B6
Cadover Bridge 5 F4
Cadoxton 18 E5
Cadoxton-Juxta-Neath 18 A2
Cadwell 32 E5
Cadwst 37 K2
Cadzow 75 F5
Cae Ddafydd 37 F1
Caeathro 46 D6
Caehopkin 27 H7
Caenby 52 D4
Caenby Corner 52 C4
Caer Llan 19 H1
Caerau Bridgend 18 B2
Caerau Cardiff 18 E4
Caerdeon 37 F4
Caerfarchell 16 A3
Caergeiliog 46 B5
Caergwrle 48 C7
Caerhun 47 F5
Caer-Lan 27 H7
Caerleon 19 G2
Caernarfon 46 C6

Caernarfon Castle (Castles & Town Walls of King Edward in Gwynedd) Gwyn LL55 2AY **46** C6
Caerphilly **18** E3
Caersws **37** K6
Caerwedros **26** C3
Caerwent **19** H2
Caerwys **47** K5
Caethle Farm **37** F6
Caggan **84** C9
Caggle Street **28** C7
Caim **46** E4
Caio **17** K2
Cairisiadar **88** F4
Cairminis **88** F9
Cairnbaan **73** G1
Cairnbulg **85** Q4
Cairncross *Angus* **81** L3
Cairncross *ScBord* **77** G4
Cairncurran **74** B4
Cairndoon **64** D7
Cairndow **79** N9
Cairness **85** Q4
Cairneyhill **75** J2
Cairngorm Mountain *High* PH22 1RB **84** D10
Cairnie **85** J6
Cairnorrie **85** N6
Cairnryan **64** A4
Cairnsmore **64** E4
Caister-on-Sea **45** K4
Caistor **52** E2
Caistor St. Edmund **45** G5
Caistron **70** E3
Caithness Crystal Visitor Centre *Norf* PE30 4NE **44** A4
Cake Street **44** E6
Cakebole **29** H1
Calbourne **11** F6
Calceby **53** G5
Calcoed **47** K5
Calcot **21** K4
Calcott *Kent* **25** H5
Calcott *Shrop* **38** D4
Calcotts Green **29** G7
Calcutt **20** E2
Caldarvan **74** C2
Caldbeck **60** E3
Caldbergh **57** F1
Caldecote *Cambs* **33** G3
Caldecote *Cambs* **42** E7
Caldecote *Herts* **33** F5
Caldecote *N'hants* **31** H3
Caldecote *Warks* **41** F6
Caldecott *N'hants* **32** C2
Caldecott *Oxon* **21** H2
Caldecott *Rut* **42** B6
Calder Bridge **60** B6
Calder Grove **51** F1
Calder Vale **55** J5
Calderbank **75** F4
Calderbrook **49** J1
Caldercruix **75** G4
Calderglen **74** E5
Calderglen Country Park *SLan* G75 0QZ **119** H8
Caldermill **74** E6
Caldey Island **16** E6
Caldicot **19** H3
Caldwell *Derbys* **40** E4
Caldwell *ERenf* **74** C5
Caldwell *NYorks* **62** B5
Caldy **48** B4
Calebreck **60** E3
Caledrhydiau **26** D3
Calford Green **33** K4
Calfsound **89** E4
Calgary **78** E5
Califer **84** E5
California *Falk* **75** H3
California *Norf* **45** K4
California *Suff* **35** G4
California Country Park *W'ham* RG40 4HT **22** A5
Calke **41** F3
Calke Abbey *Derbys* DE73 7LE **41** F3
Calkaille **82** G5
Callaly **71** F3
Callander **80** B10
Callanish (Calanais) **88** H4
Callaughton **39** F6
Callerton Lane End **71** G7
Calliburn **66** B1
Calligarry **82** G10
Callington **4** D4
Callingwood **40** D3
Callisterhall **69** H5
Callow **28** D5
Callow End **29** H4
Callow Hill *Wilts* **20** D3
Callow Hill *Worcs* **29** G1
Callow Hill *Worcs* **30** B2
Callows Grave **28** E2
Calmore **10** E3
Calmsden **20** D1
Calne **20** C4
Calow **51** G5
Calshot **11** F4
Calstock **4** E4
Calstone Wellington **20** D5
Calthorpe **45** F2
Calthwaite **61** F2
Calton *NYorks* **56** E4
Calton *Staffs* **50** D7
Calveley **48** E7
Calver **50** E5
Calver Hill **28** C4
Calverhall **39** F2
Calverleigh **7** H4
Calverley **57** H6
Calvert **31** H6
Calverton *MK* **31** J5
Calverton *Notts* **41** J1
Calvine **80** C4
Calvo **60** C1
Cam **20** A2
Camasnacroise **79** K5
Camastianavaig **82** F7
Camault Muir **83** R6
Camb **89** P3
Camber **14** E6
Camberley **22** B5
Camberwell **23** G4
Camblesforth **58** C7
Cambo **71** F5
Cambois **71** J5
Camborne **2** D5
Camborne & Redruth Mining District (Cornwall & West Devon Mining Landscape) *Corn* **2** D4
Cambourne **33** G3
Cambridge *Cambs* **33** H3
Cambridge *Glos* **20** A1
Cambridge American Military Cemetery & Memorial *Cambs* CB23 7PH **33** G3
Cambridge International Airport **33** H3
Cambridge University Botanic Garden *Cambs* CB2 1JF **33** H3
Cambus **75** G1
Cambusbarron **75** F1
Cambuskenneth **75** G1
Cambuslang **74** E4
Cambusnethan **75** G5
Camden Town **23** F3
Camel Hill **8** E2
Camel Trail *Corn* PL27 7AL **3** G1
Cameley **19** K6
Camelford **4** B2
Camelon **75** G2
Camelsdale **12** B3
Camer **24** C5
Cameron House **74** B2
Camer's Green **29** G5
Camerton *B&NESom* **19** K6
Camerton *Cumb* **60** B3
Camerton *ERid* **59** J7
Camghouran **80** A5
Camis Eskan **74** B2
Cammachmore **85** P11
Cammeringham **52** C4
Camp Hill *Pembs* **16** E4
Camp Hill *Warks* **41** F6
Campbeltown (Ceann Loch Chille Chiarain) **66** B1
Campbeltown Airport **66** A1
Camperdown **71** H6
Camperdown Country Park *Dundee* DD2 4TF **81** J7
Camps **75** K4
Camps End **33** K4
Camps Heath **45** K6
Campsall **51** H1
Campsea Ashe **35** H3
Campton **32** E5
Camptown **70** B2
Camquhart **73** H2
Camrose **16** C3
Camserney **80** D6
Camstraddan House **74** B1
Camus-luinie **83** K8
Camusrory **83** J11
Camusteel **82** H6
Camusterrach **82** H6
Camusvrachan **80** B6
Canada **10** D3
Canaston Bridge **16** D4
Candlesby **53** H6
Candy Mill **75** J6
Cane End **21** K4
Canewdon **24** E2
Canfield End **33** J6
Canford Bottom **10** B4
Canford Cliffs **10** B6
Canford Heath **10** B5
Canford Magna **10** B5
Canham's Green **34** E2
Canisbay **87** R2
Canley **30** E1
Cann **9** H2
Cann Common **9** H2
Canna **82** C10
Cannard's Grave **19** K7
Cannich (Canaich) **83** P7
Canning Town **23** H3
Cannington **8** B1
Cannock **40** B5
Cannock Wood **40** C4
Cannon Hall Country Park *SYorks* S75 4AT **50** E2
Cannon Hall Farm *SYorks* S75 4AT **50** E2
Cannon Hill Park *WMid* B13 8RD **103** H6
Cannop **29** F7
Canon Bridge **28** D4
Canon Frome **29** F4
Canon Pyon **28** D4
Canonbie **69** J6
Canons Ashby **31** G3
Canon's Town **2** C5
Canterbury **15** G2
Canterbury Cathedral (Canterbury Cathedral, St Augustine's Abbey, & St Martin's Church) *Kent* CT1 2EH **15** G2
Canterbury Tales, The *Kent* CT1 2TG **123** Canterbury
Cantley *Norf* **45** H5
Cantley *SYorks* **51** J2
Cantlop **38** E5
Canton **18** E4
Cantraydoune **84** B6
Cantsfield **56** B2
Canvey Island **24** D3
Canwell Hall **40** D5
Canwick **52** C6
Canworthy Water **4** C1
Caol **79** N3
Caolas **78** B6
Caolasnacon **79** N4
Capel *Kent* **23** K7
Capel *Surr* **22** E7
Capel Bangor **37** F7
Capel Betws Lleucu **27** F3
Capel Carmel **36** A3
Capel Celyn **37** H1
Capel Coch **46** C4
Capel Curig **47** F7
Capel Cynon **17** G1
Capel Dewi *Carmar* **17** H3
Capel Dewi *Cere* **17** H1
Capel Dewi *Cere* **37** F7
Capel Garmon **47** G7
Capel Gwyn *Carmar* **17** H3
Capel Gwyn *IoA* **46** B5
Capel Gwynfe **27** G6
Capel Hendre **17** J4
Capel Isaac **17** J3
Capel Iwan **17** F2
Capel le Ferne **15** H4
Capel Llanilltern **18** D3
Capel Mawr **46** C5
Capel Parc **46** C4
Capel Seion **27** F1
Capel St. Andrew **35** H4
Capel St. Mary **34** E5
Capel St. Silin **26** E3
Capel Tygwydd **17** F1
Capeluchaf **36** D1
Capelulo **47** F5
Capel-y-ffin **28** B5
Capel-y-graig **46** D6
Capenhurst **48** C5
Capernwray **55** J2
Capheaton **71** F5
Caplaw **74** C5
Capon's Green **35** G2
Cappercleuch **69** H1
Capplegill **69** G3
Capstone **24** D5
Capton *Devon* **5** J5
Capton *Som* **7** J2
Caputh **80** F6
Car Colston **42** A1
Caradon Mining District (Cornwall & West Devon Mining Landscape) *Corn* **3** K1
Caradon Town **4** C3
Carbellow **68** B1
Carbeth **74** D3
Carbis Bay **2** C5
Carbost *High* **82** D7
Carbost *High* **82** E6
Carbrain **75** F3
Carbrooke **44** D5
Carburton **51** J5
Carco **68** C2
Carcroft **51** H1
Cardenden **76** A1
Cardeston **38** C4
Cardew **60** E2
Cardiff (Caerdydd) **18** E4
Cardiff Airport **18** D5
Cardiff Bay Visitor Centre *Cardiff* CF10 4PA **95** C6
Cardiff Castle & Museum CF10 3RB **124** Cardiff
Cardiff Millennium Stadium CF10 1GE **124** Cardiff
Cardigan (Aberteifi) **16** E1
Cardinal's Green **33** K4
Cardington *Bed* **32** D4
Cardington *Shrop* **38** E6
Cardinham **4** B4
Cardonald **74** D4
Cardoness **65** F5
Cardow **84** F6
Cardrona **76** A7
Cardross **74** B3
Cardurnock **60** C1
Careby **42** D4
Careston **81** L5
Carew **16** D5
Carew Cheriton **16** D5
Carew Newton **16** D5
Carey **28** E5
Carfin **75** F5
Carfrae **76** D4
Carfraemill **76** D5
Cargate Green **45** H4
Cargen **65** K3
Cargenbridge **65** K3
Cargill **80** G7
Cargo **60** E1
Cargreen **4** E4
Carham **77** F7
Carhampton **7** J1
Carharrack **2** E4
Carie *P&K* **80** B5
Carie *P&K* **80** B7
Carines **2** E3
Carinish (Cairinis) **88** C2
Carisbrooke **11** F6
Carisbrooke Castle & Museum *IoW* PO30 1XY **11** F6
Cark **55** G2
Carkeel **4** E4
Carland Cross **3** F3
Carlatton **61** G1
Carlby **42** D4
Carlecotes **50** D2
Carleen **2** D6
Carleton *Cumb* **60** F1
Carleton *Cumb* **61** G4
Carleton *Lancs* **55** G5
Carleton *NYorks* **56** E5
Carleton *WYorks* **57** K7
Carleton Fishery **67** F5
Carleton Forehoe **44** E5
Carleton Rode **45** F6
Carleton St. Peter **45** H5
Carlin How **63** H5
Carlisle **60** F1
Carlisle Cathedral *Cumb* CA3 8TZ **124** Carlisle
Carlisle Park, Morpeth *N'umb* NE61 1YD **71** G5
Carloggas **3** F2
Carlops **75** K5
Carloway (Càrlabhagh) **88** H3
Carlton *Bed* **32** C3
Carlton *Cambs* **33** K3
Carlton *Leics* **41** F5
Carlton *Notts* **41** J1
Carlton *NYorks* **57** F1
Carlton *NYorks* **58** C1
Carlton *NYorks* **58** C7
Carlton *Stock* **62** D4
Carlton *Suff* **35** H2
Carlton *SYorks* **51** F1
Carlton *WYorks* **57** J7
Carlton Colville **45** K6
Carlton Curlieu **41** J6
Carlton Green **33** K3
Carlton Husthwaite **57** K2
Carlton in Lindrick **51** H4
Carlton Miniott **57** J1
Carlton Scroop **42** C1
Carlton-in-Cleveland **63** F6
Carlton-le-Moorland **52** C7
Carlton-on-Trent **52** B6
Carluke **75** G5
Carlyon Bay **4** A5
Carmacoup **68** C1
Carmarthen (Caerfyrddin) **17** H3
Carmel *Carmar* **17** J4
Carmel *Flints* **47** K5
Carmel *Gwyn* **46** C7
Carmel *IoA* **46** B4
Carmichael **75** H7
Carmunnock **74** D5
Carmyle **74** E4
Carmyllie **81** L6
Carn **72** A5
Carn Brea Village **2** D4
Carnaby **59** H3
Carnach **83** L8
Carnassarie **79** K10
Carnbee **81** L10
Carnbo **80** F10
Carnduncan **72** A4
Carnforth **55** H2
Carnhedryn **16** B3
Carnhell Green **2** D5
Carnkie *Corn* **2** D5
Carnkie *Corn* **2** E5
Carnmore **72** B6
Carno **37** J6
Carnoch *High* **83** N5
Carnoch *High* **83** P7
Carnoch *High* **84** C6
Carnock **75** J2
Carnon Downs **3** F4
Carnoustie **81** L7
Carntyne **74** E4
Carnwath **75** H6
Carnyorth **2** A5
Carol Green **30** D1
Carperby **57** F1
Carr **51** H3
Carr Hill **51** J3
Carr Houses **48** C2
Carr Shield **61** K2
Carr Vale **51** G5
Carradale **73** G7
Carrbridge **84** D8
Carrefour Selous **3** J7
Carreglefn **46** B4
Carreg-wen **17** F1
Carrhouse **51** K2
Carrick **73** H2
Carrick Castle **73** K1
Carriden **75** J2
Carrine **66** A3
Carrington *GtMan* **49** G3
Carrington *Lincs* **53** G7
Carrington *Midlo* **76** B4
Carroch **68** B4
Carrog *Conwy* **37** G1
Carrog *Denb* **38** A1
Carron *A&B* **73** H1
Carron *Falk* **75** G2
Carron *Moray* **84** G6
Carron Bridge **75** F2
Carronbridge **68** D4
Carronshore **75** G2
Carrow Hill **19** H2
Carrutherstown **69** G6
Carrville **62** D2
Carry **73** H4
Carsaig **78** G8
Carscreugh **64** C4
Carse **73** F4
Carsegowan **64** E5
Carseriggan **64** D4
Carsethorn **65** K5
Carshalton **23** F5
Carshalton Beeches **23** F5
Carsie **80** G6
Carsington **50** E7
Carsington Water *Derbys* DE6 1ST **50** E7
Carsluith **64** E5
Carsphairn **67** K4
Carstairs **75** H6
Carstairs Junction **75** H6
Carswell Marsh **21** G2
Carter's Clay **10** E2
Carterton **21** F1
Carterway Heads **62** A1
Carthew **4** A5
Carthorpe **57** J1
Cartington **71** F3
Cartland **75** G6
Cartmel **55** G2
Cartmel Fell **55** H1
Cartworth **50** D2
Carway **17** H5
Cascades Adventure Pool *Devon* EX33 1NZ **6** C2
Cascob **28** B2
Cashel Farm **74** C1
Cashes Green **20** B1
Cashlie **79** R6
Cashmoor **9** J3
Cassencarie **64** E5
Cassington **31** F7
Cassop **62** D3
Castell **47** F6
Castell Gorfod **17** F3
Castell Howell **17** H1
Castellau **18** D3
Castell-y-bwch **19** F2
Casterton **56** B2
Castle Acre **44** C4
Castle Ashby **32** B3
Castle Bolton **62** A7
Castle Bromwich **40** D7
Castle Bytham **42** C4
Castle Caereinion **38** A5
Castle Camps **33** K4
Castle Carrock **61** G1
Castle Cary **9** F1
Castle Combe **20** B4
Castle Donington **41** G3
Castle Douglas **65** H4
Castle Drogo *Devon* EX6 6PB **7** F6
Castle Eaton **20** E2
Castle Eden **62** E3
Castle Eden Dene National Nature Reserve *Dur* SR8 1NJ **62** E3
Castle End **30** D1
Castle Frome **29** F4
Castle Gate **2** B5
Castle Goring **12** E6
Castle Green **22** C5
Castle Gresley **40** E4
Castle Heaton **77** H6
Castle Hedingham **34** B5
Castle Hill *Kent* **23** K7
Castle Hill *Suff* **35** F4
Castle Howard *NYorks* YO60 7DA **58** D3
Castle Kennedy **64** B5
Castle Levan **74** A3
Castle Madoc **27** K5
Castle Morris **16** C2
Castle O'er **69** H4
Castle Rising **44** A3
Castle Semple Water Country Park *Renf* PA12 4HJ **74** B5
Castlebay (Bàgh a'Chaisteil) **88** A9
Castlebythe **16** D3
Castlecary **75** F3
Castlecraig *High* **84** C4
Castlecraig *ScBord* **75** K6
Castlefairn **68** C5
Castleford **57** K7
Castlemartin **16** C6
Castlemilk *D&G* **69** G6
Castlemilk *Glas* **74** E5
Castlemorton **29** G5
Castlerigg **60** D4
Castleside **62** A2
Castlesteads **70** A7
Castlethorpe **31** J4
Castleton *A&B* **73** G2
Castleton *Derbys* **50** D4
Castleton *GtMan* **49** H1
Castleton *Newport* **19** F3
Castleton *NYorks* **63** G6
Castletown *Dorset* **9** F7
Castletown *High* **87** P3
Castletown *IoM* **54** B7
Castletown *T&W* **62** D1
Castleweary **69** K3
Castlewigg **64** E6
Castley **57** H5
Caston **44** D6
Castor **42** E6
Castramont **65** F4
Caswell **17** J7
Cat & Fiddle Inn **50** C5
Catacol **73** H6
Catbrain **19** J3
Catbrook **19** J1
Catchall **2** B6
Catcleugh **70** C3
Catcliffe **51** G4
Catcott **8** C1
Caterham **23** G6
Catfield **45** H3
Catford **23** G4
Catforth **55** H6
Cathays **18** E4
Cathcart **74** D4
Cathedine **28** A6
Catherine-de-Barnes **40** D7
Catherington **11** H3
Catherston Leweston **8** C5
Catherton **29** F1
Cathkin **74** E5
Catisfield **11** G4
Catlodge **84** A11
Catlowdy **69** K6
Catmere End **33** H5
Caton *Devon* **5** H3
Caton *Lancs* **55** J3
Caton Green **55** J3
Cator Court **5** G3
Catrine **67** K1
Catsfield **14** C6
Catsfield Stream **14** C6
Catshaw **50** E2
Catshill **29** J1
Cattadale **72** B4
Cattal **57** K4
Cattawade **34** E5
Catterall **55** J5
Catterick **62** C7
Catterick Bridge **62** C7
Catterick Garrison **62** B7
Catterlen **61** F3
Catterline **81** P3
Catterton **58** B5
Catteshall **22** C7
Catthorpe **31** G1
Cattishall **34** C2
Cattistock **8** E4
Catton *Norf* **45** G4
Catton *NYorks* **57** J2
Catton *N'umb* **61** K1
Catton Hall **40** E4
Catwick **59** H5
Catworth **32** D1
Caudle Green **29** J7
Caudwell's Mill & Craft Centre *Derbys* DE4 2EB **50** E6
Caulcott *CenBeds* **32** D4
Caulcott *Oxon* **31** G6
Cauldhame *Stir* **74** E1
Cauldhame *Stir* **80** D10
Cauldon **40** C1
Caulkerbush **65** K5
Caulside **69** K5
Caundle Marsh **9** F3
Caunsall **40** A7
Caunton **51** K6
Causeway End *D&G* **64** E4
Causeway End *Essex* **33** K7
Causeway End *Lancs* **48** D1
Causewayhead *Cumb* **60** C1
Causewayhead *Stir* **75** G1
Causey **62** C1
Causey Arch Picnic Area *Dur* NE16 5EG **117** A7
Causey Park **71** G4
Cautley **61** H7
Cavendish **34** C4
Cavendish Bridge **41** G3
Cavenham **34** B2
Cavens **65** K5
Cavers **70** A2
Caversfield **31** G6
Caversham **22** A4
Caverswall **40** B1
Cawdor **84** C5
Cawkeld **59** F4
Cawkwell **53** F4
Cawood **58** B6
Cawsand **4** E6
Cawston *Norf* **45** F3
Cawston *Warks* **31** F1
Cawthorn **58** D1
Cawthorne **50** E2
Cawthorpe **42** D3
Cawton **58** C2
Caxton **33** G3
Caxton Gibbet **33** F2
Caynham **28** E1
Caythorpe *Lincs* **42** C1
Caythorpe *Notts* **41** J1
Cayton **59** G1
CC2000 *Pembs* SA67 8DD **16** D4
Ceallan **88** C3
Ceann a' Bhàigh *Na H-E. Siar* **88** B2
Ceann a' Bhàigh *Na H-E. Siar* **88** F9
Cearsiadair **88** J6
Cedig **37** J3
Cefn (Pontcysyllte Aqueduct & Canal) *Denb/Powys/Wrex* **38** B1
Cefn Berain **47** H6
Cefn Canol **38** B2
Cefn Cantref **27** K6
Cefn Coch *Denb* **47** K7
Cefn Coch *Powys* **37** K5
Cefn Cribwr **18** B3
Cefn Cross **18** B3
Cefn Einion **38** B7
Cefn Hengoed **18** E2
Cefn Llwyd **37** F7
Cefn Rhigos **18** C1
Cefn-brith **47** H7
Cefn-caer-Ferch **36** D1
Cefn-coch **38** A3
Cefn-coed-y-cymmer **18** D1
Cefn-ddwysarn **37** J2
Cefndeuddwr **37** G3
Cefneithin **17** J4
Cefn-gorwydd **27** J4
Cefn-gwyn **38** A7
Cefn-mawr **38** B1
Cefnpennar **18** D1
Cefn-y-bedd **48** C7
Cefn-y-pant **16** E3
Ceidio **46** C4
Ceidio Fawr **36** B2
Ceint **46** C5
Cellan **17** K1
Cellarhead **40** B1
Cemaes **46** B3
Cemmaes **37** H5
Cemmaes Road (Glantwymyn) **37** H5
Cenarth **17** F1
Cennin **36** D1
Centre for Life *T&W* NE1 4EP **129** Newcastle upon Tyne
Ceramica *Stoke* ST6 3DS **108** D3
Ceres **81** K9
Cerist **37** J7
Cerne Abbas **9** F4
Cerney Wick **20** D2
Cerrigceinwen **46** C5
Cerrigydrudion **37** J1
Cessford **70** C1
Ceunant **46** D6
Chaceley **29** H5
Chacewater **2** E4
Chackmore **31** H5
Chacombe **31** F4
Chad Valley **40** C7
Chadderton **49** J2
Chadderton Fold **49** H2
Chaddesden **41** F2
Chaddesley Corbett **29** H1
Chaddleworth **21** H4
Chadlington **30** E6
Chadshunt **30** E3
Chadstone **32** B3
Chadwell *Leics* **42** A3
Chadwell *Shrop* **39** G4
Chadwell St. Mary **24** C4
Chadwick End **30** D1
Chaffcombe **8** C3
Chafford Hundred **24** C4
Chagford **7** F7
Chailey **13** G5
Chainhurst **14** C3
Chalbury **10** B4
Chalbury Common **10** B4
Chaldon **23** G6
Chaldon Herring (East Chaldon) **9** G6
Chale **11** F7
Chale Green **11** F6
Chalfont Common **22** D2
Chalfont St. Giles **22** C2
Chalfont St. Peter **22** D2
Chalford *Glos* **20** B1
Chalford *Wilts* **20** B7
Chalgrove **21** K2
Chalk **24** C4
Chalk End **33** K7
Challaborough **5** G6
Challacombe **6** E1
Challoch **64** D4
Challock **15** F2
Chalmington **8** E4
Chalton *CenBeds* **32** D6
Chalton *Hants* **11** J3
Chalvey **22** C4
Chalvington **13** J6
Champany **75** J3
Chancery **26** E1
Chandler's Cross **22** D2
Chandler's Ford **11** F2
Channel Islands **3** G7
Channel's End **32** E3
Chantry *Som* **20** A7
Chantry *Suff* **35** F4
Chapel **76** A1
Chapel Allerton *Som* **19** H6
Chapel Allerton *WYorks* **57** J6
Chapel Amble **3** G1
Chapel Brampton **31** J2
Chapel Chorlton **40** A2
Chapel Cleeve **7** J1
Chapel Cross **13** K4
Chapel End **32** D4
Chapel Green *Warks* **31** F2
Chapel Green *Warks* **40** E7
Chapel Haddlesey **58** B7
Chapel Hill *Aber* **85** Q7
Chapel Hill *Lincs* **53** F7
Chapel Hill *Mon* **19** J1
Chapel Hill *NYorks* **57** J5
Chapel Knapp **20** B5
Chapel Lawn **28** C1
Chapel Leigh **7** K3
Chapel Milton **50** C4
Chapel of Garioch **85** M8
Chapel Rossan **64** B6
Chapel Row *Essex* **24** D1
Chapel Row *WBerks* **21** J5
Chapel St. Leonards **53** J5
Chapel Stile **60** E6
Chapel Town **3** F3
Chapeldonan **67** F3
Chapelend Way **34** B5
Chapel-en-le-Frith **50** C4
Chapelgate **43** H3
Chapelhall **75** F4
Chapelhill *P&K* **80** F7
Chapelhill *P&K* **80** H8
Chapelknowe **69** J6
Chapel-le-Dale **56** C2
Chapelthorpe **51** F1
Chapelton *Angus* **81** M6
Chapelton *Devon* **6** D3
Chapelton *SLan* **74** E6
Chapeltown *B'burn* **49** G1
Chapeltown *Cumb* **69** K6
Chapeltown *Moray* **84** G8
Chapeltown *SYorks* **51** F3
Chapmans Well **6** B6
Chapmanslade **20** B7
Chapmore End **33** G7
Chappel **34** C6
Charaton **4** D4
Chard **8** C4
Chard Junction **8** C4
Chardleigh Green **8** C3
Chardstock **8** C4
Charfield **20** A2
Charing **14** E3
Charing Cross **10** C3
Charing Heath **14** E3
Charingworth **30** C5
Charlbury **30** E7
Charlcombe **20** A5
Charlecote **30** D3
Charles **6** E2
Charles Tye **34** E3
Charlesfield **70** A1
Charleshill **22** B7
Charleston **81** J6
Charlestown *Aberdeen* **85** P10
Charlestown *Corn* **4** A5
Charlestown *Derbys* **50** C3
Charlestown *Dorset* **9** F7
Charlestown *Fife* **75** J2
Charlestown *GtMan* **49** H2
Charlestown *High* **83** J3
Charlestown *High* **84** A6
Charlestown *WYorks* **56** E7
Charlestown *WYorks* **57** G6
Charlestown (Cornwall & West Devon Mining Landscape) *Corn* **3** H3
Charlestown of Aberlour (Aberlour) **84** G6
Charlesworth **50** C3
Charlinch **8** B1
Charlotteville **22** D7
Charlton *GtLon* **23** H4
Charlton *Hants* **21** G7
Charlton *Herts* **32** E6
Charlton *N'hants* **31** G5
Charlton *N'umb* **70** D5
Charlton *Oxon* **21** H3
Charlton *Som* **19** K6
Charlton *Som* **19** K7
Charlton *Tel&W* **38** E4
Charlton *Wilts* **9** J2
Charlton *Wilts* **20** C3
Charlton *Wilts* **20** E6
Charlton *Worcs* **30** B4
Charlton *WSuss* **12** B5
Charlton Abbots **30** B6
Charlton Adam **8** E2
Charlton Down **9** F5
Charlton Horethorne **9** F2
Charlton Kings **29** J6
Charlton Mackrell **8** E2
Charlton Marshall **9** H4
Charlton Musgrove **9** G2
Charlton on the Hill **9** H4
Charlton-All-Saints **10** C2
Charlton-on-Otmoor **31** G7
Charltons **63** G5
Charlwood **23** F7
Charminster **9** F5
Charmouth **8** C5
Charndon **31** H6
Charney Bassett **21** G2
Charnock Richard **48** E1
Charsfield **35** G3
Chart Corner **14** C3
Chart Sutton **14** D3
Charter Alley **21** J6
Charterhouse **19** H6
Charterville Allotments **21** G1
Chartham **15** G2
Chartham Hatch **15** G2
Chartridge **22** C1
Chartwell *Kent* TN16 1PS **23** H6
Charvil **22** A4
Charwelton **31** G3
Chase End Street **29** G5
Chase Terrace **40** C5
Chasetown **40** C5
Chastleton **30** D6
Chasty **6** B5
Chatburn **56** C5
Chatcull **39** G2
Chatelherault Country Park *SLan* ML3 7UE **119** K8
Chatham **24** D5
Chatham Green **34** B7
Chatham Historic Dockyard *Med* ME4 4TZ **24** D5
Chathill **71** G1
Chatsworth Farmyard & Adventure Playground *Derbys* DE45 1PP **50** E5
Chatsworth House *Derbys* DE45 1PP **50** E5
Chattenden **24** D4
Chatteris **43** G7
Chattisham **34** E4
Chatto **70** C2
Chatton **71** F1
Chaul End **32** D6
Chavey Down **22** B5
Chawleigh **7** F4
Chawley **21** H1
Chawston **32** E3
Chawton **11** J1
Chazey Heath **21** K4
Cheadle *GtMan* **49** H4
Cheadle *Staffs* **40** C1
Cheadle Heath **49** H4
Cheadle Hulme **49** H4
Cheam **23** F5
Cheapside **22** C5
Chearsley **31** J7
Chebsey **40** A3
Checkendon **21** K3
Checkley *ChesE* **39** G1
Checkley *Here* **28** E5
Checkley *Staffs* **40** C2
Checkley Green **39** G1
Chedburgh **34** B3
Cheddar **19** H6
Cheddar Caves & Gorge *Som* BS27 3QF **19** H6
Cheddington **32** C7
Cheddleton **49** J7
Cheddon Fitzpaine **8** B2
Chedglow **20** C2
Chedgrave **45** H6
Chedington **8** D4
Chediston **35** H1
Chediston Green **35** H1
Chedworth **30** B7
Chedzoy **8** C1
Cheesden **49** H1
Cheeseman's Green **15** F4
Cheetham Hill **49** H2
Cheglinch **6** D1
Cheldon **7** F4
Chelford **49** H5
Chellaston **41** F2
Chells **33** F6
Chelmarsh **39** G7
Chelmondiston **35** G5
Chelmorton **50** D6
Chelmsford **24** D1
Chelmsford Cathedral *Essex* CM1 1TY **24** D1
Chelmsley Wood **40** D7
Chelsea **23** F4
Chelsfield **23** H5
Chelsham **23** G6
Chelston Heath **7** K3
Chelsworth **34** D4
Cheltenham **29** J6
Chelveston **32** C2
Chelvey **19** H5
Chelwood **19** K5
Chelwood Common **13** H4
Chelwood Gate **13** H4
Chelworth **20** C2
Cheney Longville **38** D7
Chenies **22** D2
Chepstow (Cas-gwent) **19** J2
Cherhill **20** D4
Cherington *Glos* **20** C2
Cherington *Warks* **30** D5
Cheriton *Devon* **7** F1
Cheriton *Hants* **11** G2
Cheriton *Kent* **15** G4
Cheriton *Pembs* **16** C6
Cheriton *Swan* **17** H6
Cheriton Bishop **7** F6
Cheriton Cross **7** F6
Cheriton Fitzpaine **7** G5
Cherrington **39** F3
Cherry Burton **59** F5
Cherry Green **33** J6
Cherry Hinton **33** H3
Cherry Willingham **52** D5
Chertsey **22** D5
Cheselbourne **9** G5
Chesham **22** C1
Chesham Bois **22** C1
Cheshire Farm Ice Cream, Tattenhall *ChesW&C* CH3 9NE **48** D7
Cheshunt **23** G1
Cheslyn Hay **40** B5
Chessington **22** E5
Chessington World of Adventures *GtLon* KT9 2NE **22** E5
Chestall **40** C4
Chester **48** D6
Chester Cathedral *ChesW&C* CH1 2HU **124** Chester
Chester Moor **62** C2
Chester Zoo *ChesW&C* CH2 1LH **48** D5
Chesterblade **19** K7
Chesterfield *Derbys* **51** F5
Chesterfield *Staffs* **40** D5
Chester-le-Street **62** C1
Chesters *ScBord* **70** B1
Chesters *ScBord* **70** B2
Chesters Roman Fort (Frontiers of the Roman Empire) *N'umb* NE46 4EU **70** E7
Chesterton *Cambs* **33** H2
Chesterton *Cambs* **42** E6
Chesterton *Oxon* **31** G6
Chesterton *Shrop* **39** G6
Chesterton *Staffs* **40** A1
Chesterton *Warks* **30** E3
Chesterton Green **30** E3
Chestfield **25** H5
Cheston **5** G5
Cheswardine **39** G3
Cheswick **77** J6
Cheswick Buildings **77** J6
Cheswick Green **30** C1
Chetnole **9** F4
Chettiscombe **7** H4
Chettisham **43** J7
Chettle **9** J3
Chetton **39** F6
Chetwode **31** H6
Chetwynd Aston **39** G4
Chetwynd Park **39** G3
Cheveley **33** K2
Chevening **23** H6
Cheverell's Green **32** D7
Chevin Forest Park *WYorks* LS21 3JL **57** H5
Chevington **34** B3
Chevithorne **7** H4
Chew Magna **19** J5
Chew Moor **49** F2
Chew Stoke **19** J5
Chewton Keynsham **19** K5
Chewton Mendip **19** J6
Chichacott **6** E6
Chicheley **32** C4
Chichester **12** B6
Chichester Cathedral *WSuss* PO19 1PX **12** B6
Chickerell **9** F6
Chickering **35** G1
Chicklade **9** J1
Chickney **33** J6
Chicksands **32** E5
Chidden **11** H3
Chidden Holt **11** H3
Chiddingfold **12** C3
Chiddingly **13** J5
Chiddingstone **23** H7
Chiddingstone Causeway **23** J7
Chiddingstone Hoath **23** H7
Chideock **8** D5
Chidham **11** J4
Chidswell **57** H7
Chieveley **21** H4
Chignall Smealy **33** K7
Chignall St. James **24** C1
Chigwell **23** H2
Chigwell Row **23** H2
Chilbolton **21** G7
Chilcomb **11** G2
Chilcombe **8** E5
Chilcompton **19** K6
Chilcote **40** E4
Child Okeford **9** H3
Childer Thornton **48** C5
Childerditch **24** C3
Childrey **21** G3
Child's Ercall **39** F3
Childs Hill **23** F3
Childswickham **30** B5
Childwall **48** D4
Childwick Green **32** E7
Chilfrome **8** E5
Chilgrove **12** B5
Chilham **15** F2
Chilhampton **10** B1
Chilla **6** C5
Chillaton **6** C7
Chillenden **15** H2
Chillerton **11** F6
Chillesford **35** H3
Chilley **5** H5
Chillingham **71** F1
Chillington *Devon* **5** H6
Chillington *Som* **8** C3
Chilmark **9** J1
Chilson *Oxon* **30** E7
Chilson *Som* **8** C4
Chilsworthy *Corn* **4** E3
Chilsworthy *Devon* **6** B5
Chilthorne Domer **8** E3
Chilton *Bucks* **31** H7
Chilton *Devon* **7** G5
Chilton *Dur* **62** C4
Chilton *Oxon* **21** H3
Chilton *Suff* **34** C4
Chilton Candover **11** G1
Chilton Cantelo **8** E2
Chilton Foliat **21** G4
Chilton Polden **8** C1
Chilton Street **34** B4
Chilton Trinity **8** B1
Chilvers Coton **41** F6
Chilwell **41** H2
Chilworth *Hants* **11** F3
Chilworth *Surr* **22** D7
Chimney **21** G1
Chimney Street **34** B4
Chineham **21** K6
Chingford **23** G2
Chinley **50** C4
Chinley Head **50** C4
Chinnor **22** A1
Chipchase Castle **70** D6
Chipley **7** K3
Chipnall **39** G2
Chippenham *Cambs* **33** K2
Chippenham *Wilts* **20** C4
Chipperfield **22** D1
Chipping *Herts* **33** G5
Chipping *Lancs* **56** B5
Chipping Campden **30** C5
Chipping Hill **34** C7
Chipping Norton **30** E6
Chipping Ongar **23** J1
Chipping Sodbury **20** A3
Chipping Warden **31** F4
Chipstable **7** J3
Chipstead *Kent* **23** H6
Chipstead *Surr* **23** F6
Chirbury **38** B6
Chirk (Pontcysyllte Aqueduct & Canal) *Denb/Powys/Wrex* **38** B2
Chirk (Y Waun) **38** B2
Chirk Green **38** B2
Chirmorrie **64** C3
Chirnside **77** G5
Chirnsidebridge **77** G5
Chirton *T&W* **71** J7
Chirton *Wilts* **20** D6
Chisbury **21** F5
Chiscan **66** A2
Chiselborough **8** D3
Chiseldon **20** E4
Chiserley **57** F7
Chislehampton **21** J2
Chislehurst **23** H4
Chislet **25** J5
Chiswell Green **22** E1
Chiswick **23** F4
Chiswick End **33** G4
Chisworth **49** J3
Chithurst **12** B4
Chittering **33** H1
Chitterne **20** C7
Chittlehamholt **6** E3
Chittlehampton **6** E3
Chittoe **20** C5
Chivelstone **5** H7
Chivenor **6** D2
Chobham **22** C5
Choicelee **77** F5
Cholderton **21** F7
Cholesbury **22** C1
Chollerford **70** E6
Chollerton **70** E6
Cholmondeston **49** F7
Cholsey **21** J3
Cholstrey **28** D3
Cholwell *B&NESom* **19** K6
Cholwell *Devon* **4** E3
Chop Gate **63** F7
Choppington **71** H5
Chopwell **62** B1
Chorley *ChesE* **48** E7
Chorley *Lancs* **48** E1
Chorley *Shrop* **39** F7
Chorley *Staffs* **40** C4
Chorleywood **22** D2
Chorlton **49** G7
Chorlton Lane **38** D1
Chorlton-cum-Hardy **49** H3
Chowley **48** D7
Chrishall **33** H5
Chrishall Grange **33** H4
Chrisswell **74** A3
Christ Church *Oxon* OX1 1DP **134** Oxford
Christchurch *Cambs* **43** H6
Christchurch *Dorset* **10** C5
Christchurch *Glos* **28** E7
Christchurch *Newport* **19** G3
Christian Malford **20** C4
Christleton **48** D6
Christmas Common **22** A2
Christon **19** G6
Christon Bank **71** H1
Christow **7** G7
Chryston **74** E3
Chudleigh **5** J3
Chudleigh Knighton **5** J3
Chulmleigh **6** E4
Chunal **50** C3
Church **56** C7
Church Aston **39** G4
Church Brampton **31** J2
Church Brough **61** J5
Church Broughton **40** E2
Church Charwelton **31** G3
Church Common **35** H3
Church Crookham **22** B6
Church Eaton **40** A4
Church End *Cambs* **33** G1
Church End *Cambs* **43** F7
Church End *Cambs* **43** G5
Church End *CenBeds* **32** C5
Church End *CenBeds* **32** C5
Church End *CenBeds* **32** C6
Church End *CenBeds* **32** D5
Church End *CenBeds* **32** E3
Church End *CenBeds* **32** E5
Church End *ERid* **59** G4
Church End *Essex* **33** J4
Church End *Essex* **34** B6
Church End *Glos* **29** H5
Church End *Hants* **21** K6
Church End *Herts* **32** E7
Church End *Herts* **33** H6
Church End *Lincs* **43** F2
Church End *Lincs* **53** H3
Church End *Warks* **40** E6
Church End *Wilts* **20** D4
Church Enstone **30** E6
Church Fenton **58** B6
Church Green **7** K6
Church Gresley **40** E4
Church Hanborough **31** F7
Church Hill *ChesW&C* **49** F6
Church Hill *Derbys* **51** G6
Church Houses **63** G7
Church Knowle **9** J6
Church Laneham **52** B5
Church Langley **23** H1
Church Langton **42** A6
Church Lawford **31** F1
Church Lawton **49** H7
Church Leigh **40** C2
Church Lench **30** B3
Church Mayfield **40** D1
Church Minshull **49** F6
Church Norton **12** B7
Church of St. Mary-in-Castro *Kent* CT16 1HU **125** Dover
Church Preen **38** E6
Church Pulverbatch **38** D5
Church Stoke **38** B6
Church Stowe **31** H3
Church Street *Essex* **34** B4
Church Street *Kent* **24** D4
Church Stretton **38** D6
Church Town *Leics* **41** F4
Church Town *Surr* **23** G6
Church Village **18** D3
Church Warsop **51** H6
Church Westcote **30** D6
Church Wilne **41** G2
Churcham **29** G7
Churchdown **29** H7
Churchend *Essex* **25** G2
Churchend *Essex* **33** K6
Churchend *SGlos* **20** A2
Churchfield **40** C6
Churchgate **23** G1
Churchgate Street **33** H7
Churchill *Devon* **6** D1
Churchill *Devon* **8** B4
Churchill *NSom* **19** H6
Churchill *Oxon* **30** D6
Churchill *Worcs* **29** H1
Churchill *Worcs* **29** J3
Churchill Museum & Cabinet War Rooms *GtLon* SW1A 2AQ **132** D5
Churchingford **8** B3
Churchover **41** H7
Churchstanton **7** K4
Churchstow **5** H6
Churchtown *Devon* **6** E1
Churchtown *IoM* **54** D4
Churchtown *Lancs* **55** H5
Churchtown *Mersey* **48** C1
Churnsike Lodge **70** B6
Churston Ferrers **5** K5
Churt **12** B3
Churton **48** D7
Churwell **57** H7
Chute Cadley **21** G6
Chute Standen **21** G6
Chwilog **36** D2
Chyandour **2** B5
Chysauster **2** B5
Cilan Uchaf **36** B3
Cilcain **47** K6
Cilcennin **26** E2
Cilcewydd **38** B5
Cilfrew **18** A1
Cilfynydd **18** D2
Cilgerran **16** E1
Cilgwyn *Carmar* **27** G5
Cilgwyn *Pembs* **16** D2
Ciliau Aeron **26** E3
Cille Bhrighde **88** B7
Cille Pheadair **88** B7
Cilmaengwyn **18** A1
Cilmery **27** K3
Cilrhedyn **17** F2
Cilrhedyn Bridge **16** D2
Cilsan **17** J3
Ciltalgarth **37** H1
Cilwendeg **17** F2
Cilybebyll **18** A1
Cilycwm **27** G4
Cimla **18** A2
Cinderford **29** F7
Cippenham **22** C3
Cippyn **16** E1
Cirbhig **88** G3
Cirencester **20** D1
City *GtLon* **23** G3
City *VGlam* **18** C4
City Airport **23** H3
City Dulas **46** C4
City of Bath *B&NESom* **20** A5
City of Coventry Stadium (Coventry City F.C.) *WMid* CV6 6GE **104** D2
Clabhach **78** C5
Clachaig **73** K2
Clachan *A&B* **73** F5
Clachan *A&B* **79** K6
Clachan *A&B* **79** N9
Clachan *High* **82** F7
Clachan of Campsie **74** E3
Clachan of Glendaruel **73** H2
Clachaneasy **64** D3
Clachanmore **64** A6
Clachan-Seil **79** J9
Clachbreck **73** F3
Clachnabrain **81** J4
Clachtoll **86** C7
Clackmannan **75** H1
Clackmarras **84** G5
Clacton Pier *Essex* CO15 1QX **35** F7
Clacton-on-Sea **35** F7
Cladich (An Cladach) **79** M8
Cladswell **30** B3
Claggan **79** J6
Claigan **82** C5
Claines **29** H3
Clambers Play Centre *ESuss* TN34 1LD **128** Hastings
Clandown **19** K6
Clanfield *Hants* **11** J3
Clanfield *Oxon* **21** F1
Clannaborough Barton **7** F5
Clanville **21** G7
Claonaig **73** G5
Clapgate **33** H6
Clapham *Bed* **32** D3
Clapham *Devon* **7** G7
Clapham *GtLon* **23** F4
Clapham *NYorks* **56** C3
Clapham *WSuss* **12** D6
Clapham Green **32** D3
Clapham Hill **25** H5
Clappers **77** H5
Clappersgate **60** E6
Clapton *Som* **8** D4
Clapton *Som* **19** K6
Clapton-in-Gordano **19** H4
Clapton-on-the-Hill **30** C7
Clapworthy **6** E3
Clara Vale **71** G7
Clarach **37** F7
Clarbeston **16** D3
Clarbeston Road **16** D3
Clarborough **51** K4
Clare **34** B4
Clare Castle Country Park *Suff* CO10 8NJ **34** B4
Clarebrand **65** H4
Claremont Landscape Garden *Surr* KT10 9JG **22** E5
Clarencefield **69** F7
Clarilaw **70** A2
Clark's Green **12** E3
Clarkston **74** D5
Clashban **84** A1
Clashmore **84** B2
Clashnessie **86** C6
Clatford **20** E5
Clathy **80** E8
Clatt **85** K8
Clatter **37** J6
Clattercote **31** F4
Clatterford **11** F6
Clatterford End **23** J1
Clatterin Brig **81** M3
Clatteringshaws **65** F3
Clatworthy **7** J2
Claughton *Lancs* **55** J3
Claughton *Lancs* **55** J5
Claughton *Mersey* **48** C4
Clavelshay **8** B1
Claverdon **30** C2
Claverham **19** H5
Clavering **33** H5
Claverley **39** G6
Claverton **20** A5
Claverton Down **20** A5
Clawdd-côch **18** D4
Clawdd-newydd **47** J7
Clawfin **67** K3
Clawthorpe **55** J2
Clawton **6** B6
Claxby **52** E3
Claxby Pluckacre **53** G6
Claxby St. Andrew **53** H5
Claxton *Norf* **45** H5
Claxton *NYorks* **58** C3
Claxton Grange **62** E4
Clay Common **45** J7
Clay Coton **31** G1
Clay Cross **51** F6
Clay End **33** G6
Clay Hill **19** K4
Claybrooke Magna **41** G7
Claybrooke Parva **41** G7
Claydene **23** H7
Claydon *Oxon* **31** F3
Claydon *Suff* **35** F4
Claygate *Kent* **14** C3
Claygate *Surr* **22** E5
Claygate Cross **23** K6
Clayhanger *Devon* **7** J3
Clayhanger *WMid* **40** C5
Clayhidon **7** K4
Clayhill *ESuss* **14** D5
Clayhill *Hants* **10** E4
Clayhithe **33** J2
Claypit Hill **33** G3
Claypits **20** A1
Claypole **42** B1
Claythorpe **53** H5
Clayton *Staffs* **40** A1
Clayton *SYorks* **51** G2
Clayton *WSuss* **13** F5
Clayton *WYorks* **57** G6
Clayton Green **55** J7
Clayton West **50** E1
Clayton-le-Moors **56** C6
Clayton-le-Woods **55** J7
Clayworth **51** K4
Cleadale **78** F2
Cleadon **71** J7
Clearbrook **5** F4
Clearwell **19** J1
Cleasby **62** C5
Cleat **89** D9
Cleatlam **62** B5
Cleatop **56** D3
Cleator **60** B5
Cleator Moor **60** B5
Cleckheaton **57** G7
Clee St. Margaret **38** E7
Cleedownton **38** E7
Cleehill **28** E1
Cleestanton **28** E1
Cleethorpes **53** G2
Cleethorpes Discovery Centre *NELincs* DN35 0AG **53** G2
Cleeton St. Mary **29** F1
Cleeve *NSom* **19** H5
Cleeve *Oxon* **21** K3
Cleeve Hill **29** J6
Cleeve Prior **30** B4
Cleghorn **75** G6
Clehonger **28** D5
Cleigh **79** K8
Cleish **75** J1
Cleland **75** F5
Clement's End **32** D7
Clemsfold **12** E3
Clench Common **20** E5
Clenchwarton **43** J3
Clennell **70** E3
Clent **29** J1

Eastheath 22 B5
Easthope 38 E6
Easthorpe Essex 34 D6
Easthorpe Leics 42 B4
Easthorpe Notts 51 K7
Easthouses 76 B4
Eastington 45 H2
Eastington Devon 7 F5
Eastington Glos 20 B3
Eastington Glos 30 C7
Eastleach Martin 21 F1
Eastleach Turville 21 F1
Eastleigh Devon 6 C3
Eastleigh Hants 11 F3
Eastling 14 E2
Eastmoor Derbys 51 F5
Eastmoor Norf 44 B5
Eastnor 29 G5
Eastoft 52 B1
Eastoke 11 J5
Easton Cambs 32 E1
Easton Cumb 60 D1
Easton Cumb 69 K6
Easton Devon 7 F7
Easton Dorset 9 F7
Easton Hants 11 G1
Easton IoW 10 E6
Easton Lincs 42 C3
Easton Norf 45 F4
Easton Som 19 J7
Easton Suff 35 G3
Easton Wilts 20 B4
Easton Grey 20 B3
Easton Maudit 32 B3
Easton-on-the-Hill 42 D5
Easton Royal 21 F5
Easton-in-Gordano 19 J4
Eastrea 17 J5
Eastriggs 69 H7
Eastrington 58 D6
Eastry 15 J2
East-the-Water 6 C3
Eastville 19 K4
Eastwell 42 A3
Eastwick 33 H7
Eastwood Notts 41 G1
Eastwood SYorks 50 B2
Eastwood S'end 24 E3
Eastwood WYorks 56 E7
Eastwood End 43 H6
Eathorpe 30 E2
Eaton ChesE 49 H6
Eaton ChesW&C 48 E6
Eaton GtMan 49 G1
Eaton Kent 14 E3
Eaton Leics 42 A3
Eaton Norf 45 G5
Eaton Notts 51 K5
Eaton Oxon 21 H1
Eaton Shrop 38 C7
Eaton Shrop 38 B3
Eaton Bishop 28 D5
Eaton Bray 32 C6
Eaton Constantine 39 F5
Eaton Ford 32 E3
Eaton Hall 48 D6
Eaton Hastings 21 F2
Eaton upon Tern 39 F3
Eaves Green 40 E7
Eavestone 57 H3
Ebberston 58 E1
Ebbesborne Wake 9 J2
Ebbw Vale (Glyn Ebwy) 18 E1
Ebchester 62 B1
Ebdon 19 G5
Ebford 7 H7
Ebley 20 B1
Ebnal 38 D1
Ebrington 30 C4
Ebsworthy Town 6 D6
Ecchinswell 21 H6
Ecclaw 77 H4
Ecclefechan 69 G6
Eccles GtMan 49 G3
Eccles Kent 24 D5
Eccles ScBord 77 F6
Eccles Green 28 C4
Eccles Road 44 E6
Ecclesfield 51 F3
Ecclesgreig 81 N4
Eccleshall 40 A3
Eccleshill 57 G6
Ecclesmachan 75 J3
Eccles-on-Sea 45 J3
Eccleston ChesW&C 48 D6
Eccleston Lancs 48 E1
Eccleston Mersey 48 D3
Eccup 57 H5
Echt 85 M10
Eckford 70 C1
Eckington Derbys 51 G5
Eckington Worcs 29 J4
Ecton N'hants 32 B2
Ecton Staffs 50 C7
Edale 50 D4
Eday 89 E4
Eday Airfield 89 E4
Edburton 13 F5
Edderside 60 C2
Edderton 84 B2
Eddington 21 G5
Eddleston 76 A6
Eden Camp NYorks YO17 6RT 58 E2
Eden Park 23 G5
Eden Vale 62 E3
Edenbridge 23 H7
Edenfield 49 G1
Edenhall 61 G3
Edenham 42 D3
Edensor 50 E5
Edentaggart 74 B1
Edenthorpe 51 J2
Edern 36 B2
Edgarley 8 E1
Edgbaston 40 C7
Edgcott Bucks 31 H6
Edgcott Som 7 G2
Edgcumbe 2 E5
Edge Glos 20 B1
Edge Shrop 38 C5
Edge End 28 E7
Edge Green ChesW&C 48 D7
Edge Green GtMan 48 E3
Edge Green Norf 44 E7
Edgebolton 38 E3
Edgefield 44 E2
Edgefield Street 44 E2
Edgeley 38 E1
Edgerley 38 C4
Edgerton 50 D1
Edgeworth 20 C1
Edginswell 5 J4
Edgmond 39 G4
Edgmond Marsh 39 G3
Edgton 38 C7
Edgware 23 F2
Edgworth 49 G1
Edinample 80 B8
Edinbane 82 D5
Edinburgh 76 A3
Edinburgh Airport 75 K3
Edinburgh Castle Edin EH1 2NG 126 E4

Edinburgh Zoo Edin EH12 6TS 120 C2
Edinchip 80 A8
Edingale 40 E4
Edingley 51 J7
Edingthorpe 45 H2
Edingthorpe Green 45 H2
Edington Som 8 C1
Edingthread 19 J2
Edington Wilts 20 C6
Edistone 6 A3
Edith Weston 42 C5
Edithmead 19 G7
Edlaston 40 D1
Edlesborough 32 C7
Edlingham 71 G3
Edlington 53 F5
Edmondbyers 62 A1
Edmondsham 10 B3
Edmondsley 62 C2
Edmondstown 18 D2
Edmondthorpe 42 B4
Edmonstone 89 E5
Edmonton Corn 3 G1
Edmonton GtLon 23 G2
Edmundbyers 62 A1
Ednam 77 F7
Ednaston 40 E1
Edney Common 24 C1
Edrom 77 G5
Edstaston 38 E2
Edstone 29 J2
Edvin Loach 29 F3
Edwalton 41 H2
Edwardstone 34 D4
Edwinsford 17 K2
Edwinstowe 51 J6
Edworth 33 F4
Edwyn Ralph 29 F3
Edzell 81 M4
Efail Isaf 18 D3
Efail-fâch 18 A2
Efailnewydd 36 C2
Efailwen 16 E3
Efenechtyd 47 K7
Effingham 22 E6
Effirth 89 M7
Efflinch 40 D4
Efford 7 G5
Egbury 21 H6
Egdean 12 C4
Egdon 29 J3
Egerton GtMan 49 G1
Egerton Kent 14 E3
Egerton Forstal 14 D3
Egerton Green 48 E7
Eggbuckland 4 E5
Eggborough 58 B7
Eggerness 64 E6
Eggesford Barton 6 E4
Eggington 32 C6
Egginton 40 E3
Egglescliffe 62 E5
Eggleston 61 L3
Egham 22 D4
Egham Wick 22 C4
Egleton 42 B5
Eglingham 71 G2
Eglinton 74 B6
Egloshayle 4 A3
Egloskerry 4 C2
Eglwys Cross 38 D1
Eglwys Fach 37 F6
Eglwys Nunydd 18 B3
Eglwysbach 47 G5
Eglwys-Brewis 18 D5
Eglwyswrw 16 E2
Egmanton 51 K6
Egmere 44 D2
Egremont 60 B5
Egton 63 J6
Egton Bridge 63 J6
Eight Ash Green 34 D6
Eilanreach 83 J9
Eigg 78 F2
Eildon 76 D7
Eilean Donan Castle High IV40 8DX 83 J8
Eilean Shona 78 H3
Einacleit 88 G5
Eisingrug 37 F2
Eisteddfa Gurig 37 G7
Elan Village 27 J2
Elberton 19 K3
Elborough 19 G6
Elburton 5 F5
Elcombe 20 E3
Elder Street 33 J5
Eldernell 43 G6
Eldersfield 29 G5
Eldersie 74 C4
Eldon 62 C4
Eldrick 67 G5
Eldroth 56 C3
Eldwick 57 G5
Elerch 37 F7
Elford N'umb 71 H1
Elford Staffs 40 D4
Elford Closes 33 J1
Elgin 84 G4
Elgol 82 F9
Elham 15 G3
Elie 81 L1
Elilaw 70 E3
Elim 46 B4
Eling Hants 10 E3
Eling WBerks 21 J4
Eling Tide Mill Hants SO40 9HF 92 A3
Eliock 68 D3
Elishaw 70 D4
Elkesley 51 J5
Elkington 31 H1
Elkstone 29 J7
Elland 57 G7
Elland Upper Edge 57 G7
Ellary 73 F3
Ellastone 40 D1
Ellemford 77 F4
Ellenabeich 79 J9
Ellenborough 60 B3
Ellenhall 40 A3
Ellen's Green 12 D3
Ellerbeck 62 E7
Ellerby 63 H5
Ellerdine 39 F3
Ellerdine Heath 39 F3
Ellerker 59 F7
Ellerton ERid 58 D5
Ellerton NYorks 62 C7
Ellerton Shrop 39 G3
Ellesborough 22 B1
Ellesmere 38 C2
Ellesmere Park 49 G3
Ellesmere Port 48 D5
Ellingham Hants 10 C4
Ellingham N'umb 71 G1
Ellingham Norf 45 H6
Ellingstring 57 G1
Ellington Cambs 32 E1
Ellington N'umb 71 H4
Ellington Thorpe 32 E1
Elliot 81 M7
Elliot's Green 20 A7

Ellisfield 21 K7
Ellishadder 82 E4
Ellistown 41 G4
Ellon 85 P7
Ellonby 60 F3
Ellough 45 J7
Ellough Moor 45 J7
Elloughton 59 F7
Ellwood 19 J1
Elm 43 H5
Elm Park 23 J3
Elmbridge 29 J2
Elmdon Essex 33 H5
Elmdon WMid 40 D7
Elmdon Heath 40 D7
Elmers End 23 G5
Elmer's Green 48 D2
Elmesthorpe 41 G6
Elmhurst 40 D4
Elmley Castle 29 J4
Elmley Lovett 29 H2
Elmore 29 G7
Elmore Back 29 G7
Elmscott 6 A4
Elmsett 34 E4
Elmstead Essex 34 M4
Elmstead GtLon 23 H4
Elmstead Market 34 E6
Elmstone 25 J5
Elmstone Hardwicke 29 J6
Elmswell ERid 59 F4
Elmswell Suff 34 D2
Elmton 51 H5
Elphin 86 E8
Elphinstone 76 B3
Elrick 85 N10
Elrig 64 D6
Elrigbeag 79 N9
Elsdon 70 E4
Elsecar 51 F2
Elsenham 33 J6
Elsfield 31 G7
Elsham 52 D1
Elsing 44 E4
Elslack 56 E5
Elson Hants 11 H4
Elson Shrop 38 C2
Elsrickle 75 J6
Elstead 22 C7
Elsted 22 C7
Elsthorpe 42 D3
Elstob 62 D4
Elston Lancs 55 F6
Elston Notts 42 A1
Elstone 6 E4
Elstow 32 D4
Elstree 22 E2
Elstronwick 59 J6
Elswick 55 H6
Elsworth 33 G2
Elterwater 60 E6
Eltham 23 H4
Eltisley 33 F3
Elton Cambs 42 D6
Elton ChesW&C 48 D5
Elton Derbys 50 E6
Elton Glos 29 G7
Elton GtMan 49 G1
Elton Here 28 D1
Elton Notts 42 A2
Elton Stock 62 E5
Elton Green 48 D5
Elvanfoot 68 E2
Elvaston 41 G2
Elveden 34 C1
Elvingston 76 C3
Elvington Kent 15 H2
Elvington York 58 D5
Elwick Hart 62 E3
Elwick N'umb 77 K7
Elworth 49 G6
Elworthy 7 J2
Ely Cambs 33 J1
Ely Cardiff 18 E4
Emberton 32 B4
Emberton Country Park MK MK46 5FJ 32 B3
Embleton Cumb 60 C3
Embleton Hart 62 E4
Embleton N'umb 71 H1
Embo 84 C1
Embo Street 84 C1
Emborough 19 K6
Embsay 57 F4
Embsay Steam Railway NYorks BD23 6AF 57 F4
Emerson Park 23 J3
Emery Down 10 D4
Emley 50 E1
Emmington 22 A1
Emneth 43 H5
Emneth Hungate 43 J5
Empingham 42 C5
Empshott 11 J1
Empshott Green 11 J1
Emsworth 11 J4
Enborne 21 H5
Enborne Row 21 H5
Enchmarsh 38 E6
Enderby 41 H6
Endmoor 55 J1
Endon 49 J7
Endon Bank 49 J7
Enfield 23 G2
Enfield Wash 23 G2
Enford 20 E6
Engine Common 19 K3
Englefield 21 K4
Englefield Green 22 C4
Englesea-brook 49 G7
English Bicknor 28 E7
English Frankton 38 D3
Enham Alamein 21 G7
Enmore 8 B1
Enmore Green 9 H2
Ennerdale Bridge 60 B5
Enniscaven 3 G3
Enochdhu 80 F4
Ensay 78 E6
Ensdon 38 D4
Ensis 6 D3
Enson 40 A3
Enstone 30 E6
Enterkinfoot 68 D3
Enterpen 62 E6
Eòlaigearraidh 88 B8
Eorabus 78 E7
Eòropaidh 88 L1
Epney 29 G7
Epperstone 41 J1
Epping 23 H1
Epping Green Essex 23 H1
Epping Green Herts 23 F1
Epping Upland 23 H1
Eppleby 62 B5
Eppleworth 59 G6
Epsom 23 F5
Epwell 31 F4
Epworth 52 B2
Epworth Turbary 51 K2
Erbistock 38 C1
Erbusaig 82 H8
Erchless Castle 83 Q6
Erdington 40 D6
Eredine 79 L10

Eriboll 86 G4
Ericstane 69 F3
Eridge Green 13 J3
Eriff 67 K3
Erines 73 G3
Erisey Barton 2 E7
Eriskay (Eiriosgaigh) 88 B7
Eriswell 34 B1
Erith 23 J4
Erlestoke 20 C6
Ermington 5 G5
Ernesettle 4 E4
Erpingham 45 F2
Erringden Grange 56 E7
Errogie (Earagaidh) 83 R8
Errol 80 H8
Erskine 74 C3
Ervie 66 D7
Erwarton 35 G5
Erwood 27 K4
Eryholme 62 D6
Eryrus 48 B7
Escart 73 G4
Escart Farm 73 G5
Escomb 62 B4
Escrick 58 C5
Esgair 17 G3
Esgairgeiliog 37 G5
Esgyryn 47 G5
Esh 62 B2
Esh Winning 62 B2
Esher 22 E5
Eshott 71 H4
Eshton 56 E4
Eskadale 83 Q7
Eskbank 76 B4
Eskdale Green 60 C6
Eskdalemuir 69 H4
Eskham 53 G3
Esknish 72 B4
Esperley Lane Ends 62 B4
Espley Hall 71 G4
Esprick 55 H6
Essendine 42 D4
Essendon 23 F1
Essich 84 A7
Essington 40 B5
Eston 63 F5
Etal 77 H7
Etchilhampton 20 D5
Etchingham 14 C5
Etchinghill Kent 15 G4
Etchinghill Staffs 40 C4
Etherdwick Grange 59 J6
Etherley Dene 62 B4
Ethie Mains 81 M6
Eton 22 C4
Eton Wick 22 C4
Etteridge 84 A11
Ettiley Heath 49 G6
Ettington 30 D4
Etton ERid 59 F5
Etton Peter 42 E5
Ettrick 69 H2
Ettrickbridge 69 J1
Ettrickhill 69 H2
Etwall 40 E2
Eudon George 39 F7
Euston 34 C1
Euston 34 C1
Evanston 18 D3
Evanton 84 A4
Evedon 42 D1
Evelix 84 B1
Evenjobb 28 B2
Evenley 31 G5
Evenlode 30 D6
Evenwood 62 B4
Evenwood Gate 62 B4
Everbay 89 F5
Evercreech 9 F1
Everdon 31 G3
Everingham 58 E5
Everleigh 21 F6
Everley 59 F1
Eversholt 32 C5
Evershot 8 E4
Eversley 22 A5
Eversley Cross 22 A5
Everthorpe 59 F6
Everton CenBeds 33 F3
Everton Hants 10 D5
Everton Mersey 48 C3
Everton Notts 51 J3
Evertown 69 J6
Eves Corner 25 F2
Evesbatch 29 F4
Evesham 30 B4
Evesham Country Park Shopping & Garden Centre Worcs WR11 4TP 30 B4
Evie 89 C5
Evington 41 J5
Ewart Newtown 77 H7
Ewden Village 50 E3
Ewell 23 F5
Ewell Minnis 15 H3
Ewelme 21 K2
Ewen 20 D2
Ewenny 18 C4
Ewerby 42 E1
Ewerby Thorpe 42 E1
Ewhurst 22 D7
Ewhurst Green ESuss 14 C5
Ewhurst Green Surr 12 D3
Ewloe 48 C6
Ewloe Green 48 B6
Ewood 56 B7
Ewood Bridge 56 C7
Eworthy 6 C6
Ewshot 22 B7
Ewyas Harold 28 C6
Exbourne 6 E5
Exbridge 7 H3
Exbury 11 F4
Exbury Gardens Hants SO45 1AZ 92 C8
Exceat 13 J7
Exebridge 7 H3
Exelby 57 H1
Exeter 7 H6
Exeter Cathedral Devon EX1 1HS 125 Exeter
Exeter International Airport 7 H6
Exford 7 G2
Exfords Green 38 D5
Exhall Warks 30 C3
Exhall Warks 41 F7
Exlade Street 21 K3
Exminster 7 H7
Exmoor Int. Dark Sky Reserve Devon/Som 7 F1
Exmouth 7 J7
Exnaboe 89 M11
Exning 33 K2
Explore-At-Bristol BS1 5DB 123 Bristol
Exton Devon 7 H7
Exton Hants 11 H2
Exton Rut 42 C4
Exton Som 7 H2
Exwick 7 H6

Eyam 50 E5
Eydon 31 G4
Eye Here 28 D2
Eye Peter 43 F5
Eye Suff 35 F1
Eye Green 43 F5
Eyemouth 77 H4
Eyeworth 33 F4
Eyhorne Street 14 D2
Eyke 35 H3
Eynesbury 32 E3
Eynort 82 D8
Eynsford 23 J5
Eynsham 21 H1
Eype 8 D5
Eyre 82 E5
Eythorne 15 H3
Eyton Here 28 D2
Eyton Shrop 38 C7
Eyton on Severn 38 E5
Eyton upon the Weald Moors 39 F4
Eywood 28 C3

F

Faccombe 21 G6
Faceby 62 E6
Fachwen 46 D6
Facit 49 H1
Faddiley 48 E7
Fadmoor 58 C1
Faebait 83 Q5
Faifley 74 D3
Fail 67 J1
Failand 19 J4
Failford 67 J1
Failsworth 49 H2
Fair Isle 89 K10
Fair Isle Airstrip 89 K10
Fair Oak Devon 7 J4
Fair Oak Hants 11 F3
Fair Oak Hants 21 K5
Fairbourne 37 F4
Fairburn 57 K7
Fairfield Derbys 50 C5
Fairfield Derbys 40 B1
Fairfield Kent 14 E5
Fairfield Mersey 48 B4
Fairfield Stock 62 E5
Fairfield Worcs 29 J1
Fairfield Halls, Croydon GtLon CR9 1DG 101 B12
Fairford 20 E1
Fairgirth 65 J5
Fairhaven 55 G7
Fairhill 75 F5
Fairholm 75 F5
Faircot 12 E1
Fairlands Valley Park Herts SG2 0BL 33 F6
Fairley 85 N10
Fairlie 74 A5
Fairlight 14 D6
Fairlight Cove 14 D6
Fairmile Devon 7 J6
Fairmile Surr 22 E5
Fairmilehead 76 A4
Fairnington 70 B1
Fairoak 39 G2
Fairseat 24 C5
Fairstead 34 B7
Fairwarp 13 H4
Fairwater 18 E4
Fairy Cross 6 C3
Fairyhill 17 H6
Fakenham 44 D3
Fakenham Magna 34 D1
Fala 76 C4
Fala Dam 76 C4
Falahill 76 B5
Faldingworth 52 D4
Falfield Devon 7 F4
Falfield Glos 19 K2
Falin-Wnda 17 G1
Falkenham 35 G5
Falkirk 75 G3
Falkirk Wheel Falk FK1 4RS 75 G3
Falkland 80 H10
Falla 70 C2
Fallgate 51 F6
Fallin 75 G1
Falmer 13 G6
Falmouth 3 F5
Falsgrave 59 G1
Falstone 70 C5
Fanagmore 86 D5
Fancott 32 D6
Fangdale Beck 63 F7
Fangfoss 58 D4
Fankerton 75 F2
Fanmore 78 F6
Fanner's Green 33 K7
Fans 76 E6
Far Cotton 31 J3
Far Forest 29 G1
Far Gearstones 56 C1
Far Green 20 A1
Far Moor 48 E2
Far Oakridge 20 C1
Far Royds 57 H6
Far Sawrey 60 E7
Farcet 43 F6
Farden 28 E1
Fareham 11 G4
Farewell 40 C4
Farforth 53 G5
Faringdon 21 F2
Farington 55 J7
Farlam 61 G1
Farlary 87 K9
Farleigh NSom 19 J5
Farleigh Surr 23 G5
Farleigh Hungerford 20 B6
Farleigh Wallop 21 K7
Farlesthorpe 53 H5
Farleton Cumb 55 J1
Farleton Lancs 55 J3
Farley Derbys 50 E6
Farley Shrop 38 C5
Farley Staffs 40 C1
Farley Wilts 10 D2
Farley Green Suff 34 B3
Farley Green Surr 22 D7
Farley Hill 22 A5
Farleys End 29 G7
Farlington 58 C3
Farlow 39 F7
Farm Town 41 F4
Farmborough 19 K5
Farmcote 30 B6
Farmington 30 C7
Farmoor 21 H1
Farmtown 85 K5
Farnborough GtLon 23 H5
Farnborough Hants 22 B6
Farnborough Street 22 B6
Farnborough Warks 31 F4
Farnborough WBerks 21 H3
Farncombe 22 C7

Farndish 32 C2
Farndon ChesW&C 48 D7
Farndon Notts 51 K7
Farne Islands 77 K7
Farnell 81 M5
Farnham Dorset 9 J3
Farnham Essex 33 H6
Farnham NYorks 57 J3
Farnham Suff 35 H2
Farnham Surr 22 B7
Farnham Common 22 C3
Farnham Green 33 H6
Farnham Royal 22 C3
Farningham 23 J5
Farnley NYorks 57 H5
Farnley WYorks 57 H6
Farnley Tyas 50 D1
Farnsfield 51 J7
Farnworth GtMan 49 G2
Farnworth Halton 48 E4
Farr High 84 A7
Farr High 87 K3
Farr High 84 C10
Farraline 83 R8
Farrington 7 K6
Farrington Gurney 19 K6
Farsley 57 H6
Farther Q3
Farthing Corner 24 E5
Farthing Green 14 D3
Farthinghoe 31 H3
Farthorpe 53 F5
Fartown 50 D1
Farway 7 K6
Fascadale 78 G3
Faslane 74 A2
Fasnakyle 83 P8
Fassfern 79 M3
Fatfield 62 D1
Faugh 61 G1
Fauldhouse 75 H4
Faulkbourne 34 B7
Faulkland 20 A6
Fauls 38 E2
Faulston 10 B2
Faversham 25 G5
Fawdington 57 K2
Fawdon 71 F2
Fawfieldhead 50 C6
Fawkham Green 23 J5
Fawler 30 E7
Fawley Bucks 22 A3
Fawley Hants 11 F4
Fawley WBerks 21 G3
Fawley Chapel 28 E6
Faxfleet 58 E7
Faxton 31 J1
Faygate 13 F3
Fazakerley 48 C3
Fazeley 40 E5
Fearby 57 G1
Fearn 84 C4
Fearnan 80 C6
Fearnbeg 82 H5
Fearnhead 49 F3
Fearnmore 82 H4
Fearnoch A&B 73 H2
Fearnoch A&B 73 J3
Featherstone Staffs 40 B5
Featherstone WYorks 57 K7
Featherstone Castle 70 B7
Feckenham 30 B2
Feering 34 C6
Feetham 61 L7
Feizor 56 C3
Felbridge 13 G3
Felbrigg 45 G2
Felcourt 23 G7
Felden 22 D1
Felhampton 38 D7
Felindre Carmar 17 G2
Felindre Carmar 17 J2
Felindre Carmar 17 K2
Felindre Carmar 27 G6
Felindre Cere 26 E3
Felindre Powys 28 A6
Felindre Powys 28 B5
Felindre Swan 17 K5
Felinfach Cere 26 E3
Felinfach Powys 27 K5
Felinfoel 17 J5
Felingwmisaf 17 J3
Felingwmuchaf 17 J3
Felixkirk 57 K1
Felixstowe 35 H5
Felixstowe Ferry 35 H5
Felkington 77 H6
Felldownhead 6 B7
Felling 71 H7
Felmersham 32 C3
Felmingham 45 G3
Felpham 12 C7
Felsham 34 D3
Felsted 33 K6
Feltham 22 E4
Felthamhill 22 E4
Felthorpe 45 F4
Felton Here 28 E4
Felton N'umb 71 G3
Felton NSom 19 J5
Felton Butler 38 C4
Feltwell 44 B6
Fen Ditton 33 H2
Fen Drayton 33 G2
Fen End 30 D1
Fen Street Norf 44 D6
Fen Street Norf 44 E6
Fen Street Suff 35 F2
Fenay Bridge 50 D1
Fence 56 D6
Fence Houses 62 D1
Fencott 31 G7
Fendike Corner 53 H6
Fenham 77 J6
Fenhouses 43 F1
Feniscowles 56 B7
Feniton 7 K6
Fenn Street 24 D4
Fenni-fach 27 K6
Fenny Bentley 50 D7
Fenny Bridges 7 K6
Fenny Compton 31 F3
Fenny Drayton 41 F6
Fenny Stratford 32 B5
Fenrother 71 G4
Fenstanton 33 G2
Fenton Cambs 33 G1
Fenton Lincs 52 B5
Fenton Lincs 52 B7
Fenton Notts 51 K4
Fenton Stoke 40 A1
Fenton N'umb 77 H7
Fenwick EAyr 74 C6
Fenwick N'umb 71 F6
Fenwick N'umb 77 J6
Fenwick SYorks 51 H1
Feochaig 66 B2
Feock 3 F5
Feolin 72 D4
Feolin Ferry 72 C4
Feorlan 66 A3
Feorlin 73 H1
Ferindonald 82 G10
Feriniquarrie 82 B5
Fern 81 K4
Ferndale 18 C2
Ferndown 10 B4
Ferness 84 D6
Fernham 21 F2
Fernhill Heath 29 H3
Fernhurst 12 B4
Fernie 81 J9
Fernilea 82 D7
Fernilee 50 C5
Ferrensby 57 J3
Ferrers Centre for Arts & Crafts Leics LE65 1RU 41 F3
Ferrindonald 82 G10
Ferring 12 D6
Ferry Hill 43 G7
Ferrybridge 57 K7
Ferryden 81 N5
Ferryhill 62 C3
Ferryside (Glanyferi) 17 G4
Fersfield 44 E7
Fersit 79 Q3
Ferwig 16 E1
Feshiebridge 84 C10
Festival Park BGwent NP23 8FP 18 E1
Fetcham 22 E6
Fetlar 89 Q3
Fetterangus 85 P5
Fettercairn 81 M3
Feus of Caldhame 81 M4
Fewcott 31 G6
Fewston 57 G4
Ffairfach 17 K3
Ffald-y-brenin 17 K1
Ffarmers 17 K1
Ffawyddog 28 B7
Ffestiniog (Llan Ffestiniog) 37 G1
Ffordd-las Denb 47 K6
Fforddlas Powys 28 B5
Fforest 17 J5
Fforest-fach 17 K6
Ffostrasol 17 G1
Ffos-y-ffin 26 D2
Ffridd Uchaf 46 D7
Ffrith Denb 48 B7
Ffrith Flints 48 B7
Ffrwdgrech 27 K6
Ffynnon 17 G3
Ffynnongroyw 47 K4
Fibhig 88 H3
Fiddington Glos 29 J5
Fiddington Som 19 F7
Fiddleford 9 H3
Fiddler's Green Glos 29 J6
Fiddler's Green Here 28 E5
Fiddler's Green Norf 44 D5
Fiddler's Green Norf 44 E6
Fiddlers Hamlet 23 H1
Field 40 C2
Field Broughton 55 G1
Field Dalling 44 E2
Field Head 41 G5
Fife Keith 85 J5
Fifehead Magdalen 9 G2
Fifehead Neville 9 G3
Fifehead St. Quintin 9 G3
Fifield Oxon 30 D7
Fifield W&M 22 C4
Fifield Wilts 20 E6
Fifield Bavant 10 B2
Figheldean 20 E7
Filby 45 J4
Filey 59 H1
Filgrave 32 B4
Filham 5 G5
Filkins 21 F1
Filleigh Devon 6 E3
Filleigh Devon 7 F4
Fillingham 52 C4
Fillongley 40 E7
Filmore Hill 11 H2
Filton 19 K4
Fimber 58 E3
Finavon 81 K5
Fincham 44 A5
Finchampstead 22 A5
Finchdean 11 J3
Finchingfield 33 K5
Finchley 23 F2
Findern 41 F2
Findhorn 84 E4
Findhorn Bridge 84 C8
Findo Gask 80 E8
Findochty 85 J4
Findon Aber 85 P11
Findon WSuss 12 E6
Findon Mains 84 A4
Findon Valley 12 E6
Findrack 85 L10
Finedon 32 C1
Fingal Street 35 G1
Fingask 85 M7
Fingerpost 29 G1
Fingest 22 A2
Finghall 57 G1
Fingland Cumb 60 D1
Fingland D&G 68 C2
Fingland D&G 69 H3
Finglesham 15 J2
Fingringhoe 34 E6
Finkle Street 51 F3
Finlarig 80 A7
Finmere 31 H5
Finnart A&B 74 A1
Finnart P&K 80 A5
Finningham 34 E2
Finningley 51 J3
Finnygaud 85 L5
Finsbury 23 G3
Finstall 29 J1
Finsthwaite 55 G1
Finstock 30 E7
Finstown 89 C6
Fintry Aber 85 M5
Fintry Stir 74 E2
Finwood 30 C2
Fionnphort 78 D8
Fir Tree 62 B3
Firbank 61 H7
Firbeck 51 H4
Firby NYorks 57 H1
Firby NYorks 58 D3
Firgrove 49 J1
Firs Lane 49 F2
Firsby 53 H6
Firsdown 10 D1
Firth 89 N5
Fishbourne IoW 11 G5
Fishbourne WSuss 12 B6
Fishburn 62 D3
Fishcross 75 H1
Fisher 12 B6
Fisher's Pond 11 F2
Fisher's Row 55 H5
Fisherford 85 L7
Fishers Farm Park wSuss RH14 0EG 12 D4
Fisherstreet 12 C3
Fisherton High 84 B5
Fisherton SAyr 67 G2
Fisherton de la Mere 9 J1
Fishguard (Abergwaun) 16 C2

Fishlake 51 J1
Fishleigh Barton 6 D3
Fishley 45 J4
Fishnish 78 H6
Fishpond Bottom 8 C5
Fishponds 19 K4
Fishpool 49 G2
Fishtoft 43 G1
Fishtoft Drove 43 F1
Fishtown of Usan 81 N5
Fishwick 77 H5
Fiskavaig 82 D7
Fiskerton Lincs 52 D5
Fiskerton Notts 51 K7
Fitling 59 J6
Fittleton 20 E7
Fittleworth 12 D5
Fitton End 43 H4
Fitz 38 D4
Fitzhead 7 K3
Fitzroy 7 K3
Fitzwilliam 51 G1
Fitzwilliam Museum Cambs CB2 1RB 123 Cambridge
Five Acres 28 E7
Five Ash Down 13 H4
Five Ashes 13 J4
Five Bridges 29 F4
Five Houses 11 F6
Five Lanes 19 H2
Five Oak Green 23 K7
Five Oaks Chanl 3 K7
Five Oaks WSuss 12 C6
Five Roads 17 H5
Five Turnings 28 B1
Five Wents 14 D2
Fivehead 8 C2
Fivelanes 4 C2
Flack's Green 34 B7
Flackwell Heath 22 B3
Fladbury 29 J4
Fladdabister 89 N9
Flagg 50 D6
Flamborough 59 J2
Flamborough Cliffs Nature Reserve ERid YO15 1BJ 59 J2
Flamingo Land Theme Park NYorks YO17 6UX 58 D1
Flamingo Park, Hastings ESuss TN34 3AR 14 D7
Flamstead 32 D7
Flamstead End 23 G1
Flansham 12 C6
Flanshaw 57 J7
Flasby 56 E4
Flash 50 C6
Flashader 82 D5
Flask Inn 63 K6
Flatts Lane Woodland Country Park R&C TS6 0NN 117 E4
Flaunden 22 D1
Flawborough 42 A1
Flawith 57 K3
Flax Bourton 19 J5
Flax Moss 56 C7
Flaxby 57 J4
Flaxholme 41 F1
Flaxlands 45 F6
Flaxley 29 F7
Flaxpool 7 K2
Flaxton 58 C3
Fleckney 41 J6
Flecknoe 31 G2
Fledborough 52 B5
Fleet Hants 22 B6
Fleet Lincs 43 G3
Fleet Air Arm Museum Som BA22 8HT 8 E2
Fleet Hargate 43 G3
Fleetville 22 E1
Fleetwood 55 G5
Fleggburgh (Burgh St. Margaret) 45 J4
Flemingston 18 D4
Flemington 74 E5
Flempton 34 C2
Fleoideabhagh 88 F9
Fletcher All Saints 34 C2
Fletchersbridge 4 B4
Fletchertown 60 D2
Fletching 13 H4
Fleur-de-lis 18 E2
Flexbury 6 A5
Flexford 22 C7
Flimby 60 B3
Flimwell 14 C4
Flint (Y Fflint) 48 B5
Flint Mountain 48 B5
Flint's Green 40 E7
Flinton 59 J6
Flishinghurst 14 C4
Flitcham 44 B3
Flitholme 61 J5
Flitton 32 D5
Flitwick 32 D5
Flixborough 52 B1
Flixton GtMan 49 G3
Flixton NYorks 59 G2
Flixton Suff 45 H7
Flockton 50 E1
Flockton Green 50 E1
Flodden 77 H7
Flodigarry 82 E3
Flood's Ferry 43 G6
Flookburgh 55 G2
Flordon 45 F6
Flore 31 H2
Flotta 89 C8
Flotterton 71 F3
Flowton 34 E4
Flushdyke 57 H7
Flushing Corn 2 E6
Flushing Corn 3 F5
Fluxton 7 J6
Flyford Flavell 29 J3
Foals Green 35 G1
Fobbing 24 D3
Fochabers 84 H5
Fochriw 18 E1
Fockerby 58 D7
Fodderletter 84 F8
Fodderty 83 R5
Foddington 8 E2
Foel 37 J4
Foelgastell 17 J4
Foffarty 81 K6
Fogo 77 F6
Fogorig 77 F6
Fogwatt 84 G5
Foindle 86 D5
Folda 80 H4
Fole 40 C2
Foleshill 41 F7
Folke 9 F3
Folkestone 15 H4
Folkingham 42 D2
Folkington 13 J6
Folksworth 42 E7
Folkton 59 G2
Folla Rule 85 M7
Follifoot 57 J4
Folly Dorset 9 G4
Folly Pembs 16 C3

Folly Farm, Begelly Pembs SA68 0XA 16 E5
Folly Gate 6 D6
Fonmon 18 D5
Fonthill Bishop 9 J1
Fonthill Gifford 9 J1
Fontmell Magna 9 H3
Fontmell Parva 9 H3
Fontwell 12 C6
Font-y-gary 18 D5
Foolow 50 D5
Footherley 40 D5
Foots Cray 23 H4
Force Forge 60 E7
Force Green 23 H6
Forcett 62 B5
Forches Cross 7 F5
Ford A&B 79 K10
Ford Bucks 22 A1
Ford Devon 5 F5
Ford Devon 6 C3
Ford Devon 6 D3
Ford Glos 30 B6
Ford Mersey 48 C3
Ford Midlo 76 A4
Ford N'umb 77 H7
Ford Pembs 16 C3
Ford Plym 4 E5
Ford Shrop 38 D4
Ford Som 7 J3
Ford Staffs 50 C7
Ford Wilts 20 B4
Ford End 33 K7
Ford Green 55 H5
Ford Heath 38 D4
Ford Street 7 K4
Forda 6 D6
Fordbridge 40 D7
Fordcombe 23 J7
Fordell 75 K2
Forden (Forddun) 38 B5
Forder Green 5 H4
Fordgate 8 C1
Fordham Cambs 33 K1
Fordham Essex 34 D6
Fordham Norf 44 A6
Fordham Abbey 33 K2
Fordham Heath 34 D6
Fordhouses 40 B5
Fordingbridge 10 C3
Fordon 59 G2
Fordoun 81 N3
Ford's Green 34 E2
Fordstreet 34 D6
Fordwells 30 E7
Fordwich 15 G2
Fordyce 85 K4
Forebridge 40 B3
Foredale 56 D3
Foreland 72 A4
Foremark 41 F3
Forest Coal Pit 28 B6
Forest Gate 23 H3
Forest Green 22 E7
Forest Hall Cumb 61 G6
Forest Hall T&W 71 H7
Forest Head 61 G1
Forest Hill Oxon 21 J1
Forest Lane Head 57 J4
Forest Lodge 80 E3
Forest Row 13 H3
Forest Side 11 F6
Forest Town 51 H6
Forestburn Gate 71 F4
Forest-in-Teesdale 61 K4
Forestmill 75 H1
Forestside 11 J3
Forfar 81 K6
Forgandenny 80 F9
Forge 37 G6
Forge 84 H5
Forhill 30 B1
Formby 48 B2
Forncett End 45 F6
Forncett St. Mary 45 F6
Forncett St. Peter 45 F6
Forneth 80 F6
Fornham All Saints 34 C2
Fornham St. Martin 34 C2
Forres 84 E5
Forrest 75 G4
Forrest Lodge 67 K5
Forsbrook 40 B1
Forse 87 M5
Forsinard 87 L5
Forston 9 F5
Fort Augustus (Cille Chuimein) 83 P10
Fort Fun, Eastbourne ESuss BN22 7LQ 13 K6
Fort George 84 B5
Fort William (An Gearasdan) 79 N3
Forter 80 H4
Forteviot 80 F9
Forth 75 H5
Forthampton 29 H5
Fortingall 80 C6
Forton Hants 21 H7
Forton Lancs 55 H4
Forton Shrop 38 D4
Forton Som 8 C4
Forton Staffs 39 G3
Fortrie 85 L6
Fortrose (A'Chananaich) 84 B5
Fortuneswell 9 F7
Forty Green 22 C2
Forty Hill 23 G2
Forward Green 34 E3
Fosbury 21 G6
Foscot 30 D6
Fosdyke 43 G2
Foss 80 C5
Foss Cross 20 D1
Fossdale 61 K7
Fossebridge 30 B7
Foster Street 23 H1
Fosterhouses 51 J1
Foster's Booth 31 H3
Foston Derbys 40 D2
Foston Leics 41 J6
Foston Lincs 42 B1
Foston NYorks 58 C3
Foston on the Wolds 59 H4
Fotherby 53 G3
Fothergill 60 B3
Fotheringhay 42 D6
Foubister 89 E7
Foul Mile 13 K5
Foula 89 H9
Foula Airstrip 89 H9
Foulbog 69 H3
Foulden Norf 44 B6
Foulden ScBord 77 H5
Fouldes Island 25 G2
Foulridge 56 D5
Foulsham 44 E3
Foulstone 55 J1
Foulzie 85 M4
Fountainhall 76 C6
Four Ashes Staffs 40 A7
Four Ashes Staffs 40 B6
Four Ashes Suff 34 E1
Four Crosses Denb 37 K1

Four Crosses Powys 37 K5
Four Crosses Powys 38 B4
Four Crosses Staffs 40 B5
Four Elms 23 H7
Four Forks 8 B1
Four Gotes 43 H4
Four Lane Ends ChesW&C 48 E6
Four Lane Ends York 58 C4
Four Lanes 2 D5
Four Marks 11 H1
Four Mile Bridge 46 A5
Four Oaks ESuss 14 D5
Four Oaks Glos 29 F6
Four Oaks WMid 40 D7
Four Oaks Park 40 D6
Four Roads 17 H5
Four Throws 14 C5
Fourlane Ends 51 F7
Fourlanes End 49 H7
Fourstones 70 D7
Fovant 10 B2
Foveran 85 P8
Fowey 4 B5
Fowlis 81 J7
Fowlis Wester 80 E8
Fowlmere 33 H4
Fownhope 28 E5
Fox Hatch 23 J2
Fox Lane 22 B6
Fox Street 34 E6
Fox Up 56 D2
Foxbar 74 C4
Foxcombe Hill 21 H1
Foxcote Glos 30 B7
Foxcote Som 20 A6
Foxdale 54 B6
Foxearth 34 C4
Foxfield 55 F1
Foxham 20 C4
Foxhole 3 G3
Foxholes 59 G2
Foxhunt Green 13 J5
Foxley Here 28 D4
Foxley Norf 44 E3
Foxley N'hants 31 H3
Foxley Wilts 20 B3
Foxt 40 C1
Foxton Cambs 33 H4
Foxton Dur 62 D4
Foxton Leics 41 J6
Foxup 56 D2
Foxwist Green 49 F6
Foy 28 E6
Foyers (Foithir) 83 Q8
Fraddam 2 C5
Fraddon 3 G3
Fradley 40 D4
Fradswell 40 B2
Fraisthorpe 59 H3
Framfield 13 H4
Framingham Earl 45 G5
Framingham Pigot 45 G5
Framlingham 35 G2
Frampton Dorset 9 F5
Frampton Lincs 43 G2
Frampton Cotterell 19 K3
Frampton Mansell 20 C1
Frampton on Severn 20 A1
Frampton West End 43 F1
Framsden 35 G3
France Lynch 20 C1
Franche 29 H1
Frandley 49 F5
Frankby 48 B4
Frankfort 45 H3
Frankley 40 B7
Franksbridge 28 A3
Frankton 31 F1
Frant 13 J3
Fraserburgh 85 P4
Frating 34 E6
Fratton 11 H4
Freasley 40 E6
Freathy 4 D5
Freckenham 33 K1
Freckleton 55 H7
Freeby 42 B3
Freefolk 21 H7
Freehay 40 C1
Freeland 31 F7
Freester 89 N7
Freethorpe 45 J5
Freethorpe Common 45 J5
Freiston 43 G1
Freiston Shore 43 G1
Fremington Devon 6 D2
Fremington NYorks 62 A7
French Brothers Cruises W&M SL4 5JH 22 C4
Frenchay 19 K4
Frenchbeer 6 E7
Frenich 79 R10
Frensham 22 B7
Fresgoe 87 M3
Freshbrook 20 E3
Freshfield 48 B2
Freshford 20 A5
Freshwater 10 E6
Freshwater East 16 D6
Fressingfield 35 G1
Freston 35 F5
Fretherne 20 A1
Frettenham 45 G4
Freuchie 80 H10
Freystrop Cross 16 C4
Friars Carse 68 E5
Friar's Gate 13 H3
Friars, the Aylesford Kent ME20 7BX 14 C2
Friday Bridge 43 H5
Friday Street ESuss 13 K6
Friday Street Suff 35 H3
Friday Street Suff 35 H3
Friday Street Surr 22 E7
Fridaythorpe 58 E4
Friern Barnet 23 F2
Friesthorpe 52 D4
Frieston 42 C1
Frilford 21 H2
Frilsham 21 J4
Frimley 22 B6
Frimley Green 22 B6
Frindsbury 24 D5
Fring 44 B2
Fringford 31 H6
Frinsbury 14 D2
Frinton-on-Sea 35 G7
Friockheim 81 L6
Friog 37 F2
Frisby on the Wreake 41 J4
Friskney 53 H7
Friskney Eaudyke 53 H7
Friston ESuss 13 J7
Friston Suff 35 J2
Fritchley 51 F7

Frith 14 E2
Frith Bank 43 G1
Frith Common 29 F2
Fritham 10 D3
Frithelstock 6 C4
Frithelstock Stone 6 C4
Frithville 53 G7
Frittenden 14 D3
Frittiscombe 5 J6
Fritton Norf 45 G6
Fritton Norf 45 J5
Fritwell 31 G6
Frizinghall 57 G6
Frizington 60 B5
Frocester 20 A1
Frochas 38 B5
Frodesley 38 E5
Frodesley Lane 38 E5
Frodingham 52 B1
Frodsham 48 E5
Frog End 33 H5
Frog Pool 29 G2
Frogden 70 C1
Froggatt 50 E5
Froghall 40 C1
Frogham 10 C3
Frogland Cross 19 K3
Frogmore Devon 5 H6
Frogmore Hants 22 B6
Frogmore Herts 22 E1
Frogwell 4 D4
Frolesworth 41 H6
Frome 20 A7
Frome Market 20 B6
Frome St. Quintin 8 E4
Frome Whitfield 9 F5
Fromes Hill 29 F4
Fron Gwyn 36 C2
Fron Powys 27 K2
Fron Powys 38 A6
Fron Powys 38 B5
Fron Isaf 38 B2
Froncysyllte 38 B1
Fron-goch 37 J2
Frostenden 45 J7
Frosterley 62 A3
Froxfield 21 G5
Froxfield Green 11 J2
Fruitmarket Gallery Edin EH1 1DF 126 G4
Fryerning 24 C1
Fugglestone St. Peter 10 C1
Fulbeck 52 C7
Fulbourn 33 J3
Fulbrook 30 D7
Fulflood 11 F2
Fulford Som 8 B2
Fulford Staffs 40 B2
Fulford York 58 C5
Fulham 23 F4
Fulking 13 F5
Full Sutton 58 D4
Fullaford 6 E2
Fuller Street 34 B7
Fuller's Moor 48 D7
Fullerton 10 E1
Fulletby 53 F5
Fullwood 74 C5
Fulmer 22 D3
Fulmodeston 44 D2
Fulnetby 52 E5
Fulready 30 D4
Fulstone 50 D2
Fulstow 53 G3
Fulwell Oxon 30 E6
Fulwell T&W 62 D1
Fulwood Lancs 55 J6
Fulwood SYorks 51 F4
Fun Farm, Spalding Lincs PE12 6JU 43 F3
Fundenhall 45 F6
Fundenhall Street 45 F6
Funtington 12 B6
Funtley 11 G4
Funzie 89 Q3
Furley 8 B4
Furnace A&B 79 M10
Furnace Carmar 17 J5
Furnace Cere 37 F6
Furnace End 40 E6
Furner's Green 13 H4
Furness Vale 50 C4
Furneux Pelham 33 H6
Furnham 8 C4
Further Quarter 14 D4
Furtho 31 J4
Furze Green 45 G7
Furze Platt 22 B3
Furzehill Devon 7 F1
Furzehill Dorset 10 B4
Furzeley Corner 11 H3
Furzey Lodge 10 E4
Furzley 10 D3
Fyfett 8 B3
Fyfield Essex 23 J1
Fyfield Glos 21 F1
Fyfield Hants 21 F7
Fyfield Oxon 21 H2
Fyfield Wilts 20 E5
Fylingthorpe 63 K6
Fyning 12 B4
Fyvie 85 M7

G

Gabalfa 18 E4
Gabhsann Bho Dheas 88 K2
Gabhsann Bho Thuath 88 K2
Gabroc Hill 74 C5
Gaddesby 41 J4
Gaddesden Row 32 D7
Gadebridge 22 D1
Gadshill 24 D4
Gaer Newport 19 F3
Gaer Powys 28 A6
Gaer-fawr 19 H2
Gaerllwyd 19 H2
Gaerwen 46 C5
Gagingwell 31 F6
Gaick Lodge 80 C2
Gailes 74 B7
Gailey 40 B4
Gainford 62 B5
Gainsborough 52 B3
Gainsford End 34 B5
Gairloch (Geàrrloch) 82 H3
Gairlochy (Geàrr Lòchaidh) 79 N2
Gairney Bank 75 K1
Gaitsgill 60 E2
Galabank 76 C6
Galashiels 76 C7
Galdenoch 64 B4
Gale 49 J1
Galgate 55 H4
Galhampton 9 F2
Gallantry Bank 48 E7
Gallatown 76 A1
Gallchoille 73 F1
Gallery of Modern Art Glas G1 3AH 127 F4
Galley Common 41 F6
Galleyend 24 D1
Galleywood 24 D1
Galloway Forest Dark Sky Park D&G 67 J5
Gallowfauld 81 K6
Gallowhill 74 C4
Gallows Green 40 C1
Gallowstree Common 21 K3
Gallowstree Elm 40 A7
Galltair 83 J8
Gallt-y-foel 46 D6
Gallypot Street 13 H3
Galmington 8 B2
Galmisdale 78 F2
Galmpton Devon 5 G6
Galmpton Torbay 5 J5
Galmpton Warborough 5 J5
Galphay 57 H2
Galston 74 C7
Galtrigill 82 B5
Gamble's Green 34 B7
Gamblesby 61 H3
Gamelsby 60 D1
Gamesley 50 C3
Gamlingay 33 F3
Gamlingay Cinques 33 F3
Gamlingay Great Heath 33 F3
Gammaton 6 C3
Gammaton Moor 6 C3
Gammersgill 57 F1
Gamston Notts 41 J2
Gamston Notts 51 K5
Ganarew 28 E7
Gang 4 D4
Ganllwyd 37 G3
Gannochy 81 M3
Ganstead 59 H6
Ganthorpe 58 C2
Ganton 59 F2
Ganwick Corner 23 F2
Gaodhail 78 H7
Gappah 5 J3
Gara Bridge 5 H5
Garbat 83 Q4
Garbhallt 73 J1
Garboldisham 44 E7
Garden 74 D1
Garden City 48 C6
Garden Village 50 E3
Gardeners Green 22 B5
Gardenstown 85 M4
Garderhouse 89 M8
Gardham 59 F5
Gare Hill 20 A7
Garelochhead 74 A1
Garford 21 H2
Garforth 57 K6
Gargrave 56 E4
Gargunnock 75 F1
Gariob 73 F2
Garlic Street 45 G7
Garlies Castle 64 E4
Garlieston 64 E6
Garlinge Green 15 G2
Garlogie 85 M10
Garmelow 40 A3
Garmond 85 N5
Garmony 78 H6
Garmouth 84 H4
Garmston 39 F5
Garnant 17 K4
Garndolbenmaen 36 D1
Garneddwen 37 G5
Garnett Bridge 61 G7
Garnfadryn 36 B2
Garnswllt 17 K5
Garrabost 88 L4
Garrachra 73 J2
Garras 2 E6
Garreg 37 F1
Garreg Bank 38 B4
Garrett's Green 40 D7
Garrick 80 D9
Garrigill 61 J2
Garriston 62 B7
Garroch 67 K5
Garrochty 73 J5
Garros 82 E4
Garrow 80 D7
Garryhorn 67 K4
Garynahine (Gearraidh na h-Aibhne) 88 H4
Garsdale 56 C1
Garsdale Head 61 J7
Garsdon 20 C3
Garshall Green 40 B2
Garsington 21 J1
Garstang 55 H5
Garston 48 D4
Garswood 48 E3
Gartachoil 74 D1
Gartavaich 73 G5
Gartbreck 72 A5
Gartcosh 74 E4
Garth Bridgend 18 B2
Garth Cere 37 F7
Garth Gwyn 46 D5
Garth IoM 54 C6
Garth Powys 27 J4
Garth Wrex 38 B1
Garth Row 61 G7
Garthbrengy 27 K5
Gartheli 26 E3
Garthmyl 38 A6
Garthorpe Leics 42 B3
Garthorpe NLincs 52 B1
Garths 61 G7
Garthynty 27 G4
Gartly 85 K7
Gartmore 74 D1
Gartnagrenach 73 F5
Gartnatra 72 B4
Gartness 74 D2
Gartocharn 74 C2
Garton 59 J6
Garton-on-the-Wolds 59 F3
Gartymore 87 N8
Garvald 76 D3
Garvamore 83 R11
Garvan 79 L3
Garvard 72 B1
Garve (Clachan Ghairbh) 83 P4
Garveld 66 A3
Garvestone 44 E5
Garvie 73 J2
Garvock 74 A3
Garwald 69 H3
Garwaldwaterfoot 69 H3
Garway 28 D6
Garway Hill 28 D6
Gass 67 J3
Gastard 20 B5
Gasthorpe 44 D7
Gaston Green 33 J7
Gatcombe 11 F6
Gate Burton 52 B4
Gate Helmsley 58 C4
Gate House 72 D3
Gateacre 48 D4
Gateford 51 H4
Gateforth 58 B7
Gatehead 74 B7
Gatehouse of Fleet 65 G5
Gatelawbridge 68 E4
Gateley 44 D3
Gatenby 57 J1
Gatesgarth 60 C5
Gateshaw 70 C1
Gateshead 71 H7
Gatesheath 48 D6
Gateside Angus 81 K6
Gateside Fife 80 G10
Gateside NAyr 74 B5
Gateslack 68 D3
Gathurst 48 E2
Gatley 49 H4
Gattonside 76 D7
Gatwick Airport 23 F7
Gaufron 27 J2
Gaulby 41 J5
Gauldry 81 J8
Gauntons Bank 38 E1
Gaunt's Common 10 B4
Gaunt's Earthcott 19 K3
Gautby 52 E5
Gavinton 77 F5
Gawber 51 F2
Gawcott 31 H5
Gawsworth 49 H5
Gawthrop 56 B1
Gawthwaite 55 F1
Gay Bowers 24 D1
Gay Street 12 D4
Gaydon 30 E3
Gayhurst 32 B4
Gayle 56 D1
Gayles 62 B6
Gayton Mersey 48 B4
Gayton Norf 44 B4
Gayton N'hants 31 J3
Gayton Staffs 40 B3
Gayton le Marsh 53 H4
Gayton le Wold 53 F4
Gayton Thorpe 44 B4
Gaywood 44 A3
Gazeley 34 B2
Gearach 72 A5
Gearnsary 87 K6
Gearraidh na Monadh 88 B7
Geary 82 C4
Gedding 34 D3
Geddington 42 B7
Gedgrave Hall 35 J4
Gedling 41 J1
Gedney 43 H3
Gedney Broadgate 43 H3
Gedney Drove End 43 H3
Gedney Dyke 43 H3
Gedney Hill 43 G4
Gee Cross 49 J3
Geilston 74 B3
Geinnis 88 B4
Geisiadar 88 G4
Geldeston 45 H6
Gell Conwy 47 G6
Gell Gwyn 36 D2
Gelli 18 C2
Gelli Aur Country Park Carmar SA32 8LR 17 K3
Gelli Gynan 47 K7
Gellideg 18 D1
Gellifor 47 K6
Gelligaer 18 E2
Gellilydan 37 F2
Gellioedd 37 J1
Gelly 16 D4
Gellyburn 80 F7
Gellywen 17 F3
Gelston D&G 65 H5
Gelston Lincs 42 C1
Gembling 59 H4
Genoch 64 B5
Genoch Square 64 B5
Gentleshaw 40 C4
Geocrab 88 G8
George Green 22 D3
George Nympton 7 F3
Georgeham 6 C2
Georgetown 74 C4
Gerlan 46 E6
Germansweek 6 C6
Germoe 2 C6
Gerrans 3 F5
Gerrards Cross 22 D3
Gestingthorpe 34 C5
Geuffordd 38 B4
Geufron 37 H7
Gibbet Hill 20 A7
Gibbshill 65 H3
Gibraltar Lincs 53 J7
Gibraltar Suff 35 F3
Gibraltar Point National Nature Reserve Lincs PE24 4SU 53 J7
Gibside T&W NE16 6BG 116 A6
Giddeahall 20 B4
Giddy Green 9 H6
Gidea Park 23 J3
Gidleigh 6 E7
Giffnock 74 D5
Gifford 76 D4
Giffordland 74 A6
Giffordtown 80 H9
Giggleswick 56 D3
Gigha 72 E6
Gilberdyke 58 E7
Gilbert's End 29 H4
Gilchriston 76 C4
Gilcrux 60 C3
Gildersome 57 H7
Gildingwells 51 H4
Gilesgate Moor 62 C2
Gileston 18 D5
Gilfach 18 E2
Gilfach Goch 18 C3
Gilfachrheda 26 D3
Gilgarran 60 B4
Gill 60 F4
Gillamoor 58 C1
Gillenbie 69 G5
Gillfoot 65 K3
Gilling East 58 C2
Gilling West 62 B6
Gillingham Dorset 9 H2
Gillingham Med 24 D5
Gillingham Norf 45 J6
Gillock 87 Q4
Gillow Heath 49 H7
Gills 87 R2
Gill's Green 14 C4
Gilmanscleuch 69 J1
Gilmerton Edin 76 A4
Gilmerton P&K 80 D8
Gilmilnscroft 67 K1
Gilmonby 61 L5
Gilmorton 41 H7
Gilsland 70 B7
Gilsland Spa 70 B7
Gilson 40 D6
Gilstead 57 G6
Gilston 76 C5
Gilston Park 33 H7
Giltbrook 41 G1
Gilwern 28 B7
Gimingham 45 G2
Gin Pit 49 F2
Ginclough 49 J5
Ginger's Green 13 K5
Giosla 88 G5
Gipping 34 E2
Gipsey Bridge 43 F1
Girlsta 89 N7
Girsby 62 D6
Girtford 32 E4
Girthon 65 G5
Girton Cambs 33 H2
Girton Notts 52 B6
Girvan 67 F4
Gisburn 56 D5
Gisburn Cotes 56 D5
Gisleham 45 K7
Gislingham 34 E1
Gissing 45 F7
Gittisham 7 K6
Givons Grove 22 E6
Gladestry 28 B3
Gladsmuir 76 C3
Glaic 73 J3
Glais 18 A1
Glaisdale 63 H6
Glaister 73 H7
Glamis 81 J6
Glamis Castle Angus DD8 1QJ 81 J6
Glan Conwy 47 G7
Glanaber Terrace 37 G1
Glanaman 17 K4
Glanbran 27 H5
Glan-Denys 26 E3
Glandford 44 E1
Glan-Duar 17 J1
Glandwr 16 E3
Glan-Dwyfach 36 D1
Glangrwyney 28 B7
Glanllynfi 18 B2
Glanmule 38 A6
Glan-rhyd NPT 18 A1
Glanrhyd Pembs 16 E1
Glanton 71 F2
Glanton Pyke 71 F2
Glanvilles Wootton 9 F4
Glanwern 37 F7
Glanwydden 47 G4
Glan-y-don 47 K5
Glan-y-llyn 18 E3
Glan-y-nant 37 J7
Glan-yr-afon Gwyn 37 J1
Glan-yr-afon Gwyn 37 K1
Glan-yr-afon IoA 46 E4
Glan-y-Wern 37 F2
Glapthorn 42 D6
Glapwell 51 G6
Glasbury 28 A5
Glascoed Mon 19 G1
Glascoed Wrex 48 B7
Glascote 40 E5
Glascwm 28 A3
Glasfryn 47 H7
Glasgow (Glaschu) 74 D4
Glasgow Airport 74 C4
Glasgow Botanic Garden Glas G12 0UE 118 E3
Glasgow Prestwick Airport 67 H1
Glasgow Royal Concert Hall Glas G2 3NY 127 F3
Glasgow Science Centre Glas G51 1EA 127 A4
Glasinfryn 46 D6
Glasnacardoch 82 G11
Glasnakille 82 F9
Glaspant 17 F2
Glaspwll 37 G6
Glass Studio, Eastbourne Pier ESuss BN21 3EL 125 Eastbourne
Glassel 85 L11
Glassenbury 14 C4
Glasserton 64 E7
Glassford 75 F6
Glasshouse 29 G6
Glasshouses 57 G3
Glasslie 80 H10
Glasson Cumb 69 H7
Glasson Lancs 55 H4
Glasson Dock 55 H4
Glassonby 61 G3
Glasterlaw 81 L5
Glaston 42 B5
Glastonbury 8 D1
Glastonbury Abbey Som BA6 9EL 8 E1
Glatton 42 E7
Glazebrook 49 F3
Glazebury 49 F3
Glazeley 39 G7
Gleadless 51 F4
Gleadsmoss 49 H6
Gleaston 55 F2
Glecknabae 73 J4
Gledhow 57 J6
Gledrid 38 B2
Glemsford 34 C4
Glen D&G 65 F5
Glen D&G 65 J3
Glen Auldyn 54 D4
Glen Mona 54 D5
Glen Parva 41 H6
Glen Trool Lodge 67 J5
Glen Village 75 G3
Glen Vine 54 C6
Glenae 68 E5
Glenaladale 79 K3
Glenald 74 A1
Glenapp Castle 66 E5
Glenarm 81 J4
Glenbarr 72 E7
Glenbatrick 72 D3
Glenbeg 78 G4
Glenbervie Aber 81 N2
Glenbervie Falk 75 G2
Glenboig 75 F4
Glenborrodale 78 H4
Glenbranter 73 K1
Glenbreck 69 F1
Glenbrittle 82 E8
Glenburn 74 C4
Glencaple 65 K4
Glencarse 80 G8
Glencloy 73 J7
Glencoe (A' Chàrnaich) 79 N5
Glencoe Visitor Centre High PH49 4LA 79 N5
Glencraig 75 K1
Glencrosh 68 C5
Glendearg D&G 69 H3
Glendearg ScBord 76 D7
Glendessary 83 K11
Glendevon 80 E10
Glendoebeg 83 Q10
Glendoick 80 H8
Glendoll Lodge 80 H3
Glendoune 67 F4
Glendrissaig 67 F4
Glenduisk 67 G5
Glendye Lodge 81 M2
Gleneagles Hotel 80 E9
Gleneagles House 80 E10
Glenegedale 72 B5
Glenelg (Gleann Eilg) 83 J9
Glenfarg 80 G9
Glenfield 41 H5
Glenfinnan (Gleann Fhionnainn) 79 L2
Glengap 65 G5
Glengarnock 74 B5
Glengarrisdale 72 E1
Glengennet 67 G4
Glengrasco 82 E6
Glenhead 68 D5
Gleniffer Braes Country Park Renf PA2 8TE 118 A6
Glenkerry 69 H2
Glenkiln 73 J7
Glenkin 73 K2
Glenkindie 85 J9
Glenlair 65 H3
Glenlean 73 J2
Glenlee Angus 81 K3
Glenlee D&G 68 B5
Glenlochar 65 H4
Glenluce 64 B5
Glenmallan 74 A1
Glenmanna 68 C3
Glenmavis 75 F4
Glenmaye 54 B6
Glenmeanie 83 N5
Glenmore A&B 73 J4
Glenmore High 82 E6
Glenmore Forest Park Visitor Centre High PH22 1QY 84 D9
Glenmore Lodge 84 D10
Glenmoy 81 K4
Glenochar 68 E2
Glenprosen Village 81 J4
Glenramskill 66 B2
Glenrazie 64 D4
Glenridding 60 E5
Glenrisdell 73 G5
Glenrothes 80 H10
Glenshellish 73 K1
Glensluain 73 J1
Glentaggart 68 D1
Glentham 52 D3
Glenton 85 L8
Glentress 76 A7
Glentress 7stanes ScBord EH45 8NB 76 A6
Glentress Forest Visitor Centre ScBord EH45 8NB 76 A6
Glentrool 64 D3
Glentruan 54 D3
Glentworth 52 C4
Glenuig 78 H3
Glenurquhart 84 B4
Glenwhilly 64 B3
Glespin 68 D1
Gletness 89 N7
Glewstone 28 E6
Glinton 42 E5
Glogue 17 F2
Glooston 42 A6
Glororum 77 K7
Glossop 50 C3
Gloster Hill 71 H3
Gloucester 29 H7
Gloucester Cathedral Glos GL1 2LR 128 Gloucester
Gloucester Historic Docks Glos GL1 2ER 128 Gloucester
Gloucestershire & Warwickshire Railway Glos GL54 5DT 30 B6
Gloucestershire Airport 29 H6
Gloup 89 P2
Gloweth 2 E4
Glusburn 57 F5
Glympton 31 F6
Glyn 47 F7
Glyn Ceiriog 38 B2
Glynarthen 17 G1
Glyncoch 18 D2
Glyncorrwg 18 B2
Glyn-Cywarch 37 F2
Glynde 13 H6
Glyndebourne 13 H5
Glyndyfrdwy 38 A1
Glynneath (Glyn-Nedd) 18 B1
Glynogwr 18 C3
Glyntaff 18 D3
Gnosall 40 A3
Gnosall Heath 40 A3
Go Bananas, Colchester Essex CO1 1BX 34 D6
Goadby 42 A6
Goadby Marwood 42 A3
Goatacre 20 D4
Goathill 9 F3
Goathland 63 J6
Goathurst 8 B1
Gobernuisgeach 87 M6
Gobowen 38 C2
Godalming 22 C7
Goddard's Corner 35 G2
Goddards Green 13 F4
Godden Green 23 J6
Goddington 23 H5
Godford Cross 7 K5
Godington 31 H6
Godleybrook 40 B1
Godmanchester 33 F1
Godmanstone 9 F5
Godmersham 15 F2
Godney 19 H7
Godolphin Cross 2 D5
Godor 38 B4
Godre'r-graig 18 A1
Godshill Hants 10 C3
Godshill IoW 11 G6
Godstone 23 G6
Godstone Farm Surr RH9 8LX 23 G6
Godwick 44 D3
Goetre 19 G1
Goff's Oak 23 G1
Gogar 75 K3
Gogarth 47 F4
Goginan 37 F7
Goirtein 73 H2
Golan 36 E1
Golant 4 B5
Golberdon 4 D3
Golborne 49 F3
Golcar 50 D1
Gold Hill Cambs 43 J6
Gold Hill Dorset 9 H3
Goldcliff 19 G3
Golden Acre Park WYorks LS16 8BQ 57 H5
Golden Cross 13 J5
Golden Green 23 K7
Golden Grove 17 J4
Golden Pot 22 A7
Golden Valley Derbys 51 G7
Golden Valley Glos 29 J6
Goldenhill 49 H7
Golders Green 23 F3
Goldhanger 25 F1
Goldielea 65 K3
Golding 38 E5
Goldington 32 D3
Goldsborough NYorks 57 J4
Goldsborough NYorks 63 J5
Goldsithney 2 C5
Goldstone 39 G3
Goldthorn Park 40 B6
Goldthorpe 51 G2
Goldworthy 6 B3
Goleulong 2000 Lightship Cardiff CF10 4PA 95 C6
Golford 14 C4
Gollinglith Foot 57 G1
Golspie 84 C1
Golval 87 L3
Gomeldon 10 C1
Gomersal 57 H7
Gometra 78 E6
Gomshall 22 D7
Gonachan Cottage 74 E2
Gonalston 41 J1
Gonerby Hill Foot 42 C2
Gonfirth 89 M6
Good Easter 33 K7
Gooderstone 44 B5
Goodleigh 6 E2
Goodmanham 58 E5
Goodmayes 23 H3
Goodnestone Kent 15 H2
Goodnestone Kent 25 G5
Goodrich 28 E7
Goodrington 5 J5
Goodshaw 56 D7
Goodshaw Fold 56 D7
Goodwick (Wdig) 16 C2
Goodworth Clatford 21 G7
Goodyers End 41 F7
Goole 58 D7
Goom's Hill 30 B3
Goonbell 2 E4
Goonhavern 2 E3
Goonvrea 2 E4
Goose Green Essex 35 F6
Goose Green Essex 35 F6
Goose Green GtMan 48 E2
Goose Green Kent 23 K6
Goose Green SGlos 19 K4
Goose Pool 28 D5
Gooseham 6 A4
Goosehill Green 29 J2
Goosewell 5 F5
Goosey 21 G2
Goosnargh 55 J6
Goostrey 49 G5
Gorcott Hill 30 B2
Gorddinog 46 E5
Gordon 76 E6
Gordonbush 87 L9
Gordonstown Aber 85 K5
Gordonstown Aber 85 M7
Gore Cross 20 D6
Gore End 21 H5
Gore Pit 34 C7
Gore Street 25 J5
Gorebridge 76 B4
Gorefield 43 H4
Gorey 3 K7
Gorgie 76 A3
Gorgie City Farm Edin EH11 2LA 120 C2
Goring 21 K3
Goring Heath 21 K4
Goring-by-Sea 12 E6
Gorleston-on-Sea 45 K5
Gorllwyn 17 G2
Gornalwood 40 B6
Gorran Churchtown 3 G4
Gorran Haven 4 A6
Gors 27 F1
Gorsedd 47 K5
Gorseinon 17 J6
Gorseybank 50 E7
Gorsgoch 26 D3
Gorslas 17 J4
Gorsley 29 F6
Gorsley Common 29 F6
Gorstage 49 F5
Gorstan 83 P4
Gorsty Hill 40 D3
Gorteneorn 78 H4
Gorton 49 H3
Gosbeck 35 F3
Gosberton 43 F2
Gosberton Clough 42 E3
Goseley Dale 41 F3
Gosfield 34 B6
Gosford Here 28 E2
Gosford Oxon 31 G7
Gosforth Cumb 60 B6
Gosforth T&W 71 H7
Gosland Green 48 E7
Gosmore 32 E6
Gospel End 40 A6
Gosport 11 H5
Gossabrough 89 P4
Gossington 20 A1
Gossops Green 13 F3
Goswick 77 J6
Gotham 41 H2
Gotherington 29 J6
Gothers 3 G3
Gott 89 N8
Gotton 8 B2
Goudhurst 14 C4
Goulceby 53 F5
Gourdon 81 P3
Gourock 74 A3
Govan 74 D4
Goverton 51 K7
Goveton 5 H6
Govilon 28 B7
Gowdall 58 C7
Gowerton 17 J6
Gowkhall 75 J2
Gowkthrapple 75 F5
Gowthorpe 58 D4
Goxhill ERid 59 H5
Goxhill NLincs 59 H7
Goytre 18 A3
Gozzard's Ford 21 H2
Grabhair 88 J6
Graby 42 D3
Gradbach 49 J6
Grade 2 E7
Gradeley Green 48 E7
Graffham 12 C5
Grafham Cambs 32 E2
Grafham Surr 22 D7
Grafham Water Cambs PE28 0BH 32 E2
Grafton Here 28 D5
Grafton NYorks 57 K3
Grafton Oxon 21 F1
Grafton Shrop 38 D4
Grafton Worcs 28 E2
Grafton Worcs 29 J5
Grafton Flyford 29 J3
Grafton Regis 31 J4
Grafton Underwood 42 C7
Grafty Green 14 D3
Graianrhyd 48 B7
Graig Carmar 17 H5
Graig Conwy 47 G5
Graig Denb 47 J5
Graig-fechan 47 K7
Grain 24 E4
Grainel 72 A4
Grains Bar 49 J2
Grainsby 53 F3
Grainthorpe 53 G3
Graizelound 51 K3
Grampound 3 G4
Grampound Road 3 G3
Granborough 31 J6
Granby 42 A2
Grand Pier, Teignmouth Devon TQ14 8BB 5 K3
Grandborough 31 F2
Grandes Rocques 3 J5
Grandtully 80 E5
Grange Cumb 60 D5
Grange EAyr 74 C7
Grange High 83 P7
Grange Med 24 D5
Grange Mersey 48 B4
Grange de Lings 52 C5
Grange Hill 23 H2
Grange Moor 50 E1
Grange of Lindores 80 H9
Grange Villa 62 C1
Grangemill 50 E7
Grangemouth 75 H2
Grange-over-Sands 55 H2
Grangeston 67 G4
Grangetown Cardiff 18 E4
Grangetown R&C 63 F4
Granish 84 C9
Gransmoor 59 H4
Granston 16 B2
Grantchester 33 H3
Grantham 42 C2
Grantley 57 H2
Granton 76 A3
Granton House 69 F3
Grantown-on-Spey 84 E8
Grantsfield 28 E2
Grantshouse 77 G4
Grappenhall 49 F4
Grasby 52 D2
Grasmere 60 E6
Grass Green 34 B5
Grasscroft 49 J2
Grassendale 48 C4
Grassgarth 60 F7
Grassholme 61 L4
Grassington 57 F3
Grassmoor 51 G6
Grassthorpe 51 K6
Grateley 21 F7
Gratwich 40 C2
Gravel Hill 22 D2
Graveley Cambs 33 F2
Graveley Herts 33 F6
Gravelly Hill 40 D6
Gravels 38 C5
Graven 89 N5
Graveney 25 G5
Gravesend 24 C4
Grayingham 52 C3
Grayrigg 61 G7
Grays 24 C4
Grayshott 12 B3
Grayswood 12 C3
Grazeley 21 K5
Greasbrough 51 G3
Greasby 48 B4
Great Aberystwyth Camera Obscura Cere SY23 2DN 36 E7
Great Abington 33 J4
Great Addington 32 C1
Great Alne 30 C3
Great Altcar 48 C2
Great Amwell 33 G7
Great Asby 61 H5
Great Ashfield 34 D2
Great Ayton 63 F5
Great Baddow 24 D1
Great Bardfield 33 K5
Great Barford 32 E3
Great Barr 40 C6
Great Barrington 30 D7
Great Barrow 48 D6
Great Barton 34 C2
Great Barugh 58 D2
Great Bavington 70 E5
Great Bealings 35 G4
Great Bedwyn 21 F5
Great Bentley 35 F6
Great Bernera 88 G4
Great Billing 32 B2
Great Bircham 44 B2
Great Blakenham 35 F3
Great Bolas 39 F3
Great Bookham 22 E6
Great Bourton 31 F4
Great Bowden 42 A7
Great Bradley 33 K3
Great Braxted 34 C7
Great Bricett 34 E3
Great Brickhill 32 C5
Great Bridgeford 40 A3
Great Brington 31 H2
Great Bromley 34 E6
Great Broughton Cumb 60 B3
Great Broughton NYorks 63 F6
Great Buckland 24 C5
Great Budworth 49 F5
Great Burdon 62 D5
Great Burstead 24 C2
Great Busby 63 F6
Great Cambourne 33 G3
Great Canfield 33 J7
Great Canney 24 E1
Great Carlton 53 H4
Great Casterton 42 D5
Great Chalfield 20 B5
Great Chart 14 E3
Great Chatwell 39 G4
Great Chell 49 H7
Great Chesterford 33 J4
Great Cheverell 20 C6
Great Chishill 33 H5
Great Clacton 35 F7
Great Clifton 60 B4
Great Coates 53 F2
Great Comberton 29 J4
Great Corby 61 F1
Great Cornard 34 C4
Great Cowden 59 J5
Great Coxwell 21 F2
Great Crakehall 57 H1
Great Cransley 32 B1
Great Cressingham 44 C5
Great Crosby 48 C2
Great Crosthwaite 60 D4
Great Cubley 40 D2
Great Cumbrae 73 K5
Great Dalby 42 A4
Great Doddington 32 B2
Great Doward 28 E7
Great Dunham 44 C4
Great Dunmow 33 K6
Great Durnford 10 C1
Great Easton Essex 33 K6
Great Easton Leics 42 B6
Great Eccleston 55 H5
Great Edstone 58 D1
Great Ellingham 44 E6
Great Elm 20 A7
Great Eversden 33 G3
Great Fencote 62 C7
Great Finborough 34 E3
Great Fransham 44 C4
Great Gaddesden 32 D7
Great Gidding 42 E7
Great Givendale 58 E4
Great Glemham 35 H2
Great Glen 41 J6
Great Gonerby 42 B2
Great Gransden 33 F3
Great Green Cambs 33 F4
Great Green Norf 45 G7
Great Green Suff 34 D3
Great Green Suff 34 E1
Great Green Suff 35 F1
Great Habton 58 D2
Great Hale 42 E1
Great Hall Hants SO23 8UJ 137 Winchester
Great Hallingbury 33 J7
Great Hampden 22 B1
Great Harrowden 32 B1
Great Harwood 56 C6
Great Haseley 21 K1
Great Hatfield 59 H5
Great Haywood 40 C3
Great Heath 41 F7
Great Heck 58 B7
Great Henny 34 C5
Great Hinton 20 C6
Great Hockham 44 D6
Great Holland 35 G7
Great Horkesley 34 D5
Great Hormead 33 H5
Great Horton 57 G6
Great Horwood 31 J5
Great Houghton N'hants 31 J3
Great Houghton SYorks 51 G2
Great Hucklow 50 D5
Great Kelk 59 H4
Great Kimble 22 B1
Great Kingshill 22 B2
Great Langton 62 C7
Great Leighs 34 B7
Great Limber 52 E2
Great Linford 32 B4
Great Livermere 34 C1
Great Longstone 50 E5
Great Lumley 62 C2
Great Lyth 38 D5
Great Malvern 29 G4
Great Maplestead 34 C5
Great Marton 55 G6
Great Massingham 44 B3
Great Melton 45 F5
Great Milton 21 K1
Great Missenden 22 B1
Great Mitton 56 C6
Great Mongeham 15 J2
Great Moulton 45 F6
Great Munden 33 G6
Great Musgrave 61 J5
Great Ness 38 D4
Great Notley 34 B6
Great Nurcot 7 H2
Great Oak 19 G1
Great Oakley Essex 35 F6
Great Oakley N'hants 42 B7
Great Offley 32 E6
Great Orme Tramway Conwy LL30 2HG 47 F4
Great Ormside 61 J5
Great Orton 60 E1
Great Ouseburn 57 K3
Great Oxendon 42 A7
Great Oxney Green 24 C1
Great Palgrave 44 C4
Great Parndon 23 H1
Great Paxton 33 F2
Great Plumpton 55 G6
Great Plumstead 45 H5
Great Ponton 42 C2
Great Potheridge 6 D4
Great Preston 57 J7
Great Purston 31 G5
Great Raveley 43 F7
Great Rissington 30 C7
Great Rollright 30 E5
Great Ryburgh 44 D3
Great Ryle 71 F2
Great Ryton 38 D5
Great Saling 34 B6
Great Salkeld 61 G3
Great Sampford 33 K5
Great Sankey 48 E4
Great Saredon 40 B5
Great Saxham 34 B2
Great Shefford 21 G4
Great Shelford 33 H3
Great Smeaton 62 D6
Great Snoring 44 D2
Great Somerford 20 C3
Great St. Mary's Church Cambs CB2 3PQ 123 Cambridge
Great Stainton 62 D4
Great Stambridge 25 F2
Great Staughton 32 E2
Great Steeping 53 H6
Great Stonar 15 J2
Great Strickland 61 G4
Great Stukeley 33 F1
Great Sturton 53 F5
Great Sutton ChesW&C 48 C5
Great Sutton Shrop 38 E7
Great Swinburne 70 E6
Great Tew 30 E6
Great Tey 34 C6
Great Thorness 11 F5
Great Thurlow 33 K4
Great Torr 5 G6
Great Torrington 6 C4
Great Tosson 71 F3
Great Totham Essex 34 C7
Great Totham Essex 34 C7
Great Tows 53 F3
Great Urswick 55 F2
Great Wakering 25 F3
Great Waldingfield 34 D4
Great Walsingham 44 D2
Great Waltham 33 K7
Great Warley 23 J2
Great Washbourne 29 J5
Great Weeke 7 F7
Great Welnetham 34 C3
Great Wenham 34 E5
Great Whittington 71 F6
Great Wigborough 34 D7
Great Wigsell 14 C5
Great Wilbraham 33 J3
Great Wilne 41 G2
Great Wishford 10 B1
Great Witcombe 29 J7
Great Witley 29 G2
Great Wolford 30 D5
Great Wratting 33 K4
Great Wymondley 33 F6
Great Wyrley 40 B5
Great Wytheford 38 E4
Great Yarmouth 45 K5
Great Yeldham 34 B5
Greatford 42 D4
Greatgate 40 C1
Greatham Hants 11 J1
Greatham Hart 62 E4
Greatham WSuss 12 D5
Greatness 23 J6
Greatstone-on-Sea 15 F5
Greatworth 31 G4
Green 47 J6
Green Cross 12 B3
Green End Bed 32 E3
Green End Bucks 32 C5
Green End Cambs 33 F1
Green End Cambs 33 G1
Green End Herts 33 G5
Green End Herts 33 G6
Green End Warks 40 E7
Green Hammerton 57 K4
Green Hill 20 D3
Green Lane 30 B2
Green Moor 50 E3
Green Ore 19 J6
Green Quarter 61 F6
Green Street ESuss 14 C6
Green Street Herts 22 E2
Green Street Herts 33 H6
Green Street Worcs 29 H4
Green Street WSuss 12 E4
Green Street Green GtLon 23 H5
Green Street Green Kent 23 J4
Green Tye 33 H7
Greencroft 62 B2
Greendykes 71 F1
Greenend 30 E6
Greenfaulds 75 F3
Greenfield CenBeds 32 D5
Greenfield (Maes-Glas) Flints 47 K5
Greenfield GtMan 50 C2
Greenfield High 83 N10
Greenfield Lincs 53 H5
Greenfield Oxon 22 A2
Greenford 22 E3
Greengairs 75 F3
Greengates 57 G6
Greengill 60 C3
Greenhalgh 55 H6
Greenham 21 H5
Greenhaugh 70 C5
Greenhead 70 B7
Greenheys 49 G2
Greenhill D&G 69 G6
Greenhill GtLon 22 E3
Greenhill SYorks 51 F4
Greenhithe 23 J4
Greenholm 74 D7
Greenholme 61 G6
Greenhow Hill 57 G3
Greenigo 89 D7
Greenlands 22 A3
Greenlaw 77 F6
Greenloaning 80 D10
Greenmeadow 19 F2
Greenmoor Hill 21 K3
Greenmount 49 G1
Greenock 74 A3
Greenodd 55 G1
Greens Norton 31 H4
Greenside T&W 71 G7
Greenside WYorks 50 D1
Greenstead 34 E6
Greenstead Green 34 C6
Greensted 23 J1
Greensted Green 23 J1
Greenway Pembs 16 D2
Greenway Som 8 C2
Greenwell 61 G1
Greenwich 23 G4
Greenwood Forest Park, Y Felinheli Gwyn LL56 4QN 46 D6
Greet 30 B5
Greete 28 E1
Greetham Lincs 53 G5
Greetham Rut 42 C4
Greetland 57 F7
Gregson Lane 55 J7
Greinton 8 D1
Grenaby 54 B6
Grendon N'hants 32 B2
Grendon Warks 40 E6
Grendon Common 40 E6
Grendon Green 28 E3
Grendon Underwood 31 H6
Grenitote (Greinetobht) 88 C1
Grenofen 4 E3
Grenoside 51 F3
Greosabhagh 88 G8
Gresford 48 C7
Gresham 45 F2
Greshornish 82 D5
Gress (Griais) 88 K3
Gressenhall 44 D4
Gressingham 55 J3
Greta Bridge 62 A5
Gretna 69 J7
Gretna Green 69 J7
Gretton Glos 30 B5
Gretton N'hants 42 C6
Gretton Shrop 38 E6
Grewelthorpe 57 H2
Greygarth 57 G2
Greylake 8 C1
Greys Green 22 A3
Greysouthen 60 B4
Greystead 70 C5
Greystoke 60 F3
Greystone 56 D5
Greystones 51 F4
Greywell 22 A6
Gribthorpe 58 D6
Gribton 68 E5
Griff 41 F7
Griffithstown 19 F2
Grigghall 61 F7
Grimeford Village 49 F1
Grimesthorpe 51 F3
Grimethorpe 51 G2
Griminish (Griminis) 88 B3
Grimister 89 N3
Grimley 29 H2
Grimmet 67 H2
Grimoldby 53 G4
Grimpo 38 C3
Grimsargh 55 J6
Grimsay (Griomsaigh) 88 C3
Grimsbury 31 F4
Grimsby 53 F2
Grimscote 31 H3
Grimscott 6 A5
Grimshader (Griomsiadar) 88 K5
Grimsthorpe 42 D3
Grimston ERid 59 J6
Grimston Leics 41 J3
Grimston Norf 44 B3
Grimstone 9 F5
Grimstone End 34 D2
Grindale 59 H2
Grindiscol 89 N9
Grindle 39 G5
Grindleford 50 E5
Grindleton 56 C5
Grindley 40 C3
Grindley Brook 38 E1
Grindlow 50 D5
Grindon N'umb 77 H6
Grindon Staffs 50 C7
Grindon Stock 62 D4
Grindon T&W 62 D1
Gringley on the Hill 51 K3
Grinsdale 60 E1
Grinshill 38 E3
Grinton 62 A7
Grisdale 61 J7
Gristhorpe 59 G1
Griston 44 D6
Gritley 89 E7
Grittenham 20 D3
Grittleton 20 B4
Grizebeck 55 F1
Grizedale 60 E7
Grizedale Forest Park Cumb LA22 0QJ 60 E7
Grobister 89 F5
Groby 41 H5
Groes 47 J6
Groes-faen 18 D3
Groesffordd 36 B2
Groesffordd Marli 47 J5
Groeslon Gwyn 46 C7
Groeslon Gwyn 46 D6
Groes-lwyd 38 B4
Groes-wen 18 E3
Grogport 73 G6
Groigearraidh 88 B5
Gromford 35 H3
Gronant 47 J4
Groombridge 13 J3
Groombridge Place Gardens Kent TN3 9QG 13 J3
Grosmont Mon 28 D6
Grosmont NYorks 63 J6
Groton 34 D4
Groundistone Heights 69 K2
Grouville 3 K7
Grove Bucks 32 C6
Grove Dorset 9 F7
Grove Kent 25 J5
Grove Notts 51 K5
Grove Oxon 21 H2
Grove End 24 E5
Grove Green 14 C2
Grove Park 23 H4
Grove Town 57 K7
Grovehill 22 D1
Grovesend SGlos 19 K3
Grovesend Swan 17 J5
Gruids 86 H9
Gruline 78 G6
Grumbla 2 B6
Grundisburgh 35 G3
Gruting 89 L8
Grutness 89 N11
Gualachulain 79 N6
Guardbridge 81 K9
Guarlford 29 H4
Gubbergill 60 B7
Gubblecote 32 C7
Guernsey 3 J5
Guernsey Airport 3 H6
Guestling Green 14 D6
Guestling Thorn 14 D6
Guestwick 44 E3
Guestwick Green 44 E3
Guide 56 C7
Guide Post 71 H5
Guilden Down 38 C7
Guilden Morden 33 F4
Guilden Sutton 48 D6
Guildford 22 C7
Guildford House Gallery Surr GU1 3AJ 128 Guildford
Guildtown 80 G7
Guilsborough 31 H1
Guilsfield (Cegidfa) 38 B4
Guilthwaite 51 G4
Guisborough 63 G5
Guiseley 57 G5
Guist 44 E3
Guith 89 E4
Guiting Power 30 B6
Gulberwick 89 N9
Gullane 76 C2
Gullane Bents ELoth EH31 2AZ 76 C2
Gulval 2 B5
Gulworthy 4 E3
Gumfreston 16 E5
Gumley 41 J6
Gunby Lincs 42 C3
Gunby Lincs 53 H6
Gundleton 11 H1
Gunn 6 E2
Gunnersbury 22 E4
Gunnerside 61 L7
Gunnerton 70 E6
Gunness 52 B1
Gunnislake 4 E3
Gunnister 89 M5
Gunstone 40 A5
Gunter's Bridge 12 C4
Gunthorpe Norf 44 E2
Gunthorpe Notts 41 J1
Gunthorpe Rut 42 B5
Gunville 11 F6
Gunwalloe 2 D6
Gupworthy 7 H2
Gurnard 11 F5
Gurnett 49 J5
Gurney Slade 19 K7
Gurnos MTyd 18 D1
Gurnos Powys 18 A1
Gushmere 15 F2
Gussage All Saints 10 B3
Gussage St. Andrew 9 J3
Gussage St. Michael 9 J3
Guston 15 J3
Gutcher 89 P3
Guthram Gowt 42 E3
Guthrie 81 L5
Guyhirn 43 G5
Guy's Head 43 H3
Guy's Marsh 9 H2
Guyzance 71 H3
Gwaelod-y-garth 18 E3
Gwaenysgor 47 J4
Gwaithla 28 B3
Gwalchmai 46 B5
Gwastad 16 D3
Gwastadnant 46 E7
Gwaun-Cae-Gurwen 27 G7
Gwaynynog 47 J6
Gwbert 16 E1
Gweek 2 E6
Gwehelog 19 G1
Gwenddwr 27 K4
Gwendreath 2 E7
Gwennap 2 E4
Gwennap Mining District (Cornwall & West Devon Mining Landscape) Corn 2 E4
Gwenter 2 E7
Gwernaffield 48 B6
Gwernesney 19 H1
Gwernogle 17 J2
Gwernymynydd 48 B6
Gwern-y-Steeple 18 D4
Gwersyllt 48 C7
Gwespyr 47 K4
Gwinear 2 C5
Gwithian 2 C4
Gwredog 46 C4
Gwrhay 18 E2
Gwyddelwern 37 K1
Gwyddgrug 17 H2
Gwynfryn 48 B7
Gwystre 27 K2
Gwytherin 47 G6
Gyfelia 38 C1
Gyre 89 C7
Gyrn Goch 36 D1

H

Habberley 38 C5
Habin 12 B4
Habost (Tabost) 88 L1
Habrough 52 E1
Haccombe 5 J3
Hacconby 42 E3
Haceby 42 D2
Hacheston 35 H3
Hackbridge 23 F5
Hackenthorpe 51 G4
Hackford 44 E5
Hackforth 62 C7
Hackleton 32 B3
Hacklinge 15 J2
Hackness 63 K7
Hackney 23 G3
Hackthorn 52 C4
Hackthorpe 61 G4
Hacton 23 J3

Hockwold cum Wilton 44 B7
Hockworthy 7 J4
Hoddesdon 23 G1
Hoddlesden 56 C7
Hodgehill 49 H6
Hodgeston 16 D6
Hodnet 39 F3
Hodnetheath 39 F3
Hodsoll Street 24 C5
Hodson 20 E3
Hodthorpe 51 H5
Hoe 44 D4
Hoe Gate 11 H3
Hoff 61 H5
Hoffleet Stow 43 F2
Hoggard's Green 34 C3
Hoggeston 32 B6
Hoggrill's End 40 E6
Hogha Gearraidh 88 B1
Hoghton 56 B7
Hognaston 50 E7
Hogsthorpe 53 J5
Holbeach 43 G3
Holbeach Bank 43 G3
Holbeach Clough 43 G3
Holbeach Drove 43 G4
Holbeach Hurn 43 G3
Holbeach St. Johns 43 G4
Holbeach St. Marks 43 G2
Holbeach St. Matthew 43 H2
Holbeck 51 H5
Holbeck Woodhouse 51 H5
Holberrow Green 30 B3
Holbeton 5 G5
Holborough 24 D5
Holbrook Derbys 41 F1
Holbrook Suff 35 F5
Holbrooks 41 F7
Holburn 77 J7
Holbury 11 F4
Holcombe Devon 5 K3
Holcombe GtMan 49 G1
Holcombe Som 19 K7
Holcombe Burnell Barton 7 G6
Holcombe Rogus 7 J4
Holcot 31 J2
Holden 56 C5
Holden Gate 56 D7
Holdenby 31 H2
Holdenhurst 10 C5
Holder's Green 33 K6
Holders Hill 23 F3
Holdgate 38 E7
Holdingham 42 D1
Holditch 8 C4
Hole 7 K4
Hole Park 14 D4
Holehouse 50 C3
Hole-in-the-Wall 29 F6
Holford 7 K1
Holgate 58 B4
Holker 55 G2
Holkham 44 C1
Hollacombe Devon 6 B5
Hollacombe Devon 7 G5
Hollacombe Town 6 E4
Holland Ork 89 D2
Holland Surr 23 H6
Holland Fen 43 F1
Holland-on-Sea 35 F7
Hollandstoun 89 G2
Hollee 69 H7
Hollesley 35 H4
Hollicombe 5 K4
Hollingbourne 14 D2
Hollingbury 13 G6
Hollingrove 13 K4
Hollington Derbys 40 E2
Hollington ESuss 14 C6
Hollington Staffs 40 C2
Hollingworth 50 C3
Hollins 51 F5
Hollins Green 49 F3
Hollins Lane 55 H4
Hollinsclough 50 C6
Hollinwood GtMan 49 J2
Hollinwood Shrop 38 E2
Hollocombe 6 E4
Hollow Meadows 50 E4
Holloway 51 F7
Hollowell 31 H1
Holly Bush 38 D1
Holly End 43 H5
Holly Green 22 A1
Hollybush Caerp 18 E1
Hollybush EAyr 67 H2
Hollybush Worcs 29 G5
Hollyhurst 38 E1
Hollym 59 K7
Hollywater 12 B3
Hollywood 30 B1
Holm 69 H4
Holm of Drumlanrig 68 D4
Holmbridge 50 D2
Holmbury St. Mary 22 E7
Holmbush 13 F3
Holme Cambs 42 E7
Holme Cumb 55 J2
Holme NLincs 52 C2
Holme Notts 52 B7
Holme NYorks 57 J1
Holme WYorks 50 D2
Holme Chapel 56 D7
Holme Hale 44 C5
Holme Lacy 28 E5
Holme Marsh 28 C3
Holme next the Sea 44 B1
Holme on the Wolds 59 F5
Holme Pierrepont 41 J2
Holme St. Cuthbert 60 C2
Holme-on-Spalding-Moor 58 E6
Holmer 28 E4
Holmer Green 22 C2
Holmes 48 D1
Holmes Chapel 49 G6
Holme's Hill 13 J5
Holmesfield 51 F5
Holmeswood 48 D1
Holmewood 51 G6
Holmfield 57 F7
Holmfirth 50 D2
Holmhead D&G 68 C5
Holmhead EAyr 67 K1
Holmpton 59 K7
Holmrook 60 B6
Holmside 62 C2
Holmsleigh Green 8 B4
Holmston 67 H1
Holmwrangle 61 G2
Holne 5 H4
Holnest 9 F4
Holnicote 7 H1
Holsworthy 6 B5
Holsworthy Beacon 6 B5
Holt Dorset 10 B4
Holt Norf 44 E2
Holt Wilts 20 B5
Holt Worcs 29 H2
Holt Wrex 48 D7
Holt End Hants 11 H1
Holt End Worcs 30 B2
Holt Fleet 29 H2
Holt Heath Dorset 10 B4
Holt Heath Worcs 29 H2
Holt Wood 10 B4
Holtby 58 C4
Holton Oxon 21 K1
Holton Som 9 F2
Holton Suff 35 J1
Holton cum Beckering 52 E4
Holton Heath 9 J5
Holton le Clay 53 F2
Holton le Moor 52 D3
Holton St. Mary 34 E5
Holtspur 22 C3
Holtye 13 H3
Holtye Common 13 H3
Holway 8 B2
Holwell Dorset 9 F3
Holwell Herts 32 E5
Holwell Leics 42 A3
Holwell Oxon 21 F1
Holwell Som 20 A7
Holwick 61 L4
Holworth 9 G6
Holy Cross 29 J1
Holy Island IoA 46 A5
Holy Island (Lindisfarne) N'umb 77 K6
Holy Trinity Church Warks CV37 6BG 136 Stratford-upon-Avon
Holy Trinity Church, Skipton NYorks BD23 1NJ 56 E4
Holybourne 22 A7
Holyfield 23 G1
Holyhead (Caergybi) 46 A4
Holymoorside 51 F6
Holyport 22 B4
Holystone 70 E3
Holytown 75 F4
Holywell Cambs 33 G1
Holywell Corn 2 E3
Holywell Dorset 8 E4
Holywell ESuss 13 K7
Holywell (Treffynnon) Flints 47 K5
Holywell N'umb 71 J6
Holywell Bay Fun Park Corn TR8 5PW 2 E3
Holywell Green 50 C1
Holywell Lake 7 K3
Holywell Row 34 B1
Holywood 68 E5
Hom Green 28 E6
Homer 39 F5
Homersfield 45 G7
Homington 10 C2
Honey Hill 25 H5
Honey Street 20 E5
Honey Tye 34 D5
Honeyborough 16 C5
Honeybourne 30 C4
Honeychurch 6 E5
Honeyknowle 4 E5
Honiley 30 D1
Honing 45 H3
Honingham 45 F4
Honington Lincs 42 C1
Honington Suff 34 D1
Honington Warks 30 D4
Honiton 7 K5
Honkley 48 C7
Honley 50 D1
Hoo Med 24 D4
Hoo Suff 35 G3
Hoo Green 49 G4
Hoo Meavy 5 F4
Hood Green 51 F2
Hood Hill 51 F3
Hooe ESuss 13 K6
Hooe Plym 5 F5
Hooe Common 13 K5
Hook Cambs 43 H6
Hook ERid 58 D7
Hook GtLon 22 E5
Hook Hants 11 G4
Hook Hants 22 A6
Hook Pembs 16 C4
Hook Wilts 20 D3
Hook Green Kent 13 K3
Hook Green Kent 23 J4
Hook Green Kent 24 C4
Hook Green Kent 24 C5
Hook Norton 30 E5
Hook-a-Gate 38 D5
Hooke 8 E4
Hookgate 39 G2
Hookway 7 G6
Hookwood 23 F7
Hoole 48 D6
Hooley 23 F6
Hoop 19 J1
Hooton 48 C5
Hooton Levitt 51 H3
Hooton Pagnell 51 G2
Hooton Roberts 51 G3
Hop Farm, The Kent TN12 6PY 23 K7
Hop Pocket, The Here WR6 5BT 29 F4
Hop Pole 42 E4
Hopcrofts Holt 31 F6
Hope Derbys 50 D4
Hope Devon 5 G7
Hope Flints 48 C7
Hope Powys 38 B5
Hope Shrop 38 C5
Hope Staffs 50 D7
Hope Bagot 28 E1
Hope Bowdler 38 D6
Hope End Green 33 J6
Hope Mansell 29 F7
Hope under Dinmore 28 E3
Hopehouse 69 H2
Hopeman 84 F4
Hope's Green 24 D3
Hopesay 38 C7
Hopkinstown 18 D2
Hopley's Green 28 C3
Hopperton 57 K4
Hopsford 41 G7
Hopstone 39 G6
Hopton Derbys 50 E7
Hopton Norf 45 K6
Hopton Shrop 38 C3
Hopton Shrop 38 E3
Hopton Staffs 40 B3
Hopton Suff 34 D1
Hopton Cangeford 38 E7
Hopton Castle 28 C1
Hopton Wafers 29 F1
Hoptonheath 28 C1
Hopwas 40 D5
Hopwood 30 B1
Horam 13 J5
Horbling 42 E2
Horbury 50 E1
Horden 62 E2
Horderley 38 D7
Hordle 10 D5
Hordley 38 C2
Horeb Carmar 17 H5
Horeb Cere 17 G1
Horeb Flints 48 B7
Horfield 19 J4
Horham 35 G1
Horkesley Heath 34 D6
Horkstow 52 C1
Horley Oxon 31 F4
Horley Surr 23 F7
Horn Hill 22 D2
Hornblotton 8 E1
Hornblotton Green 8 E1
Hornby Lancs 55 J3
Hornby NYorks 62 C7
Hornby NYorks 62 D6
Horncastle 53 F6
Hornchurch 23 J3
Horncliffe 77 H6
Horndean Hants 11 J3
Horndean ScBord 77 H6
Horndon 6 D7
Horndon on the Hill 24 C3
Horne 23 G7
Horne Row 24 D1
Horner 7 G1
Horniman Museum GtLon SE23 3PQ 101 C9
Horning 45 H4
Horninghold 42 B6
Horninglow 40 E3
Horningsea 33 H2
Horningsham 20 B7
Horningtoft 44 D3
Horningtops 4 C4
Horns Cross Devon 6 B3
Horns Cross ESuss 14 D5
Horns Green 23 H6
Hornsbury 8 C3
Hornsby 61 G2
Hornsby Gate 61 G1
Hornsea 59 J5
Hornsea Freeport ERid HU18 1UT 59 J5
Hornsey 23 G3
Hornton 30 E4
Horrabridge 5 F4
Horridge 5 H3
Horringer 34 C2
Horrocks Fold 49 G1
Horse Bridge 49 J7
Horsebridge Devon 4 E3
Horsebridge Hants 10 E1
Horsebrook 40 A4
Horsecastle 19 H5
Horsehay 39 F5
Horseheath 33 K4
Horsehouse 57 F1
Horsell 22 C6
Horseman's Green 38 D1
Horsenden 22 A1
Horseshoe Green 23 H7
Horseway 43 H7
Horsey 45 J3
Horsey Corner 45 J3
Horsford 45 F4
Horsforth 57 H6
Horsham Worcs 29 G3
Horsham WSuss 12 E3
Horsham St. Faith 45 G4
Horsington Lincs 52 E6
Horsington Som 9 G2
Horsington Marsh 9 G2
Horsley Derbys 41 F1
Horsley Glos 20 B2
Horsley N'umb 70 D4
Horsley N'umb 71 F7
Horsley Cross 35 F6
Horsley Woodhouse 41 F1
Horsleycross Street 35 F6
Horsleygate 51 F5
Horsleyhill 70 A2
Horsmonden 23 K7
Horspath 21 J1
Horstead 45 G4
Horsted Keynes 13 G4
Horton Bucks 32 C7
Horton Dorset 10 B4
Horton Lancs 56 D4
Horton N'hants 32 B3
Horton SGlos 20 A3
Horton Shrop 38 D3
Horton Som 8 C3
Horton Staffs 49 J7
Horton Swan 17 H7
Horton Tel&W 39 F4
Horton Wilts 20 D5
Horton W&M 22 D4
Horton Cross 8 C3
Horton Grange 71 H6
Horton Green 38 D1
Horton Heath 11 F3
Horton in Ribblesdale 56 D2
Horton Inn 10 B4
Horton Kirby 23 J5
Horton Park Farm Surr KT19 8PT 22 E5
Horton-cum-Studley 31 H7
Horwich 49 F1
Horwich End 50 C4
Horwood 6 D3
Hoscar 48 D1
Hose 42 A3
Hoses 60 D7
Hosh 80 D8
Hosta 88 B1
Hoswick 89 N10
Hotham 58 E6
Hothfield 14 E3
Hoton 41 H3
Houbie 89 Q3
Houdston 67 F4
Hough 49 G7
Hough Green 48 D4
Hougham 42 B1
Hough-on-the-Hill 42 C1
Houghton Cambs 33 F1
Houghton Cumb 60 F1
Houghton Devon 5 G6
Houghton Hants 10 E1
Houghton Pembs 16 C5
Houghton WSuss 12 D5
Houghton Bank 62 C4
Houghton Conquest 32 D4
Houghton le Spring 62 D2
Houghton on the Hill 41 J5
Houghton Regis 32 D6
Houghton St. Giles 44 D2
Houghton-le-Side 62 C4
Houlsyke 63 H6
Hound 11 F4
Hound Green 22 A6
Houndslow 76 E6
Houndsmoor 7 K3
Houndwood 77 G4
Hounsdown 10 E3
Hounslow 22 E4
House of Marbles & Teign Valley Glass Devon TQ13 9DS 5 J3
Houses Hill 50 D1
Housesteads Roman Fort (Frontiers of the Roman Empire) N'umb NE47 6NN 70 C7
Housetter 89 M4
Housham Tye 33 J7
Houston 74 C4
Houstry 87 P6
Houstry of Dunn 87 Q4
Houton 89 C7
Hove 13 F6
Hove Edge 57 G7
Hoveringham 41 J1
Hoveton 45 H4
Hovingham 58 C2
How 61 G1
How Caple 29 F5
How End 32 D4
How Green 23 H7
How Man 60 A5
Howbrook 51 F3
Howden 58 D7
Howden Clough 57 H7
Howden-le-Wear 62 B3
Howe Cumb 55 H1
Howe High 87 R3
Howe Norf 45 G6
Howe NYorks 57 J1
Howe Green 24 D1
Howe Street Essex 33 K5
Howe Street Essex 33 K7
Howegreen 24 E1
Howell 42 E1
Howey 27 K3
Howgate Cumb 60 A4
Howgate Midlo 76 A5
Howgill Lancs 56 D5
Howgill NYorks 57 F4
Howick 71 H2
Howle 39 F3
Howle Hill 29 F6
Howlett End 33 J5
Howletts Wild Animal Park Kent CT4 5EL 15 G2
Howley 8 B4
Howmore (Tobha Mòr) 88 B5
Hownam 70 C2
Hownam Mains 70 C1
Howpasley 69 J3
Howsham NLincs 52 D2
Howsham NYorks 58 D3
Howt Green 24 E5
Howtel 77 G7
Howton 28 D6
Howwood 74 C4
Hoxne 35 F1
Hoy 89 B8
Hoylake 48 B4
Hoyland 51 F2
Hoylandswaine 50 E2
Hoyle 12 C5
Hubberholme 56 E2
Hubberston 16 B5
Hubbert's Bridge 43 F1
Huby NYorks 57 H5
Huby NYorks 58 B3
Hucclecote 29 H7
Hucking 14 D2
Hucknall 41 H1
Huddersfield 50 D1
Huddington 29 J3
Huddlesford 40 D5
Hudnall 32 D7
Hudscott 6 E3
Hudswell 62 B7
Huggate 58 E4
Hugglescote 41 G4
Hugh Town 2 C1
Hughenden Valley 22 B2
Hughley 38 E6
Hugmore 48 C7
Hugus 2 E4
Huish Devon 6 D4
Huish Wilts 20 E5
Huish Champflower 7 J3
Huish Episcopi 8 D2
Huisinis 88 E6
Hulcote 32 C5
Hulcott 32 B7
Hull 59 H7
Hulland 40 E1
Hulland Ward 40 E1
Hullavington 20 B3
Hullbridge 24 E2
Hulme 40 B1
Hulme End 50 D7
Hulme Walfield 49 H6
Hulver Street 45 J7
Hulverstone 10 E6
Humber Devon 5 J3
Humber Here 28 E3
Humber Bridge Country Park ERid HU13 0LN 59 G7
Humberside Airport 52 D1
Humberston 53 G2
Humberstone 41 J5
Humberton 57 K3
Humbie 76 C4
Humbleton Dur 62 A5
Humbleton ERid 59 J6
Humbleton N'umb 70 E1
Humby 42 D2
Hume 77 F6
Humehall 77 F6
Hummer 8 E3
Humshaugh 70 E6
Huna 87 R2
Huncoat 56 C6
Huncote 41 H6
Hundalee 70 B2
Hundall 51 F5
Hunderthwaite 61 L4
Hundleby 53 G6
Hundleton 16 C5
Hundon 34 B4
Hundred Acres 11 G3
Hundred End 55 H7
Hundred House 28 A3
Hungarton 41 J5
Hungate End 31 J4
Hungerford Hants 10 C3
Hungerford Newtown 21 G4
Hungerford Shrop 38 E7
Hungerford WBerks 21 G5
Hungerton 42 B3
Hunmanby 59 G2
Hunningham 30 E2
Hunningham Hill 30 E2
Hunny Hill 11 F6
Hunsdon 33 H7
Hunsingore 57 K4
Hunslet 57 J6
Hunsonby 61 G3
Hunspow 87 Q2
Hunstanton 44 A1
Hunstanworth 61 L2
Hunston Suff 34 D2
Hunston WSuss 12 B6
Hunsworth 57 G7
Hunt End 30 B2
Hunt House 63 J7
Huntercombe End 21 K3
Hunters Forstal 25 H5
Hunter's Inn 6 E1
Hunter's Quay 73 K3
Hunterston 73 K5
Huntford 70 B3
Huntham 8 C2
Huntingdon 33 F1
Huntingfield 35 H1
Huntingford 9 H1
Huntington Here 28 B3
Huntington Here 28 D4
Huntington Staffs 40 B4
Huntington Tel&W 39 F5
Huntington York 58 C4
Huntingtower 80 F8
Huntley 29 G7
Huntly 85 K7
Huntlywood 76 E6
Hunton Hants 21 H7
Hunton Kent 14 C3
Hunton NYorks 62 B7
Hunton Bridge 22 D1
Hunt's Cross 48 D4
Huntscott 7 H1
Huntsham 7 J3
Huntshaw 6 D3
Huntshaw Cross 6 D3
Huntshaw Water 6 D3
Huntspill 19 G7
Huntworth 8 C1
Hunwick 62 B3
Hunworth 44 E2
Hurcott Som 8 C3
Hurcott Som 8 E2
Hurdley 38 B6
Hurdsfield 49 J5
Hurley Warks 40 E6
Hurley W&M 22 B3
Hurley Bottom 22 B3
Hurlford 74 C7
Hurlston Green 48 C1
Hurn 10 C5
Hursley 11 F2
Hurst NYorks 62 A6
Hurst W'ham 22 A4
Hurst Green Essex 34 E7
Hurst Green ESuss 14 C5
Hurst Green Lancs 56 B6
Hurst Green Surr 23 G6
Hurst Wickham 13 F5
Hurstbourne Priors 21 H7
Hurstbourne Tarrant 21 G6
Hurstpierpoint 13 F5
Hurstwood 56 D6
Hurtmore 22 C7
Hurworth-on-Tees 62 D5
Hury 61 L4
Husbands Bosworth 41 J7
Husborne Crawley 32 C5
Husthwaite 58 B2
Hutcherleigh 5 H5
Huthwaite 51 G7
Huttoft 53 J5
Hutton Cumb 60 F4
Hutton Essex 24 C2
Hutton Lancs 55 H7
Hutton NSom 19 G6
Hutton ScBord 77 H5
Hutton Bonville 62 D6
Hutton Buscel 59 F1
Hutton Conyers 57 J2
Hutton Cranswick 59 G4
Hutton End 60 F3
Hutton Hang 57 G1
Hutton Henry 62 E3
Hutton Magna 62 B5
Hutton Mount 24 C2
Hutton Mulgrave 63 J6
Hutton Roof Cumb 55 J2
Hutton Roof Cumb 60 E3
Hutton Rudby 62 E6
Hutton Sessay 57 K2
Hutton Wandesley 58 B4
Hutton-le-Hole 63 H7
Huxham 7 H6
Huxham Green 8 E1
Huxley 48 E6
Huyton 48 D3
Hycemoor 54 D1
Hyde Glos 20 B1
Hyde GtMan 49 J3
Hyde End WBerks 21 J5
Hyde End W'ham 22 A5
Hyde Heath 22 C1
Hyde Lea 40 B4
Hydestile 22 C7
Hylands Park Essex CM2 8WQ 24 C1
Hyndford Bridge 75 H6
Hyndlee 70 A3
Hynish 78 A7
Hyssington 38 C6
Hythe Hants 11 F4
Hythe Kent 15 G4
Hythe End 22 D4
Hyton 54 D1

I

Iarsiadar 88 G4
Ibberton 9 G4
Ible 50 E7
Ibsley 10 C4
Ibstock 41 G4
Ibstone 22 A2
Ibthorpe 21 G6
Ibworth 21 J6
Icelton 19 G5
Ickburgh 44 C6
Ickenham 22 D3
Ickford 21 K1
Ickham 15 H2
Ickleford 32 E5
Icklesham 14 D6
Ickleton 33 H4
Icklingham 34 B1
Ickwell Green 32 E4
Icomb 30 D6
Idbury 30 D6
Iddesleigh 6 D5
Ide 7 H6
Ide Hill 23 H6
Ideford 5 J3
Iden 14 E5
Iden Green Kent 14 C4
Iden Green Kent 14 D4
Idle 57 G6
Idless 3 F4
Idlicote 30 D4
Idmiston 10 C1
Idridgehay 40 E1
Idridgehay Green 40 E1
Idrigil 82 D4
Idstone 21 F3
Iffley 21 J1
Ifield 13 F3
Ifieldwood 13 F3
Ifold 12 D3
Iford Bournem 10 C5
Iford ESuss 13 H6
Ifton 19 H3
Ifton Heath 38 C2
Ightfield 38 E2
Ightham 23 J6
Ightham Mote Kent TN15 0NT 23 J6
Iken 35 J3
Ilam 50 D7
Ilchester 8 E2
Ilderton 71 F1
Ilford 23 H3
Ilfracombe 6 D1
Ilkeston 41 G1
Ilketshall St. Andrew 45 H7
Ilketshall St. Lawrence 45 H7
Ilketshall St. Margaret 45 H7
Ilkley 57 G5
Illey 40 B7
Illidge Green 49 G6
Illington 44 D7
Illingworth 57 F7
Illogan 2 D4
Illston on the Hill 42 A6
Ilmer 22 A1
Ilmington 30 D4
Ilminster 8 C3
Ilsington Devon 5 H3
Ilsington Dorset 9 G5
Ilston 17 J6
Ilton NYorks 57 G2
Ilton Som 8 C3
Imachar 73 G6
Imber 20 C7
Immingham 52 E1
Immingham Dock 53 F1
Imperial War Museum North GtMan M17 1TZ 112 F4
Imperial War Museum, Duxford Cambs CB22 4QR 33 H4
Imperial War Museum, London GtLon SE1 6HZ 132 G6
Impington 33 H2
Ince 48 D5
Ince Blundell 48 C2
Ince-in-Makerfield 48 E2
Inch Kenneth 78 F7
Inchbare 81 M4
Inchberry 84 H5
Inchbraoch 81 N5
Inchgrundle 81 K3
Inchinnan 74 C4
Inchlaggan 83 M10
Inchmarnock 73 J5
Inchnacardoch Hotel 83 P9
Inchnadamph 86 E7
Inchree 79 M4
Inchrory 84 F10
Inchture 80 H8
Indian Queens 3 G3
Inerval 72 B6
Ingatestone 24 C2
Ingbirchworth 50 E2
Ingerthorpe 57 H3
Ingestre 40 B3
Ingham Lincs 52 C4
Ingham Norf 45 H3
Ingham Suff 34 C1
Ingham Corner 45 H3
Ingleborough 43 H4
Ingleby Derbys 41 F3
Ingleby Lincs 52 B5
Ingleby Arncliffe 62 E6
Ingleby Barwick 62 E5
Ingleby Cross 62 E6
Ingleby Greenhow 63 F6
Ingleigh Green 6 E5
Inglesbatch 20 A5
Inglesham 21 F2
Ingleton Dur 62 B4
Ingleton NYorks 56 B2
Inglewhite 55 J6
Ingliston 75 K3
Ingmire Hall 61 H7
Ingoe 71 F6
Ingoldisthorpe 44 A2
Ingoldmells 53 J6
Ingoldsby 42 D2
Ingon 30 D3
Ingram 71 F2
Ingrave 24 C2
Ingrow 57 F6
Ings 60 F7
Ingst 19 J3
Ingworth 45 F3
Inhurst 21 J5
Injebreck 54 C5
Inkberrow 30 B3
Inkersall 51 G5
Inkersall Green 51 G5
Inkpen 21 G5
Inkstack 87 Q2
Inmarsh 20 C5
Innellan 73 K4
Innerhadden 80 B5
Innerleithen 76 B7
Innerleven 81 J10
Innermessan 64 A4
Innerwick ELoth 77 F3
Innerwick P&K 80 A6
Innsworth 29 H6
Insch 85 L8
Insh 84 C10
Inskip 55 H6
Instow 6 C2
Intake 51 H2
Intech, Winchester Hants SO21 1HX 11 G2
International Centre, The, Telford Tel&W TF3 4JH 39 G5
Intwood 45 F5
Inver Aber 84 G11
Inver High 84 C2
Inver Mallie 79 N2
Inverailort 79 J2
Inveralligin 83 J5
Inverallochy 85 Q4
Inveraray (Inbhir Aora) 79 M10
Inverarish 82 F7
Inverarity 81 K6
Inverarnan (Inbhir Àirnein) 79 Q9
Inverasdale 83 J2
Inverbeg 74 B1
Inverbervie 81 P3
Inverbroom 83 M2
Invercassley 86 G9
Inverchaolain 73 J3
Invercharnan 79 N6
Inverchoran 83 N5
Invercreran 79 M6
Inverdruie 84 D9
Inveresk 76 B3
Inverey 80 F2
Inverfarigaig 83 R8
Invergarry (Inbhir Garadh) 83 P10
Invergeldie 80 C8
Invergloy 79 P2
Invergordon 84 B4
Invergowrie 81 J7
Inverharroch Farm 84 H7
Inverie 82 H10
Inverinan 79 L9
Inverinate 83 K8
Inverkeilor 81 M6
Inverkeithing 75 K2
Inverkeithny 85 L6
Inverkip 74 A3
Inverkirkaig 86 C8
Inverlael 83 M2
Inverlauren 74 B2
Inverlochlarig 79 R9
Inverlussa 72 E2
Invermoriston (Inbhir Moireasdan) 83 Q9
Invernaver 87 K3
Inverneil 73 G2
Inverness (Inbhir Nis) 84 A6
Inverness Airport 84 B5
Invernoaden 73 K1
Inveroran Hotel (Inbhir Orain) 79 P6
Inverroy 79 P2
Inversnaid Hotel 79 Q10
Inverugie 85 R6
Inveruglas (Inbhir Dhubhghlais) 79 Q10
Inverurie 85 M8
Invervar 80 B6
Invervegain 73 J3
Inwardleigh 6 D6
Inworth 34 C7
Iochdar (Eochar) 88 B4
Iona 78 D8
Iping 12 B4
Ipplepen 5 J4
Ipsden 21 K3
Ipstones 40 C1
Ipswich 35 F4
Irby 48 B4
Irby Hill 48 B4
Irby in the Marsh 53 H6
Irby upon Humber 52 E2
Irchester 32 C2
Irchester Country Park N'hants NN29 7DL 32 C2
Ireby Cumb 60 D3
Ireby Lancs 56 B2
Ireland 89 M10
Ireland's Cross 39 G1
Ireleth 55 F2
Ireshopeburn 61 K3
Irlam 49 G3
Irnham 42 D3
Iron Acton 19 K3
Iron Cross 30 B3
Ironbridge 39 F5
Ironbridge Gorge Tel&W TF8 7DQ 39 F5
Irons Bottom 23 F7
Ironville 51 G7
Irstead 45 H3
Irthington 69 K7
Irthlingborough 32 C1
Irton 59 G1
Irvine 74 B7
Isauld 87 M3
Isbister Ork 89 B5
Isbister Shet 89 M3
Isbister Shet 89 P6
Isfield 13 H5
Isham 32 B1
Isington 22 A7
Island of Stroma 87 R2
Islawr-dref 37 F4
Islay 72 A4
Islay Airport 72 B5
Islay House 72 B4
Isle Abbotts 8 C2
Isle Brewers 8 C2
Isle of Bute Discovery Centre A&B PA20 0AH 73 J4
Isle of Lewis (Eilean Leodhais) 88 J3
Isle of Man 54 C5
Isle of Man Airport 54 B7
Isle of May 76 E1
Isle of Noss 89 P8
Isle of Sheppey 25 F4
Isle of Walney 54 E3
Isle of Whithorn 64 E7
Isle of Wight 11 F6
Isle of Wight (Sandown Airport) 11 G6
Isle of Wight Pearl IoW PO30 4DD 11 F6
Isle of Wight Zoo IoW PO36 8QB 11 H6
Isleham 33 K1
Isleornsay (Eilean Iarmain) 82 G9
Isles of Scilly (Scilly Isles) 2 C1
Isleworth 22 E4
Isley Walton 41 G3
Islibhig 88 E5
Islip N'hants 32 C1
Islip Oxon 31 G7
Isombridge 39 F4
Istead Rise 24 C5
Itchen 11 F3
Itchen Abbas 11 G1
Itchen Stoke 11 G1
Itchen Valley Country Park Hants SO30 3HQ 92 D2
Itchingfield 12 E4
Itchington 19 K3
Itteringham 45 F2
Itton Devon 6 E6
Itton Mon 19 H2
Itton Common 19 H2
Ivegill 60 F2
Ivelet 61 L7
Iver 22 D3
Iver Heath 22 D3
Iveston 62 B1
Ivetsey Bank 40 A4
Ivinghoe 32 C7
Ivinghoe Aston 32 C7
Ivington 28 D3
Ivington Green 28 D3
Ivy Hatch 23 J6
Ivy Todd 44 C5
Ivybridge 5 G5
Ivychurch 15 F5
Iwade 25 F5
Iwerne Courtney (Shroton) 9 H3
Iwerne Minster 9 H3
Ixworth 34 D1
Ixworth Thorpe 34 D1

J

Jack Hill 57 G4
Jackfield 39 F5
Jacksdale 51 G7
Jackton 74 D5
Jacobstow 4 B1
Jacobstowe 6 D5
Jacobswell 22 C6
James Hamilton Heritage Park SLan G74 5LB 119 G7
James Pringle Weavers of Inverness High IV2 4RB 84 A6
Jameston 16 D6
Jamestown D&G 69 J4
Jamestown WDun 74 B2
Janefield 84 B5
Janetstown 87 R4
Jarrow 71 J7
Jarvis Brook 13 J4
Jasper's Green 34 B6
Jawcraig 75 G3
Jayes Park 22 E7
Jaywick 35 F7
Jealott's Hill 22 B4
Jeater Houses 62 E7
Jedburgh 70 B1
Jeffreyston 16 D5
Jemimaville 84 B4
Jephson Gardens Warks CV32 4ER 104 C8
Jericho 49 H1
Jersay 75 G4
Jersey 3 J7
Jersey Airport 3 J7
Jersey Marine 18 A2
Jerviswood 75 G6
Jesmond 71 H7
Jevington 13 J6
Jinney Ring Craft Centre, The Worcs B60 4BU 29 J2
Jockey End 32 D7
Jodrell Bank 49 G5
Jodrell Bank Observatory & Arboretum ChesE SK11 9DL 49 G5
John Muir Country Park ELoth EH42 1UW 76 E3
John o' Groats 87 R2
Johnby 60 F3
John's Cross 14 C5
Johnshaven 81 N4
Johnson Street 45 H4
Johnston 16 C4
Johnstone 74 C4
Johnstone Castle 74 C4
Johnstonebridge 69 F4
Johnstons Cashmere Visitor Centre Moray IV30 4AF 84 G4
Johnstown Carmar 17 G4
Johnstown Wrex 38 C1
Joppa 67 J2
Jordans 22 C2
Jordanston 16 C2
Jorvik YO1 9WT 137 York
Jorvik Glass NYorks YO60 7DA 58 D3
Joy's Green 29 F7
Jumpers Common 10 C5
Juniper Green 75 K4
Juniper Hill 31 G5
Jura 72 D2
Jura House 72 C4
Jurby East 54 C4
Jurby West 54 C4

K

Kaber 61 J5
Kaimes 76 A4
Kames A&B 73 H3
Kames EAyr 68 B1
Kea 3 F4
Keadby 52 B1
Keal Cotes 53 G6
Kearsley 49 G2
Kearstwick 56 B2
Kearton 62 A7
Keasden 56 C3
Keckwick 48 E4
Keddington 53 G4
Keddington Corner 53 G4
Kedington 34 B4
Kedleston 41 F1
Keelby 52 E1
Keele 40 A1
Keeley Green 32 D4
Keelham 57 F6
Keenthorne 8 B1
Keeres Green 33 J7
Keeston 16 B4
Keevil 20 C6
Kegworth 41 G3
Kehelland 2 D4
Keighley 57 F5
Keighley & Worth Valley Railway WYorks BD22 8NJ 57 F5
Keil 66 A3
Keilhill 85 M5
Keillmore 72 E2
Keills 72 C4
Keils 72 D4
Keinton Mandeville 8 E1
Keir House 75 F1
Keir Mill 68 D4
Keisby 42 D3
Keisley 61 J4
Keiss 87 R3
Keith 85 J5
Kelbrook 56 E5
Kelby 42 D1
Keld Cumb 61 G5
Keld NYorks 61 K6
Keldholme 58 D1
Keldy Castle 63 H7
Kelfield NLincs 52 B2
Kelfield NYorks 58 B6
Kelham 51 K7
Kella 54 C4
Kellacott 6 C7
Kellan 78 G6
Kellas Angus 81 K7
Kellas Moray 84 F5
Kellaton 5 H7
Kellaways 20 C4
Kelleth 61 H6
Kelleythorpe 59 G4
Kelling 44 E1
Kellington 58 B7
Kelloe 62 D3
Kelloholm 68 C2
Kelly Corn 4 A3
Kelly Devon 6 B7
Kelly Bray 4 D3
Kelmarsh 31 J1
Kelmscott 21 F2
Kelsale 35 H2
Kelsall 48 E6
Kelsay 72 A5
Kelshall 33 G5
Kelsick 60 D1
Kelso 77 F7
Kelstedge 51 F6
Kelstern 53 F3
Kelston 20 A5
Kelton 65 K3
Kelton Hill (Rhonehouse) 65 H5
Kelty 75 K1
Kelvedon 34 C7
Kelvedon Hatch 23 J2
Kelvin Hall Glas G3 8AW 127 A2
Kelvingrove Art Gallery & Museum Glas G3 8AG 127 B2
Kelvinside 74 D4
Kelynack 2 A6
Kemacott 6 E1
Kemback 81 K9
Kemberton 39 G5
Kemble 20 C2
Kemerton 29 J5
Kemeys Commander 19 G1
Kemeys Inferior 19 G2
Kemnay 85 M9
Kemp Town 13 G6
Kempe's Corner 15 F3
Kempley 29 F6
Kempley Green 29 F6
Kemps Green 30 C1
Kempsey 29 H4
Kempsford 20 E2
Kempshott 21 J7
Kempston 32 D4
Kempston Hardwick 32 D4
Kempston West End 32 C4
Kempton 38 C7
Kemsing 23 J6
Kemsley 25 F5
Kenardington 14 E4
Kenchester 28 D4
Kencot 21 F1
Kendal 61 G7
Kenderchurch 28 D6
Kendleshire 19 K4
Kenfig 18 A3
Kenfig Hill 18 B3
Kenidjack 2 A5
Kenilworth 30 D1
Kenilworth Castle Warks CV8 1NE 104 A6
Kenknock 79 R7
Kenley GtLon 23 G5
Kenley Shrop 38 E5
Kenmore High 82 H5
Kenmore P&K 80 C6
Kenn Devon 7 H7
Kenn NSom 19 H5
Kennacraig 73 G4
Kennall Vale (Cornwall & West Devon Mining Landscape) Corn 2 E5
Kennards House 4 C2
Kennavay 88 H8
Kenneggy Downs 2 C6
Kennerleigh 7 G5
Kennet 75 H1
Kennethmont 85 K8
Kennett 33 K2
Kennford 7 H7
Kenninghall 44 E7
Kennington Kent 15 F3
Kennington Oxon 21 J1
Kennoway 81 J10
Kenny 8 C3
Kennyhill 33 K1
Kennythorpe 58 D3
Kenovay 78 A6
Kensaleyre 82 E5
Kensington 23 F3
Kensington Palace GtLon W8 4PX 99 G7
Kenstone 38 E3
Kensworth 32 D7
Kent & East Sussex Railway Kent TN30 6HE 14 D5
Kent Street ESuss 14 C6
Kent Street Kent 23 K6
Kentallen (Ceann an t-Sàilein) 79 M5
Kentchurch 28 D6
Kentford 34 B2
Kentisbeare 7 J5
Kentisbury 6 E1
Kentisbury Ford 6 E1
Kentish Town 23 F3
Kentmere 61 F6
Kenton Devon 7 H7
Kenton Suff 35 F2
Kenton T&W 71 H7
Kenton Corner 35 G2
Kentra 78 H4
Kents Bank 55 G2
Kent's Green 29 G6
Kent's Oak 10 E2
Kenwick 38 D2
Kenwood House GtLon NW3 7JR 98 H4
Kenwyn 3 F4
Kenyon 49 F3
Keoldale 86 F3
Keose (Ceòs) 88 J5
Keppoch A&B 74 B3
Keppoch High 83 J8
Keprigan 66 A2
Kepwick 62 E7
Keresley 41 F7
Kernborough 5 H6
Kerrera 79 K8
Kerridge 49 J5
Kerris 2 B6
Kerry (Ceri) 38 A6
Kerrycroy 73 K4
Kerry's Gate 28 C5
Kerrysdale 83 J3
Kersall 51 K6
Kersey 34 E4
Kersey Vale 34 E4
Kershopefoot 69 K5
Kerswell 7 J5
Kerswell Green 29 H4
Kerthen Wood 2 C5
Kesgrave 35 G4
Kessingland 45 K7
Kessingland Beach 45 K7
Kestle 3 G4
Kestle Mill 3 F3
Keston 23 H5
Keswick Cumb 60 D4
Keswick Norf 45 G5
Keswick Norf 45 H2
Ketley Bank 39 F4
Ketsby 53 G5
Kettering 32 B1
Ketteringham 45 F5
Kettins 80 H7
Kettle Corner 14 C2
Kettlebaston 34 D3
Kettlebrook 40 E5
Kettleburgh 35 G2
Kettlehill 81 J10
Kettleholm 69 G6
Kettleness 63 J5
Kettleshulme 49 J5
Kettlesing 57 H4
Kettlesing Bottom 57 H4
Kettlesing Head 57 H4
Kettlestone 44 D2
Kettlethorpe 52 B5
Kettletoft 89 F4
Kettlewell 56 E2
Ketton 42 C5
Kevingtown 23 H5
Kew 22 E4
Kewstoke 19 G5
Kexbrough 51 F2
Kexby Lincs 52 B4
Kexby York 58 D4
Key Green 49 H6
Keyham 41 J5
Keyhaven 10 E5
Keyingham 59 J7
Keymer 13 G5
Keynsham 19 K5
Key's Toft 53 H7
Keysoe 32 D2
Keysoe Row 32 D2
Keyston 32 D1
Keyworth 41 J2
Kibblesworth 62 C1
Kibworth Beauchamp 41 J6
Kibworth Harcourt 41 J6
Kidbrooke 23 H4
Kiddemore Green 40 A5
Kidderminster 29 H1
Kiddington 31 F6
Kidlington 31 F7
Kidmore End 21 K4
Kidnal 38 D1
Kidsdale 64 E7
Kidsgrove 49 H7
Kidstones 56 E1
Kidwelly (Cydweli) 17 H5
Kiel Crofts 79 L7
Kielder 70 B4
Kielder Forest N'umb NE48 1ER 70 B4
Kielder Water N'umb NE48 1BX 70 C5
Kilbarchan 74 C4
Kilbeg 82 G10
Kilberry 73 F4
Kilbirnie 74 B5
Kilbride A&B 73 J4
Kilbride A&B 79 K8
Kilbride Farm 73 H4
Kilbridemore 73 J1
Kilburn Derbys 41 F1
Kilburn GtLon 23 F3
Kilburn NYorks 58 B2
Kilby 41 J6
Kilchattan Bay 73 K5
Kilchenzie 66 A1
Kilcheran 79 K7
Kilchiaran 72 A4
Kilchoan A&B 79 J9
Kilchoan High 78 F4
Kilchoman 72 A4
Kilchrenan (Cill Chrèanain) 79 M8
Kilchrist 66 A2
Kilconquhar 81 K10
Kilcot 29 F6
Kilcoy 83 R5
Kilcreggan 74 A2
Kildale 63 G6
Kildary 84 B3
Kildavie 66 B2
Kildonan 66 E1
Kildonan Lodge 87 M7
Kildonnan 78 F2
Kildrochet House 64 A5
Kildrummy 85 J9
Kildwick 57 F5
Kilfinan 73 H3
Kilfinnan 83 N11
Kilgetty 16 E5
Kilgwrrwg Common 19 H2
Kilham ERid 59 G3
Kilham N'umb 77 G7
Kilkenneth 78 A6
Kilkenny 30 B7
Kilkerran A&B 66 B2
Kilkerran SAyr 67 H3
Kilkhampton 6 A4
Killamarsh 51 G4
Killay 17 K6
Killean 72 E6
Killearn 74 D2
Killellan 66 A2
Killen 84 A5
Killerby 62 B4
Killerton Devon EX5 3LE 7 H5
Killerton 7 H5
Killichonan 80 A5
Killiechonate 79 P2
Killiechronan 78 G6
Killiecrankie 80 E4
Killilan 83 K7
Killimster 87 R4
Killin 80 A7
Killinallan 72 B3
Killinghall 57 H4
Killington Cumb 56 B1
Killington Devon 6 E1
Killingworth 71 H6
Killochyett 76 C6
Killocraw 72 E7
Killundine 78 G6
Kilmacolm 74 B4
Kilmahog 80 B10
Kilmalieu 79 K5
Kilmaluag 82 E3
Kilmany 81 J8
Kilmarie 82 F9
Kilmarnock 74 C7
Kilmartin 73 G1
Kilmaurs 74 C6
Kilmelford 79 K9
Kilmeny 72 B4
Kilmersdon 19 K6
Kilmeston 11 G2
Kilmichael 66 A1
Kilmichael Glassary 73 G1
Kilmichael of Inverlussa 73 F2
Kilmington Devon 8 B5
Kilmington Wilts 9 G1
Kilmington Common 9 G1
Kilmorack 83 Q6
Kilmore (A' Chille Mhòr) A&B 79 K8
Kilmore High 82 G10
Kilmory A&B 73 F2
Kilmory A&B 73 F3
Kilmory High 78 G3
Kilmory High 82 D10
Kilmory NAyr 66 D1
Kilmuir High 84 A6
Kilmuir High 84 B3
Kilmun 73 K2
Kiln Green Here 29 F7
Kiln Green W'ham 22 B4
Kiln Pit Hill 62 A1
Kilnave 72 A3
Kilncadzow 75 G6
Kilndown 14 C4
Kilnhurst 51 G3
Kilninian 78 F6
Kilninver 79 K8
Kilnsea 53 H1
Kilnsey 56 E3
Kilnwick 59 F5
Kilnwick Percy 58 E4
Kiloran 72 B1
Kilpatrick 66 D1
Kilpeck 28 D5
Kilphedir 87 M8
Kilpin 58 D7
Kilpin Pike 58 D7
Kilrenny 81 L10
Kilsby 31 G1
Kilspindie 80 H8
Kilstay 64 B7
Kilsyth 75 F3
Kiltarlity 83 R6
Kilton Notts 51 H5
Kilton R&C 63 G5
Kilton Som 7 K1
Kilton Thorpe 63 G5
Kiltyrie 80 B7
Kilvaxter 82 D4
Kilve 7 K1
Kilverstone 44 C7
Kilvington 42 B1
Kilwinning 74 B6
Kimberley Norf 44 E5
Kimberley Notts 41 H1
Kimberworth 51 G3
Kimble Wick 22 B1
Kimblesworth 62 C2
Kimbolton Cambs 32 D2
Kimbolton Here 28 E2
Kimbridge 10 E2
Kimcote 41 H7
Kimmeridge 9 J7
Kimmerston 77 H7
Kimpton Hants 21 F7
Kimpton Herts 32 E7
Kinbrace 87 L6
Kinbuck 80 C10
Kincaple 81 K9
Kincardine Fife 75 H2
Kincardine High 84 A2
Kincardine O'Neil 85 K11
Kinclaven 80 G7
Kincorth 85 P10
Kincraig 84 C10
Kindallachan 80 E5
Kineton Glos 30 B6
Kineton Warks 30 E3
Kineton Green 40 D7
Kinfauns 80 G8
King Sterndale 50 C5
Kingarth 73 J5
Kingcoed 19 H1
Kingerby 52 D3
Kingham 30 D6
Kingholm Quay 65 K3
Kinghorn 76 A2
Kinglassie 76 A1
Kingoodie 81 J8
King's Acre 28 D4
King's Bank 14 D5
King's Bromley 40 D4
Kings Caple 28 E6

Olgrinmore 87 P4
Oliver 69 G1
Oliver's Battery 11 F2
Ollaberry 89 M4
Ollerton ChesE 49 G5
Ollerton Notts 51 J6
Ollerton Shrop 39 F3
Olmstead Green 33 K4
Olney 32 B3
Olrig House 87 P3
Olton 40 D7
Olveston 19 K3
Olympia GtLon W14 8UX 99 G7
Ombersley 29 H2
Ompton 51 J6
Onchan 54 C6
Onecote 50 C7
Onehouse 34 E3
Ongar Hill 43 J3
Ongar Street 28 C2
Onibury 28 D1
Onich (Omhanaich) 79 M4
Onllwyn 27 H7
Onneley 39 G1
Onslow Green 33 K7
Onslow Village 22 C7
Opinan High 82 H3
Opinan High 83 J1
Orange Lane 77 F6
Orbliston 84 H5
Orbost 82 C6
Orby 53 H6
Orcadia 73 K4
Orchard 73 K2
Orchard Portman 8 B2
Orcheston 20 D7
Orcop 28 D6
Orcop Hill 28 D6
Ord 82 G9
Ordhead 85 L9
Ordie 85 J10
Ordiequish 84 H5
Ordsall 51 J5
Ore 14 D6
Oreham Common 13 F5
Oreston 5 F5
Oreton 29 F1
Orford Suff 35 J4
Orford Warr 49 F3
Organford 9 J5
Orgreave 40 D4
Orkney Islands 89 C5
Orlestone 14 E4
Orleton Here 28 D2
Orleton Worcs 29 F2
Orleton Common 28 D2
Orlingbury 32 B1
Ormesby 63 F5
Ormesby St. Margaret 45 J4
Ormesby St. Michael 45 J4
Ormidale 73 J2
Ormiston 76 C4
Ormsaigmore 78 F4
Ormsary 73 F3
Ormskirk 48 D2
Oronsay 72 B2
Orpington 23 H5
Orrell GtMan 48 E2
Orrell Mersey 48 C3
Orrisdale 54 C4
Orroland 65 H6
Orsett 24 C3
Orsett Heath 24 C3
Orslow 40 A4
Orston 42 A1
Orton Cumb 61 H6
Orton N'hants 32 B1
Orton Longueville 42 E6
Orton Rigg 60 E1
Orton Waterville 42 E6
Orton-on-the-Hill 41 F5
Orwell 33 G3
Osbaldeston 56 B6
Osbaldwick 58 C4
Osbaston Leics 41 G5
Osbaston Shrop 38 C3
Osbaston Tel&W 38 E4
Osbaston Hollow 41 G5
Osborne 11 G5
Osborne House IoW PO32 6JX 11 G5
Osbournby 42 D2
Oscroft 48 E6
Ose 82 D6
Osgathorpe 41 G4
Osgodby Lincs 52 D3
Osgodby NYorks 58 C6
Osgodby NYorks 59 G1
Oskaig 82 F7
Osleston 40 E2
Osmaston Derby 41 F2
Osmaston Derbys 40 D1
Osmington 9 G6
Osmington Mills 9 G6
Osmondthorpe 57 J6
Osmotherley 62 E7
Osnaburgh (Dairsie) 81 K9
Ospringe 25 G5
Ossett 57 H7
Ossett Street Side 57 H7
Ossington 51 K6
Ostend 25 F2
Osterley 22 E4
Osterley Park & House GtLon TW7 4RB 99 C8
Oswaldkirk 58 C2
Oswaldtwistle 56 C7
Oswestry 38 B3
Oteley 38 D2
Otford 23 J6
Otham 14 C2
Otherton 40 B4
Othery 8 C1
Otley Suff 35 G3
Otley WYorks 57 H5
Otter 73 H3
Otter Ferry 73 H2
Otterbourne 11 F2
Otterburn NYorks 56 D4
Otterburn N'umb 70 D4
Otterburn Camp 70 D4
Otterden Place 14 E2
Otterham 4 B1
Otterham Quay 24 E5
Otterhampton 19 F7
Ottershaw 22 D5
Otterswick 89 P4
Otterton 7 J7
Otterton Mill Devon EX9 7HG 7 J7
Otterwood 11 F4
Ottery St. Mary 7 J6
Ottinge 15 G3
Ottringham 59 J7
Oughterby 60 D1
Oughtershaw 56 D1
Oughterside 60 C2
Oughtibridge 51 F3
Oulston 58 B2
Oulton Cumb 60 D1
Oulton Norf 45 F3
Oulton Staffs 39 G3
Oulton Staffs 40 B2
Oulton Suff 45 K6
Oulton WYorks 57 J7
Oulton Broad 45 K6
Oulton Grange 40 B2
Oulton Street 45 F3
Oultoncross 40 B2
Oundle 42 D7
Our Dynamic Earth Edin EH8 8AS 126 J4
Ousby 61 H3
Ousdale 87 N7
Ousden 34 B3
Ousefleet 58 E7
Ouston Dur 62 C1
Ouston N'umb 71 F6
Out Newton 59 K7
Out Rawcliffe 55 H5
Out Skerries Airstrip 89 Q5
Outcast 55 G2
Outchester 77 K7
Outgate 60 E7
Outhgill 61 J6
Outlands 39 G3
Outlane 50 C1
Outwell 43 J5
Outwood Surr 23 G7
Outwood WYorks 57 J7
Outwoods 39 G4
Ouzlewell Green 57 J7
Ovenden 57 F7
Over Cambs 33 G1
Over ChesW&C 49 F6
Over SGlos 19 J3
Over Burrows 40 E2
Over Compton 8 E3
Over Dinsdale 62 D5
Over End 50 E5
Over Green 40 D6
Over Haddon 50 E6
Over Hulton 49 F2
Over Kellet 55 J3
Over Kiddington 31 F6
Over Monnow 28 D7
Over Norton 30 E6
Over Peover 49 G5
Over Silton 62 E7
Over Stowey 7 K2
Over Stratton 8 D3
Over Tabley 49 G4
Over Wallop 10 D1
Over Whitacre 40 E6
Over Winchendon (Upper Winchendon) 31 J7
Over Worton 31 F6
Overbury 29 J5
Overcombe 9 F6
Overgreen 51 F5
Overleigh 8 D1
Overpool 48 C5
Overseal 40 E4
Overslade 31 F1
Oversland 15 F2
Oversley Green 30 B3
Overstone 32 B2
Overstrand 45 G1
Overthorpe 31 F4
Overton ChesW&C 48 E5
Overton Hants 21 J7
Overton Lancs 55 H4
Overton NYorks 58 B4
Overton Shrop 28 E1
Overton Swan 17 H7
Overton (Owrtyn) Wrex 38 C1
Overton WYorks 50 E1
Overton Bridge 38 C1
Overtown Lancs 56 B2
Overtown NLan 75 G5
Overtown Swin 20 E4
Overy 21 J2
Oving Bucks 31 J6
Oving WSuss 12 C6
Ovingdean 13 G6
Ovingham 71 F7
Ovington Dur 62 B5
Ovington Essex 34 B4
Ovington Hants 11 G1
Ovington Norf 44 D5
Ovington N'umb 71 F7
Ower Hants 10 E3
Ower Hants 11 F4
Owermoigne 9 G6
Owler Bar 50 E5
Owlpen 20 A2
Owl's Green 35 G2
Owlswick 22 A1
Owmby 52 D2
Owmby-by-Spital 52 D4
Owslebury 11 G2
Owston 42 A5
Owston Ferry 52 B2
Owstwick 59 J6
Owthorpe 41 J2
Oxborough 44 B5
Oxcliffe Hill 55 H3
Oxcombe 53 G5
Oxen End 33 K6
Oxen Park 55 G1
Oxenhall 29 G6
Oxenholme 61 G7
Oxenhope 57 F6
Oxenpill 19 H7
Oxenton 29 J5
Oxenwood 21 G6
Oxford 21 J1
Oxford Cathedral Oxon OX1 1AB 134 Oxford
Oxford Story Oxon OX1 3AJ 134 Oxford
Oxford University Museum of Natural History Oxon OX1 3PW 134 Oxford
Oxhey 22 E2
Oxhill 30 E4
Oxley 40 B5
Oxley Green 34 D7
Oxley's Green 13 K4
Oxnam 70 C2
Oxnead 45 G3
Oxnop Ghyll 61 L7
Oxshott 22 E5
Oxspring 50 E2
Oxted 23 G6
Oxton Mersey 48 C4
Oxton Notts 51 J7
Oxton ScBord 76 C5
Oxwich 17 H7
Oxwich Green 17 H7
Oxwick 44 D3
Oykel Bridge 86 F9
Oyne 85 L8
Ozleworth 20 A2

P

Pabbay 88 E9
Packington 41 F4
Packwood 30 C1
Padanaram 81 K5
Padarn Country Park Gwyn LL55 4TY 46 D6
Padbury 31 J5
Paddington 23 F3
Paddlesworth 15 G3
Paddock 14 E2
Paddock Wood 23 K7
Paddockhole 69 H5
Paddolgreen 38 E2
Padeswood 48 B6
Padfield 50 C3
Padiham 56 C6
Padside 57 G4
Padstow 3 G1
Padworth 21 K5
Paganhill 20 B1
Pagham 12 B7
Paglesham Churchend 25 F2
Paglesham Eastend 25 F2
Paignton 5 J4
Paignton & Dartmouth Steam Railway Torbay TQ4 6AF 5 J5
Paignton Pier Torbay TQ4 6BW 5 J4
Paignton Zoo Torbay TQ4 7EU 5 J5
Pailton 41 G7
Paine's Corner 13 K4
Painscastle 28 A4
Painshawfield 71 F7
Painswick 20 B1
Pairc Shiaboist 88 H3
Paisley 74 C4
Pakefield 45 K6
Pakenham 34 D2
Palace of Holyroodhouse Edin EH8 8DX 126 J4
Palace of Westminster (Palace of Westminster & Westminster Abbey inc. St Margaret's Church) GtLon SW1A 0AA 101 A7
Palacerigg Country Park NLan G67 3HU 119 M1
Pale 37 J2
Palehouse Common 13 H5
Palestine 21 F7
Paley Street 22 B4
Palgowan 67 H5
Palgrave 35 F1
Pallinsburn House 77 G7
Palmarsh 15 G4
Palmers Cross 22 D7
Palmers Green 23 G2
Palmerstown 18 E4
Palnackie 65 J5
Palnure 64 E4
Palterton 51 G6
Pamber End 21 K6
Pamber Green 21 K6
Pamber Heath 21 K5
Pamington 29 J5
Pamphill 9 J4
Pampisford 33 H4
Panborough 19 H7
Panbride 81 L7
Pancrasweek 6 A5
Pancross 18 D5
Pandy Gwyn 37 F5
Pandy Mon 28 C6
Pandy Powys 37 J5
Pandy Wrex 38 A2
Pandy Tudur 47 G6
Pandy'r Capel 47 J7
Panfield 34 B6
Pangbourne 21 K4
Pannal 57 J4
Pannal Ash 57 H4
Panshanger 33 F7
Pant 38 B3
Pant Glas 36 D1
Pant Gwyn 37 H3
Pant Mawr 37 H7
Pantasaph 47 K5
Panteg 19 G2
Pantglas 37 G6
Pantgwyn Carmar 17 J3
Pantgwyn Cere 17 F1
Pant-lasau 17 K5
Panton 52 E5
Pant-pastynog 47 J6
Pantperthog 37 G5
Pant-y-dwr 27 J1
Pantyffordd 48 B7
Pant-y-ffridd 38 A5
Pantyffynnon 17 K4
Pantygasseg 19 F2
Pantygelli 28 C7
Pantymwyn 47 K6
Panxworth 45 H4
Papa Stour 89 K6
Papa Stour Airstrip 89 K6
Papa Westray 89 D2
Papa Westray Airfield 89 D2
Papcastle 60 C3
Papple 76 D3
Papplewick 51 H7
Papworth Everard 33 F2
Papworth St. Agnes 33 F2
Par 4 A5
Paradise Park, Newhaven ESuss BN9 0DH 13 H6
Paradise Wildlife Park, Broxbourne Herts EN10 7QA 23 G1
Parbold 48 D1
Parbrook Som 8 E1
Parbrook WSuss 12 D4
Parc 37 H2
Parcllyn 26 B3
Parcrhydderch 27 F3
Parc-Seymour 19 H2
Parc-y-rhôs 17 J1
Pardshaw 60 B4
Parham 35 H2
Parish Holm 68 C1
Park 85 K5
Park Close 56 D5
Park Corner ESuss 13 J3
Park Corner Oxon 21 K3
Park End N'umb 70 D6
Park End Staffs 49 G7
Park End Worcs 29 G1
Park Gate Hants 11 G4
Park Gate Worcs 29 J1
Park Gate WYorks 50 E1
Park Gate WYorks 57 G5
Park Green 35 F2
Park Hill 51 F4
Park Lane 38 D2
Park Langley 23 G5
Park Rose Pottery & Leisure Park ERid YO15 3QF 59 H3
Park Street 22 E1
Parkend Cumb 60 E3
Parkend Glos 19 K1
Parker's Green 23 K7
Parkeston 35 G5
Parkfield Corn 4 D4
Parkfield SGlos 19 K4
Parkfield WMid 40 B6
Parkgate ChesW&C 48 B5
Parkgate D&G 69 F5
Parkgate Kent 14 D4
Parkgate Surr 23 F7
Parkgate SYorks 51 G3
Parkham 6 B3
Parkham Ash 6 B3
Parkhead 74 E4
Parkhill Angus 81 M6
Parkhill P&K 80 G6
Parkhouse 19 H1
Parkhurst 11 F5
Parkmill 17 J7
Parkneuk 81 N3
Parkside 48 C7
Parkstone 10 B5
Parkway 8 E2
Parley Cross 10 B5
Parley Green 10 C5
Parracombe 6 E1
Parrog 16 D2
Parson Cross 51 F3
Parson Drove 43 G5
Parsonage Green 33 K7
Parsonby 60 C3
Partick 74 D4
Partington 49 G3
Partney 53 H6
Parton Cumb 60 A4
Parton D&G 65 G3
Partridge Green 12 E5
Parwich 50 D7
Paslow Wood Common 23 J1
Passenham 31 J5
Passfield 12 B3
Passingford Bridge 23 J2
Paston 45 H2
Paston Street 45 H2
Pasturefields 40 B3
Patchacott 6 C6
Patcham 13 G6
Patchetts Green 22 E2
Patching 12 D6
Patchole 6 E1
Patchway 19 K3
Pateley Bridge 57 G3
Path of Condie 80 F9
Pathe 8 C1
Pathfinder Village 7 G6
Pathhead Aber 81 N4
Pathhead EAyr 68 B2
Pathhead Fife 76 A1
Pathhead Midlo 76 B4
Pathlow 30 C3
Patmore Heath 33 H6
Patna 67 J2
Patney 20 D6
Patrick 54 B5
Patrick Brompton 62 C7
Patrington 59 K7
Patrington Haven 59 K7
Patrishow 28 B6
Patrixbourne 15 G2
Patterdale 60 E5
Pattingham 40 A6
Pattishall 31 H3
Pattiswick 34 C6
Paul 2 B6
Paulerspury 31 J4
Paull 59 H7
Paull Holme 59 H7
Paul's Green 2 D5
Paulton 19 K6
Paultons Park Hants SO51 6AL 10 E3
Pauperhaugh 71 F4
Pave Lane 39 G4
Pavenham 32 C3
Pavilion Gardens, Buxton Derbys SK17 6XN 50 C5
Pawlett 19 G7
Pawston 77 G7
Paxford 30 C5
Paxhill Park 13 G4
Paxton 77 H5
Payden Street 14 E2
Payhembury 7 J5
Paythorne 56 D4
Peacehaven 13 H6
Peacemarsh 9 H2
Peachley 29 H3
Peak Dale 50 C5
Peak Forest 50 D5
Peakirk 42 E5
Pean Hill 25 H5
Pear Tree 41 F2
Pearson's Green 23 K7
Peartree 33 F7
Peartree Green Essex 23 J2
Peartree Green Here 29 F5
Pease Pottage 13 F3
Peasedown St. John 20 A6
Peasehill 41 G1
Peaseland Green 44 E4
Peasemore 21 H4
Peasenhall 35 H2
Peaslake 22 D7
Peasley Cross 48 E3
Peasmarsh ESuss 14 D5
Peasmarsh Surr 22 C7
Peaston 76 C4
Peastonbank 76 C4
Peathill 85 P4
Peathrow 62 B4
Peatling Magna 41 H6
Peatling Parva 41 H7
Peaton 38 E7
Pebble Coombe 23 F6
Pebmarsh 34 C5
Pebworth 30 C4
Pecket Well 56 E7
Peckforton 48 E7
Peckham 23 G4
Peckleton 41 G5
Pedham 45 H4
Pedmore 40 B7
Pedwell 8 D1
Peebles 76 A6
Peel IoM 54 B5
Peel Lancs 55 G6
Peening Quarter 14 D5
Peggs Green 41 G4
Pegsdon 32 E5
Pegswood 71 H5
Pegwell 25 K5
Peinchorran 82 F7
Pelaw 71 H7
Pelcomb 16 C4
Pelcomb Bridge 16 C4
Pelcomb Cross 16 C4
Peldon 34 D7
Pellon 57 F7
Pelsall 40 C5
Pelton 62 C1
Pelutho 60 C2
Pelynt 4 C5
Pemberton 48 E2
Pembrey (Pen-bre) 17 H5
Pembrey Country Park Carmar SA16 0EJ 17 H5
Pembridge 28 C3
Pembroke (Penfro) 16 C5
Pembroke Dock (Doc Penfro) 16 C5
Pembury 23 K7
Penallt 28 E7
Penally 16 E6
Penalt 28 E6
Penare 3 G4
Penarran 38 A7
Penarth 18 E4
Pen-bont Rhydybeddau 37 F7
Penboyr 17 G2
Penbryn 26 B3
Pencader 17 H2
Pen-cae 26 D3
Pen-cae-cwm 47 H6
Pencaenewydd 36 D1
Pencaitland 76 C4
Pencarnisiog 46 B5
Pencarreg 17 J1
Pencarrow 4 B2
Pencelli 27 K6
Pen-clawdd 17 J6
Pencoed 18 C3
Pencombe 28 E3
Pencoyd 28 E6
Pencraig Here 28 E6
Pencraig Powys 37 K3
Pendeen 2 A5
Penderyn 18 C1
Pendine (Pentywyn) 17 F5
Pendlebury 49 G2
Pendleton 56 C6
Pendock 29 G5
Pendoggett 4 A3
Pendomer 8 E3
Pendoylan 18 D4
Penegoes 37 G5
Penelewey 3 F4
Pen-ffordd 16 D3
Pengam 18 E2
Penge 23 G4
Pengenffordd 28 A6
Pengorffwysfa 46 C3
Pengover Green 4 C4
Pen-groes-oped 19 G1
Pengwern 47 J5
Penhale 2 D7
Penhallow 3 F3
Penhalvean 2 E5
Penhelig 37 F6
Penhill 20 E3
Penhow 19 H2
Penhurst 13 K5
Peniarth 37 F5
Penicuik 76 A4
Peniel 17 H3
Penifiler 82 E6
Peninver 66 B1
Penisa'r Waun 46 D6
Penisarcwm 37 K4
Penishawain 27 K5
Penistone 50 E2
Penjerrick 2 E5
Penketh 48 E4
Penkill 67 G4
Penkridge 40 B4
Penlean 4 C1
Penley 38 D2
Penllech 36 B2
Penllergaer 17 K6
Pen-llyn IoA 46 B4
Penllyn VGlam 18 C4
Pen-lôn 46 C6
Penmachno 47 F7
Penmaen 17 J7
Penmaenan 47 F5
Penmaenmawr 47 F5
Penmaenpool (Llyn Penmaen) 37 F4
Penmark 18 D5
Penmon 46 E4
Penmorfa 36 E1
Penmynydd 46 D5
Penn Bucks 22 C2
Penn WMid 40 A6
Penn Street 22 C2
Pennal 37 F5
Pennal-isaf 37 G5
Pennan 85 N4
Pennance 2 E4
Pennant Cere 26 E2
Pennant Powys 37 H6
Pennant Melangell 37 K3
Pennar 16 C5
Pennard 17 J7
Pennerley 38 C6
Penninghame 64 D4
Pennington Cumb 55 F2
Pennington Hants 10 E5
Pennington Flash Country Park GtMan WN7 3PA 49 F3
Pennington Green 49 F2
Pennorth 28 A6
Pennsylvania 20 A4
Penny Bridge 55 G1
Pennycross 4 E5
Pennygate 45 H3
Pennyghael 78 G8
Pennyglen 67 G2
Pennygown 78 H6
Pennymoor 7 G4
Penny's Green 45 F6
Pennyvenie 67 J3
Penparc Cere 17 F1
Penparc Pembs 16 B2
Penparcau 36 E7
Penpedairheol 19 G1
Penpethy 4 A2
Penpillick 4 A5
Penpol 3 F5
Penpoll 4 B5
Penponds 2 D5
Penpont D&G 68 D4
Penpont Powys 27 J6
Penprysg 18 C3
Penquit 5 G5
Penrherber 17 F2
Penrhiw 17 F1
Penrhiwceiber 18 D2
Penrhiwgoch 17 J4
Penrhiw-llan 17 G1
Penrhiw-pâl 17 G1
Penrhiwtyn 18 A2
Penrhos Gwyn 36 C2
Penrhos IoA 46 A4
Penrhos Mon 28 D7
Penrhos Powys 27 H7
Penrhos-garnedd 46 D5
Penrhyn Bay (Bae Penrhyn) 47 G4
Penrhyn Castle Gwyn LL57 4HN 46 E5
Penrhyn-coch 37 F7
Penrhyndeudraeth 37 F2
Penrhyn-side 47 G4
Penrhys 18 D2
Penrice 17 H7
Penrith 61 G4
Penrose Corn 3 F1
Penrose Corn 4 C2
Penruddock 60 F4
Penryn 2 E5
Pensarn Carmar 17 H4
Pensarn Conwy 47 H5
Pen-sarn Gwyn 36 D1
Pen-sarn Gwyn 36 E3
Pensax 29 G2
Pensby 48 B4
Penselwood 9 G1
Pensford 19 K5
Pensham 29 J4
Penshaw 62 D1
Penshurst 23 J7
Pensilva 4 C4
Pensnett 40 B7
Penston 76 C3
Pentewan 4 A6
Pentir 46 D6
Pentire 2 E2
Pentireglaze 3 G1
Pentlepoir 16 E5
Pentlow 34 C4
Pentlow Street 34 C4
Pentney 44 B4
Penton Mewsey 21 G7
Pentonville 23 G3
Pentraeth 46 D5
Pentre Powys 28 B2
Pentre Powys 37 K3
Pentre Powys 37 K7
Pentre Powys 38 A7
Pentre Powys 38 B6
Pentre RCT 18 C2
Pentre Shrop 28 C1
Pentre Shrop 38 C4
Pentre Wrex 38 B1
Pentre Wrex 38 C1
Pentre Berw 46 C5
Pentre Ffwrndan 48 B5
Pentre Galar 16 E2
Pentre Gwenlais 17 K4
Pentre Gwynfryn 36 E3
Pentre Halkyn 48 B5
Pentre Isaf 47 G6
Pentre Llanrhaeadr 47 J6
Pentre Maelor 38 C1
Pentre Meyrick 18 C4
Pentre Poeth 17 K6
Pentre Saron 47 J6
Pentre-bach Cere 17 J1
Pentrebach MTyd 18 D1
Pentre-bach Powys 27 J5
Pentrebach RCT 18 D3
Pentrebach Swan 17 J5
Pentre-bont 47 F7
Pentre-bwlch 38 A1
Pentrecagal 17 G1
Pentre-celyn Denb 47 K7
Pentre-celyn Powys 37 H5
Pentre-chwyth 17 K6
Pentreclwydau 18 B1
Pentre-cwrt 17 G2
Pentre-Dolau-Honddu 27 J4
Pentredwr Denb 38 A1
Pentre-dwr Swan 17 K6
Pentrefelin Carmar 17 K3
Pentrefelin Cere 17 K1
Pentrefelin Conwy 47 G5
Pentrefelin Gwyn 36 E2
Pentrefelin Powys 38 A3
Pentrefoelas 47 G7
Pentregat 26 C3
Pentreheyling 38 B6
Pentre-llwyn-llŵyd 27 J3
Pentre-llyn 27 F1
Pentre-llyn-cymmer 47 H7
Pentre-piod 37 H2
Pentre-poeth 19 F3
Pentre'r beirdd 38 A4
Pentre'r Felin 47 G6
Pentre'r-felin 27 J5
Pentre-tafarn-y-fedw 47 G6
Pentre-ty-gwyn 27 H5
Pentrich 51 F7
Pentridge 10 B3
Pentwyn Caerp 18 E1
Pen-twyn Caerp 19 F1
Pentwyn Cardiff 19 F3
Pen-twyn Mon 19 J1
Pentwyn-mawr 18 E2
Pentyrch 18 E3
Penuwch 26 E2
Penwithick 4 A5
Penwood 21 H5
Penwortham 55 J7
Penwortham Lane 55 J7
Penwyllt 27 H7
Pen-y-banc 17 K3
Pen-y-bont Carmar 17 G3
Pen-y-bont Carmar 27 G5
Penybont Powys 28 A2
Pen-y-bont Powys 37 K4
Pen-y-bont Powys 38 B3
Penybontfawr 37 K3
Penybryn Caerp 18 E2
Pen-y-bryn Gwyn 37 F4
Pen-y-bryn Pembs 16 E1
Pen-y-bryn Wrex 38 B1
Pen-y-cae Powys 27 H7
Penycae Wrex 38 B1
Pen-y-cae-mawr 19 H2
Pen-y-cefn 47 K5
Pen-y-clawdd 19 H1
Pen-y-coedcae 18 D3
Penycwm 16 B3
Pen-y-Darren 18 D1
Pen-y-fai 18 B3
Pen-y-ffordd Flints 47 K4
Penyffordd Flints 48 C6
Penyffridd 46 D7
Pen-y-gaer 28 A6
Pen-y-garn Carmar 17 J2
Pen-y-garn Cere 37 F7
Penygarn Torfaen 19 F1
Penygarnedd 38 A3
Pen-y-garreg 27 K4
Pen-y-Graig Gwyn 36 B2
Penygraig RCT 18 D2
Penygroes Carmar 17 J4
Penygroes Gwyn 46 C7
Pen-y-Gwryd Hotel 46 E7
Pen-y-lan 18 E4
Pen-y-Mynydd Carmar 17 H5
Penymynydd Flints 48 C6
Pen-y-parc 48 B6
Pen-y-Park 28 B4
Pen-yr-englyn 18 C2
Pen-yr-heol Mon 28 D7
Penyrheol Swan 17 J6
Pen-y-sarn 46 C3
Pen-y-stryt 47 K7
Penywaun 18 C1
Penzance 2 B5
Penzance Heliport 2 B5
People's Palace & Winter Gardens Glas G40 1AT 127 H6
Peopleton 29 J3
Peover Heath 49 G5
Peper Harow 22 C7
Peplow 39 F3
Pepper Arden 62 C6
Pepper's Green 33 K7
Perceton 74 B6
Perham Down 21 F7
Periton 7 H1
Perivale 22 E3
Perkins Beach 38 C5
Perkin's Village 7 J6
Perlethorpe 51 J5
Perran Downs 2 C5
Perranarworthal 2 E5
Perranporth 2 E3
Perranuthnoe 2 C6
Perranzabuloe 2 E3
Perrott's Brook 20 D1
Perry Barr 40 C6
Perry Crofts 40 E5
Perry Green Essex 34 C6
Perry Green Herts 33 H7
Perry Green Wilts 20 C3
Perry Street 24 C4
Perrymead 20 A5
Pershall 40 A3
Pershore 29 J4
Pert 81 M4
Pertenhall 32 D2
Perth (Peairt) 80 G8
Perthcelyn 18 D2
Perthy 38 C2
Perton 40 A6
Pestalozzi Children's Village 14 C6
Peter Tavy 5 F3
Peterborough 42 E6
Peterculter 85 N10
Peterhead 85 R6
Peterlee 62 E2
Peter's Green 32 E7
Peters Marland 6 C4
Petersfield 11 J2
Petersfinger 10 C2
Peterstone Wentlooge 19 F3
Peterston-super-Ely 18 D4
Peterstow 28 E6
Petham 15 G2
Petrockstowe 6 D5
Pett 14 D6
Pettaugh 35 F3
Petteril Green 61 F2
Pettinain 75 H6
Pettistree 35 G3
Petton Devon 7 J3
Petton Shrop 38 D3
Petts Wood 23 H5
Petty France 20 A3
Pettycur 76 A2
Pettymuick 85 P8
Petworth 12 C4
Pevensey 13 K6
Pevensey Bay 13 K6
Pewsey 20 E5
Pheasant's Hill 22 A3
Philham 6 A3
Philharmonic Hall, Liverpool Mersey L1 9BP 131 E5
Philiphaugh 69 K1
Phillack 2 C5
Philleigh 3 F5
Philpstoun 75 J3
Phocle Green 29 F6
Phoenix Green 22 A6
Photographer's Gallery, The, London GtLon WC2H 7HY 132 E3
Pibsbury 8 D2
Pica 60 B4
Piccadilly Corner 45 G7
Pickerells 23 J1
Pickering 58 D1
Pickering Nook 62 B1
Picket Piece 21 G7
Picket Post 10 C4
Pickford Green 40 E7
Pickhill 57 J1
Picklescott 38 D6
Pickmere 49 F5
Pickney 7 K3
Pickstock 39 G3
Pickston 80 E8
Pickup Bank 56 C7
Pickwell Devon 6 C1
Pickwell Leics 42 A4
Pickworth Lincs 42 D2
Pickworth Rut 42 C4
Picton ChesW&C 48 D5
Picton NYorks 62 E6
Piddinghoe 13 H6
Piddington Bucks 22 B2
Piddington N'hants 32 B3
Piddington Oxon 31 H7
Piddlehinton 9 G5
Piddletrenthide 9 G5
Pidley 33 G1
Piece Hall, Halifax WYorks HX1 1RE 114 C5
Piercebridge 62 C5
Pierowall 89 D3
Pigdon 71 G5
Pike Hill 56 D6
Pikehall 50 D7
Pikeshill 10 D4
Pilgrims Hatch 23 J2
Pilham 52 B3
Pill 19 J4
Pillaton Corn 4 D4
Pillaton Staffs 40 B4
Pillerton Hersey 30 D4
Pillerton Priors 30 D4
Pilleth 28 B2
Pilley Hants 10 E5
Pilley SYorks 51 F2
Pilling 55 H5
Pilling Lane 55 G5
Pillowell 19 K1
Pilning 19 J3
Pilsbury 50 D6
Pilsdon 8 D5
Pilsgate 42 D5
Pilsley Derbys 50 E5
Pilsley Derbys 51 G6
Pilson Green 45 H4
Piltdown 13 H4
Pilton Devon 6 D2
Pilton N'hants 42 D7
Pilton Rut 42 C5
Pilton Som 19 J7
Pilton Swan 17 H7
Pilton Green 17 H7
Pimhole 49 H1
Pimlico 22 D1
Pimperne 9 J4
Pin Mill 35 G5
Pinchbeck 43 F3
Pinchbeck Bars 43 F3
Pinchbeck West 43 F3
Pincheon Green 51 J1
Pinchinthorpe 63 F5
Pindon End 31 J4
Pinehurst 20 E3
Pinfold 48 C1
Pinged 17 H5
Pinhay 8 C5
Pinhoe 7 H6
Pinkneys Green 22 B3
Pinley Green 30 D2
Pinminnoch 67 F4
Pinmore 67 G4
Pinn 7 K7
Pinner 22 E3
Pinner Green 22 E2
Pinvin 29 J4
Pinwherry 67 F5
Pinxton 51 G7
Pipe & Lyde 28 E4
Pipe Gate 39 G1
Pipe Ridware 40 C4
Pipehill 40 C5
Piperhall 73 J5
Piperhill 84 C5
Pipers Pool 4 C2
Pipewell 42 B7
Pippacott 6 D2
Pipton 28 A5
Pirbright 22 C6
Pirnmill 73 G6
Pirton Herts 32 E5
Pirton Worcs 29 H4
Pisgah 80 C10
Pishill 22 A3
Pistyll 36 C1
Pitagowan 80 D4
Pitcairngreen 80 F8
Pitcaple 85 M8
Pitch Green 22 A1
Pitch Place Surr 12 B3
Pitch Place Surr 22 C6
Pitchcombe 20 B1
Pitchcott 31 J6
Pitchford 38 E5
Pitcombe 9 F1
Pitcot 18 B4
Pitcox 76 E3
Pitcur 80 H7
Pitlessie 81 J10
Pitlochry 80 E5
Pitman's Corner 35 F2
Pitmedden 85 N8
Pitminster 8 B3
Pitmunie 85 L9
Pitney 8 D2
Pitscottie 81 K9
Pitsea 24 D3
Pitsford 31 J2
Pitsford Hill 7 J2
Pitstone 32 C7
Pitt Devon 7 J4
Pitt Hants 11 F2
Pitt Rivers Museum Oxon OX1 3PP 134 Oxford
Pittendreich 84 F4
Pittentrail 87 K9
Pittenweem 81 L10
Pitteuchar 76 A1
Pittington 62 D2
Pitton Swan 17 H7
Pitton Wilts 10 D1
Pittville 29 J6
Pity Me 62 C2
Pityme 3 G1
Pixey Green 35 G1
Pixley 29 F5
Place Newton 58 E2
Plaidy 85 M5
Plain Dealings 16 D4
Plainfield 70 E3
Plains 75 F4
Plainsfield 7 K2
Plaish 38 E6
Plaistow GtLon 23 G3
Plaistow WSuss 12 D3
Plaitford 10 D3
Plaitford Green 10 D2
Plas 17 H3
Plas Gwynant 46 E7
Plas Isaf 37 K1
Plas Llwyd 47 H5
Plas Llwyngwern 37 G5
Plas Llysyn 37 J6
Plas Nantyr 38 A2
Plashett 17 F5
Plasisaf 47 H6
Plas-rhiw-Saeson 37 J5
Plastow Green 21 J5
Plas-yn-Cefn 47 J5
Platt 23 K6
Platt Bridge 49 F2
Platt Lane 38 E2
Platt's Heath 14 D2
Plawsworth 62 C2
Plaxtol 23 K6
Play Hatch 22 A4
Playden 14 E5
Playford 35 G4
Playing Place 3 F4
Playley Green 29 G5
Plealey 38 D5
Plean 75 G2
Pleasant Valley 33 J5
Pleasington 56 B7
Pleasley 51 H6
Pleasleyhill 51 H6
Pleasure Beach, Great Yarmouth Norf NR30 3EH 45 K5
Pleasure Island Theme Park NELincs DN35 0PL 53 G2
Pleasurewood Hills Theme Park Suff NR32 5DZ 45 K6
Pleck Dorset 9 G3
Pleck WMid 40 B6
Pledgdon Green 33 J6
Pledwick 51 F1
Plemstall 48 D5
Plenmeller 70 C7
Pleshey 33 K7
Plessey Woods Country Park N'umb NE22 6AN 71 H5
Plockton (Am Ploc) 83 J7
Plomer's Hill 22 B2
Plot Gate 8 E1
Plough Hill 41 F6
Plowden 38 C7
Ploxgreen 38 C5
Pluckley 14 E3
Pluckley Thorne 14 E3
Plucks Gutter 25 J5
Plumbland 60 C3
Plumbley 51 G4
Plumley 49 G5
Plumpton Cumb 61 F3
Plumpton ESuss 13 G5
Plumpton N'hants 31 H4
Plumpton Green 13 G5
Plumpton Head 61 G3
Plumstead GtLon 23 H4
Plumstead Norf 45 F2
Plumtree 41 J2
Plungar 42 A2
Plush 9 G4
Plusha 4 C2
Plushabridge 4 D3
Plwmp 26 C3
Plym Bridge 5 F5
Plymouth 4 E5
Plymouth Pavilions PL1 3LF 134 Plymouth
Plympton 5 F5
Plymstock 5 F5
Plymtree 7 J5
Pockley 58 C1
Pocklington 58 E5
Pockthorpe 44 E3
Pocombe Bridge 7 G6
Pode Hole 43 F3
Podimore 8 E2
Podington 32 C2
Podmore 39 G2
Podsmead 29 H7
Poffley End 30 E7
Point Clear 34 E7
Pointon 42 E2
Polapit Tamar 6 B7
Polbae 64 C3
Polbain 86 B8
Polbathic 4 D5
Polbeth 75 J4
Poldean 69 G4
Pole Moor 50 C1
Polebrook 42 D7
Polegate 13 J6
Polesden Lacey Surr RH5 6BD 22 E6
Polesworth 40 E5
Polgooth 3 G3
Polgown 68 C3
Poling 12 D6
Poling Corner 12 D6
Polkemmet Country Park WLoth EH47 0AD 75 H4
Polkerris 4 A5
Polla 86 F4
Pollardras 2 D5
Pollington 51 J1
Polloch 79 J4
Pollok 74 D4
Pollokshaws 74 D4
Pollokshields 74 D4
Polmassick 3 G4
Polmont 75 H3
Polnoon 74 D5
Polperro 4 C5
Polruan 4 B5
Polsham 19 J7
Polstead 34 D5
Polstead Heath 34 D4
Poltalloch 73 G1
Poltimore 7 H6
Polton 76 A4
Poltross Burn Milecastle (Frontiers of the Roman Empire) Cumb CA8 7BJ 70 B7
Polwarth 77 F5
Polyphant 4 C2
Polzeath 3 G1
Pomphlett 5 F5
Pond Street 33 H5
Ponders End 23 G2
Pondersbridge 43 F6
Ponsanooth 2 E5
Ponsonby 60 B6
Ponsongath 2 E7
Ponsworthy 5 H3
Pont Aber 27 G6
Pont Aberglaslyn 36 E1
Pont ar Hydfer 27 H6
Pont Crugnant 37 H6
Pont Cyfyng 47 F7
Pont Dolgarrog 47 F6
Pont Pen-y-benglog 46 E6
Pont Rhyd-sarn 37 H3
Pont Rhyd-y-cyff 18 B3
Pont Walby 18 B1
Pont yr Alwen 47 H7
Pontamman 17 K4
Pontantwn 17 H4
Pontardawe 18 A1
Pontarddulais 17 J5
Pont-ar-gothi 17 J3
Pont-ar-llechau 27 G6
Pontarsais 17 H3
Pontblyddyn 48 B6
Pontbren Llwyd 18 C1
Pontcysyllte Aqueduct (Pontcysyllte Aqueduct & Canal) Denb/Powys/Wrex 38 B1
Pontefract 57 K7
Ponteland 71 G6
Ponterwyd 37 G7
Pontesbury 38 D5
Pontesbury Hill 38 C5
Pontesford 38 D5
Pontfadog 38 B2
Pontfaen Pembs 16 D2
Pont-faen Powys 27 J5
Pontgarreg 26 C3
Ponthir 19 G2
Ponthirwaun 17 F1
Pontllanfraith 18 E2
Pontlliw 17 K5
Pontllyfni 46 C7
Pontlottyn 18 E1
Pontneddfechan 18 C1
Pontrhydfendigaid 27 G2
Pontrhydyfen 18 A2
Pont-rhyd-y-groes 27 G1
Pontrhydyrun 19 F2
Pontrilas 28 C6
Pontrobert 38 A4
Pont-rug 46 D6
Ponts Green 13 K5
Pontshill 29 F6
Pont-siân 17 H1
Pontsticill 27 K7
Pontwelly 17 H2
Pontyates (Pont-iets) 17 H5
Pontyberem 17 J4
Pont-y-blew 38 C2
Pontybodkin 48 B7
Pontyclun 18 D3
Pontycymer 18 C2
Pontygwaith 18 D2
Pontymister 19 F2
Pontymoel 19 F1
Pont-y-pant 47 F7
Pontypool 19 F1
Pontypridd 18 D2
Pont-y-rhyl 18 C3
Pontywaun 19 F2
Pooksgreen 10 E3
Pool Corn 2 D4
Pool WYorks 57 H5
Pool Bank 55 H1
Pool Green 40 C5
Pool Head 28 E3
Pool of Muckhart 80 F10
Pool Quay 38 B4
Pool Street 34 B5
Poole 10 B5
Poole Keynes 20 C2
Poole Pottery Poole BH15 1HJ 91 B6
Poolend 49 J7
Poolewe 83 J2
Pooley Bridge 61 F4
Pooley Street 44 E7
Poolfold 49 H7
Poolhill 29 G6
Poolsbrook 51 G5
Poolthorne Farm 52 D2
Pope Hill 16 C4
Popeswood 22 B5
Popham 21 J7
Poplar 23 G3
Porchfield 11 F5
Porin 83 P5
Poringland 45 G5
Porkellis 2 D5
Porlock 7 G1
Porlock Weir 7 G1
Port Appin (Port na h-Apainn) 79 L6
Port Askaig 72 C4
Port Bannatyne 73 J4
Port Carlisle 69 H7
Port Charlotte 72 A5
Port Clarence 63 F4
Port Driseach 73 H3
Port e Vullen 54 D4
Port Ellen 72 B6
Port Elphinstone 85 M9
Port Erin 54 A7
Port Erroll 85 Q7
Port Eynon 17 H7
Port Gaverne 4 A2
Port Glasgow 74 B3
Port Henderson 82 H3
Port Isaac 4 A2
Port Logan 64 A6
Port Lympne Wild Animal Park Kent CT21 4PD 15 G4
Port Mòr 78 F3
Port Mulgrave 63 H5
Port nan Long 88 C1
Port of Menteith 80 A10
Port of Ness (Port Nis) 88 L1
Port o'Warren 65 J5
Port Penrhyn 46 D5
Port Quin 3 G1
Port Ramsay 79 K6
Port Solent 11 H4
Port St. Mary 54 B7
Port Sunlight 48 C4
Port Talbot 18 A3
Port Tennant 17 K6
Port Wemyss 72 A5
Port William 64 D6
Portachoillan 73 F5
Portavadie 73 H4
Portbury 19 J4
Portchester 11 H4
Portencross 73 K6
Portesham 9 F6
Portessie 85 J4
Portfield 12 B6
Portfield Gate 16 C4
Portgate 6 C7
Portgordon 84 H4
Portgower 87 N8
Porth Corn 3 F2
Porth RCT 18 D2
Porth Colmon 36 A2
Porth Navas 2 E6
Porthallow Corn 2 E6
Porthallow Corn 4 C5
Porthcawl 18 B4
Porthcothan 3 F1
Porthcurno 2 A6
Porthgain 16 B2
Porthill 40 A1
Porthkerry 18 D5
Porthkerry Country Park VGlam CF62 3BY 18 D5
Porthleven 2 D6
Porthmadog 36 E2
Porthmeor 2 B5
Portholland 3 G4
Porthoustock 3 F6
Porthpean 4 A5
Porthtowan 2 D4
Porthyrhyd Carmar 17 J4
Porthyrhyd Carmar 27 G5
Porth-y-waen 38 B3
Portincaple 74 A1
Portington 58 D6
Portinscale 60 D4
Portishead 19 H4
Portknockie 85 J4
Portland Basin Museum GtMan OL7 0QA 113 L4
Portlethen 85 P11
Portloe 3 G5
Portlooe 4 C5
Portmahomack 84 D2
Portmeirion 36 E2
Portmeirion Village Gwyn LL48 6ET 36 E2
Portmellon 4 A6
Portmore 10 E5
Port-na-Con 86 G3
Portnacroish (Port na Croise) 79 L6
Portnaguran (Port nan Giùran) 88 L4
Portnahaven 72 A5
Portnalong 82 D7
Portnaluchaig 78 H2
Portobello 76 B3
Porton 10 C1
Portpatrick 64 A5
Portreath 2 D4
Portreath Harbour (Cornwall & West Devon Mining Landscape) Corn 2 D4
Portree (Port Righ) 82 E6
Portscatho 3 F5
Portsea 11 H4
Portskerra 87 L3
Portskewett 19 J3
Portslade 13 F6
Portslade-by-Sea 13 F6
Portslogan 64 A5
Portsmouth 11 H5
Portsmouth Guildhall PO1 2AB 135 Portsmouth
Portsmouth Historic Dockyard PO1 3LJ 135 Portsmouth
Portsonachan 79 M8
Portsoy 85 K4
Portway Here 28 C4
Portway Here 28 D4
Portway Here 28 D5
Portway Worcs 30 B1
Portwrinkle 4 D5
Portyerrock 64 E7
Posenhall 39 F5
Poslingford 34 B4
Postbridge 5 G3
Postcombe 22 A2
Postling 15 G4
Postwick 45 G5
Potsgrove 32 C6
Pott Row 44 B3
Pott Shrigley 49 J5
Potten End 22 D1
Potter Brompton 59 F2
Potter Heigham 45 J4
Potter Street 23 H1
Pottergate Street 45 F6
Potterhanworth 52 D6
Potterhanworth Booths 52 D6
Potteries Museum & Art Gallery ST1 3DW 136 Stoke-on-Trent
Potterne 20 C6
Potterne Wick 20 D6
Potternewton 57 J6
Potters Bar 23 F1
Potters Crouch 22 E1
Potter's Green 41 F7
Potters Marston 41 G6
Potterspury 31 J4
Potterton 57 K6
Pottle Street 20 B7
Potto 62 E6
Potton 33 F4
Pott's Green 34 D6
Poughill Corn 6 A5
Poughill Devon 7 G5
Poulshot 20 C6
Poulton 20 E1
Poulton-le-Fylde 55 G6
Pound Bank 29 G4
Pound Green ESuss 13 J4
Pound Green Suff 34 B3
Pound Green Worcs 29 G1
Pound Hill 13 F3
Pound Street 21 H5
Poundbury 9 F5
Poundffald 17 J6
Poundfield 13 J3
Poundgate 13 H4
Poundland 67 F5
Poundon 31 H6
Poundsbridge 23 J7
Poundsgate 5 H3
Poundstock 4 C1
Povey Cross 23 F7
Pow Green 29 G4
Powburn 71 F2
Powderham 7 H7
Powderham Castle Devon EX6 8JQ 7 H7

St. David's Hall CF10 1AH 124 Cardiff
St. Day 2 E4
St. Decumans 7 J1
St. Dennis 3 G3
St. Denys 11 F3
St. Dogmaels (Llandudoch) 16 E1
St. Dogwells 16 C3
St. Dominick 4 E4
St. Donats 18 C5
St. Edith's Marsh 20 C5
St. Endellion 3 G1
St. Enoder 3 F3
St. Erme 3 F4
St. Erney 4 D5
St. Erth 2 C5
St. Erth Praze 2 C5
St. Ervan 3 F1
St. Eval 3 F2
St. Ewe 3 G4
St. Fagans 18 E4
St. Fagans National History Museum Cardiff CF5 6XB 95 A5
St. Fergus 85 Q5
St. Fillans 80 B8
St. Florence 16 D5
St. Gennys 4 B1
St. George Bristol 19 K4
St. George Conwy 47 H5
St. Georges NSom 19 G5
St. George's Tel&W 39 G4
St. George's VGlam 18 D4
St. George's Hall, Liverpool Mersey L1 1JJ 131 D3
St. Germans 4 D5
St. Giles' Cathedral Edin EH1 1RE 126 F4
St. Giles in the Wood 6 D4
St. Giles on the Heath 6 B6
St. Harmon 27 J1
St. Helen Auckland 62 B4
St. Helena 45 F4
St. Helen's ESuss 14 D6
St. Helens IoW 11 H6
St. Helens Mersey 48 E3
St. Helier Chanl 3 J7
St. Helier GtLon 23 F5
St. Hilary Corn 2 C5
St. Hilary VGlam 18 D4
St. Hill 13 G3
St. Ibbs 32 E6
St. Illtyd 19 F1
St. Ippollitts 32 E6
St. Ishmael 17 G5
St. Ishmael's 16 B5
St. Issey 3 G1
St. Ive 4 D4
St. Ives Cambs 33 G1
St. Ives Corn 2 C4
St. Ives Dorset 10 C4
St. James South Elmham 45 H7
St. John Chanl 3 J6
St. John Corn 4 E5
St. John the Baptist Church, Cirencester Glos GL7 2NX 20 D1
St. John's GtLon 23 G4
St. John's IoM 54 B5
St. John's Surr 22 C6
St. John's Worcs 29 H3
St. John's Chapel Devon 6 D3
St. John's Chapel Dur 61 K3
St. John's Fen End 43 J4
St. John's Hall 62 A3
St. John's Highway 43 J4
St. John's Kirk 75 H7
St. John's Town of Dalry 68 B5
St. Judes 54 C4
St. Just 2 A5
St. Just in Roseland 3 F5
St Just Mining District (Cornwall & West Devon Mining Landscape) Corn 2 A5
St. Katherines 85 M7
St. Keverne 2 E6
St. Kew 4 A3
St. Kew Highway 4 A3
St. Keyne 4 C4
St. Lawrence Corn 4 A4
St. Lawrence Essex 25 F1
St. Lawrence IoW 11 G7
St. Lawrence's Church, Eyam Derbys S32 5QH 50 E5
St. Leonards Bucks 22 C1
St. Leonards Dorset 10 C4
St. Leonards ESuss 14 D7
St. Leonards Grange 11 F5
St. Leonard's Street 23 K6
St. Levan 2 A6
St. Lythans 18 E4
St. Mabyn 4 A3
St. Madoes 80 G8
St. Margaret South Elmham 45 H7
St. Margarets Here 28 C5
St. Margarets Herts 33 G7
St. Margarets Wilts 21 F5
St. Margaret's at Cliffe 15 J3
St Margaret's Church (Palace of Westminster & Westminster Abbey inc. St Margaret's Church) GtLon SW1P 3JX 101 A7
St. Margaret's Hope 89 D8
St. Mark's 54 B6
St. Martin Chanl 3 J5
St. Martin Chanl 3 K7
St. Martin Corn 2 E6
St. Martin Corn 4 C5
St. Martin-in-the-Fields Church GtLon WC2N 4JH 132 E3
St. Martin's IoS 2 C1
St. Martins P&K 80 G7
St. Martin's Shrop 38 C2
St Martin's Church (Canterbury Cathedral, St Augustine's Abbey, & St Martin's Church) Kent CT1 1PW 15 G2
St. Mary 3 J6
St. Mary Bourne 21 H6
St. Mary Church 18 D4
St. Mary Cray 23 H5
St. Mary Hill 18 C4
St. Mary Hoo 24 D4
St. Mary in the Marsh 15 F5
St. Mary Magdalene Chapel, Sandringham Norf PE35 6EH 44 A3
St. Mary the Virgin Church, Holy Island N'umb TD15 2RX 77 K6
St. Mary the Virgin Church, Oxford Oxon OX1 4AH 21 J1
St. Mary the Virgin Church, Rye ESuss TN31 7HE 14 E6
St. Marychurch 5 K4
St. Mary's IoS 2 C1
St. Mary's Ork 89 D7
St. Mary's Airport 2 C1
St. Mary's Bay 15 F5
St. Mary's Church, Whitby NYorks YO22 4JT 63 K5
St. Mary's Grove 19 H5
St. Maughans Green 28 D7
St. Mawes 3 F5
St. Mawgan 3 F2
St. Mellion 4 D4
St. Mellons 19 F3
St. Merryn 3 F1
St. Mewan 3 G3
St. Michael Church 8 C1
St. Michael Penkevil 3 F4
St. Michael Caerhays 3 G4
St. Michael South Elmham 45 H7
St. Michaels Fife 81 K8
St. Michaels Kent 14 D4
St. Michaels Worcs 28 E2
St. Michael's Church, Hathersage Derbys S32 1AJ 50 E4
St. Michael's Mount Corn TR17 0HT 2 C5
St. Michael's on Wyre 55 H5
St. Mildred's Church, Whippingham IoW PO32 6LP 11 G5
St. Minver 3 G1
St. Monans 81 L10
St. Mungo Museum of Religious Life & Art Glas G4 0RH 127 H3
St. Mungo's Cathedral Glas G4 0QZ 127 J3
St. Neot 4 B4
St. Neots 32 E2
St. Newlyn East 3 F3
St. Nicholas Pembs 16 C2
St. Nicholas VGlam 18 D4
St. Nicholas at Wade 25 J5
St. Ninians 75 F1
St. Osyth 35 F7
St. Ouen 3 J7
St. Owen's Cross 28 E6
St. Paul's Cathedral, London GtLon EC4M 8AD 132 J2
St. Paul's Cray 23 H5
St. Paul's Walden 32 E6
St. Peter 3 J7
St. Peter Port 3 J5
St. Peter's 25 K5
St. Petrox 16 C6
St. Pinnock 4 C4
St. Quivox 67 H1
St. Ruan 2 E7
St. Sampson 3 J5
St. Saviour Chanl 3 H5
St. Saviour Chanl 3 K7
St. Stephen 3 G3
St. Stephens Corn 4 E5
St. Stephens Corn 6 B7
St. Stephens Herts 22 E1
St. Teath 4 A2
St. Thomas 7 H6
St. Tudy 4 A3
St. Twynnells 16 C6
St. Veep 4 B5
St. Vigeans 81 M6
St. Wenn 3 G2
St. Weonards 28 D6
St. Winnow 4 B5
St. Winwaloe's Church, Gunwalloe Corn TR12 7QE 2 D6
Saintbury 30 C5
Salachail 79 M5
Salcombe 5 H7
Salcombe Regis 7 K7
Salcott 34 D7
Sale 49 G3
Sale Green 29 J3
Saleby 53 H5
Salehurst 14 C5
Salem Carmar 17 K3
Salem Cere 37 F7
Salem Gwyn 46 D7
Salen A&B 78 G6
Salen High 78 H4
Salendine Nook 50 D1
Salesbury 56 B6
Saleway 29 J3
Salford CenBeds 32 C5
Salford GtMan 49 H3
Salford Oxon 30 D6
Salford Priors 30 B3
Salfords 23 F7
Salhouse 45 H4
Saline 75 J1
Salisbury 10 C2
Salisbury Cathedral Wilts SP1 2EF 135 Salisbury
Salkeld Dykes 61 G3
Sallachy High 83 K9
Sallachy High 86 H9
Salle 45 F3
Sally Lunn's House B&NESom BA1 1NX 121 Bath
Salmonby 53 G5
Salperton 30 B6
Salph End 32 D3
Salsburgh 75 G4
Salt 40 B3
Salt Hill 22 C3
Salt Holme 62 E4
Saltaire 57 G6
Saltaire WYorks BD17 7EF 57 G6
Saltash 4 E5
Saltburn 84 B3
Saltburn-by-the-Sea 63 G4
Saltburn Cliff Lift R&C TS12 1DP 63 G4
Saltby 42 B3
Saltcoats Cumb 60 B7
Saltcoats NAyr 74 A6
Saltcotes 55 G7
Saltdean 13 G6
Salterbeck 60 A4
Salterforth 56 D5
Saltergate 63 J7
Salterswall 49 F6
Saltfleet 53 H3
Saltfleetby All Saints 53 H3
Saltfleetby St. Clements 53 H3
Saltfleetby St. Peter 53 H4
Saltford 19 K5
Salthaugh Grange 59 J7
Salthouse 44 E1
Saltley 40 C7
Saltmarshe 58 D7
Saltney 48 C6
Salton 58 D1
Saltrens 6 C3
Saltwell Park, Gateshead T&W NE9 5AX 116 C5
Saltwick 71 G6
Saltwood 15 G4
Salvington 12 E6
Salwarpe 29 H2
Salwayash 8 D5
Sambourne 30 B2
Sambrook 39 G3
Samlesbury 56 B6
Sampford Arundel 7 K4
Sampford Brett 7 J1
Sampford Courtenay 6 E5
Sampford Moor 7 K4
Sampford Peverell 7 J4
Sampford Spiney 5 F3
Samuelston 76 C3
Sanaigmore 72 A3
Sancreed 2 B6
Sancton 59 F6
Sand 19 H7
Sand Hutton 58 C4
Sandaig A&B 78 A6
Sandaig High 82 H10
Sandal Magna 51 F1
Sanday 89 F3
Sanday Airfield 89 F3
Sandbach 49 G6
Sandbank 73 K2
Sandbanks 10 B6
Sandend 85 K4
Sanderstead 23 G5
Sandford Cumb 61 J5
Sandford Devon 7 G5
Sandford Dorset 9 J6
Sandford IoW 11 G6
Sandford NSom 19 H6
Sandford Shrop 38 C3
Sandford Shrop 38 E2
Sandford SLan 75 F6
Sandford Orcas 9 F2
Sandford St. Martin 31 F6
Sandford-on-Thames 21 J1
Sandgarth 89 E6
Sandgate 15 H4
Sandgreen 65 F5
Sandhaven 85 P4
Sandhead 64 A5
Sandhills Dorset 8 E4
Sandhills Dorset 9 F3
Sandhills Surr 12 C3
Sandhills WYorks 57 J6
Sandhoe 70 E7
Sandholme ERid 58 E6
Sandholme Lincs 43 G2
Sandhurst BrackF 22 B5
Sandhurst Glos 29 H6
Sandhurst Kent 14 C5
Sandhurst Cross 14 C5
Sandhutton 57 J1
Sandiacre 41 G2
Sandilands 53 J4
Sandiway 49 F5
Sandleheath 10 C3
Sandleigh 21 H1
Sandling 14 C2
Sandlow Green 49 G6
Sandness 89 K7
Sandon Essex 24 D1
Sandon Herts 33 G5
Sandon Staffs 40 B3
Sandown 11 H6
Sandplace 4 C5
Sandquoy 89 G3
Sandridge Devon 5 J5
Sandridge Herts 32 E7
Sandridge Wilts 20 C5
Sandringham 44 A3
Sandrocks 13 G4
Sandsend 63 J5
Sandside 55 H1
Sandsound 89 M8
Sandtoft 51 K2
Sanduck 7 F7
Sandway 14 D2
Sandwell 40 C7
Sandwell Park Farm WMid B71 4BG 102 F3
Sandwell Valley Country Park WMid B71 4BG 102 F3
Sandwich 15 J2
Sandwick Cumb 60 F5
Sandwick Shet 89 N10
Sandwick (Sanndabhaig) Na H-E. Siar 88 K4
Sandwith 60 A5
Sandy Carmar 17 H5
Sandy CenBeds 32 E4
Sandy Bank 53 F7
Sandy Haven 16 B5
Sandy Lane Wilts 20 C5
Sandy Lane Wrex 38 D1
Sandy Lane WYorks 57 G6
Sandy Way 11 F6
Sandycroft 48 C6
Sandygate Devon 5 J3
Sandygate IoM 54 C4
Sandyhills 65 J5
Sandylands 55 H3
Sandypark 7 F7
Sandyway 28 D6
Sangobeg 86 G3
Sankyn's Green 29 G2
Sannaig 72 D4
Sannox 73 J6
Sanquhar 68 C3
Santon Bridge 60 C6
Santon Downham 44 C7
Sant-y-Nyll 18 D4
Sapcote 41 G6
Sapey Common 29 G2
Sapiston 34 D1
Sapperton Derbys 40 D2
Sapperton Glos 20 C1
Sapperton Lincs 42 D2
Saracen's Head 43 G3
Sarclet 87 R5
Sardis 16 C5
Sarisbury 11 G4
Sark 3 K6
Sark Dark Sky Island Channel Islands 3 K6
Sarn Bridgend 18 C3
Sarn Powys 38 B6
Sarn Bach 36 C3
Sarn Meyllteyrn 36 B2
Sarnau Carmar 17 G4
Sarnau Cere 26 C3
Sarnau Gwyn 37 J2
Sarnau Powys 27 K5
Sarnau Powys 38 B4
Sarnesfield 28 C3
Saron Carmar 17 G2
Saron Carmar 17 K4
Saron Gwyn 46 C7
Saron Gwyn 46 D6
Sarratt 22 D2
Sarre 25 J5
Sarsden 30 D6
Sartfield 54 C4
Satley 62 B2
Satron 61 L7
Satterleigh 6 E3
Satterthwaite 60 E7
Sauchen 85 L9
Saucher 75 G1
Sauchrie 67 H2
Saughall 48 C5
Saughall Massie 48 B4
Saughtree 70 A4
Saul 20 A1
Saundby 51 K4
Saundersfoot 16 E5
Saunderton 22 A1
Saunton 6 C2
Sausthorpe 53 G6
Savalbeg 86 H9
Saverley Green 40 B2
Savile Town 57 H7
Sawbridge 31 G2
Sawbridgeworth 33 H7
Sawdon 59 F1
Sawley Derbys 41 G2
Sawley Lancs 56 C5
Sawley NYorks 57 H3
Sawston 33 H4
Sawtry 42 E7
Saxby Leics 42 B4
Saxby Lincs 52 D4
Saxby All Saints 52 C1
Saxelbye 41 J3
Saxham Street 34 E2
Saxilby 52 B5
Saxlingham 44 E2
Saxlingham Green 45 G6
Saxlingham Nethergate 45 G6
Saxlingham Thorpe 45 G6
Saxmundham 35 H2
Saxon Street 33 K3
Saxondale 41 J2
Saxtead 35 G2
Saxtead Green 35 G2
Saxtead Little Green 35 G2
Saxthorpe 45 F2
Saxton 57 K6
Sayers Common 13 F5
Scackleton 58 C2
Scadabhagh 88 G8
Scaftworth 51 J3
Scagglethorpe 58 E2
Scaitcliffe 56 C7
Scalasaig 72 B1
Scalby ERid 58 E7
Scalby NYorks 59 G1
Scaldwell 31 J1
Scale Houses 61 G2
Scaleby 69 K7
Scalebyhill 69 K7
Scales Cumb 55 F2
Scales Cumb 60 E4
Scalford 42 A3
Scaling 63 H5
Scaliscro 88 G5
Scalloway 89 M9
Scalpay (Eilean Scalpaigh) 88 H8
Scamblesby 53 F5
Scampston 58 E2
Scampton 52 C5
Scaniport 84 A7
Scapa 89 D7
Scapegoat Hill 50 C1
Scarborough 59 G1
Scarborough Sea Life & Marine Sanctuary NYorks YO12 6RP 59 G1
Scarcewater 3 G3
Scarcliffe 51 G6
Scarcroft 57 J5
Scardroy 83 N5
Scarfskerry 87 Q2
Scargill 62 A5
Scarinish 78 B6
Scarisbrick 48 C1
Scarning 44 D4
Scarrington 42 A1
Scarrowhill 61 G1
Scarth Hill 48 D2
Scarthingwell 57 K6
Scartho 53 F2
Scaur D&G 65 J3
Scaur (Kippford) D&G 65 J5
Scawby 52 C2
Scawby Brook 52 C2
Scawton 58 B1
Scayne's Hill 13 G4
Scethrog 28 A6
Schaw 67 J1
Scholar Green 49 H7
Scholes SYorks 51 F3
Scholes WYorks 50 D2
Scholes WYorks 57 G7
Scholes WYorks 57 J6
School Green 49 F6
School House 8 C4
Schoose 60 B4
Sciberscross 87 K9
Science Museum GtLon SW7 2DD 99 G7
Scilly Isles (Isles of Scilly) 2 C1
Scissett 50 E1
Scleddau 16 C2
Sco Ruston 45 G3
Scofton 51 J4
Scole 35 F1
Scone 80 G8
Sconser 82 F7
Scopwick 52 D7
Scoraig 83 L1
Scorborough 59 G5
Scorrier 2 E4
Scorriton 5 H4
Scorton Lancs 55 J5
Scorton NYorks 62 C6
Scot Hay 39 G1
Scotby 60 F1
Scotch Corner 62 C6
Scotch Whisky Heritage Centre Edin EH1 2NE 126 F4
Scotforth 55 H4
Scothern 52 D5
Scotland 42 D2
Scotland End 30 E5
Scotland Street 34 D5
Scotland Street School Museum of Education Glas G5 8QB 127 C6
Scotlandwell 80 G10
Scotnish 73 F2
Scots' Gap 71 F5
Scotston 80 E6
Scotstoun 74 D4
Scotstown 79 K4
Scott Willoughby 42 D2
Scotter 52 B2
Scotterthorpe 52 B2
Scottish Exhibition & Conference Centre (S.E.C.C.) Glas G3 8YW 127 B4
Scottish National Gallery of Modern Art Edin EH4 3DR 126 B4
Scottish National Portrait Gallery Edin EH2 1JD 126 F3
Scottish Parliament Edin EH99 1SP 126 H4
Scottish Seabird Centre, North Berwick ELoth EH39 4SS 76 D2
Scottish Wool Centre, Aberfoyle Stir FK8 3UQ 80 A10
Scottlethorpe 42 D3
Scotton Lincs 52 B3
Scotton NYorks 57 J4
Scotton NYorks 62 B7
Scottow 45 G3
Scoughall 76 E2
Scoulton 44 D5
Scounslow Green 40 C3
Scourie 86 D5
Scousburgh 89 M11
Scouthead 49 J2
Scrabster 87 P2
Scrafield 53 G6
Scrainwood 70 E3
Scrane End 43 G1
Scraptoft 41 J5
Scratby 45 K4
Scrayingham 58 D4
Scredington 42 D1
Scremby 53 H6
Scremerston 77 J6
Screveton 42 A1
Scriven 57 J4
Scronkey 55 H5
Scrooby 51 J3
Scropton 40 D2
Scrub Hill 53 F7
Scruton 62 D7
Sculthorpe 44 C2
Scunthorpe 52 B1
Scurlage 17 H7
Sea 8 C3
Sea Life Centre, Blackpool FY1 5AA 121 Blackpool
Sea Life Centre, Brighton B&H BN2 1TB 123 Brighton
Sea Life Centre, Great Yarmouth Norf NR30 3AH 45 K5
Sea Life London Aquarium GtLon SE1 7PD 132 F5
Sea Life Sanctuary, Hunstanton Norf PE36 5BH 44 A1
Sea Mills 19 J4
Sea Palling 45 J3
Seaborough 8 D4
Seaburn 71 K7
Seacombe 48 C3
Seacroft Lincs 53 J6
Seacroft WYorks 57 J6
Seadyke 43 G2
Seafield A&B 73 F2
Seafield SAyr 67 H1
Seafield WLoth 75 J4
Seaford 13 H7
Seaforth 48 C3
Seagrave 41 J4
Seagry Heath 20 C3
Seaham 62 E2
Seaham Grange 62 E1
Seahouses 77 K7
Seal 23 J6
Sealand 48 C6
Seale 22 B7
Sea-Life Adventure S'end SS1 2ER 24 E3
Sealyham 16 C3
Seamer NYorks 59 G1
Seamer NYorks 62 E5
Seamill 73 K6
SeaQuarium, Rhyl Denb LL18 3AF 47 J4
Searby 52 D2
Seasalter 25 G5
Seascale 60 B6
Seathorne 53 J6
Seathwaite Cumb 60 D5
Seathwaite Cumb 60 D7
Seatle 55 G1
Seatoller 60 D5
Seaton Corn 4 D5
Seaton Cumb 60 B3
Seaton Devon 8 B5
Seaton Dur 62 D1
Seaton ERid 59 H5
Seaton N'umb 71 J6
Seaton Rut 42 C6
Seaton Burn 71 H6
Seaton Carew 63 F4
Seaton Delaval 71 J6
Seaton Junction 8 B5
Seaton Ross 58 D5
Seaton Sluice 71 J6
Seaton Tramway Devon EX12 2NQ 8 B5
Seatown 8 D5
Seave Green 63 F6
Seaview 11 H5
Seaville 60 C1
Seavington St. Mary 8 D3
Seavington St. Michael 8 D3
Seawick 35 F7
Sebastopol 19 F2
Sebergham 60 E2
Seckington 40 E5
Second Coast 83 K1
Sedbergh 61 H7
Sedbury 19 J2
Sedbusk 61 K7
Seddington 32 E4
Sedgeberrow 30 B5
Sedgebrook 42 B2
Sedgefield 62 D4
Sedgeford 44 B2
Sedgehill 9 H2
Sedgemere 30 D1
Sedgley 40 B6
Sedgwick 55 J1
Sedlescombe 14 C6
Sedlescombe Street 14 C6
Seend 20 C5
Seend Cleeve 20 C5
Seer Green 22 C2
Seething 45 H6
Sefton 48 C2
Seghill 71 H6
Seifton 38 D7
Seighford 40 A3
Seil 79 J9
Seilebost 88 F8
Seion 46 D6
Seisdon 40 A6
Seisiadar 88 L4
Seized! Revenue & Customs Uncovered Mersey L3 4AQ 131 B5
Selattyn 38 B2
Selborne 11 J1
Selby 58 C6
Selham 12 C4
Selhurst 23 G5
Selkirk 69 K1
Sellack 28 E6
Sellafield 60 B6
Sellafield Visitors Centre Cumb CA20 1PG 60 B6
Sellafirth 89 P3
Sellindge 15 G4
Selling 15 F2
Sells Green 20 C5
Selly Oak 40 C7
Selmeston 13 J6
Selsdon 23 G5
Selsey 12 B7
Selsfield Common 13 G3
Selside Cumb 61 G7
Selside NYorks 56 C2
Selsley 20 B1
Selstead 15 H3
Selston 51 G7
Selworthy 7 H1
Semblister 89 M7
Semer 34 D4
Semington 20 B5
Semley 9 H2
Send 22 D6
Send Marsh 22 D6
Senghenydd 18 E2
Sennen 2 A6
Sennen Cove 2 A6
Sennybridge 27 J6
Senwick 65 G6
Sequer's Bridge 5 G5
Serlby 51 J4
Serpentine Gallery GtLon W2 3XA 99 G7
Serrington 10 B1
Sessay 57 K2
Setchey 44 A4
Setley 10 E4
Settiscarth 89 C6
Settle 56 D3
Settrington 58 E2
Seven Ash 7 K2
Seven Bridges 20 E2
Seven Kings 23 H3
Seven Sisters 18 B1
Seven Springs 29 J7
Sevenhampton Glos 30 B6
Sevenhampton Swin 21 F2
Sevenoaks 23 J6
Sevenoaks Weald 23 J6
Severn Beach 19 J3
Severn Stoke 29 H4
Severn Valley Railway Shrop DY12 1BG 29 H1
Sevick End 32 D3
Sevington 15 F3
Sewards End 33 J5
Sewardstone 23 G2
Sewerby 59 H3
Sewerby Hall & Gardens ERid YO15 1EA 59 J3
Seworgan 2 E5
Sewstern 42 B3
Seymour Villas 6 C1
Sezincote 30 C5
Sgarasta Mhòr 88 F8
Sgiogarstaigh (Skigersta) 88 L1
Shabbington 21 K1
Shackerley 40 A5
Shackerstone 41 F5
Shackleford 22 C7
Shadfen 71 H5
Shadforth 62 D2
Shadingfield 45 J7
Shadoxhurst 14 E4
Shadsworth 56 C7
Shadwell Norf 44 D7
Shadwell WYorks 57 J5
Shaftenhoe End 33 H5
Shaftesbury 9 H2
Shafton 51 F1
Shakespeare's Birthplace Warks CV37 6QW 136 Stratford-upon-Avon
Shakespeare's Globe Theatre GtLon SE1 9DT 132 J3
Shalbourne 21 G5
Shalcombe 10 E6
Shalden 21 K7
Shalden Green 22 A7
Shaldon 5 K3
Shalfleet 11 F6
Shalford Essex 34 B6
Shalford Surr 22 D7
Shalford Green 34 B6
Shallowford Devon 7 F1
Shallowford Staffs 40 A3
Shalmsford Street 15 F2
Shalstone 31 H5
Shalunt 73 J3
Shambellie 65 K4
Shamley Green 22 D7
Shandon 74 A2
Shandwick 84 C3
Shangton 42 A6
Shankend 70 A3
Shankhouse 71 H6
Shanklin 11 G6
Shanklin Chine IoW PO37 6BW 11 G6
Shannochie 66 D1
Shantron 74 B2
Shap 61 G5
Shapinsay 89 E6
Shapwick Dorset 9 J4
Shapwick Som 8 D1
Sharcott 20 E6
Shard End 40 D7
Shardlow 41 G2
Shareshill 40 B5
Sharlston 51 F1
Sharlston Common 51 F1
Sharnal Street 24 D4
Sharnbrook 32 C3
Sharneyford 56 D7
Sharnford 41 G6
Sharnhill Green 9 G4
Sharoe Green 55 J6
Sharow 57 J2
Sharp Street 45 H3
Sharpenhoe 32 D5
Sharperton 70 E3
Sharpham House 5 J5
Sharpness 19 K1
Sharpthorne 13 G3
Sharrington 44 E2
Shatterford 39 G7
Shatterling 15 H2
Shaugh Prior 5 F4
Shave Cross 8 D5
Shavington 49 G7
Shaw GtMan 49 J2
Shaw Swin 20 E3
Shaw WBerks 21 H5
Shaw Wilts 20 B5
Shaw Green Herts 33 F5
Shaw Green NYorks 57 H4
Shaw Mills 57 H3
Shaw Side 49 J2
Shawbost (Siabost) 88 H3
Shawbury 38 E3
Shawell 41 H7
Shawfield GtMan 49 H1
Shawfield Staffs 50 C6
Shawford 11 F2
Shawforth 56 D7
Shawhead 65 J3
Shawtonhill 74 E6
Sheanachie 66 B2
Shearington 69 F7
Shearsby 41 J6
Shebbear 6 C5
Shebdon 39 G3
Shebster 87 N3
Shedfield 11 G3
Sheen 50 D6
Sheepridge 50 D1
Sheepscombe 29 H7
Sheepstor 5 F4
Sheepwash Devon 6 C5
Sheepwash N'umb 71 H5
Sheepway 19 H4
Sheepy Magna 41 F5
Sheepy Parva 41 F5
Sheering 33 J7
Sheerness 25 F4
Sheet 11 J2
Sheffield 51 F4
Sheffield Botanic Gardens SYorks S10 2LN 109 B4
Sheffield Bottom 21 K5
Sheffield Green 13 H4
Sheffield Park Garden ESuss TN22 3QX 13 H4
Shefford 32 E5
Shefford Woodlands 21 G4
Sheigra 86 D3
Sheinton 39 F5
Shelderton 28 D1
Sheldon Derbys 50 D6
Sheldon Devon 7 K5
Sheldon WMid 40 D7
Sheldwich 15 F2
Sheldwich Lees 15 F2
Shelf Bridgend 18 C3
Shelf WYorks 57 G7
Shelfanger 45 F7
Shelfield Warks 30 C2
Shelfield WMid 40 C5
Shelfield Green 30 C2
Shelford 41 J1
Shellbrook 41 F4
Shellbrook Hill 38 C1
Shelley Essex 23 J1
Shelley Suff 34 E5
Shelley WYorks 50 E1
Shellingford 21 G2
Shellow Bowells 24 C1
Shelsley Beauchamp 29 G2
Shelsley Walsh 29 G2
Shelswell 31 H5
Shelthorpe 41 H4
Shelton Bed 32 D2
Shelton Norf 45 G6
Shelton Notts 42 A1
Shelton Shrop 38 D4
Shelve 38 C6
Shelwick 28 E4
Shelwick Green 28 E4
Shenfield 24 C2
Shenington 30 E4
Shenley 22 E1
Shenley Brook End 32 B5
Shenley Church End 32 B5
Shenleybury 22 E1
Shenmore 28 C5
Shennanton 64 D4
Shenstone Staffs 40 D5
Shenstone Worcs 29 H1
Shenstone Woodend 40 D5
Shenton 41 F5
Shenval 84 G8
Shepeau Stow 43 G4
Shephall 33 F6
Shepherd's Bush 23 F4
Shepherd's Green 22 A3
Shepherd's Patch 20 A1
Shepherdswell (Sibertswold) 15 H3
Shepley 50 D2
Shepperdine 19 K2
Shepperton 22 D5
Shepreth 33 G4
Shepreth Wildlife Park Cambs SG8 6PZ 33 G4
Shepshed 41 G4
Shepton Beauchamp 8 D3
Shepton Mallet 19 K7
Shepton Montague 9 F1
Shepway 14 C2
Sheraton 62 E3
Sherborne Dorset 9 F3
Sherborne Glos 30 C7
Sherborne St. John 21 K6
Sherbourne 30 D2
Sherbourne Street 34 D4
Sherburn Dur 62 D2
Sherburn NYorks 59 F2
Sherburn Hill 62 D2
Sherburn in Elmet 57 K6
Shere 22 D7
Shereford 44 C3
Sherfield English 10 D2
Sherfield on Loddon 21 K6
Sherford Devon 5 H6
Sherford Som 8 B2
Sheriff Hutton 58 C3
Sheriffhales 39 G4
Sheringham 45 F1
Sheringham Park Norf NR26 8TL 45 F1
Sherington 32 B4
Shernal Green 29 J2
Shernborne 44 B2
Sherramore 83 R11
Sherrington 9 J1
Sherston 20 B3
Sherwood 41 H1
Sherwood Forest Country Park Notts NG21 9HN 51 J6
Sherwood Forest Fun Park Notts NG21 9QA 51 J6
Sherwood Green 6 D3
Sherwood Pines Forest Park Notts NG21 9JL 51 J6
Shetland Islands 89 L7
Shevington 48 E2
Shevington Moor 48 E1
Sheviock 4 D5
Shide 11 G6
Shiel Bridge (Drochaid Sheile) 83 K9
Shieldaig High 83 J3
Shieldaig High 83 J5
Shieldhill 75 G3
Shielfoot 78 H3
Shifford 21 G1
Shifnal 39 G5
Shilbottle 71 G3
Shildon 62 C4
Shillingford Devon 7 H3
Shillingford Oxon 21 J2
Shillingford Abbot 7 H7
Shillingford St. George 7 H7
Shillingstone 9 H3
Shillington 32 E5
Shillmoor 70 D3
Shilstone 6 D5
Shilton Oxon 21 F1
Shilton Warks 41 G7
Shimpling Norf 45 F7
Shimpling Suff 34 C3
Shimpling Street 34 C3
Shincliffe 62 C2
Shiney Row 62 D1
Shinfield 22 A5
Shingay 33 G4
Shingham 44 B5
Shingle Street 35 H4
Shinness Lodge 86 H8
Shipbourne 23 J6
Shipbrookhill 49 F5
Shipdham 44 D5
Shipham 19 H6
Shiphay 5 J4
Shiplake 22 A4
Shiplake Row 22 A4
Shipley N'umb 71 G2
Shipley Shrop 40 A6
Shipley WSuss 12 E4
Shipley WYorks 57 G6
Shipley Bridge Devon 5 G4
Shipley Bridge Surr 23 G7
Shipley Common 41 G1
Shipley Country Park Derbys DE75 7GX 106 E2
Shipmeadow 45 H6
Shippea Hill 44 A7
Shippon 21 H2
Shipston on Stour 30 D4
Shipton Glos 30 B7
Shipton NYorks 58 B4
Shipton Shrop 38 E6
Shipton Bellinger 21 F7
Shipton Gorge 8 D5
Shipton Green 12 B6
Shipton Moyne 20 B3
Shipton Oliffe 30 B7
Shipton Solers 30 B7
Shipton-on-Cherwell 31 F7
Shiptonthorpe 58 E5
Shipton-under-Wychwood 30 D7
Shirburn 21 K2
Shirdley Hill 48 C1
Shire Hall Gallery, Stafford Staffs ST16 2LD 40 B3
Shire Oak 40 C5
Shirebrook 51 H6
Shirecliffe 51 F3
Shiregreen 51 F3
Shirehampton 19 J4
Shiremoor 71 J6
Shirenewton 19 H2
Shireoaks 51 H4
Shirl Heath 28 D3
Shirland 51 F7
Shirley Derbys 40 E1
Shirley GtLon 23 G5
Shirley Hants 10 C5
Shirley Soton 11 F3
Shirley WMid 30 C1
Shirley Heath 30 C1
Shirley Warren 10 E3
Shirleywich 40 B3
Shirrell Heath 11 G3
Shirwell 6 D2
Shirwell Cross 6 D2
Shiskine 66 D1
Shittlehope 62 A3
Shobdon 28 C2
Shobley 10 C4
Shobrooke 7 G5
Shocklach 38 D1
Shocklach Green 38 D1
Shoeburyness 25 F3
Sholden 15 J2
Sholing 11 F3
Shoot Hill 38 D4
Shooter's Hill 23 H4
Shop Corn 3 F1
Shop Corn 6 A4
Shop Corner 35 G5
Shopnoller 7 K2
Shore 49 J1
Shoreditch 23 G3
Shoreham 23 J5
Shoreham-by-Sea 13 F6
Shoresdean 77 H6
Shoreswood 77 H6
Shorley 11 G2
Shorncote 20 D2
Shorne 24 C4
Shorne Ridgeway 24 C4
Short Cross 38 B5
Short Green 44 E7
Short Heath Derbys 41 F4
Short Heath WMid 40 C6
Shortacombe 6 D7
Shortbridge 13 H4
Shortfield Common 22 B7
Shortgate 13 H5
Shortgrove 33 J5
Shorthampton 30 E6
Shortlands 23 G5
Shortlanesend 3 F4
Shorton 5 J4
Shorwell 11 F6
Shoscombe 20 A6
Shotatton 38 C3
Shotesham 45 G6
Shotgate 24 D2
Shotley N'hants 42 C6
Shotley Suff 35 G5
Shotley Bridge 62 A1
Shotley Gate 35 G5
Shotleyfield 62 A1
Shottenden 15 F2
Shottermill 12 B3
Shottery 30 C3
Shotteswell 31 F4
Shottisham 35 H4
Shottle 51 F7
Shottlegate 41 F1
Shotton Dur 62 D4
Shotton Dur 62 E3
Shotton Flints 48 C6
Shotton N'umb 71 H6
Shotton Colliery 62 D2
Shotts 75 G4
Shotwick 48 C5
Shouldham 44 A5
Shouldham Thorpe 44 A5
Shoulton 29 H3
Shover's Green 13 K3
Shrawardine 38 C4
Shrawley 29 H2
Shreding Green 22 D3
Shrewley 30 D2
Shrewsbury 38 D4
Shrewton 20 D7
Shri Venkateswara (Balaji Temple of the United Kingdom) WMid B69 3DU 102 E3
Shrine of Our Lady of Walsingham (Anglican) Norf NR22 6EF 44 D2
Shripney 12 C6
Shrivenham 21 F3
Shropham 44 D6
Shroton (Iwerne Courtney) 9 H3
Shrub End 34 D6
Shucknall 28 E4
Shudy Camps 33 K4
Shugborough Estate Staffs ST17 0XB 40 B3
Shurdington 29 J7
Shurlock Row 22 B4
Shurnock 30 B2
Shurton 19 F7
Shustoke 40 E6
Shut Heath 40 A3
Shute Devon 7 G5
Shute Devon 8 B5
Shutford 30 E4
Shuthonger 29 H5
Shutlanger 31 J4
Shutt Green 40 A5
Shuttington 40 E5
Shuttlewood 51 G5
Shuttleworth 49 G1
Siadar Iarach 88 J2
Sibbaldbie 69 G5
Sibbertoft 41 J7
Sibdon Carwood 38 D7
Sibertswold (Shepherdswell) 15 H3
Sibford Ferris 30 E5
Sibford Gower 30 E5
Sible Hedingham 34 B5
Sibley's Green 33 K6
Sibsey 53 G7
Sibson Cambs 42 D6
Sibson Leics 41 F5
Sibster 87 R4
Sibthorpe 42 A1
Sibton 35 H2
Sibton Green 35 H1
Sicklesmere 34 C2
Sicklinghall 57 J5
Sidbury Devon 7 K6
Sidbury Shrop 39 F7
Sidcot 19 H6
Sidcup 23 H4
Siddal 57 G7
Siddington ChesE 49 H5
Siddington Glos 20 D2
Sidemoor 29 J1
Sidestrand 45 G2
Sidford 7 K6
Sidlesham 12 B7
Sidley 14 C7
Sidlow 23 F7
Sidmouth 7 K7
Sigford 5 H3
Sigglesthorne 59 H5
Sigingstone 18 C4
Signet 30 D7
Silbury Hill (Stonehenge, Avebury & Associated Sites) Wilts SN8 1QH 20 E5
Silchester 21 K5
Sileby 41 J4
Silecroft 54 E1
Silfield 45 F6
Silian 26 E3
Silk Willoughby 42 D1
Silkstead 11 F2
Silkstone 50 E2
Silkstone Common 50 E2
Silksworth 62 D1
Sill Field 55 J1
Silloth 60 C1
Sills 70 D3
Silpho 63 K7
Silsden 57 F5
Silsoe 32 D5
Silver End CenBeds 32 E4
Silver End Essex 34 C6
Silver Green 45 G6
Silver Street Kent 24 E5
Silver Street Som 8 E1
Silverburn 76 A4
Silvercraigs 73 G2
Silverdale Lancs 55 H2
Silverdale Staffs 40 A1
Silvergate 45 F3
Silverhill 14 C6
Silverlace Green 35 H3
Silverley's Green 35 G1
Silverstone 31 H4
Silverton 7 H5
Silvington 29 F1
Simister 49 H2
Simmondley 50 C3
Simonburn 70 D6
Simonsbath 7 F2
Simonside 71 J7
Simonstone Bridgend 18 C3
Simonstone Lancs 56 C6
Simprim 77 G6
Simpson 32 B5
Sinclair's Hill 77 G5
Sinclairston 67 J2
Sinderby 57 J1
Sinderhope 61 K1
Sindlesham 22 A5
Sinfin 41 F2
Singdean 70 A3
Singleborough 31 J5
Singleton Lancs 55 G6
Singleton WSuss 12 B5
Singlewell 24 C4
Singret 48 C7
Sinkhurst Green 14 D3
Sinnahard 85 J9
Sinnington 58 D1
Sinton Green 29 H2
Sipson 22 D4
Sirhowy 28 A7
Sirhowy Valley Country Park Caerp NP11 7BD 95 C1
Sisland 45 H6
Sissinghurst 14 C4
Sissinghurst Castle Garden Kent TN17 2AB 14 D4
Siston 19 K4
Sithney 2 D6
Sittingbourne 25 F5
Siulaisiadar 88 L4
Six Ashes 39 G7
Six Hills 41 J3
Six Mile Bottom 33 J3
Six Roads End 40 D3
Sixhills 52 E4
Sixmile 15 G3
Sixpenny Handley 10 B3
Sizewell 35 J2
Skail 87 K5
Skaill Ork 89 B6
Skaill Ork 89 E7
Skara Brae (Heart of Neolithic Orkney) Ork KW16 3LR 89 B6
Skares Aber 85 L7
Skares EAyr 67 K2
Skateraw 77 F3
Skeabost 82 E6
Skeeby 62 C6
Skeffington 42 A5
Skeffling 53 G1
Skegby 51 H6
Skegness 53 J6
Skegness Water Leisure Park Lincs PE25 1JF 53 J6
Skelberry 89 N6
Skelbo 84 B1
Skelbo Street 84 B1
Skelbrooke 51 H1
Skeld (Easter Skeld) 89 M8
Skeldon 67 H2
Skeldyke 43 G2
Skellingthorpe 52 C5
Skellister 89 N7
Skellow 51 H1
Skelmanthorpe 50 E1
Skelmersdale 48 D2
Skelmorlie 73 K4
Skelpick 87 K4
Skelton Cumb 60 F3
Skelton ERid 58 D7
Skelton NYorks 62 A6
Skelton (Skelton-in-Cleveland) R&C 63 G5
Skelton York 58 B4
Skelton-in-Cleveland (Skelton) 63 G5
Skelton-on-Ure 57 J3
Skelwick 89 D3
Skelwith Bridge 60 E6
Skendleby 53 H6
Skendleby Psalter 53 H5
Skenfrith 28 D6
Skerne 59 G4
Skeroblingarry 66 B1
Skerray 87 J3
Skerton 55 H3
Sketchley 41 G6
Sketty 17 K6
Skewen 18 A2
Skewsby 58 C2
Skeyton 45 G3
Skeyton Corner 45 G3
Skidbrooke 53 H3
Skidbrooke North End 53 H3
Skidby 59 G6
Skilgate 7 H3
Skillington 42 B3
Skinburness 60 C1
Skinflats 75 H2
Skinningrove 63 H5
Skipness 73 G5
Skippool 55 G5
Skipsea 59 H4
Skipsea Brough 59 H4
Skipton 56 E4
Skipton Castle NYorks BD23 1AW 56 E4
Skipton-on-Swale 57 J2
Skipwith 58 C6
Skirbeck 43 G1
Skirbeck Quarter 43 G1
Skirethorns 56 E3
Skirlaugh 59 H6
Skirling 75 J7
Skirmett 22 A3
Skirpenbeck 58 D4
Skirwith Cumb 61 H3
Skirwith NYorks 56 C2
Skirza 87 R3
Skittle Green 22 A1
Skomer Island 16 A5
Skullomie 87 J3
Skyborry Green 28 B1
Skye 82 E7
Skye Green 34 C6
Skyreholme 57 F3
Slack Derbys 51 F6
Slack WYorks 56 E7
Slackhall 50 C4
Slackhead 85 J4
Slad 20 B1
Slade Devon 6 D1
Slade Devon 7 K5
Slade Pembs 16 C4
Slade Swan 17 H7
Slade Green 23 J4
Slade Hooton 51 H4
Sladesbridge 4 A3
Slaggyford 61 H1
Slaidburn 56 C4
Slaithwaite 50 C1
Slaley 61 L1
Slamannan 75 G3
Slapton Bucks 32 C6
Slapton Devon 5 J6
Slapton N'hants 31 H4
Slatepit Dale 51 F6
Slattadale 83 J3
Slaugham 13 F4
Slaughden 35 J3
Slaughterford 20 B4
Slawston 42 A6
Sleaford Hants 12 B3
Sleaford Lincs 42 D1
Sleagill 61 G5
Sleap 38 D3
Sledge Green 29 H5
Sledmere 59 F3
Sleights 63 J6
Slepe 9 J5
Slerra 6 B3
Slickly 87 Q3
Sliddery 66 D1
Sligachan 82 E8
Slimbridge 20 A1
Slimbridge Wildfowl & Wetlands Trust Glos GL2 7BT 20 A1
Slindon Staffs 40 A2
Slindon WSuss 12 C6
Slinfold 12 E3
Sling 19 J1
Slingsby 58 C2
Slip End CenBeds 32 D7
Slip End Herts 33 F5
Slipper Chapel, Houghton St. Giles Norf NR22 6AL 44 D2
Slipton 32 C1
Slitting Mill 40 C4
Slochd 84 C8
Slockavullin 73 G1
Slogarie 65 G4
Sloley 45 G3
Slongaber 65 J3
Sloothby 53 H5
Slough 22 C3
Slough Green Som 8 B2
Slough Green WSuss 13 F4
Sluggan 84 C8
Slyne 55 H3
Smailholm 76 E7
Small Dole 13 F5
Small Hythe 14 D4
Smallbridge 49 J1
Smallbrook 7 G6
Smallburgh 45 H3
Smallburn 68 B1
Smalldale 50 C5
Smalley 41 G1
Smallfield 23 G7
Smallford 22 E1
Smallridge 8 B4
Smallthorne 49 H7
Smallworth 44 E7
Smannell 21 G7
Smardale 61 J6
Smarden 14 D3
Smaull 72 A4
Smeatharpe 7 K4
Smeeth 15 F4
Smeeton Westerby 41 J6
Smestow 40 A6
Smethwick 40 C7
Smethwick Green 49 H6
Smirisary 78 H3
Smisby 41 F4
Smith End Green 29 G3
Smithfield 69 K7
Smithies 51 F2
Smithincott 7 J4
Smith's End 33 G5
Smith's Green Essex 33 J6
Smith's Green Essex 33 K4
Smithton 84 B6
Smithy Green 49 G5
Smockington 41 G7
Smugglers Adventure ESuss TN34 3HY 128 Hastings
Smyrton 67 F5
Smythe's Green 34 D7
Snailbeach 38 C5
Snailwell 33 K2
Snainton 59 F1
Snaith 58 C7

Snape NYorks 57 H1
Snape Suff 35 H3
Snape Green 48 C1
Snape Watering 35 H3
Snarestone 41 F5
Snarford 52 D4
Snargate 14 E5
Snave 15 F5
Sneachill 29 J3
Snead 38 C6
Snead's Green 29 H2
Sneath Common 45 F7
Sneaton 63 J6
Sneatonthorpe 63 K6
Snelland 52 D4
Snellings 60 A6
Snelston 40 D1
Snetterton 44 D6
Snettisham 44 A2
Snipeshill 25 F5
Snitter 71 F3
Snitterby 52 C3
Snitterfield 30 D3
Snitterton 50 E6
Snittlegarth 60 D3
Snitton 28 E1
Snodhill 28 C4
Snodland 24 D5
Snow End 33 H5
Snow Street 44 E7
Snowden Hill 50 E2
Snowdon Mountain Railway Gwyn LL55 4TY 46 D7
Snowshill 30 B5
Snowshill Manor Glos WR12 7JU 30 B5
Soar Cardiff 18 D3
Soar Carmar 17 K3
Soar Devon 5 H7
Soay 82 E9
Soberton 11 H3
Soberton Heath 11 H3
Sockbridge 61 G4
Sockburn 62 D6
Sodom 47 J5
Sodylt Bank 38 C2
Softley 62 A4
Soham 33 J1
Soham Cotes 33 J1
Soldon 6 B4
Soldon Cross 6 B4
Soldridge 11 H1
Sole Street Kent 15 F3
Sole Street Kent 24 C5
Soleburn 64 A4
Solihull 30 C1
Solihull Lodge 30 B1
Sollas (Solas) 88 C1
Sollers Dilwyn 28 D3
Sollers Hope 29 F5
Sollom 48 D1
Solomon's Tump 29 G7
Solsgirth 75 H1
Solva 16 A3
Solwaybank 69 J6
Somerby Leics 42 A4
Somerby Lincs 52 D2
Somercotes 51 G7
Somerford 40 B5
Somerford Keynes 20 D2
Somerley 12 B7
Somerleyton 45 J6
Somersal Herbert 40 D2
Somersby 53 G5
Somerset House GtLon WC2R 1LA 132 F3
Somersham Cambs 33 G1
Somersham Suff 34 E4
Somerton Newport 19 G3
Somerton Oxon 31 F6
Somerton Som 8 D2
Somerton Suff 34 C3
Sompting 12 E6
Sompting Abbotts 12 E6
Sonning 22 A4
Sonning Common 22 A3
Sonning Eye 22 A4
Sontley 38 C1
Sookholme 51 H6
Sopley 10 C5
Sopworth 20 B3
Sorbie 64 E6
Sordale 87 P3
Sorisdale 78 D4
Sorn 67 K1
Sornhill 74 D7
Sortat 87 Q3
Sotby 53 F5
Sots Hole 52 E6
Sotterley 45 J7
Soudley 39 G3
Soughton 48 B6
Soulbury 32 B6
Soulby 61 J5
Souldern 31 G5
Souldrop 32 C2
Sound ChesE 39 F1
Sound Shet 89 N8
Sourhope 70 D1
Sourin 89 D4
Sourton 6 D6
Soutergate 55 F1
South Acre 44 C4
South Acton 22 E4
South Alkham 15 H3
South Allington 5 H7
South Alloa 75 G1
South Ambersham 12 C4
South Anston 51 H4
South Ascot 22 C5
South Baddesley 10 E5
South Balloch 67 H4
South Bank 63 F4
South Barrow 9 F2
South Bellsdyke 75 H2
South Benfleet 24 D3
South Bersted 12 C6
South Bockhampton 10 C5
South Boisdale 88 B7
South Bowood 8 D5
South Brent 5 G4
South Brentor 6 C7
South Brewham 9 G1
South Broomhill 71 H4
South Burlingham 45 H5
South Cadbury 9 F2
South Cairn 66 D7
South Carlton 52 C5
South Cave 59 F6
South Cerney 20 D2
South Chard 8 C4
South Charlton 71 G1
South Cheriton 9 F2
South Church 62 C4
South Cliffe 58 E6
South Clifton 52 B5
South Cockerington 53 G4
South Common 13 G5
South Cornelly 18 B3
South Corrigills 73 J7
South Cove 45 J7
South Creake 44 C2
South Crosland 50 D1
South Croxton 41 J4
South Croydon 23 G5
South Dalton 59 F5
South Darenth 23 J4
South Dell (Dail Bho Dheas) 88 K1
South Duffield 58 C6
South Elkington 53 F4
South Elmsall 51 G1
South End Bucks 32 B6
South End Cumb 55 F3
South End Hants 10 C3
South End NLincs 59 H7
South Erradale 82 H3
South Fambridge 24 E2
South Fawley 21 G3
South Ferriby 59 F7
South Field 59 G7
South Godstone 23 G7
South Gorley 10 C3
South Green Essex 24 C2
South Green Essex 34 E7
South Green Norf 44 E4
South Green Suff 35 F1
South Gyle 75 K3
South Hall 73 J3
South Hanningfield 24 D2
South Harefield 22 D3
South Harting 11 J3
South Hayling 11 J5
South Hazelrigg 77 J7
South Heath 22 C1
South Heighton 13 H6
South Hetton 62 D2
South Hiendley 51 F1
South Hill 4 D3
South Hinksey 21 J1
South Hole 6 A4
South Holme 58 C2
South Holmwood 22 E7
South Hornchurch 23 J3
South Hourat 74 A5
South Huish 5 G6
South Hykeham 52 C6
South Hylton 62 D1
South Kelsey 52 D3
South Killingholme 52 E1
South Kilvington 57 K1
South Kilworth 41 J7
South Kirkby 51 G1
South Kirkton 85 M10
South Knighton 5 J3
South Kyme 42 E1
South Lakes Wild Animal Park Cumb LA15 8JR 55 F2
South Lancing 12 E6
South Leigh 21 G1
South Leverton 51 K4
South Littleton 30 B4
South Lopham 44 E7
South Luffenham 42 C5
South Malling 13 H5
South Marston 20 E3
South Middleton 70 E1
South Milford 57 K6
South Milton 5 G6
South Mimms 23 F1
South Molton 7 F3
South Moor 62 B1
South Moreton 21 J3
South Mundham 12 B6
South Muskham 51 K7
South Newbald 59 F6
South Newington 31 F5
South Newton 10 B1
South Normanton 51 G7
South Norwood 23 G5
South Nutfield 23 G7
South Ockendon 23 J3
South Ormsby 53 G5
South Ossett 50 E1
South Otterington 57 J1
South Owersby 52 D3
South Oxhey 22 E2
South Park 23 F7
South Perrott 8 D4
South Petherton 8 D3
South Petherwin 6 B7
South Pickenham 44 C5
South Pool 5 H6
South Queensferry (Queensferry) 75 K3
South Radworthy 7 F2
South Rauceby 42 D1
South Raynham 44 C3
South Reston 53 H4
South Ronaldsay 89 D9
South Ruislip 22 E3
South Runcton 44 A5
South Scarle 52 B6
South Shields 71 J7
South Shields Museum & Art Gallery T&W NE33 2JA 116 G3
South Somercotes 53 H3
South Somercotes Fen Houses 53 H3
South Stainley 57 J3
South Stoke B&NESom 20 A5
South Stoke Oxon 21 K3
South Stoke WSuss 12 D6
South Street ESuss 13 G5
South Street GtLon 23 H6
South Street Kent 24 C5
South Street Kent 24 E5
South Street Kent 25 H5
South Tawton 6 E6
South Thoresby 53 H5
South Tidworth 21 F7
South Tottenham 23 G3
South Town Devon 7 H7
South Town Hants 11 H1
South Uist (Uibhist a Deas) 88 B5
South View 21 K6
South Walsham 45 H4
South Warnborough 22 A7
South Weald 23 J2
South Weston 22 A2
South Wheatley Corn 4 C1
South Wheatley Notts 51 K4
South Wigston 41 H6
South Willingham 52 E4
South Wingfield 51 F7
South Witham 42 C4
South Wonston 11 F1
South Woodham Ferrers 24 E2
South Wootton 44 A3
South Wraxall 20 B5
South Yardle 40 D7
South Zeal 6 E6
Southall 22 E3
Southam Glos 29 J6
Southam Warks 31 F2
Southampton 11 F3
Southampton Airport 11 F3
Southbar 74 C4
Southborough GtLon 23 H5
Southborough Kent 23 J7
Southbourne Bourne 10 C5
Southbourne WSuss 11 J4
Southbrook 7 J6
Southburgh 44 E5
Southburn 59 F4
Southchurch 25 F3
Southcott Devon 6 D6
Southcott Wilts 20 E6
Southcourt 32 B7
Southdean 70 B3
Southdene 48 D3
Southease 13 H6
Southend A&B 66 A3
Southend Bucks 22 A3
Southend (Bradfield Southend) WBerks 21 J4
Southend Wilts 20 E4
Southend Airport 24 E3
Southend Pier S'end SS1 1EE 24 E3
Southend-on-Sea 24 E3
Southerfield 60 C2
Southerly 6 D7
Southern Green 33 G5
Southerndown 18 B4
Southerness 65 K5
Southery 44 A6
Southfield 76 A1
Southfields 23 F4
Southfleet 24 C4
Southgate Cere 36 E7
Southgate GtLon 23 G2
Southgate Norf 44 A2
Southgate Norf 45 F3
Southgate Swan 17 J7
Southill 32 E4
Southington 21 J7
Southleigh 8 B5
Southmarsh 9 G1
Southminster 25 F2
Southmoor 21 G2
Southoe 32 E2
Southolt 35 F2
Southorpe 42 D5
Southowram 57 G7
Southport 48 C1
Southport Pier Mersey PR8 1QX 48 C1
Southrepps 45 G2
Southrey 52 E6
Southrop 20 E1
Southrope 21 K7
Southsea Ports 11 H5
Southsea Wrex 48 B7
Southtown Norf 45 K5
Southtown Ork 89 D8
Southwaite Cumb 61 F2
Southwaite Cumb 61 J6
Southwark Cathedral GtLon SE1 9DA 101 B7
Southwater 12 E4
Southwater Street 12 E4
Southway 19 J7
Southwell Dorset 9 F7
Southwell Notts 51 J7
Southwell Minster Notts NG25 0HD 51 K7
Southwick D&G 65 K5
Southwick Hants 11 H4
Southwick N'hants 42 D6
Southwick Som 19 G7
Southwick T&W 62 D1
Southwick Wilts 20 B6
Southwick WSuss 13 F6
Southwold 35 K1
Southwood 8 E1
Sowden 7 H7
Sower Carr 55 G5
Sowerby NYorks 57 K1
Sowerby WYorks 57 F7
Sowerby Bridge 57 F7
Sowerby Row 60 E2
Sowerhill 7 G3
Sowley Green 34 B3
Sowood 50 C1
Sowton 7 H6
Spa Common 45 G2
Spa Complex NYorks YO11 2HD 135 Scarborough
Spadeadam 70 A6
Spalding 43 F3
Spaldington 58 D6
Spaldwick 32 E1
Spalford 52 B6
Spanby 42 D2
Sparham 44 E4
Spark Bridge 55 G1
Sparkford 9 F2
Sparkhill 40 C7
Sparkwell 5 F5
Sparrow Green 44 D4
Sparrowpit 50 C4
Sparrow's Green 13 K3
Sparsholt Hants 11 F1
Sparsholt Oxon 21 G3
Spartylea 61 K2
Spath 40 C2
Spaunton 58 D1
Spaxton 8 B1
Spean Bridge (Drochaid an Aonachain) 79 P2
Spean Bridge Woollen Mill High PH34 4EP 79 P2
Spear Hill 12 E5
Speddoch 68 D5
Speedwell 19 K4
Speen Bucks 22 B2
Speen WBerks 21 H5
Speeton 59 H2
Speke 48 D4
Speldhurst 23 J7
Spellbrook 33 H7
Spelsbury 30 E6
Spen Green 49 H6
Spencers Wood 22 A5
Spennithorne 57 G1
Spennymoor 62 C3
Spernall 30 B2
Spetchley 29 H3
Spetisbury 9 J4
Spexhall 45 H7
Spey Bay 84 H4
Spilsby 53 G6
Spindlestone 77 K7
Spinkhill 51 G5
Spinnaker Tower PO1 3TN 135 Portsmouth
Spinningdale 84 A2
Spirthill 20 C4
Spital High 87 P4
Spital W&M 22 C4
Spital in the Street 52 C3
Spitalbrook 23 G1
Spitfire & Hurricane Memorial, R.A.F. Manston Kent CT12 5DF 25 K5
Spithurst 13 H5
Spittal D&G 64 D5
Spittal D&G 64 E4
Spittal ELoth 76 C3
Spittal N'umb 77 J5
Spittal Pembs 16 C3
Spittal of Glenmuick 81 J2
Spittal of Glenshee 80 G4
Spittalfield 80 G6
Spixworth 45 G4
Splayne's Green 13 H4
Splott 19 F4
Spofforth 57 J4
Spondon 41 G2
Spooner Row 44 E6
Spoonley 39 F2
Sporle 44 C4
Sportsman's Arms 47 H7
Spott 76 E3
Spratton 31 J1
Spreakley 22 B7
Spreyton 6 E6
Spriddlestone 5 F5
Spridlington 52 D4
Spring Grove 22 E4
Spring Vale 11 H5
Springburn 74 E4
Springfield A&B 73 J3
Springfield D&G 69 J7
Springfield Fife 81 J9
Springfield Moray 84 E5
Springfield WMid 40 C7
Springhill Staffs 40 B5
Springhill Staffs 40 C5
Springholm 65 J4
Springkell 69 H6
Springside 74 B7
Springthorpe 52 B4
Springwell 62 C1
Sproatley 59 H6
Sproston Green 49 G6
Sprotbrough 51 H2
Sproughton 35 F4
Sprouston 77 F7
Sprowston 45 G4
Sproxton Leics 42 B3
Sproxton NYorks 58 C1
Sprytown 6 C7
Spurlands End 22 B2
Spurstow 48 E7
Spyway 8 E5
Square Point 65 H3
Squires Gate 55 G6
Srannda 88 F9
Sròndoire 73 G3
Sronphadruig Lodge 80 C3
S.S. Great Britain Bristol BS1 6TY 96 C5
Stableford Shrop 39 G6
Stableford Staffs 40 A2
Stacey Bank 50 E3
Stackhouse 56 D3
Stackpole 16 C6
Stacksteads 56 D7
Staddiscombe 5 F5
Staddlethorpe 58 E7
Staden 50 C5
Stadhampton 21 K2
Staffield 61 G2
Staffin 82 E4
Stafford 40 B3
Stagden Cross 33 K7
Stagsden 32 C4
Stagshaw Bank 70 E7
Stain 87 R3
Stainburn Cumb 60 B4
Stainburn NYorks 57 H5
Stainby 42 C3
Staincross 51 F1
Staindrop 62 B4
Staines-upon-Thames 22 D4
Stainfield Lincs 42 D3
Stainfield Lincs 52 E5
Stainforth NYorks 56 D3
Stainforth SYorks 51 J1
Staining 55 G6
Stainland 50 C1
Stainsacre 63 K6
Stainsby Derbys 51 G6
Stainsby Lincs 53 G5
Stainton Cumb 55 J1
Stainton Cumb 61 F4
Stainton Dur 62 A5
Stainton Middl 62 E5
Stainton NYorks 62 B7
Stainton SYorks 51 H3
Stainton by Langworth 52 D5
Stainton le Vale 52 E3
Stainton with Adgarley 55 F2
Staintondale 63 K7
Stair Cumb 60 D4
Stair EAyr 67 J1
Stairfoot 51 F2
Staithes 63 H5
Stake Pool 55 H5
Stakeford 71 H5
Stakes 11 H4
Stalbridge 9 G3
Stalbridge Weston 9 G3
Stalham 45 H3
Stalham Green 45 H3
Stalisfield Green 14 E2
Stalling Busk 56 E1
Stallingborough 52 E1
Stallington 40 B2
Stalmine 55 G5
Stalybridge 49 J3
Stambourne 34 B5
Stamford Lincs 42 D5
Stamford N'umb 71 H2
Stamford Bridge ChesW&C 48 D6
Stamford Bridge ERid 58 D4
Stamfordham 71 F6
Stanah 55 G5
Stanborough 33 F7
Stanborough Park Herts AL8 6XF 33 F7
Stanbridge Dorset 10 B4
Stanbridge Earls 10 E2
Stanbridge CenBeds 32 C6
Stanbury 57 F6
Stand 75 F4
Standalone Farm, Letchworth Garden City Herts SG6 4JN 33 F5
Standburn 75 H3
Standedge Tunnel & Visitor Centre WYorks HD7 6NQ 50 C1
Standeford 40 B5
Standen 14 D4
Standen Street 14 D4
Standerwick 20 B6
Standford 12 B3
Standford Bridge 39 G3
Standish Glos 20 B1
Standish GtMan 48 E1
Standlake 21 G1
Standon Hants 11 F2
Standon Herts 33 G6
Standon Staffs 40 A2
Standon Green End 33 G7
Stane 75 G5
Stanecastle 74 B7
Stanfield 44 D3
Stanford CenBeds 32 E4
Stanford Kent 15 G4
Stanford Shrop 38 C4
Stanford Bishop 29 F3
Stanford Bridge 29 G2
Stanford Dingley 21 J4
Stanford End 22 A5
Stanford in the Vale 21 G2
Stanford on Avon 31 G1
Stanford on Soar 41 H3
Stanford on Teme 29 G2
Stanford Rivers 23 J1
Stanford-le-Hope 24 C3
Stanfree 51 G5
Stanghow 63 G5
Stanground 43 F6
Stanhoe 44 C2
Stanhope Dur 61 L3
Stanhope ScBord 69 G1
Stanion 42 C7
Stanklyn 29 H1
Stanley Derbys 41 G1
Stanley Dur 62 B1
Stanley Notts 51 G6
Stanley P&K 80 G7
Stanley Staffs 49 J7
Stanley Wilts 20 C4
Stanley WYorks 57 J7
Stanley Common 41 G1
Stanley Crook 62 B3
Stanley Gate 48 D2
Stanley Hill 29 F4
Stanleygreen 38 E2
Stanlow ChesW&C 48 D5
Stanlow Shrop 39 G6
Stanmer 13 G5
Stanmore GtLon 22 E2
Stanmore WBerks 21 H4
Stannersburn 70 C5
Stanningfield 34 C3
Stannington N'umb 71 H6
Stannington SYorks 51 F4
Stansbatch 28 C2
Stansfield 34 B3
Stanshope 50 D7
Stanstead 34 C4
Stanstead Abbotts 33 G7
Stansted 24 C5
Stansted Mountfitchet 33 J6
Stanton Derbys 40 E4
Stanton Glos 30 B5
Stanton N'umb 71 G4
Stanton Staffs 40 D1
Stanton Suff 34 D1
Stanton by Bridge 41 F3
Stanton by Dale 41 G2
Stanton Drew 19 J5
Stanton Fitzwarren 20 E2
Stanton Harcourt 21 H1
Stanton Hill 51 G6
Stanton in Peak 50 E6
Stanton Lacy 28 D1
Stanton Lees 50 E6
Stanton Long 38 E6
Stanton Prior 19 K5
Stanton St. Bernard 20 D5
Stanton St. John 21 J1
Stanton St. Quintin 20 C4
Stanton Street 34 D2
Stanton under Bardon 41 G4
Stanton upon Hine Heath 38 E3
Stanton Wick 19 K5
Stanton-on-the-Wolds 41 J2
Stanwardine in the Fields 38 D3
Stanwardine in the Wood 38 D3
Stanway Essex 34 D6
Stanway Glos 30 B5
Stanway Green Essex 34 D6
Stanway Green Suff 35 G1
Stanwell 22 D4
Stanwell Moor 22 D4
Stanwick 32 C1
Stanwix 60 F1
Stanydale 89 L7
Staoinebrig 88 B5
Stapeley 39 F1
Stapenhill 40 E3
Staple Kent 15 H2
Staple Som 7 K1
Staple Cross 7 J3
Staple Fitzpaine 8 B3
Staplecross 14 C5
Staplefield 13 F4
Stapleford Cambs 33 H3
Stapleford Herts 33 G7
Stapleford Leics 42 B4
Stapleford Lincs 52 B7
Stapleford Notts 41 G2
Stapleford Wilts 10 B1
Stapleford Abbotts 23 H2
Stapleford Tawney 23 J2
Staplegrove 8 B2
Staplehay 8 B2
Staplehurst 14 C3
Staplers 11 G6
Staplestreet 25 G5
Stapleton Cumb 70 A6
Stapleton Here 28 C2
Stapleton Leics 41 G6
Stapleton NYorks 62 C5
Stapleton Shrop 38 D5
Stapleton Som 8 D2
Stapley 7 K4
Staploe 32 E2
Staplow 29 F4
Star Pembs 17 F2
Star Som 19 H6
Starbotton 56 E2
Starcross 7 H7
Stareton 30 E1
Starkholmes 51 F7
Starling 49 G1
Starling's Green 33 H5
Starr 67 J4
Starston 45 G7
Startforth 62 A5
Startley 20 C3
Statham 49 F4
Stathe 8 C2
Stathern 42 A2
Station Town 62 E3
Staughton Green 32 E2
Staughton Highway 32 E2
Staunton Glos 28 E7
Staunton Glos 29 G6
Staunton Harold Hall 41 F3
Staunton Harold Reservoir Derbys DE73 8DN 41 F3
Staunton in the Vale 42 B1
Staunton on Arrow 28 C2
Staunton on Wye 28 C4
Staveley Cumb 61 F7
Staveley Derbys 51 G5
Staveley NYorks 57 J3
Staveley-in-Cartmel 55 G1
Staverton Devon 5 H4
Staverton Glos 29 H6
Staverton N'hants 31 G2
Staverton Wilts 20 B5
Staverton Bridge 29 H6
Stawell 8 C1
Stawley 7 J3
Staxigoe 87 R4
Staxton 59 G2
Staylittle (Penffordd-las) 37 H6
Staynall 55 G5
Staythorpe 51 K7
Stean 57 F2
Steane 31 G5
Stearsby 58 C2
Steart 19 F7
Stebbing 33 K6
Stebbing Green 33 K6
Stechford 40 D7
Stedham 12 B4
Steel Cross 13 J3
Steel Green 54 E2
Steele Road 70 A4
Steen's Bridge 28 E3
Steep 11 J2
Steep Marsh 11 J2
Steeple Dorset 9 J6
Steeple Essex 25 F1
Steeple Ashton 20 C6
Steeple Aston 31 F6
Steeple Barton 31 F6
Steeple Bumpstead 33 K4
Steeple Claydon 31 H6
Steeple Gidding 42 E7
Steeple Langford 10 B1
Steeple Morden 33 F4
Steeraway 39 F5
Steeton 57 F5
Stella 71 G7
Stelling Minnis 15 G3
Stembridge 8 D2
Stemster 87 P3
Stenalees 4 A5
Stenhill 7 J4
Stenhousemuir 75 G2
Stenigot 53 F4
Stenness 89 L5
Stenscholl 82 E4
Stenson 41 F3
Stenton 76 E3
Stepaside Pembs 16 E5
Stepaside Powys 37 K7
Stepney 23 G3
Steppingley 32 D5
Stepps 74 E4
Sternfield 35 H2
Sterridge 6 D1
Stert 20 D6
Stetchworth 33 K3
Stevenage 33 F6
Stevenston 74 A6
Steventon Hants 21 J7
Steventon Oxon 21 H2
Steventon End 33 K4
Stevington 32 C3
Stewart Park Middl TS7 8AR 117 D4
Stewartby 32 D4
Stewarton D&G 64 E6
Stewarton EAyr 74 C6
Stewkley 32 B6
Stewley 8 C3
Stewton 53 G4
Steyning 12 E5
Steynton 16 C5
Stibb 6 A4
Stibb Cross 6 C4
Stibb Green 21 F5
Stibbard 44 D3
Stibbington 42 D6
Stichill 77 F7
Sticker 3 G3
Stickford 53 G6
Sticklepath Devon 6 E6
Sticklepath Som 8 C3
Stickling Green 33 H5
Stickney 53 G7
Stiff Street 24 E5
Stiffkey 44 D1
Stifford's Bridge 29 G4
Stileway 19 H7
Stilligarry (Stadhlaigearraidh) 88 B5
Stillingfleet 58 B5
Stillington NYorks 58 B3
Stillington Stock 62 D4
Stilton 42 E7
Stinchcombe 20 A2
Stinsford 9 G5
Stirchley Tel&W 39 G5
Stirchley WMid 40 C7
Stirling (Sruighlea) 75 F1
Stirling Castle Stir FK8 1EJ 75 F1
Stirling Visitor Centre Stir FK8 1EH 75 F1
Stirton 56 E4
Stisted 34 C6
Stitchcombe 21 F5
Stithians 2 E5
Stivichall 30 E1
Stixwould 52 E6
Stoak 48 D5
Stobo 75 K7
Stoborough 9 J6
Stoborough Green 9 J6
Stobwood 75 H5
Stock 24 C2
Stock Green 29 J3
Stock Lane 21 F4
Stock Wood 30 B3
Stockbridge Hants 10 E1
Stockbridge WSuss 12 B6
Stockbury 24 E5
Stockcross 21 H5
Stockdale 2 E5
Stockdalewath 60 E2
Stockerston 42 B6
Stockgrove Country Park CenBeds LU7 0BA 32 C6
Stocking Green Essex 33 J5
Stocking Green MK 32 B4
Stocking Pelham 33 H6
Stockingford 41 F6
Stockland Cardiff 18 E4
Stockland Devon 8 B4
Stockland Bristol 19 F7
Stockleigh English 7 G5
Stockleigh Pomeroy 7 G5
Stockley 20 D5
Stocklinch 8 C3
Stockport 49 H3
Stocksbridge 50 E3
Stocksfield 71 F7
Stockton Here 28 E2
Stockton Norf 45 H6
Stockton Shrop 38 B5
Stockton Shrop 39 G6
Stockton Tel&W 39 G4
Stockton Warks 31 F2
Stockton Wilts 9 J1
Stockton Heath 49 F4
Stockton on Teme 29 G2
Stockton on the Forest 58 C4
Stockton-on-Tees 62 E5
Stockwell 29 J7
Stockwell Heath 40 C3
Stockwood Bristol 19 K5
Stockwood Dorset 8 E4
Stodday 55 H4
Stodmarsh 25 J5
Stody 44 E2
Stoer 86 C7
Stoford Som 8 E3
Stoford Wilts 10 B1
Stogumber 7 J2
Stogursey 19 F7
Stoke Devon 6 A3
Stoke Hants 11 J4
Stoke Hants 21 H6
Stoke Med 24 E4
Stoke Plym 4 E5
Stoke WMid 30 E1
Stoke Abbott 8 D4
Stoke Albany 42 B7
Stoke Ash 35 F1
Stoke Bardolph 41 J1
Stoke Bishop 19 J4
Stoke Bliss 29 F2
Stoke Bruerne 31 J3
Stoke by Clare 34 B4
Stoke Canon 7 H6
Stoke Charity 11 F1
Stoke Climsland 4 D3
Stoke D'Abernon 22 E6
Stoke Doyle 42 D7
Stoke Dry 42 B6
Stoke Edith 29 F4
Stoke Farthing 10 B2
Stoke Ferry 44 B6
Stoke Fleming 5 J6
Stoke Gabriel 5 J5
Stoke Gifford 19 K4
Stoke Golding 41 F6
Stoke Goldington 32 B4
Stoke Green 22 C3
Stoke Hammond 32 B6
Stoke Heath Shrop 39 F3
Stoke Heath Worcs 29 J2
Stoke Holy Cross 45 G5
Stoke Lacy 29 F4
Stoke Lyne 31 G6
Stoke Mandeville 32 B7
Stoke Newington 23 G3
Stoke on Tern 39 F3
Stoke Orchard 29 J6
Stoke Pero 7 G1
Stoke Poges 22 C3
Stoke Pound 29 J2
Stoke Prior Here 28 E3
Stoke Prior Worcs 29 J2
Stoke Rivers 6 E2
Stoke Rochford 42 C3
Stoke Row 21 K3
Stoke St. Gregory 8 C2
Stoke St. Mary 8 B2
Stoke St. Michael 19 K7
Stoke St. Milborough 38 E7
Stoke sub Hamdon 8 D3
Stoke Talmage 21 K2
Stoke Trister 9 G2
Stoke Villice 19 J5
Stoke Wake 9 G4
Stoke-by-Nayland 34 D5
Stokeford 9 H6
Stokeham 51 K5
Stokeinteignhead 5 K3
Stokenchurch 22 A2
Stokenham 5 J6
Stoke-on-Trent 40 A1
Stokesay 38 D7
Stokesby 45 J4
Stokesley 63 F6
Stolford 19 F7
Ston Easton 19 K6
Stonar Cut 25 K5
Stondon Massey 23 J1
Stone Bucks 31 J7
Stone Glos 19 K2
Stone Kent 14 E5
Stone Kent 23 J4
Stone Som 8 E1
Stone Staffs 40 B2
Stone SYorks 51 H4
Stone Worcs 29 H1
Stone Allerton 19 H6
Stone Cross Dur 62 A5
Stone Cross ESuss 13 J4
Stone Cross ESuss 13 K6
Stone Cross Kent 13 J3
Stone Cross Kent 15 F4
Stone House 56 C1
Stone Street Kent 23 J6
Stone Street Suff 34 D5
Stone Street Suff 45 H7
Stonea 43 H6
Stonebridge ESuss 13 J4
Stonebridge NSom 19 G6
Stonebridge Warks 40 E7
Stonebroom 51 G7
Stonecross Green 34 C3
Stonefield A&B 73 G3
Stonefield Staffs 40 B2
Stonegate ESuss 13 K4
Stonegate NYorks 63 H6
Stonegrave 58 C2
Stonehaugh 70 C6
Stonehaven 81 P2
Stonehenge (Stonehenge, Avebury & Associated Sites) Wilts SP4 7DE 20 E7
Stonehill 22 C5
Stonehouse ChesW&C 48 E5
Stonehouse D&G 65 J4
Stonehouse Glos 20 B1
Stonehouse N'umb 61 H1
Stonehouse Plym 4 E5
Stonehouse SLan 75 F6
Stoneleigh Surr 23 F5
Stoneleigh Warks 30 E1
Stoneley Green 49 F7
Stonely 32 E2
Stoner Hill 11 J2
Stones 56 E7
Stones Green 35 F6
Stones of Stenness (Heart of Neolithic Orkney) Ork KW16 3JZ 89 C6
Stonesby 42 B3
Stonesfield 30 E7
Stonestreet Green 15 F4
Stonethwaite 60 D5
Stoney Cross 10 D3
Stoney Middleton 50 E5
Stoney Stanton 41 G6
Stoney Stoke 9 G1
Stoney Stratton 9 F1
Stoney Stretton 38 C5
Stoneyburn 75 H4
Stoneyford 7 J7
Stoneygate 41 J5
Stoneyhills 25 F2
Stoneykirk 64 A5
Stoneywood 85 N9
Stonham Aspal 35 F3
Stonham Barns Suff IP14 6AT 35 F3
Stonnall 40 C5
Stonor 22 A3
Stonton Wyville 42 A6
Stony Houghton 51 G6
Stony Stratford 31 J4
Stonybreck 89 K10
Stoodleigh Devon 6 E2
Stoodleigh Devon 7 H4
Stopham 12 D5
Stopsley 32 E6
Stoptide 3 G1
Storeton 48 C4
Stormontfield 80 G8
Stornoway (Steòrnabhagh) 88 K4
Stornoway Airport 88 K4
Storridge 29 G4
Storrington 12 D5
Storrs 50 E4
Storth 55 H1
Storwood 58 D5
Stotfold 33 F5
Stottesdon 39 F7
Stoughton Leics 41 J5
Stoughton Surr 22 C6
Stoughton WSuss 11 J3
Stoughton Cross 19 H7
Stoul 82 H11
Stoulton 29 J4
Stour Provost 9 G2
Stour Row 9 H2
Stourbridge 40 A7
Stourhead Wilts BA12 6QD 9 G1
Stourpaine 9 H4
Stourport-on-Severn 29 H1
Stourton Staffs 40 A7
Stourton Warks 30 D5
Stourton Wilts 9 G1
Stourton Caundle 9 G3
Stoven 45 J7
Stow Lincs 52 B4
Stow ScBord 76 C6
Stow Bardolph 44 A5
Stow Bedon 44 D6
Stow cum Quy 33 J2
Stow Longa 32 E1
Stow Maries 24 E2
Stow Pasture 52 B4
Stowbridge 43 J5
Stowe Glos 19 J1
Stowe Shrop 28 C1
Stowe Staffs 40 D4
Stowe Landscape Gardens Bucks MK18 5DQ 31 H5
Stowe-by-Chartley 40 C3
Stowehill 31 H3
Stowell Glos 30 B7
Stowell Som 9 F2
Stowey 19 J6
Stowford Devon 6 C7
Stowford Devon 6 E3
Stowford Devon 7 K7
Stowlangtoft 34 D2
Stowmarket 34 E3
Stow-on-the-Wold 30 C6
Stowting 15 G3
Stowupland 34 E3
Straad 73 J4
Stracathro 81 M4
Strachan 85 L11
Strachur (Clachan Strachur) 79 M10
Stradbroke 35 G1
Stradishall 34 B3
Stradsett 44 A5
Stragglethorpe 52 C7
Straight Soley 21 G4
Straiton Edin 76 A4
Straiton SAyr 67 H3
Straloch 80 F4
Stramshall 40 C2
Strands 54 E1
Strang 54 C6
Strangford 28 E6
Stranraer 64 A4
Strata Florida 27 G2
Stratfield Mortimer 21 K5
Stratfield Saye 21 K5
Stratfield Turgis 21 K6
Stratford CenBeds 32 E4
Stratford Glos 29 H5
Stratford GtLon 23 G3
Stratford St. Andrew 35 H3
Stratford St. Mary 34 E5
Stratford sub Castle 10 C1
Stratford Tony 10 B2
Stratford-upon-Avon 30 D3
Stratford-upon-Avon Butterfly Farm Warks CV37 7LS 136 Stratford-upon-Avon
Strath 87 Q4
Strathan High 83 K11
Strathan High 86 C7
Strathaven 75 F6
Strathblane 74 D3
Strathcanaird 86 D9
Strathcarron 83 K6
Strathclyde Country Park NLan ML1 3ED 119 K6
Strathdon 84 H9
Strathkinness 81 K9
Strathmiglo 80 H9
Strathpeffer (Strath Pheofhair) 83 Q5
Strathtay 80 E5
Strathwhillan 73 J7
Strathy 87 L3
Strathyre 80 A9
Stratton Corn 6 A5
Stratton Dorset 9 F5
Stratton Glos 20 D1
Stratton Audley 31 H6
Stratton Hall 35 G5
Stratton St. Margaret 20 E3
Stratton St. Michael 45 G6
Stratton Strawless 45 G3
Stratton-on-the-Fosse 19 K6
Stravanan 73 J5
Strawberry Hill 22 E4
Stream 7 J2
Streat 13 G5
Streatham 23 F4
Streatham Ice and Leisure Centre GtLon SW16 6HX 101 A10
Streatham Vale 23 F4
Streatley CenBeds 32 D6
Streatley WBerks 21 J3
Street Devon 7 K7
Street Lancs 55 J4
Street NYorks 63 H6
Street Som 8 C4
Street Som 8 D1
Street Ashton 41 G7
Street Dinas 38 C2
Street End 12 B7
Street Gate 62 C1
Street Houses 58 B5
Street Lane 41 F1
Street on the Fosse 9 F1
Streethay 40 D4
Streethouse 57 J7
Streetlam 62 D7
Streetly 40 C6
Streetly End 33 K4
Strefford 38 D7
Strelley 41 H1
Strensall 58 C3
Strensham 29 J4
Stretcholt 19 F7
Strete 5 J6
Stretford GtMan 49 G3
Stretford Here 28 D3
Stretford Here 28 E3
Strethall 33 H5
Stretham 33 J1
Strettington 12 B6
Stretton ChesW&C 48 D7
Stretton Derbys 51 F6
Stretton Rut 42 C4
Stretton Staffs 40 A4
Stretton Staffs 40 E3
Stretton Warr 49 F4
Stretton en le Field 41 F4
Stretton Grandison 29 F4
Stretton Heath 38 C4
Stretton Sugwas 28 D4
Stretton under Fosse 41 G7
Stretton Westwood 38 E6
Stretton-on-Dunsmore 31 F1
Stretton-on-Fosse 30 D5
Stribers 55 G1
Strichen 85 P5
Strines 49 J4
Strinesdale Countryside Area GtMan OL4 2JJ 113 L1
Stringston 7 K1
Strixton 32 C2
Stroat 19 J2
Stromeferry 83 J7
Stromemore 83 J7
Stromness 89 B7
Stronaba 79 P2
Stronachlachar 79 R9
Strone A&B 73 K2
Strone High 79 N2
Strone High 83 R8
Stronlonag 73 K2
Stronmilchan (Srón nam Mialchon) 79 N8
Stronsay 89 F5
Stronsay Airfield 89 F5
Strontian (Srón an t-Sithein) 79 K4
Strood 24 D5
Strood Green Surr 23 F7
Strood Green WSuss 12 D4
Strood Green WSuss 12 E3
Stroquhan 68 D5
Stroud Glos 20 B1
Stroud Hants 11 J2
Stroud Common 22 D7
Stroud Green Essex 24 E2
Stroud Green Glos 20 B1
Stroude 22 D5
Stroul 74 A2
Stroxton 42 C2
Struan 80 C4
Strubby Lincs 52 E5
Strubby Lincs 53 H4
Strumpshaw 45 H5
Struy 83 Q7
Stryd y Facsen 46 B4
Stryt-cae-rhedyn 48 B6
Stryt-issa 38 B1
Stuart & Waterford Crystal Factory Shop, Crieff P&K PH7 4HQ 80 D8
Stuart Line Cruises, Exmouth Devon EX8 1EJ 7 J7
Stuartfield 85 P6
Stub Place 60 B7
Stubber's Green 40 C5
Stubbington 11 G4
Stubbins 49 G1
Stubbs Green 45 H6
Stubhampton 9 J3
Stubley 51 F5
Stubshaw Cross 48 E2
Stubton 42 B1
Stuck A&B 73 J4
Stuck A&B 73 K1
Stuckbeg 74 A1
Stuckreoch 73 J1
Stuckton 10 C3
Stud Green 22 B4
Studdon 61 K1
Studfold 56 D2
Studham 32 D7
Studholme 60 D1
Studland 10 B6
Studland & Godlingston Heath NNR Dorset BH19 3AX 10 B6
Studley Warks 30 B2
Studley Wilts 20 C4
Studley Common 30 B2
Studley Green 22 A2
Studley Roger 57 H2
Studley Royal Park & Ruins of Fountains Abbey NYorks HG4 3DY 57 H3
Stuggadhoo 54 C6
Stump Cross Essex 33 J4
Stump Cross Lancs 55 J6
Stuntney 33 J1
Stunts Green 13 K5
Sturbridge 40 A2
Sturgate 52 B4
Sturmer 33 K4
Sturminster Common 9 G3
Sturminster Marshall 9 J4
Sturminster Newton 9 G3
Sturry 25 H5
Sturton by Stow 52 B4
Sturton le Steeple 51 K4
Stuston 35 F1
Stutton NYorks 57 K5
Stutton Suff 35 F5
Styal 49 H4
Styrrup 51 J3
Suardail 88 K4
Succoth 79 P10
Succothmore 79 N10
Suckley 29 G3
Suckley Green 29 G3
Suckley Knowl 29 G3
Sudborough 42 C7
Sudbourne 35 J3
Sudbrook Lincs 42 C1
Sudbrook Mon 19 J3
Sudbrooke 52 D5
Sudbury Derbys 40 D2
Sudbury GtLon 22 E3
Sudbury Suff 34 C4
Sudbury Hall Derbys DE6 5HT 40 D2
Sudden 49 H1
Sudgrove 20 C1
Suffield Norf 45 G2
Suffield NYorks 63 K7
Sugarloaf 14 E4
Sugnall 39 G2
Sugwas Pool 28 D4
Suie Lodge Hotel 79 R8
Sulby IoM 54 C4
Sulby IoM 54 C6
Sulgrave 31 G4
Sulham 21 K4
Sulhamstead 21 K5
Sullington 12 D5
Sullom 89 M5
Sullom Voe Oil Terminal 89 M5
Sully 18 E5
Sumburgh 89 M11
Sumburgh Airport 89 M11
Summer Bridge 57 H3
Summer Isles 86 B9
Summer Lodge 61 L7
Summercourt 3 F3
Summerfield Norf 44 B2
Summerfield Worcs 29 H1
Summerhill 48 C7
Summerhouse 62 C5
Summerlands 55 J1
Summerleaze 19 H3
Summertown 21 J1
Summit 49 J1
Sun Green 49 J3
Sunadale 73 G6
Sunbiggin 61 H6
Sunbury 22 E5
Sundaywell 68 D5
Sunderland Cumb 60 C3
Sunderland Lancs 55 H4
Sunderland T&W 62 D1
Sunderland Bridge 62 C3
Sunderland Museum & Winter Gardens T&W SR1 1PP 136 Sunderland
Sundhope 69 J1
Sundon Park 32 D6
Sundown Adventure Land Notts DN22 0HX 51 K5
Sundridge 23 H6
Sundrum Mains 67 J1
Sunhill 20 E1
Sunk Island 53 F1
Sunningdale 22 C5
Sunninghill 22 C5
Sunningwell 21 H1
Sunniside Dur 62 B3
Sunniside T&W 62 C1
Sunny Bank 60 D7
Sunny Brow 62 B3
Sunnylaw 75 F1
Sunnyside N'umb 70 E7
Sunnyside SYorks 51 G3
Sunnyside WSuss 13 G3
Sunton 21 F6
Sunwick 77 G5
Surbiton 22 E5
Surfleet 43 F3
Surfleet Seas End 43 F3
Surlingham 45 H5
Sustead 45 F2
Susworth 52 B2
Sutcombe 6 B4
Sutcombemill 6 B4
Suton 44 E6
Sutors of Cromarty 84 C4
Sutterby 53 G5
Sutterton 43 F2
Sutton Cambs 33 H1
Sutton CenBeds 33 F4
Sutton Devon 5 H6
Sutton Devon 7 F5
Sutton GtLon 23 F5
Sutton Kent 15 J3
Sutton Lincs 52 B7
Sutton Norf 45 H3
Sutton Notts 42 A2
Sutton Notts 51 J4
Sutton Oxon 21 H1
Sutton Pembs 16 C4
Sutton Peter 42 D6
Sutton Shrop 38 C3
Sutton Shrop 38 E4
Sutton Shrop 39 F2
Sutton Shrop 39 G7
Sutton Staffs 39 G3
Sutton Suff 35 H4
Sutton SYorks 51 H1
Sutton WSuss 12 C5
Sutton Abinger 22 E7
Sutton at Hone 23 J4
Sutton Bank National Park Centre NYorks YO7 2EH 58 B1
Sutton Bassett 42 A7
Sutton Benger 20 C4
Sutton Bingham 8 E3
Sutton Bonington 41 H3
Sutton Bridge 43 H3
Sutton Cheney 41 G5
Sutton Coldfield 40 D6
Sutton Courtenay 21 J2
Sutton Crosses 43 H3
Sutton Grange 57 H2
Sutton Green Oxon 21 H1
Sutton Green Surr 22 D6
Sutton Green Wrex 38 D1
Sutton Holms 10 B4
Sutton Howgrave 57 J2
Sutton in Ashfield 51 G7
Sutton in the Elms 41 H6
Sutton Ings 59 H6
Sutton Lane Ends 49 J5
Sutton le Marsh 53 J4
Sutton Leach 48 E3
Sutton Maddock 39 G5
Sutton Mallet 8 C1
Sutton Mandeville 9 J2
Sutton Montis 9 F2
Sutton on Sea 53 J4
Sutton on the Hill 40 E2
Sutton on Trent 51 K6
Sutton Poyntz 9 G6
Sutton Scarsdale 51 G6
Sutton Scotney 11 F1
Sutton St. Edmund 43 G4
Sutton St. James 43 H4
Sutton St. Nicholas 28 E4
Sutton upon Derwent 58 D5
Sutton Valence 14 D3
Sutton Veny 20 B7
Sutton Waldron 9 H3
Sutton Weaver 48 E5
Sutton Wick B&NESom 19 J6
Sutton Wick Oxon 21 H2
Sutton-in-Craven 57 F5
Sutton-on-Hull 59 H6
Sutton-on-the-Forest 58 B3
Sutton-under-Brailes 30 E5
Sutton-under-Whitestonecliffe 57 K1
Swaby 53 G5
Swadlincote 41 F4
Swaffham 44 C5
Swaffham Bulbeck 33 J2
Swaffham Prior 33 J2
Swafield 45 G2
Swainbost (Suainebost) 88 L1
Swainby 62 E6
Swainshill 28 D4
Swainsthorpe 45 G5
Swainswick 20 A5
Swalcliffe 30 E5
Swalecliffe 25 H5
Swallow 52 E2
Swallow Beck 52 C6
Swallow Falls Conwy LL24 0DW 47 F7
Swallowcliffe 9 J2
Swallowfield 22 A5
Swallows Cross 24 C2
Swampton 21 H6
Swan Green ChesW&C 49 G5
Swan Green Suff 35 G1
Swan Street 34 C6
Swanage 10 B7
Swanage Railway Dorset BH19 1HB 10 B7
Swanbach 39 F1
Swanbourne 32 B6
Swanbridge 18 E5
Swancote 39 G6
Swanland 59 F7
Swanlaws 70 C2
Swanley 23 J5
Swanley Village 23 J5
Swanmore Hants 11 G3
Swanmore IoW 11 G5
Swannington Leics 41 G4
Swannington Norf 45 F4
Swanscombe 24 C4
Swansea (Abertawe) 17 K6
Swansea Museum SA1 1SN 136 Swansea
Swanston 76 A4
Swanton Abbot 45 G3
Swanton Morley 44 E4
Swanton Novers 44 E2
Swanton Street 14 D2
Swanwick Derbys 51 G7
Swanwick Hants 11 G4
Swanwick Green 38 E1
Swarby 42 D1
Swardeston 45 G5
Swarkestone 41 F3
Swarland 71 G3
Swarraton 11 G1
Swarthmoor 55 F2
Swaton 42 E2
Swavesey 33 G2
Sway 10 D5
Swayfield 42 C3
Swaythling 11 F3
Swaythorpe 59 G3
Sweetham 7 G6
Sweethay 8 B2
Sweetshouse 4 A4
Sweffling 35 H2

Swell 8 C2
Swepstone 41 F4
Swerford 30 E5
Swettenham 49 H6
Swffryd 19 F2
Swift's Green 14 D3
Swiftsden 14 C5
Swilland 35 F3
Swillington 57 J6
Swimbridge 6 E3
Swimbridge Newland 6 D2
Swinbrook 30 D7
Swincliffe 57 H4
Swincombe 6 E1
Swindon 56 D4
Swinderby 52 B6
Swindon Staffs 40 A6
Swindon Swin 20 E3
Swindon Village 29 J6
Swine 59 H6
Swinefleet 58 D7
Swineford 19 K5
Swineshead Bed 32 D2
Swineshead Lincs 43 F1
Swineshead Bridge 43 F1
Swineside 57 F1
Swiney 87 Q6
Swinford Leics 31 G1
Swinford Oxon 21 H1
Swingate 41 H1
Swingfield Minnis 15 H3
Swingleton Green 34 D4
Swinhoe 71 H1
Swinhope 53 F3
Swinithwaite 57 F1
Swinscoe 40 D1
Swinside Hall 70 C2
Swinstead 42 D3
Swinton GtMan 49 G2
Swinton NYorks 57 H2
Swinton NYorks 58 D2
Swinton ScBord 77 G6
Swinton SYorks 51 G3
Swinton Quarter 77 G6
Swintonmill 77 G6
Swithland 41 H4
Swordale 83 R4
Swordland 82 H11
Swordly 87 K3
Sworton Heath 49 F4
Swyddffynnon 27 F2
Swyncombe 21 K2
Swynnerton 40 A2
Swyre 8 E6
Sychnant 27 J1
Syde 29 J7
Sydenham GtLon 23 G4
Sydenham Oxon 22 A1
Sydenham Damerel 4 E3
Syderstone 44 C2
Sydling St. Nicholas 9 F5
Sydmonton 21 H6
Sydney 49 G7
Syerston 42 A1
Sykehouse 51 J1
Sykes 56 B4
Sylen 17 J5
Symbister 89 P6
Symington SAyr 74 B7
Symington SLan 75 H7
Symonds Yat 28 E7
Symondsbury 8 D5
Synod Inn (Post-mawr) 26 D3
Syre 87 J5
Syreford 30 B6
Syresham 31 H4
Syston Leics 41 J4
Syston Lincs 42 C1
Sytchampton 29 H2
Sywell 32 B2
Sywell Country Park N'hants NN6 0QX 32 B2

T

Tableyhill 49 G5
Tachbrook Mallory 30 E2
Tackley 31 F6
Tacolneston 45 F6
Tadcaster 57 K5
Tadden 9 J4
Taddington Derbys 50 D5
Taddington Glos 30 B5
Taddiport 6 C4
Tadley 21 K5
Tadlow 33 F4
Tadmarton 30 E5
Tadpole Bridge 21 G1
Tadworth 23 F6
Tafarnaubach 28 A7
Tafarn-y-bwlch 16 D2
Tafarn-y-Gelyn 47 K6
Taff Merthyr Garden Village 18 E2
Taff's Well (Ffynnon Taf) 18 E3
Tafolwern 37 H5
Taibach NPT 18 A3
Tai-bach Powys 38 A3
Taicynhaeaf 37 F4
Tain 84 B2
Tai'n Lôn 36 D1
Tai'r Bull 27 J6
Tairgwaith 27 G7
Tai'r-heol 18 E2
Tairlaw 67 J3
Tai'r-ysgol 17 K6
Takeley 33 J6
Takeley Street 33 J6
Talachddu 27 K5
Talacre 47 K4
Talardd 37 H3
Talaton 7 J6
Talbenny 16 B4
Talbot Green 18 D3
Talbot Village 10 B5
Talerddig 37 J5
Talgarreg 26 D3
Talgarth 28 A5
Taliesin 37 F6
Talisker 82 D7
Talke Pits 49 H7
Talkin 61 G1
Talkin Tarn Country Park Cumb CA8 1HN 61 G1
Talla Linnfoots 69 G1
Talladale 83 K3
Talland 4 C5
Tallarn Green 38 D1
Tallentire 60 C3
Talley (Talyllychau) 17 K2
Tallington 42 D5
Talmine 86 H3
Talog 17 G3
Tal-sarn 26 E3
Talsarnau 37 F2
Talskiddy 3 G2
Talwrn IoA 46 C5
Talwrn Wrex 38 B1
Talwrn Wrex 38 C1
Talybont Cere 37 F7
Tal-y-bont Conwy 47 F6
Tal-y-bont Gwyn 36 E3
Tal-y-bont Gwyn 46 E5
Talybont-on-Usk 28 A6
Tal-y-Cae 46 E6
Tal-y-cafn 47 F5
Tal-y-coed 28 D7
Talygarn 18 D3
Tal-y-llyn Gwyn 37 G5
Talyllyn Powys 28 A6
Talysarn 46 C7
Tal-y-wern 37 H5
Tamar Valley Mining District (Cornwall & West Devon Mining Landscape) Corn/Devon 4 E3
Tamavoid 74 D1
Tamerton Foliot 4 E4
Tamworth 40 E5
Tamworth Green 43 G1
Tan Office Green 34 B3
Tandem 50 D1
Tandridge 23 G6
Tanerdy 17 H3
Tanfield 62 B1
Tanfield Lea 62 B1
Tang 57 H4
Tang Hall 58 C4
Tangiers 16 C4
Tangley 21 G6
Tangmere 12 C6
Tangy 66 A1
Tank Museum, Bovington Dorset BH20 6JG 9 H6
Tankersley 51 F2
Tankerton 25 H5
Tan-lan 37 F1
Tannach 87 R5
Tannadice 81 K5
Tannington 35 G2
Tannochside 74 E4
Tansley 51 F7
Tansley Knoll 51 F7
Tansor 42 D6
Tantobie 62 B1
Tanton 63 F5
Tanworth in Arden 30 C1
Tan-y-fron 47 H6
Tan-y-graig 36 C2
Tanygrisiau 37 F1
Tan-y-groes 17 F1
Tan-y-pistyll 37 K3
Tan-yr-allt 47 J4
Taobh Siar 88 G7
Tapeley 6 C3
Taplow 22 C3
Tapton Grove 51 G5
Taransay (Tarasaigh) 88 F7
Tarbert A&B 72 E2
Tarbert A&B 72 E5
Tarbert A&B 73 G4
Tarbert Na H-E. Siar 88 G7
Tarbet (An Tairbeart) A&B 79 Q10
Tarbet High 82 H11
Tarbet High 86 D5
Tarbock Green 48 D4
Tarbolton 67 J1
Tarbrax 75 J5
Tardebigge 29 J2
Tardy Gate 55 J7
Tarfside 81 K3
Tarland 85 J10
Tarleton 55 H7
Tarlscough 48 D1
Tarlton 20 C2
Tarnbrook 55 J4
Tarnock 19 G6
Tarporley 48 E6
Tarr 7 K2
Tarrant Crawford 9 J4
Tarrant Gunville 9 J3
Tarrant Hinton 9 J3
Tarrant Keyneston 9 J4
Tarrant Launceston 9 J4
Tarrant Monkton 9 J4
Tarrant Rawston 9 J4
Tarrant Rushton 9 J4
Tarrel 84 C2
Tarring Neville 13 H6
Tarrington 29 F4
Tarrnacraig 73 H7
Tarskavaig 82 F10
Tarves 85 N7
Tarvie (Tairbhidh) High 83 Q5
Tarvie P&K 80 F4
Tarvin 48 D6
Tarvin Sands 48 D6
Tasburgh 45 G6
Tasley 39 F6
Taston 30 E6
Tate Britain GtLon SW1P 4RG 101 A7
Tate Liverpool Mersey L3 4BB 131 B5
Tate Modern GtLon SE1 9TG 132 H4
Tate St. Ives Corn TR26 1TG 2 C4
Tatenhill 40 E3
Tathall End 32 B4
Tatham 56 B3
Tathwell 53 G4
Tatsfield 23 H6
Tattenhall 48 D7
Tattenhoe 32 B5
Tatterford 44 C3
Tattersett 44 C2
Tattershall 53 F7
Tattershall Bridge 52 E7
Tattershall Thorpe 53 F7
Tattingstone 35 F5
Tatton Park ChesE WA16 6QN 49 G4
Tatworth 8 C4
Tauchers 84 H6
Taunton 8 B2
Tavelty 85 M9
Taverham 45 F4
Tavernspite 16 E4
Tavistock 4 E3
Tavistock (Cornwall & West Devon Mining Landscape) Devon 4 E3
Taw Bridge 6 E5
Taw Green 6 E6
Tawstock 6 D3
Taxal 50 C5
Tayburn 74 D6
Tayinloan 72 E6
Taylors Cross 6 A4
Taynish 73 F2
Taynton Glos 29 G6
Taynton Oxon 30 D7
Taynuilt (Taigh an Uillt) 79 M7
Tayock 81 M5
Tayovullin 72 A3
Tayport 81 K8
Tayvallich 73 F2
Tea Green 32 E6
Tealby 52 E3
Tealing 81 K7
Team Valley 71 H7
Teanamachar 88 B2
Teangue 82 G10
Tebay 61 H6
Tebworth 32 C6
Techniquest Cardiff CF10 5BW 95 C6
Tedburn St. Mary 7 G6
Teddington Glos 29 J5
Teddington GtLon 22 E4
Tedstone Delamere 29 F3
Tedstone Wafre 29 F3
Teeton 31 H1
Teffont Evias 9 J1
Teffont Magna 9 J1
Tegryn 17 F2
Tehidy Country Park Corn TR14 0HA 2 D4
Teifi Marshes Nature Reserve Pembs SA43 2TB 16 E1
Teigh 42 B4
Teign Village 7 G7
Teigngrace 5 J3
Teignmouth 5 K3
Telford 39 F5
Telford Wonderland Tel&W TF3 4AY 39 G5
Telham 14 C6
Tellisford 20 B6
Telscombe 13 H6
Telscombe Cliffs 13 H6
Templand 69 F5
Temple Corn 4 B3
Temple Midlo 76 B5
Temple Balsall 30 D1
Temple Bar 26 E3
Temple Cloud 19 K6
Temple End 33 K3
Temple Ewell 15 H3
Temple Grafton 30 C3
Temple Guiting 30 B6
Temple Herdewyke 30 E3
Temple Hirst 58 C7
Temple Newsam WYorks LS15 0AE 115 M3
Temple Normanton 51 G6
Temple Sowerby 61 H4
Templecombe 9 G2
Templeton Devon 7 G4
Templeton Pembs 16 E4
Templeton Bridge 7 G4
Tempsford 32 E3
Ten Mile Bank 44 A6
Tenbury Wells 28 E2
Tenby (Dinbych-y-pysgod) 16 E5
Tendring 35 F6
Tendring Green 35 F6
Tenterden 14 D4
Terally 64 B6
Terling 34 B7
Tern 39 F4
Ternhill 39 F2
Terregles 65 K3
Terriers 22 B2
Terrington 58 C2
Terrington St. Clement 43 J3
Terrington St. John 43 J4
Terry's Green 30 C1
Teston 14 C2
Testwood 10 E3
Tetbury 20 B2
Tetbury Upton 20 B2
Tetchill 38 C2
Tetcott 6 B6
Tetford 53 G5
Tetney 53 G2
Tetney Lock 53 G2
Tetsworth 21 K1
Tettenhall 40 A5
Tettenhall Wood 40 A6
Tetworth 33 F3
Teversal 51 G6
Teversham 33 H3
Teviot Smokery and Water Gardens ScBord TD5 8LE 70 C1
Teviothead 69 K3
Tewin 33 F7
Tewkesbury 29 H5
Tewkesbury Abbey Glos GL20 5RZ 29 H5
Teynham 25 F5
Thackley 57 G6
Thainston 81 M3
Thakeham 12 E5
Thame 22 A1
Thames Ditton 22 E5
Thames Haven 24 D3
Thamesmead 23 H3
Thanington 15 G2
Thankerton 75 H7
Tharston 45 F6
Thatcham 21 J5
Thatto Heath 48 E3
Thaxted 33 K5
The Apes Hall 43 J6
The Bage 28 B4
The Balloch 80 D9
The Bar 12 E4
The Birks 85 M10
The Bog 38 C6
The Bourne 22 B7
The Bratch 40 A6
The Broad 28 D2
The Bryn 19 G1
The Burf 29 H2
The Butts 20 A7
The Camp 20 C1
The Chequer 38 D1
The City Bucks 22 A2
The City Suff 45 H7
The Common Wilts 10 D1
The Common Wilts 20 D3
The Craigs 83 Q1
The Cronk 54 C4
The Delves 40 C6
The Den 74 B5
The Den & The Glen Aber AB12 5FT 85 N11
The Dicker 13 J6
The Down 39 F6
The Drums 81 J4
The Eaves 19 K1
The English Lake District Cumb 60 D5
The Flatt 70 A6
The Folly 32 E7
The Forge 28 C3
The Forstal ESuss 13 J3
The Forstal Kent 15 F4
The Forth Bridge Fife/ WLoth EH30 9SF 75 K2
The Grange Lincs 53 J5
The Grange Shrop 38 C2
The Grange Surr 23 G7
The Green Cumb 54 E1
The Green Essex 34 B7
The Green Flints 48 B6
The Green Wilts 9 H1
The Grove 29 H4
The Haven 12 D3
The Headland 63 F3
The Heath 40 C2
The Herberts 18 C4
The Hermitage 23 F6
The Hill 54 E1
The Holme 57 H4
The Howe 54 A7
The Isle 38 D4
The Laurels 45 H6
The Leacon 14 E4
The Lee 22 B1
The Leigh 29 H6
The Lhen 54 C3
The Lodge 73 K1
The Lodge Visitor Centre Stir FK8 3SX 80 A10
The Luxulyan Valley (Cornwall & West Devon Mining Landscape) Corn 3 H3
The Marsh 38 C6
The Moor ESuss 14 D6
The Moor Kent 14 C5
The Mumbles 17 K7
The Murray 74 E5
The Mythe 29 H5
The Narth 19 J1
The Node 33 F7
The Oval 20 A5
The Port of Hayle (Cornwall & West Devon Mining Landscape) Corn 2 C5
The Quarter 14 D3
The Reddings 29 J6
The Rhos 16 D4
The Rookery 49 H7
The Rowe 40 A2
The Sale 40 D4
The Sands 22 B7
The Shoe 20 B4
The Slade 21 J4
The Smithies 39 F6
The Stocks 14 E5
The Swillett 22 D2
The Thrift 33 G5
The Vauld 28 E4
The Wern 48 B7
The Wyke 39 G5
Theakston 57 J1
Thealby 52 B1
Theale Som 19 H7
Theale WBerks 21 K4
Thearne 59 G6
Theberton 35 J2
Thedden Grange 11 H1
Theddingworth 41 J7
Theddlethorpe All Saints 53 H4
Theddlethorpe St. Helen 53 H4
Thelbridge Barton 7 F4
Thelbridge Cross 7 F4
Thelnetham 34 E1
Thelveton 45 F7
Thelwall 49 F4
Themelthorpe 44 E3
Thenford 31 G4
Therfield 33 G5
Thermae Bath Spa B&NESom BA1 1SJ 121 Bath
Thetford Lincs 42 E4
Thetford Norf 44 C7
Thetford Forest Park Norf IP27 0TJ 44 C7
Thethwaite 60 E2
Theydon Bois 23 H2
Theydon Garnon 23 H2
Theydon Mount 23 H2
Thickwood 20 B4
Thimbleby Lincs 53 F6
Thimbleby NYorks 62 E7
Thingley 20 B5
Thirkleby 57 K2
Thirlby 57 K1
Thirlestane 76 D6
Thirn 57 H1
Thirsk 57 K1
Thirston New Houses 71 G4
Thirtleby 59 H6
Thistleton Lancs 55 H6
Thistleton Rut 42 C4
Thistley Green 33 K1
Thixendale 58 E3
Thockrington 70 E6
Tholomas Drove 43 G5
Tholthorpe 57 K3
Thomas Chapel 16 E5
Thomas Close 60 F2
Thompson 44 D6
Thomshill 84 G5
Thong 24 C4
Thongsbridge 50 D2
Thoralby 57 F1
Thoresby 51 J5
Thoresthorpe 53 H5
Thoresway 52 E3
Thorganby Lincs 53 F3
Thorganby NYorks 58 C5
Thorgill 63 H7
Thorington 35 J1
Thorington Street 34 E5
Thorlby 56 E4
Thorley 33 H7
Thorley Houses 33 H6
Thorley Street Herts 33 H7
Thorley Street IoW 10 E6
Thormanby 57 K2
Thornaby-on-Tees 62 E5
Thornage 44 E2
Thornborough Bucks 31 J5
Thornborough NYorks 57 H2
Thornbury Devon 6 B5
Thornbury Here 29 F3
Thornbury SGlos 19 K2
Thornbury WYorks 57 G6
Thornby 31 H1
Thorncliff 50 E1
Thorncliffe 50 C7
Thorncombe 8 C4
Thorncombe Street 22 D7
Thorncote Green 32 E4
Thorncross 11 F6
Thorndon 35 F2
Thorndon Country Park Essex CM13 3RZ 24 C2
Thorndon Cross 6 D6
Thorne 51 J1
Thorne St. Margaret 7 J3
Thorner 57 J6
Thorney Bucks 22 D4
Thorney Notts 52 B5
Thorney Peter 43 F5
Thorney Som 8 D2
Thorney Close 62 D1
Thorney Hill 10 D5
Thornfalcon 8 B2
Thornford 9 F3
Thorngrafton 70 C7
Thorngrove 8 C1
Thorngumbald 59 J7
Thornham 44 B1
Thornham Magna 35 F1
Thornham Parva 35 F1
Thornhaugh 42 D5
Thornhill Cardiff 18 E3
Thornhill Cumb 60 B6
Thornhill Derbys 50 D4
Thornhill D&G 68 D4
Thornhill Soton 11 F3
Thornhill Stir 80 B10
Thornhill WYorks 50 E1
Thornhill Lees 50 E1
Thornholme 59 H3
Thornicombe 9 H4
Thornley Dur 62 B3
Thornley Dur 62 D3
Thornley Gate 61 K1
Thornliebank 74 D5
Thornroan 85 N7
Thorns 34 B3
Thorns Green 49 G4
Thornsett 50 C4
Thornthwaite Cumb 60 D4
Thornthwaite NYorks 57 G4
Thornton Bucks 31 J5
Thornton ERid 58 D5
Thornton Fife 76 A1
Thornton Lancs 55 G5
Thornton Leics 41 G5
Thornton Lincs 53 F6
Thornton Mersey 48 C2
Thornton Middl 62 E5
Thornton N'umb 77 H6
Thornton Pembs 16 C5
Thornton WYorks 57 G6
Thornton Bridge 57 K2
Thornton Curtis 52 D1
Thornton Heath 23 G5
Thornton Hough 48 C4
Thornton in Lonsdale 56 B2
Thornton le Moor 52 D3
Thornton Park 77 H6
Thornton Rust 56 E1
Thornton Steward 57 G1
Thornton Watlass 57 H1
Thorntonhall 74 D5
Thornton-in-Craven 56 E5
Thornton-le-Beans 62 E7
Thornton-le-Clay 58 C3
Thornton-le-Dale 58 E1
Thornton-le-Moor 57 J1
Thornton-le-Moors 48 D5
Thornton-le-Street 57 K1
Thorntonloch 77 F3
Thornwood 23 H1
Thornylee 76 C7
Thoroton 42 A1
Thorp Arch 57 K5
Thorpe Derbys 50 D7
Thorpe ERid 59 F5
Thorpe Lincs 53 H4
Thorpe Norf 45 J6
Thorpe Notts 51 K7
Thorpe NYorks 57 F3
Thorpe Surr 22 D5
Thorpe Abbotts 45 F7
Thorpe Acre 41 H3
Thorpe Arnold 42 A3
Thorpe Audlin 51 G1
Thorpe Bassett 58 E2
Thorpe Bay 25 F3
Thorpe by Water 42 B6
Thorpe Constantine 40 E5
Thorpe Culvert 53 H6
Thorpe End 45 G4
Thorpe Green Essex 35 F6
Thorpe Green Lancs 55 J7
Thorpe Green Suff 34 D3
Thorpe Hall 58 B2
Thorpe Hesley 51 F3
Thorpe in Balne 51 H1
Thorpe in the Fallows 52 C4
Thorpe Langton 42 A6
Thorpe Larches 62 D4
Thorpe le Street 58 E5
Thorpe Malsor 32 B1
Thorpe Mandeville 31 G4
Thorpe Market 45 G2
Thorpe Morieux 34 D3
Thorpe on the Hill Lincs 52 C6
Thorpe on the Hill WYorks 57 J7
Thorpe Park Surr KT16 8PN 22 D5
Thorpe Row 44 D5
Thorpe Salvin 51 H4
Thorpe Satchville 42 A4
Thorpe St. Andrew 45 G5
Thorpe St. Peter 53 H6
Thorpe Street 34 E1
Thorpe Thewles 62 E4
Thorpe Tilney Dales 52 E7
Thorpe Underwood NYorks 57 K4
Thorpe Underwood N'hants 42 A7
Thorpe Waterville 42 D7
Thorpe Willoughby 58 B6
Thorpefield 57 K2
Thorpe-le-Soken 35 F6
Thorpeness 35 J2
Thorpland 44 A5
Thorrington 34 E6
Thorverton 7 H5
Thrandeston 35 F1
Thrapston 32 C1
Threapland 56 E3
Threapwood 38 D1
Threapwood Head 40 C1
Three Ashes 19 K7
Three Bridges 13 F3
Three Burrows 2 E4
Three Chimneys 14 D4
Three Cocks (Aberllynfi) 28 A5
Three Counties Showground Worcs WR13 6NW 29 G4
Three Crosses 17 J6
Three Cups Corner 13 K4
Three Hammers 4 C2
Three Holes 43 J5
Three Leg Cross 13 K3
Three Legged Cross 10 B4
Three Mile Cross 22 A5
Three Oaks 14 D6
Threehammer Common 45 H4
Threekingham 42 D2
Threemilestone 2 E4
Threlkeld 60 E4
Threshfield 56 E3
Threxton Hill 44 C5
Thrigby 45 J4
Thrigby Hall Wildlife Gardens Norf NR29 3DR 45 J4
Thringarth 61 L4
Thringstone 41 G4
Thrintoft 62 D7
Thriplow 33 H4
Throapham 51 H4
Throckenholt 43 G5
Throcking 33 G5
Throckley 71 G7
Throckmorton 29 J4
Throop 9 H5
Throphill 71 G5
Thropton 71 F3
Througham 20 C1
Throwleigh 6 E6
Throwley 14 E2
Throws 33 K6
Thrumpton Notts 41 H2
Thrumpton Notts 51 K5
Thrumster 87 R5
Thrunton 71 F2
Thrupp Glos 20 B1
Thrupp Oxon 21 F2
Thrupp Oxon 31 F7
Thrushelton 6 C7
Thrussington 41 J4
Thruxton Hants 21 F7
Thruxton Here 28 D5
Thrybergh 51 G3
Thrybergh Country Park SYorks S65 4NU 109 G1
Thulston 41 G2
Thunder Bridge 50 D1
Thundergay 73 G6
Thundersley 24 E3
Thundridge 33 G7
Thurcaston 41 H4
Thurcroft 51 G4
Thurdon 6 A4
Thurgarton Norf 45 F2
Thurgarton Notts 41 J1
Thurgoland 50 E2
Thurlaston Leics 41 H6
Thurlaston Warks 31 F1
Thurlbear 8 B2
Thurlby Lincs 42 E4
Thurlby Lincs 52 C6
Thurlby Lincs 53 H5
Thurleigh 32 D3
Thurlestone 5 G6
Thurloxton 8 B1
Thurlstone 50 E2
Thurlton 45 J6
Thurlwood 49 H7
Thurmaston 41 J5
Thurnby 41 J5
Thurne 45 J4
Thurnham 14 D2
Thurning Norf 44 E3
Thurning N'hants 42 D7
Thurnscoe 51 G2
Thursby 60 E1
Thursden 56 E6
Thursford 44 D2
Thursford Collection Norf NR21 0AS 44 D2
Thursley 12 C3
Thurso (Inbhir Theòrsa) 87 P3
Thurstaston 48 B4
Thurston 34 D2
Thurston Clough 49 J2
Thurstonfield 60 E1
Thurstonland 50 D1
Thurton 45 H5
Thurvaston Derbys 40 D2
Thurvaston Derbys 40 E2
Thuxton 44 E5
Thwaite NYorks 61 K7
Thwaite Suff 35 F2
Thwaite Head 60 E7
Thwaite St. Mary 45 H6
Thwaites 57 F5
Thwaites Brow 57 F5
Thwing 59 G2
Tibbermore 80 F8
Tibberton Glos 29 G6
Tibberton Tel&W 39 F3
Tibberton Worcs 29 J3
Tibbie Shiels Inn 69 H1
Tibenham 45 F6
Tibshelf 51 G6
Tibthorpe 59 F4
Ticehurst 13 K3
Tichborne 11 G1
Tickencote 42 C5
Tickenham 19 H4
Tickford End 32 B4
Tickhill 51 H3
Ticklerton 38 D6
Ticknall 41 F3
Tickton 59 G5
Tidbury Green 30 C1
Tidcombe 21 F6
Tiddington Oxon 21 K1
Tiddington Warks 30 D3
Tiddleywink 20 B4
Tidebrook 13 K4
Tideford 4 D5
Tideford Cross 4 D4
Tidenham 19 J2
Tidenham Chase 19 J2
Tideswell 50 D5
Tidmarsh 21 K4
Tidmington 30 D5
Tidpit 10 B3
Tidworth 21 F7
Tiers Cross 16 C4
Tiffield 31 H3
Tigerton 81 L4
Tighnabruaich 73 H3
Tigley 5 H4
Tilbrook 32 D2
Tilbury 24 C4
Tilbury Green 34 B4
Tile Hill 30 D1
Tilehurst 21 K4
Tilford 22 B7
Tilgate 13 F3
Tilgate Forest Row 13 F3
Tilgate Park WSuss RH10 5PQ 13 F3
Tillathrowie 85 J7
Tillers' Green 29 F5
Tilley 38 E3
Tillicoultry 75 H1
Tillietudlem 75 G6
Tillingham 25 F1
Tillington Here 28 D4
Tillington WSuss 12 C4
Tillington Common 28 D4
Tillyfourie 85 L9
Tilmanstone 15 J2
Tiln 51 K4
Tilney All Saints 43 J4
Tilney Fen End 43 J4
Tilney High End 43 J4
Tilney St. Lawrence 43 J4
Tilshead 20 D7
Tilstock 38 E2
Tilston 48 D7
Tilstone Fearnall 48 E6
Tilsworth 32 C6
Tilton on the Hill 42 A5
Tiltups End 20 B2
Timberland 52 E7
Timberland Dales 52 E6
Timbersbrook 49 H6
Timberscombe 7 H1
Timble 57 G4
Timewell 7 H3
Timperley 49 G4
Timsbury B&NESom 19 K6
Timsbury Hants 10 E2
Timsgearraidh 88 F4
Timworth 34 C2
Timworth Green 34 C2
Tincleton 9 G5
Tindale 61 H1
Tindon End 33 K5
Tingewick 31 H5
Tingley 57 H7
Tingrith 32 D5
Tingwall 89 D5
Tingwall (Lerwick Airport) 89 N8
Tinhay 6 B7
Tinney 4 C1
Tinshill 57 H6
Tinsley 51 G3
Tinsley Green 13 F3
Tintagel 4 A2
Tintagel Castle Corn PL34 0HE 4 A2
Tintern Parva 19 J1
Tintinhull 8 D3
Tintwistle 50 C3
Tinwald 69 F5
Tinwell 42 D5
Tippacott 7 F1
Tipps End 43 H6
Tiptoe 10 D5
Tipton 40 B6
Tipton St. John 7 J6
Tiptree 34 C7
Tiptree Heath 34 C7
Tiptree Museum Essex CO5 0RF 34 D7
Tirabad 27 H4
Tiree 78 A6
Tiree Airport 78 B6
Tirley 29 H6
Tirphil 18 E1
Tirril 61 G4
Tir-y-dail 17 K4
Tisbury 9 J2
Tisman's Common 12 D3
Tissington 50 D7
Titchberry 6 A3
Titchfield 11 G4
Titchmarsh 32 D1
Titchwell 44 B1
Titchwell Marsh Norf PE31 8BB 44 B1
Tithby 41 J2
Titley 28 C2
Titlington 71 G2
Titmore Green 33 F6
Titsey 23 H6
Titson 6 A5
Tittensor 40 A2
Tittesworth Reservoir & Visitor Centre Staffs ST13 8TQ 49 J7
Tittleshall 44 C3
Tiverton ChesW&C 48 E6
Tiverton Devon 7 H4
Tivetshall St. Margaret 45 F7
Tivetshall St. Mary 45 F7
Tivington 7 H1
Tixall 40 B3
Tixover 42 C5
Toab Ork 89 E7
Toab Shet 89 M11
Tobermory 78 G5
Toberonochy 79 J10
Tobson 88 G4
Tockenham 20 D4
Tockenham Wick 20 D3
Tockholes 56 B7
Tockington 19 K3
Tockwith 57 K4
Todber 9 G2
Toddington CenBeds 32 D6
Toddington Glos 30 B5
Todenham 30 D5
Todhills 69 J7
Todlachie 85 L9
Todmorden 56 E7
Todwick 51 G4
Toft Cambs 33 G3
Toft Lincs 42 D4
Toft Shet 89 N5
Toft Hill 62 B4
Toft Monks 45 J6
Toft next Newton 52 D4
Toftrees 44 C3
Tofts 87 R3
Toftwood 44 D4
Togston 71 H3
Tokavaig 82 G9
Tokers Green 21 K4
Tolastadh a'Chaolais 88 G4
Tolastadh Bho Thuath 88 L3
Toll Bar 51 H2
Toll of Birness 85 Q7
Tolland 7 K2
Tollard Farnham 9 J3
Tollard Royal 9 J3
Tollcross 74 E4
Toller Down Gate 8 E4
Toller Fratrum 8 E5
Toller Porcorum 8 E5
Toller Whelme 8 E4
Tollerton Notts 41 J2
Tollerton NYorks 58 B3
Tollesbury 34 D7
Tollesby 63 F5
Tolleshunt D'Arcy 34 D7
Tolleshunt Knights 34 D7
Tolleshunt Major 34 C7
Tolpuddle 9 G5
Tolvah 84 C11
Tolworth 22 E5
Tomatin 84 C8
Tombreck 84 A7
Tomchrasky (Tom Chrasgaidh) 83 N9
Tomdoun 83 M10
Tomich High 83 P8
Tomich High 84 B3
Tomintoul 85 F9
Tomnavoulin 84 G8
Ton Pentre 18 C2
Tondu 18 B3
Tonedale 7 K3
Tonfanau 36 E5
Tong Kent 14 D3
Tong Shrop 39 G5
Tong WYorks 57 H6
Tong Norton 39 G5
Tong Street 57 G6
Tonge 41 G3
Tongham 22 B7
Tongland 65 G5
Tongue 86 H4
Tongwynlais 18 E3
Tonmawr 18 B2
Tonna 18 A2
Tonwell 33 G7
Tonypandy 18 C2
Tonyrefail 18 D3
Toot Baldon 21 J1
Toot Hill 23 J1
Toothill Hants 10 E3
Toothill Swin 20 E3
Tooting Graveney 23 F4
Top End 32 D2
Top of Hebers 49 H2
Topcliffe 57 K2
Topcroft 45 G6
Topcroft Street 45 G6
Toppesfield 34 B5
Toppings 49 G1
Toprow 45 F6
Topsham 7 H7
Topsham Bridge 5 H5
Torbeg Aber 84 H11
Torbeg NAyr 66 D1
Torbothie 75 G5
Torbryan 5 J4
Torcross 5 J6
Tordarroch 84 A7
Tore (An Todhar) 84 A5
Torfrey 4 B5
Torksey 52 B5
Torlum 88 B3
Torlundy (Tòrr Lunndaidh) 79 N3
Tormarton 20 A4
Tormisdale 72 A5
Tormore 73 G7
Tornagrain 84 B5
Tornahaish 84 G10
Tornaveen 85 L10
Torness 83 R8
Toronto 62 B3
Torpenhow 60 D3
Torphichen 75 H3
Torphins 85 L10
Torpoint 4 E5
Torquay 5 K4
Torquhan 76 C6
Torr 5 F5
Torrance 74 E3
Torrance House 74 E5
Torre Som 7 J1
Torre Torbay 5 K4
Torridon 83 J5
Torrin 82 F8
Torrisdale A&B 73 F7
Torrisdale High 87 J3
Torrish 87 M8
Torrisholme 55 H3
Torroble 86 H9
Torry Aber 85 J6
Torry Aberdeen 85 P10
Torryburn 75 J2
Torsonce 76 C6
Torthorwald 69 F6
Tortington 12 C6
Torton 29 H1
Tortworth 20 A2
Torvaig 82 E6
Torver 60 D7
Torwood 75 G2
Torworth 51 J4
Tosberry 6 A3
Toscaig 82 H7
Toseland 33 F2
Tosside 56 C4
Tostock 34 D2
Totaig 82 B5
Tote Hill 12 B4
Totegan 87 L3
Totford 11 G1
Totham Hill 34 C7
Tothill 53 H4
Totland 10 E6
Totley 51 F5
Totnes 5 J4
Toton 41 H2
Totscore 82 D4
Tottenham 23 G2
Tottenhill 44 A4
Tottenhill Row 44 A4
Totteridge Bucks 22 B2
Totteridge GtLon 23 F2
Totternhoe 32 C6
Tottington GtMan 49 G1
Tottington Norf 44 C6
Totton 10 E3
Toulton 7 K2
Tournaig 83 J2
Tovil 14 C2
Tow Law 62 B3
Towan Cross 2 E4
Toward 73 K4
Towcester 31 H4
Towednack 2 B5
Tower Bridge Exhibition GtLon SE1 2UP 101 B7
Tower End 44 A4
Tower of London GtLon EC3N 4AB 101 B7
Towersey 22 A1
Towie 85 J9
Towiemore 84 H6
Town End Cambs 43 H6
Town End Cumb 55 H1
Town End Mersey 48 D4
Town Green Lancs 48 D2
Town Green Norf 45 H4
Town of Lowton 49 F3
Town Row 13 J3
Town Street 44 B7
Town Yetholm 70 D1
Towneley Hall Art Gallery & Museum Lancs BB11 3RQ 56 D6
Townfield 61 L2
Townhead D&G 65 G6
Townhead SYorks 50 D2
Townhead of Greenlaw 65 H4
Townhill Fife 75 K2
Townhill Swan 17 K6
Towns End 21 J6
Towns Green 49 F6
Townshend 2 C5
Towthorpe ERid 59 F3
Towthorpe York 58 C4
Towton 57 K6
Towyn 47 H5
Toynton All Saints 53 G6
Toynton Fen Side 53 G6
Toynton St. Peter 53 H6
Toy's Hill 23 H6
Trabboch 67 J1
Traboe 2 E6
Tradespark 84 C5
Trafford Centre 49 G3
Trafford Park 49 G3
Trago Mills, Newton Abbot Devon TQ12 6JD 5 J3
Trallong 27 J6
Trallwn 17 K6
Tram Inn 28 D5
Tranent 76 C3
Tranmere 48 C4
Trantlebeg 87 L4
Trantlemore 87 L4
Tranwell 71 G5
Trap 17 K4
Trap Street 49 H6
Traprain 76 D3
Trap's Green 30 C2
Traquair 76 B7
Trawden 56 E6
Trawscoed (Crosswood) 27 F1
Trawsfynydd 37 G2
Trealaw 18 D2
Treales 55 H6
Trearddur 46 A5
Tre-Aubrey 18 D4
Trebah Garden Corn TR11 5JZ 2 E6
Trebanog 18 D2
Trebanos 18 A1
Trebarrow 4 C1
Trebartha 4 C3
Trebarvah 2 E5
Trebarwith 4 A2
Trebeath 4 C2
Trebetherick 3 G1
Trebister 89 N9
Tre-boeth 17 K6
Treborough 7 J2
Trebudannon 3 F2
Trebullett 4 D3
Treburley 4 D3
Treburrick 3 F1
Trebyan 4 A4
Trecastle 27 H6
Trecott 6 E5
Trecrogo 6 B7
Trecwn 16 C2
Trecynon 18 C1
Tredaule 4 C2
Tredavoe 2 B6
Treddiog 16 B3
Tredegar 18 E1
Tredington Glos 29 J6
Tredington Warks 30 D4
Tredinnick Corn 3 G2
Tredinnick Corn 4 C5
Tredogan 18 D5
Tredomen 28 A5
Tredrissi 16 D1
Tredunnock 19 G2
Tredustan 28 A5
Tredworth 29 H7
Treen Corn 2 A6
Treen Corn 2 B5
Treesmill 4 A5
Treeton 51 G4
Trefasser 16 B2
Trefdraeth 46 C5
Trefecca 28 A5
Trefechan 18 D1
Trefeglwys 37 J6
Trefenter 27 F2
Treffgarne 16 C3
Treffynnon 16 B3
Trefgarn Owen 16 B3
Trefil 28 A7
Trefilan 26 E3
Trefin 16 B2
Treflach 38 B3
Trefnanney 38 B4
Trefnant 47 J5
Trefonen 38 B3
Trefor Gwyn 36 C1
Trefor IoA 46 B4
Treforest 18 D3
Treforest Industrial Estate 18 E3
Trefriw 47 F6
Tregadillett 4 C2
Tregaian 46 C5
Tregare 28 D7
Tregarland 4 C5
Tregarne 2 E6
Tregaron 27 F3
Tregarth 46 E6
Tregaswith 3 F2
Tregavethan 2 E4
Tregear 3 F3
Tregeare 4 C2
Tregeiriog 38 A2
Tregele 46 B3
Tregidden 2 E6
Tregiskey 4 A6
Treglemais 16 B3
Tregolds 3 F1
Tregole 4 B1
Tregonetha 3 G2
Tregonning & Gwinear Mining Districts with Trewavas (Cornwall & West Devon Mining Landscape) Corn 2 C5
Tregony 3 G4
Tregoodwell 4 B2
Tregoss 3 G2
Tregowris 2 E6
Tregoyd 28 A5
Tregrehan Mills 4 A5
Tre-groes 17 H1
Treguff 18 D4
Tregullon 4 A4
Tregunnon 4 C2
Tregurrian 3 F2
Tregynon 37 K6
Trehafod 18 D2
Trehan 4 E5
Treharris 18 D2
Treherbert 18 C2
Tre-hill 18 D4
Trekenner 4 D3
Treknow 4 A2
Trelan 2 E7
Trelash 4 B1
Trelassick 3 F3
Trelawnyd 47 J5
Trelech 17 F2
Treleddyd-fawr 16 A3
Trelewis 18 E2
Treligga 4 A2
Trelights 3 G1
Trelill 4 A3
Trelissick Corn TR3 6QL 3 F5
Trelissick 3 F5
Trelleck 19 J1
Trelleck Grange 19 H1
Trelogan 47 K4
Trelowla 4 C5
Trelystan 38 B5
Tremadog 36 E1
Tremail 4 B2
Tremain 17 F1
Tremaine 4 C2
Tremar 4 C4
Trematon 4 D5
Tremeirchion 47 J5
Tremethick Cross 2 B5
Tremore 4 A4
Trenance Corn 3 F2
Trenance Corn 3 G1
Trenarren 4 A6
Trench Tel&W 39 F4
Trench Wrex 38 C2
Trencreek 3 F2
Trenear 2 D5
Treneglos 4 C2
Trenewan 4 B5
Trengune 4 B1
Trent 8 E3
Trent Port 52 B4
Trent Vale 40 A1
Trentham 40 A1
Trentishoe 6 E1
Trenwheal 2 D5
Treoes 18 C4
Treorchy 18 C2
Treowen 19 F2
Trequite 4 A3
Tre'r-ddol 37 F6
Trerhyngyll 18 D4
Tre-Rhys 16 E1
Trerulefoot 4 D5
Tresaith 26 B3
Tresco 2 B1
Trescott 40 A6
Trescowe 2 C5
Tresean 2 E3
Tresham 20 A2
Tresillian 3 F4
Tresinney 4 B2
Tresinwen 16 B1
Treskinnick Cross 4 C1
Treslea 4 B4
Tresmeer 4 C2
Tresowes Green 2 D6
Tresparrett 4 B1
Tresparrett Posts 4 B1
Tressait 80 D4
Tresta Shet 89 M7
Tresta Shet 89 Q3
Treswell 51 K5
Trethewey 2 A6
Trethomas 18 E3
Trethurgy 4 A5
Tretio 16 A3
Tretire 28 E6
Tretower 28 A6
Treuddyn 48 B7
Trevadlock 4 C3
Trevalga 4 A1
Trevalyn 48 C7
Trevanson 3 G1
Trevarnon 2 C5
Trevarrack 2 B5
Trevarren 3 G2
Trevarrian 3 F2
Trevarrick 3 G4
Trevaughan Carmar 16 E4
Tre-vaughan Carmar 17 G3
Treveighan 4 A3
Trevellas 2 E3
Trevelmond 4 C4
Trevenen 2 D6
Treverva 2 E5
Trevescan 2 A6
Trevethin 19 F1
Trevigro 4 D4
Treviscoe 3 G3
Trevivian 4 B2
Trevone 3 F1
Trevor 38 B1
Trewalder 4 A2
Trewarmett 4 A2
Trewarthenick 3 G4
Trewassa 4 B2
Trewellard 2 A5
Trewen Corn 4 C2
Trewen Here 28 E7
Trewennack 2 D6
Trewent 16 D6
Trewern 38 B4
Trewethern 4 A3
Trewidland 4 C5
Trewilym 16 E1
Trewint Corn 4 B1
Trewint Corn 4 C2
Trewithian 3 F5
Trewoon 3 G3
Treworga 3 F4
Treworlas 3 F5
Trewornan 3 G1
Treworthal 3 F5
Tre-wyn 28 C6
Treyarnon 3 F1
Treyford 12 B5
Trezaise 3 G3
Triangle 57 F7
Trickett's Cross 10 B4
Triermain 70 A7
Trimdon 62 D3
Trimdon Colliery 62 D3
Trimdon Grange 62 D3
Trimingham 45 G2
Trimley Lower Street 35 G5
Trimley St. Martin 35 G5
Trimley St. Mary 35 G5
Trimsaran 17 H5
Trimstone 6 C1
Trinafour 80 C4
Trinant 19 F2
Tring 32 C7
Trinity Angus 81 M4
Trinity Chanl 3 K7
Trinity Edin 76 A3
Trisant 27 G1
Triscombe Som 7 H2
Triscombe Som 7 K2
Trislaig 79 M3
Trispen 3 F3
Tritlington 71 H4
Trochry 80 F7
Troedyraur 17 G1
Troedyrhiw 18 D1
Trofarth 47 G5
Trondavoe 89 M5
Troon Corn 2 D5
Troon SAyr 74 B7
Tropical Butterfly House, Wildlife & Falconry Centre SYorks S25 4EQ 109 H5
Tropical World, Roundhay WYorks LS8 2ER 115 L1
Troston 34 C1
Troswell 4 C1
Trottiscliffe 24 C5
Trotton 12 B4
Trough Gate 56 D7
Troughend 70 D4
Troustan 73 J3
Troutbeck Cumb 60 E4
Troutbeck Cumb 60 F6
Troutbeck Bridge 60 F6
Trow Green 19 J1
Troway 51 F5
Trowbridge Cardiff 19 F3
Trowbridge Wilts 20 B6
Trowell 41 G2
Trowle Common 20 B6
Trowley Bottom 32 D7
Trows 76 E7
Trowse Newton 45 G5
Troy 57 H6
Trudernish 72 C5
Trudoxhill 20 A7
Trull 8 B2
Trumpan 82 C4
Trumpet 29 F5
Trumpington 33 H3
Trumps Green 22 C5
Trunch 45 G2
Trunnah 55 G5
Truro 3 F4
Truro Cathedral Corn TR1 2AF 3 F4
Truscott 6 B7
Trusham 7 G7
Trusley 40 E2
Trusthorpe 53 J4
Truthan 3 F3
Trysull 40 A6
Tubney 21 H2
Tuckenhay 5 J5
Tuckhill 39 G7
Tuckingmill 2 D4
Tuddenham Suff 34 B1
Tuddenham Suff 35 F4
Tudeley 23 K7
Tudeley Hale 23 K7
Tudhoe 62 C3
Tudweiliog 36 B2
Tuesley 22 C7
Tuffley 29 H7
Tufton Hants 21 H7
Tufton Pembs 16 D3
Tugby 42 A5
Tugford 38 E7
Tughall 71 H1
Tulleys Farm WSuss RH10 4PE 13 G3
Tullibardine Distillery P&K PH4 1QG 80 E10
Tullibody 75 G1
Tullich A&B 79 M9
Tullich High 84 C3
Tullich Muir 84 B3
Tullie House Museum & Art Gallery Cumb CA3 8TP 124 Carlisle
Tulliemet 80 F5
Tulloch 84 A3
Tullochgorm 73 H1
Tullybelton 80 F7
Tullynessle 85 K9
Tulse Hill 23 G4
Tumble (Y Tymbl) 17 J4
Tumby 53 F7
Tumby Woodside 53 F7

Tummel Bridge 80 C5
Tunbridge Wells 13 J3
Tundergarth Mains 69 G5
Tunga 88 K4
Tungate 45 G3
Tunley 19 K6
Tunstall ERid 59 K6
Tunstall Kent 24 E5
Tunstall Lancs 56 B2
Tunstall Norf 45 J5
Tunstall NYorks 62 C7
Tunstall Stoke 49 H2
Tunstall Suff 35 H3
Tunstall T&W 62 D1
Tunstead GtMan 50 C2
Tunstead Norf 45 H3
Tunstead Milton 50 C4
Tunworth 21 K7
Tupholme 52 E6
Tupsley 28 E5
Tupton 51 F6
Tur Langton 42 A6
Turbiskill 73 F2
Turgis Green 21 K6
Turkdean 30 C7
Turleigh 20 B5
Turn 49 H1
Turnastone 28 C5
Turnberry 67 G3
Turnchapel 4 E5
Turnditch 40 E1
Turner's Green 30 C2
Turners Hill 13 G3
Turners Puddle 9 H5
Turnford 23 G1
Turnworth 9 H4
Turriff 85 M6
Turton Bottoms 49 G1
Turvey 32 C3
Turville 22 A2
Turville Heath 22 A2
Turweston 31 H5
Tutankhamun Exhibition, Dorchester Dorset DT1 1UW 9 F5
Tutbury 40 E3
Tutbury Castle Staffs DE13 9JF 40 E3
Tutnall 29 J1
Tutshill 19 J2
Tuttington 45 G3
Tutts Clump 21 J4
Tutwell 4 D3
Tuxford 51 K5
Twatt Ork 89 B5
Twatt Shet 89 M7
Twechar 75 F3
Tweedbank 76 D7
Tweedmouth 77 H5
Tweedsmuir 69 F1
Twelve Oaks 13 K4
Twelveheads 2 E4
Twemlow Green 49 G6
Twenty 42 E3
Twickenham 22 E4
Twigworth 29 H6
Twineham 13 F4
Twineham Green 13 F4
Twinhoe 20 A6
Twinstead 34 C5
Twiss Green 49 F3
Twiston 56 D5
Twitchen Devon 7 F2
Twitchen Shrop 28 C1
Twitton 23 J6
Twizell House 71 G1
Two Bridges Devon 5 G3
Two Bridges Glos 19 K1
Two Dales 50 E6
Two Gates 40 E5
Two Mills 48 C5
Twycross 41 F5
Twycross Zoo Leics CV9 3PX 41 F5
Twyford Bucks 31 H6
Twyford Derbys 41 F3
Twyford Dorset 9 H3
Twyford Hants 11 F2
Twyford Leics 42 A4
Twyford Norf 44 E3
Twyford Oxon 31 F5
Twyford W'ham 22 A4
Twyford Common 28 E5
Twyn Shôn-Ifan 18 E2
Twynholm 65 G5
Twyning 29 H5
Twynllanan 27 G6
Twyn-yr-odyn 18 E4
Twyn-y-Sheriff 19 H1
Twywell 32 C1
Ty Croes 46 B5
Tyberton 28 C5
Tycroes 17 K4
Tycrwyn 38 A4
Tydd Gote 43 H4
Tydd St. Giles 43 H4
Tydd St. Mary 43 H4
Tye Common 24 C2
Tye Green Essex 23 H1
Tye Green Essex 33 J6
Tye Green Essex 33 K7
Tye Green Essex 34 B6
Tyersal 57 G6
Ty-hen 36 A2
Tyldesley 49 F2
Tyle-garw 18 D3
Tyler Hill 25 H5
Tylers Green Bucks 22 C2
Tyler's Green Essex 23 J1
Tylorstown 18 D2
Tylwch 27 J1
Ty-Mawr Conwy 37 J1
Ty-mawr Denb 38 A1
Ty-nant Conwy 37 J1
Ty-nant Gwyn 37 J3
Tyndrum (Taigh an Droma) 79 Q7
Tyne Green Country Park N'umb NE46 3RY 70 E7
Tyneham 9 H6
Tynehead 76 B5
Tynemouth 71 J7
Tynewydd 18 C2
Tyninghame 76 E3
Tynron 68 D4
Tyntesfield 19 J4
Tyn-y-cefn 37 K1
Ty'n-y-coedcae 18 E3
Tyn-y-cwm 37 H7
Tyn-y-ffridd 38 A2
Ty'n-y-garn 18 B3
Tyn-y-gongl 46 D4
Tynygraig Cere 27 F2
Tyn-y-graig Powys 27 K4
Ty'n-y-groes 47 F5
Tyringham 32 B4
Tyseley 40 D7
Tythegston 18 B4
Tytherington ChesE 49 J5
Tytherington SGlos 19 K3
Tytherington Som 20 A7
Tytherington Wilts 20 C7
Tytherton Lucas 20 C4
Tyttenhanger 22 E1
Ty-uchaf 37 J3
Tywardreath 4 A5
Tywardreath Highway 4 A5
Tywyn 36 E5

U
Uachdar 88 C3
Ubberley 40 B1
Ubbeston Green 35 H1
Ubley 19 J6
Uckerby 62 C6
Uckfield 13 H4
Uckinghall 29 H5
Uckington 29 J6
Uddingston 74 E4
Uddington 75 G7
Udimore 14 D6
Udley 19 H5
Udny Green 85 N8
Udston 75 F5
Udstonhead 75 F6
Uffcott 20 E4
Uffculme 7 J4
Uffington Lincs 42 D5
Uffington Oxon 21 G3
Uffington Shrop 38 E4
Ufford Peter 42 D5
Ufford Suff 35 G3
Ufton 30 E2
Ufton Green 21 K5
Ufton Nervet 21 K5
Ugborough 5 G5
Ugford 10 B1
Uggeshall 45 J7
Ugglebarnby 63 J6
Ugley 33 J6
Ugley Green 33 J6
Ugthorpe 63 H5
Uig A&B 73 K2
Uig (Uige High) 82 D4
Uiskevagh (Uisgebhagh) 88 C3
Ulbster 87 R5
Ulcat Row 60 F4
Ulceby Lincs 53 H5
Ulceby NLincs 52 E1
Ulceby Cross 53 H5
Ulceby Skitter 52 E1
Ulcombe 14 D3
Uldale 60 D3
Uldale House 61 J7
Uley 20 A2
Ulgham 71 H4
Ullapool (Ullapul) 83 M1
Ullenhall 30 C2
Ullenwood 29 J7
Ulleskelf 58 B6
Ullesthorpe 41 H7
Ulley 51 G4
Ulley Reservoir Country Park SYorks S26 3XL 109 F4
Ullingswick 28 E4
Ullinish 82 D7
Ullock 60 B4
Ullswater Steamers Cumb CA11 0US 60 E5
Ulpha Cumb 55 H1
Ulpha Cumb 60 C7
Ulrome 59 H4
Ulsta 89 N4
Ulting 24 E1
Ulva 78 F7
Ulverston 55 F2
Ulwell 10 B6
Ulzieside 68 C3
Umberleigh 6 E3
Unapool 86 E6
Underbarrow 61 F7
Undercliffe 57 G6
Underhill 23 F2
Underling Green 14 C3
Underriver 23 J6
Underwood Newport 19 G3
Underwood Notts 51 G7
Underwood Plym 5 F5
Undley 44 A7
Undy 19 H3
Unifirth 89 L7
Union Mills 54 C6
Union Street 14 C4
University of Glasgow Visitor Centre Glas G12 8QQ 127 B1
Unst 89 Q1
Unst Airport 89 Q2
Unstone 51 F5
Unstone Green 51 F5
Unsworth 49 H2
Unthank Cumb 61 F3
Unthank Derbys 51 F5
Up Cerne 9 F4
Up Exe 7 H5
Up Hatherley 29 J6
Up Holland 48 E2
Up Marden 11 J3
Up Mudford 8 E3
Up Nately 21 K6
Up Somborne 10 E1
Up Sydling 9 F4
Upavon 20 E6
Upchurch 24 E5
Upcott Devon 6 C6
Upcott Devon 6 D2
Upcott Here 28 C3
Upcott Som 7 H3
Upend 33 K3
Upgate 45 F4
Upgate Street Norf 44 E6
Upgate Street Norf 45 G6
Uphall Dorset 8 E4
Uphall WLoth 75 J3
Uphall Station 75 J4
Upham Devon 7 G5
Upham Hants 11 G2
Uphampton Here 28 C2
Uphampton Worcs 29 H2
Uphempston 5 J4
Uphill 19 G6
Uplands Glos 20 B1
Uplands Swan 17 K6
Uplawmoor 74 C5
Upleadon 29 G6
Upleatham 63 G5
Uplees 25 G5
Uploders 8 E5
Uplowman 7 J4
Uplyme 8 C5
Upminster 23 J3
Upottery 8 B4
Upper Affcot 38 D7
Upper Ardroscadale 73 J4
Upper Arley 39 G7
Upper Arncott 31 H7
Upper Astley 38 E4
Upper Aston 40 A6
Upper Astrop 31 G5
Upper Basildon 21 K4
Upper Bayble (Pabail Uarach) 88 L4
Upper Beeding 12 E5
Upper Benefield 42 C7
Upper Bentley 29 J2
Upper Berwick 38 D4
Upper Boat 18 E3
Upper Boddam 85 L7
Upper Boddington 31 F3
Upper Borth 37 F7
Upper Boyndlie 85 P4
Upper Brailes 30 E5
Upper Breakish 82 G8
Upper Breinton 28 D4
Upper Broadheath 29 H3
Upper Broughton 41 J3
Upper Brynamman 27 G7
Upper Bucklebury 21 J5
Upper Burgate 10 C3
Upper Caldecote 32 E4
Upper Canada 19 G6
Upper Catesby 31 G3
Upper Catshill 29 J1
Upper Chapel 27 K4
Upper Cheddon 8 B2
Upper Chicksgrove 9 J1
Upper Chute 21 F6
Upper Clatford 21 G7
Upper Coberley 29 J7
Upper Colwall 29 G4
Upper Cotton 40 C1
Upper Cound 38 E5
Upper Cumberworth 50 E2
Upper Cwmbran 19 F2
Upper Dean 32 D2
Upper Denby 50 E2
Upper Denton 70 B7
Upper Derwent Reservoirs Derbys S33 0AQ 50 D4
Upper Diabaig 83 J4
Upper Dicker 13 J5
Upper Dovercourt 35 G5
Upper Dunsforth 57 K3
Upper Dunsley 32 C7
Upper Eastern Green 40 E7
Upper Egleton 29 F4
Upper Elkstone 50 C7
Upper End 50 C5
Upper Enham 21 G7
Upper Farringdon 11 J1
Upper Framilode 29 G7
Upper Froyle 22 A7
Upper Godney 19 H7
Upper Gornal 40 B6
Upper Gravenhurst 32 E5
Upper Green Essex 33 H5
Upper Green Essex 33 J5
Upper Green Mon 28 C7
Upper Green WBerks 21 G5
Upper Grove Common 28 E6
Upper Hackney 50 E6
Upper Halliford 22 D5
Upper Halling 24 C5
Upper Hambleton 42 C5
Upper Harbledown 15 G2
Upper Hardres Court 15 G2
Upper Hartfield 13 H3
Upper Hatton 40 A2
Upper Hayesden 23 J7
Upper Hayton 38 E7
Upper Heath 38 E7
Upper Heaton 50 D1
Upper Hellesdon 45 G4
Upper Helmsley 58 C4
Upper Hengoed 38 B2
Upper Hergest 28 B3
Upper Heyford N'hants 31 H3
Upper Heyford Oxon 31 F6
Upper Hill Here 28 D3
Upper Hill SGlos 19 K2
Upper Horsebridge 13 J5
Upper Howsell 29 G4
Upper Hulme 50 C6
Upper Inglesham 21 F2
Upper Kilchattan 72 B1
Upper Killay 17 J6
Upper Knockando 84 F6
Upper Lambourn 21 G3
Upper Langford 19 H6
Upper Langwith 51 H6
Upper Leigh 40 C2
Upper Ley 29 G7
Upper Loads 51 F6
Upper Longdon 40 C4
Upper Longwood 39 F5
Upper Ludstone 40 A6
Upper Lybster 87 Q6
Upper Lydbrook 29 F7
Upper Lyde 28 D4
Upper Lye 28 C2
Upper Maes-coed 28 C5
Upper Midhope 50 E3
Upper Milton 30 D7
Upper Minety 20 D2
Upper Moor 29 J4
Upper Morton 19 K2
Upper Nash 16 D5
Upper Newbold 51 F5
Upper North Dean 22 B2
Upper Norwood 23 G4
Upper Oddington 30 D6
Upper Padley 50 E5
Upper Pennington 10 D5
Upper Pollicott 31 J7
Upper Poppleton 58 B4
Upper Quinton 30 C4
Upper Ratley 10 E2
Upper Rissington 30 D6
Upper Rochford 29 F2
Upper Sanday 89 E7
Upper Sapey 29 F2
Upper Scolton 16 C3
Upper Seagry 20 C3
Upper Shelton 32 C4
Upper Sheringham 45 F1
Upper Shuckburgh 31 F2
Upper Siddington 20 D2
Upper Skelmorlie 74 A4
Upper Slaughter 30 C6
Upper Soudley 29 F7
Upper Staploe 32 E3
Upper Stoke 45 G5
Upper Stondon 32 E5
Upper Stowe 31 H3
Upper Street Hants 10 C3
Upper Street Norf 35 F1
Upper Street Norf 45 H4
Upper Street Suff 35 F5
Upper Strensham 29 J5
Upper Sundon 32 D6
Upper Swanmore 11 G3
Upper Swell 30 C6
Upper Tean 40 C2
Upper Thurnham 55 H4
Upper Tillyrie 82 D3
Upper Tooting 23 F4
Upper Town Derbys 50 E6
Upper Town Derbys 50 E7
Upper Town Here 28 E4
Upper Town NSom 19 J5
Upper Tysoe 30 E4
Upper Upham 21 F4
Upper Upnor 24 D4
Upper Vobster 20 A7
Upper Wardington 31 F4
Upper Waterhay 20 D2
Upper Weald 32 B5
Upper Weedon 31 H3
Upper Welson 28 B3
Upper Weston 20 A5
Upper Whiston 51 G4
Upper Wick 29 H3
Upper Wield 11 H1
Upper Winchendon (Over Winchendon) 31 J7
Upper Witton 40 C6
Upper Woodford 10 C1
Upper Woolhampton 21 J5
Upper Wootton 21 J6
Upper Wraxall 20 B4
Upper Wyche 29 G4
Upperby 60 F1
Uppermill 49 J2
Upperthong 50 D2
Upperton 12 C4
Uppertown 51 F6
Uppingham 42 B5
Uppington 38 E5
Upsall 57 K1
Upsettlington 77 G6
Upshire 23 H1
Upstreet 25 J5
Upthorpe 34 D1
Upton Bucks 31 J7
Upton Cambs 32 E1
Upton ChesW&C 48 D6
Upton Corn 4 C3
Upton Corn 6 A5
Upton Devon 5 H6
Upton Devon 7 J5
Upton Dorset 9 G6
Upton Dorset 9 J5
Upton ERid 59 H4
Upton Hants 10 E3
Upton Hants 21 G6
Upton Leics 41 F6
Upton Lincs 52 B4
Upton Mersey 48 B4
Upton Norf 45 H4
Upton Notts 51 K5
Upton Notts 51 K7
Upton N'hants 31 J2
Upton Oxon 21 J3
Upton Oxon 30 D7
Upton Pembs 16 D5
Upton Peter 42 E5
Upton Slo 22 C4
Upton Som 7 H3
Upton Som 8 D2
Upton Wilts 9 H1
Upton WYorks 51 G1
Upton Bishop 29 F6
Upton Cheyney 19 K5
Upton Country Park Poole BH17 7BJ 91 B5
Upton Cressett 39 F6
Upton Crews 29 F6
Upton Cross 4 C3
Upton End 32 E5
Upton Grey 21 K7
Upton Hellions 7 G5
Upton Lovell 20 C7
Upton Magna 38 E4
Upton Noble 9 G1
Upton Park 23 H3
Upton Pyne 7 H6
Upton Scudamore 20 B7
Upton Snodsbury 29 J3
Upton St. Leonards 29 H7
Upton upon Severn 29 H4
Upton Warren 29 J2
Upwaltham 12 C5
Upware 33 J1
Upwell 43 J5
Upwey 9 F6
Upwick Green 33 H6
Upwood 43 F7
Urafirth 89 M5
Urbis, Manchester GtMan M4 3BG 133 E2
Urchfont 20 D6
Urdimarsh 28 E4
Urgha 88 G8
Urlay Nook 62 E5
Urmston 49 G3
Urpeth 62 C1
Urquhart 84 G4
Urquhart Castle High IV63 6XJ 83 R8
Urra 63 F6
Ushaw Moor 62 C2
Usher Hall Edin EH1 2EA 126 E5
Usk (Brynbuga) 19 G1
Usselby 52 D3
Usworth 62 D1
Utley 57 F5
Uton 7 G6
Utterby 53 G3
Uttoxeter 40 C2
Uwchmynydd 36 A3
Uxbridge 22 D3
Uyeasound 89 P2
Uzmaston 16 C4

V
Valley (Y Fali) 46 A5
Valley Truckle 4 B2
Valleyfield D&G 65 G5
Valleyfield Fife 75 J2
Valsgarth 89 Q1
Vange 24 D3
Vardre 17 K5
Varteg (Y Farteg) 19 F1
Vatersay (Bhatarsaigh) 88 A9
Vatten 82 C6
Vaul 78 B6
Vaynor 27 K7
Vaynor Park 38 A5
Veaullt 28 A3
Veensgarth 89 N8
Velindre Pembs 16 D2
Velindre Powys 28 A5
Vellow 7 J2
Venn 5 H6
Venn Ottery 7 J6
Venngreen 6 B4
Vennington 38 C5
Venny Tedburn 7 G6
Venterdon 4 D3
Ventnor 11 G7
Ventnor Botanic Gardens IoW PO38 1UL 11 G7
Venton 5 F5
Venue Cymru, Llandudno Conwy LL30 1BB 47 F4
Vernham Dean 21 G6
Vernham Street 21 G6
Vernolds Common 38 D7
Verwood 10 B4
Veryan 3 G5
Veryan Green 3 G4
Vickerstown 54 E3
Victoria 3 G2
Victoria & Albert Dundee 125 Dundee
Victoria & Albert Museum GtLon SW7 2RL 99 G7
Victoria Art Gallery B&NESom BA2 4AT 121 Bath
Vidlin 89 N6
Viewpark 75 F4
Vigo 40 C5
Vigo Village 24 C5
Villavin 6 D4
Vindobala (Frontiers of the Roman Empire) N'umb 71 G7
Vindolandia (Chesterholm) (Frontiers of the Roman Empire) N'umb NE47 7JN 70 C7
Vinehall Street 14 C5
Vine's Cross 13 J5
Viney Hill 19 K1
Virginia Water 22 C5
Virginstow 6 B6
Virley 34 D7
Vobster 20 A7
Voe 89 N6
Vogrie Country Park Midlo EH23 4NU 120 H5
Voirrey Embroidery Mersey CH63 6JA 110 A6
Volks Electric Railway B&H BN2 1EN 13 G6
Vowchurch 28 C5
Voy 89 B6
Vron Gate 38 C5

W
Waberthwaite 60 C7
Wackerfield 62 B4
Wacton 45 F6
Wadbister 89 N8
Wadborough 29 J4
Waddesdon 31 J7
Waddesdon Manor Bucks HP18 0JH 31 J7
Waddeton 5 J5
Waddicar 48 C3
Waddingham 52 C3
Waddington Lancs 56 C5
Waddington Lincs 52 C6
Waddingworth 52 E5
Waddon Devon 5 J3
Waddon GtLon 23 G5
Wadebridge 3 G1
Wadeford 8 C3
Wadenhoe 42 D7
Wadesmill 33 G7
Wadhurst 13 K3
Wadshelf 51 F5
Wadworth 51 H3
Wadworth Hill 59 J7
Waen Denb 47 H6
Waen Denb 47 K6
Waen Aberwheeler 47 J6
Waen-fâch 38 B4
Waen-wen 46 D6
Wainfleet All Saints 53 H7
Wainfleet Bank 53 H7
Wainfleet St. Mary 53 H7
Wainford 45 H6
Waingroves 41 G1
Wainhouse Corner 4 B1
Wainscott 24 D4
Wainstalls 57 F7
Waitby 61 J6
Wakefield 57 J7
Wakehurst Place WSuss RH17 6TN 13 G3
Wakerley 42 C6
Wakes Colne 34 C6
Walberswick 35 J1
Walberton 12 C6
Walbottle 71 G7
Walcot Lincs 42 D2
Walcot Lincs 52 E7
Walcot NLincs 58 E7
Walcot Shrop 38 C7
Walcot Tel&W 38 E4
Walcot Green 45 F7
Walcote Leics 41 H7
Walcote Warks 30 C3
Walcott 45 H2
Walcott Dales 52 E7
Walden 57 F1
Walden Head 56 E1
Walden Stubbs 51 H1
Walderslade 24 D5
Walderton 11 J3
Walditch 8 D5
Waldley 40 D2
Waldridge 62 C2
Waldringfield 35 G4
Waldron 13 J5
Wales 51 G4
Wales Millennium Centre Cardiff CF10 5AL 95 C6
Walesby Lincs 52 E3
Walesby Notts 51 J5
Waleswood 51 G4
Walford Here 28 C1
Walford Here 28 E6
Walford Shrop 38 D3
Walford Staffs 40 A2
Walford Heath 38 D4
Walgherton 39 F1
Walgrave 32 B1
Walhampton 10 E5
Walk Mill 56 D6
Walkden 49 G2
Walker 71 H7
Walker Art Gallery, Liverpool Mersey L3 8EL 131 D3
Walker Fold 56 B5
Walkerburn 76 B7
Walkeringham 51 K3
Walkerith 51 K3
Walkern 33 F6
Walker's Green 28 E4
Walkford 10 D5
Walkhampton 5 F4
Walkingham Hill 57 J3
Walkington 59 F6
Walkwood 30 B2
Wall Corn 2 D5
Wall N'umb 70 E7
Wall Staffs 40 D5
Wall End 55 F1
Wall Heath 40 A7
Wall Houses 71 F7
Wall under Heywood 38 E6
Wallace Collection, London GtLon W1U 3BN 132 A2
Wallacehall 69 H6
Wallacetown 67 G3
Wallasey 48 B3
Wallaston Green 16 C5
Wallend 24 E4
Waller's Green 29 F5
Wallingford 21 K3
Wallington GtLon 23 F5
Wallington Hants 11 G4
Wallington Herts 33 F5
Wallington Wrex 38 D1
Wallingwells 51 H4
Wallis 16 D3
Wallisdown 10 B5
Walliswood 12 E3
Walls 89 L8
Wallsend 71 J7
Wallyford 76 B3
Walmer 15 J2
Walmer Bridge 55 H7
Walmersley 49 H1
Walmley 40 D6
Walmsgate 53 G5
Walpole 35 H1
Walpole Cross Keys 43 J4
Walpole Highway 43 J4
Walpole Marsh 43 H4
Walpole St. Andrew 43 J4
Walpole St. Peter 43 J4
Walrond's Park 8 C2
Walrow 19 G7
Walsall 40 C6
Walsall Arboretum Illuminations WMid WS1 2AB 102 F1
Walsall Wood 40 C5
Walsden 56 E7
Walsgrave on Sowe 41 F7
Walsham le Willows 34 E1
Walshford 57 K4
Walsoken 43 H4
Walston 75 J6
Walsworth 32 E5
Walter's Ash 22 B2
Walterston 18 D4
Walterstone 28 C6
Waltham Kent 15 G3
Waltham NELincs 53 F2
Waltham Abbey 23 G1
Waltham Chase 11 G3
Waltham Cross 23 G1
Waltham on the Wolds 42 A3
Waltham St. Lawrence 22 B4
Walthamstow 23 G3
Walton Bucks 32 B7
Walton Cumb 70 A7
Walton Derbys 51 F6
Walton Leics 41 H7
Walton Mersey 48 C3
Walton MK 32 B5
Walton Peter 42 E5
Walton Powys 28 B3
Walton Shrop 28 D1
Walton Som 8 D1
Walton Staffs 40 A2
Walton Suff 35 H5
Walton Tel&W 38 E4
Walton Warks 30 D3
Walton WYorks 51 F1
Walton WYorks 57 K5
Walton Cardiff 29 J5
Walton East 16 D3
Walton Elm 9 G3
Walton Hall Gardens Warr WA4 6SN 111 M5
Walton Highway 43 H4
Walton Lower Street 35 G5
Walton on the Hill 23 F6
Walton on the Naze 35 G6
Walton on the Wolds 41 H4
Walton Park D&G 65 H3
Walton Park NSom 19 H4
Walton West 16 B4
Walton-in-Gordano 19 H4
Walton-le-Dale 55 J7
Walton-on-Thames 22 E5
Walton-on-the-Hill 40 B3
Walton-on-Trent 40 E4
Walwen Flints 47 K5
Walwen Flints 48 B5
Walwick 70 E6
Walworth 62 C5
Walworth Gate 62 C4
Walwyn's Castle 16 B4
Wambrook 8 B4
Wanborough Surr 22 C7
Wanborough Swin 21 F3
Wandel 68 E1
Wandon 71 F1
Wandon End 32 E6
Wandsworth 23 F4
Wandylaw 71 G1
Wangford Suff 35 J1
Wangford Suff 44 B7
Wanlip 41 H4
Wanlockhead 68 D2
Wannock 13 J6
Wansbeck Riverside Park N'umb NE63 8TX 71 H5
Wansford ERid 59 G4
Wansford Peter 42 D6
Wanshurst Green 14 C3
Wanstrow 20 A7
Wanswell 19 K1
Wantage 21 G3
Wapley 20 A4
Wappenbury 30 E2
Wappenham 31 H4
Warblebank 60 D2
Warbleton 13 K5
Warblington 11 J4
Warborough 21 J2
Warboys 43 G7
Warbreck 55 G6
Warbstow 4 C1
Warburton 49 F4
Warcop 61 J5
Ward End 40 D7
Ward Green 34 E2
Warden Kent 25 G4
Warden N'umb 70 E7
Warden Hill 29 J6
Warden Street 32 E4
Wardington 31 F4
Wardle ChesE 49 F7
Wardle GtMan 49 J1
Wardley GtMan 49 G2
Wardley Rut 42 B5
Wardley T&W 71 J7
Wardlow 50 D5
Wardsend 49 J4
Wardy Hill 43 H7
Ware Herts 33 G7
Ware Kent 25 J5
Wareham 9 J6
Warehorne 14 E4
Waren Mill 77 K7
Warenford 71 G1
Warenton 77 K7
Wareside 33 G7
Waresley Cambs 33 F3
Waresley Worcs 29 H1
Warfield 22 B4
Wargrave Mersey 48 E3
Wargrave W'ham 22 A4
Warham Here 28 D5
Warham Norf 44 D1
Wark N'umb 70 D6
Wark N'umb 77 G7
Warkleigh 6 E3
Warkton 32 B1
Warkworth N'hants 31 F4
Warkworth N'umb 71 H3
Warland 56 E7
Warleggan 4 B4
Warley Essex 23 J2
Warley WMid 40 C7
Warley Town 57 F7
Warlingham 23 G6
Warmfield 57 J7
Warmingham 49 G6
Warminghurst 12 E5
Warmington N'hants 42 D6
Warmington Warks 31 F4
Warminster 20 B7
Warmlake 14 D3
Warmley 19 K4
Warmley Hill 19 K4
Warmsworth 51 H2
Warmwell 9 G6
Warndon 29 H3
Warners End 22 D1
Warnford 11 H2
Warnham 12 E3
Warningcamp 12 D6
Warninglid 13 F4
Warren ChesE 49 H5
Warren Pembs 16 C6
Warren House 6 E7
Warren Row 22 B3
Warren Street 14 E2
Warrenby 63 F4
Warren's Green 33 F6
Warrington MK 32 B3
Warrington Warr 49 F4
Warsash 11 F4
Warslow 50 C7
Warsop Vale 51 H6
Warter 58 E4
Warthill 58 C4
Wartling 13 K6
Wartnaby 42 A3
Warton Lancs 55 H7
Warton Lancs 55 J2
Warton N'umb 71 F3
Warton Warks 40 E5
Warton Bank 55 H7
Warwick 30 D2
Warwick Bridge 61 F1
Warwick Castle Warks CV34 4QU 104 A8
Warwick Wold 23 G6
Warwick-on-Eden 61 F1
Wasbister 89 C4
Wasdale Head 60 C6
Waseley Hills Country Park Worcs B45 9AT 102 D8
Wash 50 C4
Wash Common 21 H5
Washall Green 33 H5
Washaway 4 A4
Washbourne 5 H5
Washbrook 19 H7
Washfield 7 H4
Washfold 62 A6
Washford Som 7 J1
Washford Warks 30 B2
Washford Pyne 7 G4
Washingborough 52 D5
Washington T&W 62 D1
Washington WSuss 12 E5
Washmere Green 34 D4
Wasing 21 J5
Waskerley 62 A2
Wasperton 30 D3
Wasps Nest 52 D6
Wass 58 B2
Wat Tyler Country Park Essex SS16 4UH 24 D3
Watchet 7 J1
Watchfield Oxon 21 F2
Watchfield Som 19 G7
Watchgate 61 G7
Watcombe 5 K4
Watendlath 60 D5
Water 56 D7
Water Eaton MK 32 B5
Water Eaton Oxon 31 G7
Water End Bed 32 E4
Water End CenBeds 32 D5
Water End ERid 58 D6
Water End Essex 33 J4
Water End Herts 23 F1
Water End Herts 32 D7
Water Newton 42 E6
Water Orton 40 D6
Water Stratford 31 H5
Water Yeat 55 F1
Waterbeach 33 H2
Waterbeck 69 H6
Watercombe 9 G6
Waterend 22 A2
Waterfall 50 C7
Waterfoot ERenf 74 D5
Waterfoot Lancs 56 D7
Waterford 33 G7
Watergate 4 B2
Waterhead Cumb 60 E6
Waterhead D&G 68 C5
Waterheath 45 J6
Waterhouses Dur 62 B2
Waterhouses Staffs 50 C7
Wateringbury 23 K6
Waterlane 20 C1
Waterloo Derbys 51 G6
Waterloo GtMan 49 J2
Waterloo High 82 G8
Waterloo Mersey 48 C3
Waterloo NLan 75 G5
Waterloo Norf 45 G4
Waterloo Pembs 16 C5
Waterloo P&K 80 F7
Waterloo Poole 10 B5
Waterloo Cross 7 J4
Waterloo Port 46 C6
Waterlooville 11 H3
Watermead Country Park Leics LE7 4PF 105 E4
Watermeetings 68 E2
Watermillock 60 F4
Watermouth Castle Devon EX34 9SL 6 D1
Waterperry Oxon OX33 1JZ 21 K1
Waterperry 21 K1
Waterrow 7 J3
Waters Upton 39 F4
Watersfield 12 D5
Watershed Mill Visitor Centre, Settle NYorks BD24 9LR 56 D3
Watersheddings 49 J2
Waterside Aber 85 Q8
Waterside Bucks 22 C1
Waterside B'burn 56 C7
Waterside EAyr 67 J3
Waterside EAyr 74 C6
Waterside EDun 74 E3
Watersmeet House Devon EX35 6NT 7 F1
Waterstock 21 K1
Waterston 16 C5
Waterthorpe 51 G4
Waterworld, Hanley ST1 5PU 136 Stoke-on-Trent
Watford Herts 22 E2
Watford N'hants 31 H2
Watford Park 18 E3
Wath NYorks 57 G3
Wath NYorks 57 J2
Wath Brow 60 B5
Wath upon Dearne 51 G2
Watley's End 19 K3
Watlington Norf 44 A4
Watlington Oxon 21 K2
Watnall 41 H1
Watten 87 Q4
Wattisfield 34 E1
Wattisham 34 E3
Watton Dorset 8 D5
Watton ERid 59 G5
Watton Norf 44 D5
Watton at Stone 33 G7
Watton Green 44 D5
Watton's Green 23 J2
Wattston 75 F4
Wattstown 18 D2
Wattsville 19 F2
Waun Fawr 37 F7
Waun y Clyn 17 H5
Waunarlwydd 17 K6
Waunclunda 17 K2
Waunfawr 46 D7
Waun-Lwyd 18 E1
Wavendon 32 C5
Waverbridge 60 D2
Waverton ChesW&C 48 D6
Waverton Cumb 60 D2
Wavertree 48 C4
Wawne 59 G6
Waxham 45 J3
Waxholme 59 K7
Way Gill 56 E3
Way Village 7 G4
Way Wick 19 G5
Wayford 8 D4
Waytown 8 D5
Weacombe 7 K1
Weald 21 G1
Weald & Downland Open Air Museum WSuss PO18 0EU 12 B5
Weald Country Park Essex CM14 5QS 23 J2
Wealdstone 22 E2
Weardley 57 H5
Weare 19 H6
Weare Giffard 6 C3
Wearhead 61 K3
Wearne 8 D2
Weasenham All Saints 44 C3
Weasenham St. Peter 44 C3
Weathercote 56 C2
Weatheroak Hill 30 B1
Weaverham 49 F5
Weaverthorpe 59 F2
Webheath 30 B2
Webton 28 D5
Weddington 41 F6
Wedhampton 20 D6
Wedmore 19 H7
Wednesbury 40 B6
Wednesfield 40 B6
Weedon 32 B7
Weedon Bec 31 H3
Weedon Lois 31 H4
Weeford 40 D5
Week Devon 7 F4
Week Devon 5 H4
Week Devon 7 F4
Week Som 7 H2
Week Orchard 6 A5
Week St. Mary 4 C1
Weeke 11 F1
Weekley 42 B7
Weel 59 G6
Weeley 35 F6
Weeley Heath 35 F6
Weem 80 D5
Weeping Cross 40 B3
Weethley 30 B3
Weeting 44 B7
Weeton ERid 59 K7
Weeton Lancs 55 G6
Weeton NYorks 57 H5
Weetwood 57 H6
Weir Essex 24 E3
Weir Lancs 56 D7
Weir Quay 4 E4
Weirbrook 38 C3
Weisdale 89 M7
Welbeck Abbey 51 H5
Welborne 44 E5
Welbourn 52 C7
Welburn NYorks 58 C1
Welburn NYorks 58 D3
Welbury 62 D6
Welby 42 C2
Welches Dam 43 H7
Welcombe 6 A4
Weldon 42 C7
Welford N'hants 41 J7
Welford WBerks 21 H4
Welford-on-Avon 30 C3
Welham Leics 42 A6
Welham Notts 51 K4
Welham Green 23 F1
Well Hants 22 A7
Well Lincs 53 H5
Well NYorks 57 H1
Well End Bucks 22 B3
Well End Herts 23 F2
Well Hill 23 H5
Well Street 23 K6
Well Town 7 H5
Welland 29 G4
Wellesbourne 30 D3
Wellhill 84 D4
Wellhouse WBerks 21 J4
Wellhouse WYorks 50 C1
Welling 23 H4
Wellingborough 32 B2
Wellingham 44 C3
Wellingore 52 C7
Wellington Cumb 60 B6
Wellington Here 28 D4
Wellington Som 7 K3
Wellington Tel&W 39 F4
Wellington Heath 29 G4
Wellington Marsh 28 D4
Wellow B&NESom 20 A6
Wellow IoW 10 E6
Wellow Notts 51 J6
Wells 19 J7
Wells Cathedral Som BA5 2UE 19 J7
Wells Green 40 D7
Wellsborough 41 F5
Wells-next-the-Sea 44 D1
Wellstyle Green 33 K7
Wellwood 75 J2
Welney 43 J6
Welsh Bicknor 28 E7
Welsh End 38 E2
Welsh Frankton 38 C2
Welsh Hook 16 C3
Welsh Mountain Zoo Conwy LL28 5UY 47 G5
Welsh Newton 28 D7
Welsh St. Donats 18 D4
Welshampton 38 D2
Welshpool (Y Trallwng) 38 B5
Welton B&NESom 19 K6
Welton Cumb 60 E2
Welton ERid 59 F7
Welton Lincs 52 D4
Welton N'hants 31 G2
Welton le Marsh 53 H6
Welton le Wold 53 F4
Welwick 59 K7
Welwyn 33 F7
Welwyn Garden City 33 F7
Wem 38 E3
Wembdon 8 B1
Wembley GtLon HA9 0WS 98 E5
Wembley 22 E3
Wembley Park 22 E3
Wembury 5 F6
Wembworthy 6 E4
Wemyss Bay 73 K4
Wenallt Cere 27 F1
Wenallt Gwyn 37 J1
Wendens Ambo 33 J5
Wendlebury 31 G7
Wendling 44 D4
Wendover 22 B1
Wendover Dean 22 B1
Wendron 2 D5
Wendron Mining District (Cornwall & West Devon Mining Landscape) Corn 2 D5
Wendy 33 G4
Wenfordbridge 4 A3
Wenhaston 35 J1
Wenlli 47 G6
Wennington Cambs 33 F1
Wennington GtLon 23 J3
Wennington Lancs 56 B2
Wensley Derbys 50 E6
Wensley NYorks 57 F1
Wensleydale Cheese Visitor Centre, Hawes NYorks DL8 3RN 56 D1
Wentbridge 51 G1
Wentnor 38 C6
Wentworth Cambs 33 H1
Wentworth SYorks 51 F3
Wenvoe 18 E4
Weobley 28 C3
Weobley Marsh 28 D3
Weoley Castle 40 C7
Wepham 12 D6
Wepre 48 B6
Wepre Country Park Flints CH5 4HL 48 B6
Wereham 44 A5
Wergs 40 A5
Wern Gwyn 36 E2
Wern Powys 28 A7
Wern Powys 38 B4
Wern Shrop 38 B2
Wernffrwd 17 J6
Wern-olau 17 J6
Wernrheolydd 28 C7
Wern-y-cwrt 19 G1
Werrington Corn 6 B7
Werrington Peter 42 E5
Werrington Staffs 40 B1
Wervil Grange 26 C3
Wervin 48 D5
Wesham 55 H6
Wessington 51 F7
West Aberthaw 18 D5
West Acre 44 B4
West Acton 22 E3
West Allerdean 77 H6
West Alvington 5 H6
West Amesbury 20 E7
West Anstey 7 G3
West Ashby 53 F5
West Ashford 6 D2
West Ashling 12 B6
West Ashton 20 B6
West Auckland 62 B4
West Ayton 59 F1
West Bagborough 7 K2
West Barkwith 52 E4
West Barnby 63 J5
West Barns 76 E3
West Barsham 44 D2
West Bay 8 D5
West Beckham 45 F2
West Benhar 75 G4
West Bergholt 34 D6
West Bexington 8 E6
West Bilney 44 B4
West Blatchington 13 F6
West Boldon 71 J7
West Bourton 9 G2
West Bowling 57 G6
West Brabourne 15 F3
West Bradford 56 C5
West Bradley 8 E1
West Bretton 50 E1
West Bridgford 41 H2
West Bromwich 40 C6
West Buckland Devon 6 E2
West Buckland Som 7 K3
West Burrafirth 89 L7
West Burton NYorks 57 F1
West Burton WSuss 12 C5
West Butsfield 62 A2
West Butterwick 52 B2
West Byfleet 22 D5
West Caister 45 K4
West Calder 75 J4
West Camel 8 E2
West Carbeth 74 D3
West Carr Houses 51 K2
West Cauldcoats 74 E6
West Chaldon 9 G6
West Challow 21 G3
West Charleton 5 H6
West Chevington 71 H4
West Chiltington 12 D5
West Chiltington Common 12 D5
West Chinnock 8 D3
West Chisenbury 20 E6
West Clandon 22 D6
West Cliffe 15 J3
West Clyne 87 L9
West Coker 8 E3
West Compton Dorset 8 E5
West Compton Som 19 J7
West Cowick 58 C7
West Cross 17 K7
West Crudwell 20 C2
West Curry 4 C1
West Curthwaite 60 E2
West Dean Wilts 10 D2
West Dean WSuss 12 B5
West Deeping 42 E5
West Derby 48 C3
West Dereham 44 A5
West Ditchburn 71 G1
West Down 6 D1
West Drayton GtLon 22 D4
West Drayton Notts 51 K5
West Edington 71 G5
West Ella 59 G7
West End Bed 32 C3
West End BrackF 22 B4
West End Caerp 19 F2
West End Cambs 43 H6
West End ERid 59 G3
West End Hants 11 F3
West End Herts 23 F1
West End Kent 25 H5
West End Lancs 55 H3
West End Lincs 53 G3
West End Norf 44 D5
West End Norf 45 J4
West End NSom 19 H5
West End NYorks 57 G4
West End Oxon 21 H5
West End Oxon 21 J3
West End SLan 75 H6
West End Suff 45 J7
West End Surr 22 C5
West End Surr 22 E5
West End Wilts 9 J2
West End Wilts 9 J2
West End Wilts 20 C4
West End Green 21 K5
West Farleigh 14 C2
West Farndon 31 G3
West Felton 38 C3
West Firle 13 H6
West Fleetham 71 G1
West Flotmanby 59 G2
West Garforth 57 J6
West Ginge 21 H3
West Glen 73 H3
West Grafton 21 F5
West Green GtLon 23 G3
West Green Hants 22 A6
West Grimstead 10 D2
West Grinstead 12 E4
West Haddlesey 58 B7
West Haddon 31 H1
West Hagbourne 21 J3
West Hagley 40 B7
West Hall 70 A7
West Hallam 41 G1
West Halton 59 F7
West Ham 23 G3
West Handley 51 F5
West Hanney 21 H2
West Hanningfield 24 D2
West Hardwick 51 G1
West Harnham 10 C2
West Harptree 19 J6
West Harrow 22 E3
West Harting 11 J2
West Hatch Som 8 B2
West Hatch Wilts 9 J2
West Head 43 J5
West Heath ChesE 49 H6
West Heath GtLon 23 H4
West Heath Hants 21 J6
West Heath Hants 22 B6
West Heath WMid 30 B1
West Helmsdale 87 N8
West Hendon 23 F3
West Hendred 21 H3
West Heslerton 59 F2
West Hewish 19 G5
West Hill Devon 7 J6
West Hill ERid 59 H3
West Hill NSom 19 H4
West Hoathly 13 G3
West Holme 9 H6
West Horndon 24 C3
West Horrington 19 J7
West Horsley 22 D6
West Horton 77 J7
West Hougham 15 H4
West Howe 10 B5
West Howetown 7 H2
West Huntspill 19 G7
West Hyde 22 D2
West Hythe 15 G4
West Ilsley 21 H3
West Itchenor 11 J4
West Keal 53 G6
West Kennet Long Barrow (Stonehenge, Avebury & Associated Sites) Wilts SN8 1QH 20 E5
West Kennett 20 E5
West Kilbride 74 A6
West Kingsdown 23 J5
West Kington 20 B4
West Kington Wick 20 B4
West Kirby 48 B4
West Knapton 58 E2
West Knighton 9 G6
West Knoyle 9 H1
West Kyloe 77 J6
West Lambrook 8 D3
West Langdon 15 J3
West Langwell 87 J9
West Lavington Wilts 20 D6
West Lavington WSuss 12 B4
West Layton 62 B6
West Leake 41 H3
West Learmouth 77 G7
West Lees 62 E6
West Leigh Devon 5 H5
West Leigh Devon 6 E5
West Leigh Som 7 K2
West Leith 32 C7
West Lexham 44 C4
West Lilling 58 C3
West Linton 75 K5
West Liss 11 J2
West Littleton 20 A4
West Lockinge 21 H3
West Looe 4 C5
West Lulworth 9 H6
West Lutton 59 F3
West Lydford 8 E1
West Lyn 7 F1
West Lyng 8 C2
West Lynn 44 A4
West Mains 77 J6
West Malling 23 K6
West Malvern 29 G4
West Marden 11 J3
West Markham 51 K5
West Marsh 53 F2
West Marton 56 D4
West Melbury 9 H2
West Melton 51 G2
West Meon 11 H2
West Meon Hut 11 H2
West Mersea 34 E7
West Midland Safari Park & Leisure Park Worcs DY12 1LF 29 H1
West Milton 8 D5
West Minster 25 F4
West Molesey 22 E5
West Monkton 8 B2
West Moors 10 B4
West Morden 9 J5
West Morriston 76 E6
West Morton 57 F5
West Mostard 61 J7
West Mudford 8 E2
West Ness 58 C2
West Newbiggin 62 D5
West Newton ERid 59 H6
West Newton Norf 44 A3
West Norwood 23 G4
West Ogwell 5 J3
West Orchard 9 H3
West Overton 20 E5
West Panson 6 B6
West Park 48 E3
West Parley 10 B5
West Peckham 23 K6
West Pelton 62 C1
West Pennard 8 E1
West Pentire 2 E2
West Perry 32 E2
West Porlock 7 G1
West Prawle 5 H7
West Preston 12 D6
West Pulham 9 G4
West Putford 6 B4
West Quantoxhead 7 K1
West Raddon 7 G5
West Rainton 62 D2
West Rasen 52 D4
West Raynham 44 C3
West Retford 51 J4
West Rounton 62 E6
West Row 33 K1
West Rudham 44 C3
West Runton 45 F1
West Saltoun 76 C4
West Sandford 7 G5
West Sandwick 89 N4
West Scrafton 57 F1
West Shepton 19 K7
West Somerset Railway Som TA24 5BG 7 K2
West Somerton 45 J4
West Stafford 9 G6
West Stockwith 51 K3
West Stoke 12 B6
West Stonesdale 61 K6
West Stoughton 19 H7
West Stour 9 G2
West Stourmouth 25 J5
West Stow 34 C1
West Stow Country Park Suff IP28 6HG 34 B1
West Stowell 20 E5
West Stratton 21 J7
West Street Kent 14 E2
West Street Med 24 D4
West Street Suff 34 D1
West Tanfield 57 H2
West Taphouse 4 B4
West Tarbert 73 G4
West Tarring 12 E6
West Thirston 71 G3
West Thorney 11 J4

West Thurrock 23 J4
West Tilbury 24 C4
West Tisted 11 H2
West Tofts Norf 44 C6
West Tofts P&K 80 G7
West Torrington 52 E4
West Town B&NESom 19 J5
West Town Hants 11 J5
West Town NSom 19 H5
West Town Som 8 E1
West Tytherley 10 D2
West Walton 43 H4
West Wellow 10 D3
West Wembury 5 F6
West Wemyss 76 B1
West Wick 19 G5
West Wickham Cambs 33 K4
West Wickham GtLon 23 G5
West Williamston 16 D5
West Winch 44 A4
West Winterslow 10 D1
West Wittering 11 J5
West Witton 57 F1
West Woodburn 70 D5
West Woodhay 21 G5
West Woodlands 20 A7
West Worldham 11 J1
West Worlington 7 F4
West Worthing 12 E6
West Wratting 33 K3
West Wycombe 22 B2
West Yatton 20 B4
West Yell 89 N4
West Youlstone 6 A4
Westbere 25 H5
Westborough 42 B1
Westbourne Bourne 10 B5
Westbourne WSuss 11 J4
Westbourne Green 23 F3
Westbrook Kent 25 K4
Westbrook WBerks 21 H4
Westbrook Wilts 20 C5
Westbury Bucks 31 H5
Westbury Shrop 38 C5
Westbury Wilts 20 B6
Westbury Leigh 20 B6
Westbury on Trym 19 J4
Westbury-on-Severn 29 G7
Westbury-sub-Mendip 19 J7
Westby Lancs 55 G6
Westby Lincs 42 C3
Westcliff-on-Sea 24 E3
Westcombe 9 F1
Westcot 21 G3
Westcott Bucks 31 J7
Westcott Devon 7 J5
Westcott Surr 22 E7
Westcott Barton 31 F6
Westcourt 21 F5
Westcroft 32 B5
Westdean 13 J7
Westdowns 4 A2
Westend Town 20 A4
Wester Balgedie 80 G10
Wester Dechmont 75 J3
Wester Gruinards 83 R1
Wester Hailes 76 A4
Wester Quarff 89 N9
Wester Skeld 89 L8
Westerdale High 87 P4
Westerdale NYorks 63 G6
Westerfield 35 F4
Westergate 12 C6
Westerham 23 H6
Westerhope 71 G7
Westerleigh 19 K4
Westerloch 87 R4
Westerton Aber 85 M11
Westerton Dur 62 C3
Westerton P&K 80 D9
Westerwick 89 L8
Westfield Cumb 60 A4
Westfield ESuss 14 D6
Westfield High 87 N3
Westfield NLan 75 F3
Westfield Norf 44 D5
Westfield WLoth 75 H3
Westfield WYorks 57 H7
Westfield Sole 24 D5
Westgate Dur 61 L3
Westgate NLincs 51 K2
Westgate Norf 44 D1
Westgate N'umb 71 G6
Westgate Hill 57 H7
Westgate on Sea 25 K4
Westhall Aber 85 L8
Westhall Suff 45 J7
Westham Dorset 9 F7
Westham ESuss 13 K6
Westham Som 19 H7
Westhampnett 12 B6
Westhay Devon 8 C4
Westhay Som 19 H7
Westhead 48 D2
Westhide 28 E4
Westhill 85 N10
Westhope Here 28 D3
Westhope Shrop 38 D7
Westhorp 31 G3
Westhorpe Lincs 43 F2
Westhorpe Notts 51 J7
Westhorpe Suff 34 E2
Westhoughton 49 F2
Westhouse 56 B2
Westhouses 51 G7
Westhumble 22 E6
Westing 89 P2
Westlake 5 G5
Westlands 40 A1
Westlea 20 E3
Westleigh Devon 6 C3
Westleigh Devon 7 J4
Westleigh GtMan 49 F2
Westleton 35 J2
Westley Shrop 38 C5
Westley Suff 34 C2
Westley Heights 24 C3
Westley Waterless 33 K3
Westlington 31 J7
Westlinton 69 J7
Westloch 76 A5
Westmancote 29 J5
Westmarsh 25 J5
Westmeston 13 G5
Westmill 33 G6
Westminster 23 F4
Westminster Abbey (Palace of Westminster & Westminster Abbey inc. St Margaret's Church) GtLon SW1P 3PA 101 A7
Westminster Cathedral GtLon SW1P 2QW 132 C6
Westmuir 81 J5
Westness 89 C5
Westnewton Cumb 60 C2
Westnewton N'umb 77 H7
Westoe 71 J7
Weston B&NESom 20 A5
Weston ChesE 49 G7
Weston Devon 7 K5
Weston Devon 7 K7
Weston Dorset 9 F7
Weston Halton 48 E4
Weston Hants 11 J2
Weston Here 28 C3
Weston Herts 33 F5
Weston Lincs 43 F3
Weston Notts 51 K6
Weston NYorks 57 G5
Weston N'hants 31 G4
Weston Shrop 28 C1
Weston Shrop 38 E3
Weston Shrop 38 E6
Weston Soton 11 F3
Weston Staffs 40 B3
Weston WBerks 21 H4
Weston Bampfylde 9 F2
Weston Beggard 28 E4
Weston by Welland 42 A6
Weston Colville 33 K3
Weston Corbett 21 K7
Weston Coyney 40 B1
Weston Favell 31 J2
Weston Green Cambs 33 K3
Weston Green Norf 45 F4
Weston Heath 39 G4
Weston Hills 43 F3
Weston in Arden 41 F7
Weston Jones 39 G3
Weston Longville 45 F4
Weston Lullingfields 38 D3
Weston Park Staffs TF11 8LE 40 A4
Weston Patrick 21 K7
Weston Point 48 D4
Weston Rhyn 38 B2
Weston Subedge 30 C4
Weston Town 20 A7
Weston Turville 32 B7
Weston under Penyard 29 F6
Weston under Wetherley 30 E2
Weston Underwood Derbys 40 E1
Weston Underwood MK 32 B3
Westonbirt 20 B3
Westonbirt - The National Arboretum Glos GL8 8QS 20 B3
Westoning 32 D5
Weston-in-Gordano 19 H4
Weston-on-Avon 30 C3
Weston-on-the-Green 31 G7
Weston-on-Trent 41 G3
Weston-super-Mare 19 G5
Weston-under-Lizard 40 A4
Westonzoyland 8 C1
Westow 58 D3
Westpoint Arena, Clyst St. Mary Devon EX5 1DJ 7 H6
Westport A&B 66 A1
Westport Som 8 C2
Westra 18 E4
Westray 89 D3
Westray Airfield 89 D2
Westridge Green 21 J4
Westrigg 75 H4
Westruther 76 E6
Westry 43 G6
Westvale 48 D3
Westville 41 H1
Westward 60 D2
Westward Ho! 6 C3
Westwell Kent 14 E3
Westwell Oxon 21 F1
Westwell Leacon 14 E3
Westwick Cambs 33 H2
Westwick Dur 62 A5
Westwick Norf 45 G3
Westwick NYorks 57 J3
Westwood Devon 7 J6
Westwood Peter 42 E6
Westwood SLan 74 E5
Westwood Wilts 20 B6
Westwood Heath 30 D1
Westwoodside 51 K3
Wetham Green 24 E5
Wetheral 61 F1
Wetherby 57 K5
Wetherden 34 E2
Wetherden Upper Town 34 E2
Wetheringsett 35 F2
Wethersfield 34 B5
Wethersta 89 M6
Wetherup Street 35 F2
Wetley Abbey 40 B1
Wetley Rocks 40 B1
Wettenhall 49 F6
Wettenhall Green 49 F6
Wetton 50 D7
Wetwang 59 F4
Wetwood 39 G2
Wexcombe 21 F6
Wexham Street 22 C3
Weybourne Norf 45 F1
Weybourne Surr 22 B7
Weybread 45 G7
Weybread Street 35 G1
Weybridge 22 D5
Weycroft 8 C4
Weyhill 21 G7
Weymouth 9 F7
Weymouth Sea Life Adventure Park & Marine Sanctuary Dorset DT4 7SX 9 F6
Whaddon Bucks 32 B5
Whaddon Cambs 33 G4
Whaddon Glos 29 H7
Whaddon Glos 29 J6
Whaddon Wilts 10 C2
Whaddon Wilts 20 B5
Whaddon Gap 33 G4
Whale 61 G4
Whaley 51 H5
Whaley Bridge 50 C4
Whaley Thorns 51 H5
Whalley 56 C6
Whalsay 89 P6
Whalsay Airport 89 P6
Whalton 71 G5
Wham 56 C3
Whaplode 43 G3
Whaplode Drove 43 G4
Whaplode St. Catherine 43 G4
Wharfe 56 C3
Wharles 55 H6
Wharley End 32 C4
Wharncliffe Side 50 E3
Wharram le Street 58 E3
Wharram Percy 58 E3
Wharton ChesW&C 49 F6
Wharton Here 28 E3
Whashton 62 B6
Whatcote 30 E4
Whateley 40 E6
Whatfield 34 E4
Whatley 20 A7
Whatlington 14 C6
Whatsole Street 15 G3
Whatstandwell 51 F7
Whatton 42 A2
Whauphill 64 E6
Whaw 61 L6
Wheal Peevor (Cornwall & West Devon Mining Landscape) Corn 2 E4
Wheatacre 45 J6
Wheatcroft 51 F7
Wheatenhurst 20 A1
Wheatfield 21 K2
Wheathampstead 32 E7
Wheathill Shrop 39 F7
Wheathill Som 8 E1
Wheatley Hants 11 J1
Wheatley Oxon 21 K1
Wheatley WYorks 57 F7
Wheatley Hill 62 D3
Wheatley Lane 56 D6
Wheatley Park 51 H2
Wheaton Aston 40 A4
Wheddon Cross 7 H2
Wheelerstreet 22 C7
Wheelock 49 G7
Wheelock Heath 49 G7
Wheelton 56 B7
Wheen 81 J3
Wheldale 57 K7
Wheldrake 58 C5
Whelford 20 E2
Whelley 48 E2
Whelpley Hill 22 C1
Whelpo 60 E3
Whelston 48 B5
Whenby 58 C3
Whepstead 34 C3
Wherstead 35 F4
Wherwell 21 G7
Wheston 50 D5
Whetley Cross 8 D4
Whetsted 23 K7
Whetstone GtLon 23 F2
Whetstone Leics 41 H6
Whicham 54 E1
Whichford 30 E5
Whickham 71 H7
Whiddon 6 C5
Whiddon Down 6 E6
Whifflet 75 F4
Whigstreet 81 K6
Whilton 31 H2
Whim 76 A5
Whimble 6 B5
Whimple 7 J6
Whimpwell Green 45 H3
Whin Lane End 55 G5
Whinburgh 44 E5
Whinlatter Forest Cumb CA12 5TW 60 C4
Whinny Hill 62 D5
Whinnyfold 85 Q7
Whippingham 11 G5
Whipsnade 32 D7
Whipsnade Zoo CenBeds LU6 2LF 32 D7
Whipton 7 H6
Whirlow 51 F4
Whisby 52 C6
Whissendine 42 B4
Whissonsett 44 D3
Whisterfield 49 H5
Whistley Green 22 A4
Whiston Mersey 48 D3
Whiston N'hants 32 B2
Whiston Staffs 40 A4
Whiston Staffs 40 C1
Whiston SYorks 51 G3
Whiston Cross 39 G5
Whiston Eaves 40 C1
Whitacre Fields 40 E6
Whitacre Heath 40 E6
Whitbeck 54 E1
Whitbourne 29 G3
Whitburn T&W 71 K7
Whitburn WLoth 75 H4
Whitby ChesW&C 48 C5
Whitby NYorks 63 J5
Whitby Abbey NYorks YO22 4JT 63 K5
Whitby Lifeboat Museum NYorks YO21 3PU 63 J5
Whitbyheath 48 C5
Whitchurch B&NESom 19 K5
Whitchurch Bucks 32 B6
Whitchurch Cardiff 18 E3
Whitchurch Devon 4 E3
Whitchurch Hants 21 H7
Whitchurch Here 28 E7
Whitchurch Pembs 16 A3
Whitchurch Shrop 38 E1
Whitchurch Warks 30 D4
Whitchurch Canonicorum 8 C5
Whitchurch Hill 21 K4
Whitchurch-on-Thames 21 K4
Whitcombe 9 G6
Whitcott Keysett 38 B7
White Ball 7 J4
White Colne 34 C6
White Coppice 49 F1
White Cross Corn 3 F3
White Cross Devon 7 J6
White Cross Here 28 D4
White Cross Wilts 9 G1
White Cube GtLon N1 6PB 100 B6
White End 29 H6
White Hill 9 H1
White Houses 51 K5
White Kirkley 62 A3
White Lackington 9 G5
White Ladies Aston 29 J3
White Lund 55 H3
White Mill 17 H3
White Moor 41 F1
White Notley 34 B7
White Ox Mead 20 A6
White Pit 53 G5
White Post Farm Centre, Farnsfield Notts NG22 8HL 51 J7
White Rocks 28 D6
White Roding 33 J7
White Waltham 22 B4
Whiteash Green 34 B5
Whitebirk 56 C7
Whitebridge (An Drochaid Bhàn) 83 Q9
Whitebrook 19 J1
Whiteburn 76 D6
Whitecairn 64 C5
Whitecairns 85 P9
Whitecastle 75 J6
Whitechapel 55 J5
Whitechurch 16 E2
Whitecote 57 H6
Whitecraig 76 B3
Whitecroft 19 K1
Whitecrook 64 B5
Whitecross Corn 2 C5
Whitecross Corn 3 G1
Whitecross Dorset 8 D5
Whitecross Falk 75 H3
Whitefield Devon 7 F2
Whitefield Dorset 9 J5
Whitefield GtMan 49 H2
Whitegate 49 F6
Whitehall Devon 7 K4
Whitehall Hants 22 A6
Whitehall Ork 89 F5
Whitehall WSuss 12 E4
Whitehaven 60 A5
Whitehill Hants 11 J1
Whitehill Kent 14 E2
Whitehill Midlo 76 B4
Whitehill NAyr 74 A5
Whitehills 85 L4
Whitehouse A&B 73 G4
Whitehouse Aber 85 L9
Whitehouse Common 40 D6
Whitekirk 76 D2
Whitelackington 8 C3
Whitelaw 77 G5
Whitelees 74 B7
Whiteley 11 G4
Whiteley Bank 11 G6
Whiteley Green 49 J5
Whiteley Village 22 D5
Whiteleys 64 A5
Whitemans Green 13 G4
Whitemire 84 D5
Whitemoor 3 G3
Whiteness 89 M8
Whiteoak Green 30 E7
Whiteparish 10 D2
Whiterashes 85 N8
Whiteshill 20 B1
Whiteside N'umb 70 C7
Whiteside WLoth 75 H4
Whitesmith 13 J5
Whitestaunton 8 B3
Whitestone A&B 73 F7
Whitestone Devon 7 G6
Whitestreet Green 34 D5
Whiteway 29 J7
Whitewell Lancs 56 B5
Whitewell Wrex 38 D1
Whiteworks 5 G3
Whitewreath 84 G5
Whitfield Here 28 D5
Whitfield Kent 15 J3
Whitfield N'hants 31 H5
Whitfield N'umb 61 J1
Whitfield SGlos 19 K2
Whitford Devon 8 B5
Whitford (Chwitffordd) Flints 47 K5
Whitgift 58 E7
Whitgreave 40 A3
Whithorn 64 E6
Whiting Bay 66 E1
Whitkirk 57 J6
Whitland (Hendy-Gwyn) 17 F4
Whitland Abbey 17 F4
Whitleigh 4 E4
Whitletts 67 H1
Whitley NYorks 58 B7
Whitley Read 22 A5
Whitley Wilts 20 B5
Whitley WMid 30 E1
Whitley Bay 71 J6
Whitley Chapel 61 L1
Whitley Heath 40 A3
Whitley Lower 50 E1
Whitley Row 23 H6
Whitlock's End 30 C1
Whitminster 20 A1
Whitmore Dorset 10 B4
Whitmore Staffs 40 A1
Whitnage 7 J4
Whitnash 30 E2
Whitnell 19 F7
Whitney-on-Wye 28 B4
Whitrigg Cumb 60 D1
Whitrigg Cumb 60 D3
Whitsbury 10 C3
Whitsome 77 G5
Whitson 19 G3
Whitstable 25 H5
Whitstone 4 C1
Whittingham 71 F2
Whittingslow 38 D7
Whittington Derbys 51 F5
Whittington Glos 30 B6
Whittington Lancs 56 B2
Whittington Norf 44 B6
Whittington Shrop 38 C2
Whittington Staffs 40 A7
Whittington Staffs 40 D5
Whittington Worcs 29 H3
Whittlebury 31 H4
Whittle-le-Woods 55 J7
Whittlesey 43 F6
Whittlesford 33 H4
Whittlestone Head 49 G1
Whitton GtLon 22 E4
Whitton NLincs 59 F7
Whitton N'umb 71 F3
Whitton Powys 28 B2
Whitton Shrop 28 E1
Whitton Stock 62 D4
Whitton Suff 35 F4
Whittonditch 21 F4
Whittonstall 62 A1
Whitway 21 H6
Whitwell Derbys 51 H5
Whitwell Herts 32 E6
Whitwell IoW 11 G7
Whitwell NYorks 62 C7
Whitwell Rut 42 C5
Whitwell Street 45 F3
Whitwell-on-the-Hill 58 D3
Whitwick 41 G4
Whitwood 57 K7
Whitworth 49 H1
Whitworth Art Gallery, Manchester GtMan M15 6ER 113 H5
Whixall 38 E2
Whixley 57 K4
Whorlton Dur 62 B5
Whorlton NYorks 62 E6
Whygate 70 C6
Whyle 28 E2
Whyteleafe 23 G6
Wibdon 19 J2
Wibsey 57 G6
Wibtoft 41 G7
Wichenford 29 G2
Wichling 14 E2
Wick Bourne 10 C5
Wick Devon 7 K5
Wick (Inbhir Ùige) High 87 R4
Wick SGlos 20 A4
Wick Som 8 E1
Wick Som 19 F7
Wick VGlam 18 C4
Wick Wilts 10 C2
Wick Worcs 29 J4
Wick WSuss 12 D6
Wick John O'Groats Airport 87 R4
Wick Hill Kent 14 D3
Wick Hill W'ham 22 A5
Wick St. Lawrence 19 G5
Wicken Cambs 33 J1
Wicken N'hants 31 J5
Wicken Bonhunt 33 H5
Wickenby 52 D4
Wicker Street Green 34 D4
Wickersiack 61 H5
Wickersley 51 G3
Wicketwood Hill 41 J1
Wickford 24 D2
Wickham Hants 11 G3
Wickham WBerks 21 G4
Wickham Bishops 34 C7
Wickham Heath 21 H5
Wickham Market 35 H3
Wickham Skeith 34 E2
Wickham St. Paul 34 C5
Wickham Street Suff 34 B3
Wickham Street Suff 34 E2
Wickhambreaux 15 H2
Wickhambrook 34 B3
Wickhamford 30 B4
Wickhampton 45 J5
Wicklewood 44 E5
Wickmere 45 F2
Wicksteed Park N'hants NN15 6NJ 32 B1
Wickstreet 13 J6
Wickwar 20 A3
Widcombe 20 A5
Widdington 33 J5
Widdop 56 E6
Widdrington 71 H4
Widdrington Station 71 H4
Wide Open 71 H6
Widecombe in the Moor 5 H3
Widegates 4 C5
Widemouth Bay 6 A5
Widewall 89 D8
Widford Essex 24 C1
Widford Herts 33 H7
Widford Oxon 30 D7
Widgham Green 33 K3
Widmer End 22 B2
Widmerpool 41 J3
Widnes 48 E4
Widworthy 8 B5
Wigan 48 E2
Wigan Pier GtMan WN3 4EU 48 E2
Wiganthorpe 58 C2
Wigborough 8 D3
Wiggaton 7 K6
Wiggenhall St. Germans 43 J4
Wiggenhall St. Mary Magdalen 43 J4
Wiggenhall St. Mary the Virgin 43 J4
Wiggenhall St. Peter 44 A4
Wiggens Green 33 K4
Wigginton Herts 32 C7
Wigginton Oxon 30 E5
Wigginton Shrop 38 C2
Wigginton Staffs 40 E5
Wigginton York 58 C4
Wigglesworth 56 D4
Wiggonby 60 E1
Wiggonholt 12 D5
Wighill 57 K5
Wighton 44 D2
Wightwizzle 50 E3
Wigley 10 E3
Wigmore Here 28 D2
Wigmore Med 24 E5
Wigsley 52 B5
Wigsthorpe 42 D7
Wigston 41 J6
Wigston Parva 41 G7
Wigthorpe 51 H4
Wigtoft 43 F2
Wigton 60 D2
Wigtown 64 E5
Wike 57 J5
Wilbarston 42 B7
Wilberfoss 58 D4
Wilburton 33 H1
Wilby Norf 44 E6
Wilby N'hants 32 B2
Wilby Suff 35 G1
Wilcot 20 E5
Wilcott 38 C4
Wilcrick 19 H3
Wilday Green 51 F5
Wildboarclough 49 J6
Wilde Street 34 B1
Wilden Bed 32 D3
Wilden Worcs 29 H1
Wildhern 21 G6
Wildhill 23 F1
Wildmoor 29 J1
Wildsworth 52 B3
Wilford 41 H2
Wilkesley 39 F1
Wilkhaven 84 D2
Wilkieston 75 K4
Wilksby 53 F6
Willand Devon 7 J4
Willand Som 7 K4
Willaston ChesE 49 F7
Willaston ChesW&C 48 C5
Willaston Shrop 38 E2
Willen 32 B4
Willen Lakeside Park MK MK15 9HQ 97 E3
Willenhall WMid 30 E1
Willenhall WMid 40 B6
Willerby ERid 59 G7
Willerby NYorks 59 G2
Willersey 30 C5
Willersley 28 C4
Willesborough 15 F3
Willesborough Lees 15 F3
Willesden 23 F3
Willesleigh 6 D2
Willesley 20 B3
Willett 7 K2
Willey Shrop 39 F6
Willey Warks 41 G7
Willey Green 22 C6
William's Green 34 D4
Williamscot 31 F4
Williamson Park, Lancaster Lancs LA1 1UX 55 H3
Willian 33 F5
Willimontswick 70 C7
Willingale 23 J1
Willingdon 13 J6
Willingham 33 H1
Willingham by Stow 52 B4
Willingham Green 33 K3
Willington Bed 32 E3
Willington Derbys 40 E3
Willington Dur 62 B3
Willington Kent 14 C2
Willington T&W 71 J7
Willington Warks 30 D5
Willington Corner 48 E6
Willisham 34 E3
Willitoft 58 D6
Williton 7 J1
Willoughbridge 39 G1
Willoughby Lincs 53 H5
Willoughby Warks 31 G2
Willoughby Waterleys 41 H6
Willoughby-on-the-Wolds 41 J3
Willoughton 52 C3
Willow Green 49 F5
Willows Farm Village Herts AL2 1BB 22 E1
Willows Green 34 B7
Willsbridge 19 K4
Willslock 40 C2
Willsworthy 6 D7
Willtown 8 C2
Wilmcote 30 C3
Wilmington B&NESom 19 K5
Wilmington Devon 8 B4
Wilmington ESuss 13 J6
Wilmington Kent 23 J4
Wilmslow 49 H4
Wilnecote 40 E5
Wilney Green 44 E7
Wilpshire 56 B6
Wilsden 57 F6
Wilsford Lincs 42 D1
Wilsford Wilts 10 C1
Wilsford Wilts 20 E6
Wilsham 7 F1
Wilshaw 50 D2
Wilsill 57 G3
Wilsley Green 14 C4
Wilsley Pound 14 C4
Wilson 41 G3
Wilstead 32 D4
Wilsthorpe ERid 59 H3
Wilsthorpe Lincs 42 D4
Wilstone 32 C7
Wilton Cumb 60 B5
Wilton Here 28 E6
Wilton NYorks 58 E1
Wilton R&C 63 F5
Wilton ScBord 69 K2
Wilton Wilts 10 B1
Wilton Wilts 21 F5
Wilton House Wilts SP2 0BJ 10 B1
Wiltown 7 K4
Wimbish 33 J5
Wimbish Green 33 K5
Wimblebury 40 C4
Wimbledon 23 F4
Wimbledon All England Lawn Tennis & Croquet Club GtLon SW19 5AG 99 G10
Wimblington 43 H6
Wimborne Minster Dorset BH21 1HT 91 B3
Wimborne Minster 10 B4
Wimborne St. Giles 10 B3
Wimbotsham 44 A5
Wimpole 33 G3
Wimpole Home Farm Cambs SG8 0BW 33 G4
Wimpole Lodge 33 G4
Wimpstone 30 D4
Wincanton 9 G2
Winceby 53 G6
Wincham 49 F5
Winchburgh 75 J3
Winchcombe 30 B6
Winchelsea 14 E6
Winchelsea Beach 14 E6
Winchester 11 F2
Winchester Cathedral Hants SO23 9LS 137 Winchester
Winchet Hill 14 C3
Winchfield 22 A6
Winchmore Hill Bucks 22 C2
Winchmore Hill GtLon 23 G2
Wincle 49 J6
Wincobank 51 F3
Windermere 60 F7
Windermere Lake Cruises Cumb LA12 8AS 60 E7
Winderton 30 E4
Windle Hill 48 C5
Windlehurst 49 J4
Windlesham 22 C5
Windley 41 F1
Windmill 50 D5
Windmill Hill ESuss 13 K5
Windmill Hill Som 8 C3
Windmill Hill (Stonehenge, Avebury & Associated Sites) Wilts SN4 9NW 20 D4
Windmill Hill Worcs 29 J4
Windrush 30 C7
Windsor 22 C4
Windsor Castle W&M SL4 1NJ 137 Windsor
Windsor Green 34 C3
Windy Nook 71 H7
Windygates 81 J10
Windy-Yett 74 C5
Wineham 13 F4
Winestead 59 J7
Winewall 56 E6
Winfarthing 45 F7
Winford IoW 11 G6
Winford NSom 19 J5
Winforton 28 B4
Winfrith Newburgh 9 H6
Wing Bucks 32 B6
Wing Rut 42 B5
Wingate 62 D3
Wingates GtMan 49 F2
Wingates N'umb 71 F4
Wingerworth 51 F6
Wingfield CenBeds 32 D6
Wingfield Suff 35 G1
Wingfield Wilts 20 B6
Wingfield Green 35 G1
Wingham 15 H2
Wingham Well 15 H2
Wingmore 15 G3
Wingrave 32 B7
Winkburn 51 K7
Winkfield 22 C4
Winkfield Row 22 B4
Winkhill 50 C7
Winkleigh 6 E5
Winksley 57 H2
Winkton 10 C5
Winlaton 71 G7
Winlaton Mill 71 G7
Winmarleigh 55 H5
Winnard's Perch 3 G2
Winnersh 22 A4
Winnington 49 F5
Winscombe 19 H6
Winsford ChesW&C 49 F6
Winsford Som 7 H2
Winsham Devon 6 C2
Winsham Som 8 C4
Winshill 40 E3
Winsh-wen 17 K6
Winskill 61 G3
Winslade 21 K7
Winsley 20 B5
Winslow 31 J6
Winson 20 D1
Winsor 10 E3
Winster Cumb 60 F7
Winster Derbys 50 E6
Winston Dur 62 B5
Winston Suff 35 F2
Winston Green 35 F2
Winstone 20 C1
Winswell 6 C4
Winter Gardens NSom BS23 1AJ 137 Weston-super-Mare
Winter Gardens FY1 1HW 121 Blackpool
Winterborne Came 9 G6
Winterborne Clenston 9 H4
Winterborne Herringston 9 F6
Winterborne Houghton 9 H4
Winterborne Kingston 9 H5
Winterborne Monkton 9 F6
Winterborne Stickland 9 H4
Winterborne Whitechurch 9 H4
Winterborne Zelston 9 H5
Winterbourne SGlos 19 K3
Winterbourne Stoke 20 D7
Winterbourne WBerks 21 H4
Winterbourne Abbas 9 F5
Winterbourne Bassett 20 E4
Winterbourne Dauntsey 10 C1
Winterbourne Earls 10 C1
Winterbourne Gunner 10 C1
Winterbourne Monkton 20 E4
Winterbourne Steepleton 9 F6
Winterbrook 21 K3
Winterburn 56 E4
Wintercleugh 68 E2
Winteringham 59 F7
Winterley 49 G7
Wintersett 51 F1
Wintershill 11 G3
Winterslow 10 D1
Winterton 52 C1
Winterton-on-Sea 45 J4
Winthorpe Lincs 53 J6
Winthorpe Notts 52 B7
Winton Bourne 10 B5
Winton Cumb 61 J5
Wintringham 58 E2
Winwick Cambs 42 E7
Winwick N'hants 31 H1
Winwick Warr 49 F3
Wirksworth 50 E7
Wirksworth Moor 51 F7
Wirral Country Park Mersey CH61 0HN 48 B4
Wirswall 38 E1
Wisbech 43 H5
Wisbech St. Mary 43 H5
Wisborough Green 12 D4
Wiseton 51 K4
Wishaw NLan 75 F5
Wishaw Warks 40 D6
Wisley 22 D6
Wispington 53 F5
Wissett 35 H1
Wissington 34 D5
Wistanstow 38 D7
Wistanswick 39 F3
Wistaston 49 F7
Wiston Pembs 16 D4
Wiston SLan 75 H7
Wiston WSuss 12 E5
Wistow Cambs 43 F7
Wistow NYorks 58 B6
Wiswell 56 C6
Witcham 33 H1
Witchampton 9 J4
Witchburn 66 B1
Witchford 33 J1
Witcombe 8 D2
Witham 34 C7
Witham Friary 20 A7
Witham on the Hill 42 D4
Withcall 53 F4
Withcote 42 A5
Withdean 13 G6
Witherenden Hill 13 K4
Witherhurst 13 K4
Witheridge 7 G4
Witherley 41 F6
Withern 53 H4
Withernsea 59 K7
Withernwick 59 H5
Withersdale Street 45 G7
Withersfield 33 K4
Witherslack 55 H1
Witherslack Hall 55 H1
Withiel 3 G2
Withiel Florey 7 H2
Withielgoose 4 A4
Withington Glos 30 B7
Withington GtMan 49 H3
Withington Here 28 E4
Withington Shrop 38 E4
Withington Staffs 40 C2
Withington Green 49 H5
Withington Marsh 28 E4
Withleigh 7 H4
Withnell 56 B7
Withnell Fold 56 B7
Withybrook Som 19 K7
Withybrook Warks 41 G7
Withycombe 7 J1
Withycombe Raleigh 7 J7
Withyham 13 H3
Withypool 7 G2
Witley 12 C3
Witnesham 35 F3
Witney 21 G1
Wittering 42 D5
Wittersham 14 D5
Witton Angus 81 L3
Witton Norf 45 H5
Witton Worcs 29 H2
Witton Bridge 45 H2
Witton Gilbert 62 C2
Witton Park 62 B3
Witton-le-Wear 62 B3
Wiveliscombe 7 J3
Wivelsfield 13 G4
Wivelsfield Green 13 G5
Wivenhoe 34 E6
Wiveton 44 E1
Wix 35 F6
Wixford 30 B3
Wixhill 38 E3
Wixoe 34 B4
Woburn 32 C5
Woburn Safari Park CenBeds MK17 9QN 97 H5
Woburn Sands 32 C5
Wokefield Park 21 K5
Woking 22 D6
Wokingham 22 B5
Wolborough 5 J3
Wold Newton ERid 59 G2
Wold Newton NELincs 53 F3
Woldingham 23 G6
Wolds Village, Bainton ERid YO25 9EF 59 F4
Wolfelee 70 A3
Wolferlow 29 F2
Wolferton 44 A3
Wolfhampcote 31 G2
Wolfhill 80 G7
Wolfpits 28 B3
Wolf's Castle 16 C3
Wolfsdale 16 C3
Woll 69 K1
Wollaston N'hants 32 C2
Wollaston Shrop 38 C4
Wollaston WMid 40 A7
Wollaton 41 H2
Wollerton 39 F3
Wollescote 40 B7
Wolsingham 62 A3
Wolston 31 F1
Wolsty 60 C1
Wolvercote 21 J1
Wolverhampton 40 B6
Wolverhampton Art Gallery WMid WV1 1DU 102 B1
Wolverley Shrop 38 D2
Wolverley Worcs 29 H1
Wolvers Hill 19 G5
Wolverton Hants 21 J6
Wolverton MK 32 B4
Wolverton Warks 30 D2
Wolverton Wilts 9 G1
Wolverton Common 21 J6
Wolvesnewton 19 H2
Wolvey 41 G7
Wolvey Heath 41 G7
Wolviston 62 E4
Womaston 28 B2
Wombleton 58 C1
Wombourne 40 A6
Wombwell 51 F2
Womenswold 15 H2
Womersley 51 H1
Wonastow 28 D7
Wonersh 22 D7
Wonford 7 H6
Wonson 6 E7
Wonston 11 F1
Wooburn 22 C3
Wooburn Green 22 C3
Wood Bevington 30 B3
Wood Burcote 31 H4
Wood Dalling 44 E3
Wood Eaton 40 A4
Wood End Bed 32 D2
Wood End Bed 32 D4
Wood End Bucks 31 J5
Wood End Herts 33 G6
Wood End Warks 30 C1
Wood End Warks 40 E6
Wood End Warks 40 E7
Wood End WMid 40 B5
Wood Enderby 53 F6
Wood Green GtLon 23 G2
Wood Green Norf 45 G6
Wood Green Animal Shelter, Godmanchester Cambs PE29 2NH 33 F2
Wood Lane 38 D2
Wood Norton 44 E3
Wood Seats 51 F3
Wood Stanway 30 B5
Wood Street 45 H3
Wood Street Village 22 C6
Woodacott 6 B5
Woodale 57 F2
Woodall 51 G4
Woodbastwick 45 H4
Woodbeck 51 K5
Woodborough Notts 41 J1
Woodborough Wilts 20 E6
Woodbridge Devon 7 K6
Woodbridge Dorset 9 G3
Woodbridge Suff 35 G4
Woodbury Devon 7 J7
Woodbury Som 19 J7
Woodbury Salterton 7 J7
Woodchester 20 B1
Woodchurch Kent 14 E4
Woodchurch Mersey 48 B4
Woodcombe 7 H1
Woodcote Oxon 21 K3
Woodcote Tel&W 39 G4
Woodcote Green 29 J1
Woodcott 21 H6
Woodcroft 19 J2
Woodcutts 9 J3
Woodditton 33 K3
Woodeaton 31 G7
Woodend Cumb 60 C7
Woodend N'hants 31 H4
Woodend WSuss 12 B6
Woodend Green 33 J6
Woodfalls 10 C2
Woodfield Oxon 31 G6
Woodfield SAyr 67 H1
Woodfoot 61 H5
Woodford Corn 6 A4
Woodford Devon 5 H5
Woodford Glos 19 K2
Woodford GtLon 23 H2
Woodford GtMan 49 H4
Woodford N'hants 32 C1
Woodford Som 7 J2
Woodford Bridge 23 H2
Woodford Green 23 H2
Woodford Halse 31 G3
Woodgate Devon 7 K4
Woodgate Norf 44 E4
Woodgate WMid 40 B7
Woodgate Worcs 29 J2
Woodgate WSuss 12 C6
Woodgreen 10 C3
Woodhall Invcly 74 B3
Woodhall NYorks 61 L7
Woodhall Hills 57 G6
Woodhall Spa 52 E6
Woodham Bucks 31 H7
Woodham Dur 62 C4
Woodham Surr 22 D5
Woodham Ferrers 24 D2
Woodham Mortimer 24 E1
Woodham Walter 24 E1
Woodhaven 81 K8
Woodhead Aber 85 M7
Woodhead Staffs 40 C1
Woodhenge (Stonehenge, Avebury & Associated Sites) Wilts SP4 7AR 20 E7
Woodhey 48 C4
Woodhey Green 48 E7
Woodhill Shrop 39 G7
Woodhill Som 8 C2
Woodhorn 71 H5
Woodhouse Cumb 55 J1
Woodhouse Leics 41 H4
Woodhouse SYorks 51 G4
Woodhouse WYorks 57 G7
Woodhouse WYorks 57 H6
Woodhouse WYorks 57 J7
Woodhouse Down 19 K3
Woodhouse Eaves 41 H4
Woodhouse Green 49 J6
Woodhouses GtMan 49 H2
Woodhouses Staffs 40 C5
Woodhouses Staffs 40 D4
Woodhuish 5 K5
Woodhurst 33 G1
Woodingdean 13 G6
Woodington 10 E2
Woodland Devon 5 H4
Woodland Dur 62 A4
Woodland Kent 15 G3
Woodland Head 7 F6
Woodlands Dorset 10 B4
Woodlands Hants 10 E3
Woodlands NYorks 57 J4
Woodlands Shrop 39 G7
Woodlands Som 7 K1
Woodlands Leisure Park, Dartmouth Devon TQ9 7DQ 5 J5
Woodlands Park 22 B4
Woodlands St. Mary 21 G4
Woodlane 40 D3
Woodleigh 5 H6
Woodlesford 57 J7
Woodley GtMan 49 J3
Woodley W'ham 22 A4
Woodmancote Glos 20 A2
Woodmancote Glos 20 D1
Woodmancote Glos 29 J6
Woodmancote WSuss 11 J4
Woodmancote WSuss 13 F5
Woodmancott 21 J7
Woodmansey 59 G6
Woodmansterne 23 F6
Woodmanton 7 J7
Woodmill 40 D3
Woodminton 10 B2
Woodmoor 38 B5
Woodnesborough 15 J2
Woodnewton 42 D6
Woodperry 31 G7
Woodplumpton 55 J6
Woodrising 44 D5
Woodrow 9 G3
Wood's Corner 13 K5
Woods Eaves 28 B4
Wood's Green 13 K3
Woodseaves Shrop 39 F2
Woodseaves Staffs 39 G3
Woodsend 21 F4
Woodsetts 51 H4
Woodsford 9 G5
Woodside BrackF 22 C4
Woodside CenBeds 32 D7
Woodside Cumb 60 B3
Woodside D&G 69 F6
Woodside GtLon 23 G5
Woodside Hants 10 E5
Woodside Herts 23 F1
Woodside NAyr 74 B5
Woodside P&K 80 G7
Woodside Shrop 38 C7
Woodside WMid 40 B7
Woodside Animal Farm & Leisure Park CenBeds LU1 4DG 32 D7
Woodside Green 14 E2
Woodstock Oxon 31 F7
Woodstock Pembs 16 D3
Woodthorpe Derbys 51 G5
Woodthorpe Leics 41 H4
Woodthorpe Lincs 53 H4
Woodthorpe SYorks 51 F4
Woodthorpe York 58 B5
Woodton 45 G6
Woodtown 6 C3
Woodvale 48 C1
Woodville 41 F4
Woodwall Green 39 G2
Woodwalton 43 F7
Woodwick 89 C5
Woodworth Green 48 E7
Woodyates 10 B3
Woofferton 28 E2
Wookey 19 J7
Wookey Hole 19 J7
Wookey Hole Caves & Papermill Som BA5 1BB 19 J7
Wool 9 H6
Woolacombe 6 C1
Woolage Green 15 H3
Woolage Village 15 H2
Woolaston 19 J1
Woolaston Slade 19 J1
Woolavington 19 G7
Woolbeding 12 B4
Woolcotts 7 H2
Wooldale 50 D2
Wooler 70 E1
Woolfardisworthy Devon 6 B3
Woolfardisworthy Devon 7 G5
Woolfold 49 G1
Woolfords Cottages 75 J5
Woolgarston 9 J6
Woolgreaves 51 F1
Woolhampton 21 J5
Woolhope 29 F5
Woolland 9 G4
Woollard 19 K5
Woollaton 6 C4
Woollensbrook 23 G1
Woolley B&NESom 20 A5
Woolley Cambs 32 E1
Woolley Corn 6 A4
Woolley Derbys 51 F6
Woolley WYorks 51 F1
Woolley Green Wilts 20 B5
Woolley Green W&M 22 B4
Woolmer Green 33 F7
Woolmere Green 29 J2
Woolmersdon 8 B1
Woolpit 34 D2
Woolpit Green 34 D2
Woolscott 31 F2
Woolsgrove 7 F5
Woolstaston 38 D6
Woolsthorpe 42 B2
Woolsthorpe by Colsterworth 42 C3
Woolston Devon 5 H6
Woolston Shrop 38 C3
Woolston Shrop 38 D7
Woolston Soton 11 F3
Woolston Warr 49 F4
Woolston Green 5 H4
Woolstone Glos 29 J5
Woolstone MK 32 B5
Woolstone Oxon 21 F3
Woolton 48 D4
Woolton Hill 21 H5
Woolverstone 35 F5
Woolverton 20 A6
Woolwich 23 H4
Woonton 28 C3
Wooperton 71 F2
Woore 39 G1
Wootten Green 35 G1
Wootton Bed 32 D4
Wootton Hants 10 D5
Wootton IoW 11 G5
Wootton Kent 15 H3
Wootton NLincs 52 D1
Wootton N'hants 31 J3
Wootton Oxon 21 H1
Wootton Oxon 31 F7
Wootton Shrop 28 D1
Wootton Shrop 38 C3
Wootton Staffs 40 A3
Wootton Staffs 40 D1
Wootton Bassett 20 D3
Wootton Bridge 11 G5
Wootton Common 11 G5
Wootton Courtenay 7 H1
Wootton Fitzpaine 8 C5
Wootton Green 32 D4
Wootton Rivers 20 E5
Wootton St. Lawrence 21 J6
Wootton Wawen 30 C2
Worcester 29 H3
Worcester Cathedral Worcs WR1 2LH 137 Worcester
Worcester Park 23 F5
Worcester Woods Country Park Worcs WR5 2LG 29 H3
Wordsley 40 A7
Wordwell 34 C1
Worfield 39 G6
Worgret 9 J6
Workhouse End 32 E3
Workington 60 B4
Worksop 51 H5
Worlaby Lincs 53 G5
Worlaby NLincs 52 D1
World Museum Liverpool Mersey L3 8EN 131 D3
World of Beatrix Potter Attraction Cumb LA23 3BX 60 F7
World of James Herriot, Thirsk NYorks YO7 1PL 57 K1
World's End Bucks 22 B1
Worlds End Hants 11 H3
World's End WBerks 21 H4
Worlds End WMid 40 D7
Worle 19 G5
Worleston 49 F7
Worlingham 45 J7
Worlington 33 K1
Worlingworth 35 G2
Wormald Green 57 J3
Wormbridge 28 D5
Wormegay 44 A4
Wormelow Tump 28 D5
Wormhill 50 D5
Wormingford 34 D5
Worminghall 21 K1
Wormington 30 B5
Worminster 19 J7
Wormit 81 J8
Wormleighton 31 F3
Wormley Herts 23 G1
Wormley Surr 12 C3
Wormley West End 23 G1
Wormshill 14 D2
Wormsley 28 D4
Worplesdon 22 C6
Worrall 51 F3
Worsbrough 51 F2
Worsley 49 G2
Worstead 45 H3
Worsted Lodge 33 J3
Worsthorne 56 D6
Worston 56 C5
Worswell 5 F6
Worth Kent 15 J2
Worth WSuss 13 F3
Worth Matravers 9 J7
Wortham 34 E1
Worthen 38 C5
Worthenbury 38 D1
Worthing Norf 44 D4
Worthing WSuss 12 E6
Worthington 41 G3
Worting 21 K6
Wortley Glos 20 A2
Wortley SYorks 51 F3
Wortley WYorks 57 H6
Worton NYorks 56 E1
Worton Wilts 20 C6
Wortwell 45 G7
Wothersome 57 K5
Wotherton 38 B5
Wotter 5 F4
Wotton 22 E7
Wotton Underwood 31 H7
Wotton-under-Edge 20 A2
Woughton on the Green 32 B5
Wouldham 24 D5
W.R. Outhwaite & Son Ropemakers NYorks DL8 3NT 56 D1
Wrabness 35 F5
Wrafton 6 C2
Wragby 52 E5
Wragholme 53 G3
Wramplingham 45 F5
Wrangaton 5 G5
Wrangle 53 H7
Wrangle Lowgate 53 H7
Wrangway 7 K4
Wrantage 8 C2
Wrawby 52 D2
Wraxall NSom 19 H4
Wraxall Som 9 F1
Wray 56 B3
Wray Castle 60 E6
Wrays 23 F7
Wraysbury 22 D4
Wrayton 56 B2
Wrea Green 55 G6
Wreay Cumb 60 F2
Wreay Cumb 60 F4
Wrecclesham 22 B7
Wrekenton 62 C1
Wrelton 58 D1
Wrenbury 38 E1
Wrench Green 59 F1
Wreningham 45 F6
Wrentham 45 J7
Wrenthorpe 57 J7
Wrentnall 38 D5
Wressle ERid 58 D6
Wressle NLincs 52 C2
Wrestlingworth 33 F4
Wretham 44 D6
Wretton 44 A5
Wrexham (Wrecsam) 38 C1
Wrexham Arts Centre Wrex LL11 1AU 48 C7
Wrexham Industrial Estate 38 C1
Wribbenhall 29 G1
Wrightington Bar 48 E1
Wrightpark 74 E1
Wright's Green 33 J7
Wrinehill 39 G1
Wrington 19 H5
Writhlington 19 K6
Writtle 24 C1
Wrockwardine 39 F4
Wroot 51 K2
Wrose 57 G6
Wrotham 23 K6
Wrotham Heath 23 K6
Wrotham Hill Park 24 C5
Wrottesley 40 A5
Wroughton 20 E3
Wroxall IoW 11 G7
Wroxall Warks 30 D1
Wroxeter 38 E5
Wroxham 45 H4
Wroxham Barns Norf NR12 8QU 45 H3
Wroxton 31 F4
Wstrws 17 G1
WWT London Wetland Centre GtLon SW13 9WT 99 F8
Wyaston 40 D1
Wyberton 43 G1

Collins

Published by Collins
An imprint of HarperCollins Publishers
Westerhill Road, Bishopbriggs, Glasgow G64 2QT

www.harpercollins.co.uk

Copyright © HarperCollins Publishers Ltd 2018

Collins® is a registered trademark of HarperCollins Publishers Limited

Mapping generated from Collins Bartholomew digital databases

Contains Ordnance Survey data © Crown copyright and database right (2018)

The grid on the maps is the National Grid used on Ordnance Survey mapping

Please note that roads and other facilities which are under construction at the time of going to press,
and are due to open before the end of 2018, are shown in this atlas as open. Roads due to open
during 2019 or begin construction before the end of June 2019 are shown as 'proposed or under construction'.

Printed in China

Paperback   ISBN 978 0 00 827268 5   10 9 8 7 6 5 4 3 2 1
Spiral   ISBN 978 0 00 827269 2   10 9 8 7 6 5 4 3 2 1

e-mail: roadcheck@harpercollins.co.uk   facebook.com/collinsref   @collins_ref

© Natural England copyright. Contains Ordnance Survey data © Crown copyright and database right (2015)

Information for the alignment of the Wales Coast Path provided by © Natural Resources Wales.
All rights reserved. Contains Ordnance Survey data. Ordnance Survey licence number 100019741.
Crown copyright and database right (2013).

Information for the alignment of several Long Distance Trails in Scotland provided by © walkhighlands

Information on fixed speed camera locations provided by PocketGPSWorld.com

With thanks to the Wine Guild of the United Kingdom for help with researching vineyards.

For the latest information on Blue Flag award beaches visit www.blueflag.global

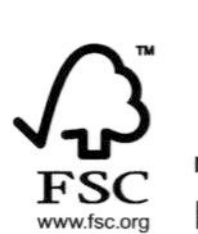

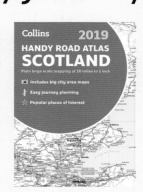
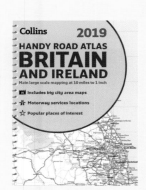
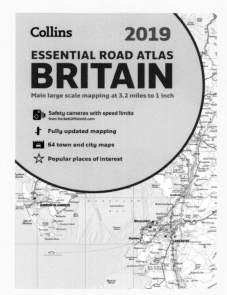
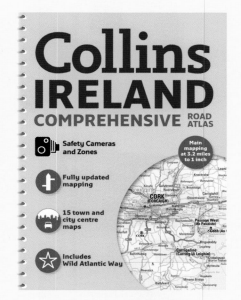

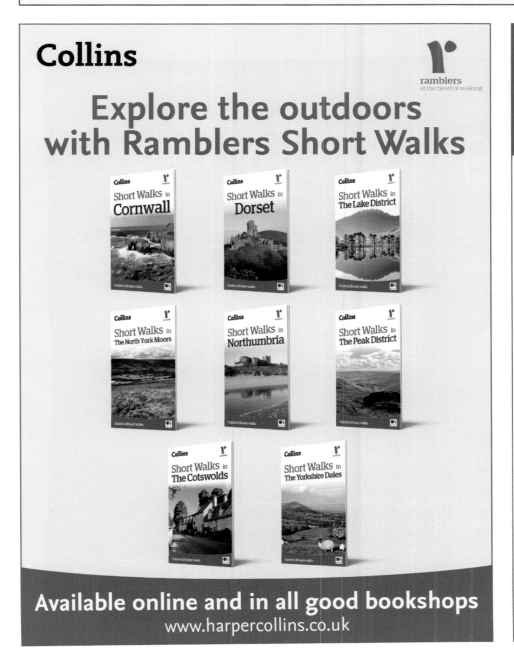

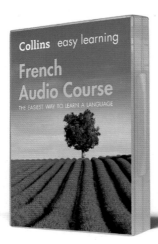 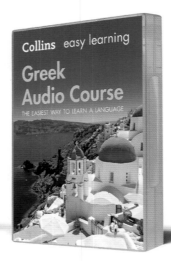

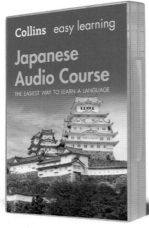   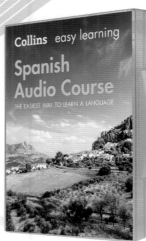